Advanced Organic Chemistry

FOURTH EDITION

Part A: Structure and Mechanisms

D0207332

Advanced Organic Chemistry

PART A: Structure and Mechanisms
PART B: Reactions and Synthesis

Advanced Organic Chemistry

FOURTH EDITION

Part A: Structure and Mechanisms

FRANCIS A. CAREY
and RICHARD J. SUNDBERG

University of Virginia
Charlottesville, Virginia

 Springer

Library of Congress Cataloging-in-Publication Data

Carey, Francis A., 1937–
    Advanced organic chemistry/Frank A. Carey and Richard J. Sundberg.—4th ed.
        p. cm.
    Includes bibliographical references and index.
    Contents: pt. A. Structure and mechanisms
    ISBN 00-306-46242-7 (hard: pt. A)—ISBN 0-306-46243-5 (pbk.: pt. A)
     1. Chemistry, Organic.   I. Sundberg, Richard J., 1938–   II. Title.

    QD251.2 .C36 2000
    547—dc21

                                                                                00-027456

Library of Congress Cataloging-in-Publication Data

Carey, Francis A., 1937–
    Advanced organic chemistry/Frank A. Carey and Richard J. Sundberg.—4th ed.
        p. cm.
    Includes bibliographical references and index.
    Contents: pt. A. Structure and mechanisms
    ISBN 00-306-46242-7 (hard: pt. A)—ISBN 0-306-46243-5 (pbk.: pt. A)
     1. Chemistry, Organic. I. Sundberg, Richard J., 1938– II. Title.

    QD251.2 .C36 2000
    547—dc21   00-027456

A C.I.P. record for this book is available from the Library of Congress

ISBN: 0-306-46242-7 (Hardbound)
ISBN: 0-306-46243-5 (Paperback)

©2000, 1990, 1984, 1977 Springer Science+Business Media, Inc.

Printed in the United States of America

10 9 8 7 6 5 4 3

springeronline.com

Preface to the Fourth Edition

The goal of this text is to build on the foundation of introductory organic chemistry to provide students and other readers a deeper understanding of structure and mechanism and the relationships between them. We have provided specific data and examples with which to illustrate the general principles that are discussed. Our purpose is to solidify the student's understanding of the basic concepts, but also to illustrate the way specific structural changes influence mechanism and reactivity.

The first three chapters discuss fundamental bonding theory, stereochemistry, and conformation, respectively. Chapter 4 discusses the means of study and description of reaction mechanisms. Chapter 9 focuses on aromaticity and aromatic stabilization and can be used at an earlier stage of a course if an instructor desires to do so. The other chapters discuss specific mechanistic types, including nucleophilic substitution, polar additions and eliminations, carbon acids and enolates, carbonyl chemistry, aromatic substitution, concerted reactions, free-radical reactions, and photochemistry.

Both the language of valence bond theory and of molecular orbital theory are used in discussing structural effects on reactivity and mechanism. Our intent is to illustrate both approaches to interpretation. A decade has passed since the publication of the Third Edition. That decade has seen significant developments in areas covered by the text. Perhaps most noteworthy has been the application of computational methods to a much wider range of problems of structure and mechanism. We have updated the description of computational methods and have included examples throughout the text of application of computational methods to specific reactions.

References to the primary literature are provided for specific issues of structure, reactivity, and mechanism. These have been chosen to illustrate the topic of discussion and, of course, cannot be comprehensive. The examples and references chosen do not imply any priority of concept or publication. References to general reviews which can provide a broader coverage of the various topics are usually given.

The problems at the end of each chapter represent application of concepts to new structures and circumstances, rather than review of material explicitly presented in the text. The level of difficulty is similar to that of earlier editions, and we expect that many will present a considerable challenge to students. Some new problems have been added in this edition. References to the literature material upon which the problems are based are given at the end of the book.

The companion volume, Part B, has also been revised to reflect the continuing development of synthetic methodology. Part B emphasizes synthetic application of organic reactions. We believe that the material in Part A and Part B will provide advanced undergraduate and beginning graduate students with a background which will permit them to understand, analyze, and apply the primary and review literature in organic chemistry. We hope that this new edition will continue to serve students and teachers in fostering both an understanding of the structural and mechanistic foundations of organic chemistry and a broad knowledge of the most fundamental reaction types in organic chemistry. We welcome comments and suggestions which can improve the text or correct errors.

F. A. Carey
R. J. Sundberg

Charlottesville, Virginia

Contents of Part A

Chapter 6. Polar Addition and Elimination Reactions

Chapter 7. Carbanions and Other Nucleophilic Carbon Species

Chapter 8. Reactions of Carbonyl Compounds

Contents of Part B

Chemical Bonding and Structure

Introduction

Organic chemistry is a broad field which intersects with such diverse areas as biology, medicine and pharmacology, polymer technology, agriculture, and petroleum engineering. At the heart of organic chemistry are fundamental concepts of molecular structure and reactivity of carbon-containing compounds. The purpose of this text is to explore this central core, which is concerned with how the structures of organic compounds are related to reactivity. Reactivity, in turn, determines the methods that can be used for synthesis. Understanding of *structure*, *reactivity*, and *synthesis* can be used within organic chemistry or applied to other fields, such as those named above, which require contributions from organic chemistry. *Structure* includes the description of bonding in organic molecules and the methods for determining, analyzing, and predicting molecular structure. Dynamic aspects of structure, such as conformational equilibria, are also included. Stereochemistry is also a crucial aspect of structure in organic chemistry. *Reactivity* pertains to the aspects of a given structure that determine its chemical transformations. Is the molecule electron-rich or electron-poor? Is it easily reduced or oxidized? What is the distribution of the most reactive electrons? Which bonds are weakest and therefore most likely to engage in reactions? Unlike structure, which is an inherent property of the molecule, reactivity usually describes an interaction with other molecules. Understanding reactivity includes describing the mechanisms, that is, the stepwise process by which reactions occur. Reactivity also encompasses the *stereochemical aspects* of the transformation. *Synthesis* encompasses those activities which are directed toward finding methods that convert existing substances into different compounds. Synthesis involves the control of reactivity to achieve specified transformations. It involves the choice of reagents, catalysts, and reaction conditions that will accomplish a given transformation within the required parameters. In various circumstances, the limiting parameters might include yield, purity of product, stereochemical control, availability or cost of reagents, or safety and environmental consequences. Structure, reactivity, and synthesis are all interrelated.

Synthesis is built on knowledge of both structure and reactivity, and understanding reactivity ultimately rests on detailed knowledge about molecular structure. A firm grounding in the principles of structure and chemical bonding is therefore an essential starting point for fuller appreciation of reactivity and synthesis. In this first chapter, we will discuss the ideas that have proven most useful to organic chemists for describing and organizing facts, concepts, and theories about the structure of organic molecules.

Structural formulas serve as key devices for communication of chemical information, and it is important to recognize the symbolic relationship between structural formulas and molecular structure. The current system of structural formulas arose largely as a result of research done in the last half of the 19th century. Elemental analyses, interrelation of various compounds, and systematic investigation of the reactivity of various "functional groups" permitted chemists to correctly deduce much information about molecular structure. For most molecules, it became possible to draw conclusions as to which atoms were directly connected (*constitution*). Lines drawn between atoms were used to represent direct connections or bonds. It was recognized that the various elements formed a characteristic numbers of bonds. The capacity of an element to form bonds was called valence, and the number of bonds a given element could form was called its *valence number*. These structural deductions predated modern electronic concepts of atomic and molecular structure and the nature of the forces that bind atoms together in molecules. Nevertheless, structural formulas proved to be readily adaptable to description of chemical bonding in terms of electron-pair bonds since the bonds came to symbolize the shared pair of electrons.

The precise description of molecular structure specifies nuclear positions with respect to other nuclei in the molecule and the distribution of the electrons associated with the nuclei. Because chemical properties are primarily determined by the outer shell of valence electrons, chemists focus attention primarily on these electrons. Spectroscopic methods and diffraction methods, especially X-ray crystal structure determination, have provided a large amount of information about atomic positions and bond lengths. Dynamic aspects of molecular structure involving such issues as alternative molecular shapes arising by bond rotations (*conformations*) can also be characterized by spectroscopic methods, especially nuclear magnetic resonance (NMR) spectroscopy. These *experimental methods* for structure determination have been joined by *computational methods*. Computational approaches for calculating molecular structures are based on systematic searching for the most stable arrangement of the atoms having a particular bonding pattern (molecular connectivity). Computational methods can be based on observed relationships between energy and structure (*molecular mechanics*) or on theoretical descriptions of bonding based on quantum chemistry.

Theories of molecular structure attempt to describe the nature of chemical bonding both qualitatively and quantitatively. To be useful to chemists, the bonding theories must provide insight into the properties and reactivity of molecules. The structural theories and concepts that are most useful in organic chemistry are the subject of this chapter. Our goal is to be able to relate molecular structure, as depicted by structural formulas and other types of structural information, such as bond lengths and electronic distributions, to the chemical reactivity and physical properties of molecules.

1.1. Valence Bond Approach to Chemical Bonding

The idea put forth by G. N. Lewis in 1916 that chemical bonding results from a sharing of electron pairs between two atoms was a fundamental advance in bonding theory.[1] The concept of valence is related to the number of electrons available to each atom and, for the second-row elements, to the "octet rule," that is, to the stability associated with four pairs of electrons. Lewis's proposal was put on the sound ground of quantum mechanics by Heitler and London's treatment of the hydrogen molecule in 1927. This treatment marked the beginning of what we now know as *valence bond theory*.[2] A central feature of this theory was the conclusion that most of the binding energy between the two atoms at the most stable internuclear separation results from sharing of the electrons between the two nuclei. This conclusion arose in a direct way from the Heitler–London calculations. If electron 1 were constrained to be associated only with nucleus 1, and electron 2 with nucleus 2, then the calculated binding energy was only a small fraction of the experimentally determined bond energy. If this constraint was removed so that the electrons were indistinguishable and permitted to interact equally with both nuclei, the calculated potential energy curve exhibited a deep minimum at the equilibrium internuclear distance. The bonding energy associated with this minimum corresponded quite well with the experimental bond energy. The covalent bond represented by a line in the simple notation H—H then takes on more precise meaning. It symbolizes the presence of *two bonding electrons* in the region between the two nuclei. The region of space occupied by an electron is called an *orbital*. In the H_2 molecule, the bonding arises from the two electrons in an orbital formed by overlap of the spherically symmetrical $1s$ atomic orbital of each hydrogen atom, as shown in Fig. 1.1. Similarly, the bonding orbitals of other molecules arise from the atomic orbitals of the constituent atoms.

Application of valence bond theory to more complex molecules involves writing as many plausible Lewis structures as possible that correspond to the correct molecular connectivity. Valence bond theory assumes that the actual molecule is a hybrid of these "canonical forms." As a simple example, the hydrogen chloride molecule is considered to be a hybrid of the limiting canonical forms H—Cl, H^+Cl^-, and H^-Cl^+. In mathematical terms, the molecule can be represented as the weighted combination of the contributing structures. Unfortunately, the extension of this approach to larger molecules results in a large number of canonical structures, which makes both conceptual and computational interpretation difficult. For example, more than 175 individual structures, most with charge separation, can be written for benzene.[3] For this reason, qualitative concepts which arise from the valence bond treatment of simple molecules are applied to larger molecules. The key ideas that are used to adapt the concepts of valence bond theory to complex molecules are *hybridization* and *resonance*. In this qualitative form, valence bond theory describes molecules in terms of orbitals that are mainly localized between two atoms. The shapes of these orbitals are assumed to be similar to those of orbitals described by more quantitative

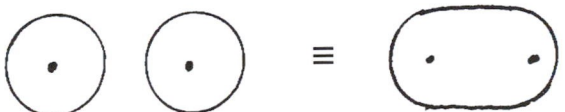

Fig. 1.1. Representation of σ bond of H_2 formed by overlap of $1s$ orbitals.

1. G. N. Lewis, *J. Am. Chem. Soc.* **38**:762 (1916).
2. W. Heitler and F. London, *Z. Phys.* **44**:455 (1927). For a historical review, see M. Simonetta, in *Structural Chemistry and Molecular Biology*, A. Rich and N. Davidson, eds., W. H. Freeman, San Francisco, 1968, pp. 769–782.
3. C. Amovilli, R. D. Harcourt, and R. McWeeny, *Chem. Phys. Lett.* **187**:494 (1991).

treatment of simpler molecules. The properties of complex molecules are regarded as derived from the combination of the properties of the constituent bonds. This conceptual approach is in accord with a large body of chemical knowledge which indicates that structure and reactivity of similar bonds and groups are relatively constant in different molecules.

1.1.1. Orbital Hybridization

The concepts of *directed valence* and *orbital hybridization* were developed by Linus Pauling soon after the description of the hydrogen molecule by the valence bond theory. These concepts were applied to an issue of specific concern to organic chemistry, the tetrahedral orientation of the bonds to tetracoordinate carbon.[4] Pauling reasoned that because covalent bonds require mutual overlap of orbitals, stronger bonds would result from better overlap. Orbitals that possess directional properties, such as p orbitals, should therefore be more effective than spherically symmetric s orbitals.

The electronic configuration of a carbon atom in its ground state is $1s^2 2s^2 2p^2$, and is not consistent with a simple rationalization of the tetrahedral bonding at carbon. Pauling suggested that four atomic orbitals ($2s$, $2p_x$, $2p_y$, $2p_z$) are replaced by a set of four equivalent *hybrid orbitals*, designated sp^3. The approximate shapes of these orbitals are shown in Fig. 1.2. Notice particularly that the probability distribution is highly directional for the sp^3 orbitals, with the region of greatest probability concentrated to one side of the nucleus.

Orbital hybridization has two important consequences. First, four bonds, rather than two, can be formed to carbon. Second, the highly directional sp^3 orbitals provide for more effective overlap and stronger bonds. Thus, although an isolated carbon atom with one electron in each of four equivalent sp^3-hybridized orbitals would be of higher energy than the ground state, the energy required in a formal sense to promote two electrons from a $2s$ orbital to sp^3 orbitals is more than compensated for by the formation of four bonds rather than two. In addition, each of the bonds is stronger owing to the directional properties of the hybrid orbitals. Tetrahedral geometry is predicted by the mathematical description of hybridization. Methane is found experimentally to be a perfect tetrahedron, with each H−C−H bond angle equal to 109.5°. The valence bond representation of methane in Fig. 1.3 shows the orbital overlaps that give rise to four equivalent C−H bonds. These bonds, in which the electron density is cylindrically symmetric about the internuclear axis are called σ bonds.

The hybridization concept can also be applied to molecules containing double and triple bonds. The descriptive valence bond approach to the bonding in ethylene and

$2s \qquad 2p \qquad 2sp^3$

Fig. 1.2. Cross section of angular dependence of orbitals.

4. L. Pauling, *J. Am. Chem. Soc.* **53**:1367 (1931).

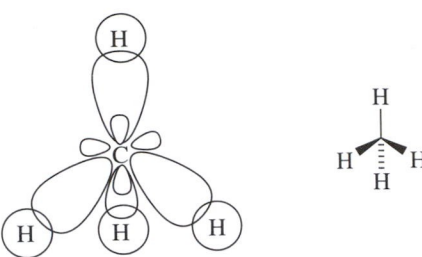

Fig. 1.3. Valence bond structural representation of methane resulting from overlap of H $1s$ orbitals with four equivalent sp^3 orbitals of carbon.

acetylene and their congeners is analogous to that for methane. In ethylene (Fig. 1.4), each carbon bears three ligands and is sp^2 hybridized. Three sp^2 orbitals are generated from the $2s$ and two of the $2p$ orbitals. The three sp^2 orbitals are coplanar and orthogonal to the remaining $2p$ orbital. A bond is formed between the two carbon atoms by overlap of an sp^2 orbital of each. The four hydrogens are bonded by σ bonds involving hydrogen $1s$ orbitals and the remaining two sp^2 hybrid orbitals. Additional bonding between the two carbon atoms is portrayed as resulting from overlap of the unhybridized p orbitals on each carbon atom, each of which contains one electron. This overlap is somewhat less effective than that of a σ bond and corresponds to a π bond. The electron distribution in a π bond is concentrated above and below the plane of the σ framework. The molecule is planar, and the plane defined by the nuclei is a nodal plane for the π system. The hybridization at each carbon atom of acetylene is sp, and the two carbon atoms are considered as bonded by a σ bond and two π bonds, as shown in Fig. 1.4.

The concept of hybrid orbitals is deeply ingrained in the thinking of organic chemists, as widely reflected in texts and the research literature. However, Pauling and others recognized that there was a different conceptual starting point in which multiple bonds can be represented as bent bonds.[4]

ethylene   acetylene

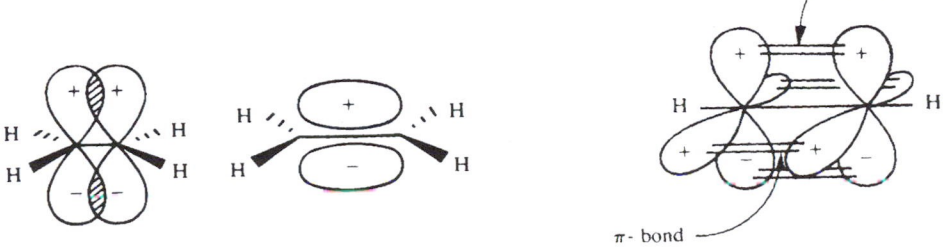

Fig. 1.4. The π bond in ethylene and the π bonds in acetylene.

It has been shown that description of bonding based on the bent-bond concept can be just as successful in describing molecular structure as the hybridization concept.[5] We will, however, use the hybridization terminology.

The relation between the number of ligands on carbon (its coordination number), hybridization, and molelcular geometry is summarized in Table 1.1. Unless all the ligands on a particular carbon atom are identical, there will be deviations from the perfectly symmetrical structures implied by the hybridization schemes. For example, whereas methane and carbon tetrachloride are tetrahedral with bond angles of 109.5°, the C−C−C angle in cyclohexane is 111.5°. The H−C−H angle in formaldehyde is 118° rather than 120°. Benzene, however, is a regular hexagon with 120° bond angles.

Large deviations in bond angles from the normal values are found in cyclopropane, cyclobutane, and other molecules containing three- and four-membered rings. These molecules are less stable than molecules with larger rings, and the difference in energy is referred to as *angle strain*. Because the three carbon atoms of a cyclopropane ring are required by symmetry to be at the vertices of an equilateral triangle, the internuclear angles are 60°. This arrangement represents a serious distortion of the normal tetrahedral bond angle and engenders unique chemical and physical properties. To develop a valence bond model of the bonding in cyclopropane, it is assumed than the carbon atoms will adopt the hybridization that produces the most stable bonding arrangement.[6] The orbitals used for forming the carbon–carbon bonds in cyclopropane can overlap more effectively if they have more p character than normal sp^3 bonds, since additional p character corresponds to a reduced bond angle. Consequently, the orbitals used for bonding to hydrogen must have increased s character. This adjustment in hybridization can be described quantitatively by assignment of numerical values to the "percent s character" in the C−H bonds. Values of 33% and 17%, respectively, have been suggested for the C−H and C−C bonds of cyclopropane, on the basis of NMR measurements.[7] The picture of the bonding in cyclopropane indicates that the region of maximum orbital overlap would not correspond to the internuclear axis. The C−C bonds are described as "bent bonds" (Fig. 1.5).

The change in hybridization is associated with a change in electronegativity. The greater the s character of a particular carbon orbital, the greater is its electronegativity. As a result, carbon atoms that are part of strained rings are more electronegative than normal toward hydrogen.[8] Figure 1.6 shows some calculated charges for cyclopropane and other

Table 1.1. Dependence of Structure on Hybridization of Carbon

Number of ligands	Hybridization	Geometry	Examples
4	sp^3	Tetrahedral	Methane, cyclohexane, methanol, carbon tetrachloride
3	sp^2	Trigonal	Ethylene, formaldehyde, benzene methyl cation, carbonate ion
2	sp	Linear	Acetylene, carbon dioxide, hydrogen cyanide, allene (C-2)

5. P. A. Shultz and R. P. Messmer, *J. Am. Chem. Soc.* **115**:10925, 10938, 10943 (1993).
6. For a review of various descriptions of the bonding in cyclopropane, see A. de Meijer, *Angew. Chem. Int. Ed. Engl.* **18**:809 (1979); K. B. Wiberg, in *The Chemistry of the Cyclopropyl Group*, Z. Rappoport, ed., John Wiley & Sons, New York, Chapter 1, 1987; B. Rozsondai, in *The Chemistry of the Cyclopropyl Group*, Vol. 2, Z. Rappoport (ed.), John Wiley & Sons, New York, Chapter 3, 1995.
7. F. J. Weigert and J. D. Roberts, *J. Am. Chem. Soc.* **89**:5962 (1967).
8. K. B. Wiberg, R. F. W. Bader, and C. D. H. Lau, *J. Am. Chem. Soc.* **109**:1001 (1987).

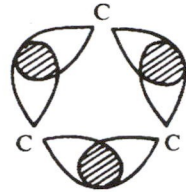

Fig. 1.5. Bent bonds in cyclopropane.

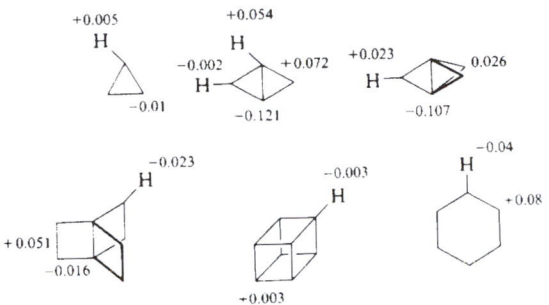

Fig. 1.6. Charge distributions in strained cyclic hydro-
carbons in comparison with cyclohexane. Data are from
K. B. Wiberg, R. F. W. Bader, and C. D. H. Lau, *J. Am.
Chem. Soc.* **109**:1001 (1987).

strained hydrocarbons in comparison with the unstrained reference cyclohexane. Notice
that the greater the distortion from the normal tetrahedral angle, the greater is the negative
charge on carbon. This relative electronegativity is reflected in both the acidity and the
NMR chemical shifts of hydrogens attached to strained ring systems.

Even more drastic distortions from ideal geometry are found when several small rings
are assembled into bicyclic and tricyclic molecules. The synthesis of such highly strained
molecules is not only a challenge to the imagination and skill of chemists, but also
provides the opportunity to test bonding theories by probing the effects of unusual bonding
geometry on the properties of molecules. One series of such molecules is the *propellanes*.[9]
The structures of some specific propellanes and the strain energies of the molecules are
shown in Fig. 1.7. Each of the molecules in Fig. 1.7 has been synthesized, and some of
their physical and structural properties have been analyzed. In the propellanes with small
rings, the bridgehead must be severely flattened to permit bonding. In order to attain this
geometry, the hybridization at the bridgehead carbons must change as the size of the
bridges decreases. Whereas the hybridization at the bridgehead carbons in [4.4.4]propel-
lane can be approximately the normal sp^3, in [2.2.2]propellane the flattening of the
bridgehead must result in a change to approximately sp^2 hybridization, with the central
bond between the two bridgehead carbons being a σ bond formed by overlap of two p
orbitals. The distortion is still more extreme in [1.1.1]propellane, in which the bridgehead

9. K. B. Wiberg, *Acc. Chem. Res.* **17**:379 (1984); K. B. Wiberg, *Chem. Rev.* **89**:975 (1989).
10. J. E. Jackson and L. C. Allen, *J. Am. Chem. Soc.* **106**:591 (1984); K. B. Wiberg, R. F. W. Bader, and C. D. H.
Lau, *J. Am. Chem. Soc.* **109**:985 (1987); K. B. Wiberg, R. F. W. Bader, and C. D. H. Lau, *J. Am. Chem. Soc.*
109:1001 (1987).

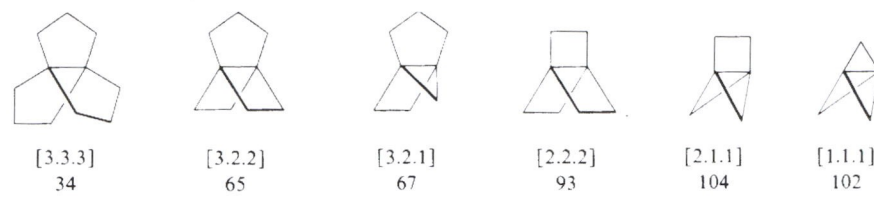

| [3.3.3] | [3.2.2] | [3.2.1] | [2.2.2] | [2.1.1] | [1.1.1] |
| 34 | 65 | 67 | 93 | 104 | 102 |

Fig. 1.7. Strain energies of some propellanes in kcal/mol.

carbon is an "inverted carbon" with all four bonds to one side. The resulting bond has quite special characteristics and is not adequately described as a localized bond.[10]

[1.1.1]propellane   [2.2.2]propellane   [4.4.4]propellane

The distortion of the bond angles in propellanes leads both to strain and to unusual chemical reactivity. [3.2.1]Propellane, for example, is found to have a strain energy of 67 kcal/mol, as compared to 27 kcal/mol for cyclopropane. The molecule is exceptionally reactive and undergoes a variety of reactions involving the cleavage of the central bond under mild conditions. For example, it undergoes bromination instantaneously at $-50°C$.[11] The strain should be less in [3.3.3]propellane, in which a smaller distortion at the bridgehead carbons is required to permit bonding. This is reflected by the lower strain energy of 34 kcal/mol. On the other hand, the smaller bridges lead to increased strain. [2.2.1]Propellane conforms to the expectation that it would be highly reactive. It can be observed when isolated in solid argon at 45 K but decomposes at temperatures higher than this and cannot be isolated as a pure substance.[12] [1.1.1]Propellane is a surprisingly stable substance. Although the strain is comparable to that of [3.1.1]- and [2.2.1]propellane, the relief of strain on rupture of the center bond is quite small, leading to greater thermal stability.[13] The "inverted" center bond is largely p in character, and, as a result, there is a considerable charge density external to the ring. This permits a variety of radical and electrophilic addition reactions to occur.[14]

11. K. B. Wiberg and G. J. Burgmaier, *J. Am. Chem. Soc.* **94**:7396 (1972).
12. K. B. Wiberg, C. M. Breneman, and T. J. LePage, *J. Am. Chem. Soc.* **112**:61 (1990); A. Gobbi and G. Frenking, *J. Am. Chem. Soc.* **116**:9275 (1994).
13. K. B. Wiberg and F. H. Walker, *J. Am. Chem. Soc.* **104**:5239 (1982); C. Y. Zhao, Y. Zhang, and X. Z. You, *J. Phys. Chem. A*, **101**, 5174 (1997).
14. K. B. Wiberg and S. T. Waddell, *J. Am. Chem. Soc.* **112**:2194 (1990); D. S. Toops and M. R. Barbachyn, *J. Org. Chem.* **58**:6505 (1993).

1.1.2. Resonance

A second concept that makes valence bond theory useful for the structural description of complex molecules is *resonance theory*. Resonance theory is an extension of valence bond theory which recognizes that, for many molecules, more than one Lewis structure can be written. Its usefulness in organic chemistry lies in its being a convenient way of depicting electron delocalization. Resonance theory is particularly useful in describing conjugated compounds and reactive intermediates. Arguments based on resonance theory are usually made in a qualitative way, although a mathematical treatment can be applied.[15] The elements of resonance theory that are necessary for qualitative applications can be summarized as follows:

a. Whenever alternative Lewis structures can be written for a molecule that differ only in assignment of electrons among the nuclei, with the nuclear positions being constant for all the structures, then the molecule is not adequately represented by a single Lewis structure but has weighted properties of all of the alternative Lewis structures.

b. All structures are restricted to the maximum number of valence electrons that is appropriate for each atom, that is, two for hydrogen and eight for the first-row elements.

c. Some individual Lewis structures are more stable than others. The structures that approximate the actual molecule most closely are those that incorporate the following features: maximum number of covalent bonds, minimum separation of unlike charges, and placement of any negative charges on the most electronegative atom (or any positive charge on the most electropositive atom). Stated in another way, the most favorable (lowest-energy) resonance structure makes the greatest contribution to the true (hybrid) structure.

d. In most cases, the delocalization of electrons, as represented by the writing of alternative Lewis structures, is associated with enhanced stability relative to a single localized structure. This is not always true, however, since molecules and ions are known in which electron delocalization produces an increase in energy relative to a localized model.

Some fundamental structure–stability relationships can be employed to illustrate the use of resonance concepts. The allyl cation is known to be a particularly stable carbocation. This stability can be understood by recognizing that the positive charge is delocalized between two carbon atoms, as represented by the two equivalent resonance structures. The delocalization imposes a structural requirement. The *p* orbitals on the three contiguous carbon atoms must all be aligned in the same direction to permit electron delocalization. As a result, there is an energy barrier to rotation about the carbon–carbon

---

15. For a classical presentation of resonance theory, see G. W. Wheland, *Resonance Theory in Organic Chemistry*, John Wiley & Sons, New York, 1955. Models of molecular structure based on mathematical description of valence bond theory have been developed: F. W. Bodrowicz and W. A. Goddard III, in *Modern Theoretical Chemistry, Methods of Electronic Structure Theory*, Vol. 3, H. F. Schaefer III, ed., Plenum Press, New York, 1977, Chapter 4; A. Voter and W. A. Goddard III, *Chem. Phys.* **57**:253 (1981); N. D. Epiotis, *Unified Valence Bond Theory of Electronic Structure*, Springer-Verlag, Berlin, 1983; D. J. Klein and N. Trinajstic, eds., *Valence Bond Theory and Chemical Structure*, Elsevier, Amsterdam, 1990; D. L. Cooper, J. Gerratt, and M. Raimondi, *Chem. Rev.* **91**:929 (1991).

bonds in the allyl cation. The most stable geometry is planar, and the barrier to rotation is estimated to be 36–38 kcal/mol.[16]

resonance interaction
disrupted by rotation

Another important and familiar example of resonance is the stabilization of the enolate anions formed by deprotonation of carbonyl compounds. This can be illustrated by considering the relative acidity of 2-methylpropene (isobutene) and 2-propanone (acetone). The relative acidity is indicated by the pK values, which are ~ 25 and ~ 45, respectively. The difference of ~ 20 pK units (20 powers of 10) shows that it is much easier for a proton to be removed from acetone than from isobutene. The main reason for the difference in acidity is the difference in stability of the two conjugate bases. A resonance-stabilized anion is generated in each case, but one of the contributing structures for the acetone anion has a negative charge on oxygen. In both resonance structures for the anion of isobutene, the negative charge is on carbon. Because oxygen is a more electronegative element than carbon, application of resonance theory leads to the conclusion that the acetone anion will be more stable than the isobutene anion and that acetone will therefore be more acidic.

Another example of the effect of resonance is in the relative acidity of carboxylic acids as compared to alcohols. Carboxylic acids derived from saturated hydrocarbons have pK_a values near 5, whereas saturated alcohols have pK_a values in the range 16–18. This implies that the carboxylate anion can accept negative charge more readily than an oxygen on a saturated carbon chain. This can be explained in terms of stabilization of the negative charge by resonance.[17]

Resonance stabilization of
carboxylate anion.

Alcoholate anions have negative
charge localized on a single oxygen.

One of the structural implications of the delocalization of the negative charge in carboxylate anions, the identity of the two C—O bond lengths, intermediate between those of single and double bonds, has been verified by many crystal structure determinations.

Dynamic structural characteristics can also be interpreted in terms of resonance. There is a substantial barrier to rotation about the C—N single bonds in carboxamides. A frequently observed consequence is the nonidentity of NMR peaks due to the *syn* and *anti*

16. K. B. Wiberg, C. M. Breneman, and T. J. LePage, *J. Am. Chem. Soc.* **112**:61 (1990); A. Gobbi and G. Frenking, *J. Am. Chem. Soc.* **116**:9275 (1994).
17. F. G. Bordwell and A. V. Satish, *J. Am. Chem. Soc.* **116**:8885 (1994); P. C. Hiberty and C. P. Byrman, *J. Am. Chem. Soc.* **117**:9875 (1995); J. D. da Motta Neto and M. A. C. Nascimento, *J. Phys. Chem.* **100**:15105 (1996).

substituents on nitrogen. The barrier is in the range of 15–20 kcal/mol.[18] A planar structure is imposed by the necessity of *p*-orbital overlap for delocalization. Other structural parameters of amides, such as bond lengths and bond force constants, are also consistent with the resonance model.[19]

Resonance concepts are especially important in systems in which two or more double bonds are in conjugation. Resonance structures permit a description of the interaction between the bonds. Carbonyl compounds having a carbon–carbon double bond adjacent to the carbonyl group provide a good example of how structural features can be related to resonance interactions. While only a single uncharged stucture can be drawn, two structures with charges can be drawn. The structure with a negative charge on oxygen is far more important because of the higher electronegativity of oxygen relative to carbon. The structure with a positive charge on oxygen is unfavorable and would make only a minor contribution.

Some of the structural features of this class of compounds which are in accord with the resonance picture are as follows:

a. The $C=O$ bond is not as strong as in saturated carbonyl compounds. This is revealed by the infrared stretching frequency, which comes at lower energy (typically $1690\ cm^{-1}$ versus $1730\ cm^{-1}$ for saturated compounds).

b. Carbon-13 NMR spectroscopy also reveals that the β carbon is less shielded (lower electron density) than is the case for a simple alkene. This results from the delocalization of π electrons from this carbon to the carbonyl oxygen.

c. The chemical reactivity of the double bond is also affected by the presence of the conjugated carbonyl group. Simple alkenes are not very reactive toward nucleophiles. In contrast, double bonds adjacent to carbonyl groups do react with nucleophiles. The partial positive charge depicted by the resonance structure makes the β carbon subject to nucleophilic attack.

18. B. M. Pinto, in *Acyclic Organonitrogen Stereodynamics*, J. B. Lambert and Y. Takeuchi, eds., VCH Publishers, New York, 1992, pp. 149–175.

19. The issue of resonance in carboxamides has recently been reexamined. Most of the structural consequences predicted by resonance are consistent with a stabilizing interaction which results in a net shift of electron density from nitrogen to carbon and oxygen; A. J. Bennet, V. Somayaji, R. S. Brown, and B. D. Santarsiero, *J. Am. Chem. Soc.* **113**:7563 (1991); A. Greenberg, T. D. Thomas, C. R. Bevilacqua, M. Coville, D. Ji, J.-C. Tsai, and G. Wu, *J. Org. Chem.* **57**:7093 (1992).

Also, the carbonyl group stabilizes the negative charge that develops at the α carbon as a result of nucleophilic attack. All of these effects are summed up by saying that the carbonyl group acts as an *electron-withdrawing group* toward the double bond, as is depicted in the most important of the charged resonance structures.

There are also substituents that can act as electron-releasing groups through resonance. Among familiar examples are alkoxy and amino groups in vinyl ethers and enamines, respectively.

We might wonder how important such resonance effects are since the resonance structures have the unfavorable feature of charge separation and place a positive charge on electronegative atoms. Several approaches have been taken to determining the importance of these interactions. Most are based on molecular orbital theory (see Sections 1.3 and 1.6). Typically, the results reveal only small ground-state structural changes, but parameters which reflect reactivity do indicate that these molecules are very much enhanced in reactivity toward electrophiles.[20] Carbon and proton NMR shifts are also upfield in vinyl ethers and enamines, as implied by the charged resonance structures. The most detailed analyses suggest that there is indeed a π polarization of the type indicated, but that it is compensated by a σ polarization in the opposite sense.[21]

The most impressive example of resonance stabilization is benzene, in which the delocalization is responsible for a stabilization of 30–36 kcal/mol, the *resonance energy* of benzene.

cyclohexatriene   benzene

We will return to the aromatic stabilization of benzene in more detail in Chapter 9, but substituted benzenes provide excellent examples of how proper use of the resonance concept can be valuable in predicting reactivity. Many substituents can be readily classified

20. A. R. Katritzky and M. Karelson, *Tetrahedron Lett.* **31**:2987 (1990); K. B. Wiberg, R. E. Rosenberg, and P. R. Rablen, *J. Am. Chem. Soc.* **113**:2890 (1991).
21. K. B. Wiberg and P. R. Rablen, *J. Am. Chem. Soc.* **115**:9234 (1993).

as electron-releasing or electron-withdrawing simply by noting whether the substituent can donate or accept electrons by π-orbital interaction with the ring.

Electron-withdrawing substituents

Electron-releasing substituents

We will address this issue further in Chapter 10, where the polar effects of the substituents on both the σ and π electrons will be considered. For the case of electrophilic aromatic substitution, where the energetics of interaction of an approaching electrophile with the π system determines both the rate of reaction and position of substitution, simple resonance arguments are extremely useful.

1.2. Bond Energy, Polarity, and Polarizability

1.2.1. Bond Energies

Of the various geometric parameters associated with molecular shape, the one most nearly constant from molecule to molecule and most nearly independent of substituent effects is bond length. Bond lengths to carbon depend strongly on the hybridization of the carbon involved but are little influenced by other factors. Table 1.2 lists the interatomic distances for some of the most common bonds in organic molecules. The near constancy of bond lengths from molecule to molecule reflects the fact that the properties of individual bonds are, to a good approximation, independent of the remainder of the molecule.

Table 1.3 gives some bond-energy data. Part A includes bond energies for some simple diatomic molecules and generalized values for some of the types of bonds found

Table 1.2. Bond Lengths (Å)[a]

sp^3	C−H	1.09	sp^3–sp^3	C−C	1.54	sp^2–sp^2	C−C	1.46	C−O	1.42
sp^2	C−H	1.086	sp^3–sp^2	C−C	1.50	sp^2–sp^2	C=C	1.34	C=O	1.22
sp	C−H	1.06	sp^3–sp	C−C	1.47	sp–sp	C≡C	1.20		

a. From experimental values tabulated for simple molecules by M. J. S. Dewar and W. Thiel, *J. Am. Chem. Soc.* **99**:4907 (1977).

most often in organic molecules. The assumption that bond energies are independent of the remainder of the molecule is a rather rough one. Part B of Table 1.3 lists some specific $C-H$, $C-C$, and other bond energies. It is apparent that some are substantially different from the generalized values. For example, the CH_2-H bond dissociation energies listed for propene and toluene are 85 kcal/mol, which is significantly less than for a $C-H$ bond in ethane (98 kcal/mol). The reason for the relative weakness of these particular bonds is that the allyl and benzyl radicals that are produced by the bond dissociations are stabilized by resonance.

$$H_2C=CH-CH_2\bullet \longleftrightarrow \bullet H_2C-CH=CH_2$$

A similar explanation lies behind the diminished strength of the sp^3-sp^3 carbon–carbon bond in ethylbenzene. The general trend toward weaker $C-C$ bonds with increased substitution that can be recognized in Table 1.3 reflects the increased stability of substituted radicals relative to primary radicals.

The bond energies in Table 1.3 refer to *homolytic* bond dissociation to uncharged radical fragments. Many reactions involve heterolytic bond cleavages. The energy for heterolytic cleavage of $C-H$ or $C-C$ bonds in the gas phase is very high, largely because of the energy required for charge separation. However, in solution, where stabilization of the ions by solvation becomes possible, heterolytic bond dissociation can become energetically feasible. Heterolytic bond dissociations are even more sensitive to structural changes than homolytic ones. Table 1.4 gives a series of comparable ionization energies for cleavage of H^- from hydrocarbons and Cl^- from chlorides in the gas phase. Besides

Table 1.3. Bond Energies (kcal/mol)

A. Some common bond energies[a]					
H–H	103	C–H	98	C=C	145
C–C	81	N–H	92	C≡C	198
O–O	34	O–H	109	N≡N	225
Cl–Cl	57	Cl–H	102	C=O	173
Br–Br	45	Br–H	87	C–O	79
I–I	36	I–H	71	C–N	66

B. Some specific bond dissociation energies[b]					
H_3C-H	104	H_3C-CH_3	88	H_3C-F	108
CH_3CH_2-H	98	$H_5C_2-CH_3$	85	H_3C-Cl	84
$H_2C=CH-H$	110[c]	$(CH_3)_2CH-CH_3$	83	H_3C-Br	70
$HC\equiv C-H$	131	$PhCH_2-CH_3$	70	H_3C-I	56
$H_2C=CHCH_2-H$	85	$H_5C_2-C_2H_5$	82	H_3C-OH	91
$PhCH_2-H$	85	$(CH_3)_2CH-CH(CH_3)_2$	78		
H_2N-H	103	$H_2C=CH_2$	171[c]		
CH_3NH-H	92	$HC\equiv CH$	228[c]		
CH_3O-H	102				

a. From Table 1, G. J. Janz, *Thermodynamic Properties of Organic Compounds*, Academic Press, New York, 1967.
b. Except where noted, from J. A. Kerr, *Chem. Rev.* **66**:465 (1966).
c. K. M. Ervin, S. Gronert, S. E. Barlow, M. K. Gilles, A. G. Harrison, V. M. Bierbaum, C. H. DePuy, W. C. Lineberger, and G. B. Ellison, *J. Am. Chem. Soc.* **112**:5750 (1990).

**Table 1.4. Heterolytic Bond Dissociation Energies for
Some C—H and C—Cl Bonds**[a]

R	$R-H \rightarrow R^+ + H^-$	$R-Cl \rightarrow R^+ + Cl^-$
	ΔE_{C-H} (kcal/mol)	ΔE_{C-Cl} (kcal/mol)
CH_3	312.2	227.1
CH_3CH_2	272.6	190.3
$(CH_3)_2CH$	249.9	171.0
$CH_2=CH$	290.2	
$CH_2=CHCH_2$	255.3	

a. Data from C. G. Screttas, *J. Org. Chem.* **45**:333 (1980).

noting the higher energies in comparison with the homolytic bond dissociation energies, one can see that branching decreases the energy requirement for heterolytic bond cleavage more dramatically than for homolytic cleavage. These bond-energy relationships are consistent with the familiar order of carbocation stability, *tert* > *sec* > *pri* > methyl. Note also that the heterolytic bond dissociation energies for allyl and vinyl bonds reflect the stability of allyl cations and the instability of vinyl cations.

Smaller, but nevertheless significant, differences in energies of organic molecules also result from less obvious differences in structure. Table 1.5 gives the heats of formation of some hydrocarbons. These energy values represent the heat evolved on formation of the compound from its constituent elements under standard conditions. The heats of formation therefore permit precise comparison of the stability of *isomeric compounds*. The more negative the heat of formation, the greater is the stability. Direct comparison of compounds having different elemental composition is not meaningful, because the total number of bonds formed is then different.

Part A of Table 1.5 shows all the acyclic C_4–C_6 and some of the C_8 hydrocarbons. A general trend is discernible in the data. Branched-chain hydrocarbons are more stable than straight-chain hydrocarbons. For example, ΔH_f° for *n*-octane is -49.82 kcal/mol, whereas the most highly branched isomer possible, 2,2,3,3-tetramethylbutane, is the most stable of the octanes, with ΔH_f° of -53.99 kcal/mol. Similar trends are observed in the other series.

Part B of Table 1.5 gives heats of formation for the C_4, C_5, and some of the C_6 alkenes. A general relationship is also observed for the alkenes. The more highly substituted the double bond, the more stable is the compound. There are also other factors that enter into alkene stability. *trans*-Alkenes are usually more stable than *cis*-alkenes, probably largely because of increased nonbonded repulsion in the *cis* isomer.[22]

1.2.2. Electronegativity and Polarity

Another fundamental property of chemical bonds is *polarity*. In general, it is to be expected that the distribution of the pair of electrons in a covalent bond will favor one of the two atoms. The tendency of an atom to attract electrons is called *electronegativity*. There are a number of different approaches to assigning electronegativity, and most are numerically scaled to a definition originally proposed by Pauling.[23] Part A of Table 1.6

22. For a theoretical discussion of this point, see N. D. Epiotis, R. L. Yates, and F. Bernardi, *J. Am. Chem. Soc.* **97**:5961 (1975).

23. For leading references, see G. Simons, M. E. Zandler, and E. R. Talaty, *J. Am. Chem. Soc.* **98**:7869 (1976).

Table 1.5. Standard Enthalpies of Formation of Some Hydrocarbons (kcal/mol)[a]

A. Saturated hydrocarbons			
C_4		C_8	
Butane	-30.15	Octane	-49.82
2-Methylpropane	-32.15	2-Methylheptane	-51.50
		3-Methylheptane	-50.82
C_5		4-Methylheptane	-50.69
Pentane	-35.00	2,2,-Dimethylhexane	-53.71
2-Methylbutane	-36.90	2,3-Dimethylhexane	-51.13
2,2-Dimethylpropane	-36.97	2,4-Dimethylhexane	-52.44
		3,3-Dimethylhexane	-52.61
C_6		2,2,3-Trimethylpentane	-52.61
Hexane	-39.96	2,2,4-Trimethylpentane	-53.57
2-Methylpentane	-41.66	2,2,3,3-Tetramethylbutane	-53.99
3-Methylpentane	-41.02		
2,3-Dimethylbutane	-42.49		
2,2-Dimethylbutane	-44.35		

B. Alkenes			
C_4		C_6	
1-Butene	-0.03	1-Hexene	-9.96
trans-2-Butene	-2.67	*trans*-2-Hexene	-12.56
cis-2-Butene	-1.67	*cis*-2-Hexene	-11.56
2-Methylpropene	-4.04	*trans*-3-Hexene	-12.56
		cis-3-Hexene	-11.56
C_5		2-Methyl-1-pentene	-13.56
1-Pentene	-5.00	3-Methyl-pentene	-11.02
trans-2-Pentene	-7.59	4-Methyl-1-pentene	-11.66
cis-2-Pentene	-6.71	2-Methyl-2-pentene	-14.96
2-Methyl-1-butene	-8.68	3- Methyl-2-pentene	-14.32
3-Methyl-1- butene	-6.92	2,3-Dimethyl-1-butene	-14.78
2- Methyl-2-butene	-10.17	3,3-Dimethyl-1-butene	-14.25
		2,3-Dimethyl-2-butene	-15.91

a. From F. D. Rossini, K. S. Pitzer, R. L. Arnett, R. M. Braun, and G. C. Pimentel, *Selected Values of Physical and Thermodynamic Properties of Hydrocarbons and Related Compounds*, Carnegie Press, Pittsburgh, 1953.

gives the original Pauling values and also a more recent set based on theoretical calculation of electron distributions. The concept of electronegativity can also be expanded to include functional groups. Part B of Table 1.6 gives some values which are scaled to be numerically consistent with elemental electronegativities, as well as a set based on theoretical electron distribution calculations. These electronegativity values can serve to convey a qualitative impression of the electron-attracting capacity of these groups.

The unequal distribution of electron density in covalent bonds produces a *bond dipole*, the magnitude of which is expressed by the dipole moment, having the units of charge times distance.[24] Bonds with significant bond dipoles are described as being polar. The bond and group dipole moments of some typical substituents are shown in Table 1.7.

24. For more detailed discussion of dipole moments, see L. E. Sutton, in *Determination of Organic Structures by Physical Methods*, Vol. 1, E. A. Braude and F. C. Nachod, eds., Academic Press, New York, 1955, Chapter 9; V. I. Minkin, O. A. Osipov, and Y. A. Zhdanov, *Dipole Moments in Organic Chemistry*, Plenum Press, New York, 1970.

Table 1.6. Atomic and Group Electronegativities

A. Atomic electronegativities[a]

H 2.1	C 2.5; **2.35**	N 3.0; **3.16**	O 3.5; **3.52**	F 4.0; **4.00**
	Si 1.8; **1.64**	P 2.1; **2.11**	S 2.5; **2.52**	Cl 3.0; **2.84**
		As 2.0; **1.99**	Se 2.4; **2.40**	Br 2.8; **2.52**
				I 2.5

B. Empirical electronegativities for some organic functional groups[b]

CH_3	2.3; **2.55**	H	2.28; **1.20**	F	3.95; **4.00**
CH_2Cl	2.75; **2.61**	NH_2	3.35; **3.12**	Cl	3.03; **3.05**
$CHCl_2$	2.8; **2.66**	$^+NH_3$	3.8; **3.21**	Br	2.80; **2.75**
CCl_3	3.0; **2.70**	NO_2	3.4; **3.22**	I	2.28;
CF_3	3.35; **2.71**	OH	3.7; **3.55**		
Ph	3.0; **2.58**				
$CH{=}CH_2$	3.0; **2.58**				
$C{\equiv}CH$	3.3; **2.66**				
$C{\equiv}N$	3.3; **2.69**				

a. From L. Pauling, *The Nature of the Chemical Bond*, 3rd edition, Cornell University Press, Ithaca, New York, 1960. Boldface values from G. Simons, M. E. Zandler, and E. R. Talaty, *J. Am. Chem. Soc.* **98**:7869 (1976).
b. From P. R. Wells, *Prog. Phys. Org. Chem.* **6**:111 (1968). Boldface values from R. J. Boyd and S. L. Boyd, *J. Am. Chem. Soc.* **114**:1652 (1992).

It is possible to estimate with a fair degree of accuracy the dipole moment of a molecule as the vector sum of the component bond dipoles. A qualitative judgment of bond polarity can be made by comparing the electronegativities of the bound atoms or groups. The larger the difference in electronegativity, the greater will be the bond dipole.

For most purposes, hydrocarbon groups can be considered to be nonpolar. There are, however, small dipoles associated with C—H bonds and bonds between carbons of different hybridization or substitution pattern. For normal sp^3 carbon, the carbon is found to be slightly negatively charged relative to hydrogen.[25] The electronegativity order for hybridized carbon orbitals is $sp > sp^2 > sp^3$. Scheme 1.1 lists the dipole moments of some hydrocarbons and some other organic molecules.

Electronegativity is a fundamental characteristic of atoms that is transferred into functional groups. Electronegativity correlates strongly with position in the periodic table.

Table 1.7. Bond and Group Dipoles for Some Organic Functional Groups[a]

Bond moments[b]		Bond moments[b]		Group moments[b]	
C—H	0.4	C—N	0.22	MeO	1.3
C—F	1.41	C—O	0.74	NH_2	1.2
C—Cl	1.46	C=O	2.3	CO_2H	1.7
C—Br	1.38	$C{\equiv}N$	3.5	COMe	2.7
C—I	1.19			NO_2	3.1
				CN	4.0

a. From C. P. Smyth, *Dielectric Behavior and Structure*, McGraw-Hill Book Company, New York, 1955, pp. 244, 253.
b. In e.s. units $\times 10^{18}$.

25. K. B. Wiberg, R. F. W. Bader, and C. D. H. Lau, *J. Am. Chem. Soc.* **109**:1001 (1987).

Scheme 1.1. Dipole Moments for Some Organic Compounds[a]

A. Hydrocarbons

$H_2C=CHCH_2CH_3$ $HC\equiv CCH_2CH_3$

$$\begin{array}{cc} CH_3 & CH_3 \\ & C=C \\ H & H \end{array}$$

$$\begin{array}{cc} CH_3 & H \\ & C=C \\ H & CH_3 \end{array}$$

0.34   0.800   0.253   0
                                                 (symmetry)

$$\begin{array}{c} CH_3 \\ H_2C=C \\ CH_3 \end{array}$$

$CH_3C\equiv CCH_3$ $H_2C=CHCH_3$ $HC\equiv CCH_3$

0.503   0   0.366   0.781
           (symmetry)

B. Substituted molecules

$$\begin{array}{cc} H & H \\ & C=C \\ Cl & Cl \end{array}$$

$$\begin{array}{cc} H & Cl \\ & C=C \\ Cl & H \end{array}$$

$$\begin{array}{cc} H & Cl \\ & C=C \\ H & Cl \end{array}$$

1.90   0   1.34   1.63
        (symmetry)

CH_3CN CH_3NO_2 CH_3OCH_3 CH_3OH CH_3CO_2H $CH_3\overset{O}{\underset{||}{C}}CH_3$ $CH_3\overset{O}{\underset{||}{S}}CH_3$

3.92   3.46   1.30   1.70   1.74   2.88   3.96

a. Units are in debye. Data are from *Handbook of Chemistry and Physics*, 78th edition, CRC Press, Inc., Boca Raton, Florida, 1997.

Electronegativity increases going to the right in any row of the periodic table and decreases going down any column. Because electronegativity is such an important property in relation to chemical reactivity, there has been much effort to relate it to other atomic properties. Mulliken pointed out that there is a relationship between ionization energy and electron affinity and the numerical electronegativity scale.[26] The Mulliken electronegativity can be expressed as

$$\chi = \frac{I+A}{2}$$

where I is ionization potential and A is electron affinity. Ionization energy measures the energy required to remove an electron from an atom, whereas electron affinity is the energy released upon addition of an electron to an element. Both of these energies reflect the degree of attraction the atomic nucleus has for electrons and correlate with position in the periodic table. The metallic atoms readily release electrons, whereas the halogen atoms are avid electron acceptors.

The polarity of covalent bonds between carbon and substituents is the basis of important structure–reactivity relationships in organic chemistry. The effects of polar bonds are generally considered to be transmitted in two ways. Successive polarization through bonds is called the *inductive effect*. It is expected that such an effect would diminish as the number of intervening bonds increases.

$$\overset{\delta^-}{X}\diagdown\underset{\delta\delta^+}{\diagup}\overset{\delta^+}{\diagdown}\underset{}{\diagup}\overset{\delta\delta\delta^+}{\diagdown}Y$$

26. R. S. Mulliken, *J. Chem. Phys.* **2**:782 (1934); R. S. Mulliken, *J. Chem. Phys.* **3**:573 (1935).

The second component is called a *field effect* and is attributed to through-space interactions of the electric dipoles resulting from polar bonds.

In Chapter 4, we will discuss the relative importance of inductive effects and field effects on reactivity. Generally, field effects appear to be the dominant mechanism for the transmission of electrostatic effects of polar bonds to other parts of a molecule.

One of the most extensively explored series of substituent effects involves the acidity of carboxylic acids. The pK_a values of some derivatives of acetic acid are presented in Table 1.8. These data illustrate that substitution of hydrogen by a more electronegative atom or group increases the equilibrium constant for ionization, that is, makes the derivative a stronger acid. The highly electronegative fluorine atom causes a larger increase in acidity than the somewhat less electronegative chlorine atom. A slight acid-weakening effect of a methyl substituent is observed when propanoic acid is compared with acetic acid. The pK_a data refer to measurements in solution, and some care must be taken in interpreting the changes in acidity solely on the basis of electronegativity. In fact, measurements in the gas phase show that propanoic acid is a slightly stronger acid than acetic acid.[27] In the case of acid dissociation in solution, both the proton and the carboxylate anion are strongly solvated, and this greatly favors the dissociation process in solution relative to the gas phase. We can discuss trichloroacetic acid, in which polar $C-Cl$ bonds have replaced $C-H$ bonds as an example.

very little dipolar stabilization of the anion

strong dipolar stabilization of the anion

Table 1.8. Acidity of Substituted Acetic Acids

	H_2O $(pK_a)^a$	Gas phase $(\Delta G)^b$
$(CH_3)_3CCO_2H$	5.0	337.6
$(CH_3)_2CHCO_2H$	4.9	339.0
$CH_3CH_2CO_2H$	4.9	340.4
CH_3CO_2H	4.8	341.5
HCO_2H	3.8	338.3
FCH_2CO_2H		331.6
$ClCH_2CO_2H$	2.7	329.0
F_3CCO_2H		317.4
Cl_3CCO_2H	0.7	
$NCCH_2CO_2H$	2.6	
$O_2NCH_2CO_2H$	1.3	

a. From *Stability Constants and Stability Constants, Supplement No. 1*, Special Publications 17 (1964) and 25 (1971), The Chemical Society, London.
b. ΔG for $AH \rightarrow H^+ + A^-$ in kcal/mol at $300\,K$; from G. Caldwell, R. Renneboog, and P. Kebarle, *Can. J. Chem.* **67**:611 (1989).

27. R. Yamdagni and P. Kebarle, *J. Am. Chem. Soc.* **95**:4050 (1973).

For the acid dissociation equilibrium,

$$RCO_2H + H_2O \rightleftharpoons RCO_2^- + H_3O^+$$

dissociation places a negative charge on the carboxylate residue and increases the electron density at the carboxyl group carbon and oxygen atoms. For acetic acid, where $R = CH_3$, this increase occurs adjacent to the carbon on the methyl group, which bears a very small negative charge. In the case of trichloroacetic acid, $R = CCl_3$, the corresponding carbon is somewhat positive as a result of the $C-Cl$ bond dipoles (inductive effect). The cumulative effect of the three $C-Cl$ bond dipoles establishes a strong dipole (field effect) associated with the CCl_3 group. As a result, the development of negative charge is more favorable, and trichloroacetic acid is a stronger acid than acetic acid.

It is always important to keep in mind the *relative* nature of substituent effects. Thus, the effect of the chlorine atoms in the case of trichloroacetic acid is primarily to stabilize the dissociated anion. The acid is more highly dissociated than in the unsubstituted case because there is a more favorable energy *difference* between the parent acid and the anion. It is the energy differences, not the absolute energies, that determine the equilibrium constant for ionization. As we will discuss more fully in Chapter 4, there are other mechanisms by which substituents affect the energy of reactants and products. The detailed understanding of substituent effects will require that we separate polar effects from these other factors.

It can be seen from the data in Table 1.8 that alkyl substituents slightly decrease the acidity of carboxylic acids in aqueous solution. This is a general effect and is attributable to two factors. One is a steric effect. The increasing steric bulk somewhat destabilizes the carboxylate anion by decreasing the effectiveness of solvation.[28] Alkyl groups also have an inductive/field effect that decreases acidity. This results from the weak electron-donating character of the alkyl substituent which causes acetic acid and other alkanoic acids to be weaker acids than formic acid both in solution and in the gas phase. The alkyl group donates charge density to the carbonyl carbon, thereby reducing the acidity of the carboxyl hydrogen.[29]

The second column in Table 1.8 shows the free energy for dissociation of some of the same acids in the gas phase. The effects of strongly electron-withdrawing groups are still evident. There is frequently a large difference in the energy of chemical processes in the gas and solution phases because of the importance of solvation. In the gas phase, alkyl groups enhance acidity. This is attributed to the greater polarizability of the larger substituents. In the case of molecules in the gas phase, any stabilization of the negative charge must be accomplished by internal charge redistribution.[30] With larger substituents, the larger molecular volume and the larger number of atoms that can participate lead to a greater stabilization.

1.2.3. Polarizability—Hardness and Softness

Another property that is closely related to electronegativity and position in the periodic table is polarizability. Polarizability is related to the size of atoms and ions and the

28. K. Bowden and R. G. Young, *Can. J. Chem.* **47**:2775 (1969).
29. A. D. Headley, M. E. McMurry, and S. D. Starnes, *J. Org. Chem.* **59**:1863 (1994); M. R. F. Siggel and T. D. Thomas, *J. Am. Chem. Soc.* **114**:5795 (1992).
30. G. Caldwell, R. Renneboog, and P. Kebarle, *Can. J. Chem.* **67**:611 (1989).

ease with which the electron cloud can be distorted. There is a mathematical correlation between polarizability and atomic volume, and polarizability is also related to the LUMO–HOMO energy gap. (See Section 1.6 if the terms HOMO and LUMO are unfamiliar.)

hard   soft

Qualitative relationship between orbital
energies, hardness, and softness

Numerical measures of polarizability analogous to electronegativity have been defined.[31] Hardness η is numerically expressed as

$$\eta = \tfrac{1}{2}(I - A)$$

where I is the ionization potential and A is the electron affinity. Softness σ is the opposite of hardness and is numerically expressed as

$$\sigma = 1/\eta$$

These two properties are closely related to the HOMO and LUMO energies of molecules and ions. The larger the HOMO–LUMO gap, the greater is the hardness. Numerically, hardness is approximately equal to half the energy gap, as defined above for atoms. In general, chemical reactivity increases as LUMO energies decrease and HOMO energies increase. The implication is that softer chemical species, those with smaller HOMO–LUMO gaps, will tend to be more reactive than harder ones. In qualitative terms, this can be described as the ability of nucleophiles or bases to donate electrons more readily to electrophiles and acids to begin the process of bond formation. Interactions between harder chemical entities are more likely to be dominated by electrostatic interactions. Table 1.9 gives some hardness values for atoms and common small molecules and ions.

Figure 1.8 shows the I–A gap (2η) for several radicals. Note that there is a correlation with electronegativity and position in the periodic table. The halogen anions and radicals become progressively softer from fluorine to iodine. Across the second row, softness decreases from carbon to fluorine. Highly electronegative atoms tend to be hard, whereas less electronegative atoms are softer. Polarizability is also a function of atomic number. Larger atoms are softer than smaller atoms of similar electronegativity. The charge on an atom also influences polarizability. Metal cations, for example, become harder as the oxidation number increases. Values for the hardness of some metal cations that are frequently of interest in organic chemistry are included in Table 1.9. A useful precept for understanding Lewis acid–base interactions is that hard acids prefer hard bases, and soft acids prefer soft bases. The hard–hard interactions are dominated by electrostatic attraction, whereas soft–soft interactions are dominated by mutual polarization.[32]

31. R. G. Parr and R. G. Pearson, *J. Am. Chem. Soc.* **105**:7512 (1983).
32. R. G. Pearson, *J. Am. Chem. Soc.* **85**:3533 (1963); T. L. Ho, *Hard and Soft Acids and Bases in Organic Chemistry*, Academic Press, New York, 1977; W. B. Jensen, *The Lewis Acid–Base Concept*, Wiley-Interscience, New York, 1980, Chapter 8.

Table 1.9. Hardness of Some Atoms, Acids, and Bases^a

Atom	η	Acid	η_A	Base	η_B
H	6.4	H^+	∞	H^-	6.8
Li	2.4	Li^+	35.1	F^-	7.0
C	5.0	Mg^{2+}	32.5	Cl^-	4.7
N	7.3	Na^+	21.1	Br^-	4.2
O	6.1	Ca^{2+}	19.7	I^-	3.7
F	7.0	Al^{3+}	45.8	CH_3^-	4.0
Na	2.3	Cu^+	6.3	NH_2^-	5.3
Si	3.4	Cu^{2+}	8.3	OH^-	5.6
P	4.9	Fe^{2+}	7.3	SH^-	4.1
S	4.1	Fe^{3+}	13.1	CN^-	5.3
Cl	4.7	Hg^{2+}	7.7		
		Pb^{2+}	8.5		
		Pd^{2+}	6.8		

a. From R. G. Parr and R. G. Pearson, *J. Am. Chem. Soc.* **105:**7512 (1983).

Another useful generalization is the principle of maximum hardness.[33] This states that molecular arrangements that maximize hardness are preferred. Electronegativity and hardness determine the extent of electron transfer between two molecular fragments in a reaction. This can be approximated numerically by the expression

$$\Delta N = \frac{\chi_X - \chi_Y}{2(\eta_X + \eta_Y)}$$

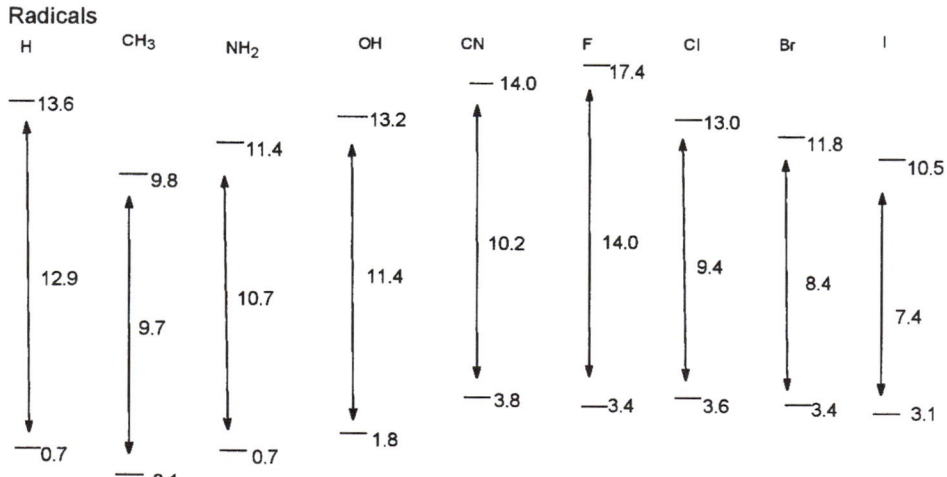

Fig. 1.8. Ionization energy (I) and electron affinity (A) gaps in eV for radicals. [Adapted from R. G. Pearson, *J. Am. Chem. Soc.* **110:**7684 (1988).]

33. R. G. Pearson, *Acc. Chem. Res.* **26:**250 (1993); R. G. Parr and Z. Zhou, *Acc. Chem. Res.* **26:**256 (1993); R. G. Pearson, *J. Org. Chem.* **54:**1423 (1989); R. G. Parr and J. L. Gazquez, *J. Phys. Chem.* **97:**3939 (1993).

where χ is absolute electronegativity and η is hardness for the reacting species. For example, one can calculate the values for each of the four halogen atoms reacting with the methyl radical to form the corresponding methyl halide.

$$X\cdot + CH_3\cdot \longrightarrow CH_3X$$

X	χ_X	η_X	ΔN	η_{CH_3X}
F·	10.4	7.0	0.23	9.4
Cl·	8.3	4.7	0.17	7.5
Br·	7.6	4.2	0.14	5.8
I·	6.8	3.7	0.10	4.7

According to this analysis, the C−X bond will be successively both more polar and harder in the order I < Br < Cl < F. This result is in agreement with both the properties and the reactivities of the methyl halides. When comparable bonds are considered, reacting species of greater hardness result in a larger net charge transfer, which adds an increment to the exothermicity of bond formation. That is, bonds formed between two harder reactants will be stronger than those between two softer reactants.[34] This is an example of a general relationship that recognizes that there is an increment to bond strength resulting from added ionic character. Indeed, quite accurate estimates of bond strength can be made by methods that add an increment due to the electronegativity difference between atoms.[35]

The concepts of electronegativity, hardness, and polarizability are all interrelated. For the kind of qualitative applications we will make in discussing reactivity, the concept that initial interactions between reacting molecules can be dominated by either partial electron transfer by bond formation (soft reactants) or by electrostatic interaction (hard reactants) is a useful generalization.

The ideas about bond length, bond energies, polarity, and polarizability discussed in this section are very useful because of the relative constancy of these properties from molecule to molecule. Thus, data obtained from simple well-characterized molecules can provide a good guide to the properties of substances whose structures are known but which have yet to be studied in detail. Organic chemists have usually discussed this transferability of properties in terms of valence bond theory. Thus, the properties are thought of as characteristic of the various types of atoms and bonds. The properties of the molecule are thought of as the sum of the properties of the bonds. This has been a highly fruitful conceptual approach in organic chemistry. As we shall see in the next section, there is an alternative description of molecules which is also highly informative and useful.

1.3. Molecular Orbital Theory and Methods

Another broad approach to the description of molecular structure that is of importance in organic chemistry is molecular orbital theory. Molecular orbital (MO) theory pictures electrons as being distributed among a set of molecular orbitals of discrete

34. P. K. Chattaraj, A. Cedillo, R. G. Parr, and E. M. Arnett, *J. Org. Chem.* **60**:4707 (1995).
35. R. R. Reddy, T. V. R. Rao, and R. Viswanath, *J. Am. Chem. Soc.* **111**:2914 (1989).

energies. In contrast to the orbitals described by valence bond theory, which are usually concentrated between two specific atoms, molecular orbitals can extend over the entire molecule. MO theory is based on the Schrödinger equation,

$$H\psi = E\psi$$

in which ψ is a wave function describing an orbital, H is the Hamiltonian operator, and E is the energy of an electron in a particular orbital. The wave function describes the interaction of the electron with the other electrons and nuclei of the molecule. The total electronic energy is the sum of the individual electron energies:

$$E = \int \psi H \psi \, d\tau \qquad \text{when} \int \psi^2 \, d\tau = 1$$

In order to make the mathematics tractable, approximations must be made. The choice of approximations has produced a variety of MO methods, the judicious application of which can provide valuable insight into questions of bonding, structure, dynamics, and reactivity. The discussion that follows will not be sufficiently detailed or complete for the reader to understand how the calculations are performed or the details of the approximations. Instead, the nature of the information that is obtained will be described, and the ways in which organic chemists have applied the results of MO theory will be illustrated. Several excellent books are available which provide detailed treatment of various aspects of MO methods.[36]

There is a trade-off between the accuracy of the calculation and the amount of computation required. In general, the more severe the approximations, the more limited is the range of applicability of the particular calculation. An organic chemist who wishes to make use of the results of MO calclulations must therefore make a judgment about the suitability of the various methods to the particular problem. The rapid increases that have occurred in computer speed and power have made the application of sophisticated methods practical for increasingly larger molecules.

Mathematically, the molecular orbitals are treated as linear combinations of atomic orbitals, so that the wave function, ψ, is expressed as a sum of individual atomic orbitals multiplied by appropriate weighting factors (atomic coefficients):

$$\psi = c_1\phi_1 + c_2\phi_2 + \cdots + c_n\phi_n$$

The coefficients indicate the contribution of each atomic orbital to the molecular orbital. This method of representing the molecular orbital wave function in terms of combinations of atomic orbital wave functions is known as the *linear combination of atomic orbitals* approximation (LCAO). The combination of atomic orbitals chosen is called the *basis set*.

36. M. J. S. Dewar, *The Molecular Orbital Theory of Organic Chemistry*, McGraw-Hill, New York, 1969; W. T. Borden, *Modern Molecular Orbital Theory for Organic Chemists*, Prentice-Hall, Englewood Cliffs, New Jersey, 1975; H. E. Zimmerman, *Quantum Mechanics for Organic Chemists*, Academic Press, New York, 1975; I. G. Csizmadia, *Theory and Practice of MO Calculations on Organic Molecules*, Elsevier, Amsterdam, 1976; M. J. S. Dewar and R. C. Dougherty, *The PMO Theory of Organic Chemistry*, Plenum Press, New York, 1975; T. A. Albright, J. K. Burdett, and M.-H. Whangbo, *Orbital Interactions in Chemistry*, John Wiley & Sons, New York, 1985; W. G. Richards and D. L. Cooper, *Ab Initio Molecular Orbital Calculations for Chemists*, 2nd ed., Clarendon Press, Oxford, U.K., 1983; W. J. Hehre, L. Radom, P. Schleyer, and J. Pople, *Ab Initio Molecular Orbital Theory*, Wiley-Interscience, New York, 1986.

A minimum basis set for molecules containing C, H, O, and N would consist of $2s$, $2p_x$, $2p_y$, and $2p_z$ orbitals for each C, N, and O and a $1s$ orbital for each hydrogen. The basis sets are mathematical expressions describing the properties of the atomic orbitals.

Two main streams of computational techniques branch out from this point. These are referred to as *ab initio* and *semiempirical* calculations. In both *ab initio* and semiempirical treatments, mathematical formulations of the wave functions which describe hydrogen-like orbitals are used. Examples of wave functions that are commonly used are Slater-type orbitals (abbreviated STO) and Gaussian-type orbitals (GTO). There are additional variations which are designated by additions to the abbreviations. Both *ab initio* and semiempirical calculations treat the linear combination of orbitals by iterative computations that establish a self-consistent electrical field (SCF) and minimize the energy of the system. The minimum-energy combination is taken to describe the molecule.

The various semiempirical methods differ in the approximations which are made concerning repulsions between electrons in different orbitals. The approximations are then corrected for by "parameterization," whereby parameters are included in the protocol to adjust the results to match more accurate calculations or experimental data. The reliability and accuracy of the semiempirical methods have evolved, and the increasing power of computers has permitted wider application of the more accurate methods. The earliest semiempirical methods to be applied extensively to organic molecules included the extended Hückel theory[37] (EHT) and the CNDO (complete neglect of differential overlap) methods.[38] These methods gave correct representations of the shapes and trends in charge distribution in the various molecular orbitals but were only roughly reliable in describing molecular geometry. These methods tend to make large errors in calculation of total energies of molecules. Improved semiempirical calculations give better representations of charge distributions, molecular geometry, and ground-state total energies. Among these methods are MINDO-3,[39] MNDO,[40] AM1,[41] and PM3.[42] (The acronyms refer to titles of the methods.) There are differences among the methods in the ranges of compounds for which the results are satisfactory.

Ab initio calculations are iterative procedures based on *self-consistent field* (SCF) methods. Normally, calculations are approached by the Hartree–Fock closed-shell approximation, which treats a single electron at a time interacting with an aggregate of all the other electrons. Self-consistency is achieved by a procedure in which a set of orbitals is assumed, and the electron–electron repulsion is calculated; this energy is then used to calculate a new set of orbitals, which in turn are used to calculate a new repulsive energy. The process is continued until convergence occurs and self-consistency is achieved.[43]

The individual *ab initio* calculations are further identified by abbreviations for the basis set orbitals that are used. These abbreviations include, for example, STO-3G,[44] 4-31G[45] and 6-31G.[46] In general, the *ab initio* calculations make fewer assumptions than semiempirical methods, and therefore the computations are more demanding.

37. R. Hoffmann, *J. Chem. Phys.* **39**:1397 (1963).
38. J. A. Pople and G. A. Segal, *J. Chem. Phys.* **44**:3289 (1966).
39. R. C. Bingham, M. J. S. Dewar, and D. H. Lo, *J. Am. Chem. Soc.* **97**:1285, 1294, 1302 (1975).
40. M. J. S. Dewar and W. Thiel, *J. Am. Chem. Soc.* **99**:4907 (1977).
41. M. J. S. Dewar, E. G. Zoebisch, E. F. Healy, and J. J. P. Stewart, *J. Am. Chem. Soc.* **109**:3902 (1985).
42. J. J. P. Stewart, *J. Comput. Chem.* **10**:209, 221 (1989).
43. C. C. J. Roothaan, *Rev. Mod. Phys.* **23**:69 (1951); R. Pariser and R. G. Parr, *J. Chem. Phys.* **21**:767 (1953); J. A. Pople, *J. Phys. Chem.* **61**:6 (1957).
44. W. J. Hehre, R. F. Stewart, and J. A. Pople, *J. Chem. Phys.* **51**:2657 (1971).
45. R. Ditchfield, W. J. Hehre, and J. A. Pople, *J. Chem. Phys.* **54**:724 (1971).
46. W. J. Hehre, R. Ditchfield, and J. A. Pople, *J. Chem. Phys.* **56**:2257 (1972).

Another distinguishing aspect of MO methods is the extent to which they deal with *electron correlation*. The Hartree–Fock approximation does not deal with correlation between individual electrons, and the results are expected to be in error because of this, giving energies above the exact energy. MO methods that include electron correlation have been developed. The calculations are usually done using Møller–Plesset perturbation theory and are designated MP calculations.[47]

Among the most widely used *ab initio* methods are those referred to as G1[48] and G2.[49] These methods incorporate large basis sets including *d* and *f* orbitals, called 6-311**. The calculations also have extensive configuration interaction terms at the Møller–Plesset fourth order (MP4) and further terms referred to as quadratic configuration interaction (QCISD).[50] Finally, there are systematically applied correction terms calibrated by exact energies from small molecules.

Current MO methods give excellent results on ground-state molecular geometry and charge distribution. They can also give excellent agreement with experimental data in the calculation of relative molecular energies. The calculation of these quantities normally refers to isolated molecules; i.e., the calculated values correspond to those for the molecule in the gas phase. The energy changes associated with dynamic processes can be studied by calculation of molecular energy as a function of molecular distortion. Much effort is also being devoted to the description of reaction processes. This is an especially formidable task because information about solvent participation and exact separation of reacting molecules is imprecise. In cases where good assumptions about such variables can be made, *ab initio* MO calculations can give good estimates of the energy changes associated with chemical reactions.

The results of all types of MO calculations include the energy of each MO, the total electronic energy of the molecule relative to the separated atoms, and the coefficients of the atomic orbitals (AOs) contributing to each MO. Such information may be applied directly to a number of physical and chemical properties. The total electronic energy obtained by summing the energies of the occupied orbitals gives the calculated molecular energy. Comparison of isomeric molecules permits conclusions about the relative stabilities of the compounds. Conclusions about molecular stability can be checked by comparison with thermodynamic data when they are available. Conformational effects can be probed by calculating the total energy as a function of molecular geometry. The minimum energy should correspond to the most favorable conformation. Most calculations are done on single molecules, not an interacting array of molecules. Thus, the results are most comparable to the situation in the gas phase, where intermolecular forces are weak. The types of data that are obtained by MO calculations are illustrated in the following paragraphs.

The coefficients for the AOs that comprise each MO may be related to the electron density at each atom by the equation

$$q_r = \sum n_j c_{jr}{}^2$$

47. C. Møller and M. S. Plesset, *Phys. Rev.* **46:**618 (1934); J. S. Brinkley and J. A. Pople, *Int. J. Quantum Chem.* **9S:**229 (1975); K. Raghavachari and J. B. Anderson, *J. Phys. Chem.* **100:**12960 (1996).
48. J. A. Pople, M. Head-Gordon, D. J. Fox, K. Raghavachari, and L. A. Curtiss, *J. Chem. Phys.* **90:**5622 (1989); M. Head-Gordon, *J. Phys. Chem.* **100:**13213 (1996).
49. L. A. Curtiss, K. Raghavachari, G. W. Trucks, and J. A. Pople, *J. Chem. Phys.* **94:**7221 (1991); L. A. Curtiss, K. Raghavachari, and J. A. Pople, *J. Chem. Phys.* **98:**1293 (1993).
50. J. A. Pople, M. Head-Gordon, and K. Raghavachari, *J. Chem. Phys.* **87:**5968 (1987).

which gives the electron density at atom r as the sum over all the occupied MOs of the product of the number of electrons in each orbital and the *square* of the coefficient at atom r for each orbital. To illustrate, the coefficients calculated for the methyl cation by the CNDO/2 method are given in Table 1.10. There are seven MOs generated from the three hydrogen $1s$ and the carbon $2s$, $2p_x$, $2p_y$, and $2p_z$ atomic orbitals. The electron densities are calculated from the coefficients of ψ_1, ψ_2, and ψ_3 only, because these are the occupied orbitals for the six-electron system. The carbon atom is calculated to have 3.565 electrons (exclusive of the $1s$ electrons), and each hydrogen atom is calculated to have 0.812 electrons. Because neutral carbon has four valence electrons, its net charge in the methyl cation is $+0.435$. Each hydrogen atom has a charge of $+0.188$. The total charge is $0.435 + 3(0.188) = 1.000$ electron. A sample calculation of the hydrogen electron density from the orbital coefficients follows:

$$q_H = 2(0.3528)^2 + 2(0.0999)^2 + 2(0.5210)^2 = 0.812$$

Further examination of Table 1.10 reveals that the lowest unoccupied molecular orbital, ψ_4, is a pure p orbital, localized on carbon, since the coefficients are zero for all but the $2p_z$ orbital. The MO picture is in agreement with the usual qualitative hybridization picture for the methyl cation.

MO methods can also be used to calculate heats of formation (ΔH_f) of molecules or heats of reaction (ΔH) by comparing the heats of formation of reactants and products. The total stabilization energy calculated for even a small hydrocarbon, relative to the separated nuclei and electrons, is very large (typically, 50,000–100,000 kcal/mol for C_2 and C_4 compounds, respectively). The energy differences that are of principal chemical interest, such as ΔH for a reaction, are likely to be in the range of 0–50 kcal/mol. A very small error, relative to the total energy, in a MO calculation becomes an enormous error in a calculated ΔH. Fortunately, the absolute errors for compounds of similar structure are likely to be comparable so that the errors will cancel in calculation of the *energy differences* between related molecules. Calculation of heats of formation and heats of reaction is frequently done on the basis of *isodesmic reactions*[51] in order to provide for maximum cancellation of errors in total binding energies. An isodesmic reaction is defined as a process in which the number of formal bonds of each type is kept constant; that is, the numbers of $C-H$, $C=C$, $C=O$, etc., bonds on each side of the equation are identical.[52] Although the reaction may not correspond to any real chemical process, the calculation can

Table 1.10. Coefficients of Wave Functions Calculated for Methyl Cation by the CNDO/2 Approximation[a]

Orbital	C_{2s}	C_{2p_x}	C_{2p_y}	C_{2p_z}	H	H	H
ψ_1	0.7915	0.0000	0.0000	0.0000	0.3528	0.3528	0.3528
ψ_2	0.0000	0.1431	0.7466	0.0000	0.0999	0.4012	-0.5011
ψ_3	0.0000	0.7466	-0.1431	0.0000	0.5210	-0.3470	-0.1740
ψ_4	0.0000	0.0000	0.0000	1.0000	0.0000	0.0000	0.0000
ψ_5	-0.6111	0.0000	0.0000	0.0000	0.4570	0.4570	0.4570
ψ_6	0.0000	0.5625	-0.3251	0.0000	-0.5374	0.5377	-0.0003
ψ_7	0.0000	0.3251	0.5625	0.0000	-0.3106	-0.3101	0.6207

a. The orbital energies (eigenvalues) are not given. The lowest-energy orbital is ψ_1; the highest-energy orbital, ψ_7.

51. W. J. Hehre, R. Ditchfield, L. Radom, and J. A. Pople, *J. Am. Chem. Soc.* **92**:4796 (1970).
52. D. A. Ponomarev and V. V. Takhistov, *J. Chem. Educ.* **74**:201 (1997).

Table 1.11. Calculated and Experimental ΔH Values for Some Isodemic Reactions

Reaction	Calculated ΔH (4-31G)[a] (kcal/mol)	Calculated ΔH (6-31G)[b] (kcal/mol)	Experimental ΔH (kcal/mol)
$CH_3CH_2CH_3 + CH_4 \rightarrow 2CH_3CH_3$	1.0	0.8	2.6
$\triangle + 3CH_4 \rightarrow 2CH_3CH_3 + CH_2{=}CH_2$	-58	-50.4	-43.9
$H_2C{=}C{=}O + CH_4 \rightarrow CH_2{=}CH_2 + H_2C{=}O$	12.8	13.3	15.0
$CH_3CN + CH_4 \rightarrow CH_3CH_3 + HCN$	12.0	11.7	14.4

a. Data from W. J. Hehre, R. Ditchfield, L. Radom, and J. A. Pople, *J. Am. Chem. Soc.* **92**:4796 (1970).
b. From W. J. Hehre, L. Radom, P. v. R. Schleyer, and J. Pople, *Ab Initio Molecular Orbital Theory*, John Wiley & Sons, New York, 1986, pp. 299–305.

provide a test of the reliability of the computational methods because of the comparability of ΔH_f data. The "experimental" ΔH for a reaction can be obtained by summation of tabulated ΔH_f for reactants and products. Table 1.11 gives some ΔH values calculated at the 4-31G and 6-31G level of theory for some isodesmic reactions.

The relative merits of various MO methods have been discussed in the literature.[53] In general, the *ab initio* type of calculations will be more reliable, but the semiempirical calculations are faster in terms of computer time. The complexity of calculation also increases rapidly as the number of atoms in the molecule increases. The choice of a method is normally made on the basis of evidence that the method is adequate for the problem at hand and the availability of appropriate computer programs and equipment. Results should be subjected to critical evaluation by comparison with experimental data or checked by representative calculations using higher-level methods. Table 1.12 lists some reported deviations from experimental ΔH_f for some small hydrocarbons. The extent of deviation gives an indication of the accuracy of the various types of MO calculations in this application.

The use of MO methods to probe the relationship between structure and energy can be illustrated by a study of CH_3^+, $CH_3\cdot$, and CH_3^-. The study employed *ab initio* calculations and the 4-31G basis set and was aimed at exploring the optimum geometry and resistance to deformation in each of these reaction intermediates.[54] Figure 1.9 is a plot of the calculated energy as a function of deformation from planarity for the three species. Whereas CH_3^+ and $CH_3\cdot$ are found to have minimum energy at $\theta = 0°$, that is, when the molecule is planar, CH_3^- is calculated to have a nonplanar equilibrium geometry. This calculated result is in good agreement with a variety of experimental observations which will be discussed in later chapters where these intermediates are discussed in more detail.

53. J. A. Pople, *J. Am. Chem. Soc.* **97**:5306 (1975); W. J. Hehre, *J. Am. Chem. Soc.* **97**:5308 (1975); T. A. Halgren, D. A. Kleier, J. H. Hall, Jr., L. D. Brown, and W. N. Lipscomb, *J. Am. Chem. Soc.* **100**:6595 (1978); M. J. S. Dewar and G. P. Ford, *J. Am. Chem. Soc.* **101**:5558 (1979); W. J. Hehre, *Acc. Chem. Res.* **9**:399 (1976); M. J. S. Dewar, E. G. Zoebisch, E. F. Healy, and J. J. P. Stewart, *J. Am. Chem. Soc.* **107**:3902 (1985); J. N. Levine, *Quantum Chemistry*, 3rd ed., Allyn and Bacon, Boston, 1983, pp. 507–512; W. Hehre, L. Radom, P. v. R. Schleyer, and J. A. Pople, *Ab Initio Molecular Orbital Calculations*, John Wiley & Sons, New York, 1986, Chapter 6; B. H. Besler, K. M. Merz, Jr., and P. Kollman, *J. Comput. Chem.* **11**:431 (1990); M. Sana and M. T. Nguyen, *Chem. Phys. Lett.* **196**:390 (1992).
54. E. D. Jemmiss, V. Buss, P. v. R. Schleyer, and L. C. Allen, *J. Am. Chem. Soc.* **98**:6483 (1976).

Table 1.12. Deviations of Calculated ΔH_f Values from Experimental ΔH_f Data[a]

Hydrocarbon	MNDO	AM1	PM3[b]	3-21G	6-31G	G2[c]	MP4/QCI[d]
Methane	5.9	9.0	4.9	−0.9	−0.5	0.7	−0.6
Ethane	0.3	2.6	2.1	0.2	1.9	−0.2	−0.7
Ethene	3.1	4.0	4.2	−1.6	−2.4	0.3	0.1
Allene	−1.6	0.6	1.5	−2.6	−6.8	0.0[e]	
1,3-Butadiene	2.7	3.6	5.0	−4.7	−2.9[f]	0.5[g]	
Cyclopropane	−1.5	5.1	3.5		−2.4		
Cyclobutane	−18.7	0.2	−10.6				
Cyclopentane	−12.0	−10.5	−5.6		−6.1[f]		
Cyclohexane	−5.3	−9.0	−1.5		−9.1[f]		
Benzene	1.5	2.2	3.6			4.0[h]	

a. Except as noted, energy (kcal/mol) comparisons are from M. J. S. Dewar, E. G. Zoebisch, E. F. Healy, and J. J. P. Stewart, *J. Am. Chem. Soc.* **107**:3902 (1985).
b. J. J. P. Stewart, *J. Comput. Chem.* **10**:221 (1989).
c. J. A. Pople, M. Head-Gordon, D. J. Fox, K. Raghavachari, and L. A. Curtiss, *J. Chem. Phys.* **90**:5622 (1989); L. A. Curtiss, K. Raghavachari, G. W. Trucks, and J. A. Pople, *J. Chem. Phys.* **94**:7221 (1991).
d. M. Sana and M. T. Nguyen, *Chem. Phys. Lett.* **196**:390 (1992).
e. D. W. Rogers and F. W. McLafferty, *J. Phys. Chem.* **99**:1375 (1993).
f. M. Selmi and J. Tomasi, *J. Phys. Chem.* **99**:5894 (1995).
g. M. N. Glukhovtsev and S. Laiter, *Theor. Chim. Acta* **92**:327 (1995).
h. A. Nicolaides and L. Radom, *J. Phys. Chem.* **98**:3092 (1994).

Substituent effects on intermediates can also be analyzed by MO methods. Take, for example, methyl cations where adjacent substituents with lone pairs of electrons can form π bonds, as can be expressed in either valence bond or MO terminology.

Fig. 1.9. Total energy as a function of distortion from planarity for methyl cation, methyl radical, and methyl anion. [Reproduced from *J. Am. Chem. Soc.* **98**:6483 (1976) by permission of the American Chemical Society.]

Table 1.13. Calculated Stabilization Resulting from Substituents on the Methyl Cation

Substituent	Stabilization in kcal/mol			
	4-31G[a]	6-31G*[b,c]	MP2/6-31G**[c]	MP3/6-31G*[d]
F	2.1	14.16		21.5
OH	48.0	53.77		62.7
NH_2	93.0	87.33		97.8
CH_3	30.0	30.2	41.5	34.1
$CH=CH_2$		53.9	66.1	
$CH=O$		−5.2	0.2	
CN		−13.0	−4.3	
NO_2		−40.7	−22.3	

a. Y. Apeloig, P. v. R. Schleyer, and J. A. Pople, *J. Am. Chem. Soc.* **99**:1291 (1977).
b. F. Bernardi, A. Bottoni, and A. Venturini, *J. Am. Chem. Soc.* **108**:5395 (1986).
c. X. Creary, Y.-X. Wang, and Z. Jiang, *J. Am. Chem. Soc.* **117**:3044 (1995).
d. Y. Apeloig, in *Heteroatom Chemistry*, E. Block, ed., VCH Press, Weinheim, 1990, p. 27.

An *ab initio* study using 4-31G basis set orbitals gave the stabilization energies shown in Table 1.13.[55] The table shows the calculated stabilization, relative to the methyl cation, that results from electron release by the substituent. The π-donor effects of the fluorine, oxygen, and nitrogen atoms are partially counterbalanced by the inductive electron withdrawal through the σ bond. In the case of the oxygen and nitrogen substitutents, the π-donor effect is dominant, and these substituents strongly stabilize the carbocation. For the fluorine substituent, the balance is much closer and the overall stabilization is calculated to be quite small. We will return to the case of the methyl group and its stabilizing effect on the methyl cation a little later.

In the case of the methyl anion, stabilization will result from electron-accepting substituents. Table 1.14 gives some stabilization energies calculated for a range of substituents.[56] Those substituents that have a low-lying π orbital capable of accepting electrons from the carbon $2p_z$ orbital (BH_2, $C\equiv N$, NO_2, and $CH=O$) are strongly stabilizing. Electronegative substituents without π-acceptor capacity reveal a smaller stabilization of the methyl anion. The order is $F > OH > NH_2$, which parallels the ability of these substituents to act as σ-electron acceptors. The strong effect of the trifluoromethyl group is a combination of both σ- and π-bond effects.

The substituent stabilization effects calculated for the methyl cation and the methyl anion refer to the gas phase, where no solvation effects are present, and therefore are substantially larger, in terms of energy, than would be the case in solution, where solvation contributes to stabilization and attenuates the substituent effects.

The allyl carbocation is an example of an intermediate whose structure has been extensively investigated by MO methods. The hybridization/resonance approach discussed earlier readily rationalizes some of the most prominent features of the allyl carbocation. The resonance structures suggest a significant stabilization and imply that the molecule would be planar in order to maximize the overlap of the π system.

55. Y. Apeloig, P. v. R. Schleyer, and J. A. Pople, *J. Am. Chem. Soc.* **99**:1291 (1977).
56. A. Pross, D. J. DeFrees, B. A. Levi, S. K. Pollack, L. Radom, and W. J. Hehre, *J. Org. Chem.* **46**:1693 (1981).

Table 1.14. Calculated Stabilization of Methyl Anion by Substituents

Substituent	Stabilization in kcal/mol		
	4-31G[a]	6-31G*[b]	QC150[c]
BH_2	68	61.4	54.0
CH_3	2	1.4	−3.2
NH_2	5	3.3	−0.5
F	25	14.6	10.4
$CH=CH_2$	38		
$CH=O$	72		
CN	61		
CF_3	57		
NO_2	98		

a. A. Pross, D. J. DeFrees, B. A. Levi, S. K. Pollack, L. Radom, and W. J. Hehre, *J. Org. Chem.* **46**:1693 (1981).
b. G. W. Spitznagel, T. Clark, J. Chandrasekhar, and P. v. R. Schleyer, *J. Comput. Chem.* **3**:363 (1982).
c. A. M. El-Nahas and P. v. R. Schleyer, *J. Comput. Chem.* **15**:596 (1994).

These structural effects are also found by MO calculations. Calculations at the MP4/6-311++G** level have been performed on the allyl cation and indicate a rotation barrier of 36–38 kcal/mol.[57]

1.4. Hückel Molecular Orbital Theory

Before computers enabled elaborate MO calculations to be performed routinely, it was essential that greatly simplifying approximations be applied to the molecules of interest to organic chemists. The most useful of these approximations were those incorporated in Hückel molecular orbital (HMO) theory for treatment of conjugated systems. HMO theory is based on the assumption that the π system can be treated independently of the σ framework in conjugated planar molecules and that it is the π system that is of paramount importance in determining the chemical and spectroscopic properties of conjugated polyenes and aromatic compounds. The basis for treating the σ and π systems as independent of each other is their orthogonality. The σ skeleton of a planar conjugated system lies in the nodal plane of the π system and does not interact with it. Because of its simplicity, HMO theory has been extremely valuable in the application of MO concepts to organic chemistry. It provides a good qualitative description of the π molecular orbitals in both cyclic and acyclic conjugated systems. In favorable cases such as aromatic ring systems, it provides a quite thorough analysis of the relative stability of related structures.

In the HMO approximation, the π-electron wave function is expressed as a linear combination of the p_z atomic orbitals (for the case in which the plane of the molecule coincides with the x–y plane). Minimizing the total π-electron energy with respect to the coefficients leads to a series of equations from which the atomic coefficients can be extracted. Although the mathematical operations involved in solving the equation are not

57. K. B. Wiberg, C. M. Breneman, and T. J. LePage, *J. Am. Chem. Soc.* **112**:61 (1990); A. Gobbi and G. Frenking, *J. Am. Chem. Soc.* **116**:9275 (1994).

difficult, we will not describe them in detail but will instead concentrate on the interpretation of the results of the calculations. For many systems, the Hückel MO energies and atomic coefficients have been tabulated.[58]

The most easily obtained information from such calculations is the relative orderings of the energy levels and the atomic coefficients. Solutions are readily available for a number of frequently encountered delocalized systems, which we will illustrate by referring to some typical examples. Consider, first, *linear polyenes* of formula C_nH_{n+2} such as 1,3-butadiene, 1,3,5-hexatriene, and so forth. The energy levels for such compounds are given by the expression

$$E = \alpha + m_j\beta$$

where

$$m_j = 2\cos\frac{j\pi}{n+1} \qquad \text{for } j = 1, 2, \ldots, n$$

and n is the number of carbon atoms in the conjugated chain. This calculation generates a series of MOs with energies expressed in terms of the quantities α and β, which symbolize the *Coulomb integral* and *resonance integral*, respectively. The Coulomb integral, α, is related to the binding of an electron in a $2p$ orbital, and this is taken to be a constant for all carbon atoms but will vary for heteroatoms as a result of the difference in electronegativity. The resonance integral, β, is related to the energy of an electron in the field of two or more nuclei. In the Hückel method, β is assumed to be zero when nuclei are separated by distances greater than the normal bonding distance. The approximation essentially assumes that the electron is affected only by nearest-neighbor nuclei. Both α and β are negative numbers and represent unspecified units of energy.

The coefficient corresponding to the contribution of the $2p$ AO of atom r to the jth MO is given by

$$c_{rj} = \left(\frac{2}{n+1}\right)^{1/2}\left(\sin\frac{rj\pi}{n+1}\right)$$

Carrying out the numerical operations for 1,3,5-hexatriene gives the results shown in Table 1.15. Because the molecule is a six-π-electron system, ψ_1, ψ_2, and ψ_3 are all doubly

Table 1.15. Energy Levels and Coefficients for HMOs of 1,3,5-Hexatriene

π-Orbital: ψ_j	m_j	c_1	c_2	c_3	c_4	c_5	c_6
ψ_1	1.802	0.2319	0.4179	0.5211	0.5211	0.4179	0.2319
ψ_2	1.247	0.4179	0.5211	0.2319	-0.2319	-0.5211	-0.4179
ψ_3	0.445	0.5211	0.2319	-0.4179	-0.4179	0.2319	0.5211
ψ_4	-0.445	0.5211	-0.2319	-0.4179	0.4179	0.2319	-0.5211
ψ_5	-1.247	0.4179	-0.5211	0.2319	0.2319	-0.5211	0.4179
ψ_6	-1.802	0.2319	-0.4179	0.5211	-0.5211	0.4179	-0.2319

58. C. A. Coulson and A. Streitwieser, Jr., *Dictionary of π-Electron Calculations*, W. H. Freeman, San Francisco, 1965; E. Heilbronner and P. A. Straub, *Hückel Molecular Orbitals*, Springer-Verlag, Berlin, 1966.

occupied, giving a total π-electron energy of $6\alpha + 6.988\beta$. The general solution for this system is based on the assumption that the electrons are delocalized. If this assumption were not made and the molecule were considered to be composed of alternating single and double bonds, the total π-electron energy would have been $6\alpha + 6\beta$, identical to that for three ethylene units. The differences between the electron energy calculated for a system of delocalized electrons and that calculated for alternating single and double bonds is referred to as the *delocalization energy* and is a measure of the extra stability afforded a molecule containing delocalized electrons compared to a molecule containing localized bonds. The calculated delocalization energy (DE) for 1,3,5-hexatriene is 0.988β. The value of β (as expressed in conventional energy units) is not precisely defined. One of the frequently used values is 18 kcal/mol, which is based on the value of 36 kcal/mol for the resonance energy of benzene, for which the calculated DE is 2β.

Inspection of the coefficients and a familiarity with the way they translate into symmetry properties of orbitals can be used in an extremely powerful way to aid in understanding a number of aspects of the properties of conjugated unsaturated compounds. Such considerations apply particularly well to the class of reactions classified as *concerted*, which will be described in detail in Chapter 11. It can be seen in Table 1.15 that the coefficients are all of like sign in the lowest-energy orbital, ψ_1, and that the number of times that a sign *change* occurs in the wave function increases with the energy of the orbital. A change in sign of the coefficients of the AOs on adjacent atoms corresponds to an antibonding interaction between the two atoms, and a node exists between them. Thus, ψ_1 has no nodes, ψ_2 has one, ψ_3 has two, and so on up to ψ_6, which has five nodes and no bonding interactions in its combination of AOs. A diagrammatic view of the bonding and antibonding interactions among the AOs of 1,3,5-hexatriene is presented in Fig. 1.10. Notice that for the bonding orbitals ψ_1, ψ_2, and ψ_3, there are more bonding interactions than antibonding interactions, whereas the opposite is true of the antibonding orbitals.

The success of simple HMO theory in dealing with the relative stabilities of cyclic conjugated polyenes is impressive. Simple resonance arguments do not explain the unique stability of benzene as compared with the elusive and unstable nature of cyclobutadiene. (Two apparently analogous resonance structures can be drawn in each case.) This contrast in stability is readily explained by Hückel's rule, which states that a species will be strongly stabilized (aromatic) if it is composed of a planar monocyclic array of atoms, each of which contributes a p orbital to the π system, when the number of electrons in the π system is $4n + 2$, where n is an integer. By this criterion, benzene, with six π electrons, is aromatic, whereas cyclobutadiene, with four, is not. An understanding of the theoretical basis for Hückel's rule can be gained by examining the results of HMO calculations. For cyclic polyenes, the general solution for the energy levels is

$$E = \alpha + m_j\beta$$

where

$$m_j = 2\cos\frac{2j\pi}{n} \qquad \text{for } j = 0, \pm1, \pm2, \ldots, \begin{cases} \pm(n-1)/2 \text{ for } n \text{ odd} \\ \pm n/2 \text{ for } n \text{ even} \end{cases}$$

and n is the number of carbon atoms in the ring. This solution gives the energy level diagrams for cyclobutadiene and benzene shown in Fig. 1.11.

The total π-electron energy of benzene is $6\alpha + 8\beta$, corresponding to a DE of 2β. Cyclobutadiene is predicted to have a triplet ground state (for a square geometry) and zero

DE, since the π-electron energy is $4\alpha + 4\beta$, the same as that for two independent double bonds. Thus, at this level of approximation, HMO theory predicts no stabilization for cyclobutadiene from delocalization and furthermore predicts that the molecule will have unpaired electrons, which would lead to very high reactivity. In addition, cyclobutadiene would suffer angle strain, which is not present in benzene. The extreme instability of cyclobutadiene is then understandable. Higher-level MO calculations modify this picture somewhat and predict that cyclobutadiene will be a rectangular molecule, as will be discussed in Chapter 9. These calculations, nevertheless, agree with simple HMO theory in

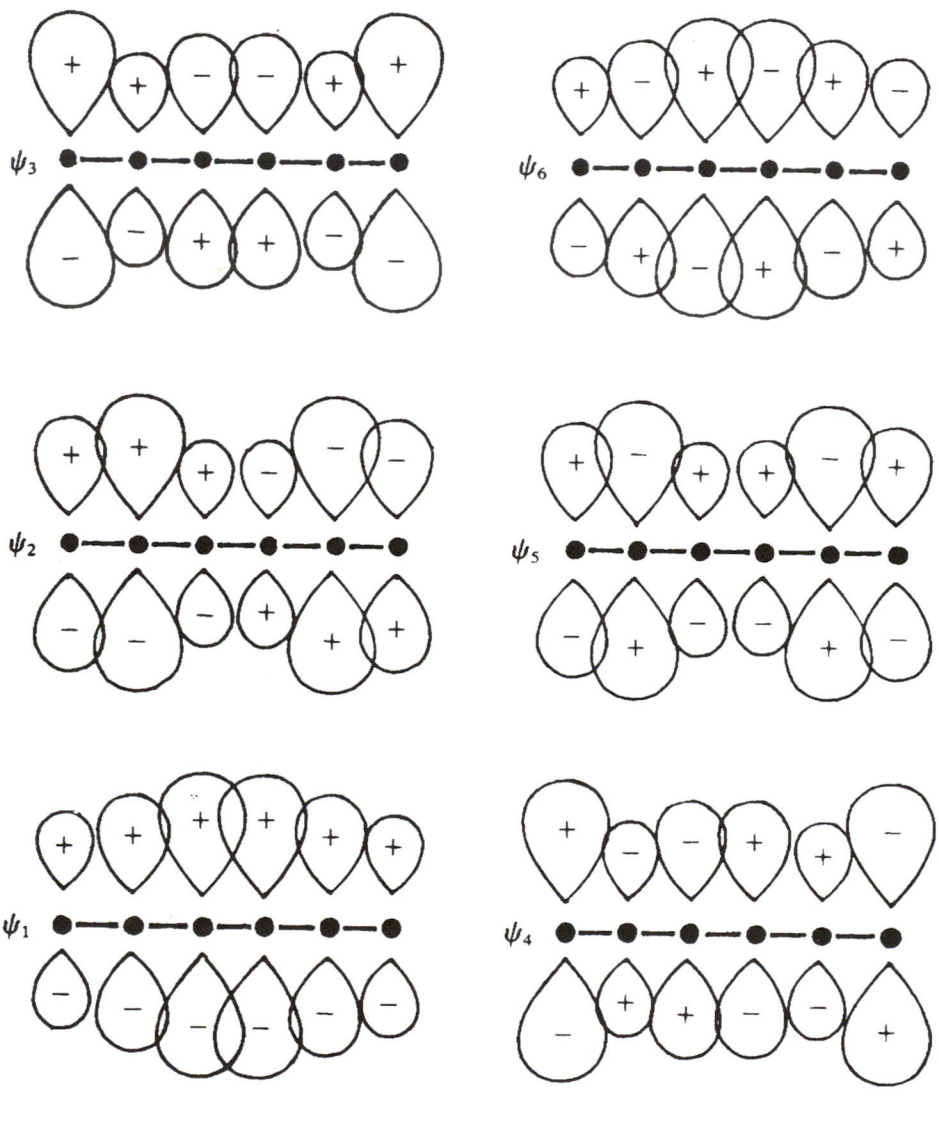

bonding orbitals   antibonding orbitals

Fig. 1.10. Graphic representation of π-molecular orbitals of 1,3,5-hexatriene as combinations of $2p$ AOs. The sizes of the orbitals are roughly proportional to the coefficients of the Hückel wave functions.

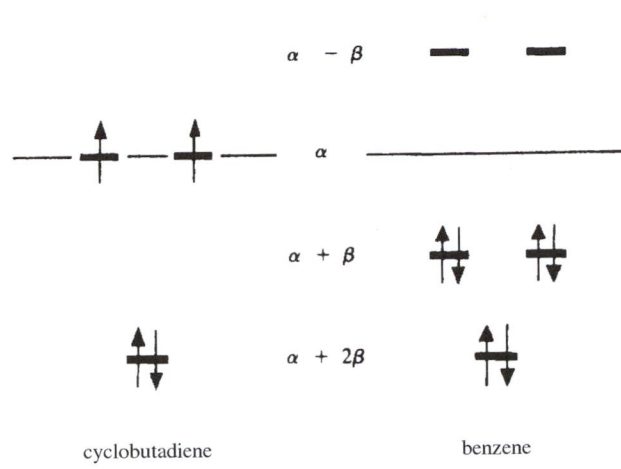

cyclobutadiene   benzene

Fig. 1.11. Energy level diagrams for cyclobutadiene and benzene.

concluding that there will be no stabilization of butadiene resulting from delocalization of the π electrons of the conjugated double bonds.

    A useful mnemonic device for quickly setting down the HMOs for cyclic systems is *Frost's circle*.[59] If a regular polygon of n sides is inscribed in a circle of diameter 4β with one corner at the lowest point, the points at which the corners of the polygon touch the circle define the energy levels. The energy levels obtained for benzene and cyclobutadiene with Frost's circle are shown in Fig. 1.12.

    The energy level diagrams for charged C_3H_3 and C_5H_5 systems are readily constructed and are presented in Fig. 1.13. Cyclopropenyl cation has a total of two π electrons, which occupy the bonding HMO, and a total π-electron energy of $2\alpha + 4\beta$. This gives a DE of 2β and is indicative of a stabilized species. Addition of two more π electrons to the system to give cyclopropenide anion requires population of higher-energy antibonding orbitals and results in a net destabilization of the molecule. The opposite is true for the C_5H_5 case, where the anionic species is stabilized more than the cation and the cation is predicted to have unpaired electrons.

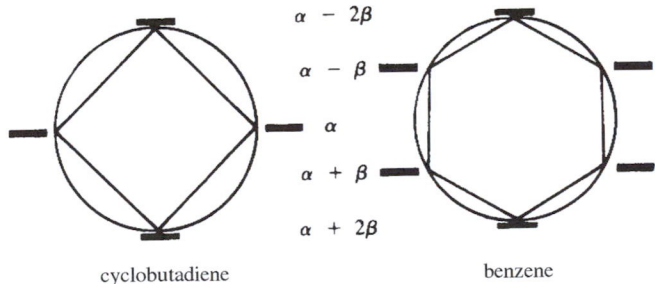

cyclobutadiene   benzene

Fig. 1.12. Energy level diagrams for cyclobutadiene and benzene, illustrating the application of Frost's circle.

59. A. A. Frost and B. Musulin, *J. Chem. Phys.* **21**:572 (1953).

Fig. 1.13. Energy level diagrams for C_3H_3 and C_5H_5 systems.

Monocyclic conjugated polyenes are referred to as *annulenes*, and there exists ample experimental evidence to support the conclusions based on application of HMO theory to neutral and charged annulenes. The relationship between stability and structure in cyclic conjugated systems will be explored more fully in Chapter 9.

While Hückel's $4n + 2$ rule applies only to monocyclic systems, HMO theory is applicable to many other systems. HMO calculations of fused-ring systems are carried out in much the same way as for monocyclic species and provide energy levels and atomic coefficients for the systems. The incorporation of heteroatoms is also possible. Because of the underlying assumption of orthogonality of the σ and π systems of electrons, HMO theory is restricted to planar molecules.

Although the Hückel method has now been supplanted by more complete treatments for theoretical analysis of organic reactions, the pictures of the π orbitals of both linear and cyclic conjugated polyene systems that it provides are correct as to symmetry and the relative energy of the orbitals. In many reactions where the π system is the primary site of reactivity, these orbitals correctly describe the behavior of the systems. For that reason, the reader should develop a familiarity with the qualitative description of the π orbitals of typical linear polyenes and conjugated cyclic hydrocarbons. These orbitals will be the basis for further discussion in Chapters 9 and 11.

1.5. Qualitative Application of Molecular Orbital Theory

As with valence bond theory, the full mathematical treatment of MO theory is too elaborate to apply to many situations. It is important to be able to develop qualitative approaches based on the fundamental concepts of MO theory that can be applied without the need for detailed calculations. A key tool for this type of analysis is a molecular orbital energy diagram. The construction of an approximate energy level diagram can be accomplished without recourse to detailed calculations by keeping some basic principles in mind. These principles can be illustrated by referring to some simple examples. Consider, first, diatomic species formed from atoms in which only the $1s$ orbitals are involved in the bonding scheme. The two $1s$ orbitals can combine in either a bonding or an antibonding manner to give two molecular orbitals, as indicated in Fig. 1.14.

The number of molecular orbitals (bonding + nonbonding + antibonding) is equal to the sum of the atomic orbitals in the basis set from which they are generated. The bonding combination is characterized by a positive overlap in which the coefficients are of like sign, while the antibonding combination is characterized by a negative overlap with coefficients of opposite sign.

A

B

Fig. 1.14. Graphic description of combination of two $1s$ orbitals to give two molecular orbitals for H_2 (A) and HHe^+ (B).

Orbitals are occupied by a maximum of two electrons, beginning with the orbital of lowest energy (the Aufbau principle). The number of electrons is determined by the number of electrons present on the interacting atoms. The orbitals in Fig. 1.14A could be applied to systems such as H_2^+ (one electron), H_2 (two electrons), He_2^+ (three electrons), or He_2 (four electrons). A reasonable conclusion would be that H_2 would be the most stable of these diatomic species because it has the largest net number of electrons in the bonding orbital (two). The He_2 molecule has no net bonding because the antibonding orbital contains two electrons and cancels the bonding contribution of the occupied bonding orbital. Both H_2^+ and He_2^+ have one more electron in bonding orbitals than in antibonding orbitals. These species have been determined to have bond energies of 61 and 60 kcal/mol, respectively. The bond energy of H_2, for comparison, is 103 kcal/mol.

A slight adjustment in the energy level diagram allows it to be applied to heteronuclear diatomic species such as HHe^+. The diagram that results from this slight modification is shown in Fig. 1.14B. Rather than being a symmetrical diagram, this diagram shows the He $1s$ level to be lower than the H $1s$ level, owing to the increased nuclear charge on helium. Calculations for the HHe^+ ion indicate a bond energy of 43 kcal/mol.[60]

The second-row elements, including carbon, oxygen, and nitrogen, involve p atomic orbitals as well as $2s$ orbitals. An example of a heteronuclear diatomic molecule involving these elements is carbon monoxide, CO. The carbon monoxide molecule has 14 electrons, and the orbitals for each atom are $1s$, $2s$, $2p_x$, $2p_y$, and $2p_z$. For most chemical purposes, the carbon $1s$ and oxygen $1s$ electrons are ignored. This simplification is valid because the energy gap between the $1s$ and $2s$ levels is large and the effect of the $1s$ levels on the valence electrons is very small. The 10 valence electrons are distributed among eight MOs generated by combining the four valence AOs from carbon with the four from oxygen, as illustrated in Fig. 1.15.

Figure 1.16 illustrates in a qualitative way the interactions between the AOs that give rise to the MOs. Each pair of AOs leads to a bonding and an antibonding combination. The $2s$ orbitals give the σ and σ^* orbitals. The $2p_x$ and $2p_y$ combinations form MOs that are π in character. The $2p_z$ combination gives a σ-type orbital labeled σ' as well as the corresponding antibonding orbital. The lower five orbitals are each doubly occupied, accounting for the 10 valence-shell electrons in the molecule. Of these five occupied orbitals, one is antibonding; thus, the net number of bonding electrons is six, consistent with the triple bond found in the Lewis structure for carbon monoxide. The shapes of the

60. H. H. Michels, *J. Chem. Phys.* **44**:3834 (1966).

Fig. 1.15. Energy levels in the carbon monoxide molecule. (Adapted from H. B. Gray and G. P. Haight, *Basic Principles of Chemistry*, W. A. Benjamin, New York, 1967, p. 289.)

molecular orbitals can also be depicted as in Fig. 1.17. Here the nodes in the MOs are represented by a change from solid to dashed lines, and the sizes of the lobes are scaled to represent the distribution of the orbital. One gains from these pictures an impression of the distortion of the bonding orbital toward oxygen as a result of the greater electronegativity of the oxygen atom.

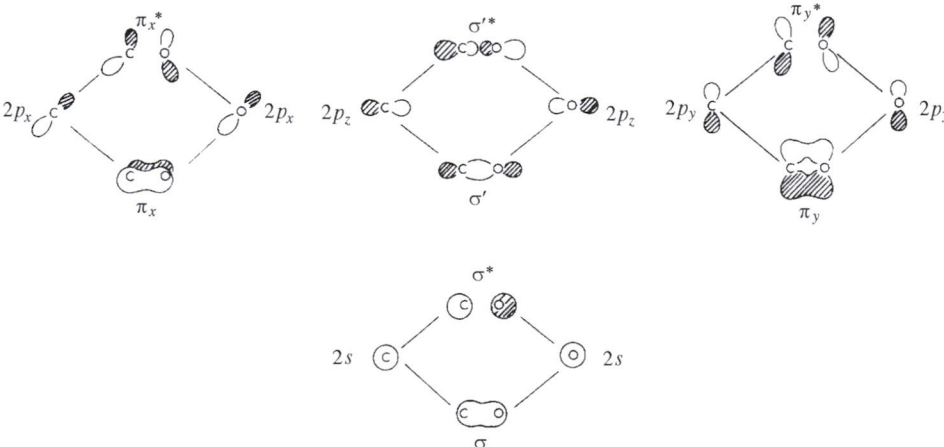

Fig. 1.16. Interaction of atomic orbitals of carbon and oxygen leading to molecular orbitals of carbon monoxide.

σ'* E=0.2755

π$_x^*$ E=0.1268

π$_y^*$ E=0.1268

σ' E=−0.5544

π$_x$ E=−0.6395

π$_y$ E=−0.6395

σ* E=−0.8038

σ E=−1.5210

Fig. 1.17. Representation of the molecular orbitals of carbon monoxide. Energies are given in atomic units (1 a.u. 27.21 eV). (From W. L. Jorgensen and L. Salem, *The Organic Chemist's Book of Orbitals*, Academic Press, New York, 1973. Reproduced with permission.)

Just as we were able to state some guiding rules for application of resonance theory, it is possible to state some conditions by which to test the correctness of an MO energy level diagram derived by qualitative considerations.

a. The total number of MOs must equal the number of AOs from which they were constructed.
b. The symmetry of the MOs must conform to the symmetry of the molecule. That is, if a molecule possesses a plane of symmetry, for example, *all* the MOs must be either symmetric (unchanged) or antisymmetric (unchanged except for sign) with respect to that plane.
c. AOs that are orthogonal to one another do not interact. Thus, two different carbon $2p$ orbitals will not contribute to the same MO.
d. The energies of similar AOs (s or p) are lower for elements of higher electronegativity. MOs will reflect these differences in energy.
e. The relative energy of MOs in a molecule increases with the number of nodes in the orbital.

By applying these rules and recognizing the elements of symmetry present in the molecule, it is possible to construct MO diagrams for more complex molecules. In the succeeding paragraphs, the MO diagrams of methane and ethylene are constructed on the basis of these kinds of considerations.

To provide a basis for comparison, Fig. 1.18 gives the results of an *ab initio* calculation on the methane molecule.[61] This particular calculation used as a basis set the $1s$, $2s$, and three $2p$ orbitals of carbon and the $1s$ orbitals of the four hydrogens. The lowest MO is principally $1s$ in character. A significant feature of this and other MO calculations of methane is that, unlike a picture involving localized bonds derived from sp^3 hybrid carbon orbitals, there are not four equivalent orbitals. We can obtain an understanding of this feature of the MO picture by a qualitative analysis of the origin of the methane MOs. We will consider the orbitals to be derived from the carbon $2s$, $2p_x$, $2p_y$, and $2p_z$ orbitals and ignore the carbon $1s$ orbital. The most convenient frame of reference for the tetrahedral methane molecule is a cube with hydrogen atoms at alternate corners and the carbon atom centered in the cube, as shown in Fig. 1.19. This orientation of the molecule reveals that methane possesses three twofold symmetry axes, one each along the x, y, and z axes. Because of this molecular symmetry, the proper MOs of methane must possess symmetry with respect to these same axes. There are two possibilities: the orbital may be unchanged by $180°$ rotation about the axis (symmetric), or it may be transformed into an orbital of identical shape but opposite sign by the symmetry operation (antisym-

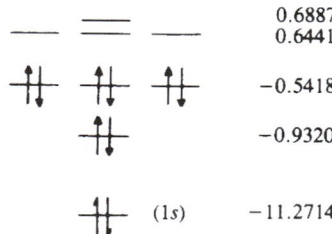

Fig. 1.18. Molecular orbital energy diagram for methane. Energies are in atomic units.

61. W. E. Palke and W. N. Lipscomb, *J. Am. Chem. Soc.* **88**:2384 (1966).

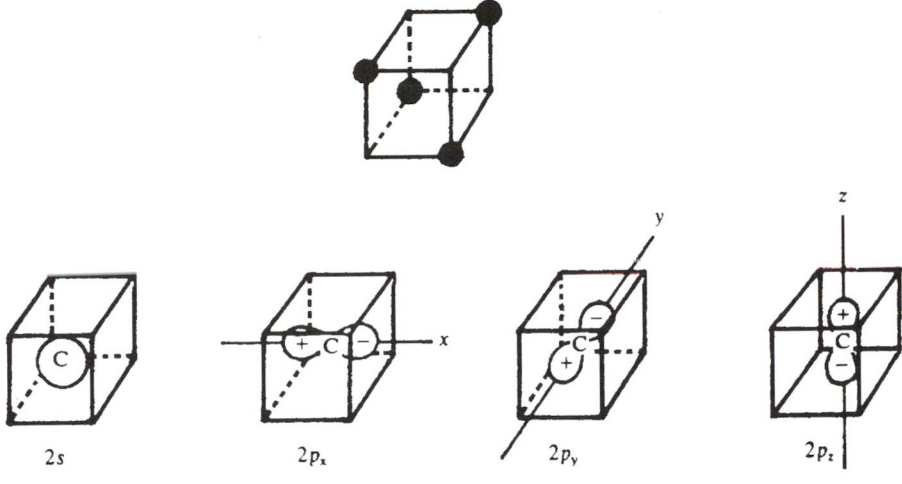

Fig. 1.19. Atomic orbitals of carbon relative to methane in a cubic frame of reference.

metric). The carbon $2s$ orbital is symmetric with respect to each axis, but the three $2p$ orbitals are each antisymmetric with respect to two of the axes and symmetric with respect to one. The combinations that give rise to MOs that meet these symmetry requirements are shown in Fig. 1.20.

The bonding combination of the carbon $2s$ orbital with the four $1s$ hydrogen orbitals leads to an MO that encompasses the entire molecule and has no nodes. Each of the MOs derived from a carbon $2p$ orbital has a node at carbon. The three combinations are equivalent, but higher in energy than the MO with no nodes. The four antibonding orbitals arise from similar combinations, but with the carbon and hydrogen orbitals having opposite signs in the region of overlap. Thus, the MO diagram arising from these considerations shows one bonding MO with no nodes and three degenerate (having the same energy) MOs with one node. The diagram is given in Fig. 1.21. Note that except for

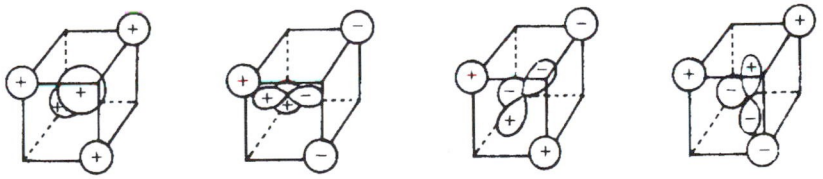

Fig. 1.20. Atomic orbital combinations giving rise to bonding molecular orbitals for methane.

three identical
orbitals (one
node)

single unique
orbital (no nodes)

Fig. 1.21. Qualitative molecular orbital diagram for methane.

inclusion of the $1s$ orbital, this qualitative picture corresponds to the calculated orbital diagram in Fig. 1.18.

A qualitative approach cannot assign energies to the orbitals. In many cases, it is, however, possible to assign an ordering of energies. The relationship between relative energy and number of nodes has already been mentioned. In general, σ-type orbitals are lower in energy than π-type orbitals because of this factor. Conversely, antibonding σ^* orbitals are higher in energy than antibonding π^* orbitals. Orbitals derived from more electronegative atoms are lower in energy than those derived from less electronegative atoms.

The process of constructing the MOs of ethylene is similar to that used for carbon monoxide, but the total number of AOs is greater, 12 instead of 8, because of the additional AOs from hydrogen. We must first define the symmetry of ethylene. Ethylene is known from experiment to be a planar molecule.

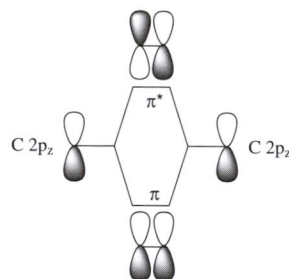

This geometry possesses three important elements of symmetry, the molecular plane and two planes that bisect the molecule. All MOs must be either symmetric or antisymmetric with respect to each of these symmetry planes. With the axes defined as in the diagram above, the orbitals arising from carbon $2p_z$ have a node in the molecular plane. These are the familiar π and π^* orbitals.

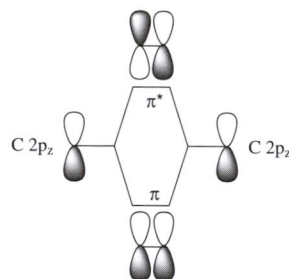

The π orbital is symmetric with respect to both the x–z plane and the y–z plane. It is antisymmetric with respect to the molecular (x–y) plane. On the other hand, π^* is antisymmetric with respect to the y–z plane.

The remaining AOs are the four H $1s$, two C $1s$, and four C $2p$ orbitals. All lie in the molecular plane. Only two combinations of the C $2s$ and H $1s$ orbitals meet the molecular symmetry requirements. One of these, σ, is bonding between all atoms, whereas σ^* is antibonding between all nearest-neighbor atoms. No other combination corresponds to the symmetry of the ethylene molecule.

Let us next consider the interaction of $2p_y$ with the four hydrogen $1s$ orbitals. There are four possibilities that conform to the molecular symmetry:

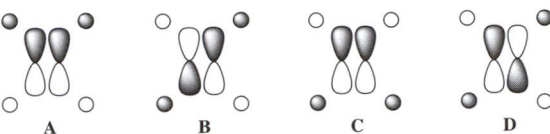

Orbital **A** is bonding between all nearest-neighbor atoms, whereas **B** is bonding within the CH_2 units but antibonding with respect to the two carbons. The orbital labeled **C** is C–C bonding but antibonding with respect to the hydrogens. Finally, orbital **D** is antibonding with respect to all nearest-neighbor atoms. Similarly, when the $2p_x$ orbitals are considered, four possible combinations arise. Notice that the nature of the overlap of the $2p_x$ orbitals is different from that in the $2p_y$ case, so that the two sets of MOs should have different energies.

The final problem in construction of a qualitative MO diagram for ethylene is the relative placement of the orbitals. There are some guidelines which are useful. First, because π-type interactions are normally weaker than σ-type, we expect the separation between σ and σ^* to be greater than that between π and π^*. Within the sets **ABCD** and **EFGH**, the orbitals can be placed in the order $A < B < C < D$ and $E < F < G < H$ on the basis that C–H bonding interactions will outweigh C–C antibonding interactions arising from relatively weak p–p overlaps. The placement of the set **ABCD** in relation to **EFGH** is not qualitatively obvious. Calculations give the results shown in Fig. 1.22.[62] Pictorial representations of the orbitals are given in Fig. 1.23.

The kind of qualitative considerations that have been used to construct the ethylene MO diagram do not give an indication of how much each AO contributes to the individual

D	——	0.89
σ^*	——	0.84
H	——	0.63
G	——	0.62
C	——	0.59
π^*	——	0.24
π	——	-0.37
B	——	-0.51
F	——	-0.56
A	——	-0.64
E	——	-0.78
σ	——	-1.0

Fig. 1.22. Ethylene molecular orbital energy levels. Energies are given in atomic units.

62. W. L. Jorgensen and L. Salem, *The Organic Chemists's Book of Orbitals*, Academic Press, New York, 1973.

MOs. This information is obtained from the coefficients provided by an MO calculation. Without these coefficients, we cannot specify the shapes of the MOs very precisely. However, the qualitative ideas do permit conclusions about the *symmetry* of the orbitals. As will be seen in Chapter 11, just knowing the symmetry of the MOs provides very useful insight into many chemical reactions.

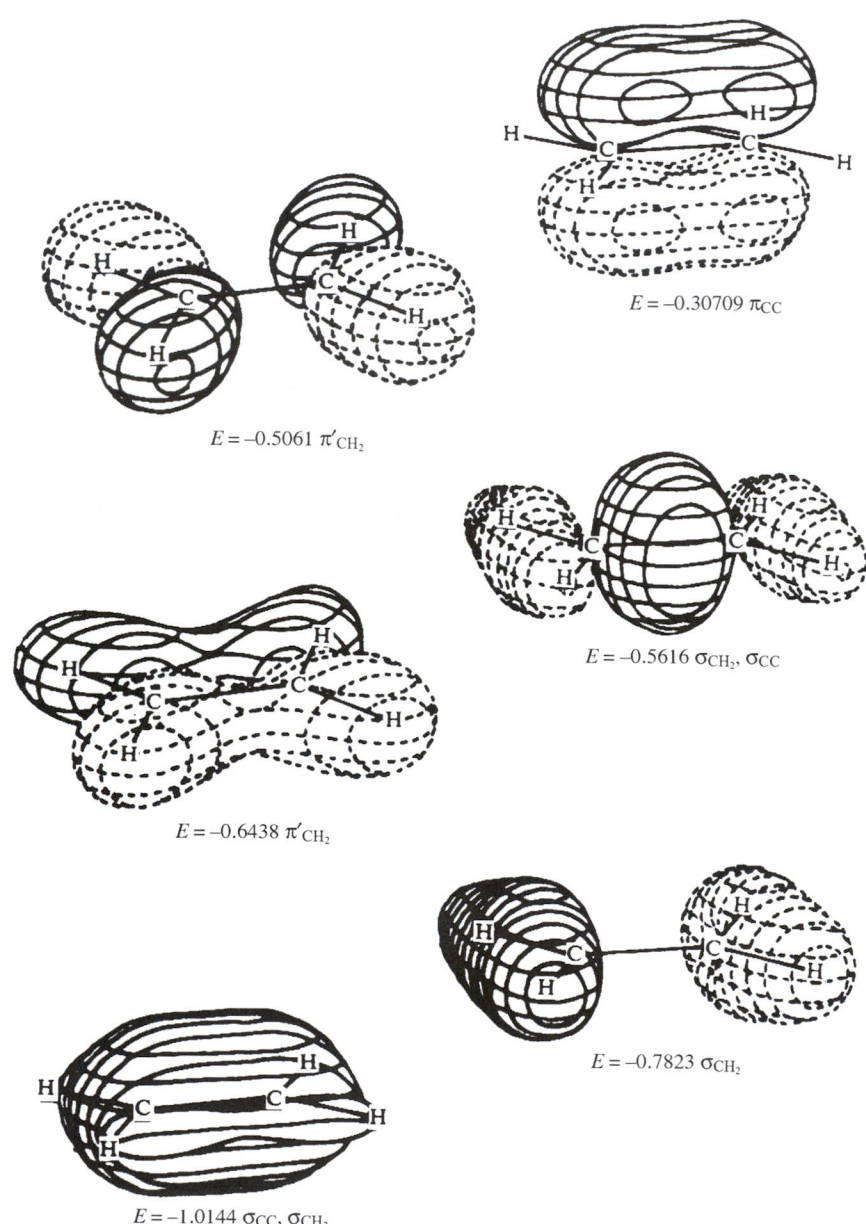

$E = -0.30709 \, \pi_{CC}$

$E = -0.5061 \, \pi'_{CH_2}$

$E = -0.5616 \, \sigma_{CH_2}, \sigma_{CC}$

$E = -0.6438 \, \pi'_{CH_2}$

$E = -0.7823 \, \sigma_{CH_2}$

$E = -1.0144 \, \sigma_{CC}, \sigma_{CH_2}$

Fig. 1.23. Representation of the molecular orbitals of ethylene. (From W. L. Jorgensen and L. Salem, *The Organic Chemist's Book of Orbitals*, Academic Press, New York, 1973. Reproduced with permission.)

$E = 0.8917\ \pi^*_{CH_2}$

$E = 0.8453\ \sigma^*_{CC},\ \sigma^*_{CH_2}$

$E = 0.6395\ \sigma^*_{CH_2},\ \sigma^*_{CC}$

$E = 0.6206\ \sigma^*_{CH_2}$

$E = 0.5868\ \sigma^*_{CH_2}$

$E = 0.2426\ \pi^*_{CC}$

Fig. 1.23. (*continued*)

1.6. Application of Molecular Orbital Theory to Reactivity

The construction of MO diagrams under the guidance of the general principles and symmetry restrictions which have been outlined can lead to useful insights into molecular structure. Now we need to consider how these structural concepts can be related to reactivity. In valence bond terminology, structure is related to reactivity in terms of substituent effects. The impact of polar and resonance effects on the electron distribution and stability of reactants, transition states, and intermediates is assessed. In MO theory, reactivity is related to the relative energies and shapes of the orbitals that are involved as the reactants are transformed to products. Reactions which can take place through relatively stable transition states and intermediates are more favorable and faster than reactions which proceed through less stable ones. The *symmetry* of the MOs is a particularly important feature of many analyses of reactivity based on MO theory. The shapes of orbitals also affect the energy of reaction processes. Orbital shapes are quantified by the atomic coefficients. The strongest interaction (bonding when the overlapping orbitals have the same sign) occurs when the orbitals on two reaction centers have high coefficients on the atoms which undergo bond formation.

The qualitative description of reactivity in MO terms must begin with a basic understanding of the MOs of the reacting systems. At this point, we have developed a familiarity with the MOs of ethylene and conjugated unsaturated systems from the discussion of Hückel MO theory and the construction of the ethylene MOs from AOs. To apply these ideas to new systems, we need to be able to understand how a change in structure will affect the MOs. One approach to this problem is called *perturbation molecular orbital theory* or PMO for short.[63] In this approach, a system under analysis is compared to another related system for which the MO pattern is known, and the MO characteristics of the new system are deduced by analyzing how the change in structure affects the MO pattern. The types of changes that can be handled in a qualitative way are substitution of atoms by other elements, with the resulting change in electronegativity and changes in connectivity that revise the pattern of direct orbital overlap. The fundamental thesis of PMO theory is that the resulting changes in the MO energies are relatively small and can be treated as adjustments on the original MO system.

Another aspect of qualitative application of MO theory is the analysis of interactions of the orbitals in reacting molecules. As molecules approach one another and reaction proceeds, there is a mutual perturbation of the orbitals. This process continues until the reaction is complete and the new product (or intermediate in a multistep reaction) is formed. PMO theory incorporates the concept of *frontier orbital control*. This concept proposes that the most important interactions will be between a particular pair of orbitals.[64] These orbitals are the highest filled orbital of one reactant (the HOMO, highest occupied molecular orbital) and the lowest unfilled (LUMO, lowest unoccupied molecular orbital) orbital of the other reactant. The basis for concentrating attention on these two orbitals is that they will be the closest in energy of the interacting orbitals. A basic postulate of PMO

63. C. A. Coulson and H. C. Longuet-Higgins, *Proc. R. Soc. London, Ser. A* **192**:16 (1947); L. Salem, *J. Am. Chem. Soc.* **90**:543 (1968); M. J. S. Dewar and R. C. Dougherty, *The PMO Theory of Organic Chemistry*, Plenum Press, New York, 1975; G. Klopman, *Chemical Reactivity and Reaction Paths*, Wiley-Interscience, New York, 1974, Chapter 4.

64. K. Fukui, *Acc. Chem. Res.* **4**:57 (1971); I. Fleming, *Frontier Orbital and Organic Chemical Reactions*, John Wiley & Sons, New York, 1976; L. Salem, *Electrons in Chemical Reactions*, John Wiley & Sons, New York, 1982, Chapter 6.

theory is that interactions are strongest between orbitals that are close in energy. Frontier orbital theory proposes that these strong initial interactions will guide the course of the reaction as it proceeds to completion. A further general feature of MO theory is that only MOs of matching symmetry can interact so as to lead to bond formation. Thus, analysis of a prospective reaction path will direct attention to the *relative energy* and *symmetry* of the interacting orbitals.

These ideas can be illustrated here by considering some very simple cases. Let us consider the fact that the double bonds of ethylene and formaldehyde have very different chemical reactivities. Formaldehyde reacts readily with nucleophiles whereas ethylene does not. The π bond in ethylene is more reactive toward electrophiles than the formaldehyde C=O π bond. We have already described the ethylene MOs in Figs. 1.22 and 1.23. How will those of formaldehyde differ? In the first place, the higher atomic number of oxygen provides two additional electrons so that in place of the CH_2 group of ethylene, the oxygen of formaldehyde has two pairs of nonbonding electrons. The key change, however, has to do with the frontier orbitals, the π (HOMO) and π^* (LUMO) orbitals. These are illustrated in Fig. 1.24. One significant difference between the two molecules is the lower energy of the π and π^* orbitals in formaldehyde. These are lower in energy than the corresponding ethylene orbitals because they are derived in part from the lower-lying (more electronegative) $2p_z$ orbital of oxygen. Because of its lower energy, the π^* orbital is a better acceptor of electrons from the HOMO of any attacking nucleophile than is the LUMO of ethylene. On the other hand, we also can see why ethylene is more reactive toward electrophiles than formaldehyde. In electrophilic attack, it is the HOMO that is involved as an electron donor to the approaching electrophile. In this case, the fact that the HOMO of ethylene lies higher in energy than the HOMO of formaldehyde will mean that electrons can be more easily attracted by the approaching electrophile. The unequal electronegativities of the oxygen and carbon atoms also distort the π-MOs. Whereas the π-MO has a symmetrical distribution in ethylene, the formaldehyde π-MO has a higher atomic coefficient at oxygen. This results in a net positive charge in the vicinity of the carbon atom, which is a favorable circumstance for approach by a nucleophilic reactant.

One principle of PMO theory is that the degree of perturbation is a function of the degree of overlap of the orbitals. Thus, in the qualitative application of MO theory, it is important to consider the shapes of the orbitals (as indicated quantitatively by their atomic coefficients) and the proximity that can be achieved within the limits of the geometry of the reacting molecules. Secondly, the strength of the interaction between orbitals depends on their relative energy. The closer the orbitals are in energy, the greater will be their mutual interaction. This principle, if used in conjunction with reliable estimates of relative orbital energies, can be of value in predicting the relative importance of various possible interactions.

Fig. 1.24. Relative energy of the π and π^* orbitals in ethylene and formaldehyde.

Let us illustrate these ideas by returning to the comparisons of the reactivity of ethylene and formaldehyde toward a nucleophilic species and an electrophilic species. The perturbations which arise as both a nucleophile and an electrophile approach are sketched in Fig. 1.25. The electrophilic species E^+ must have a low-lying empty orbital. The strongest interaction will be with the ethylene π orbital, and this leads to a stabilizing effect on the complex because the electrons are located in an orbital that is stabilized. The same electrophilic species would lie further from the π orbital of formaldehyde because the formaldehyde orbitals are shifted to lower energy. As a result, the mutual interaction with the formaldehyde HOMO will be weaker than in the case of ethylene. The conclusion is that such an electrophile will undergo a greater stabilizing attraction on approaching within bonding distance of ethylene than on approaching within bonding distance of formalde-hyde. In the case of Nu^-, a strong bonding interaction with π^* of formaldehyde is possible (Fig. 1.25D). In the case of ethylene, there may be a stronger interaction with the HOMO, but this is a destabilizing interaction because both orbitals are filled and the lowering of one orbital is canceled by the raising of the other. Thus, we conclude that a nucleophile with a high-lying HOMO will interact more favorably with formaldehyde than with ethylene.

The ideas of MO theory can be used in a slightly different way to describe substituent effects. Let us consider, for example, the effect of a π-donor substituent or a π-acceptor substituent on the MO levels of ethylene and upon the reactivity of substituted ethylenes. We can take the amino group as an example of a π-donor substituent. The nitrogen atom provides an additional $2p_z$ orbital and two electrons to the π system. The π orbitals of aminoethylene are very similar to those of an allyl anion but with some distortion because the system is no longer symmetrical. The highest charge density is on the terminal atoms, that is, the nitrogen atom and the β carbon, since the HOMO has a node at the center carbon. Furthermore, the HOMO is considerably higher in energy than the HOMO in ethylene. The HOMO in aminoethylene resembles ψ_2 of the allyl anion. It is not quite so high in energy as the allyl ψ_2 because of the greater electronegativity of the nitrogen atom, but is substantially higher than the HOMO of ethylene. Thus, we expect aminoethylene, with its high-lying HOMO, to be more reactive toward electrophiles than ethylene. Furthermore, the HOMO has the highest coefficients on the terminal atoms, so we expect an electrophile to become bonded to the β carbon or nitrogen, but not to the α carbon. On the other hand, the LUMO will now correspond to the higher-energy ψ_3 of the

Interaction of ethylene frontier orbitals with E^+ and Nu^-

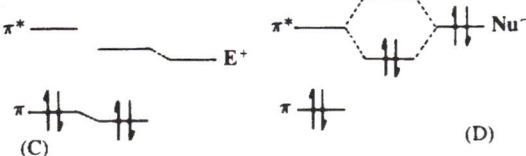

Interaction of formaldehyde frontier orbitals with E^+ and Nu^-

Fig. 1.25. PMO description of interaction of ethylene and formaldehyde with an electrophile (E^+) and a nucleophile (Nu^-).

allyl anion, so we expect aminoethylene to be even less reactive toward nucleophiles than is ethylene.

π MO energy levels for ethylene with a π-donor substituent.

An example of a π-acceptor group is the formyl group as in acrolein.

$$CH_2=CHCH=O$$

In this case, the π-MOs resemble those of butadiene. Relative to the butadiene orbitals, however, the acrolein orbitals lie somewhat lower in energy because of the effect of the more electronegative oxygen atom. This factor also increases the electron density at oxygen relative to carbon.

π MO energy levels for ethylene with a π-acceptor substituent.

The LUMO, which is the frontier orbital in reactions with nucleophiles, has a larger coefficient on the β-carbon atom, whereas the two occupied orbitals are distorted in such a way as to have larger coefficients on oxygen. The overall effect is that the LUMO is relatively low-lying and has a high coefficient on the β-carbon atom. The frontier orbital theory therefore predicts that nucleophiles will react preferentially at the β-carbon atom.

MO calculations at the 6-31G** level have been done on both acrolein and aminoethylene. The resulting MOs were used to calculate charge distributions. Figure 1.26 gives the π-electron densities calculated for butadiene, acrolein, and aminoethylene.[65] Inclusion of the hydrogen and σ orbitals leads to overall charges as shown. These charge distributions result from σ polarization which is counter to the π polarization.

Notice that the MO picture gives the same qualitative picture of the substituent effects as described by resonance structures. The amino group is pictured by resonance as an electron donor which causes a buildup of electron density at the β carbon, whereas the formyl group is an electron acceptor which diminishes electron density at the β carbon.

$$CH_2=CH-\ddot{N}H_2 \longleftrightarrow {}^-CH_2-CH=NH_2{}^+ \qquad CH_2=CH-CH=O \longleftrightarrow {}^+CH_2-CH=CH-O^-$$

65. K. B. Wiberg, R. E. Rosenberg, and P. R. Rablen, *J. Am. Chem. Soc.* **113**:2890 (1991).

Fig. 1.26. Charge distribution in butadiene, acrolein, and aminoethylene. Data are from K. B. Wiberg, R. E. Rothenberg, and P. R. Rablen, *J. Am. Chem. Soc.* **113**:2890 (1991).

The chemical reactivity of these two substituted ethylenes is in agreement with the ideas encompassed by both the MO and resonance descriptions. Enamines, as amino-substituted alkenes are called, are very reactive toward electrophilic species, and it is the β carbon that is the site of attack. For example, enamines are protonated on the β carbon. Acrolein is an electrophilic alkene, as predicted, and the nucleophile attacks the β carbon.

Both MO theory and experimental measurements provide a basis for evaluation of the energetic effects of conjugation between a double bond and adjacent substituents. Table 1.16 gives some representative values. The theoretical values ΔE are for the isodesmic reaction

$$CH_2=CH-X + CH_4 \rightarrow CH_3-X + CH_2=CH_2$$

and are calculated at the *ab initio* 4-31G level of theory. These values refer to the gas phase. The ΔH values are based on the experimentally determined thermodynamic ΔH_f of the compounds. Notice that both electron-withdrawing and electron-accepting substituents result in a net stabilization of the conjugated system. This stabilization results from the lowering in energy of the lowest-lying MO in each case. The effect on the HOMO is

Table 1.16. Stabilization Resulting from Conjugation of Ethylene with Substituents[a]

Substituent	ΔE (kcal/mol)	ΔH (kcal/mol)	Substituent	ΔE (kcal/mol)	ΔH (kcal/mol)
H	0	0	NH_2	13.3	13.3
F	6.4	6.7 (3.4)	CN	3.3	4.8
CH_3	4.3	5.4	$COCH_3$	4.0	10.5 (3.8)
OCH_3	10.9	12.3 (4.9)	CO_2CH_3	8.0	11.9 (3.4)

a. From A. Greenberg and T. A. Stevenson, *J. Am. Chem. Soc.* **107**:3488 (1985). Values in parentheses are from J. Hine and M. J. Skoglund, *J. Org. Chem.* **47**:4766 (1982) and are based on experimental equilibrium measurement values. These measurements are in solution, and the difference between the two sets of values may reflect the effect of solvent.

different for electron-withdrawing as compared with electron-accepting substituents. For donor substituents, the HOMO is raised in energy, relative to the HOMO in ethylene. For electron-accepting substituents, it is lowered relative to the HOMO in ethylene.

Frontier orbital theory also provides the basic framework for analysis of the effect that the symmetry of orbitals has upon reactivity. One of the basic tenets of MO theory is that the symmetries of two orbitals must match to permit a strong interaction between them. This symmetry requirement, when used in the context of frontier orbital theory, can be a very powerful tool for predicting reactivity. As an example, let us examine the approach of an allyl cation and an ethylene molecule and ask whether the following reaction is likely to occur.

The positively charged allyl cation would be expected to be the electron acceptor in any initial interaction with ethylene. Therefore, to consider this reaction in terms of frontier orbital theory, the question we need to answer is, "do the ethylene HOMO and allyl cation LUMO interact favorably as the reactants approach one another?" The orbitals that are involved are shown in Fig. 1.27. If we analyze a symmetrical approach, which would be necessary for the simultaneous formation of the two new bonds, we see that the symmetries of the two orbitals do not match. Any bonding interaction developing at one end would be canceled by an antibonding interaction at the other end. The conclusion that is drawn from this analysis is that this particular reaction process is not favorable. We would need to consider other modes of approach to analyze the problem more thoroughly, but this analysis indicates that simultaneous (concerted) bond formation between ethylene and an allyl cation to form a cyclopentyl cation is not possible.

Fig. 1.27. MOs for ethylene and allyl cation.

Let us now consider another reaction, this time between the allyl cation and butadiene. Again, the assumption will be made that it is the frontier π orbitals that will govern the course of the reaction. We will also be slightly more formal about the issue of symmetry. This can be done by recognizing the elements of symmetry that would be maintained as the reaction proceeds. If the reaction is to proceed in a single step, the geometry must permit simultaneous overlap of the orbitals on the carbons where new bonds are being formed. A geometry of approach that permits such a simultaneous overlap is shown below.

The allyl cation could approach from the the top (or bottom) of the *s-cis* conformation of butadiene, and the new bonds would be formed from the π orbitals. This arrangement would maintain a plane of summetry during the course of the reaction. The plane bisects butadiene between C-2 and C-3 and the allyl cation at C-2. The orbitals can be classified as symmetric (S) or antisymmetric (A) with respect to this plane. This gives rise to the MO diagram shown in Fig. 1.28. Because strong interactions will occur only between orbitals of the same symmetry, the mutual perturbation of the approaching reactants will affect the orbital energy levels as shown in the diagram. As in all such perturbations, one orbital of the interacting pair will be stabilized and the other will move to higher energy. The

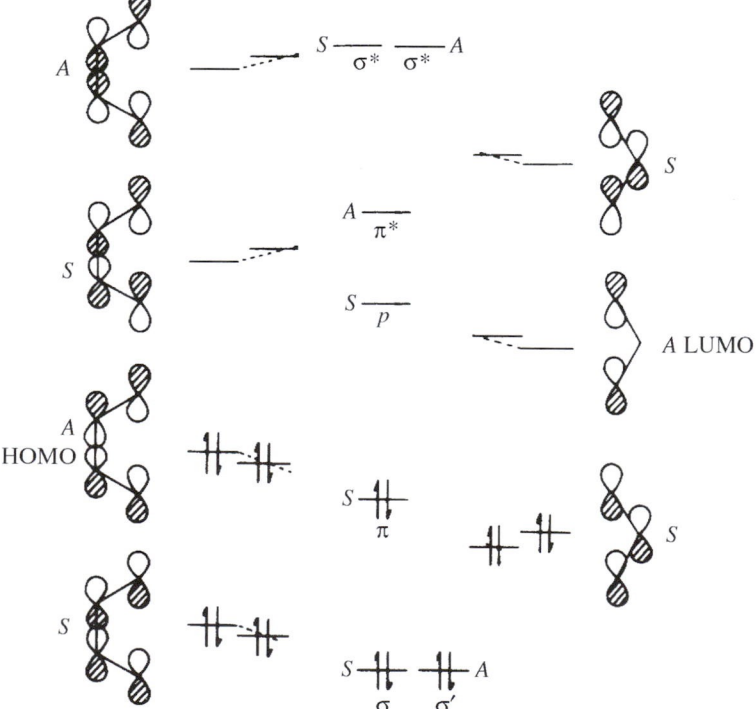

Fig. 1.28. MO diagram showing mutual perturbation of MOs of butadiene and allyl cation.

perturbed orbitals at some point on the way to the transition state are shown in the diagram. Eventually, when the reaction has proceeded to completion, a new set of orbitals belonging to the product will have been formed. These are shown in the center of the diagram, but we will be considering only the initial perturbed set. The lowest-lying π orbitals of both butadiene and the allyl cation are filled. These will interact, with one moving down in energy and the other up. These two changes in energy are partially compensating, with the total energy change being a net increase in the energy of the system. Both the HOMO and LUMO are antisymmetric and will interact strongly, but in this case, because only two electrons are involved, only the energy of the stabilized orbital will affect the total energy since the destabilized orbital is empty. This HOMO–LUMO interaction then contributes a net bonding contribution as the transition state is approached. From this analysis, we conclude that there is the possibility of a favorable bonding interaction between the two reactant species. Notice that the reaction is only *permitted* and that nothing can be said about its actual rate or position of equilibrium on the basis of the analysis given. Such matters as steric hindrance to approach of the reactants and the geometric requirements for satisfactory overlap of the orbitals could still cause the reaction to proceed with difficulty. The analysis does establish, however, that there is a pathway by which the orbitals of the reactants can interact in a way that is favorable for reaction.

A more complete analysis of interacting molecules would examine all of the involved MOs in a similar way. A *correlation diagram* would be constructed to determine which reactant orbital is transformed into which product orbital. Reactions which permit smooth transformation of the reactant orbitals to product orbitals without intervention of high-energy transition states or intermediates can be identified in this way. If no such transformation is possible, a much higher activation energy is likely since the absence of a smooth transformation implies that bonds must be broken before they can be re-formed. This treatment is more complete than the frontier orbital treatment because it focuses attention not only on the reactants but also on the products. We will describe this method of analysis in more detail in Chapter 11. The qualitative approach that has been described here is a useful and simple way to apply MO theory to reactivity problems, and we will employ it in subsequent chapters to problems in reactivity that are best described in MO terms.

It is worth noting that in the case of the reactions of ethylene and butadiene with the allyl cation, the MO description has provided a prediction that would not have been recognized by a pictorial application of valence bond terminology. Thus, we can write an apparently satisfactory description of both reactions.

It is only on considering the symmetry of the interacting orbitals that we find reason to suspect that only the second of the two reactions is possible.

1.7. Interactions between σ and π Systems—Hyperconjugation

One of the key assumptions of the Hückel approximation is the noninteracton of the π-orbital system with the σ-molecular framework. This is a good approximation for planar π molecules in which the σ framework is in the nodal plane of the π system. For other molecules, as for example when an sp^3 carbon is added as a substituent group, this approximation is no longer entirely valid. Qualitative application of MO theory can be enlightening in describing interactions between the π system and substituent groups. In valence bond theory, a special type of resonance called *hyperconjugation* is used to describe such interactions. For example, much chemical and structural evidence indicates that alkyl substituents on a carbon–carbon double bond act as electron donors to the π system. In valence bond language, "no bond" resonance structures are introduced to indicate this electronic interaction.

The MO picture of such interactions flows from the idea that individual orbitals encompass the entire molecule. Thus, while the MO description of ethylene involved no interaction between the C $2p_z$ orbitals and the H $1s$ orbitals (see p. 43 to recall this discussion), this strict separation would not exist in propene because the hydrogens of the methyl group are not in the nodal plane of the π bond. The origin of interactions of these hydrogens with the π orbital can be indicated as in Fig. 1.29, which shows propene in a geometry in which two of the hydrogen $1s$ AOs are in a position to interact with the $2p_z$ orbital of carbon 2. An *ab initio* calculation using a STO-3G basis set was carried out on propene in two distinct geometries, eclipsed and staggered.

The calculations of the optimum geometry show a slight lenghthening of the C−H bonds because of the electron release to the π system. These calculations also reveal a barrier to rotation of the methyl group of about 1.5–2.0 kcal/mol. Interaction between the hydrogens and the π system favors the eclipsed conformation to this extent.[66] Let us examine the

Fig. 1.29. Interactions between two hydrogen $1s$ orbitals and carbon $2p_z$ orbitals stabilize the eclipsed conformation of propene.

66. W. J. Hehre, J. A. Pople, and A. J. P. Devaquet, *J. Am. Chem. Soc.* **98**:664 (1976); A. Pross, L. Radom, and N. V. Riggs, *J. Am. Chem. Soc.* **102**:2253 (1980); K. B. Wiberg and E. Martin, *J. Am. Chem. Soc.* **107**:5035 (1985); A. E. Dorigo, D. W. Pratt, and K. N. Houk, *J. Am. Chem. Soc.* **109**:6589 (1987).

reason for the preference for the eclipsed conformation. This issue can be approached by analyzing the interactions between the carbon $2p_z$ orbitals and the CH_3 fragment in a little more detail. The bonding and antibonding combinations that arise from interaction of the appropriate CH_3-π and CH_3-π^* orbitals with the $2p_z$ orbitals are shown in Fig. 1.30. The strongest interaction is a repulsive one between the filled CH_3-π and C=C-π orbitals. It is this interaction which is primarily responsible for the favored eclipsed conformation. The eclipsed structure minimizes the repulsion by maximizing the separation between the hydrogens and the π bond. The second interaction is the stabilizing hyperconjugative one between CH_3-π and C=C-π^*. This is a bonding interaction because π^* is an empty orbital and can accept electron density from CH_3-π. It is this bonding interaction which transfers electron density from the methyl group to the terminal carbon of the double bond. Notice that there is a correspondence between the MO picture and the valence bond resonance structure in that both specify a net transfer of electron density from C−H bonds to the π system with a net strengthening of the bond between C-2 and C-3 but a weakening of the C(1)−C(2) π bond.

hyperconjugation

repulsive interaction between CH_3-π and π

attractive interaction between CH_3-π and π^*

One of the fundamental structural facets of organic chemistry, which has been explained most satisfactorily in MO terms, is the existence of a small barrier to rotation about single bonds. In ethane, for example, it is known that the staggered conformation is about 3 kcal/mol more stable than the eclipsed conformation so that the eclipsed conformation represents a transition state for transformation of one staggered conformation into another by rotation.

staggered

eclipsed

Valence bond theory offers no immediate qualitative explanation since the σ bond that is involved is cylindrically symmetrical. A steric argument based on repulsions between hydrogens also fails because on detailed examination of this hypothesis, it is found that the

Fig. 1.30. Interactions between CH_3-π and CH_3-π^* orbitals and carbon $2p_z$ orbitals.

hydrogens are too small and too distant from one another to account for the observed energy. MO ideas, however, succeed in correctly predicting and calculating the magnitude of the ethane rotational barrier.[67] The origin of the barrier is a repulsive interaction between the filled C−H orbitals which is maximal in the eclipsed geometry.

The interaction can be further examined by consideration of the ethane MOs.[68] Because ethane contains two carbon atoms and six hydrogens, the MOs are constructed from six H $1s$, two C $2s$, and six C $2p$ orbitals. Figure 1.31 depicts the seven bonding MOs, assuming the staggered geometry. The σ, σ', and σ_x orbitals are not affected much by the rotation of the two CH_3 groups with respect to one another because the H $1s$ orbitals all have the same sign within each CH_3 group. The other MOs, however, are of a π type, having a nodal plane derived from the nodal plane of the C $2p_z$ orbitals. The extent of the overlap in these orbitals clearly changes as the two CH_3 groups are rotated with respect to one another. Analysis of the relative magnitudes of the bonding and antibonding interactions that take place as rotation occurs indicates that the change in energy of these two pairs of MOs is the source of the ethane rotational barrier.

The interaction of the lone-pair electrons on an amine nitrogen with adjacent C−H bonds is another example of a hyperconjugative effect that can be described in MO language. The lone-pair electrons, when properly aligned with the C−H bond, lead to a

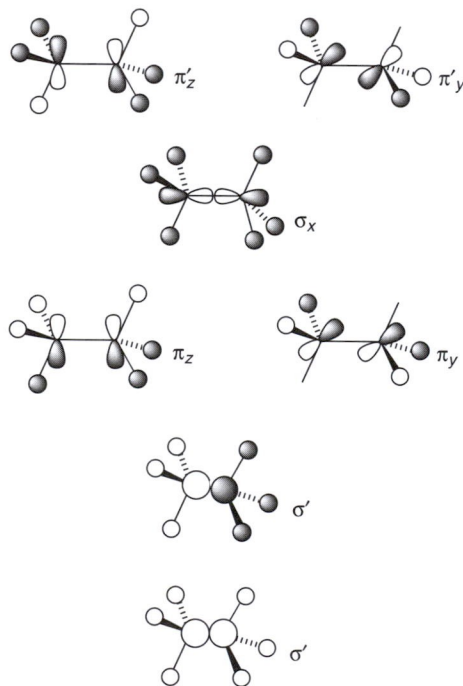

Fig. 1.31. Molecular orbitals of ethane revealing π character of π_z, π_y, π_z', and π_y' orbitals. Only the filled orbitals are shown.

67. R. M. Pitzer, *Acc. Chem. Res.* **16**:207 (1983); R. Hoffmann, *J. Chem. Phys.* **39**:1397 (1963); R. M. Pitzer and W. N. Lipscomb, *J. Chem. Phys.* **39**:1995 (1963); J. A. Pople and G. A. Segal, *J. Chem. Phys.* **43**:5136 (1956).
68. J. P. Lowe, *Prog. Phys. Org. Chem.* **6**:1 (1968); J. P. Lowe, *J. Am. Chem. Soc.* **92**:3799 (1970); J. P. Lowe, *Science* **179**:527 (1973).

donation of electron density from the lone-pair orbital to the antibonding C−H orbital. The overall effect is to weaken the C−H bond.

57

SECTION 1.8.
OTHER
QUANTITATIVE
DESCRIPTIONS OF
MOLECULAR
STRUCTURE

representation
by hyperconjugation

Electron donation from N n orbital to C−H σ^* orbital.

In acyclic structures, such effects are averaged by rotation, but in cyclic structures differences in C−H bond strengths based on the different alignments can be recognized.[69] The C−H bonds that are in an *anti* orientation to the lone pair are weaker than the C−H bonds in other orientations.

The examples that have been presented in this section illustrate the approach that is used to describe structure and reactivity effects within the framework of MO description of structure. In the chapters that follow, both valence bond theory and MO theory will be used in the discussion of structure and reactivity. Qualitative valence bond terminology is normally most straightforward for saturated systems. MO theory provides useful insights into conjugated systems and into effects that depend upon the symmetry of the molecules under discussion.

1.8. Other Quantitative Descriptions of Molecular Structure

One of the difficulties of MO computations is that the concept of the electron-pair bond disappears. The bonding between individual atoms appears as a contribution from several MOs. Thus, while MO theory is valuable in describing overall molecular characteristics such as structure, total energy, and charge distribution, it is less useful for focusing attention on individual parts of a molecule. Although the frontier orbital concept is a useful guide to reactivity, MO theory provides no unambiguous method to relate the MOs to properties of a particular atom or functional group. For this reason, many chemists have pursued theoretical descriptions of molecules that would lend themselves more readily to the concepts of transferable structural units and functional groups which arise from qualitative valence bond theory. Although we do not have space here to fully develop these approaches, we can describe some of the key ideas and illustrate specific applications.

1.8.1. Atoms in Molecules

MO calculations can provide the minimum-energy structure, total energy, and overall electron density of a given molecule. However, this information is in the form of the sum of the individual MOs and cannot be easily dissected into contributions by specific atoms or groups. How can the properties described by the MOs be related to our concept of molecules as a collection of atoms or functional groups held together by chemical bonds?

69. A. Pross, L. Radom, and N. V. Riggs, *J. Am. Chem. Soc.* **102**:2253 (1980); T. Laube and T.-K. Ha, *J. Am. Chem. Soc.* **110**:5511 (1988).

One approach is to define critical bond points and surfaces that subdivide molecules into the constituent atoms. R. F. W. Bader has developed such an approach.[70] The total electron density obtained from MO or other computational methods is partitioned among atoms. The quantitative results can be depicted qualitatively in the form of *molecular graphs*. The atoms are connected by atomic interaction lines that define maximum charge density between neighboring atoms. The atomic interaction lines correspond to chemical bonds by revealing accumulation of electron density between nuclei. The network of lines constitutes a molecular graph. In addition, the graphs locate *critical points* where electron density is either at a maximum or a minimum with respect to dislocation in any of the directions of three- dimensional space. The similarity to a classical structural formula is clear, but the molecular graphs are based on quantum-chemical theory rather than the qualitative concepts of valence bonds and hybridized orbitals. Figure 1.32 gives some representative molecular graphs and shows the bond critical points in these structures.

The subunits which can be defined can include atoms or collections of atoms corresponding to functional groups. The subunits can be represented as regions of space defined by electon density. These representations correspond well with the qualitative concepts that arise from valence bond structures. The mathematical evaluation can assign shape and charge density to atoms. Table 1.17 gives the C and H charge densities in some

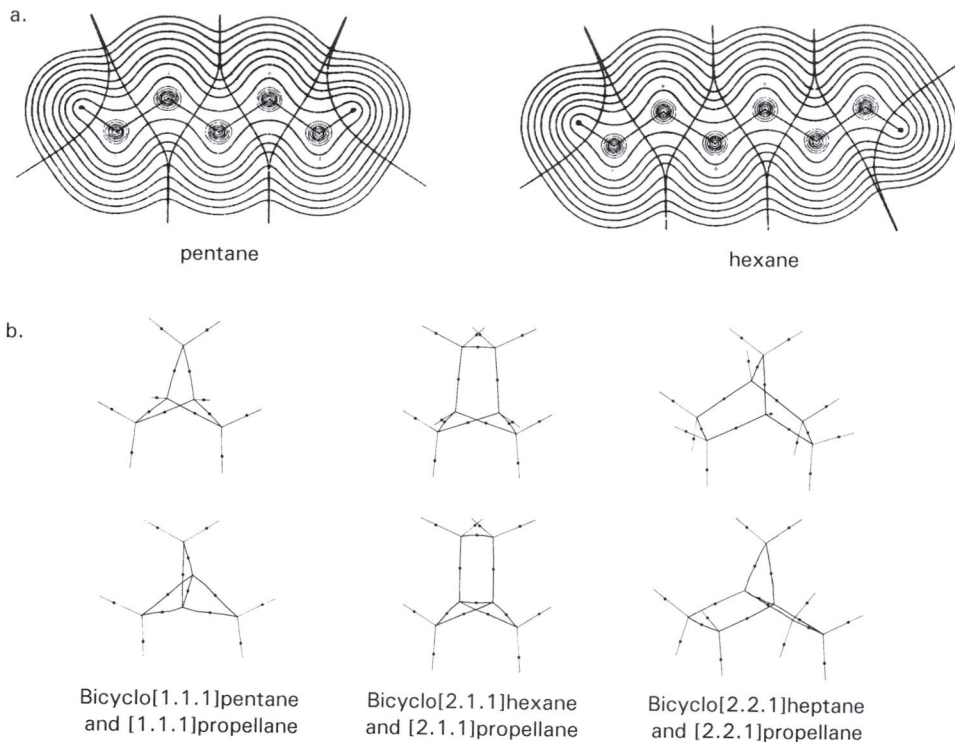

a.

pentane   hexane

b.

Bicyclo[1.1.1]pentane   Bicyclo[2.1.1]hexane   Bicyclo[2.2.1]heptane
and [1.1.1]propellane   and [2.1.1]propellane   and [2.2.1]propellane

Fig. 1.32. (a) Molecular graphs and electron density contours for pentane and hexane. Dots on bond paths represent critical points. (b) Comparison of molecular graphs for bicycloalkanes and corresponding propellanes. (Reproduced from *Chem. Rev.* **91**:893 (1991) with permission of the American Chemical Society.)

70. R. F. W. Bader, *Atoms in Molecules—A Quantum Theory*, Oxford University Press, Oxford, U.K., 1990; R. F. W. Bader, *Chem. Rev.* **91**:893 (1991).

59

SECTION 1.8.
OTHER
QUANTITATIVE
DESCRIPTIONS OF
MOLECULAR
STRUCTURE

Table 1.17. Atomic Charges in Organic Structures[a]

	$q(C)$	$q(H)$
CH_4	$+0.175$	-0.044
C_2H_6	$+0.184$	-0.061
C_2H_4	$+0.035$	-0.017
C_6H_6	$+0.020$	-0.020
C_2H_2	-0.136	$+0.136$
CH_3^+	$+0.179$	$+0.274$
$CH_3\cdot$	-0.035	$+0.012$
CH_3^-	-0.422	-0.193

a. From R. F. W. Bader, P. L. A. Popelier, and T. A. Keith, *Angew. Chem. Int. Ed. Engl.* **33**:620 (1994).

fundamental organic molecules and intermediates.[71] Figure 1.33 shows the carbon and hydrogen atoms dissected from ethane, ethylene, ethyne, and benzene. Especially noteworthy is the decreasing size of hydrogen as the carbon electronegativity changes with hybridization in the order $sp > sp^2 > sp^3$. Figure 1.34 shows the shape of carbon in CH_3^+, CH_3^-, and $CH_3\cdot$.

Figure 1.35 shows the second-row elements Li through F in their compounds with hydrogen. Note the transformation of hydrogen from "hydride" in LiH to a much smaller, protonlike entity in HF as the electronegativity of the heavier atom increases.

The lesson in these figures is that the qualitative concepts of chemical structures can be given a pictorial representation based on the quantitative application of the principles of quantum chemistry. Various, indeed all, molecular properties can, in principle, be calculated from the electronic distribution these pictures represent.

1.8.2. Electron Density Functionals

Another approach to calculating molecular geometry and energy is based on *density functional theory* (DFT).[72] DFT focuses on the electron cloud corresponding to a molecule. The energy of a molecule is uniquely specified by the electron density functional.[73] The calculation involves the construction of an expression for the electron density. The energy of the system is then expressed as

$$E = T + v_{en} + J_{ee} + v_{xc}$$

where T is the kinetic energy, v_{en} and J_{ee} are Coulombic electron–nuclear and electron–electron interactions, respectively, and v_{xc} are correlation and exchange effects. As in the Hartree–Fock MO approach, the minimization of energy should provide the most accurate description of the electronic field. The mathematical problem is to define each of the terms, with v_{xc} being the most challenging. The formulation cannot be done exactly, but various approaches have been developed and calibrated by comparison with experimental data. The methods used most frequently by chemists were developed by A. D. Becke.[74] This approach is often called the B3LYP method. The computations can be done with

71. R. F. W. Bader, P. L. A. Popelier, and T. A. Keith, *Angew. Chem. Int. Ed. Engl.* **33**:620 (1994).

72. R. G. Parr and W. Yang, *Density Functional Theory of Atoms and Molecules,* Oxford University Press, Oxford, U.K., 1989.

73. P. Hohenberg and W. Kohn, *Phys. Rev. A* **136**:864 (1964); M. Levy, *Proc. Natl. Acad. Sci. U.S.A.* **76**:6062 (1979).

74. A. D. Becke, *Phys. Rev. A* **38**:3098 (1988); A. D. Becke, *J. Chem. Phys.* **96**:2155 (1992); A. D. Becke, *J. Chem. Phys.* **97**:9173 (1992); A. D. Becke, *J. Chem. Phys.* **98**:5648 (1993).

Fig. 1.33. The C and H atoms in ethane (a), ethene (b), ethyne (c), and benzene (d), respectively. Note that the H atom is largest in ethane and smallest in ethyne. (Reproduced from *Angew. Chem. Int. Ed. Engl.* **33:**620 (1994) by permission of Wiley-VCH.)

considerably less computer time than required by the most advanced (G2) *ab initio* MO methods (see Section 1.3), and there has been considerable interest in comparing B3LYP results to MO calculations.

Table 1.18 gives total bonding energies in kilocalories per mole for some simple molecules. The B3LYP results are comparable in accuracy to G1 and G2 results. Another comparison was done with a series of cyclic hydrocarbons as the test case. The calculations were done using an isodesmic reaction scheme. The results are given in Table 1.19. Density functional calculations have also been successfully extended to functionalized molecules.[75]

75. A. St. Amant, W. D. Cornell, P. A. Kollmar, and T. A. Halgren, *J. Comput. Chem.* **16:**1483 (1995).

61

SECTION 1.8.
OTHER
QUANTITATIVE
DESCRIPTIONS OF
MOLECULAR
STRUCTURE

Fig. 1.34. The carbon atoms in the methyl cation (a), radical (b), and anion (c), respectively. (Reproduced from *Angew. Chem. Int. Ed. Engl.* **33**:620 (1994) by permission of Wiley-VCH.)

DFT turns out to be well suited to quantitative expression of some of the qualitative concepts introduced in Section 1.2, such as electronegativity, hardness, and softness.[76] The principle of maximum hardness[77] (p. 22) can be derived as a consequence of DFT, as can the concepts of hardness and softness.[78]

A very simple definition of electronegativity also finds a foundation in DFT. The definition is

$$V = n/r$$

which relates electronegativity to the number of valence-shell electrons and the effective atomic radius.[79] This measure of electronegativity is both empirically correlated and theoretically related to the Mulliken electronegativity χ, defined as $\chi = (I + A)/2$.[80] The values assigned for some of the atoms of most interest in organic chemistry are given in Table 1.20.

Concepts such as relative acidity and carbocation stability can be fundamentally related to hardness and electronegativity as defined by DFT.

$$R\text{---}H$$
$$H^- \diagup \quad \Big| {\scriptstyle -H\cdot} \quad \diagdown {\scriptstyle -H^+}$$
$$R^+ \rightleftharpoons R\cdot \rightleftharpoons R^-$$

The energy difference between R^+ and R^- can be expressed as

$$\Delta G = -23.06[E_{\text{red}}(R^+) - E_{\text{red}}(R\cdot)]$$

where the E_{red} are the reduction potentials of R^+ and $R\cdot$. This energy is approximately the same as 2χ, since it represents the HOMO–LUMO gap for $R\cdot$.[81]

76. P. W. Chattaraj and R. G. Parr, *Struct. Bonding* **80**:11 (1993); G.-H. Liu and R. G. Parr, *J. Am. Chem. Soc.* **117**:3179 (1995).

77. R. G. Parr and P. K. Chattaraj, *J. Am. Chem. Soc.* **113**:1854 (1991); T. K. Ghanty and S. K. Ghosh, *J. Phys. Chem.* **100**:12295 (1996).

78. P. K. Chattaraj, H. Lee, and R. G. Parr, *J. Am. Chem. Soc.* **113**:1855 (1991).

79. Y.-R. Luo and S. W. Benson, *J. Phys. Chem.* **92**:5255 (1988); Y.R. Luo and S. W. Benson, *J. Am. Chem. Soc.* **111**:2480 (1989); Y.R. Luo and S. W. Benson, *J. Phys. Chem.* **94**:914 (1990); Y.R. Luo and S. W. Benson, *Acc. Chem. Res.* **25**:375 (1992).

80. Y.-R. Luo and P. D. Pacey, *J. Am. Chem. Soc.* **113**:1465 (1991).

Fig. 1.35. Representations of the atoms in the second-row hydrides AH_n. In the hydridic members LiH, BeH_2, and BH_3, the A atom consists primarily of a core of decreasing radius, dressed with some residual valence density. The form and properties of the atoms undergo a marked change at methane, a nonpolar molecule: no core is visible on the C atom, and the H atoms, considerably reduced in size and population, now lie on the convex side of the interatomic surface. The increasing polarity of the remaining members is reflected in the decreasing size of the H atom and the increasing convexity of its interatomic surface. (Reproduced from *Angew. Chem. Int. Ed. Engl.* **33:**620 (1994) by permission of Wiley-VCH.)

63

SECTION 1.8.
OTHER
QUANTITATIVE
DESCRIPTIONS OF
MOLECULAR
STRUCTURE

Table 1.18. Comparison of *Ab Initio* and DFT Calculations of Atomization Energies in kcal/mol.

	Exp	G1[a]	G2[b]	DFT[c]	B3LYP[d]
H_2O	219.3	218.4	219.3	217.0	220.7
NH_3	276.7	275.4	276.6	276.8	279.7
CH_4	392.5	393.1	393.3	393.5	393.0
C_2H_6	666.3	667.6	666.6	668.7	667.5
C_2H_4	531.9	533.5	531.4	534.3	532.1
C_2H_2	388.9	391.2	386.6	389.0	391.0
CH_3OH	480.8	482.6	481.9	480.8	481.3
$CH_2=O$	357.2	362.2	358.7	357.9	356.7

a. J. A. Pople, M. Head-Gordon, D. J. Fox, K. Raghavachari, and L. A. Curtiss, *J. Chem. Phys.* **90**:5622 (1989).
b. L. A. Curtiss, J. E. Carpenter, K. Raghavachari, and J. A. Pople, *J. Chem. Phys.* **96**:9030 (1992).
c. A. D. Becke, *J. Chem. Phys.* **98**:5648 (1993).
d. C. W. Bauschlicher, Jr., and H. Partridge, *Chem. Phys. Lett.* **240**:533 (1995).

Table 1.19. Comparison of *Ab Initio* and DFT Enthalpies with Experimental Values in kcal/mol for the Isodemic Reaction:[a]

$$C_nH_m + (3n - m)CH_4 \rightarrow (2n - m/2)C_2H_6$$

Structure	ΔH_f (exp)	*Ab Initio*[b]	DFT
	6.99	5.79	5.12
	11.51	11.87	9.94
	13.12	5.39	5.14
	10.43	12.37	8.99
	14.21	3.16	1.47
	21.73	6.41	3.00
	−7.66	12.23	9.11

a. A. Fortunelli and M. Selmi, *Chem. Phys. Lett.* **223**:390 (1994)
b. Gaussian 92.

81. E. M. Arnett and R. T. Ludwig, *J. Am. Chem. Soc.* **117**:6627 (1995).

Table 1.20. Electronegativity of Some Atoms by the Definition n/r^a

Atom	Single bond	Double bond	Triple bond
C	5.18	6.00	6.63
N	6.67	8.1	
O	8.11	9.7	
F	9.92		
Cl	7.04		
Br	6.13		
I	5.25		

a. Based on covalent radii given by W. Gordy and R. L. Cook, in *Microwave Molecular Spectroscopy*, John Wiley & Sons, New York, 1984.

The same concept can be applied to carbon radicals bound to atoms other than hydrogen.[82]

$$R^+ + X^- \quad \xleftarrow{} \quad R{-}X \quad \xrightarrow{} \quad R^- + X^+$$
$$\searrow_{+e\bullet} \quad \downarrow_{+e\bullet} \quad \nearrow_{-e\bullet}$$
$$R\bullet + X^-$$

The energy for the electron transfer

$$2R\bullet \rightarrow R^+ + R^-$$

is given by

$$\Delta E = I - A = 2\eta \approx \text{HOMO–LUMO gap}$$

These relationships also underlie observed correlations between acidity, hydride affinity, or heterolytic bond energies and oxidation potentials. Table 1.21 gives some hardness values for simple compounds of hydrogen, carbon, and the halogens. Note particularly the trend towards greater softness in the hydrogen halides and methyl halides as the halogen becomes larger.

Table 1.21. Hardness for Some Small Molecules^a

H_2	8.7	HF	11.0	F_2		CH_4	10.3	CH_3F	9.4
H_2O	9.5	Cl	8.0	Cl_2	4.6	$CH_2{=}CH_2$	6.2	CH_3Cl	7.5
NH_3	8.2	HBr		Br_2	4.0	$HC{\equiv}CH$	7.0	CH_3Br	5.8
H_2S	6.2	HI	5.3	I_2	3.4	C_6H_6	5.3	CH_3I	4.7

a. R. G. Pearson, *J. Org. Chem.* **54:**1423 (1989).

82. P. K. Chattaraj, A. Cedillo, R. G. Parr, and E. M. Arnett, *J. Org. Chem.* **60:**4707 (1995).

83. D. J. Klein and N. Trinajstic, eds., *Valence Bond Theory and Chemical Structure*, Elsevier, Amsterdam, 1990.

84. D. L. Cooper, J. Gerratt, and M. Raimondi, *Chem. Rev.* **91:**929 (1991).

85. R. McWeeny, *Methods of Molecular Quantum Mechanics*, 2nd ed., Academic Press, New York, 1992; R. McWeeny, *NATO ASI Ser., Ser. B* **293:**325 (1992).

86. J. H. van Lenthe and G. G. Balint-Kurti, *J. Chem. Phys.* **78:**5699 (1983); J. Verbeek and J. H. van Lenthe, *THEOCHEM* **229:**115 (1991).

87. F. W. Bobrowicz and W. A. Goddard III, in *Methods in Electronic Structure*, H. F. Shaefer, ed., Plenum Press, New York, 1977.

Several methods of quantitative description of molecular structure based on the concepts of valence bond theory have been developed.[83–87] These methods employ orbitals similar to localized valence bond orbitals, but permitting modest delocalization. These orbitals allow many fewer structures to be considered and remove the need for incorporating many ionic structures, in agreement with chemical intuition. To date, these methods have not been as widely applied in organic chemistry as MO calculations. They have, however, been successfully applied to fundamental structural issues. For example, successful quantitative treatments of the structure and energy of benzene and its heterocyclic analogs have been developed.[88] It remains to be seen whether computations based on DFT and modern valence bond theory will come to rival the widely used MO programs in analysis and interpretation of structure and reactivity.

General References

T. A. Albright, J. K. Burdett, and M.-H. Whangbo, *Orbital Interactions in Chemistry*, John Wiley & Sons, New York, 1985.

R. F. W. Bader, *Atoms in Molecules; A Quantum Theory*, Clarendon Press, Oxford, U.K., 1990.

W. T. Borden, *Modern Molecular Orbital Theory for Organic Chemists*, Prentice-Hall, Englewood Cliffs, New Jersey, 1975.

M. J. S. Dewar, *The Molecular Orbital Theory of Organic Chemistry*, McGraw-Hill, New York, 1969.

M. J. S. Dewar and R. C. Dougherty, *The PMO Theory of Organic Chemistry*, Plenum Press, New York, 1975.

I. Fleming, *Frontier Orbitals and Organic Chemical Reactions*, John Wiley & Sons, New York, 1976.

W. J. Hehre, L. Radom, P. v. R. Schleyer, and J. Pople, *Ab Initio Molecular Orbital Theory*, Wiley-Interscience, New York, 1986.

R. F. Hout, W. J. Pietro, and W. J. Hehre, *A Pictorial Approach to Molecular Structure and Reactivity*, John Wiley & Sons, New York, 1984.

C. K. Ingold, *Structure and Mechanism in Organic Chemistry*, 2nd ed., Cornell University Press, Ithaca, New York, 1969.

R. G. Parr and W. Yang, *Density-Functional Theory of Atoms and Molecules*, Oxford University Press, Oxford, U.K., 1989.

L. Salem, *Electrons in Chemical Reactions*, John Wiley & Sons, New York, 1982.

P. v. R. Schleyer, ed., *Encyclopedia of Computational Chemistry*, John Wiley & Sons, New York, 1998.

R. B. Woodward and R. Hoffmann, *The Conservation of Orbital Symmetry*, Verlag Chemie, Weinheim, 1970.

H. E. Zimmerman, *Quantum Mechanics for Organic Chemists*, Academic Press, New York, 1975.

Problems

(References for these problems will be found on page 791.)

1. Use thermochemical relationships to obtain the required information.

   (a) The heats of formation of cyclohexane, cyclohexene, and benzene are, respectively, -29.5, -1.1, and $+18.9$ kcal/mol. Estimate the resonance energy of benzene using these data.

88. D. L. Cooper, S. C. Wright, J. Gerratt, and M. Raimondi, *J. Chem. Soc., Perkin Trans. 2* **1989**:255; D. L. Cooper, S. C. Wright, J. Gerratt, P. A. Hyams, and M. Raimondi, *J. Chem. Soc., Perkin Trans. 2* **1989**:719; C. Amovilli, R. D. Harcourt, and R. McWeeny, *Chem. Phys. Lett.* **187**:494 (1991); J. Gerratt, D. L. Cooper, P. B. Karadakov, and M. Raimondi, *Chem. Soc. Rev.* **26**:87 (1997).

(b) Calculate ΔH for the air oxidation of benzaldehyde to benzoic acid given that the heats of formation of benzaldehyde and benzoic acid are -8.8 and $-70.1\,\text{kcal/mol}$, respectively.

(c) Using the appropriate heats of formation in Table 1.5, calculate the heat of hydrogenation of 2-methyl-1-pentene.

2. Suggest an explanation for the following observations:

(a) The dipole moment of the hydrocarbon calicene has been estimated to be as large as 5.6 D.

calicene

(b) The measured dipole moment of *p*-nitroaniline (6.2 D) is larger than the value calculated using empirical group moments (5.2 D).

(c) The dipole moment of furan is smaller than and in the opposite direction from that of pyrrole.

0.71 D   1.80 D

3. Predict the energetically preferred site of protonation for each of the following molecules and explain the basis of your prediction.

(a)
$$C_6H_5CH{=}N{-}C_6H_5$$

(b)
$$CH_3{-}C\overset{\displaystyle O}{\underset{\displaystyle NHCH_3}{\big|}}$$

(c)

(d)
$$-NH_2$$

4. What physical properties, such as absorption spectra, bond length, dipole moment, etc., could be examined to obtain evidence of resonance interactions in the following molecules? What deviations from "normal" physical properties would you expect to find?

(a)   (b) C_6H_5 C_6H_5 (c)

5. Certain $C{-}H$ bonds have significantly lower bond dissociation energies than do the "normal" $C{-}H$ bonds in saturated hydrocarbons. Offer a structural rationalization of the lowered bond energy in each of the following compounds, relative to the saturated

hydrocarbon C—H bond taken as a reference. (The bond dissociation energies are given in kcal/mol.)

(a) [benzene ring]—CH₂—H (85) versus CH₃—H (103)

(b) HOCH₂—H (92) versus CH₃—H (103)

(c) CH₃C—H
         ‖
         O
    (88) versus CH₃CH₂—H (98)

6. (a) Carboxamides have substantial rotational barriers on the order of 20 kcal/mol for the process

$$
\begin{array}{ccc}
\text{R—C(=O)—N(—CH}_3^*)\text{—CH}_3 & \rightleftharpoons & \text{R—C(=O)—N(—CH}_3)\text{—CH}_3^*
\end{array}
$$

Develop a structural explanation for the existence of this barrier in both resonance and molecular orbital terminology.

(b) In the gas phase the rotational barrier of *N,N*-dimethylformamide is about 19.4 kcal/mol, which is about 1.5 kcal/mol less than in solution. Is this change consistent with the ideas you presented in (a)? Explain.

(c) Explain the relative rates of alkaline hydrolysis of the following pairs of carboxamides.

PhC(=O)—N◁ 10^5> PhCNH₂   [bicyclic structure] 10^3> Ph—N[piperidinone]

7. Construct a qualitative MO diagram showing how the π-molecular orbitals in the following molecules are modified by the addition of the substituent:

(a) vinyl fluoride, compared to ethylene
(b) acrolein, compared to ethylene
(c) acrylonitrile, compared to ethylene
(d) benzyl cation, compared to benzene
(e) propene, compared to ethylene
(f) fluorobenzene, compared to benzene

8. The data below give the stabilization calculated in kcal/mol by MO methods for the reaction:

$$\text{CH}_3\text{—X} + \text{CH}_2\text{=CH}_2 \longrightarrow \text{CH}_4 + \text{CH}_2\text{=CH—X}$$

X	ΔH
F	− 6.4
OCH₃	− 10.9
NH₂	− 13.3

(a) Demonstrate that this indicates that there is a stabilizing interaction between the substituent and a carbon–carbon double bond.

(b) Draw resonance structures showing the nature of the interaction.

(c) Construct a qualitative MO diagram which rationalizes the existence of a stabilizing interaction.

(d) Both in resonance and molecular orbital terminology, explain the order of the stabilization $N > O > F$.

9. Construct a qualitative MO diagram for the H-bridged ethyl cation by analyzing the interaction of the ethylene MOs given in Fig. 1.23 with a proton approaching the center of the ethylene molecule from a direction perpendicular to the molecular plane. Indicate which ethylene orbitals will be lowered by this interaction and which will be raised or left relatively unchanged. Assume that the hydrogens of ethylene are slightly displaced away from the direction of approach of the proton.

10. In the Hückel treatment, atomic orbitals on nonadjacent atoms are assumed to have no interaction. They are neither bonding nor antibonding. The concept of *homoconjugation* suggests that such orbitals may interact, especially in rigid structures which direct orbitals toward one another. Consider, for example, bicyclo[2.2.1]hepta-2,5-diene:

(a) Construct the MO diagram according to simple Hückel theory and assign energies to the orbitals.

(b) Extend the MO description by allowing a significant interaction between the C-2 and C-6 and between the C-3 and C-5 orbitals. Construct a qualitative MO diagram by treating the interaction as a perturbation on the orbitals shown for (a).

11. (a) Sketch the nodal properties of the highest occupied molecular orbital of pentadienyl cation ($CH_2{=}CHCH{=}CHCH_2{}^+$).

(b) Two of the π-MOs of pentadienyl are given below. Specify which one is of lower energy, and classify each as to whether it is bonding, nonbonding, or antibonding. Explain your reasoning.

$$1 \ \ 2 \ \ 3 \ \ 4 \ \ 5$$
$$o{-}o{-}o{-}o{-}o$$

$$\psi_x = 0.50\phi_1 + 0.50\phi_2 - 0.50\phi_4 - 0.50\phi_5$$
$$\psi_y = 0.58\phi_1 - 0.58\phi_3 + 0.58\phi_5$$

12. The diagrams below give STO-3G calculated charge densities for the benzyl cation and its α-formyl and α-cyano derivatives. Analyze the effect of these substituents on

the charge density.

13. Calculate the energy levels and coefficients for 1,3-butadiene using Hückel MO theory.

14. (a) Estimate from HMO theory the delocalization energy, expressed in units of β, of cyclobutadienyl dication ($C_4H_4^{2+}$).
    (b) Estimate, in units of β, the energy associated with the long-wavelength UV-VIS absorption of 1,3,5,7-octatetraene. Does it appear at longer or shorter wavelengths than the corresponding absorption for 1,3,5-hexatriene?

15. Addition of methylmagnesium bromide to 2-methylcyclohexanone followed by iodine-catalyzed dehydration of the resulting alcohol gave three alkenes in the ratio $A:B:C = 3:31:66$. Each isomer gave a mixture of *cis*- and *trans*-1,2-dimethylcyclohexane on catalytic hydrogenation. When the alkene mixture is heated with a small amount of sulfuric acid, the ratio $A:B:C$ is changed to $0.0:15:85$. Assign structures to A, B, and C.

16. The propellanes are highly reactive substances which readily undergo reactions involving rupture of the central bond. It has been suggested that the polymerization of propellanes occurs by a dissociation of the central bond:

Somewhat surprisingly perhaps, it has been found that [1.1.1]propellane is considerably *less reactive* than [2.2.1]propellane. Use the theoretically calculated enthalpy data below to estimate the bond dissociation energy of the central bond in each of the three propellanes shown. How might this explain the relative reactivity of the [1.1.1]- and [2.2.1]propellanes?

Enthalpy for addition of hydrogen to give the corresponding bicycloalkane

[2.2.1]propellane + H_2 → bicyclo[2.2.1]heptane, $\quad \Delta H = -99$ kcal/mol

[2.1.1]propellane + H_2 → bicyclo[2.1.1]hexane, $\quad \Delta H = -73$ kcal/mol

[1.1.1]propellane + H_2 → bicyclo[1.1.1]pentane, $\quad \Delta H = -39$ kcal/mol

Assume that the bond dissociation energy of the bridgehead hydrogens in each bicycloalkane is 104 kcal/mol. Indicate and discuss any other assumptions you have made.

17. Examine the following thermochemical data pertaining to hydrogenation of unsaturated eight-membered ring hydrocarbons to give cyclooctane:

Unsaturated ring hydrocarbon	$-\Delta H$ (kcal/mol)
cis,cis,cis,cis-1,3,5,7-Cyclooctatetraene	97.96
cis,cis,cis-1,3,5-Cyclooctatriene	76.39
cis,cis,cis-1,3,6-Cyclooctatriene	79.91
cis,cis-1,5-Cyclooctadiene	53.68
cis,cis-1,4-Cyclooctadiene	52.09
cis,cis-1,3-Cyclooctadiene	48.96
trans-Cyclooctene	32.24
cis-Cyclooctene	22.98

(a) Discuss the differences observed in each isomeric series of compounds, and offer an explanation for these differences.

(b) Comment on whether the conjugation present in cyclooctatetraene has a stabilizing or destabilizing effect on the C=C bonds.

18. Cyclic amines such as piperidine and its derivatives show substantial differences in the properties of the axial C-2 and C-6 versus the equatorial C-2 and C-6 C−H bonds.

The axial C−H bonds are *weaker* than the equatorial C−H bonds as can be demonstrated by a strongly shifted C−H stretching frequency in the IR spectrum. Axial C-2 and C-6 methyl groups *lower* the ionization potential of the lone-pair electrons on nitrogen substantially more than do equatorial C-2 or C-6 methyl groups. Discuss the relationship between these observations and provide a rationalization in terms of qualitative MO theory.

19. (a) The strain energy of spiropentane (62.5 kcal/mol) is considerably greater than twice that of cyclopropane (27.5 kcal/mol). Suggest an explanation.

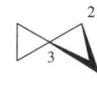

(b) The fractional *s* character in bonds to carbon in organic molecules may be estimated by its relation to $^{13}C-^{13}C$ coupling constants, as determined by NMR. Estimate the fractional *s* character of C-1 in its bond to C-3 of spiropentane, given the following information:

$$s_{1(3)} = \frac{J_{^{13}C-^{13}C}}{Ks_{3(1)}}$$

where K is a constant equal to 550 Hz, the $^{13}C-^{13}C$ coupling constant J between C-1 and C-3 is observed to be 20.2 Hz, and $s_{3(1)}$ is the s character at C-3 in its bond to C-1.

20. Predict which direction will be favored in the thermodynamic sense for each of the following reactions:

(a) CH_4 + $\rightleftharpoons$ CH_3^- +

(b) CH_4 + $N\equiv CCH_2^+$ $\rightleftharpoons$ CH_3^+ + $CH_3C\equiv N$

(c) CHF_3 + CH_3^- $\rightleftharpoons$ CF_3^- + CH_4

21. Predict which compound would give the faster (k) or more complete (K) reaction. Explain the basis for your prediction.

(a) $-I$ + Ag^+ $\xrightarrow{k}$

or

$-I$ + Ag^+ $\xrightarrow{k}$

(b) Ph_3CH + $^-NH_2$ $\xrightleftharpoons{K}$ Ph_3C^- + NH_3

or

+ $^-NH_2$ $\xrightleftharpoons{K}$ + NH_3

(c) $-OH$ + H^+ $\xrightleftharpoons{K}$ + H_2O

or

OH + H^+ $\xrightleftharpoons{K}$ + H_2O

22. Computational comparison of structures of the benzyl cation (**A**) and singlet phenylcarbene (**B**) indicates a much greater double-bond character for the exocyclic

bond in **A** than in **B**.

A   **B**

Can you provide a rationalization of this difference in terms of both valence bond-resonance and PMO considerations? Explain.

23. The ionization potentials of some substituted norbornadienes have been measured by photoelectron spectroscopy. The values which pertain to the π orbitals are shown:

		IP (eV)	
	X	1	2
	H	8.69	9.55
	CH_3O	8.05	9.27
	CN	9.26	10.12

Use PMO theory to describe the effect of the substituents on the ionization potential. Use an MO diagram to explain the interaction of the substituents with the π bonds. Explicitly take into account the fact that the two orbitals interact and therefore cannot be treated as separate entities (see Problem 10).

24. *Ab initio* MO calculations using 4-31G orbitals indicate that the eclipsed conformation of acetaldehyde is more stable than the staggered conformation.

$E = -152.685$ au $E = -152.683$ au
H/C = O eclipsed   H/C = O staggered

Provide a rationalization of this structural effect in terms of MO theory. Construct a qualitative MO diagram for each conformation, and point out the significant differences that can account for the preference for the eclipsed conformation.

25. Interesting stabilization and structural trends have been noted using MP2/6-31G* calculations on the effect of substituents on imines. The data below give ΔE for the isodesmic reaction and show that stabilization tends to increase with χ_{BE}, the group electronegativity of the substituent. The X—N=CH$_2$ bond angle decreases with χ_{BE}.

Account for these trends.

$$X-N=CH_2 + CH_3CH=CH_2 \longrightarrow CH_3N=CH_2 + XCH=CH_2$$

Discuss these trends in relation to the interaction of the nitrogen unshared pair and the C=N electrons with the substituent.

Substituent	χ_{BE}	ΔE (kcal/mol)	$\angle XN=CH_2$ (°)
H	2.20	+ 4.1	110
CH_3	2.55	0.0	116
CH=O	2.66	+ 3.6	114
CN	2.69	− 4.8	117
CF_3·	2.71	+ 4.7	118
NO_2	3.22	− 10.0	111
OH	3.55	− 20.5	110
F	4.00	− 29.0	108
SiH_3	1.90	+ 13.2	120

Principles of Stereochemistry

Introduction

For most combinations of atoms, a number of molecular structures that differ from each other in the sequence of bonding of the atoms are possible. Each individual molecular assembly is called an *isomer*, and the *constitution* of a compound is the particular combination of bonds between atoms (molecular connectivity) which is characteristic of that structure. Propanal, allyl alcohol, acetone, 2-methyloxirane, and cyclopropanol each correspond to the molecular formula C_3H_6O, but differ in constitution and are isomers of one another.

$CH_3CH_2CH{=}O$ $CH_2{=}CHCH_2OH$ $CH_3\overset{O}{\overset{\|}{C}}CH_3$ $CH_3CH{-}CH_2$ (OH)

propanal   allyl alcohol   acetone   methyloxirane   cyclopropanol

When structures having the same constitution differ with respect to their spatial arrangement, they are *stereoisomers*. Stereoisomers are described by specifying their topology and the nature of their relationship to other stereoisomers of the same constitution. Stereoisomers differ in *configuration*, and in order to distinguish between stereoisomeric compounds, it is necessary to specify the configuration.[1,2] If two stereoisomers are nonsuperimposable mirror images, the molecules are *enantiomers*. Structures that have nonsuperimposable mirror images are called *chiral*. *Chirality* is the property of any molecule (or other object) of being nonsuperimposable on its mirror image. Samples that contain only one enantiomer are called enantiomerically pure or *homochiral*. Stereoisomers that are not enantiomers are *diastereomers*.

It is possible to obtain pure enantiomers of chiral compounds. One property of separated enantiomers is to cause the rotation of the plane of polarized light by opposite

1. The IUPAC rules and definintions for fundamental stereochemistry are given with examples in *J. Org. Chem.* **35**:2849 (1970); see also G. Krow, *Top. Stereochem.* **5**:31 (1969).
2. K. Mislow and M. Raban, *Top. Stereochem.* **1**:1 (1967); J. K. O'Loane, *Chem. Rev.* **80**:41 (1980).

but equal amounts. Samples that contain equal amounts of two enantiomers have zero net rotation and are called *racemic mixtures*. Samples that contain only one of the enantiomers are said to be *enantiomerically pure*. Samples that have an excess of one enantiomer over the other are *enantiomerically enriched* and show a net rotation of polarized light and are said to be *optically active*.

In addition to constitution and configuration, there is a third important level of structure, that of *conformation*. Conformations are discrete molecular arrangements that differ in spatial arrangement as a result of facile rotations about single bonds. Usually, conformers are in thermal equilibrium and cannot be separated. The subject of conformational interconversion will be discussed in detail in Chapter 3. A special case of stereoisomerism arises when rotation about single bonds is sufficiently restricted by steric or other factors that the different conformations can be separated. The term *atropisomer* is applied to stereoisomers that result from restricted bond rotation.[3]

In this chapter, configurational relationships will be emphasized. Both structural and dynamic aspects of stereochemical relationships will be considered. We will be concerned both with the fundamental principles of stereochemistry and the conventions which have been adopted to describe the spatial arrangements of molecules. We will consider the stereochemical consequences of chemical reactions so as to provide a basis for understanding the relationships between stereochemistry and reaction mechanism that will be encountered later in the book.

2.1. Enantiomeric Relationships

The relationship between chirality and optical activity is historically such a close one that chemists sometimes use the terms imprecisely. Optical activity refers to just one property of chiral molecules, namely, the ability to rotate plane-polarized light. Measurement of optical activity is useful both for determining the configuration of chiral molecules and for investigating the stereochemical relationship between reactants and products. The mechanics of measuring optical rotation will not be discussed here since the basic method is described in most introductory texts. Both the sign and the magnitude of optical rotation are dependent on the conditions of the measurement, including temperature, solvent, and the wavelength of the light. By convention, single-wavelength measurements are usually made at the 589-nm emission line of sodium arc lamps. This wavelength is known as the sodium D line, and optical rotations measured at this wavelength are designated $[\alpha]_D$.

Pure enantiomeric substances show rotations that are equal in magnitude but opposite in direction. Unequal mixtures of enantiomers rotate light in proportion to the composition. The relationship between optical purity and measured rotation is

$$\text{optical purity (\%)} = \frac{\alpha_{\text{mixture of enantiomers}}}{\alpha_{\text{pure enantiomer}}} \times 100$$

The optical purity is numerically equivalent to the *enantiomeric excess* (e.e.), which is defined as

$$\text{enantiomeric excess} = \text{mole fraction}_{\text{major enantiomer}} - \text{mole fraction}_{\text{minor enantiomer}} \times 100$$

---

3. M. Oki, *Top. Stereochem.* **14**:1 (1983).

Measurement of rotation as a function of wavelength is useful in structural studies aimed at determining the chirality of a molecule. This technique is called *optical rotatory dispersion* (ORD).[4] The resulting plot of rotation against wavelength is called an ORD curve. The shape of the ORD curve is determined by the configuration of the molecule and its absorption spectrum. In many cases, the ORD curve can be used to specify the configuration of a molecule by relating it to those of similar molecules of known configuration. Chiral substances also show differential absorption of circularly polarized light. This is called *circular dichroism* (CD) and is quantitatively expressed as the molecular ellipticity, θ:

$$\theta = 3330 \, (\varepsilon_L - \varepsilon_R)$$

where ε_L and ε_R are the extinction coefficients of left and right circularly polarized light, respectively. Figure 2.1 shows the ultraviolet (UV), ORD and CD spectra of an enantiomerically pure sulfonium ion salt.[5]

The molecular ellipticity is analogous to specific rotation in that two enantiomers have exactly opposite values of θ at every wavelength. Two enantiomers will thus show CD spectra having opposite signs. A compound with several absorption bands may show both

Fig. 2.1. UV absorption, ORD, and CD curves of ethyl methyl *p*-tolyl sulfonium tetrafluoroborate. [Reproduced with permission from *J. Org. Chem.* **41**:3099 (1976).]

4. P. Crabbe, *Top. Stereochem.* **1**:93 (1967); C. Djerassi, *Optical Rotatory Dispersion*, McGraw-Hill, New York, 1960; P. Crabbe, *Optical Rotatory Dispersion and Circular Dichroism in Organic Chemistry*, Holden Day, San Francisco, 1965; E. Charney, *The Molecular Basis of Optical Activity. Optical Rotatory Dispersion and Circular Dichroism*, John Wiley & Sons, New York, 1979.
5. K. K. Andersen, R. L. Caret, and D. L. Ladd, *J. Org. Chem.* **41**:3096 (1976).

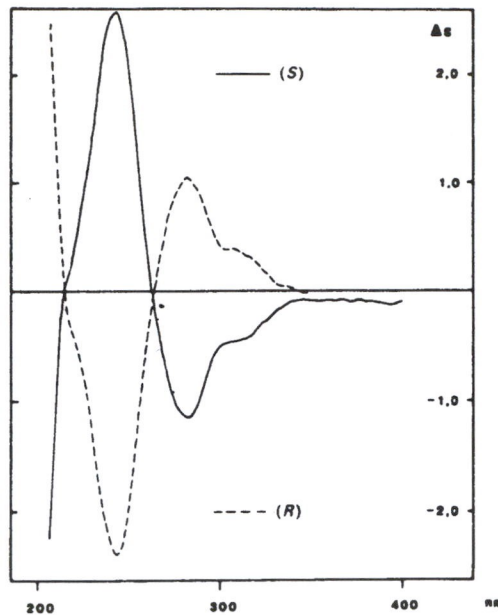

Fig. 2.2. CD spectra of (*S*)- and (*R*)-2 amino-1-phenyl-1-propanone hydrochloride. [Reproduced with permission from *Helv. Chim. Acta* **69**:1498 (1986).]

positive and negative bands. Figure 2.2 shows the CD curves for both enantiomers of 2-amino-1-phenyl-1-propanone.[6]

Although measurements of optical rotation and ORD or CD spectra have historically been the main methods for determining enantiomeric purity and assigning configuration, other analytical techniques are also available. High-performance liquid chromatography (HPLC) using chiral column packing material can resolve enantiomers on both an analytical and a preparative scale. Chiral packing materials for gas–liquid chromatography (GLC) have also been developed. Several other approaches to determining enantiomeric purity that depend upon formation of diastereomers will be discussed in Section 2.2.

Compounds in which one or more carbon atoms have four nonidentical substituents are the largest class of chiral molecules. Carbon atoms with four nonidentical ligands are referred to as asymmetric carbon atoms because the molecular environment at such a carbon atom possesses no element of symmetry. Asymmetric carbons are a specific example of a *stereogenic center*. A stereogenic center is any structural feature that gives rise to chirality in a molecule. 2-Butanol is an example of a chiral molecule and exists as two nonsuperimposable mirror images. Carbon-2 is a stereogenic center.

Ethanol is an achiral molecule. The plane defined by atoms C-1, C-2, and O is a plane of symmetry. Any carbon atom with two identical ligands contains a plane of symmetry that

6. J.-P. Wolf and H. Pfander, *Helv. Chim. Acta* **69**:1498 (1986).

includes the two nonidentical ligands. Any molecule, no matter how complex, that possesses a plane of symmetry is achiral.

There are a number of important kinds of stereogenic centers besides asymmetric carbon atoms. One example is furnished by sulfoxides with nonidentical substituents on sulfur.[7] Sulfoxides are pyramidal and maintain their configuration at room temperature. Unsymmetrical sulfoxides are therefore chiral and exist as enantiomers. Sulfonium salts with three nonidentical ligands are also chiral as a result of their pyramidal shape. Some examples of chiral derivatives of sulfur are given in Scheme 2.1.

Although unsymmetrically substituted amines are chiral, the configuration is not stable because of rapid inversion at nitrogen. The activation energy for pyramidal inversion at phosphorus is much higher than at nitrogen, and many optically active phosphines have been prepared.[8] The barrier to inversion is usually in the range of 30–35 kcal/mol so that enantiomerically pure phosphines are stable at room temperature but racemize by inversion at elevated temperatures. Asymmetrically substituted tetracoordinate phosphorus compounds such as phosphonium salts and phosphine oxides are also chiral. Scheme 2.1 includes some examples of chiral phosphorus compounds.

The chirality of a molecule is described by specifying its configuration. The system that is used is the *Cahn–Ingold–Prelog* convention, which uses the descriptors R and S. The *Fischer convention*, employing the descriptors D and L, is historically important and is still used with certain types of molecules.

The Cahn–Ingold–Prelog descriptors R and S are assigned by using the *sequence rule* to assign a priority order to the substituents on the atom to which a configuration is being assigned. The substituent atoms are assigned decreasing priority in the order of decreasing atomic number. When two or more of the substituent atoms are the same element (e.g., carbon), the assignment of priority is based on the next attached atom in those substituents. This process is continued until the order of priority of all substituents has been established. An atom that is multiply bonded is counted once for each formal bond. When the substituent priority has been established, the molecule is viewed in an orientation that places the lowest-priority substituent behind the stereogenic center. The three remaining substituents project toward the viewer. The remaining substituents have one of two possible arrangements. The substituents decrease in priority in either a clockwise manner or in a counterclockwise manner. In the former case, the configuration R (for *rectus*) is assigned. If the priority decreases in the counterclockwise sense, the atom is of S (for *sinister*) configuration.

The configuration of the 2-butanol enantiomer shown below is established as S as follows. The highest-priority atom bonded to the asymmetric carbon is O; the lowest is H. The remaining two atoms are each C, and the choice as to which of these is of higher priority is made by comparing their ligands. The methyl group has (H, H, H), while the ethyl group has (C, H, H); therefore, the ethyl group is of higher priority than the methyl

7. For reviews of chiral sulfoxides, see M. Cinquini, F. Cozzi, and F. Montanari, *Stud. Org. Chem.* **19**:355 (1985); M. R. Barbachy and C. R. Johnson, in *Asymmetric Synthesis*, Vol. 4, J. D. Morrison and J. W. Scott, eds., Academic Press, New York, 1984, Chapter 2.

8. D. Valentine, Jr., in *Asymmetric Synthesis*, Vol. 4, J. D. Morrison and J. W. Scott, eds., Academic Press, New York, 1984, Chapter 3.

Scheme 2.1. Chiral Compounds with Stereogenic Centers at Sulfur and Phosphorus

1[a]

$[a]_D = +92.4°$
R-enantiomer

2[b]

$[a]_D = -15.8°$
R-enantiomer

3[c]

$[a]_D = +172.4°$
S-enantiomer

4[d]

$[a]_D = +17°$
R-enantiomer

5[e]

$[a]_D = 16.8°$
S-enantiomer

6[f]

$[a]_D = +101.6°$
S-enantiomer

7[g]

$[a]_D = -86.9°$
R-enantiomer

a. C. R. Johnson and D. McCants, Jr., *J. Am. Chem. Soc.* **87**:5404 (1965).
b. K. K. Andersen, R. L. Caret, and D. L. Ladd, *J. Org. Chem.* **41**:3096 (1976).
c. C. R. Johnson and C. W. Schroeck, *J. Am. Chem. Soc.* **95**:7418 (1973); C. R. Johnson, C. W. Schroeck, and J. R. Shanklin, *J. Am. Chem. Soc.* **95**:7424 (1973).
d. O. Korpiun, R. A. Lewis, J. Chickos, and K. Mislow, *J. Am. Chem. Soc.* **90**:4842 (1968).
e. L. Horner, H. Winkler, A. Rapp, A. Mentrup, H. Hoffman, and P. Beck, *Tetrahedron Lett.* **1961**:161.
f. W.-D. Balzer, *Chem. Ber.* **102**:3546 (1969).
g. L. Horner and M. Jordan, *Phosphorus and Sulfur* **8**:225 (1980).

group. The complete priority list is: OH > C_2H_5 > CH_3 > H. When viewed from the side opposite the lowest-priority ligand, the remaining groups appear in order of decreasing priority in counterclockwise fashion, and the configuration is S:

Some other examples of assignment of configuration are illustrated below.

When a stereogenic center is tricoordinate, as is the case for sulfoxides, sulfonium salts, and phosphines, then a "phantom atom" of atomic number zero is taken to occupy the lowest-priority site of a presumed tetrahedral atom. Application of the sequence rule in the usual manner allows the configurations of the enantiomers of phenyl *p*-tolyl sulfoxide and allylmethylphenylphosphine shown in Scheme 2.1 to be assigned as *R* and *S*, respectively.

Glyceraldehye is the point of reference for describing the configuration of carbohydrates and other natural substances in accordance with the *Fischer convention*. The two enantiomers were originally arbitrarily assigned the configurations D and L as shown below. Subsequently, a determination of the configuration of sodium rubidium tartrate by X-ray crystallography and the relationship of this material to D glyceraldehyde established that the original arbitrary assignments were the correct ones.

D-(+)-glyceraldehyde   L-(−)-glyceraldehyde

In the Fischer convention, the configurations of other molecules are described by the descriptors D and L, which are assigned by comparison with the reference molecule glyceraldehyde. In employing the Fischer convention, it is convenient to use *projection formulas*. These are planar representations defined in such a way as to convey three-dimensional structural information. The molecule is oriented with the major carbon chain aligned vertically in such a manner that the most oxidized terminal carbon is at the top. The vertical bonds at each carbon are directed back, away from the viewer, and the horizontal bonds are directed toward the viewer. The D and L forms of glyceraldehyde are shown below with the equivalent Fischer projection formulas.

D   L

The assignment of the configuration of any other chiral molecule in the Fischer convention is done by comparison with D- and L-glyceraldehyde. The molecule is aligned with the chain vertical and the most oxidized carbon at the top, as specified by the Fischer convention. The stereogenic center with the highest number (at the lowest position in the Fischer projection) is compared with C-2 of glyceraldehyde. If the configuration is that of D-glyceraldehye, the molecule is assigned the D-configuration, whereas if it is like that of L-glyceraldehyde, it is assigned the L-configuration. This is illustrated below with several carbohydrates.

D-ribose   L-fucose   D-mannose   L-fructose

The amino acids found in proteins have the L-configuration, as illustrated for alanine, serine, and leucine.

$$
\begin{array}{ccc}
\text{CO}_2\text{H} & \text{CO}_2\text{H} & \text{CO}_2\text{H} \\
\text{H}_2\text{N}\!-\!\!\!-\!\!\!-\!\text{H} & \text{H}_2\text{N}\!-\!\!\!-\!\!\!-\!\text{H} & \text{H}_2\text{N}\!-\!\!\!-\!\!\!-\!\text{H} \\
\text{CH}_3 & \text{CH}_2\text{OH} & \text{CH}_2\text{CH(CH}_3)_2 \\
\text{L-alanine} & \text{L-serine} & \text{L-leucine}
\end{array}
$$

At the present time, use of the Fischer convention is almost entirely restricted to carbohydrates, amino acids, and biologically important molecules of closely related structural types. The problem with more general use is that there are no adequate rules for deciding whether a chiral atom is "like" D-glyceraldehyde or L-glyceraldehyde when the structures are not closely similar to the reference molecules. This relationship is clear for carbohydrates and amino acids.

The property of chirality is determined by overall molecular topology, and there are many molecules that are chiral even though they do not possess an asymmetrically substituted atom. The examples in Scheme 2.2 include allenes (entries 1 and 2) and spiranes (entries 7 and 8). Entries 3 and 4 are examples of separable chiral atropisomers in which the barrier to rotation results from steric restriction of rotation of the bond between the aryl rings. The chirality of E-cyclooctene and Z,E-cyclooctadiene is also dependent on restricted rotation. Manipulation of a molecular model will illustrate that each of these molecules can be converted into its enantiomer by a rotational process by which the ring is turned "inside-out."

There is no direct relationship between the configurational descriptors R and S or D and L and the sign of rotation of the molecule. R or S molecules can have either $+$ or $-$ signs for rotation, as can D or L molecules. Thus, even though a configuration can be specified on the basis of these conventions, additional information is necessary to establish which molecule of an enantiomeric pair possesses the specified configuration. Determination of the *absolute configuration* establishes the configuration of each enantiomer. There are several approaches to this problem. One is to establish a direct structural relationship to a molecule of known configuration by chemical transformation.[9] This is the way in which most of the reference molecules whose absolute configurations are known were initially assigned. The existence of a base of molecules whose absolute configurations are known has permitted the development of correlations based on the CD and ORD curves of certain types of chromophores. When chromophores are located close to stereogenic centers, the spectroscopic properties are affected in a predictable way so that the sign and shape of the ORD or CD curve can be a reliable basis for configurational assignment.[10] While routine X-ray crystal structure determination does not provide the absolute configuration of the molecule, special analysis of the diffraction data does allow assignment of absolute configuration.[11] These methods are important, for example, in assigning the absolute configuration of new natural products.

9. For a review of chemical methods for determining absolute configuration, see *Stereochemistry, Fundamentals and Methods*, Vol. 3, H. B. Kagan, ed., G. Thieme, Stuttgart, 1977.

10. K. Nakanishi and N. Harada, *Circular Dichroism Spectroscopy: Exciton Coupling in Organic Stereochemistry*, University Science Books, Mill Valley, California, 1983; D. N. Kirk, *Tetrahedron* **42**:777 (1986).

11. D. Rogers, *Acta Crystallogr.. Sect. A* **37**:734 (1981).

Scheme 2.2. Examples of Chiral Molecules Lacking Asymmetric Atoms

1[a]

R-(−)-Glutinic acid

2[b]

R-(−)-1,3-Dimethylallene

3[c]

S-(+)-1,1′-Binaphthyl

4[d]

R-(+)-2,2′-Diamino-6,6′-dimethylbiphenyl

5[e]

R-(−)-*E*-Cyclooctene

6[f]

Z,*E*-1,3-cyclooctadiene

7[g]

S-(+)-Spiro[3,3]-hepta-1,5-diene

8[h]

R-(+)-1,1′-Spirobiindan

9[i]

10[j]

a. W. C. Agosta, *J. Am. Chem. Soc.* **86**:2638 (1964).
b. W. L. Waters, S. S. Linn, and M. C. Caserio, *J. Am. Chem. Soc.* **90**:6741 (1968).
c. P. A. Browne, M. M. Harris, R. Z. Mazengo, and S. Singh, *J. Chem. Soc., C* **1971**:3990.
d. L. H. Pignolet, R. P. Taylor, and W. DeW. Horrocks, Jr., *J. Chem. Soc. Chem. Commun.* **1968**:1443.
e. A. C. Cope and A. S. Mehta, *J. Am. Chem. Soc.* **86**:1268 (1964).
f. R. Isaksson, J. Rochester, J. Sandstrom, and L.-G. Wirstrand, *J. Am. Chem. Soc.* **107**:4074 (1985).
g. L. A. Hulshof, M. A. McKervey, and H. Wynberg, *J. Am. Chem. Soc.* **96**:3906 (1974).
h. J. H. Brewster and R. T. Prudence, *J. Am. Chem. Soc.* **95**:1217 (1973); R. K. Hill and D. A. Cullison, *J. Am. Chem. Soc.* **95**:1229 (1973).
i. E. Vogel, W. Tückmantel, K. Schlögl, M. Widhalm, E. Kraka, and D. Cremer, *Tetrahedron Lett.* **25**:4925 (1984).
j. N. Harada, H. Uda, T. Nozoe, Y. Okamoto, H. Wakabayashi, and S. Ishikawa, *J. Am. Chem. Soc.* **109**:1661 (1987).

2.2. Diastereomeric Relationships

Diastereomers include all stereoisomers that are not related as an object and its mirror image. Consider the four structures in Fig. 2.3. These structures represent the four stereoisomers of 2,3,4-trihydroxybutanal. The configurations of C-2 and C-3 are indicated. Each stereogenic center is designated R or S by application of the sequence rule. Each of the four structures is stereoisomeric with respect to any of the others. The $2R,3R$ and $2S,3S$ isomers are enantiomeric, as are the $2R,3S$ and $2S,3R$ pair. The $2R,3S$ isomer is diastereomeric with the $2S,3S$ and $2R,3R$ isomers because they are stereoisomers but not enantiomers. Any given structure can have only one enantiomer. All other stereo-isomers of that molecule are diastereomeric. The relative configuration of diastereomeric molecules is frequently specified using the terms *syn* and *anti*. The molecules are represented as extended chains. Diastereomers with substituents on the same side of the extended chain are *syn* stereoisomers, whereas those with substituents on opposite sides are *anti* stereoisomers.

Diastereoisomers differ in both physical properties and chemical reactivity. They generally have different melting points, boiling points, solubility, chromatographic mobility, and so on. The specific rotations of diastereomeric molecules differ in both magnitude and sign. The difference in chemical reactivity can be small, such as a difference in rate, or two diastereomers can lead to entirely different products, depending on the mechanism of the particular reaction. Because of their differing physical and chemical properties, diastereomers can be separated by methods such as crystallization or chromatography.

Sometimes the terms *erythro* and *threo* are used to specify the relative configuration of two adjacent stereogenic centers. The terms are derived fom the sugars erythrose and threose. The terms were originally defined such that a Fischer projection formula in which two adjacent substituents were on the same side was the *erythro* isomer and that in which the substituents were on opposite sides was the *threo* isomer.

Fig. 2.3. Stereoisomeric relationships in 2,3,4-trihydroxy-butanal.

Unfortunately, assignment of molecules that are not closely related to the reference molecules becomes a subjective matter of assigning which substituents are "similar." The application of the terminology to cases in which the chiral centers are not adjacent is also ambiguous. As a result, the *threo–erythro* terminology is not a general method of specifying stereochemical relationships.

Fischer projection formulas can be used to represent molecules with several stereogenic centers and are commonly used for carbohydrates. For other types of structures, a more common practice is to draw the molecule in an extended conformation with the main chain horizontal. In this arrangement, each tetrahedral carbon has two additional substituents, one facing out and one in. The orientation is specified with solid wedged bonds for substituents facing out and with dashed bonds for substituents that point in.

representation of 3,4-dimethyldecane-5,6-diol.

Since the main chain in this representation is in an entirely staggered conformation, whereas in the Fischer projection formulas the conformation represented is completely eclipsed, an *anti* relationship between two adjacent substituents in an extended structure corresponds to being on the same side in a Fischer projection formula (*erythro*) whereas a *syn* relationship corresponds to being on opposite sides in the Fischer projection (*threo*).

Since chirality is a property of a molecule as a whole, the specific juxtaposition of two or more stereogenic centers in a molecule may result in an achiral molecule. For example, there are three stereoisomers of tartaric acid (2,3-dihydroxybutanedioic acid). Two of these are chiral and optically active but the third is not.

D-tartaric acid   L-tartaric acid   *meso*-tartaric acid

Plane of symmetry in
eclipsed conformation
of *meso*-tartaric acid

Center of symmetry in
anti staggered conformation
of *meso*-tartaric acid

The reason that the third stereoisomer is achiral is that the substituents on the two asymmetric carbons are located with respect to each other in such a way that a molecular plane of symmetry exists. Compounds that incorporate asymmetric atoms but are nevertheless achiral are called *meso* forms. This situation occurs whenever pairs of stereogenic centers are disposed in the molecule in such a way as to create an element of symmetry. A particularly striking example is the antibiotic nonactin.[12] (Work Problem 2.24 to convince yourself that nonactin is a *meso* form.)

nonactin

Incorporation of stereogenic centers into cyclic structures produces special stereochemical circumstances. Except in the case of cyclopropane, the lowest-energy conformation of the rings is not planar. Most cyclohexane derivatives adopt a chair conformation. For example, the two conformers of *cis*-1,2-dimethylcyclohexane are both chiral. However, the two conformers are enantiomeric so the conformational change leads to racemization. Because the barrier to this conformational change is low (10 kcal/mol), the two enantiomers are rapidly interconverted.

rotate 120°

enantiomeric

While individual conformers of *cis*-1,2-dimethylcyclohexane
are chiral, the two conformers are enantiomeric.

Certain dimethylcycloalkanes contain a plane of symmetry. For example, both chair conformers of *cis*-1,3-dimethylcyclohexane possess a plane of symmetry bisecting the molecule through C-2 and C-5. The *trans* isomer does not have any element of symmetry and is chiral.

identical

enantiomeric

enantiomeric

cis-1,3-Dimethylcyclohexane has a plane
of symmetry through C-2 and C-5 and is achiral.

identical

trans-1,3-Dimethylcyclohexane is chiral.

One simple test for chirality of substituted cycloalkanes is to represent the ring in planar form. If the planar form is achiral because of a symmetry element, the compound will not

12. J. Dominguez, J. D. Dunitz, H. Gerlach, and V. Prelog, *Helv. Chim. Acta* **45**:129 (1962); H. Gerlach and V. Prelog, *Liebigs Ann. Chem.* **669**:121 (1963); B. T. Kilbourn, J. D. Dunitz, L. A. R. Pioda, and W. Simon, *J. Mol. Biol.* **30**:559 (1967).

exist as an enantiomerically biased sample, even if individual conformers may be chiral. Ring sizes 3–6 are classified in this way in Scheme 2.3.

Since the presence of a plane of symmetry in a molecule ensures that it will be achiral, one approach to classification of stereoisomers as chiral or achiral is to examine the molecule for symmetry elements. There are other elements of symmetry in addition to planes of symmetry that ensure that a molecule will be superimposable on its mirror image. The *trans,cis,cis* and *trans,trans,cis* stereoisomers of 1,3-dibromo-*trans*-2,4-dimethylcyclobutane are illustrative. This molecule does not possess a plane of symmetry, but the mirror images are superimposable, as illustrated below. This molecule possesses a *center of symmetry*. A center of symmetry is a point from which any line drawn through the molecule encounters an identical environment in either direction from the center of symmetry.

trans, cis, cis-1,3-Dibromo-2,4-dimethylcyclobutane has a plane of symmetry. The mirror images are superimposable.

trans, trans, cis-1,3-Dibromo-2,4-dimethylcyclobutane has a center of symmetry. The mirror images are superimposable.

Scheme 2.3. Chiral and Achiral Disubstituted Cycloalkanes

A. Achiral structures

B. Chiral structures

Because diastereoisomers have different physical and chemical properties, they can be separated by a range of chemical and physical methods. The process of *resolution* is the separation of a racemic mixture. Separation is frequently effected by converting the enantiomers into a mixture of diastereomers by reaction with a pure enantiomer of a second reagent, the *resolving agent*.[13] Because the two resulting products will be diastereomeric, they can be separated. The separated diastereomers can then be reconverted to the pure enantiomers by reversing the initial chemical transformation. An example of this method is shown in Scheme 2.4 for the resolution of a racemic carboxylic acid by way of a diastereomeric salt resulting from reaction with an enantiomerically pure amine. The *R*-acid, *R*-amine and *S*-acid, *R*-amine salts are separated by fractional recrystallization. The resolved acids are regenerated by reaction with a strong acid, which liberates the carboxylic acid from the amine salt.

Although the traditional method of separating the diastereomeric compounds generated in a resolution procedure is fractional crystallization, chromatographic procedures are now common and convenient.[14] Diastereomeric compounds exhibit different adsorption

Scheme 2.4. Resolution of 2-Phenyl-3-methylbutanoic Acid[a]

a. C. Aaron, D. Dull, J. L. Schmiegel, D. Jaeger, Y. Ohashi, and M. S. Mosher, *J. Org. Chem.* **32**:2797 (1967).

13. For reviews of resolution methods, see S. H. Wilen, *Top. Stereochem.* **6**:107 (1971); S. H. Wilen, A. Collet, and J. Jacques, *Tetrahedron* **33**:2725 (1977); A. Collet, M. J. Brienne, and J. Jacques, *Chem. Rev.* **80**:215 (1980); J. Jacques, A. Collet, and S. H. Wilen, *Enantiomers, Racemates and Resolutions*, Wiley-Interscience, New York, 1981.

14. G. Subramanian, ed., *A Practical Approach to Chiral Separations by Liquid Chromatography*, VCH Publishers, Weinheim, 1994; S. G. Allenmark, *Chromatographic Enantioseparation; Methods and Applications*, Ellis Horwood, New York, 1991.

on achiral materials and can be separated by column chromatography or by taking advantage of the greater separation powers of HPLC.

Separation of enantiomers by physical or chemical methods requires the use of a chiral material, reagent, or catalyst. Both natural materials, such as polysaccharides and proteins, and solids that have been synthetically modified to incorporate chiral structures have been developed for use in separation of enantiomers by HPLC. The use of a chiral stationary phase makes the interactions between the two enantiomers with the adsorbent nonidentical and thus establishes a different rate of elution through the column. The interactions typically include hydrogen bonding, dipolar interactions, and $\pi-\pi$ interactions. These attractive interactions may be disturbed by steric repulsions, and frequently the basis of enantioselectivity is a better steric fit for one of the two enantiomers.[15,16]

The potential for use of chiral natural materials such as cellulose for separation of enantiomers has long been recognized, but development of efficient materials occurred relatively recently. Several acylated derivatives of cellulose are effective chiral stationary phases. Benzoate esters and aryl carbamates are particularly useful. These materials are commercially available on a silica support and under the trademark Chiralcel. Figure 2.4 shows the resolution of γ-phenyl-γ-butyrolactone with the use of acetylated cellulose as the adsorbent material.

silica surface

Synthetic chiral adsorbents are usually prepared by tethering a chiral molecule to a silica surface. The attachment to the silica is through alkylsiloxy bonds. A study which demonstrates the technique reports the resolution of a number of aromatic compounds on a 1- to 8-g scale. The adsorbent is a silica that has been derivatized with a chiral reagent. Specifically, hydroxyl groups on the silica surface are covalently bound to a derivative of R-phenylglycine. A medium-pressure chromatography apparatus is used. The racemic mixture is passed through the column, and, when resolution is successful, the separated enantiomers are isolated as completely resolved fractions.[17] Scheme 2.5 shows some other examples of chiral stationary phases.

Another means of resolution depends on the difference in rates of reaction of two enantiomers with a chiral reagent. The transition-state energies for reaction of each enantiomer with one enantiomer of a chiral reagent will be different. This is because the transition states and intermediates (R-substrate...R-reactant) and (S-substrate...R-reactant) are diastereomeric. *Kinetic resolution* is the term used to describe the separation of enantiomers based on different reaction rates with an enantiomerically pure reagent.

15. D. R. Taylor and K. Maher, *J. Chromatogr. Sci.* **30**:67 (1992).
16. Y. Okamoto and Y. Kaida, *J. Chromatogr., A* **666**:403 (1994); K. Oguni, H. Oda, and A. Ichida, *J. Chromatogr.* **694**:91 (1995).
17. W. H. Pirkle and J. M. Finn, *J. Org. Chem.* **47**: 4037 (1982). For other examples of chiral HPLC adsorbents, see W. H. Pirkle and M. H. Hyun, *J. Org. Chem.* **49**: 3043 (1984); W. H. Pirkle, T. C. Pochapsky, G. S. Mahler, D. E. Corey, D. S. Reno, and D. M. Alessi, *J. Org. Chem.* **51**: 4991 (1986).

Fig. 2.4. Preparative chromatographic resolution of 5 g of γ-phenyl-γ-butyrolactone on 480 g of CTA I (column 5 cm × 60 cm). [Reproduced from *Helv. Chim. Acta* **70**:1569 (1987) by permission of Verlag Helvetica Chimica Acta, A.G.]

Scheme 2.5. Chiral Stationary Phases for HPLC Separation of Enantiomers

Tether   O—Si—(CH₂)ₙR

1^a $n = 3$, R =

2^b $n = 3$, R =

3^c $n = 10$, R =

a. W. H. Pirkle, D. W. House, and J. M. Finn, *J. Chromatogr.* **192**:143 (1980).
b. W. H. Pirkle and M. H. Hyun, *J. Chromatogr.* **322**:287 (1985).
c. W. H. Pirkle, C. J. Welch, and B. Lamm, *J. Org. Chem.* **57**:3854 (1997); W. H. Pirkle and C. J. Welch, *J. Liq. Chromatogr.* **15**:1947 (1992).

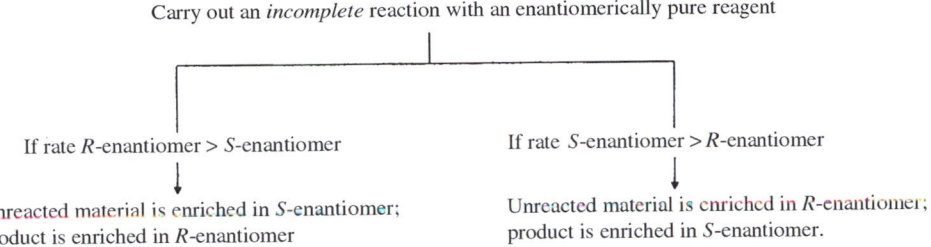

S-enantiomer + *R*-enantiomer

Carry out an *incomplete* reaction with an enantiomerically pure reagent

If rate *R*-enantiomer > *S*-enantiomer

Unreacted material is enriched in *S*-enantiomer;
product is enriched in *R*-enantiomer

If rate *S*-enantiomer > *R*-enantiomer

Unreacted material is enriched in *R*-enantiomer;
product is enriched in *S*-enantiomer.

Fig. 2.5. Basis of kinetic resolution.

Figure 2.5 summarizes the basis of kinetic resolution. Because the separation is based on differing rates of reaction, the degree of resolution that can be achieved depends on both the magnitude of the rate difference and the extent of reaction. The greater the difference in the two rates, the higher the enantiomeric purity of both the reacted and the unreacted enantiomer. The extent of enantiomeric purity can be controlled by controlling the degree of conversion. As the degree of conversion increases, the enantiomeric purity of the *unreacted* enantiomer becomes very high.[18] The relationship between the relative rate of reaction, extent of conversion, and enantiomeric purity of the unreacted enantiomer is shown in Fig. 2.6. Of course, the high conversion required for high enantiomeric purity when the relative reactivity difference is low has a serious drawback. The *yield* of the *unreacted* substrate is low if the overall conversion is high. Thus, with relative reactivity

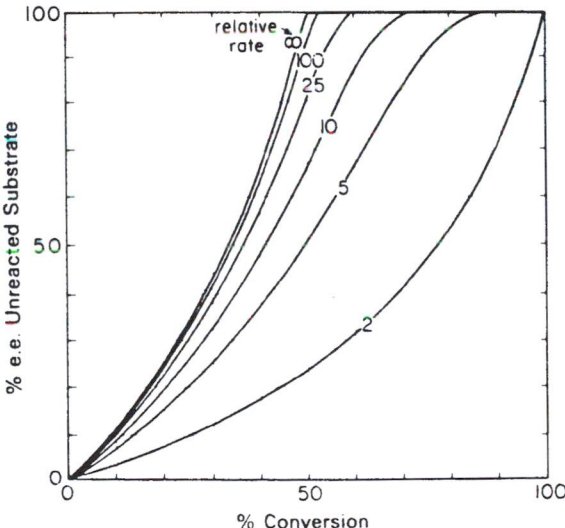

Fig. 2.6. Dependence of enantiomeric excess on relative rate of reaction and extent of conversion with a chiral reagent in kinetic resolution. [Reproduced from *J. Am. Chem. Soc.* **103**:6237 (1981) by permission of the American Chemical Society.]

18. V. S. Martin, S. S. Woodard, T. Katsuki, Y. Yamada, M. Ikeda, and K. B. Sharpless, *J. Am. Chem. Soc.* **103**:6237 (1981).

differences of < 10, high enantiomeric purity can be achieved only at the expense of low yield. Scheme 2.6 gives some specific examples of kinetic resolution procedures.

Preparation of enantiomerically enriched materials by use of chiral catalysts is also based on differences in transition-state energies. While the reactant is part of a complex or intermediate containing a chiral catalyst, it is in a chiral environment. The intermediates and complexes containing each enantiomeric reactant and a homochiral catalyst are diastereomeric and differ in energy. This energy difference can then control selection between the stereoisomeric products of the reaction. If the reaction creates a new stereogenic center in the reactant molecule, there can be a preference for formation of one enantiomer over the other.

Enzymes constitute a particularly important group of enantioselective catalysts.[19] Enzymes are highly efficient and selective catalysts and can carry out a variety of transformations. Because the enzymes are derived from L-amino acids, they are homochiral, and usually one enantiomer of a reactant is much more reactive than the other. The reason is that the interaction of the enzyme with one enantiomer is diastereomeric to its interaction with the other. Because enzyme catalysis is usually based on a specific fit to an "active site," the degree of selection between the two enantiomers is often very high. Enzyme-catalyzed reactions can therefore be used to resolve organic compounds. The most completely characterized enzymes that are available are those which catalyze hydrolysis of esters and amides (esterases, lipases, peptidases, acylases) and those which oxidize alcohols to ketones or aldehydes (dehydrogenases). Purified enzymes can be used, or the reaction can be done by incubating the reactant with an organism (yeast, for example) that produces an appropriate enzyme during fermentation. Scheme 2.7 gives some specific examples of enzymic resolutions.

The differing physical properties of diastereomers are also the basis for a particularly sensitive method for assessing the enantiomeric purity of compounds. Although, in principle, enantiomeric purity can be determined by measuring the optical rotation, this method is reliable only if the rotation of the pure compound is accurately known. This is never the case for a newly prepared material and is often uncertain for previously prepared compounds. If a derivative of a chiral compound is prepared in which a new chiral center is introduced, the two enantiomers will give different diastereomers. Because these will have different physical properties, their relative amounts can be determined. NMR spectroscopy is a convenient means of detecting and quantitating the two diastereomeric products. A pure enantiomer will give only a single spectrum, but a partially resolved material will show two overlapping spectra in the ratio of the two diastereomeric derivatives. The most widely used derivatizing reagent for the NMR method is a compound known as *Mosher's reagent*.[20] One reason that this compound is particularly useful is that the aromatic ring usually induces markedly different chemical shifts in the two diastereomeric products that are formed.

Mosher's reagent

19. J. B. Jones, *Tetrahedron* **42**:3351 (1986); J. B. Jones, in *Asymmetric Synthesis*, Vol. 5, J. D. Morrison, ed., Academic Press, New York, 1985, Chapter 9; G. M. Whitesides and C.-H. Wong, *Angew. Chem. Int. Ed. Engl.* **24**:617 (1985).
20. J. A. Dale, D. L. Dull, and H. S. Mosher, *J. Org. Chem.* **34**:2543 (1969).

1[a]

ratio of $R,S : S,S$ = 1.75 with 0.25 equiv anhydride

2[b]

Ti(OR)$_4$
t-BuOOH
(+) diisopropyl tartrate

37% yield recovered, 95% e.e.

3[c]

$(PhOCHCO)_2$ + CH_3CH—$\langle\rangle$—N → $PhOCHCOCH$—$\langle\rangle$—N + $PhOCHCO_2H$

1 equiv

S-isomer
1 equiv

92% R,S ester
8% S,S ester

84% e.e. S

4[d]

racemic + L-valine $\xrightarrow{H^+}$

27% yield, 96% d.e.

5[e]

S-2,2′-bis(diphenylphosphino)-1,1′-binaphthyl complex of Ru(OAc)$_2$

H$_2$

48% recovery, 96% e.e.

6[f]

racemic + Cl$_3$CC(CH$_3$)$_2$OC(O) ... $\xrightarrow{ZnCl_2}$

30% conversion

38% e.e. R

+

91% e.e. S

7[g]

Ti(O-i-Pr)$_4$, R-binaphthol
t-BuOOH

31% yield, 97% e.e.

a. Y. Hiraki and A. Tai, *Bull. Chem. Soc. Jpn.* **57**:1570 (1984).
b. S. Miyano, L. D. Lu, S. M. Viti, and K. B. Sharpless, *J. Org. Chem.* **48**:3608 (1983).
c. U. Salz and C. Rüchardt, *Chem. Ber.* **117**:3457 (1984).
d. P. Stead, H. Marley, M. Mahmoudian, G. Webb, D. Noble, Y. T. Ip, E. Piga, T. Rossi, S. Roberts, and M. J. Dawson, *Tetrahedron Asymmetry* **7**:2247 (1996).
e. M. Kitamura, I. Kasahara, K. Manabe, R. Noyori, and H Takaya *J. Org. Chem.* **53**:708 (1988).
f. E. Vedejs and X. Chen, *J. Am. Chem. Soc.* **118**:1809 (1996).
g. N. Komatsu, M. Hashizume, T. Sugita, and S. Uemura, *J. Org. Chem.* **58**:7624 (1993).

Scheme 2.7. Examples of Enzymatic Resolutions

a. B. Cambou and A. M. Klibanov, *J. Am. Chem. Soc.* **106:**2687 (1984).
b. A. Sharma, A. S. Pawar, and S. Chattopadhyay, *Synth. Commun.* **26:**19 (1996).
c. Y. Terao, K. Tsuji, M. Murata, K. Achiwa, T. Nishio, N. Watanabe, and K. Seto, *Chem. Pharm. Bull.* **37:**1653 (1989).
d. H. Ziffer, K. Kawai, M. Kasai, M. Imuta, and C. Froussios, *J. Org. Chem.* **48:**3017 (1983).
e. G. Caron and R. J. Kazlauskas, *J. Org. Chem.* **56:** 7251 (1991).
f. J. M. Roper and D. P. Bauer, *Synthesis* **1983:**1041.
g. N. Kosui, M. Waki, T. Kato, and N. Izumiya, *Bull. Chem. Soc. Jpn.* **55:**918 (1982).
h. M. Kalesse and M. Eh, *Tetrahedron Lett.* **37:**1767 (1996).
i. G. Fantin, M. Fogagnolo, A. Medici, P. Pedrini, S. Poli, and M. Sinigaglia, *Tetrahedron Lett.* **34:**883 (1993).
j. U. Wandel, M. Mischitz, W. Kroutil, and K. Faber, *J. Chem. Soc., Perkin Trans. 1* **1995:**735.

Changes in NMR spectra can also be observed as the result of formation of noncovalent complexes between enantiomeric molecules and another chiral reagent. This is the basis of the use of *chiral shift reagents* to determine the enantiomeric purity of chiral substances.[21] Several of the lanthanide elements have the property of forming strong complexes with alcohols, ketones, and other functional groups having Lewis base character. If the lanthanide ion is in a chiral environment as the result of an enantiomerically pure ligand, two diastereomeric complexes are formed. The lanthanide elements induce large NMR shifts, and, as a result, shifted spectra are seen for the two complexed enantiomers. The relative intensities of the two spectra correspond to the ratio of enantiomers present in the sample. Figure 2.7 shows the NMR spectrum of an unequal mixture of the two enantiomers of 1-phenylethylamine in the presence of a europium shift reagent.[22]

Geometric isomers of alkenes are diastereomeric, since they are stereoisomers but not enantiomeric. The specification of the geometry of double bonds as *cis* and *trans* suffers from the same ambiguity as specifying configuration by the Fischer convention; that is, it requires a subjective judgment about the "similarity" of groups. The sequence rule is the basis for an unambiguous method for assignment of alkene geometry.[23] The four substituents on the double bond are taken in pairs. The sequence rules are used to determine if the higher-priority groups on the atoms forming the double bond are on the same side or opposite sides of the double bond. If the higher-priority groups are on the

Fig. 2.7. NMR spectrum of 1-phenylethylamine in the presence of a chiral shift reagent, showing differential chemical shift of methine and methyl signals and indicating ratio of *R*- to *S*-enantiomers. [Reproduced from *J. Am. Chem. Soc.* **93**:5914 (1971) by permission of the American Chemical Society.]

21. G. R. Sullivan, *Top. Stereochem.* **10**:287 (1978); R. R. Fraser, in *Asymmetric Synthesis*, Vol. 1, J. D. Morrison, ed., Academic Press, New York, 1983, Chapter 9.

22. G. M. Whitesides and D. W. Lewis, *J. Am. Chem. Soc.* **93**:5914 (1971); M. D. McCreary, D. W. Lewis, D. L. Wernick, and G. M. Whitesides, *J. Am. Chem. Soc.* **96**:1038 (1974).

23. J. E. Blackwood, C. L. Gladys, K. L. Loening, A. E. Petrarca, and J. E. Rush, *J. Am. Chem. Soc.* **90**:509 (1968).

same side, the descriptor is Z (from the German word *zusammen*, together); if they are on opposite sides, the descriptor is E (from *entgegen*, opposite). As in applying the sequence rule to stereogenic centers, if the atoms directly attached to the double bond have the same atomic number, the priorities are assigned by sequentially comparing atoms in the substituent until priority can be established. The system can also be applied to multiple bonds involving elements other than carbon, such as C=N. The Z and E descriptors have replaced *syn* and *anti* for describing the stereochemistry of oximes. As in the case of stereogenic centers, if an atom at a double bond does not have two substituents (as is the case for oximes), then a "phantom ligand" with atomic number zero is assumed and assigned the lower priority. Scheme 2.8 shows some stereoisomeric compounds named according to the sequence rule convention.

Scheme 2.8. Stereoisomeric Alkenes and Related Molecules with the Double-Bond Geometry Named According to the Sequence Rule

1^a (Z)-3-Decenoic acid (the sex pheromone of the furniture carpet beetle)

$$CH_3(CH_2)_5 \quad CH_2CO_2H$$
$$H \quad\quad H$$

2^b Methyl (2E,6E,10Z)-10,11-epoxy-3,7,11-trimethyltridecadienoate (the juvenile hormone of the tobacco hornworm)

$$CH_3CH_2 \quad\quad CH_3 \quad\quad CH_3$$
$$H_3C \quad\quad\quad\quad\quad\quad\quad\quad\quad CO_2CH_3$$
$$O$$

3^c Nitrones and oxime ethers

$$p\text{-}CH_3C_6H_4 \quad\quad O \quad\quad\quad\quad p\text{-}CH_3C_6H_4$$
$$=N^+ \quad\quad\quad\quad\quad\quad\quad\quad =N$$
$$C_6H_5 \quad CH(C_6H_5)_2 \quad\quad C_6H_5 \quad OCH(C_6H_5)_2$$
$$\quad\quad Z \quad\quad\quad\quad\quad\quad\quad\quad\quad E$$

4^d (2Z,4Z,6E,8E)-9-(3′-Furyl)-2,6-dimethylnona-2,4,6,8-tetraen-4-olide (dihydrofreelingyne)

a. H. Fukui, F. Matsumara, M. C. Ma, and W. E. Burkholder, *Tetrahedron Lett.* **1974**:3536.
b. R. C. Jennings, K. J. Judy, and D. A. Schooley, *J. Chem. Chem. Commun.* **1975**:21.
c. T. S. Dobashi and E. J. Grubbs, *J. Am. Chem. Soc.* **95**:5070 (1973).
d. C. F. Ingham and R. A. Massy-Westropp, *Aust. J. Chem.* **27**:1491 (1974).

2.3. Stereochemistry of Reactions

Up to this point, we have emphasized the stereochemical properties of molecules as objects, without concern for processes which affect the molecular shape. The term *dynamic stereochemistry* applies to the topology of processes which effect a structural change. The cases that are most important in organic chemistry are chemical reactions, conformational changes, and noncovalent complex formation. In order to understand the stereochemical aspects of a dynamic process, it is essential not only that the stereochemical relationship between starting and product states be established, but also that the spatial features of proposed intermediates and transition states must account for the observed stereochemical transformations.

In describing the stereochemical features of chemical reactions, we can distinguish between two types: stereospecific reactions and stereoselective reactions.[24] A *stereospecific reaction* is one in which stereoisomeric starting materials afford stereoisomerically different products under the same reaction conditions. A *stereoselective reaction* is one in which a single reactant has the capacity of forming two or more stereoisomeric products in a particular reaction but one is formed preferentially.

The stereochemistry of the most fundamental reaction types such as addition, substitution, and elimination are described by terms which specify the stereochemical relationship between the reactants and products. Addition and elimination reactions are classified as *syn* or *anti*, depending on whether the covalent bonds which are made or broken are on the same face or opposite faces of the plane of the double bond.

Substitution reactions at tetrahedral centers are classified as proceeding with retention or inversion of configuration or with racemization. The term *retention of configuration* applies to a process in which the relative spatial arrangement at the reaction center is the same in the reactant and the product. *Inversion of configuration* describes a process in which the substitution site in the product has a mirror-image relationship to that of the reactant. A substitution process that generates both possible enantiomers of a product from a single enantiomer of the reactant occurs with *racemization*. Such a process can result in complete or partial racemization depending on whether the product is a racemic mixture or if an excess of one enantiomer is formed. The term *epimerization* is used to describe the case of racemization of a single stereogenic center in a diastereomer, the configuration of the other centers being maintained.

While it may be convenient to use optically active reactants to probe the stereochemistry of substitution reactions, it should be emphasized that the stereochemistry of a reaction is a feature of the mechanism, not the means of determining it. Thus, it is proper to speak of a substitution process such as the hydrolysis of methyl iodide as proceeding

24. E. L. Eliel, S. H. Wilen, and L. N. Mander, *Stereochemistry of Organic Compounds*, John Wiley & Sons, New York, 1993, p. 837.

with inversion, even though the configuration is not directly discernible because of the achiral nature of the reactant and product.

Inversion of configuration at carbon
in hydrolysis of methyl iodide.

Some stereospecific reactions are listed in Scheme 2.9. Examples of stereoselective reactions are presented in Scheme 2.10. As can be seen in Scheme 2.9, the starting materials in these stereospecific processes are stereoisomeric pairs, and the products are stereoisomeric with respect to each other. Each reaction proceeds to give a single stereoisomer without contamination by the alternative stereoisomer. The stereochemical relationships between reactants and products are determined by the reaction mechanism. Detailed discussion of the mechanisms of these reactions will be deferred until later chapters, but some comments can be made here to illustrate the concept of stereospecificity.

Scheme 2.9. Stereospecific Reactions

A. Stereospecific addition to alkenes

1[a]   Epoxidation

2[b]   Addition of dibromocarbene

3[c]   Bromination

Scheme 2.9. (*continued*)

B. Nucleophilic substitution

4[d] *cis*- and *trans*-4-*t*-Butylcyclohexyl *p*-toluenesulfonate

H $OSO_2C_6H_4pCH_3$ NaSPh   PhS H

H $C(CH_3)_3$ H $C(CH_3)_3$

p-$CH_3C_6H_4SO_2O$ H   NaSPh   H SPh

H $C(CH_3)_3$ H $C(CH_3)_3$

5[e,f] S-(+)- and R-(−)-1-Methylheptyl *p*-toluenesulfonate

[e] $CH_3(CH_2)_5$ H
$-OSO_2C_6H_4CH_3$ $\xrightarrow{CH_3CO_2^-}$ CH_3CO_2
H_3C H $(CH_2)_5CH_3$ CH_3

[f] $CH_3(CH_2)_5$ CH_3
$-OSO_2C_6H_4CH_3$ $\xrightarrow{CH_3CO_2^-}$ CH_3CO_2
H H_3C $(CH_2)_5CH_3$ H

C. Elimination

6[g] Dehydrohalogenation

CH_3 Br
Ph   Ph $\xrightarrow{KOCH_2CH_3}$ H_3C Ph
H   H   Ph H
threo

CH_3 Br
Ph   H $\xrightarrow{KOCH_2CH_3}$ H_3C H
H   Ph   Ph Ph
erythro

7[h] Pyrolysis of amine oxides

H_3C CH_3
Ph   H $\xrightarrow{115°C}$ H_3C CH_3
H $\overset{+}{N}(CH_3)_2$ Ph H
    ^{-}O
threo

H_3C H
Ph CH_3 $\xrightarrow{125°C}$ H_3C H
H $\overset{+}{N}(CH_3)_2$ Ph CH_3
    ^{-}O
erythro

a. L. P. Witnauer and D. Swern, *J. Am. Chem. Soc.* **72**:3364 (1950).
b. P. S. Skell and A. Y. Garner, *J. Am. Chem. Soc.* **78**:3409 (1956).
c. A. Modro, G. H. Schmid, and K. Yates, *J. Org. Chem.* **42**:3673 (1977).
d. E. L. Eliel and R. S. Ro, *J. Am. Chem. Soc.* **79**:5995 (1957).
e. A. Streitwieser, Jr., A. C. Waiss, Jr., *J. Org. Chem.* **27**:290 (1962).
f. H. Philips, *J. Chem. Soc.* **1925**:2582.
g. D. J. Cram, F. D. Greene, and C. H. DePuy, *J. Am. Chem. Soc.* **78**:790 (1956).
h. D. J. Cram and J. E. McCarty, *J. Am. Chem. Soc.* **76**:5740 (1954).

Entries 1 and 2 in Scheme 2.9 are typical of concerted *syn* addition to alkene double bonds. On treatment with peroxyacetic acid, the *Z*-alkene affords the *cis*-oxirane, whereas the *E*-alkene affords only the *trans*-oxirane. Similarly, addition of dibromocarbene to *Z*-2-butene yields exclusively 1,1-dibromo-*cis*-2,3-dimethylcyclopropane, whereas only 1,1-dibromo-*trans*-2,3-dimethylcyclopropane is formed from *E*-2-butene. There are also numerous stereospecific *anti* additions. Entry 3 shows the *anti* stereochemistry typical of bromination of simple alkenes.

Nucleophilic substitution reactions at sp^3 carbon by direct displacement proceed with inversion of configuration at the carbon atom bearing the leaving group. Thus, *cis*-4-*t*-butylcyclohexyl *p*-toluenesulfonate is converted by thiophenoxide ion to *trans*-4-*t*-butylcyclohexyl phenyl thioether. The stereoisomeric *trans*-*p*-toluenesulfonate gives the *cis*-phenyl thioether. (entry 4). 1-Methylheptyl *p*-toluenesulfonate esters react with acetate ion to give the substitution product of inverted configuration, as can be demonstrated with the use of optically active reactant (entry 5).

Entry 6 is an example of a stereospecific elimination reaction of an alkyl halide in which the transition state requires the proton and bromide ion that are lost to be in an *anti* orientation with respect to each other. The diastereomeric *threo*- and *erythro*-1-bromo-1,2-diphenyl-propanes undergo β-elimination to produce stereoisomeric products. Entry 7 is an example of a pyrolytic elimination requiring a *syn* orientation of the proton that is removed and the nitrogen atom of the amine oxide group. The elimination proceeds through a cyclic transition state in which the proton is transferred to the oxygen of the amine oxide group.

The stereoselective reactions in Scheme 2.10 include one example that is completely stereoselective (entry 3), one that is highly stereoselective (entry 6), and others in which the stereoselectivity is modest to low (entries l, 2, 4, 5, and 7). The addition of formic acid to norbornene (entry 3) produces only the *exo* ester. Reduction of 4-*t*-butylcyclohexanone (entry 6) is typical of the reduction of unhindered cyclohexanones in that the major diastereomer produced has an equatorial hydroxyl group. Certain other reducing agents, particularly sterically bulky ones, exhibit the opposite stereoselectivity and favor the formation of the diastereomer having an axial hydroxyl group. The alkylation of 4-*t*-butylpiperidine with benzyl chloride (entry 7) provides only a slight excess of one diastereomer over the other.

We have previously seen (Scheme 2.9, entry 6), that the dehydrohalogenation of alkyl halides is a stereospecific reaction involving an *anti* orientation of the proton and the halide leaving group in the transition state. The elimination reaction is also moderately stereoselective (Scheme 2.10, entry 1) in the sense that the more stable of the two alkene isomers is formed preferentially. Both isomers are formed by *anti* elimination processes, but these processes involve stereochemically distinct hydrogens. Base-catalyzed elimination of 2-iodobutane affords three times as much *E*-2-butene as *Z*-2-butene.

A. Formation of alkenes

1[a]   Dehydrohalogenation

CH₃CH₂CHCH₃ (with I substituent) → KOC(CH₃)₃ / dimethyl sulfoxide

$$CH_3CH_2CHCH_3 \xrightarrow[\text{dimethyl sulfoxide}]{KOC(CH_3)_3}$$

(60%) + (20%) + CH₃CH₂CH=CH₂ (20%)

2[b]   Thermal elimination of amine oxide

CH₃CH₂CHCH₃ with ⁻O—N⁺(CH₃)₂

(71%) + (29%)

B. Addition of alkenes

3[c]   Addition of formic acid to norbornene

HCO₂H →

4[d]   Addition of phenoxycarbene to cyclohexene

PhOCH₂Cl / n-BuLi/ether →

(26%) + (14%)

C. Addition to carbonyl groups

5[e]

CH₃MgI →

Ph ... CH₃ / OH   syn (67%) + Ph ... CH₃ / OH   anti (33%)

6[f]

(CH₃)₃C—(cyclohexanone) → LiAlH₄

(CH₃)₃C—...OH (90%) + (CH₃)₃C—...OH (10%)

D. Formation of quaternary ammonium salts

7[g]

PhCH₂Cl / CH₃OH, 30°C →

(58%) + (42%)

a. R. A. Bartsch, G. M. Pruss, B. A. Bushaw, and K. W. Wiegers, *J. Am. Chem. Soc.* **95**:3405 (1973).
b. A. C. Cope, N. A. LeBel, H.-H. Lee, and W. R. Moore, *J. Am. Chem. Soc.* **79**:4720 (1957).
c. D. C. Kleinfelter and P. von R. Schleyer, *Org. Synth.* **V**:852 (1973).
d. U. Schöllkopf, A. Lerch, and W. Pitteroff, *Tetrahedron Lett.* **1962**:241.
e. D. J. Cram and F. A. Abd Elhafez, *J. Am. Chem. Soc.* **74**:5828 (1952).
f. E. L. Eliel and M. N. Rerick, *J. Am. Chem. Soc.* **82**:1367 (1960).
g. A. T. Bottini and M. K. O'Rell, *Tetrahedron Lett.* **1967**:423.

Moderate stereoselectivity is also seen in the addition of phenoxycarbene to cyclohexene (entry 4), in which the product ratio is apparently influenced by steric factors that favor introduction of the larger group (PhO versus H) in the less crowded *exo* position.

The addition of methylmagnesium iodide to 2-phenylpropanal is stereoselective in producing twice as much *syn*-3-phenyl-2-butanol as the *anti* isomer (entry 5). The stereoselective formation of a particular configuration at a new stereogenic center in a reaction of a chiral reactant is called *asymmetric induction*. This particular case is one in which the stereochemistry can be predicted on the basis of an empirical correlation called *Cram's rule*. The structural and mechanistic basis of Cram's rule will be discussed in Chapter 3.

Standing in contrast to stereospecific and stereoselective processes are the racemization processes which result in formation of products of both configurations. The most common mechanistic course by which organic reactions lead to racemic products is by cleavage of one of the ligands from a stereogenic center to give a planar or rapidly inverting tricoordinate intermediate, such as a carbocation or free radical. In the absence of any special solvation effects, such intermediates are achiral and produce equal quantities of the two possible enantiomeric products. Nucleophilic substitution proceeding through a carbocation intermediate is a familiar example of such a racemization process. This reaction will be discussed in detail in Chapter 5.

The term racemization can be used to describe any process that leads to formation of both configurations at a stereogenic center and is not restricted to processes which involve bond cleavage. Examples include pyramidal inversion at trivalent nitrogen, sulfur, or phosphorus. The rate of racemization of such compounds depends upon the barrier to the inversion process. For ammonia and simple amines, the barrier is very low and inversion of configuration at nitrogen is rapid at room temperature. Thus, although unsymmetrically substituted amines are chiral, the process of racemization is too rapid to allow separation of the enantiomers. Incorporation of the nitrogen into a three-membered ring raises the barrier for inversion, owing to the additional strain in the planar transition state. For aziridine, the energy barrier to pyramidal inversion is 12 kcal/mol. Although this is too

inversion barrier 8–12 kcal/mol
depending on X

Certain substituted aziridines can be isolated as enantiomers as the result of still higher barriers. Most of these compounds are *N*-chloro- or *N*-alkoxyaziridines.[26]

Whereas the barrier for pyramidal inversion is low for second-row elements, the heavier elements have much higher barriers to inversion. The preferred bonding angle at trivalent phosphorus and sulfur is about $100°$, and thus a greater distortion is required to reach a planar transition state. Typical barriers for trisubstituted phosphines are 30–35 kcal/mol, whereas for sulfoxides the barriers are about 35–45 kcal/mol. Many phosphines and sulfoxides have been isolated in enantiomerically enriched form, and they undergo racemization by pyramidal inversion only at high temperature.[27]

(Ref. 28)

inversion barrier 32.7 kcal/mol at 130°C

Molecules that are chiral as a result of barriers to conformational interconversion can be racemized if the enantiomeric conformers are interconverted. The rate of racemization will depend upon the conformational barrier. For example, *E*-cyclooctene is chiral. *E*-Cycloalkenes can be racemized by a conformational process involving reorienting of the

25. J. D. Andose, J. M. Lehn, K. Mislow, and J. Wagner, *J. Am. Chem. Soc.* **92**:4050 (1970).
26. S. J. Brois, *J. Am. Chem. Soc.* **90**:506, 508 (1968); S. J. Brois, *J. Am. Chem. Soc.* **92**:1079 (1970); V. F. Rudchenko, O. A. D'yachenko, A. B. Zolotoi, L. O. Atovmyan, I. I. Chervin, and R. G. Kostyanovsky, *Tetrahedron* **38**:961 (1982).
27. For a review of racemization via vibrational inversion at trivalent stereogenic centers, see J. B. Lambert, *Top. Stereochem.* **6**:19 (1971).
28. R. D. Baechler and K. Mislow, *J. Am. Chem. Soc.* **93**:773 (1971).

double bond. The process is represented below but is more easily seen by working with a molecular model.

Since one of the vinyl hydrogens must "slip through" the ring, the energy barrier depends upon the ring size. *E*-Cyclooctene is quite stable to thermal racemization and can be recovered with no loss of enantiomeric purity after 7 days at 61°C.[29] When the ring size is larger, passage of the double bond through the ring occurs more easily and racemization takes place more rapidly. The half-life for racemization of *E*-cyclononene is 5 min at 0°C.[30] The rate of racemization of *E*-cyclodecene is so fast that racemization occurs immediately on its release from the platinum complex employed for its resolution.[30]

The dynamic stereochemistry of biaryls is conceptually similar. The energy barrier for racemization of optically active 1,1'-binaphthyl (Scheme 2.2, entry 3, p. 83) is 21–23 kcal/mol.[31] The two rings are not coplanar in the ground state, and the racemization takes place by rotation about the 1,1'-bond.

Rotation about the 1,1'-bond is resisted by van der Waals interactions between the hydrogens shown in the structures. These hydrogens crowd each other when the two naphthyl groups are coplanar, and the racemization process requires the hydrogens to move past each other. The existence of enantiomeric substituted biphenyls also depends on steric interactions between substituents. The relationship between the rate of racemization and

29. A. C. Cope, C. R. Ganellin, H. W. Johnson, Jr., T. V. VanAuken, and H. J. S. Winkler, *J. Am. Chem. Soc.* **85**:3276 (1963); the activation energy is 35.6 kcal/mol: A. C. Cope and B. A. Pawson, *J. Am. Chem. Soc.* **87**:3649 (1965).

30. A. C. Cope, K. Banholzer, H. Keller, B. A. Pawson, J. J. Whang, and H. J. S. Winkler, *J. Am. Chem. Soc.* **87**:3644 (1965).

31. A. K. Colter and L. M. Clemens, *J. Phys. Chem.* **68**:651 (1964).

the size of the substituents has been investigated.[32] There is a correlation between the barrier to rotation and the extent of steric repulsion between substituents.[33]

2.4. Prochiral Relationships

It is frequently necessary to distinguish between identical ligands that, even though they are bonded to the same atom, may be topologically nonequivalent. Let us consider 1,3-propanediol as an example. If a process occurs in which a proton at C-2 is substituted by another ligand, say, deuterium, the two possible substitution modes generate identical products. The two protons at C-2 are therefore topologically equivalent and are termed *homotopic* ligands.

Substitution products are superimposable. There is a plane of symmetry defined by the atoms H–C(2)–D.

If a similar process occurred involving the two protons at C-1, a stereochemically different situation will result. Substitution at C-1 produces a chiral product, 1-*deuterio*-1,3-propanediol:

(R)-1-*deuterio*-1,3-propanediol   (S)-1-*deuterio*-1,3-propanediol

The two protons at C-1 are topologically nonequivalent, since substitution of one produces a product that is stereochemically distinct from that produced by substitution of the other. Ligands of this type are termed *heterotopic*, and, because the products of substitution are enantiomers, the more precise term *enantiotopic* also applies.[34] If a chiral assembly is generated when a particular ligand is replaced by a new ligand, the original assembly is *prochiral*. Both C-1 and C-3 of 1,3-propanediol are *prochiral centers*.

The sequence rule may be applied directly to the specification of heterotopic ligands in prochiral molecules using the descriptors *pro-R* and *pro-S*. The assignment is done by selecting one of the heterotopic ligands at the prochiral center and arbitrarily assigning it a higher priority than the other, without disturbing the priorities of the remaining ligands. If application of the sequence rule results in assignment of R as the configuration of the

32. F. H. Westheimer, in *Steric Effects in Organic Chemistry*, M. S. Newman, ed., John Wiley & Sons, New York, 1956, Chapter 12.

33. G. Bott, L. D. Field, and S. Sternhell, *J. Am. Chem. Soc.* **102**:5618 (1980).

34. For a more complete discussion of definitions and terminology, see E. L. Eliel, *J. Chem. Educ.* **57**:52 (1980); E. L. Eliel, *Top. Curr. Chem.* **105**:1 (1982); K. R. Hanson, *J. Am. Chem. Soc.* **88**:2731 (1966).

prochiral center, then the selected ligand is *pro-R*. If the prochiral center is S, then the selected ligand is *pro-S*. It is customary to designate prochirality in structures by a subscript R or S at the appropriate atoms. For 1,3-propanediol, the prochiral hydrogens are indicated as shown below.

Enantiotopic atoms or groups have equivalent physical properties but differ in reactivity toward chiral reagents or catalysts. Many examples of discrimination between enantiotopic groups are found among enzyme-catalyzed reactions. The enzymes *liver alcohol dehydrogenase* and *yeast alcohol dehydrogenase*, for example, distinguish between the enantiotopic C-1 hydrogens of ethanol. Ethanol is a prochiral molecule, and it has been shown that its oxidation to acetaldehyde by either enzyme results in the loss of the *pro-R* hydrogen. Both enzymes require nicotinamide adenine dinucleotide (NAD$^+$) as a coenzyme, which serves as the immediate hydrogen acceptor. Incubation of (S)-1-*deuterio*-ethanol with the enzyme–coenzyme system produces exclusively acetaldehyde-1-*d*, whereas the same treatment of (R)-1-*deuterio*-ethanol affords acetaldehyde containing no deuterium.

The enzyme-catalyzed interconversion of acetaldehyde and ethanol serves to illustrate a second important feature of prochiral relationships, that of *prochiral faces*. Addition of a fourth ligand, different from the three already present, to the carbonyl carbon of acetaldehyde will produce a chiral molecule. The original molecule presents to the approaching reagent two faces which bear a mirror-image relationship to one another and are therefore enantiotopic. The two faces may be classified as *re* (from *rectus*) or *si* (from *sinister*), according to the sequence rule. If the substituents viewed from a particular face appear clockwise in order of decreasing priority, then that face is *re*; if counterclockwise, then *si*. The *re* and *si* faces of acetaldehyde are shown below.

Reaction of an achiral reagent with a molecule exhibiting enantiotopic faces will produce equal quantities of enantiomers, and a racemic mixture will result. The achiral reagent sodium borodeuteride, for example, will produce racemic 1-*deuterio*-ethanol. Chiral reagent can discriminate between the prochiral faces, and the reaction will be enantioselective. Enzymatic reduction of acetaldehyde-1-*d* produces R-1-*deuterio*-ethanol that is enantiomerically pure.[35]

35. H. R. Levy, F. A. Loewus, and B. Vennesland, *J. Am. Chem. Soc.* **79**:2949 (1957).

Fumaric acid is converted to L-malic acid by hydration in the presence of the enzyme *fumarase*. From the structure of the substrate and the configuration of the product, it is apparent that the hydroxyl group has been added to the *si* face of one of the carbon atoms of the double bond. Each of the trigonal carbon atoms of an alkene has its face specified separately. The molecule of fumaric acid shown below is viewed from the *re–re* face.

As was the case for kinetic resolution of enantiomers, enzymes typically exhibit a high degree of selectivity toward enantiotopic reaction sites. Selective reactions of enantiotopic groups provide enantiomerically enriched products. Thus, the treatment of an achiral material containing two enantiotopic functional groups is a means of obtaining enantiomerically enriched material. Most successful examples reported to date have involved hydrolysis. Several examples are outlined in Scheme 2.11.

Most enzyme-catalyzed processes, such as the examples just discussed, are highly enantioselective, leading to products of high enantiomeric purity. Reactions with other chiral reagents exhibit a wide range of enantioselectivity. A frequent objective of the study

Scheme 2.11. Enantioselective Transformations Based on Enzyme-Catalyzed Reactions Which Differentiate Enantiotopic Substituents

a. Y.-F. Wang and C. J. Sih, *Tetrahedron Lett.* **25**:4999 (1984).
b. A. S. Gopalan and C. J. Sih, *Tetrahedron Lett.* **25**:5235 (1984).
c. D. R. Deardorff, A. J. Matthews, D. S. McMeekin, and C. L. Craney, *Tetrahedron Lett.* **27**:1255 (1986).
d. K. P. Lok, I. J. Jakovac, and J. B. Jones, *J. Am. Chem. Soc.* **107**:2521 (1985).

of such reactions is to find the best reagent and conditions to optimize the enantioselectivity of the reaction.

Chiral chemical reagents can react with prochiral centers in achiral substances to give partially or completely enantiomerically pure product. An example of such processes is the preparation of enantiomerically enriched sulfoxides from achiral sulfides with the use of chiral oxidant. The reagent must preferentially react with one of the two prochiral faces of the sulfide, that is, the enantiotopic electron pairs.

An achiral reagent cannot distinguish between these two faces. In a complex with a chiral reagent, however, the two (phantom ligand) electron pairs are in different (enantiotopic) environments. The two complexes are therefore diastereomeric and are formed and react at different rates. Two reaction systems that have been used successfully for enantioselective formation of sulfoxides are illustrated below. In the first example, the $Ti(O-i-Pr)_4-t-BuOOH$–diethyl tartrate reagent is chiral by virtue of the presence of the chiral tartrate ester in the reactive complex.[36] With simple aryl methyl sulfides, up to 90% enantiomeric purity of the product is obtained.

(Ref. 37)

A second method uses sodium periodate ($NaIO_4$) as the oxidant in the presence of the readily available protein bovine serum albumin. In this procedure, the sulfide is complexed in the chiral environment of the protein. Although the oxidant is achiral, it encounters the sulfide in a chiral environment in which the two faces of the sulfide are differentiated.

(Ref. 38)

Another important example of an enantioselective reaction mediated by a chiral catalyst is the hydrogenation of 3-substituted 2-acetamidoacrylic acid derivatives.

36. F. Di Furia, G. Modena, and R. Seraglia, *Synthesis* **1984**:325.
37. P. Pitchen, E. Dunack, M. N. Desmukh, and H. B. Kagan, *J. Am. Chem. Soc.* **106**:8188 (1984).
38. T. Sugimoto, T. Kokubu, J. Miyazaki, S. Tanimoto, and M. Okano, *J. Chem. Soc., Chem. Commun.* **1979**:402; see also S. Colonna, S. Banfi, F. Fontana, and M. Sommaruga, *J. Org. Chem.* **50**:769 (1985).

Depending on the stereoselectivity of the reaction, either the R or the S configuration can be generated at C-2 in the product. This corresponds to enantioselective synthesis of the D and L enantiomers of α-amino acids. Hydrogenation using chiral catalysts has been carefully investigated.[39] The most effective catalysts for the reaction are rhodium

Scheme 2.12. Enantioselective Reduction of 2-Acetamidoacrylic Acids by Chiral Phosphine Complexes of Rhodium

R	Chiral ligand	Configuration of product	Enantiomeric excess (%)
1[a]		S	94
2[b] $(CH_3)_2CH-$		R	100
3[c]		R	86
4[d]		R	91
5[e]		R	86

a. B. D. Vineyard, W. S. Knowles, M. J. Sabacky, G. L. Backman, and D. J. Weinkauf, *J. Am. Chem. Soc.* **99**:5946 (1977).
b. M. B. Fryzuk and B. Bosnich, *J. Am. Chem. Soc.* **99**:6262 (1977).
c. U. Hengartner, D. Valentine, Jr., K. K. Johnson, M. E. Larschied, F. Pigott, F. Scheidl, J. W. Scott, R. C. Sun, J. M. Townsend, and T. H. Williams, *J. Org. Chem.* **44**:3741 (1979).
d. K. Achiwa, *J. Am. Chem. Soc.* **98**:8265 (1976).

39. W. S. Knowles, *Acc. Chem. Res.* **16**:106 (1983); D. Valentine, Jr., and J. W. Scott, *Synthesis* **1978**:329; A. S. C. Chan, J. J. Pluth, and J. Halpern, *J. Am. Chem. Soc.* **102**:5952 (1980); J. S. Giovanetti, C. M. Kelly, and C. R. Landis, *J. Am. Chem. Soc.* **115**:4040 (1993); T. Ohta and R. Noyori, in *Catalytic Asymmetric Synthesis*, I. Ojima, ed., VCH Publishers, New York, 1993, pp. 1–39.

complexes with chiral phosphine ligands. Scheme 2.12 records some illustrative results. The details of the catalytic mechanism will be considered later (see Section 5.1 in Part B). The fundamental point is that the chiral environment at the catalytic rhodium atoms causes a preference for reaction at either the *re* or the *si* face of the reactant. The hydrogen delivered from the catalyst then establishes the configuration of the stereogenic center at C-2.

The enantioselective reduction of acetophenone (1-phenylethanone) has been extensively explored. This compound is a simple example of a prochiral ketone, and the difference between the methyl and phenyl substituents provides the basis for discrimination by chiral reagents and catalysts. Three types of reducing agents have been explored, including (a) chiral catalysts for hydrogenation, (b) chiral Lewis acids which activate the carbonyl group by complexation at oxygen and promote enantioselective hydride delivery, and (c) hydride donors with chiral ligands. For enantioselective hydrogenation, homogeneous catalysts consisting of a transition metal with chiral ligands are used. Scheme 2.13 shows some examples. The hydride transfer takes place from the metal in the catalytic complex. The chiral ligands in the assembly make the process enantioselective.

The most successful of the Lewis acid catalysts are oxazaborolidines prepared from chiral amino alcohols and boranes. These compounds lead to enantioselective reduction of acetophenone by an external reductant, usually diborane. The chiral environment established in the complex leads to facial selectivity. The most widely known example of these reagents is derived from the amino acid proline.[40] Several other examples of this type of reagent have been developed, and these will be discussed more completely in Section 5.2 of part B.

The hydride-donor class of reductants has not yet been successfully paired with enantioselective catalysts. However, a number of chiral reagents that are used in stoichiometric quantity can effect enantioselective reduction of acetophenone and other prochiral ketones. One class of reagents consists of derivatives of $LiAlH_4$ in which some of the hydrides have been replaced by chiral ligands. Section C of Scheme 2.13 shows some examples where chiral diols or amino alcohols have been introduced. Another type of reagent represented in Scheme 2.13 is chiral trialkylborohydrides.[41] Chiral boranes are quite readily available (see Section 4.9 in Part B) and easily converted to borohydrides.

40. E. J. Corey, R. K. Bakshi, S. Shibata, C.-P. Chen, and V. C. Singh, *J. Am. Chem. Soc.* **109**:7925 (1987); E. J. Corey and J. O. Link, *Tetrahedron Lett.* **33**:4141 (1992); D. J. Mathre, A. S. Thompson, A. W. Douglas, K. Hoogsteen, J. D. Carroll, E. G. Corley, and E. J. J. Grabowski, *J. Org. Chem.* **58**:2880 (1993).
41. M. M. Midland, *Chem. Rev.* **89**:1553 (1989).

Reductant   Reductant

A.   Chiral catalysts for hydrogenation

1[a]

Ru, Ph₂P, Ph₂P (BINAP)

2[b]

$[Rh(C_6H_{10})Cl]_2$

CH₃ H, H₂N, Ar, NH₂, Ar 87% o.o.
Ar = 4-methoxyphenyl

Ph, Ph, CH₃HN, NHCH₃ 67% e.e.

B.   Chiral oxazaborolidine catalysts

3[c]

BH₃, 96% e.e.

4[d]

92% e.e.

C.   Hydride donors with chiral ligands

5[e]

LiAlH₄, 20–30% e.e.

6[f]

LiAlH₄, 95% e.e.

7[g]

LiAlH₄, $(CH_3)_2NCH_2CH$—Ph 65–75% e.e.
CH₃ OH, CH₂Ph

8[h]

$LiAlH_3OC_2H_5$, 70–95% e.e.

9[i]

$PhCH_2OCH_2CH_2$ 70% e.e.

10[j]

87% e.e.

a. T. Ohkuma, H. Ooka, S. Hashiguchi, T. Ikariya, and R. Noyori, *J. Am. Chem. Soc.* **117**:2675 (1995).
b. P. Gamez, F. Fache, P. Mangeney, and M. Lemaire, *Tetrahedron Lett.* **34**:6897 (1993).
c. E. J. Corey, R. K. Bakshi, S. Shibata, C.-P. Chen, and V. K. Singh, *J. Am. Chem. Soc.* **109**:7925 (1987).
d. G. J. Quallich and T. M. Woodall, *Tetrahedron Lett.* **34**:4145 (1993).
e. S. R. Landor, B. J. Miller, and A. P. Tatchell, *J. Chem. Soc., C* **1966**:2280.
f. M. Asami and T. Mukaiyama, *Heterocycles* **12**:499 (1979).
g. S. Yamaguchi and H. S. Mosher, *J. Org. Chem.* **38**:1870 (1973); C. J. Reich, G. R. Sullivan, and H. S. Mosher, *Tetrahedron Lett.* **1973**:505.
h. R. Noyori, I. Tomino, Y. Tanimoto, and M. Nishizawa, *J. Am. Chem. Soc.* **106**:6709 (1984).
i. M. M. Midland and A. Kazubski, *J. Org. Chem.* **47**:2495 (1982).
j. H. C. Brown and G. G. Pai, *J. Org. Chem.* **50**:1384 (1985).

The concept of heterotopic atoms, groups, and faces can be extended from enantiotopic to diastereotopic types. If each of two nominally equivalent ligands in a molecule is replaced by a test group and the molecules that are generated are diastereomeric, then the ligands are *diastereotopic*. Similarly, if reaction at one face of a trigonal atom generates a molecule diastereomeric with that produced at the alternate face, the faces are diastereotopic.

As an example of a molecule with diastereotopic ligands, consider the amino acid L-phenylalanine. The two protons at C-3 are diastereotopic, since substitution of either of them would generate a molecule with two stereogenic centers. Because the existing center is S, the two diastereomers would be the $2S,3R$ and $2S,3S$ stereoisomers. As in the case of enantiotopic protons, diastereotopic protons are designated *pro-R* or *pro-S*. The enzyme *phenylalanine ammonia-lyase* catalyzes the conversion of phenylalanine to *trans*-cinnamic acid by a process involving *anti* elimination of the amino group and the 3-*pro-S* hydrogen. This stereochemical course has been demonstrated using deuterium-labeled L-phenylalanine as shown below.[42]

diastereotopic protons
in L-phenylalanine

The environments of diastereotopic groups are topologically nonequivalent. An important property of diastereotopic ligands is that they are chemically nonequivalent toward achiral as well as chiral reagents, and they can also be distinguished by physical probes, especially NMR spectroscopy. As a consequence of their nonequivalence, they experience different shielding effects and have different chemical shifts in the NMR spectrum. (Enantiotopic groups have identical chemical shifts and are not distinguishable in NMR spectra.) A clear example of this effect can be seen in the proton NMR spectra of the *cis* and *trans* isomers of 1-benzyl-2,6-dimethylpiperidine shown in Fig. 2.8.[43] The methylene protons of the benzyl group of the *cis* isomer are enantiotopic and appear as a sharp singlet. The methylene protons of the *trans* isomer are diastereotopic and appear as a four-line AB pattern.

cis-isomer-methylene protons
of benzyl group enantiotopic

trans-isomer-methylene protons
of benzyl group diastereotopic

42. R. H. Wightman, J. Staunton, A. R. Battersby, and K. R. Hanson, *J. Chem. Soc., Perkin Trans. 1* **1972**:2355.
43. R. K. Hill and T. H. Chan, *Tetrahedron* **21**:2015 (1965); for additional discussion of chemical shift nonequivalence in diastereotopic groups, see W. B. Jennings, *Chem. Rev.* **75**:307 (1975).

Fig. 2.8. Equivalent benzyl CH_2 protons in 1-benzyl-*cis*-2,6-dimethylpiperidine compared with nonequivalent protons in the *trans* isomer. [Reproduced from *Tetrahedron* **21**:2015 (1965) by permission of Elsevier Science.]

In general, any pair of hydrogens in a methylene group in a chiral molecule are diastereotopic. Whether the topological difference is detectable in the NMR spectrum is determined by the proximity to the chiral center and the particular shielding effects of the molecule.

Similarly, the two faces at a trigonal carbon in a molecule containing a stereogenic center are diastereotopic. Both chiral and achiral reactants can distinguish between these diastereotopic faces. Many examples of diastereotopic transformations of such compounds are known. One of the cases that has been examined closely is addition reactions at a trigonal center adjacent to an asymmetric carbon. Particular attention has been given to the case of nucleophilic addition to carbonyl groups.

$$
\underset{b}{\overset{a}{\underset{c''''}{\diagup}}}C-\overset{O}{\underset{R}{\overset{\parallel}{C}}}
\quad\xrightarrow[\text{2) H}^+]{\text{1) Nu}^-}\quad
\underset{b}{\overset{a}{\underset{c''''}{\diagup}}}C-\overset{OH}{\underset{R}{\overset{}{C}}}\!-\!Nu
\quad or \quad
\underset{b}{\overset{a}{\underset{c''''}{\diagup}}}C-\overset{Nu}{\underset{R}{\overset{}{C}}}\!-\!OH
$$

Such reactions are usually *diastereoselective*; that is, one of the diastereomeric products is formed in excess. For the case of the nucleophile being hydride from a metal hydride reductant or alkyl or aryl groups from organometallic reagents, the preferred stereochemistry can frequently be predicted on the basis of *Cram's rule*. This empirically based prediction is applied by considering a conformation in which the sterically most demanding of the three substituents at the adjacent stereogenic center is *anti* to the carbonyl group. The major product is the stereoisomer in which the nucleophile is added from the face of the carbonyl group occupied by the smaller of the remaining substituents.[44]

S = smallest group
M = intermediate group
L = largest group

We will discuss the structural and mechanistic basis of Cram's rule in Chapter 3. As would probably be expected, the influence of a stereogenic center on the diastereoselectivity of the reaction is diminished when the center is more remote from the reaction site.

General References

D. Ager and M. B. East, *Asymmetric Synthetic Methodology*, CRC Press, Boca Raton, Florida, 1996.

R. S. Atkinson, *Stereoselective Synthesis*, John Wiley & Sons, New York, 1995.

E. L. Eliel, *Stereochemistry of Carbon Compounds*, McGraw-Hill, New York, 1962.

E. L. Eliel, S. H. Wilen, and L. N. Mander, *Stereochemistry of Organic Compounds*, John Wiley & Sons, New York, 1993.

J. Jacques, A. Collet, and S. H. Wilen, *Enantiomers, Racemates and Resolutions*, John Wiley & Sons, New York, 1981.

K. Mislow, *Introduction to Stereochemistry*, W. A. Benjamin, New York, 1966.

J. D. Morrison, ed., *Asymmetric Synthesis*, Vols. 1–4, Academic Press, New York, 1983–1984.

G. Procter, *Asymmetric Synthesis*, Oxford University Press, New York, 1996.

A. Rodger and B. Norden, *Circular Dichroism and Linear Dichroism*, Oxford University Press, Oxford, U.K., 1997.

Problems

(References for these problems will be found on page 792.)

1. Indicate whether the relationship in each of the following pairs of compounds is identical, enantiomeric, or diastereomeric:

(a)

(b)

(c)

44. D. J. Cram and F. A. Abd Elhafez, *J. Am. Chem. Soc.* **74**:5828 (1952); D. J. Cram and K. R. Kopecky, *J. Am. Chem. Soc.* **81**:2748 (1959); E. L. Eliel, in *Asymmetric Synthesis*, Vol. 2, J. D. Morrison, ed., Academic Press, New York, 1983, Chapter 5.

(d)

and

(e)

and

(f)

and

(g)

and

2. The structure originally proposed for cordycepic acid, $[\alpha]_D = +40.3°$, has been shown to be incorrect. Suggest a reason to be skeptical about the original structure, which is given below.

3. Each reaction in the sequence shown is reported to proceed with retention of configuration, yet the starting material has the R configuration and the product has the S configuration. Reconcile this apparent contradiction.

4. Using the sequence rule, specify the configuration at each stereogenic center in the following molecules:

(a)

(b)

(c)

(d)

(e) (f) (g)

5. Draw structural formulas for each of the following compounds, clearly showing stereochemistry:

  (a) (E)-3,7-dimethyl-2,6-octadien-1-ol (geraniol)
  (b) (R)-4-methyl-4-phenyl-2-cyclohexenone
  (c) L-erythro-2-(methylamino)-1-phenylpropan-1-ol [(−)-ephedrine]
  (d) (7R,8S)-7,8-epoxy-2-methyloctadecane (the sex attractant of the female gypsy moth)
  (e) methyl (1S)-cyano-(2R)-phenylcyclopropanecarboxylate
  (f) (Z)-2-methyl-2-butenol
  (g) (E)-(3-methyl-2-pentenylidene)triphenylphosphorane

6. The racemization of medium-ring trans-cycloalkenes depends upon ring size and substitution, as indicated by the data below. Discuss these relative reactivities in terms of the structures of the cycloalkenes and the mechanism of racemization.

Ring size	n	R	$t_{1/2}$ for racemization
8	6	H	10^5 years at 25°C
9	7	H	10 s at 25°C
10	8	CH_3	3 days at 100°C
12	10	CH_3	1 day at 25°C

7. Compound **A** can be prepared in enantiomerically pure form. It is racemized by heating to 120°C with an E_a of about 30 kcal/mol. Suggest a mechanism for the racemization process.

8. Reaction of 1,3,5-tris(thiomethyl)benzene with potassium hydroxide and then with tris(2-bromoethyl)methane gives a product of formula $C_{16}H_{22}S_3$ having the following NMR signals: δ − 1.7, septet, 1H; 1.0, multiplet, 6H; 2.1, multiplet, 6H; 3.0, singlet, 6H; 7.1, singlet, 3H. Propose a structure which is consistent with the observed properties of this material and explain the basis of your proposal.

9. The substance chaetochromin A, structure **A**, has been shown by X-ray diffraction to have the absolute configuration indicated in the structure. The CD spectra of **A** and the related compounds cephalochromin (**B**) and ustilaginoidin (**C**) are shown in the figure. Deduce the absolute stereochemistry of cephalochromin and ustilaginoidin from these data, and draw perspective structures indicating the absolute configuration.

CD spectra of chaetochromin A, cephalochromin, and ustilaginoidin A (in dioxane), – · – · –, Chaetochromin A (**A**); ———, cephalochromin (**B**); - - - -, ustilaginoidin (**C**).

A
chaetochromin A

B
cephalochromin

C
ustilaginoidin

10. When partially resolved samples of 5-hydroxymethylpyrrolidin-2-one are allowed to react with benzaldehyde in the presence of an acid catalyst, two products, **A** ($C_{12}H_{13}NO_2$) and **B** ($C_{24}H_{26}N_2O_4$), are formed. The ratio **A** : **B** depends on the enantiomeric purity of the starting material. When the starting material is optically pure, only **A** is formed. When it is racemic, only **B** is formed. Partially resolved material gives both **A** and **B**. The more nearly it is enantiomerically pure, the less **B** is formed. The products **A** is optically active but **B** is achiral. Develop an explanation for these observations, including structures for **A** and **B**.

11. Give the product(s) described for each reaction. Specify all aspects of stereochemistry.

   (a) stereospecific *anti* addition of bromine to *Z*- and *E*-cinnamic acid
   (b) solvolysis of (*S*)-3-bromooctane in methanol with 6% racemization
   (c) stereospecific *syn* elimination of acetic acid from (*R,S*)-1,2-diphenylpropyl acetate
   (d) stereoselective epoxidation of bicyclo[2.2.1]hept-2-ene proceeding 94% from the *exo* direction

12. Compound **A** can be resolved to give an enantiomerically pure substance, $[\alpha]_D^{25} = -124°$. Oxidation gives the pure ketone **B**, which is optically active, $[\alpha]_D^{25} \simeq -439°$. Heating the alcohol **A** gives partial conversion (an equilibrium is established) to an isomer with $[\alpha]_D^{25} = +22°$. Oxidation of this isomer gives the enantiomer of the ketone **B**. Heating either enantiomer of the ketone leads to the racemic mixture. Explain the stereochemical relationships between these compounds.

13. Some of the compounds shown contain diastereotopic atoms or groups. Which possess this characteristic? For those that do, indicate the atoms or groups that are diastereotopic and indicate which atom or group is *pro-R* and which is *pro-S*.

   (a)

   (b)

   (c) $C_6H_5CHCO_2CH(CH_3)_2$
       $|$
       OCH_3

   (d) $(CH_3)_2CHCHCO_2H$
       $|$
       NH_2

   (e) $C_6H_5CH_2CHCNHCH_2CO_2H$ (with O double-bonded above the C)
       $|$
       NH_2

   (f) $BrCH_2CH(OC_2H_5)_2$

14. Indicate which of the following molecules are chiral and which are achiral. For each molecule that is achiral, indicate the element of symmetry that is present in the molecule.

(a)

(b)

(c)

(d)

(e)

(f)

(g)

(h)

(i)

(j)

(k)

(l)

(m)

15. Assign configurations, using the sequence rule, to each chiral center of the stereo-isomeric isocitric acids and alloisocitric acids:

$$
\begin{array}{cccc}
CO_2H & CO_2H & CO_2H & CO_2H \\
H\!-\!OH & HO\!-\!H & H\!-\!OH & HO\!-\!H \\
HO_2C\!-\!H & H\!-\!CO_2H & H\!-\!CO_2H & HO_2C\!-\!H \\
CH_2CO_2H & CH_2CO_2H & CH_2CO_2H & CH_2CO_2H
\end{array}
$$

isocitric acids   alloisocitric acids

16. The enzyme enolase catalyzes the following reaction:

$$
\begin{array}{ccc}
CO_2^- & & ^-O_2C \quad OPO_3^{2-} \\
H\!-\!OPO_3^{2-} & \rightleftharpoons & \diagdown\diagup \\
CH_2OH & & H \quad\quad H
\end{array}
\;\; + \; H_2O
$$

When $(2R,3R)$-2-phosphoglycerate-3-d was used as the substrate, the E-isomer of phosphoenolpyruvate-3-d was produced. Is the stereochemistry of elimination *syn* or *anti*?

17. An important sequence in valine biosynthesis in bacteria is

$$
\underset{\underset{HO\ \ OH}{|\ \ \ \ |}}{(CH_3)_2C\!-\!CHCO_2H} \longrightarrow (CH_3)_2CH\!-\!\overset{\overset{O}{\|}}{C}\!-\!CO_2H \longrightarrow \underset{\underset{\underset{valine}{NH_2}}{|}}{(CH_3)_2CH\!-\!CHCO_2H}
$$

The stereochemical aspects of this sequence have been examined, using a diol substrate in which one of the methyl groups has been replaced by CD_3. Given the information that labeled starting diol of configuration $2R,3R$ produces labeled valine of configuration $2S,3S$, deduce whether the C-3 hydroxyl group is replaced with overall retention or inversion of configuration.

18. 1,2-Diphenyl-1-propanol may be prepared in either of two ways:

  (a) lithium aluminum hydride reduction of 1,2-diphenyl-1-propanone
  (b) reaction of 2-phenylpropanal with phenylmagnesium bromide
Which method would you choose to prepare the anti isomer? Explain.

$$
\begin{array}{c}
CH_3 \\
| \\
Ph\diagdown\diagup\!\!-\!Ph \\
| \\
OH
\end{array}
$$

19. A mixture of tritium ($^3H = T$) labeled **A** and **B** was carried through the reaction

sequence shown:

D-Amino acid oxidase will oxidase only serine having the R configuration at C-2. Glycolate oxidase will remove only the *pro-R* hydrogen of glycolic acid. Does the product ($O=CHCO_2H$) contain tritium? Explain your reasoning.

20. Enzymatic oxidation of naphthalene by bacteria proceeds by way of the intermediate *cis*-diol shown. Which prochiral faces of C-1 and C-2 of naphthalene are hydroxylated in this process?

21. An amino acid having the constitution shown has been isolated from horse chestnuts. It is configurationally related to L-proline and has the R configuration at C-3. Write a stereochemically correct representation for this compound.

(horse-chestnut amino acid)   (proline — no stereochemistry implied)

22. (a) One of the diastereomers of 2,6-dimethylcyclohexyl benzyl ether exhibits two doublets for the benzylic protons in its NMR spectrum. Deduce the stereochemistry of this isomer.

(b) The NMR spectrum of the highly hindered molecule trimesitylmethane indicates that there are two enantiomeric species present in solution, the interconversion of which is separated by a barrier of 22 kcal/mol. Discuss the source of the observed chirality of this molecule.

23. A synthesis of the important biosynthetic intermediate mevalonic acid starts with the enzymatic hydrolysis of the diester **A** by pig liver esterase. The *pro-R* group is selectively hydrolyzed. Draw a three-dimensional structure of the product.

24. The structure of a natural product is shown without any specification of stereochemistry. It is a pure substance which gives no indication of being a mixture of stereoisomers and has zero optical rotation. It is not a racemic mixture because it does not yield separate peaks on a chiral HPLC column. When the material is completely hydrolyzed, it gives a racemic sample of the product shown. Deduce the complete stereochemical structure of the natural product from this information.

Conformational, Steric, and Stereoelectronic Effects

Introduction

The various shapes that a given molecule can attain are called *conformations*. The total energy of a molecule is directly related to its shape. Several components of the total energy can be recognized and, to some extent, attributed to specific structural features. Among the factors that contribute to total energy and have a recognizable connection with molecular structure are nonbonded repulsions, ring strain in cyclic systems, torsional strain arising from eclipsing of bonds, and destabilization resulting from distortion of bond lengths or bond angles from optimal values. Conversely, there are stabilizing interactions that have geometric constraints. Most of these can be classed as stereoelectronic effects; that is, *a particular geometric relationship is required to maximize the stabilizing interaction*. In addition, there are other interactions, such as hydrogen-bond formation and dipole–dipole interactions, for which the strength of the interaction strongly depends on the geometry of the molecule. The principles on which analysis of conformational equilibria are based were developed within a classical-mechanical framework. A molecule adopts the minimum-energy shape that is available by rotations about single bonds and adjustment of bond angles and bond lengths. Because bond angles and bond lengths vary relatively little from molecule to molecule, molecular shape is primarily determined by the rotational processes. Many molecules exhibit *strain* caused by nonideal geometry. Any molecule adopts a minimum-energy conformation at equilibrium, but the structural adjustments may not compensate entirely for nonideal bonding arrangements. *Strain energy* is the excess energy relative to an unstrained reference molecule. This chapter will focus on the sources of strain, the structural evidence of strain, and the energetic and reactivity consequences of strain.[1]

1. For a review of the concept of strain, see K. B. Wiberg, *Angew. Chem. Int. Ed. Engl.* **25**:312 (1986).

124

CHAPTER 3
CONFORMATIONAL,
STERIC, AND
STEREOELECTRONIC
EFFECTS

From a molecular orbital viewpoint, the energy of a molecule is the sum of the energy of the occupied orbitals. Calculation of total energy in different spatial arrangements reveals the energy as a function of geometry. The physical interpretation is given in terms of the effectiveness of orbital overlap. Maximum overlap between orbitals that have a bonding interaction lowers the molecular energy whereas overlap of antibonding orbitals raises the energy. The term *stereoelectronic effect* is used to encompass relationships between structure, conformation, energy, and reactivity that can be traced to geometry-dependent orbital interactions.

3.1. Strain and Molecular Mechanics

A system of analyzing the energy differences among molecules and among various conformations of a particular molecule has been developed and is based on some fundamental concepts formalized by Westheimer.[2] The method is now known by the term *molecular mechanics*. A molecule adopts that geometry which minimizes its total energy. The minimum-energy geometry will be strained to a degree that depends on the extent to which its structural parameters deviate from their ideal values. The energy for a particular kind of distortion is a function of the amount of distortion and the opposing force. The total strain energy, originally called *steric energy*, is the sum of several contributions:

$$E_{\text{steric}} = E(r) + E(\theta) + E(\phi) + E(d)$$

where $E(r)$ is the energy component associated with stretching or compression of bonds, $E(\theta)$ is the energy of bond-angle distortion, $E(\phi)$ is the torsional strain, and $E(d)$ are energy increments that result from nonbonded interactions between atoms. Molecular mechanics calculations involve summation of the force fields for each type of strain.

The mathematical expressions for the force fields are derived from classical-mechanical potential energy functions. The energy required to stretch a bond or to bend a bond angle increases as the square of the distortion. For bond stretching,

$$E(r) = 0.5k_r(r - r_0)^2$$

where k_r is the stretching force constant, r is the bond length, and r_0 is the normal bond length. For bond-angle bending,

$$E(\theta) = 0.5k_\theta(\Delta\theta)^2$$

where k_θ is the bending force constant and $\Delta\theta$ is the deviation of the bond angle from its normal value.

2. F. H. Westheimer, in *Steric Effects in Organic Chemistry*, M. S. Newman, ed., John Wiley & Sons, New York, 1956, Chapter 12; J. E. Williams, P. J. Stang, and P. v. R. Schleyer, *Annu. Rev. Phys. Chem.* **19**:531 (1968); D. B. Boyd and K. P. Lipkowitz, *J. Chem. Educ.* **59**:269 (1982); P. J. Cox, *J. Chem. Educ.* **59**:275 (1982); N. L. Allinger, *Adv. Phys. Org. Chem.* **13**:1 (1976); E. Osawa and H. Musso, *Top. Stereochem.* **13**:117 (1982); U. Burkett and N. L. Allinger, *Molecular Mechanics*, ACS Monograph 177, American Chemical Society, Washington, D.C., 1982.

The torsional strain is a sinusoidal function of the torsion angle. Torsional strain results from the barrier to rotation about single bonds as described for ethane on p. 56. For molecules with a threefold barrier such as ethane, the form of the torsional barrier is

$$E(\phi) = 0.5V_0(1 + \cos 3\phi)$$

where V_0 is the rotational energy barrier and ϕ is the torsional angle. For hydrocarbons, V_0 can be taken as being equal to the ethane barrier (2.8–2.9 kcal/mol). The potential energy diagram for rotation about the C$-$C bond of ethane is given in Fig. 3.1.

Nonbonded interaction energies are the most difficult contributions to evaluate and may be attractive or repulsive. When two uncharged atoms approach each other, the interaction between them is very small at large distances, becomes increasingly attractive as the separation approaches the sum of their van der Waals radii, but then becomes strongly repulsive as the separation becomes less than the sum of their van der Waals radii. This behavior is represented graphically by a Morse potential diagram such as in Fig. 3.2. The attractive interaction results from a mutual polarization of the electrons of the atoms. Such attractive forces are called *London forces* or *dispersion forces* and are relatively weak interactions. London forces vary inversely with the sixth power of internuclear distance and become negligible as internuclear separation increases. At distances smaller than the

Eclipsed conformations correspond to torsion angles of 0°, 120°, 240°.

Staggered conformations correspond to torsion angles of 60°, 180°, and 300°

Fig. 3.1. Potential energy as a function of torsion angle for ethane.

126

CHAPTER 3
CONFORMATIONAL,
STERIC, AND
STEREOELECTRONIC
EFFECTS

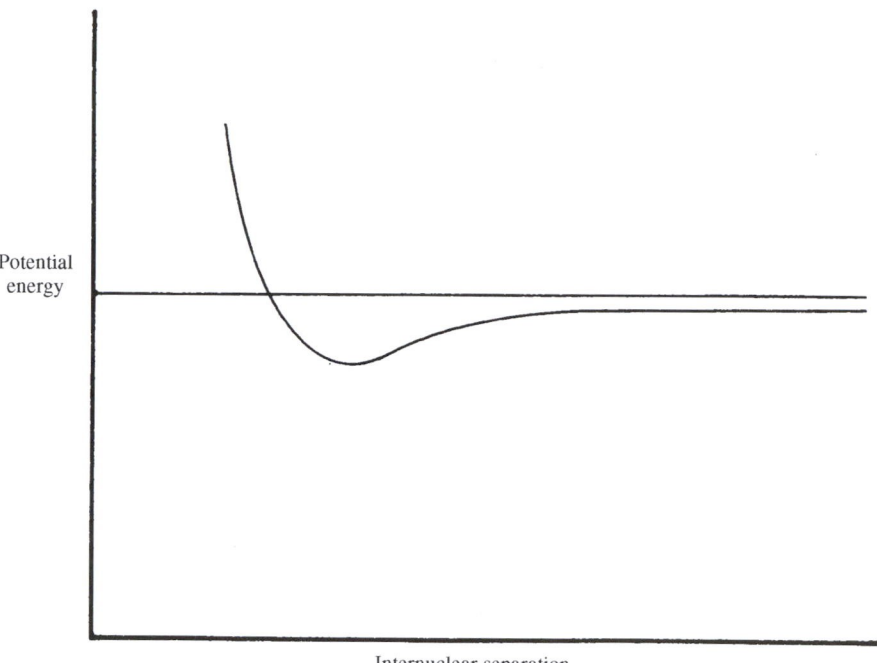

Fig. 3.2. Energy as a function of internuclear distance for nonbonded atoms.

sum of the van der Waals radii, the much stronger repulsive forces are dominant. Table 3.1 lists van der Waals radii for atoms and groups most commonly encountered in organic molecules.

The interplay between torsional strain and nonbonded interactions can be illustrated by examining the conformations of *n*-butane. The relationship between potential energy and the torsion angle for rotation about the C(2)−C(3) bond is presented in Fig. 3.3. The potential energy diagram of butane resembles that of ethane in having three maxima and three minima but differs in that one of the minima is lower than the other two, and one of the maxima is of higher energy than the other two. The minima correspond to staggered conformations. Of these, the *anti* conformation is lower in energy than the two *gauche* conformations. The energy difference between the *anti* and *gauche* conformations in butane is about 0.8 kcal/mol.[3] The maxima correspond to eclipsed conformations, with the highest-energy conformation being the one with the two methyl groups eclipsed with each

Table 3.1. Van der Waals Radii of Several Atoms and Groups (Å)[a]

H	1.20			CH$_3$	2.0		
N	1.55	P	1.80				
O	1.52	S	1.80				
F	1.47	Cl	1.75	Br	1.85	I	1.98

a. From A. Bondi, *J. Phys. Chem.* **68:**441 (1964).

3. G. J. Szasz, N. Sheppard, and D. H. Rank, *J. Chem. Phys.* **16:**704 (1948); P. B. Woller and E. W. Garbisch, Jr., *J. Am. Chem. Soc.* **94:**5310 (1972).

other. The methyl–methyl eclipsed conformation is about 2.6 kcal/mol higher in energy than the methyl–hydrogen eclipsed conformation and 6 kcal/mol higher in energy than the staggered *anti* conformation.

The rotational profile of butane can be understood as a superimposition of van der Waals forces on the ethane rotational energy profile. The two *gauche* conformations are raised in energy relative to the *anti* conformation by an energy increment resulting from the van der Waals repulsion between the two methyl groups of 0.8 kcal/mol. The eclipsed conformations all incorporate 2.8 kcal/mol of torsional strain relative to the staggered conformations, just as is true in ethane. The methyl–methyl eclipsed conformation is further strained by the van der Waals repulsion between the methyl groups. The van der Waals repulsion between methyl and hydrogen in the other eclipsed conformations is smaller. The methyl–methyl eclipsed barrier is not known precisely. The range in

Fig. 3.3. Potential energy diagram for rotation about C(2)–C(3) bond of n-butane.

128

CHAPTER 3
CONFORMATIONAL,
STERIC, AND
STEREOELECTRONIC
EFFECTS

experimental and theoretical values is between 4.0 and 6.6 kcal/mol. The most recent values are at the low end of the range.[4] The 120° and 240° eclipsed conformations are strained by 0.6 kcal/mol over and above the torsional strain, that is, by 0.3 kcal/mol for each of the two methyl–hydrogen repulsions.

The populations of the various conformations are related to the free energy difference between them by the equation

$$\Delta G° = -RT \ln K$$

For the case of butane, the equilibrium

$$gauche \rightleftharpoons anti$$

has $\Delta H° = -0.8$ kcal/mol. Because there are two enantiomeric *gauche* conformers, the free energy also reflects an entropy contribution:

$$\Delta S° = -R \ln 2$$

and

$$\Delta G° = \Delta H - T\Delta S°$$
$$= -0.8 \text{ kcal/mol} - (-RT \ln 2)$$

At 298 K, $\Delta G° = -0.8$ kcal/mol $+ 0.41$ kcal/mol $= -0.39$ kcal/mol and

$$K = [anti]/[gauche] = 1.9$$

This corresponds to a distribution of 66% *anti* and 34% *gauche*. Table 3.2 gives the relationship between free-energy difference, equilibrium constant, and percent composition of a two-component mixture.

Examples of attractive nonbonded interactions can be found in certain halogenated hydrocarbons. In 1-chloropropane, for example, the *gauche* conformation is slightly

Table 3.2. Composition–Equilibrium–Free-Energy Relationships[a]

More stable isomer (%)	Equilibrium constant (K)	Free energy $\Delta G°_{25}$ (kcal/mol)
50	1	0.0
55	1.22	−0.119
60	1.50	−0.240
65	1.86	−0.367
70	2.33	−0.502
75	3.00	−0.651
80	4.00	−0.821
85	5.67	−1.028
90	9.00	−1.302
95	19.00	−1.744
98	49.00	−2.306
99	99.00	−2.722
99.9	999.00	−4.092

a. From E. L. Eliel, *Stereochemistry of Carbon Compounds*, McGraw-Hill, New York, 1962.

4. N. L. Allinger, R. S. Grev, B. F. Yates, and H. F. Schaefer III, *J. Am. Chem. Soc.* **112**:114 (1990); W. A. Herrebout, B. J. van der Veken, A. Wang, and J. R. Durig, *J. Phys. Chem.* **99**:578 (1995).

preferred over the *anti* conformation at equilibrium. This is not solely the result of the statistical (entropy) factor of $2:1$ but also reflects a $\Delta H = 0.3 + 0.3\,\text{kcal/mol}$, which is attributed to a London attractive force. The chlorine atom and methyl group are separated by about the sum of their van der Waals radii in the *gauche* conformation.[5]

The separation of the total strain energy into component elements of bond-length strain, bond-angle strain, torsional strain, and nonbonded repulsion in a qualitative way is useful for analysis and rationalization of structural and steric effects on equilibria and reactivity. The quantitative application of the principles of molecular mechanics for calculation of ground-state geometries, heats of formation, and strain energies has been developed to a high level. Minimization of the total strain energy of a molecule, expressed by a multiparameter equation for each of the force fields, can be accomplished by iterative computation. The calculational methods have been refined to the point that geometries of hydrocarbons of moderate size can be calculated to an accuracy of 0.01 Å (1 pm) in bond length and $1-2°$ in bond angle.[6] A similar degree of accuracy can be expected from calculations on a wide variety of molecules including both the usual functional groups and certain intermediates such as carbocations.[7] In these types of systems, terms for dipole–dipole interactions and mutual polarization must be included in the parameterized equations. These computations can be done by a number of commercially available programs. The properties of the parameters used in the programs determine the range of applicability and reliability of the results.

Several systems of parameters and equations for carrying out the calculations have been developed. The most frequently used in organic chemistry is that developed by Allinger and co-workers, which is frequently referred to as MM (molecular mechanics) calculations. The most recent version is MM3.[8] The computations involve iterations to locate an energy minimum. Precautions must be taken to establish that a true ("global") minimum, as opposed to a local minimum energy, has been achieved. This can be accomplished by using a number of different initial geometries and comparing the structures of the minima that are located.

An example of the application of molecular mechanics in the investigation of chemical reactions is a study of the correlation between steric strain in a molecule and the ease of rupture of carbon–carbon bonds. For a series of hexasubstituted ethanes, it was found that there is a good correlation between the strain calculated by the molecular mechanics method and the rate of thermolysis.[9] Some of the data are shown in Table 3.3.

3.2. Conformations of Acyclic Molecules

The conformations of simple hydrocarbons can be interpreted by extensions of the principles illustrated in the analysis of rotational equilibria in ethane and butane. The staggered

5. W. E. Steinmetz, F. H. Hickernell, I. K. Mun, and L. H. Scharpen, *J. Mol. Spectrosc.* **68**:173 (1977); N. Y. Morino and K. Kuchitsu, *J. Chem. Phys.* **28**:175 (1958).
6. N. L. Allinger, M. A. Miller, F. A. VanCatledge, and J. A. Hirsch, *J. Am. Chem. Soc.* **89**:4345 (1967); N. L. Allinger, *J. Am. Chem. Soc.* **99**: 8127 (1977).
7. For a summary, see N. L. Allinger, *Adv. Phys. Org. Chem.* **13**:1 (1976).
8. N. L. Allinger, Y. H. Yuh, and J.-H. Lii, *J. Am. Chem. Soc.* **111**:8551 (1989).
9. C. Rüchardt and H.-D. Beckhaus, *Angew. Chem. Int. Ed. Engl.* **19**:429 (1980).

130

CHAPTER 3
CONFORMATIONAL,
STERIC, AND
STEREOELECTRONIC
EFFECTS

Table 3.3. Correlation between Intramolecular Strain from Molecular Mechanics (MM) Calculations and Activation Energies for Dissociation of C−C Bonds in Substituted Ethanes[a]

$$R^1-\underset{\underset{R^3}{|}}{\overset{\overset{R^2}{|}}{C}}-\underset{\underset{R^3}{|}}{\overset{\overset{R^2}{|}}{C}}-R^1 \longrightarrow 2R^1-\underset{R^3}{\overset{R^2}{C{\cdot}}}$$

R^1	R^2	R^3	$\Delta G^{\ddagger}$ (kcal/mol)	MM strain (kcal/mol)	R^1	R^2	R^3	$\Delta G^{\ddagger}$ (kcal/mol)	MM strain (kcal/mol)
H	H	H	79	0	Ph	H	CH_3	50	3.6
CH_3	H	H	69	0	Ph	H	C_2H_5	49.7	4.8
CH_3	CH_3	H	68	2.7	Ph	H	$CH(CH_3)_2$	47.4	8.4
CH_3	CH_3	CH_3	60	6.9	Ph	H	$C(CH_3)_3$	42.1	22.2
CH_3	CH_3	C_2H_5	55.3	12.3	Ph	CH_3	CH_3	37.9	18.4
CH_3	CH_3	$CH(CH_3)_2$	47.3	22.4	Ph	CH_3	C_2H_5	34.9	23.7
CH_3	CH_3	$C(CH)_3$	33.7	45.3	Ph	CH_3	$CH(CH_3)_2$	26.7	40.9

a. C. Rüchardt and H.-D. Beckhaus, *Angew. Chem. Int. Ed. Engl.* **19**:429 (1980).

conformations correspond to potential energy minima, and the eclipsed conformations to potential energy maxima. Of the staggered conformations, *anti* forms are more stable than *gauche* forms. The magnitudes of the barriers to rotation of many small organic molecules have been measured.[10] Some representative examples are listed in Table 3.4. The

Table 3.4. Rotational Energy Barriers of Compounds of the Type $CH_3 - X$[a]

Compound	Barrier height (kcal/mol)
Alkanes	
1. CH_3-CH_3	2.88
2. $CH_3-CH_2CH_3$	3.4
3. $CH_3-CH(CH_3)_2$	3.9
4. $CH_3-C(CH_3)_3$	4.7
5. CH_3-SiH_3	1.7
Haloethanes	
6. CH_3-CH_2F	3.3
7. CH_3-CH_2Cl	3.7
8. CH_3-CH_2Br	3.7
9. CH_3-CH_2I	3.2
Heteroatom substitution	
10. CH_3-NH_2	1.98
11. CH_3-NHCH_3	3.62
12. CH_3-OH	1.07
13. CH_3-OCH_3	2.7

a. Taken from the compilation of J. P. Lowe, *Prog. Phys. Org. Chem.* **6**:1 (1968); barriers are those for rotation about the bond indicated in the formula.

10. For a review, see J. P. Lowe, *Prog. Phys. Org. Chem.* **6**:1 (1968).

experimental techniques used to study rotational processes include microwave spectroscopy, electron diffraction, ultrasonic absorption, and infrared (IR) spectroscopy.[11]

Substitution of methyl groups for hydrogen atoms on one of the carbon atoms produces a regular increase of about 0.5–0.6 kcal/mol in the height of the rotational energy barrier. The barrier in ethane is 2.88 kcal/mol. In propane, the barrier is 3.4 kcal/mol, corresponding to an increase of ~0.5 kcal/mol for methyl–hydrogen eclipsing. When two methyl–hydrogen eclipsing interactions occur, as in 2-methylpropane, the barrier is raised to 3.9 kcal/mol. The increase in going to 2,2-dimethylpropane, in which the barrier is 4.7 kcal/mol, is 1.8 kcal/mol for the total of three methyl hydrogen eclipsing interactions.

Rotational barrier increases with the number of CH_3/H eclipsing interactions.

The rotational barrier in methylsilane (Table 3.4, entry 5) is significantly smaller than that in ethane (1.7 versus 2.88 kcal/mol). This reflects the decreased electron–electron repulsions in the eclipsed conformation resulting from the longer carbon–silicon bond length (1.87 Å) compared to the carbon–carbon bond length (1.54 Å) in ethane.

The haloethanes all have similar rotational barriers of 3.2–3.7 kcal/mol. The increase in the barrier height relative to ethane is probably due to a van der Waals repulsive effect. The heavier halogens have larger van der Waals radii, but this is offset by the longer bond lengths, so that the net effect is a relatively constant rotational barrier for each of the ethyl halides.

Changing the atom bound to a methyl group from carbon to nitrogen to oxygen, as in going from ethane to methylamine to methanol, produces a decrease in the rotational barrier from 2.88 to 1.98 to 1.07 kcal/mol. This closely approximates the 3 : 2 : 1 ratio of the number of H—H eclipsing interactions in these three molecules.

Rotational barrier decreases with the number of H/H eclipsing interactions.

Entries 11 and 13 in Table 3.4 present data relating the effect of methyl substitution on methanol and methylamine. The data show an increased response to methyl substitution. While the propane barrier is 3.4 kcal/mol (compared to 2.88 in ethane), the dimethylamine barrier is 3.6 kcal/mol (compared to 1.98 in methylamine) and in dimethyl ether it is 2.7 kcal/mol (compared to 1.07 in methanol). Thus, while methyl–hydrogen eclipsing raised the propane barrier by 0.5 kcal/mol, the increase for both dimethylamine and dimethyl ether is 1.6 kcal/mol. This increase in the barrier is attributed to greater van der Waals repulsions resulting from the shorter C—N and C—O bonds, relative to the C—C bond.

11. Methods for determination of rotational barriers are discussed in Ref. 10 and by E. Wyn-Jones and R. A. Pethrick, *Top. Stereochem.* **5**:205 (1969).

132

CHAPTER 3
CONFORMATIONAL,
STERIC, AND
STEREOELECTRONIC
EFFECTS

There are two families of conformations available to terminal alkenes. These are the eclipsed and bisected conformations shown below for propene. The eclipsed conformation is more stable by about 2 kcal/mol.[12]

eclipsed   bisected

The origin of the preference for the eclipsed conformation of propene can be explained in MO terms by focusing attention on the interaction between the double bond and the π component of the orbitals associated with the methyl group. The dominant interaction is a repulsive one between the filled methyl group orbitals and the filled π orbital of the double bond. This repulsive interaction is greater in the bisected conformation than in the eclipsed conformation.[13]

maximum repulsive   reduced repulsive
interaction   interaction

With more substituted terminal alkenes, additional conformations are available as indicated below for 1-butene.

A   B   C   D

Conformations **A** and **B** are of the eclipsed type, whereas **C** and **D** are bisected. It has been determined by microwave spectroscopy that the eclipsed conformations are more stable than the bisected ones and that **B** is about 0.15 kcal more stable than **A**.[14] MO calculations at the 6-31G* level have found relative energies of 0.00, −0.25, 1.75, and 1.74 kcal/mol, respectively, for **A−D**.[13]

Further substitution can introduce additional factors, especially nonbonded repulsions, which influence conformational equilibria. For example, methyl substitution at C−2, as in 2-methyl-1-butene, introduces a methyl–methyl *gauche* interaction in the conformation analogous to **B**, with the result that in 2-methyl-1-butene the two eclipsed conformations are of approximately equal energy.[15] Increasing the size of the group at

12. K. B. Wiberg and E. Martin, *J. Am. Chem. Soc.* **107**:5035 (1985); A. E. Dorigo, D. W. Pratt, and K. N. Houk, *J. Am. Chem. Soc.* **109**:6591 (1987); J. L. Broeker, R. W. Hoffmann, and K. N. Houk, *J. Am. Chem. Soc.* **113**:5006 (1991).
13. W. J. Hehre, J. A. Pople, and A. J. P. Devaquet, *J. Am. Chem. Soc.* **98**:664 (1976).
14. S. Kondo, E. Hirota, and Y. Morino, *J. Mol. Spectrosc.* **28**:471 (1968).
15. T. Shimanouchi, Y. Abe, and K. Kuchitsu, *J. Mol. Struct.* **2**:82 (1968).

C−3 increases the preference for the eclipsed conformation analogous to **B** at the expense of **A**. 4,4-Dimethyl-1-pentene exists mainly in the hydrogen-eclipsed conformation.

favored   disfavored

The preferred conformations of carbonyl compounds, like those of 1-alkenes, are eclipsed rather than bisected, as shown below for acetaldehyde and propionaldehyde. The barrier for acetaldehyde is 1.2 kcal/mol.[16] This is about one-third of the barrier in ethane, and MO calculations indicate that the origin of the barrier is largely the hydrogen–hydrogen repulsion in the conformation in which the aldehyde hydrogen is eclipsed with a hydrogen of the methyl group.[13] More detailed analysis has indicated that small adjustments in molecular geometry, including σ-bond lengthening, must be taken into account to analyze the barrier quantitatively.[17] In propionaldehyde, it is the methyl group, rather than the hydrogen, that is eclipsed with the carbonyl group in the most stable conformation. The energy difference between the two eclipsed conformations has been determined to be 0.9 kcal/mol by microwave spectroscopy.[18] A number of other aldehydes have been studied by NMR and found to have analogous rotameric compositions.[19] When the alkyl substituent becomes more sterically demanding, the hydrogen-eclipsed conformation becomes relatively more stable, as is the case with 3,3-dimethylbutanal.

Preferred conformations for acetaldehyde, propionaldehyde, and 3,3-dimethylbutanal

Ketones also favor eclipsed conformations. The preference is for the rotamer in which the alkyl group, rather than a hydrogen, is eclipsed with the carbonyl group because this conformation allows the two alkyl groups to be *anti* rather than *gauche*. Electron diffraction studies of 3-pentanone indicate the conformation shown to be the most stable rotamer, in accord with this generalization.[20]

more stable   less stable   Preferred conformation for 3-pentanone

16. R. W. Kilb, C. C. Lin, and E. B. Wilson, Jr., *J. Chem. Phys.* **26**:1695 (1957).
17. L. Goodman, T. Kundu, and J. Leszczynski, *J. Am. Chem. Soc.* **117**:2082 (1995).
18. S. S. Butcher and E. B. Wilson, Jr., *J. Chem. Phys.* **40**:1671 (1964).
19. G. J. Karabatsos and N. Hsi, *J. Am. Chem. Soc.* **87**:2864 (1965).
20. C. Romers and J. E. G. Creutzberg, *Recl. Trav. Chim. Pays-Bas* **75**:331 (1956).

134

CHAPTER 3
CONFORMATIONAL,
STERIC, AND
STEREOELECTRONIC
EFFECTS

1,3-Dienes would be expected to adopt conformations in which the double bonds are coplanar, so as to permit effective π-orbital overlap and electron delocalization. The two alternative planar conformations for 1,3-butadiene are referred to as *s-trans* and *s-cis*. In addition to the two planar conformations, there is a third conformation, referred to as the *skew* conformation, which is cisoid but not planar. Various types of studies have shown that the *s-trans* conformation is the most stable one for 1,3-butadiene.[21] A small amount of one of the *skew* conformations is also present in equilibrium with the major conformer.[22] The planar *s-cis* conformation incorporates a van der Waals repulsion between the hydrogens on C$-$1 and C$-$4. This is relieved in the *skew* conformation.

s-trans   *s-sic*   *skew*

The barrier for conversion of the skew conformation to the *s-trans* conformation is 3.9 kcal/mol. This energy maximum presumably refers to the conformation (transition state) in which the two π bonds are mutually perpendicular. Various MO calculations find the *s-trans* conformation to be 2–5 kcal/mol lower in energy than either the planar or *skew* cisoid conformations.[23] Most high-level calculations favor the skew conformation over the planar *s-cis*, but the energy differences found are quite small.[22,24]

The case of α,β-unsaturated carbonyl compounds is analogous to that of 1,3-dienes, in that stereoelectronic factors favor coplanarity of the C=C$-$C=O system. The rotamers that are important are the *s-trans* and *s-cis* conformations. Microwave data indicate that the *s-trans* form is the only conformation present in detectable amounts in acrolein (2-propenal).[25] The equilibrium distribution of *s-trans* and *s-cis* conformations of α,β-unsaturated ketones depends on the extent of van der Waals interaction between substituents.[26] Methyl vinyl ketone has minimal unfavorable van der Waals repulsions between substituents and exists predominantly as the *s-trans* conformer:

s-trans (73%)   *s-cis* (27%)

21. A. Almenningen, O. Bastiansen, and M. Traetteburg, *Acta Chem. Scand.* **12**:1221 (1958); K. K. Kuchitsu, T. Fukuyama, and Y. Morino, *J. Mol. Struct.* **1**:643 (1967); R. L. Lipnick and E. W. Garbisch, Jr., *J. Am. Chem. Soc.* **95**:6370 (1973).

22. K. B. Wiberg and R. E. Rosenberg, *J. Am. Chem. Soc.* **112**:1509 (1990).

23. A. J. P. Devaquet, R. E. Townshend, and W. J. Hehre, *J. Am. Chem. Soc.* **98**:4068 (1976); K. B. Wiberg, P. R. Rablen, and M. Marquez, *J. Am. Chem. Soc.* **114**:8654 (1992); M. Head-Gordon and J. A. Pople, *J. Phys. Chem.* **97**:1147 (1993).

24. J. Breulet, T. J. Lee, and H. F. Schaefer III, *J. Am. Chem. Soc.* **106**:6250 (1984); D. Feller and E. R. Davidson, *Theor. Chim. Acta* **68**:57 (1985).

25. E. A. Cherniak and C. C. Costain, *J. Chem. Phys.* **45**:104 (1966).

26. G. Montaudo, V. Librando, S. Caccamese, and P. Maravigna, *J. Am. Chem. Soc.* **95**:6365 (1973).

When larger alkyl groups are substituted for methyl, the mole fraction of the *s-cis* form progressively increases as the size of the alkyl group increases.[27]

	s-trans	*s-cis*
R = CH$_3$	0.7	0.3
CH$_3$CH$_2$	0.55	0.45
(CH$_3$)$_2$CH	0.3	0.7
(CH$_3$)$_3$C	0.0	1.0

An unfavorable methyl–methyl interaction destabilizes the *s-trans* conformation of 4-methyl-3-penten-2-one relative to the *s-cis* conformation, and the equilibrium favors the *s-cis* form.

s-trans (28%)   s-cis (72%)

3.3. Conformations of Cyclohexane Derivatives

The conformational analysis of compounds containing six-membered rings is particularly well understood. A major reason for the depth of study and resulting detailed knowledge has to do with the nature of the system itself. Cyclohexane and its derivatives lend themselves well to thorough analysis, because they are characterized by a small number of energy minima. The most stable conformations are separated by rotational energy barriers that are somewhat higher, and more easily measured, than rotational barriers in acyclic compounds or in other ring systems.

The most stable conformation of cyclohexane is the chair. Electron diffraction studies in the gas phase reveal a slight flattening of the chair compared with the geometry obtained when tetrahedral molecular models are used. The torsion angles are 55.9°, compared with 60° for the "ideal" chair conformation, and the axial C–H bonds are not perfectly parallel but are oriented outward by about 7°. The length of the C–C bonds is 1.528 Å, the length of the C–H bonds is 1.119 Å, and the C–C–C angles are 111.05°.[28]

structural features of cyclohexane chair conformation

27. A. Bienvenue, *J. Am. Chem. Soc.* **95**:7345 (1973).
28. H. J. Geise, H. R. Buys, and F. C. Mijlhoff, *J. Mol. Struct.* **9**:447 (1971).

136

CHAPTER 3
CONFORMATIONAL,
STERIC, AND
STEREOELECTRONIC
EFFECTS

Two other nonchair conformations of cyclohexane that have normal bond angles and bond lengths are the *twist* and the *boat* conformations.[29] Both the twist and the boat conformations are less stable than the chair. Molecular mechanics calculations indicate that the twist conformation is about 5 kcal/mol, and the boat about 6.4 kcal/mol, higher in energy than the chair.[6] A direct measurement of the chair–twist energy difference has been made using low-temperature IR spectroscopy.[30] The chair was determined to be 5.5 kcal/mol lower in enthalpy than the twist. The twist and boat conformations are more flexible than the chair but are destabilized by torsional strain. In addition, the boat conformation is further destabilized by a van der Waals repulsion between the "flagpole" hydrogens, which are separated from each other by about 1.83 Å, a distance considerably less than the sum of their van der Waals radii of 2.4 Å.

eclipsing in
boat conformation

flagpole interaction in
boat conformation

partial relief of eclipsing
in twist conformation

Interconversion of chair forms is known as *conformational inversion* and occurs by rotation about carbon–carbon bonds. For cyclohexane, the first-order rate constant for ring inversion is 10^4–10^5 s^{-1} at 300 K. The enthalpy of activation is 10.8 kcal/mol.[31] Calculation of the geometry of the transition state by molecular mechanics suggests a half-twist form lying 12.0 kcal/mol above the chair. The transition state incorporates 0.2 kcal/mol of compression energy from bond deformation, 2.0 kcal/mol of bond-angle strain, 4.4 kcal/mol of van der Waals strain, and 5.4 kcal/mol of torsional strain.[6] Figure 3.4 presents an energy diagram illustrating the process of conformational inversion in cyclohexane. The boat form is not shown in the diagram, because the chair forms can interconvert without passing through the boat. The boat lies 1–2 kcal/mol above the twist conformation and is a transition state for interconversion of twist forms.

Substitution on a cyclohexane ring does not greatly affect the rate of conformational inversion but does change the equilibrium distribution between alternative chair forms. All substituents that are axial in one chair conformation become equatorial on ring inversion, and vice versa. For methylcyclohexane, ΔG° for the equilibrium

is −1.8 kcal/mol, corresponding to a composition with 95% of the equatorial methyl conformation. Two factors contribute to the preference for the equatorial conformation. The equatorial methyl conformation corresponds to an *anti* arrangement with respect to the C(2)−C(3) and C(6)−C(5) bonds, whereas the axial methyl group is in a *gauche* relationship to these bonds. We have seen earlier that the *gauche* conformation of *n*-butane is 0.8 kcal/mol higher in energy than the *anti* conformation. In addition, there is a

29. For a review of nonchair conformations of six-membered rings, see G. M. Kellie and F. G. Riddell, *Top. Stereochem.* **8:**225 (1974).
30. M. Squillacote, R. S. Sheridan, O. L. Chapman, and F. A. L. Anet, *J. Am. Chem. Soc.* **97:**3244 (1975).
31. F. A. L. Anet and A. J. R. Bourn, *J. Am. Chem. Soc.* **89:**760 (1967).

Fig. 3.4. Energy diagram for ring inversion of cyclohexane. [For a rigorous analysis of ring inversion in cyclohexane, see H. M. Pickett and H. L. Strauss, *J. Am. Chem. Soc.* **92**:7281 (1979).]

van der Waals repulsion between the axial methyl group and the axial hydrogens at C−3 and C−5. Interactions of this type are called *1,3-diaxial interactions*.

gauche-butane arrangements
in axial methylcyclohexane

anti-butane arrangements
in equatorial methylcyclohexane

Energy differences between conformations of substituted cyclohexanes can be measured by several physical methods, as can the kinetics of the ring inversion processes. NMR spectroscopy has been especially valuable for both thermodynamic and kinetic studies.[32] In NMR terminology, the transformation of an equatorial substituent to axial and vice versa is called a *site exchange process*. Depending on the rate of the process, the difference between the chemical shifts of the nucleus at the two sites, and the field strength

32. G. Binsch, *Top. Stereochem.* **3**:97 (1968); F. G. Riddell, *Nucl. Magn. Reson.* **12**:246 (1983); J. Sandstrom, *Dynamic NMR Spectroscopy*, Academic Press, New York, 1982; J. L. Marshall, *Nuclear Magnetic Resonance*, Verlag Chemie, Deerfield Beach, Florida, 1983; M. Oki, *Applications of Dynamic NMR to Organic Chemistry*, VCH Publishers, Deerfield Beach, Florida, 1985; Y. Takeuchi and A. P. Marchand, eds., *Applications of NMR Spectroscopy in Stereochemistry and Conformational Analysis*, VCH Publishers, Deerfield Beach, Florida, 1986.

138

CHAPTER 3
CONFORMATIONAL,
STERIC, AND
STEREOELECTRONIC
EFFECTS

of the spectrometer, the spectrum will be either a weighted-average spectrum (rapid site exchange, $k > 10^5$ s^{-1}) or a superposition of the spectra of the two conformers reflecting the conformational composition (slow site exchange, $k < 10^2$ s^{-1}). At intermediate rates of exchange, broadened spectra are observed. Analysis of the temperature dependence of the spectra can lead to the activation parameters for the conformational process. Figure 3.5 illustrates the change in appearance of a simple spectrum.

For substituted cyclohexanes, the slow-exchange condition is met at temperatures below about $-50°$C. Table 3.5 presents data for the half-life for conformational equilibration of cyclohexyl chloride as a function of temperature. From these data, it can be seen that conformationally pure solutions of equatorial cyclohexyl chloride could be maintained at low temperature. This has been accomplished experimentally.[33] Crystallization of cyclohexyl chloride at low temperature affords crystals containing only the

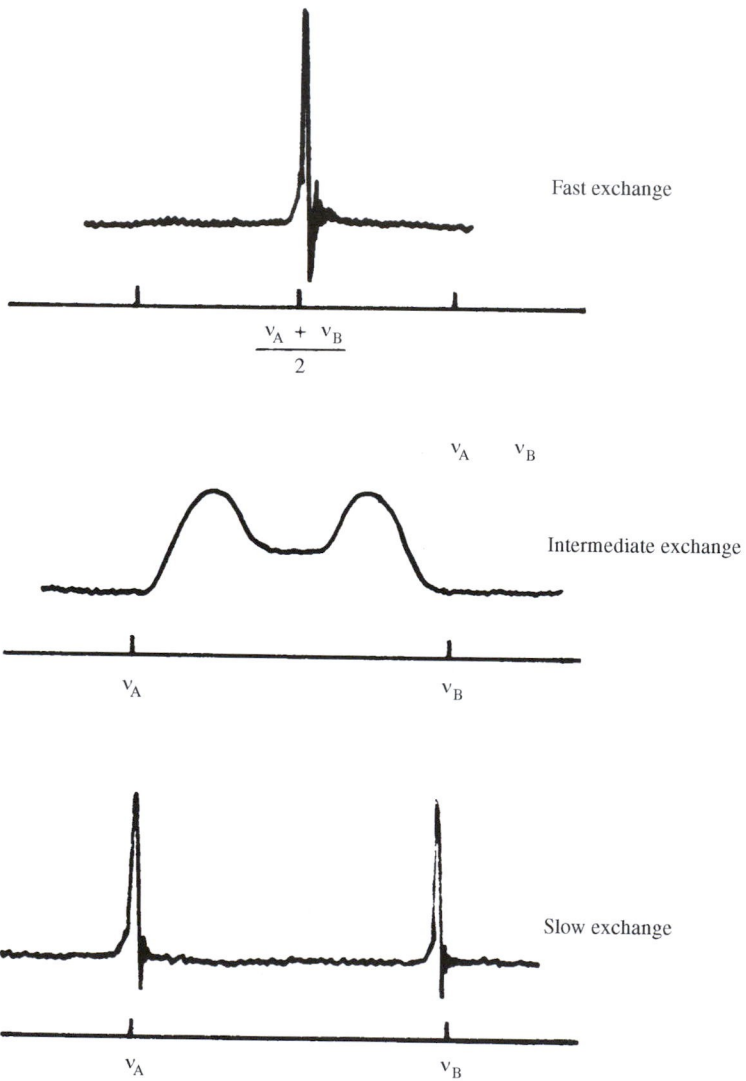

Fig. 3.5. Appearance of NMR spectra for system undergoing two-site exchange (A $\rightleftharpoons$ B).

Table 3.5. Half-life for Conformational Inversion of Cyclohexyl Chloride at Various Temperatures[a]

Temperature (°C)	Half-life
25	1.3×10^{-5} s
−60	2.5×10^{-2} s
−120	23 min
−160	22 yr

a. F. R. Jensen and C. H. Bushweller, *J. Am. Chem. Soc.* **91**:3223 (1969).

equatorial isomer. When the solid is dissolved at −150°C, the NMR spectrum of the solution exhibits only the signal characteristic of the equatorial conformer. When the solution is warmed, the conformational equilibrium is reestablished.

The free-energy difference between conformers is referred to as the *conformational free energy*. For substituted cyclohexanes, it is conventional to specify the value of $-\Delta G^\circ$ for the equilibrium

$$\text{axial} \rightleftharpoons \text{equatorial}$$

Because ΔG° will be negative when the equatorial conformation is more stable than the axial, the value of $-\Delta G^\circ$ is positive for groups which favor the equatorial position. The larger the value of $-\Delta G^\circ$, the greater is the preference for the equatorial position.

The case of cyclohexyl iodide provides an example of the use of NMR spectroscopy to determine the conformational equilibrium constant and the value of $-\Delta G^\circ$. At −80°C, the NMR spectrum of cyclohexyl iodide shows two distinct peaks in the area of the C*H*I signal as shown in Fig. 3.6.[34] The multiplet at higher field is a triplet of triplets with coupling constants of 3.5 and 12 Hz. This pattern is characteristic of a hydrogen in an axial position with two axial–axial couplings and two axial–equatorial couplings. The broader peak at lower field is characteristic of a proton at an equatorial position and reflects the four equatorial–equatorial couplings of such a proton. The relative area of the two peaks is 3.4 : 1 in favor of the conformer with the axial hydrogen. This corresponds to a $-\Delta G^\circ$ value of 0.47 kcal/mol for the iodo substituent. Conformational free-energy values for many substituent groups on cyclohexane have been determined by NMR methods; some are recorded in Table 3.6.

A second important method for measuring conformational free energies involves establishing an equilibrium between *diastereomers* differing only in the orientation of the designated substituent group. The equilibrium constant can then be determined and used to calculate the free-energy difference between the isomers. For example, *cis-* and *trans-t-*butylcyclohexanol can be equilibrated with the use of nickel catalyst in refluxing benzene

33. F. R. Jensen and C. H. Bushweller, *J. Am. Chem. Soc.* **91**:3223 (1969).
34. F. R. Jensen, C. H. Bushweller, and B. H. Beck, *J. Am. Chem. Soc.* **91**:344 (1969).

140

CHAPTER 3
CONFORMATIONAL,
STERIC, AND
STEREOELECTRONIC
EFFECTS

Fig. 3.6. NMR spectrum of cyclohexyl iodide at $-80°C$. Only the lowest-field signals are shown (100-MHz spectrum). [Reproduced from *J. Am. Chem. Soc.* **91**:344 (1969) by permission of the American Chemical Society.]

to give a mixture containing 28% *cis*-4-*t*-butylcyclohexanol and 72% *trans*-*t*-butylcyclohexanol.[35]

$$(CH_3)_3C \quad \xrightarrow[\text{benzene, 80°C}]{\text{nickel}} \quad (CH_3)_3C \quad \text{OH}$$

(28%)   (72%)

Table 3.6. Conformational Free Energies ($-\Delta G°$) for Substituent Groups[a]

Substituent	$-\Delta G°$ (kcal/mol)	Reference
$-F$	0.24–0.28	b
$-Cl$	0.53	b
$-Br$	0.48	b
$-I$	0.47	b
$-CH_3$	1.8	c
$-CH_2CH_3$	1.8	c
$-CH(CH_3)_2$	2.1	c
$-C(CH_3)_3$	> 4.5	d
$-CH=CH_2$	1.7	e
$-C\equiv CH$	0.5	f
$-C_6H_5$	2.9	e
$-CN$	0.15–0.25	b
$-O_2CCH_3$	0.71	b
$-CO_2H$	1.35	d
$-CO_2C_2H_5$	1.1–1.2	d
$-OH$ (aprotic solvents)	0.52	d
$-OH$ (protic solvents)	0.87	d
$-OCH_3$	0.60	d
$-NO_2$	1.16	b
$-HgBr$	0	b

a. For a more extensive compilation including other groups, see E. L. Eliel, S. H. Wilen, and L. N. Monder, *Stereochemistry of Organic Compounds*, John Wiley & Sons, New York, 1993, pp. 696–697.
b. F. R. Jensen and C. H. Bushweller, *Adv. Alicyclic Chem.* **3**:140 (1971).
c. N. L. Allinger and L. A. Freiberg, *J. Org. Chem.* **31**:804 (1966).
d. J. A. Hirsch, *Top Stereochem.* **1**:199 (1967).
e. E. L. Eliel and M. Manoharan, *J. Org. Chem.* **46**:1959 (1981).
f. H. J. Schneider and V. Hoppen, *J. Org. Chem.* **43**:3866 (1978).

35. E. L. Eliel and S. H. Schroeter, *J. Am. Chem. Soc.* **87**:5031 (1965).

Assuming that only conformations that have the *t*-butyl group in the equatorial position are significant, the free-energy change for the equilibration is equal to the free-energy difference between an axial and an equatorial hydroxyl group. The equilibrium constant leads to a value of $-\Delta G^\circ = 0.7\,\text{kcal/mol}$ for the hydroxyl substituent. This approach also assumes that the *t*-butyl group does not distort the ring in any way or interact directly with the hydroxyl group.

There are several other methods available for determining conformational free energies.[36] Values for many substituents in addition to those listed in Table 3.6 have been compiled.[37]

Some insight into the factors that determine the $-\Delta G^\circ$ values for various substituents can be gained by considering some representative groups. Among the halogens, fluorine has the smallest preference for an equatorial conformation. The other halogens have very similar conformational free energies. This is the result of the compensating trends in van der Waals radii and bond lengths. Although the van der Waals radius increases with atomic number, the bond length also increases, so the net effect is small. There may also be a contribution from attractive London forces, which would increase with the size of the halogen atom.

The alkyl groups methyl, ethyl, and isopropyl have similar conformational energies, with the value for the isopropyl group being only slightly larger than that for the methyl and ethyl groups. The similar values for the three substituents reflect the fact that rotation about the bond between the substituent and the ring allows the ethyl and isopropyl groups to adopt a conformation that minimizes the effect of their additional methyl substituents.

methyl substituent: R = R′ = H
ethyl substituent: R = H, R′ = CH₃
isopropyl substituent: R = R′ = CH₃

A *t*-butyl substituent experiences a strong van der Waals repulsion with the *syn*-axial hydrogens in the axial orientation which cannot be relieved by rotation about the bond to the ring. As a result, the $-\Delta G^\circ$ value for the *t*-butyl group is much larger than the values for the other alkyl groups. A value of about 5 kcal/mol has been calculated by molecular mechanics.[38] Experimental attempts to measure the $-\Delta G^\circ$ value for *t*-butyl have provided only a lower limit because very little of the axial conformation is present and the energy difference is very similar to that between the chair and twist forms of the cyclohexane ring.

The strong preference for a *t*-butyl group to occupy the equatorial position has made it a useful group for the study of conformationally biased systems. The presence of a *t*-butyl substituent will ensure that the equilibrium lies heavily to the side having the *t*-butyl group equatorial but does not stop the process of conformational inversion. It should be emphasized that "conformationally biased" is not synonymous with "conformationally

36. F. R. Jensen and C. H. Bushweller, *Adv. Alicyclic. Chem.* **3**:139 (1971).
37. E. L. Eliel, S. H. Wilen, and L. N. Mander, *Stereochemistry of Organic Compounds*, John Wiley & Sons, New York, 1993, pp. 696–697.
38. N. L. Allinger, J. A. Hirsch, M. A. Miller, I. J. Tyminski, and F. A. VanCatledge, *J. Am. Chem. Soc.* **90**:1199 (1968); B. van de Graf, J. M. A. Baas, and B. M. Wepster, *Recl. Trav. Chim. Pays-Bas* **97**:268 (1978); J. M. A. Baas, A. van Veen, and B. M. Wepster, *Recl. Trav. Chim. Pays-Bas* **99**:228 (1980).

142

CHAPTER 3
CONFORMATIONAL,
STERIC, AND
STEREOELECTRONIC
EFFECTS

locked." Because ring inversion can still occur, it is inappropriate to think of the systems as being "locked" in a single conformation.

When two or more substituents are present on a cyclohexane ring, the interactions between the substituents must be included in the analysis. The dimethylcyclohexanes provide an example in which a straightforward interpretation is in complete agreement with the experimental data. For 1,2-, 1,3-, and 1,4-dimethylcyclohexane, the free-energy change of the equilibrium for the $cis \rightleftharpoons trans$ isomerization is given below.[6]

The more stable diastereomer in each case is the one having both methyl groups equatorial. The free-energy difference favoring the diequatorial isomer is about the same for each case (about 1.9 kcal/mol) and is close to the $-\Delta G°$ value of the methyl group (1.8 kcal/mol). This implies that there are no important interactions present that are not also present in methylcyclohexane. This is reasonable since in each case the axial methyl group interacts only with the 3,5-diaxial hydrogens, just as in methylcyclohexane.

Conformations in which there is a 1,3-diaxial interaction between substituent groups larger than hydrogen are destabilized by van der Waals repulsion. Equilibration of mixtures of cis- and $trans$-1,1,3,5-tetramethylcyclohexane reveals that the cis isomer is favored by 3.7 kcal/mol.[39] This provides a value for a 1,3-diaxial methyl interaction that is 1.9 kcal/mol higher than that for the 1,3-methyl–hydrogen interaction.

The decalin (bicyclo[4.4.0]decane) ring system provides another important system for study of conformational effects in cyclohexane rings. Equilibration of the cis and $trans$ isomers reveals that the $trans$ isomer is favored by about 2.8 kcal/mol. Note that this represents a change in *configuration*, not *conformation*. The energy difference can be analyzed by noting that the cis isomer has three more *gauche* butane interactions that are

39. N. L. Allinger and M. A. Miller, *J. Am. Chem. Soc.* **83**:2145 (1961).

not present in the *trans* isomer. Assigning a value of 0.8 kcal/mol to the *gauche* interaction would predict an enthalpy difference of 2.4 kcal/mol between the two isomers.

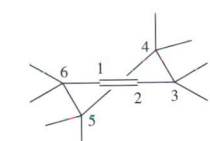

cis-decalin   *trans*-decalin

$\Delta H° = -2.8$ kcal/mol

There is an important difference between the *cis*- and *trans*-decalin systems with respect to their conformational flexibility. *trans*-Decalin, because of the nature of the ring fusion, is incapable of ring inversion. *cis*-Decalin is conformationally mobile and undergoes ring inversion at a rate only slightly slower than cyclohexane ($\Delta G^{\ddagger} = 12.3 - 12.4$ kcal/mol).[40] The *trans*-decalin system is a "conformationally locked" system and can be used to determine the difference in stability and reactivity of groups in axial or equatorial environments.

The effect of introducing sp^2-hybridized atoms into open-chain molecules was discussed earlier, and it was noted that torsional barriers in 1-alkenes and aldehydes are somewhat smaller than in alkanes. Similar effects are noted when sp^2 centers are incorporated into six-membered rings. Whereas the free-energy barrier for ring inversion in cyclohexane is 10.3 kcal/mol, it is reduced to 7.7 kcal/mol in methylenecyclohexane[41] and to 4.9 kcal/mol in cyclohexanone.[42]

The conformation of cyclohexene is described as a half-chair. Structural parameters determined on the basis of electron diffraction and microwave spectroscopy reveal that the double bond can be accommodated into the ring without serious distortion.[43]

half-chair conformation of cyclohexane

The C(1)–C(2) bond length is 1.335 Å, and the C(1)–C(2)–C(3) bond angle is 123°. The substituents at C–3 and C–6 are tilted from the usual axial and equatorial directions and are referred to as *pseudoaxial* and *pseudoequatorial*. The activation energy for ring inversion is 5.3 kcal/mol.[44] The preference for equatorial orientation of a methyl group in cyclohexene is less than in cyclohexane because of the ring distortion and the removal of one 1,3-diaxial interaction. A value of 1 kcal/mol has been suggested for the $-\Delta G°$ value for a methyl group in 4-methylcyclohexene.[45]

40. F. R. Jensen and B. H. Beck, *Tetrahedron Lett.* **1966**:4523; D. K. Dalling, D. M. Grant, and L. F. Johnson, *J. Am. Chem. Soc.* **93**:367 (1971); B. E. Mann, *J. Magn. Reson.* **21**:17 (1976).
41. J. T. Gerig, *J. Am. Chem. Soc.* **90**:1065 (1968).
42. F. R. Jensen and B. H. Beck, *J. Am. Chem. Soc.* **90**: 1066 (1968).
43. J. F. Chiang and S. H. Bauer, *J. Am. Chem. Soc.* **91**:1898 (1969); L. H. Scharpen, J. E. Wollrab, and D. P. Ames, *J. Chem. Phys.* **49**:2368 (1968).
44. F. A. L. Anet and M. Z. Haq, *J. Am. Chem. Soc.* **87**:3147 (1965).
45. B. Rickborn and S.-Y. Lwo, *J. Org. Chem.* **30**:2212 (1965).

144

CHAPTER 3
CONFORMATIONAL,
STERIC, AND
STEREOELECTRONIC
EFFECTS

It has been found that alkylidenecyclohexanes bearing alkyl groups of moderate size at C−2 tend to adopt the conformation with the alkyl group axial in order to relieve unfavorable van der Waals interactions with the alkylidene group. This results from van der Waals repulsion between the alkyl group in the equatorial position and *cis* substituents on the exocyclic double bond. The term *allylic strain* is used to designate this steric effect.[46] The repulsive energy is minimal for methylenecyclohexanes, but molecular mechanics calculations indicate that the axial conformation **B** is 2.6 kcal/mol more stable than **A** with an exocyclic isopropylidene group.[47]

1,3-Allyl strain influences the conformation of *Z*-alkenes. A 4-substituted 2-alkene will prefer conformation **C** over **D** or **E** to minimize the steric interaction with the C−1 methyl group.[48]

If R^1 and R^2 are different, the two faces of the double bond become nonequivalent, permitting stereoselective reactions at the double bond. These effects have been explored, for example, using 4-silyl-2-pentenes. Reactions such as epoxidation and hydroboration proceed by preferential addition from the face opposite the bulky silyl substituents.

46. F. Johnson, *Chem. Rev.* **68**:375 (1968); R. W. Hoffmann, *Chem. Rev.* **89**:1841 (1989).
47. N. L. Allinger, J. A. Hirsch, M. A. Miller, and I. J. Tyminski, *J. Am. Chem. Soc.* **90**:5773 (1968); P. W. Rabideau, ed., *The Conformational Analysis of Cyclohexadiene and Related Hydroaromatic Compounds*, VCH Publishers, Weinheim, 1989.
48. J. L. Broeker, R. W. Hoffmann, and K. N. Houk, *J. Am. Chem. Soc.* **113**:5006 (1991).
49. I. Fleming, A. K. Sarkar, and A. P. Thomas, *J. Chem. Soc., Chem. Commun.* **1987**:157.
50. I. Fleming and N. J. Lawrence, *Tetrahedron Lett.* **27**:2077 (1988).

By analogy with acyclic aldehydes and ketones, an alkyl group at C−2 of a cyclohexanone ring would be expected to be more stable in the equatorial than in the axial orientation. The alkyl group in the equatorial orientation is eclipsed with the carbonyl group, and this conformation corresponds to the more stable conformation of open-chain ketones. This conformation also avoids 3,5-diaxial interactions with *syn*-diaxial hydrogens as in cyclohexane. Conformational free energies for 2-alkyl substituents in cyclohexanones have been determined by equilibration studies. The conformational free energy for the methyl group is similar to that for cyclohexane whereas the values for ethyl and isopropyl are somewhat smaller.[51]

The conformational energy of an alkyl group at C−3 of cyclohexanone is substantially less than that of an alkyl group in cyclohexane because of reduced 1,3-diaxial interactions. A C−3 methyl group in cyclohexanone has a $-\Delta G°$ of 1.3–1.4 kcal/mol.[35]

The preferred conformation of 2-bromo- and 2-chlorocyclohexanones depends upon the polarity of the solvent. In solvents of low dielectric constant, the halogen substituent is more stable in the axial orientation. For example, in chloroform the Br-axial conformation of 2-bromocyclohexanone is favored by nearly 3 : 1.[52] The axial preference increases in the order F < Cl < Br < I. The equatorial halogens are eclipsed with the carbonyl group. The *α-haloketone effect*, as this phenomenon is known, is believed to be the result of dipolar and stereoelectronic interactions between the carbonyl group and the carbon–halogen bond.[53] The bond dipoles partially cancel in the conformation with an axial halogen but are additive for the equatorial halogen. The conformation with the smaller dipole moment will be favored in solvents of low dielectric constant.

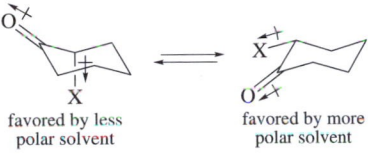

favored by less   favored by more
polar solvent   polar solvent

	Axial : equatorial ratio	
X	Cyclohexane	CHCl₃
F	56 : 44	17 : 83
Cl	77 : 23	45 : 55
Br	87 : 13	71 : 29
I	95 : 5[a]	88 : 12

a. In hexane.

The relative preference for the axial orientation for α-halocyclohexanones can also be interpreted in stereoelectronic terms. In 2-chlorocyclohexanone, the axial arrangement

51. N. L. Allinger and H. M. Blatter, *J. Am. Chem. Soc.* **83**:994 (1961); B. Rickborn, *J. Am. Chem. Soc.* **84**:2414 (1962); E. L. Eliel, N. L. Allinger, S. J. Angyal, and G. A. Morrison, *Conformational Analysis*, Interscience, New York, 1965, pp. 113–114.

52. E. A. Basco, C. Kaiser, R. Rittner, and J. B. Lambert, *J. Org. Chem.* **58**:7865 (1993).

53. J. B. Lambert, in *The Conformational Analysis of Cyclohexadiene and Related Hydroaromatic Compounds*, P. W. Rabideau, ed., VCH Publishers, Weinheim, 1989, Chapter 2.

146

CHAPTER 3
CONFORMATIONAL,
STERIC, AND
STEREOELECTRONIC
EFFECTS

places the C—Cl bond nearly perpendicular to the plane of the carbonyl group. This permits interaction between the π orbitals of the carbonyl group and the σ orbitals associated with the C—Cl bond. There are two interactions possible: $\sigma \rightarrow \pi^*$ donation and $\pi \rightarrow \sigma^*$ donation. This $\sigma-\pi$ delocalization is not possible in the case of equatorial orientation of the chlorine because the C—Cl bond then lies approximately in the nodal plane of the carbonyl group.

$\pi \longrightarrow \sigma^*$ interaction $\sigma \longrightarrow \pi^*$ interaction

3.4. Carbocyclic Rings Other Than Six-Membered

The most important structural features that influence the conformation and reactivity of cycloalkanes differ depending on whether small (cyclopropane and cyclobutane), common (cyclopentane, cyclohexane, and cycloheptane), medium (cyclooctane through cycloundecane), or large (cyclododecane and up) rings are considered. The small rings are dominated by angle strain and torsional strain. The common rings are relatively unstrained, and their conformations are most influenced by torsional factors. Medium rings exhibit conformational equilibria and chemical properties indicating that cross-ring van der Waals interactions play an important role. Large rings become increasingly flexible and possess a large number of low-energy conformations. Table 3.7 presents data on the strain energies of cycloalkanes up to cyclododecane.

The cyclopropane ring is necessarily planar, and the question of conformation does not arise. The C—C bond lengths are slightly shorter than normal at 1.5 Å, and the H—C—H angle of 115° is opened somewhat from the tetrahedral angle.[54] These structural

Table 3.7. Strain Energies of Cycloalkanes

Cycloalkane	Strain energy (kcal/mol)[a]
Cyclopropane	28.1[b]
Cyclobutane	26.3
Cyclopentane	7.3
Cyclohexane	1.4
Cycloheptane	7.6
Cyclooctane	11.9
Cyclononane	15.5
Cyclodecane	16.4
Cyclododecane	11.8

a. Estimated values taken from E. M. Engler, J. D. Andose, and P. v. R. Schleyer, *J. Am. Chem. Soc.* **95:**8005 (1973).
b. Estimated values taken from P. v. R. Schleyer, J. E. Williams, and K. R. Blanchard, *J. Am. Chem. Soc.* **92:**2377 (1970).

54. O. Bastiansen, F. N. Fritsch, and K. Hedberg, *Acta Crystallogr.* **17:**538 (1964).

features and the relatively high reactivity of cyclopropane rings are explained by the concept of "bent bonds" in which the electron density is displaced from the internuclear axis (review Section 1.1.1).

Cyclobutane adopts a puckered conformation in which substituents then occupy axial-like or equatorial-like positions.[55] 1,3-Disubstituted cyclobutanes show small energy preferences for the *cis* isomer since this places both substituents in equatorial-like positions.[56] The energy differences and the barrier to inversion are both smaller than in cyclohexane.

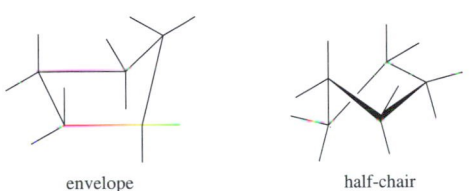

trans   *cis*

R = R′ = Br $\Delta G° = -0.4$ kcal/mol (Ref. 56a)
R = CH$_3$, R′ = CO$_2$CH$_3$ $\Delta G° = -0.3$ kcal/mol (Ref. 56b)

Cyclopentane is nonplanar, and the two minimum-energy geometries are the envelope and half-chair.[57] In the envelope conformation, one carbon atom is displaced from the plane of the other four. In the half-chair conformation, three carbons are coplanar, with one of the remaining two being above the plane and the other below. Both conformations suffer from torsional strain. The energy differences between the conformers are very small, and interconversion is rapid.[58] All of the carbon atoms rapidly move through planar and nonplanar positions. The process is called *pseudorotation*.

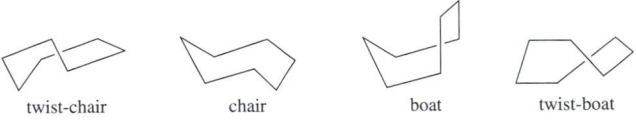

envelope   half-chair

As ring size increases, there are progressively more conformations that need to be considered. For cycloheptane, four conformations have been calculated to be particularly stable.[59] NMR investigations indicate that the twist-chair is the most stable.[60] Various cycloheptane derivatives adopt mainly twist-chair conformations.[61]

twist-chair   chair   boat   twist-boat

55. A. Almenningen, O. Bastiansen, and P. N. Skancke, *Acta Chem. Scand.* **15:**711 (1961).
56. (a) K. B. Wiberg and G. M. Lampman, *J. Am. Chem. Soc.* **88:**4429 (1966); (b) N. L. Allinger and L. A. Tushaus, *J. Org. Chem.* **30:**1945 (1965).
57. A. C. Legon, *Chem. Rev.* **80:**231 (1980); B. Fuchs, *Top. Stereochem.* **10:**1 (1978).
58. W. J. Adams, H. J. Geise, and L. S. Bartell, *J. Am. Chem. Soc.* **92:**5013 (1970); J. B. Lambert, J. J. Papay, S. A. Khan, K. A. Kappauf, and E. S. Magyar, *J. Am. Chem. Soc.* **96:** 6112 (1974).
59. J. B. Hendrickson, *J. Am. Chem. Soc.* **89:** 7036 (1967).
60. J. B. Hendrickson, R. K. Boeckman, Jr., J. D. Glickson, and E. Grunwald, *J. Am. Chem. Soc.* **95:** 494 (1973).
61. F. H. Allen, J. A. K. Howard, and N. A. Pitchford, *Acta Crystallog. Sect. B,* **49:**910 (1993).

148

CHAPTER 3
CONFORMATIONAL,
STERIC, AND
STEREOELECTRONIC
EFFECTS

The total spread in energies calculated for the four conformations is only 2.7 kcal/mol. The individual twist-chair conformations interconvert rapidly by pseudorotation.[62]

For cyclooctane, a total of 11 conformations have been suggested for consideration and their relative energies calculated. The boat-chair was calculated to be the most stable conformation.[49] This prediction was confirmed by analysis of the temperature dependence of the ^{19}F-NMR spectra of fluorocyclooctanes.[63] The activation energy for interconversion of conformers is 5–8 kcal/mol. A few of the most stable conformations are shown below.

boat-chair   boat-boat (saddle)   crown

The number of conformational possibilities for larger rings quickly becomes very large.[64] One interesting and simplifying concept has emerged. The diamond lattice is the most stable arrangement for a large array of sp^3 carbon atoms. There are also both theoretical and experimental results that show that complex polycyclic saturated hydrocarbons are most stable in diamond-type structures. Adamantane is the most familiar example of this type of structure.

adamantane

It might be anticipated that large flexible rings would adopt similar structures incorporating the chair cyclohexane conformation. Conformations for C_{10} through C_{24} cycloalkanes corresponding to diamond-lattice sections have been identified by systematic topological analysis using models or computation.[65] This type of relationship is illustrated in Fig. 3.7 for cyclodecane. Molecular mechanics computations indicate that this is indeed the minimum-energy conformation for cyclododecane.[64,66]

Studies of cyclodecane derivatives by X-ray crystallographic methods have demonstrated that the boat-chair-boat conformation is adopted in the solid state.[67] (Notice that "boat" is used here in a different sense than for cyclohexane.) As was indicated in Table 3.7 (p. 146), cyclodecane is significantly more strained than cyclohexane. Examination of the boat-chair-boat conformation reveals that the source of most of this strain is the close van der Waals contacts between two sets of three hydrogens on either side of the molecule,

62. D. F. Bocian, H. M. Pickett, T. C. Rounds, and H. L. Strauss, *J. Am. Chem. Soc.* **97**:687 (1975).

63. J. E. Anderson, E. S. Glazer, D. L. Griffith, R. Knorr, and J. D. Roberts, *J. Am. Chem. Soc.* **91**:1386 (1969); see also F. A. Anet and M. St. Jacques, *J. Am. Chem. Soc.* **88**:2585, 2586 (1966).

64. I. Kolossvary and W. C. Guida, *J. Am. Chem. Soc.* **115**:2107 (1993).

65. J. Dale, *J. Chem. Soc.* **1963**:93; M. Saunders, *Tetrahedron* **23**:2105 (1967); J. Dale, *Top. Stereochem.* **9**:199 (1976).

66. M. Saunders, *J. Comput. Chem.* **12**:645 (1991).

67. J. D. Dunitz, in *Perspectives in Structural Chemistry*, Vol. II, J. D. Dunitz and J. A. Ibers, eds., John Wiley & Sons, New York, 1968, pp. 1–70.

149

SECTION 3.5.
THE EFFECT OF
HETEROATOMS ON
CONFORMATIONAL
EQUILIBRIA

Side view   Front view

Fig. 3.7. Equivalent diamond-lattice conformations of cyclodecane (boat-chair-boat).

as indicated in the drawing below. Distortion of the molecule to twist forms relieves this interaction but introduces torsional strain.

boat-chair-boat   twist-boat-chair

As the ring size increases, the number of possible conformations increases further so that many alternative diamond-lattice conformations are available.[68]

3.5. The Effect of Heteroatoms on Conformational Equilibria

The replacement of carbon by other elements produces changes in several structural parameters and consequently affects the conformational characteristics of the molecule. In this section, we will first describe some stereochemical features of heterocyclic analogs of cycloalkanes.[69] For the purpose of elaborating conformational principles, the discussion will focus on six-membered rings, so that the properties may be considered in the context of a ring system possessing a limited number of low-energy conformations.

The most obvious changes that occur on introduction of a heteroatom into a six-membered ring have to do with bond lengths and angles. Both the carbon–oxygen and carbon–nitrogen bond lengths (1.43 and 1.47 Å, respectively) are shorter than the carbon–carbon bond length of 1.54 Å, whereas the carbon–sulfur bond length (1.82 Å) is considerably longer. The normal valence angles are somewhat smaller than tetrahedral

68. M. Saunders, *J. Am. Chem. Soc.* **109**:3150 (1987); V. L. Shannon, H. L. Strauss, R. G. Snyder, C. A. Elliger, and W. L. Mattice, *J. Am. Chem. Soc.* **111**:1947 (1989); M. Saunders, K. N. Houk, Y.-D. Wu, W. C. Still, M. Lipton, G. Chang, and W. C. Guida. *J. Am. Chem. Soc.* **112**:1419 (1990).
69. For reviews, see J. B. Lambert and S. I. Featherman, *Chem. Rev.* **75**:611 (1975); F. G. Ridell, *The Conformational Analysis of Heterocyclic Compounds*, Academic Press, New York, 1980; E. L. Eliel, *Acc. Chem. Res.* **3**:1 (1970).

150

CHAPTER 3
CONFORMATIONAL,
STERIC, AND
STEREOELECTRONIC
EFFECTS

at oxygen and nitrogen, and significantly so for sulfur, for which the normal C$-$S$-$C angle is about 100°. The six-membered heterocycles containing oxygen (tetrahydropyran), nitrogen (piperidine), and sulfur (thiane) all resemble the chair conformation of cyclohexane but are modified so as to accommodate the bond lengths and bond angles characteristic of the heteroatom. The rings are all somewhat more puckered than cyclohexane. Because of the shorter C$-$O bond distances, 2-substituents in tetrahydropyran and 1,3-dioxane rings have larger conformational free energies than in cyclohexane rings. The shorter bonds lead to stronger repulsive interaction with the 4- and 6-axial hydrogens. Table 3.8 presents $-\Delta G°$ values for several alkyl groups in tetrahydropyrans, 1,3-dioxanes, and 1,3-dithianes, along with their comparative $-\Delta G°$ values in cyclohexane.

tetrahydropyran   piperidine   thiane   1,3-dioxane   1,3-dithiane

An important feature associated with heterocyclic rings is the reduced steric repulsion for axial substituents that results from replacement of a methylene group in cyclohexane by oxygen, nitrogen, or sulfur. This effect is readily apparent in *cis*-2-methyl-5-*t*-butyl-1,3-dioxane, in which the preferred conformation has the *t*-butyl group axial and the methyl group equatorial.[70] Divalent oxygen has no substituents, so the 1,3-diaxial interaction, which is the main unfavorable interaction for axial substituents in cyclohexanes, is not present. The diaxial interactions between the methyl group and the C-4 and C-6 hydrogens outweigh any repulsions between the *t*-butyl group and the oxygen lone-pair electrons.

preferred conformation

It is consistently found that 5-alkyl substituents in 1,3-dioxane exhibit a smaller equatorial preference than they do in cyclohexane. This decreased preference is due to

Table 3.8. Comparison of Conformational Free-Energy Values for Substituents on Tetrahydropyran, 1,3-Dioxane, and 1,3-Dithiane Rings with Those for Cyclohexane

			$-\Delta G°$ (kcal/mol)			
		Tetrahydro-pyran[a]	1,3-Dioxane[b]		1,3-Dithiane[c]	
Group	Cyclohexane	2-Position	2-Position	5-Position	2-Position	5-Position
CH$_3-$	1.8	2.9	4.0	0.8	1.8	1.0
CH$_3$CH$_2-$	1.8		4.0	0.7	1.5	0.8
(CH$_3$)$_2$CH$-$	2.1		4.2	1.0	1.5	0.8
(CH$_3$)$_3$C$-$	>4.5			1.4	> 2.7	
CH$_2$=CH$-$	1.7	2.3				
CH≡C$-$	0.5	0.3				

a. E. L. Eliel, K. D. Hargrave, K. M. Pietrusiewicz, and M. Manoharan, *J. Am. Chem. Soc.* **104**:3635 (1982).
b. E. L. Eliel and M. C. Knoeber, *J. Am. Chem. Soc.* **90**:3444 (1968); F. W. Nader and E. L. Eliel, *J. Am. Chem. Soc.* **92**:3050 (1970).
c. E. L. Eliel and R. O. Hutchins, *J. Am. Chem. Soc.* **91**:2703 (1969).

70. E. L. Eliel and M. C. Knoeber, *J. Am. Chem. Soc.* **90**:3444 (1968).

decreased van der Waals repulsions in the axial orientation, since there are no hydrogens that are *syn*-axial to the 5-alkyl substituent. A 2-alkyl substituent, on the other hand, has a greater preference for the equatorial orientation in 1,3-dioxane than in cyclohexane, because the decreased C—O bond length (relative to C—C) brings an axial 2-alkyl group into closer contact with the *syn*-axial hydrogens at C-4 and C-6, resulting in an increased van der Waals repulsion. Similarly, an axial 4-alkyl substituent in a 1,3-dioxane suffers a greater van der Waals repulsion with the axial hydrogen at C-2 than it does in cyclohexane. The general point to be recognized is that the conformational free energy is a function not only of the size of the group but also of the molecular environment that it encounters.[71]

The decreased preference for the equatorial orientation of a 5-alkyl group in 1,3-dioxanes and 1,3-dithianes is evident from the data in Table 3.8. It is also interesting that the increased preference for the equatorial orientation of a 2-methyl group in 1,3-dioxane disappears in going to 1,3-dithiane. The conformational free energies of 2-alkyl substituents in 1,3-dithianes are more similar to those of cyclohexane (actually, slightly smaller) because of the longer C—S bond length compared to C—O.

When a polar substituent is present, interactions between the substituent and the ring heteroatom can become an important factor in the position of the conformational equilibrium. In some cases, the interactions are straightforward and readily assessed. For example, the preferred conformation of 5-hydroxy-1,3-dioxane has the hydroxyl group in the axial position.[72] This conformation is favored because hydrogen bonding of the hydroxyl group with the ring oxygen is possible only when the hydroxyl group is axial and serves as a stabilizing force for this conformation.

3.6. The Anomeric Effect

The incorporation of heteroatoms can result in stereoelectronic effects that have a pronounced effect on conformation and, ultimately, on reactivity. It is known from numerous examples in carbohydrate chemistry that pyranose sugars substituted with an electron-withdrawing group such as halogen or alkoxy at C-1 are often more stable when the substituent has an axial, rather than an equatorial, orientation. This tendency is not limited to carbohydrates but carries over to simpler ring systems such as 2-substituted tetrahydropyrans. The phenomenon is known as the *anomeric effect*, because it involves a substituent at the anomeric position in carbohydrate pyranose rings.[73] Scheme 3.1 lists

71. For a review of conformational analysis of dioxanes, see M. J. O. Anteunis, D. Tavernier, and F. Borremans, *Heterocycles* **4**:293 (1976).

72. J. L. Alonso and E. B. Wilson, *J. Am. Chem. Soc.* **102**:1248 (1980); N. Baggett, M. A. Bukhari, A. B. Foster, J. Lehmann, and J. M. Webber, *J. Chem. Soc.* **1963**:4157.

73. For reviews, see R. U. Lemieux, *Pure Appl. Chem.* **25**: 527 (1971); W. A. Szarek and D. Horton, eds., *Anomeric Effects*, ACS Symposium Series, No. 87, American Chemical Society, Washington, D.C., 1979; A. J. Kirby, *The Anomeric Effect and Related Stereoelectronic Effects at Oxygen*, Springer-Verlag, Berlin, 1983; P. Deslongchamps, *Stereoelectronic Effects in Organic Chemistry*, Pergamon Press, Oxford, 1983; M. L. Sinot, *Adv. Phys. Org. Chem.* **24**:113 (1988); P. R. Graczyk and M. Mikolajczyk, *Top. Stereochem.* **21**:159 (1994); E. Juraisti and G. Cuevas, *The Anomeric Effect*, CRC Press, Boca Raton, Florida, 1995; C. J. Cramer, *THEOCHEM* **370**:135 (1996).

152

CHAPTER 3
CONFORMATIONAL,
STERIC, AND
STEREOELECTRONIC
EFFECTS

Scheme 3.1. Equilibria in Compounds that Exhibit the Anomeric Effect

1[a] Glucose pentaacetate

$K = 5$ (in 50% acetic acid; acetic anhydride, $0.1\ M\ H_2SO_4$ at 25°C)
$\Delta H° = -1.4$ kcal/mol

2[b] 2-Chloro-4-methyltetrahydropyran

$K = 32$ (pure liquid at 40°C)

3[c] 2-Methoxy-6-methyltetrahydropyran

$K = 3.4$ (in carbon tetrachloride)
$K = 1.8$ (in acetonitrile)

4[d] *trans, cis, trans*-2,3,5,6-Tetrachloro-1,4-dioxane

Equilibrium constant not known in solution; crystalline form has all chlorines axial.

5[e] Tri-*O*-acetyl-β-D-xylopyranosyl chloride

The NMR spectrum in $CDCl_3$ indicates that the all-axial form is strongly favored.
The equilibrium constant is not known.

6[f] α-D-Altropyranose pentaacetate

only form present at equilibrium by NMR analysis

a. W. A. Bonner, *J. Am. Chem. Soc.* **73**:2659 (1951).
b. C. B. Anderson and D. T. Sepp, *J. Org. Chem.* **32**:607 (1967).
c. E. L. Eliel and C. A. Giza, *J. Org. Chem.* **33**:3754 (1968).
d. E. W. M. Rutten, N. Nibbering, C. H. MacGillavry, and C. Romers, *Rec. Trav. Chim.* **87**:888 (1968).
e. C. V. Holland, D. Horton, and J. S. Jewell, *J. Org. Chem.* **32**:1818 (1967).
f. B. Coxon, *Carbohydr. Res.* **1**:357 (1966).

several compounds that exhibit the anomeric effect, along with some measured equilibrium distributions. In entries 1–3, the equilibria are between diastereoisomers, while entries 4–6 illustrate the anomeric effect in conformationally mobile systems. In all cases, the more stable isomer is written on the right.

The magnitude of the anomeric effect depends on the nature of the substituent and decreases with increasing dielectric constant of the medium.[74] The effect of the substituent can be seen by comparing the related 2-chloro- and 2-methoxy-substituted tetrahydropyrans in entries 2 and 3. The 2-chloro compound exhibits a significantly greater preference for the axial orientation than the 2-methoxy compound. Entry 3 also provides data relative to the effect of solvent polarity; it is observed that the equilibrium constant is larger in carbon tetrachloride ($\varepsilon = 2.2$) than in acetonitrile ($\varepsilon = 37.5$).

Compounds in which conformational, rather than configurational, equilibria are influenced by the anomeric effect are depicted in entries 4–6. Single-crystal X-ray diffraction studies have unambiguously established that all the chlorine atoms of *trans*, *cis*, *trans*-2,3,5,6-tetrachloro-1,4-dioxane occupy axial sites in the crystal. Each chlorine in the molecule is bonded to an anomeric carbon and is subject to the anomeric effect. Equally striking is the observation that all the substituents of the tri-*O*-acetyl-β-D-xylopyranosyl chloride shown in entry 5 are in the axial orientation *in solution*. Here, no special crystal packing forces can be invoked to rationalize the preferred conformation. The anomeric effect of a single chlorine is sufficient to drive the equilibrium in favor of the conformation that puts the three acetoxy groups in axial positions.

Several structural factors have been considered as possible causes of the anomeric effect. In localized valence bond terminology, it can be recognized that there will be a dipole–dipole repulsion between the polar bonds at the anomeric carbon in the equatorial conformation. This dipole–dipole interaction is reduced in the axial conformation, and this factor probably contributes to the solvent dependence of the anomeric effect.

From the molecular orbital viewpoint, the anomeric effect is described as resulting from an interaction between the lone-pair electrons on the pyran oxygen and the σ^* orbital associated with the bond to the electronegative C-2 substituent.[75] When the C–X bond is axial, an interaction between an occupied *p*-type orbital on oxygen (lone-pair electrons) and the antibonding σ^* orbital of the C–X combination is possible. This permits delocalization of the lone-pair electrons and would be expected to shorten and strengthen the C–O bond while lengthening and weakening the C–X bond.

74. K. B. Wiberg and M. Marquez, *J. Am. Chem. Soc.* **116**:2197 (1994).

75. S. Wolfe, A. Rauk, L. M. Tel, and I. G. Csizmaida, *J. Chem. Soc. B* **1971**:136; S. O. David, O. Eisenstein, W. J. Hehre, L. Salem, and R. Hoffmann, *J. Am. Chem. Soc.* **95**:306 (1973); F. A. VanCatledge, *J. Am. Chem. Soc.* **96**:5693 (1974).

154

CHAPTER 3
CONFORMATIONAL,
STERIC, AND
STEREOELECTRONIC
EFFECTS

Studies of the temperature dependence of the ^{13}C-NMR chemical shifts of 2-methoxytetrahydropyran have determined $\Delta G°$ values ranging from 0.5 to 0.8 kcal/mol, depending on the solvent.[76] MO calculations indicate that the axial methoxy group is favored by about 1.5 kcal/mol.[77] In 2-alkoxytetrahydropyran derivatives, there is a correlation between the length of the exocyclic C−O bond and the nature of the oxygen substituent. The more electron-withdrawing the group, the longer is the bond to the oxygen. This indicates that the extent of the anomeric effect increases with the electron-accepting capacity of the exocyclic oxygen.[78]

Extent of bond lengthening increases with electron-accepting capacity of OR.

The axial–equatorial conformational equilibria for 2-fluoro- and 2-chlorotetrahydropyran have been investigated with several MO calculations, including calculations at the MP2/6-31G* level. The MP2/6-31G* calculations give values of 3.47 and 2.84 kcal/mol, respectively, for the energy favoring the axial conformer.[79] Solvent effects were also explored computationally and show the usual trend of reduced stability for the axial conformation as solvent polarity increases.

The anomeric effect is also present in acyclic systems and stabilizes conformations that allow antiperiplanar alignment of the C−X bond with a lone-pair orbital of the heteroatom. Anomeric effects are prominent in determining the conformation of acetals and α-alkoxyamines, as well as α-haloethers. MO calculations (4-31G) have found 4 kcal/mol as the difference between the two conformations shown below for methoxymethyl chloride.[80]

The synclinal conformation (*sc*) is appropriate for overlap of an oxygen nonbonded pair with the σ* C−Cl orbital. The preferred *ap* relationship, requires an antiperiplanar alignment of a lone-pair orbital with the bond to the electronegative substituent. Because of the donor–acceptor nature of the interaction it is enhanced in the order F < O < N for the donor (D) atom and N < O < F for the acceptor (A) atom.

76. H. Booth, J. M. Dixon, and S. A. Readshaw, *Tetrahedron* **48**:6151 (1992).
77. U. Salzner and P. v. R. Schleyer, *J. Org. Chem.* **59**:2138 (1994).
78. A. J. Briggs, R. Glenn, P. G. Jones, A. J. Kirby, and P. Ramaswamy, *J. Am. Chem. Soc.* **106**:6200 (1984).
79. I. Tvaroska and J. P. Carver, *J. Phys. Chem.* **98**:6452 (1994).
80. G. A. Jeffrey and J. H. Yates, *J. Am. Chem. Soc.* **101**:820 (1979).

$$CH_4 + X{-}CH_2{-}Y \longrightarrow CH_3{-}X + CH_3{-}Y$$

should measure the stabilizing effect. The results were obtained using 3-21G level calculations.[81] Calculations for a related isodesmic reaction at the MP2/6-311G(d,f) level gave much lower values (shown in parentheses) for the three geminally disubstituted derivatives.[82]

Stabilization Calculated for Anomeric Effect (kcal/mol)

X(donor)	Y(acceptor)		
	NH$_2$	OH	F
NH$_2$	10.6 (3.3)	12.7	17.6
OH		17.4 (9.0)	16.2
F			13.9 (4.5)

The case of fluoromethanol is also illustrative. There is a substantial barrier to rotation of the hydroxyl hydrogen with respect to the fluoromethyl group, with the preferred orientation having the hydroxyl hydrogen *gauche* to the fluorine.[83] This conformation is 12.6 kcal/mol more stable than that having the fluorine *anti* to the hydroxyl hydrogen.

most stable   less stable
conformation   conformation

The preference for the *gauche* arrangement is an example of the anomeric effect. An oxygen lone pair is *anti* to fluorine in the stable conformation but not in the unstable conformation.

Even molecules as simple as dimethoxymethane give evidence of anomeric effects. The preferred conformation of dimethoxymethane aligns each C$-$O bond with a lone-pair orbital of the adjacent oxygen.[84]

81. P. v. R. Schleyer, E. D. Jemmis, and G. W. Spitznagel, *J. Am. Chem. Soc.* **107**:6393 (1985).
82. G. Leroy, J.-P. Dewispelaere, H. Bbenkadour, D. R. Temsamani, and C. Wilante, *THEOCHEM* **334**:137 (1995); see also Y.-P. Chang and T.-M. Su, *THEOCHEM* **365**:183 (1996).
83. S. Wolfe, M.-H. Whangbo, and D. J. Mitchell, *Carbohydr. Res.* **69**: 1 (1979).
84. K. B. Wiberg and M. A. Murcko, *J. Am. Chem. Soc.* **111**:4821 (1989); J. R. Kneisler and N. L. Allinger, *J. Comput. Chem.* **17**:757 (1996).

156

CHAPTER 3
CONFORMATIONAL,
STERIC, AND
STEREOELECTRONIC
EFFECTS

Simple acyclic acetals can possess six distinct conformations.

The preferred conformation is **D** because it maximizes the number of antiperiplanar relationships between nonbonded electron pairs and C—O bonds while avoiding the R′–R′ van der Waals repulsions in conformations **E** and **F**.

In cyclic systems such as **1**, the dominant conformation is the one with the maximum anomeric effect. In the case of **1**, only conformation **1A** provides the preferred antiperiplanar geometry for both oxygens.[85] Antiperiplanar relationships are indicated by including lone pairs in the oxygen orbitals. Other effects, such as torsional strain and nonbonded repulsion, contribute to the conformational equilibrium, of course. Normally, a value of about 1.5 kcal/mol is assigned to the stabilization due to an optimum anomeric interaction in an acetal.

Another example of a stereoelectronic effect is observed in amines. Amines in which a C—H bond is oriented antiperiplanar to the nitrogen lone pair show a shift in the C—H bond stretching frequency that corresponds to a weakening of the bond by about 0.4 kcal/mol.[86] This effect, called a *trans lone-pair effect*, results from an interaction between the lone pair and the antibonding σ* C—H orbital. Similar effects are also present in oxygen compounds but are weaker because of the greater electronegativity of oxygen.

85. P. Deslongchamps, D. D. Rowan, N. Pothier, G. Sauve, and J. K. Saunders, *Can. J. Chem.* **59**:1132 (1981).
86. H. D. Thomas, K. Chen, and N. L. Allinger, *J. Am. Chem. Soc.* **116**:5887 (1994).

3.7. Conformational Effects on Reactivity

Conformational effects on reactivity have been particularly thoroughly studied in cyclohexane systems. The difference between an axial and an equatorial environment of a functional group can lead to significant differences in reaction rates. One of the most common ways of studying the effect of orientation on reactivity is to use an appropriately placed *t*-butyl or other large substituent to ensure that the reacting group is overwhelmingly in the equatorial or axial position. The conformationally rigid *trans*-decalin system can also be used to assess reactivity differences between functional groups in axial versus equatorial positions.

Scheme 3.2 gives some data that illustrate the differences in reactivity between groups in axial and equatorial positions. It should be noted that a group can be either more or less reactive in an axial position as compared to the corresponding equatorial position.

The effect of conformation on reactivity is intimately associated with the details of the mechanism of a reaction. The examples of Scheme 3.2 illustrate some of the ways in which substituent orientation can affect reactivity. It has been shown that oxidation of *cis*-4-*t*-butylcyclohexanol is faster than oxidation of the *trans* isomer, but the rates of acetylation are in the opposite order. Let us consider the acetylation first. The rate of the reaction will depend on the free energy of activation for the rate-determining step. For acetylation, this step involves nucleophilic attack by the hydroxyl group on the acetic anhydride carbonyl

Scheme 3.2. Effects of Functional-Group Orientation on Rates and Equilibria

Relative rate of oxidation
$(CrO_3) = 3.23^a$
Relative rate of acetylation
$= 1.00^b$

Relative rate of oxidation
$(CrO_3) = 1.00^a$
Relative rate of acetylation
$= 3.70^b$

Relative rate of saponification
$= 1.00^c$

Relative rate of saponification
$= 19.8^c$

$pK_a = 8.23^d$

$pK_a = 7.79^d$

a. E. L. Eliel, S. H. Schroeter, T. J. Brett, F. J. Biros, and J.-C. Richer, *J. Am. Chem. Soc.* **88**:3327 (1966).
b. E. L. Eliel and F. J. Biros, *J. Am. Chem. Soc.* **88**:3334 (1966).
c. E. L. Eliel, H. Haubenstock, and R. V. Acharya, *J. Am. Chem. Soc.* **83**:2351 (1961).
d. R. D. Stolow, *J. Am. Chem. Soc.* **81**:5806 (1959).

158

CHAPTER 3
CONFORMATIONAL,
STERIC, AND
STEREOELECTRONIC
EFFECTS

group to form the tetrahedral intermediates **I** and **J**. A qualitative energy diagram is given in Fig. 3.8.

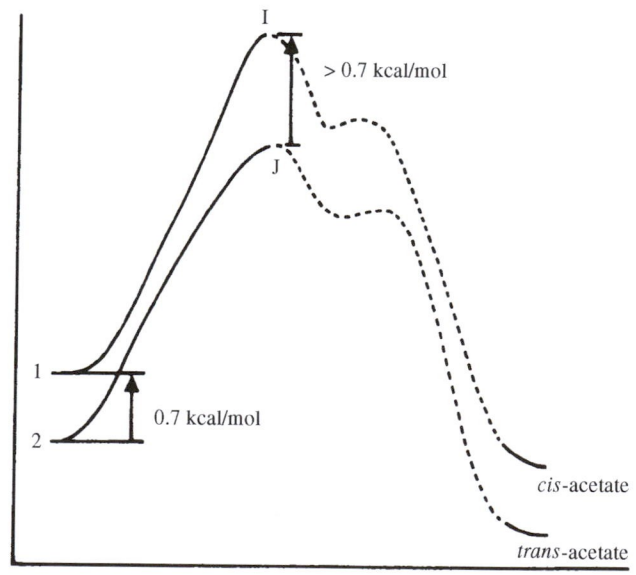

1 OH axial (*cis*)
2 OH equatorial (*trans*)

I *cis*
J *trans*

Because its hydroxyl group occupies an equatorial position, the *trans* isomer **2** is more stable than the *cis* isomer **1** by an amount equal to $-\Delta G^\circ$ for the hydroxyl group. It can be assumed that the transition state for the rate-determining step will resemble the tetrahedral intermediates **I** and **J**. Because the substituent group has become larger as the acetylating reagent is bonded to the hydroxyl group, the value of $-\Delta G^\circ$ for the substituent at the transition state should be *greater than* that for the hydroxyl group, 0.7 kcal/mol. Intermediate **I**, then, must be higher in energy than **J** by more than 0.7 kcal/mol. From this information, it can be predicted that **1** will acetylate more slowly than **2**, because a larger free energy of activation will be required. This is illustrated in Fig. 3.8. As shown by the data in Scheme 3.2, the prediction is correct.

Extensive research has established that axial cyclohexanols are more reactive than equatorial alcohols toward chromic acid oxidation.[87] The basis for this effect can be seen

Fig. 3.8. Approximate energy diagram for acetylation of *cis*- and *trans*-4-*t*-butylcyclohexanol.

87. E. L. Eliel, S. H. Schroeter, T. J. Brett, F. J. Biros, and J.-C. Richer, *J. Am. Chem. Soc.* **88**:3327 (1966); P. Mueller and J.-C. Perlberger, *J. Am. Chem. Soc.* **98**:8407 (1976).

by analyzing the free energies of activation for the reactions. The available evidence indicates that the rate-determining step is a breakdown of a chromate ester intermediate. The transition state involves partial cleavage of the C—H bond proceeding toward loss of chromium. An approximate energy diagram is given in Fig. 3.9.

The diaxial interactions that are responsible for a large portion of the conformational free energy of the hydroxyl group are relieved in the transition state as the reaction proceeds toward sp^2 hybridization at the carbon atom undergoing oxidation. Because the substituent is effectively becoming smaller as the reaction proceeds, the energy difference between the diastereomeric transition states is less than that in the reactant alcohols. Putting it another way, the 1,3-diaxial interactions are relieved in the rate-determining transition state. Under these circumstances, the higher energy *cis*-isomer is the more reactive of the two alcohols.

A similar analysis of the hydrolysis of the esters **3** and **4** is possible. From Table 3.6 (p. 140), we see that the conformational free energy of the carboethoxy group is 1.2 kcal/mol. The *cis* isomer is this much higher in energy than the *trans* isomer. The transition states resemble **M** and **N**. The substituent group changes from sp^2 to sp^3 hydbridization and increases in size as the transition state is reached. As a result, the difference in energy between **M** and **N** must be greater than 1.2 kcal/mol. As can be

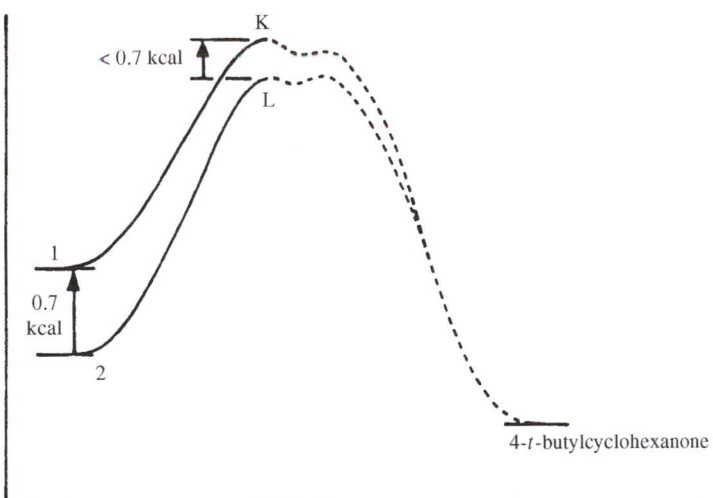

Fig. 3.9. Approximate energy diagram for oxidation of *cis*- and *trans*-4-*t*-butylcyclohexanol.

160

CHAPTER 3
CONFORMATIONAL,
STERIC, AND
STEREOELECTRONIC
EFFECTS

concluded by examining the approximate energy diagram in Fig. 3.10, the *trans* isomer with the equatorial carboethoxy group hydrolyzes significantly faster than the *cis* isomer.

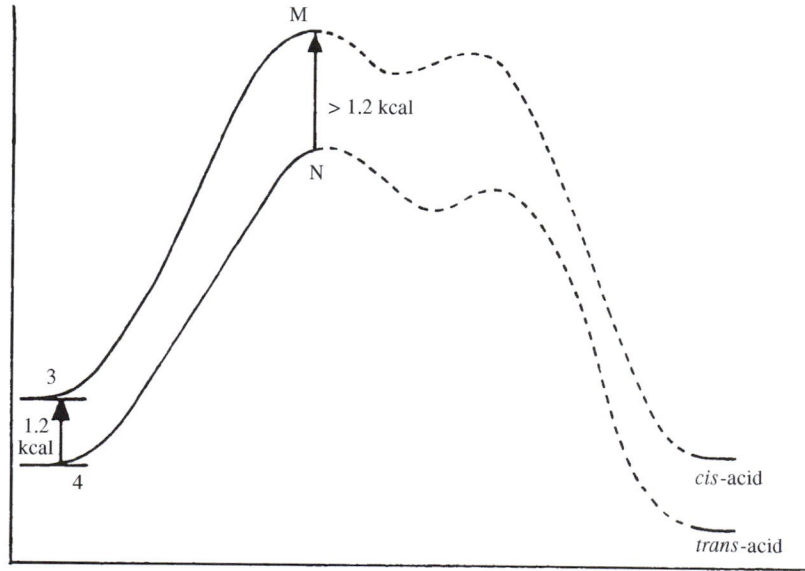

Many examples of reactivity effects that are due to the anomeric effect have been identified. For example, CrO_3 can oxidize some pyranose acetals, leading eventually to δ-ketoesters.

Isomers with equatorial 2-alkoxy groups are more reactive than those with axial 2-alkoxy groups.[88] The greater reactivity of the equatorial isomers is the result of the alignment of

Fig. 3.10. Approximate energy diagram for saponification of ethyl esters of *cis*- and *trans*-4-*t*-butylcyclohexanecarboxylic acid.

the lone pairs on both the endocyclic and the exocyclic oxygen to assist in hydrogen abstraction.

two lone-pair orbitals in
antiperiplanar arrangement

only one lone-pair orbital in
antiperiplanar arrangement

Other reagents which oxidize acetals such as ozone and N-bromosuccinimide show similar reactivity trends.[89]

Another example is the absence of oxygen exchange with solvent in the hydrolysis of gluconolactone. Simple acyclic esters usually undergo isotopic exchange at a rate that is competitive with hydrolysis. This occurs through the tetrahedral addition intermediate.

Gluconolactone shows no exchange.[90] The reason is that the tetrahedral intermediate is formed and breaks down stereoselectively. Even though proton exchange can occur in the tetrahedral intermediate, the anomeric effect leads to preferential loss of the axial oxygen.

reverse reaction favored by
two antiperiplanar orbitals

expulsion of equatorial OH favored
by only one antiperiplanar orbital

88. S. J. Angyal and K. James, *Aust. J. Chem.* **23**:1209 (1970).
89. P. Deslongchamps, C. Moreau, D. Frehel, and R. Chevenet, *Can. J. Chem.* **53**:1204 (1975) and preceding papers; C. W. McClelland, *J. Chem. Soc., Chem. Commun.* **1979**:751.
90. Y. Pocker and E. Green, *J. Am. Chem. Soc.* **95**:113 (1973).

3.8. Angle Strain and Its Effect on Reactivity

Another important factor in reactivity is angle strain. Angle strain results from the distortion of bond angles and increases the energy content of the molecule.[91] Table 3.9 gives some data on the total angle strain for some cyclic, bicyclic, and tricyclic hydrocarbons. Six-membered rings are nearly strain-free, whereas the strain energy in smaller rings increases from 6–7 kcal/mol for cyclopentane to nearly 30 kcal/mol for cyclopropane. In more complex structures, total strain increases as molecular geometry requires greater distortion from optimal bond angles.

Because of the increased ground-state energy resulting from angle strain, reactions which lead to ring opening of strained compounds often proceed much more readily than do similar reactions in unstrained systems. Furthermore, the ring strain causes qualitative changes in the nature of the bonds (hybridization), and these changes can increase reactivity.[92] For example, whereas normal saturated hydrocarbons are inert to bromine

Table 3.9. Strain Energies in some Alicylic Compounds (kcal/mol)[a]

△	27.5	⋈	63.2		68
△	55.2	▱	63.9		98
□	26.5	⬠	54.9		37
⬠	6.2		31.0		104
⬡	0.0		25.6		7.4
	−1.9		15.8		89
	1.2		9.4		67[c]
	154.7		96[b]		

a. Data from K. B. Wiberg, *Angew Chem. Int. Ed. Engl.* **25**:312 (1986).
b. D. S. Kabakoff, J.-C. G. Bünzli, J. F. M. Oth, W. B. Mammon, and J. A. Berson, *J. Am. Chem. Soc.* **97**:1510 (1975).
c. K. B. Wiberg, H. A. Connon, and W. E. Pratt, *J. Am. Chem. Soc.* **101**:6970 (1979).

91. K. B. Wiberg, *Angew. Chem. Int. Ed. Engl.* **25**:312 (1986); B. Halton, *Adv. Strain Org. Chem.* **1**:1 (1991).

in the dark, cyclopropane reacts rapidly, giving ring-opened products.[93] The major product arises from ring opening to yield a carbocation, followed by capture by bromide ion. The other products arise from rearrangement of the cationic intermediate.

$$CH_3CHCHCH_2Br + BrCH_2CHCH(CH_3)_2 + \text{other products}$$

~50%   ~30%

Kinetic and product-structure studies of the reaction of acetic acid with cyclopropanes and bicyclic compounds incorporating three-membered rings have shown that the protonation of the cyclopropane ring is followed by addition of the nucleophile at the most substituted carbon. The regioselectivity is determined by the ability of the more highly substituted carbons to sustain more of the positive charge. The extent of substitution at the incipient carbocation is the most important factor in determining the degree of reactivity. The relative rates of solvolysis of cyclopropane and its methyl, 1,1-dimethyl, 1,1,2-trimethyl, and 1,1,2,2-tetramethyl derivatives in acetic acid, demonstrate the accelerating effect of electron-donating methyl groups.[94]

products

substituents	relative rate	acetate	alkene
none	1	100	0
1-Methyl	91	67	33
1,1-Dimethyl	639	10	96
1,1,2-Trimethyl	668	25	75
1,1,2,2-Tetramethyl	2135	15	85

Bicyclo[1.1.0]butane is an example of a molecule in which severe angle strain results in decreased stability and greatly enhanced reactivity.[95] The bicyclo[1.1.0]butane ring has a strain energy of 63.9 kcal/mol, and the central bond is associated with a relatively high energy HOMO.[96] The central bond in bicyclo[1.1.0]butane is formed from nearly pure p orbitals of the two bridgehead carbons.[97] These structural features are reflected in enhanced reactivity toward electrophiles. Acid-catalyzed solvolysis gives products char-

92. A. Sella, H. Basch, and S. Hoz, *J. Am. Chem. Soc.* **118**:416 (1996).
93. J. B. Lambert and B. A. Iwanetz, *J. Org. Chem.* **37**:4082 (1972); J. B. Lambert and K. Kobayashi, *J. Org. Chem.* **41**:671 (1976); P. S. Skell, J. C. Day, and K. H. Shea, *J. Am. Chem. Soc.* **94**:1126 (1972); J. B. Lambert, W. J. Schulz, Jr., P. H. Mueller, and K. Kobayashi, *J. Am. Chem. Soc.* **106**:792 (1984).
94. K. B. Wiberg and S. R. Kass, *J. Am. Chem. Soc.* **107**:988 (1985); K. B. Wiberg, S. R. Kass, and K. C. Bishop III, *J. Am. Chem. Soc.* **107**:996 (1985).
95. S. Hoz in *The Chemistry of the Cyclopropyl Group*, Part 2, S. Patai and Z. Rappoport, eds., Wiley, Chichester, U.K., 1987, pp. 1121–1192; M. Christl, *Adv. Strain Org. Chem.* **4**:163 (1995).
96. K. B. Wiberg, G. B. Ellison, and K. S. Peters, *J. Am. Chem. Soc.* **99**:3941 (1977).
97. J. M. Schulmand and G. J. Fisanick, *J. Am. Chem. Soc.* **92**:6653 (1970); R. D. Bertrand, D. M. Grant, E. L. Allred, J. C. Hinshaw, and A. B. Strong, *J. Am. Chem. Soc.* **94**:997 (1972); D. R. Whitman and J. F. Chiang, *J. Am. Chem. Soc.* **94**:1126 (1972).

164

CHAPTER 3
CONFORMATIONAL,
STERIC, AND
STEREOELECTRONIC
EFFECTS

acteristic of the bicyclobutonium cation (see Section 5.12). The products are derived from cleavage of both the central and the peripheral bonds:

The propellanes (see pp. 7–8) represent another interesting group of molecules whose reactivity reflects the effects of distorted bond angles. As the bridges in the propellanes are made smaller the molecules become more reactive. As the structural formula implies, [1.1.1]propellane has a very unusual shape. All four bonds to the bridgehead carbons are directed to the same side of the atom.[99] [2.2.1]Propellane can be isolated in solid argon at 45 K but decomposes at higher temperatures and has not been obtained as a pure substance.[100] The trend toward decreasing stability with increasing strain makes a sharp reversal at [1.1.1]propellane. This compound is observed to decompose much more slowly than [2.2.1]propellane. To understand this situation, it must be recognized that a major factor in determining stability is the strength of the central bond toward homolytic cleavage, which provides a path for decomposition. This energy is strongly influenced by the *difference* in the strain energy of the reactant and the resulting diradical. The effect is seen in the estimates of the energy required to break the central bond in [1.1.1]-, [2.1.1]-, and [2.2.1]propellane.[101]

65 kcal/mol   30 kcal/mol   5 kcal/mol

The more strain relieved by the bond rupture, the more reactive is the molecule. The substantially increased stability of [1.1.1]propellane is due to the fact that not as much strain is relieved at the diradical stage because the diradical remains highly strained.[85]

98. (a) K. B. Wiberg and G. Szeimies, *J. Am. Chem. Soc.* **92**:571 (1970); (b) W. G. Dauben, J. H. Smith, and J. Saltiel, *J. Org. Chem.* **34**:261 (1969); (c) W. R. Moore, K. G. Taylor, P. Muller, S. S. Hall, and Z. L. F. Gaibel, *Tetrahedron Lett.* **1970**: 2365.

99. L. Hedberg and K. Hedberg, *J. Am. Chem. Soc.* **107**:7257 (1985).

100. F. H. Walker, K. B. Wiberg, and J. Michl, *J. Am. Chem. Soc.* **104**:2056 (1982); K. B. Wiberg and F. H. Walker, *J. Am. Chem. Soc.* **104**:5239 (1982).

101. K. B. Wiberg, *Angew. Chem. Int. Ed. Engl.*, **25**:312 (1985)

Another manifestation of the relatively small release of energy associated with breaking the central bond comes from MP4/6-31G* calculations on the reverse ring closure.[102]

$\Delta E = -27$ kcal/mol

Alkenes exhibit large strain energy when molecular geometry does not permit all the bonds to the two sp^2-hybridized carbons to be coplanar. An example that illustrates this point is *E*-cycloheptene:

With only five methylene units available to bridge the *trans* double bond, the molecule is highly strained and very reactive. Isolation of *E*-cycloheptene has not been possible, but evidence for its formation has been obtained by trapping experiments.[103] The alkene is generated in the presence of a reagent expected to react rapidly with it—in this case, the very reactive Diels–Alder diene 2,5-diphenyl-3,4-isobenzofuran. The adduct that is isolated has the structure and stereochemistry anticipated for that derived from *E*-cycloheptene. The lifetime of *E*-cycloheptene has been measured after generation by photoisomerization of the *Z*-isomer. The activation energy for isomerization to *Z*-cycloheptene is about 17 kcal/mol. The lifetime in pentane is on the order of minutes at 0°C.[104]

Although *E*-cyclohexene has been postulated as a reactive intermediate, it has not been observed directly. MO calculations at the 6-31G* level predict it to be 56 kcal/mol less stable than the *Z*-isomer and yield a value for the barrier to isomerization of about 15 kcal/mol.[105]

E-Cyclooctene is also significantly strained, but less so than *E*-cycloheptene. As the ring size is increased, the amount of strain decreases. The *E*-isomers of both cyclononene and cyclodecene are less stable than the corresponding *Z*-isomers, but for cycloundecene and cyclododecene, the *E*-isomers are the more stable.[106] Table 3.10 gives data concerning the relative stability of the C_7 through C_{12} cycloalkenes.

The geometry of bicyclic rings can also cause distortion of the alkene bond from coplanarity. An example is bicyclo[2.2.1]hept-1-ene:

102. W. Adcock, G. T. Binmore, A. R. Krstic, J. C. Walton, and J. Wilkie, *J. Am. Chem. Soc.* **117**:2758 (1995).
103. E. J. Corey, F. A. Carey, and R. A. E. Winter, *J. Am. Chem. Soc.* **87**:934 (1965).
104. Y. Inoue, S. Takamuku, and H. Sakurai, *J. Chem. Soc., Perkin Trans. 2* **1977**:1635; Y. Inoue, T. Ueoka, T. Kuroda, and T. Hakushi, *J. Chem. Soc. Perkin Trans. 2* **1983**:983.
105. J. Verbeek, J. H. van Lenthe, P. J. J. A. Timmermans, A. Mackor, and P. H. M. Budzelaar, *J. Org. Chem.* **52**:2955 (1987).
106. A. C. Cope, P. T. Moore, and W. R. Moore, *J. Am. Chem. Soc.* **82**:1744 (1960).

166

CHAPTER 3
CONFORMATIONAL,
STERIC, AND
STEREOELECTRONIC
EFFECTS

Table 3.10. Relative Stabilities of Z- and E-Cycloalkenes

Cycloalkene	$\Delta H° (E \rightleftharpoons Z)$ (kcal/mol)	Reference
Cycloheptene	−20.3	a
Cyclooctene	−9.7	b
Cyclononene	−2.8	b
Cyclodecene	−3.5	b
Cycloundecene	+0.1	b
Cyclododecene	+0.4	b

a. Calculated value from N. L. Allinger and J. T. Sprague, *J. Am. Chem. Soc.* **94**:5734 (1972).
b. From R. B. Turner and W. R. Meador, *J. Am. Chem. Soc.* **79**:4133 (1957); A. C. Cope, P. T. Moore, and W. R. Moore, *J. Am. Chem. Soc.* **82**:1744 (1960).

Attempts to construct a model of this molecule will show that the geometry of the bicyclic system does not permit coplanarity of the atoms bound to the sp^2 carbons. As a result of the strain, the molecule has, at most, transitory existence.[107] The absence of such "bridgehead double bonds" was noted long ago and formulated as *Bredt's rule*. As the structural basis for Bredt's rule became clear, it was evident that the prohibition against bridgehead double bonds would not be absolute.[108] When the bridges of the bicyclic system are large enough to permit planarity of the double bond, bridgehead alkenes are capable of existence. It has been proposed that the limit for unstable but isolable bridgehead alkenes is reached when the largest ring containing the double bond is at least eight-membered. Bridgehead alkenes in which the largest ring is seven-membered are expected to be capable only of short existence.[109] These proposals have subsequently been tested and verified by the development of successful synthesis of bridgehead alkenes, such as those shown in Scheme 3.3.[110] The strained double bonds in these molecules are exceptionally reactive and undergo a variety of addition reactions. The total strain in the bridgehead alkenes can be computed by molecular mechanics methods. Some of the calculated strain energies are included in Scheme 3.3. The total strain energy can be dissected to indicate that fraction of the total strain which is due to the twist of the carbon–carbon double bond. This strain proves to be a quite reliable predictor of the stability of bridgehead alkenes.[111]

3.9. Relationships between Ring Size and Rate of Cyclization

Many examples of intramolecular reactions leading to ring closure have served to establish a correlation between the rate of a reaction and the size of the ring being formed.[112]

107. R. Keese and E.-P. Krebs, *Angew. Chem. Int. Ed. Engl.* **11**:518 (1972).
108. G. Kobrich, *Angew. Chem. Int. Ed. Engl.* **12**:464 (1973).
109. J. R. Wiseman, *J. Am. Chem. Soc.* **89**:5966 (1967); J. R. Wiseman and W. A. Pletcher, *J. Am. Chem. Soc.* **92**:956 (1970).
110. For reviews of the synthesis and properties of bridgehead alkenes, see G. L. Buchanan, *Chem. Soc. Rev.* **3**:41 (1974); K. J. Shea, *Tetrahedron* **36**:1683 (1980); R. Keese, *Angew. Chem. Int. Ed. Engl.* **14**:528 (1975); G. Szeimies, in *Reactive Intermediates*, Vol. 3, R. A. Abramovitch, ed., Plenum Press, New York, 1983, Chapter 5; G. Szeimies, *Adv. Strain Org. Chem.* **2**:1 (1992).
111. W. F. Maier and P. v. R. Schleyer, *J. Am. Chem. Soc.* **103**:1891 (1981).
112. G. Illuminati and L. Mandolini, *Acc. Chem. Res.* **14**:95 (1981); L. Mandolini, *Adv. Phys. Org. Chem.* **22**:1 (1986).

Scheme 3.3. Bridgehead Alkenes[a]

167

SECTION 3.9.
RELATIONSHIPS
BETWEEN RING SIZE
AND RATE OF
CYCLIZATION

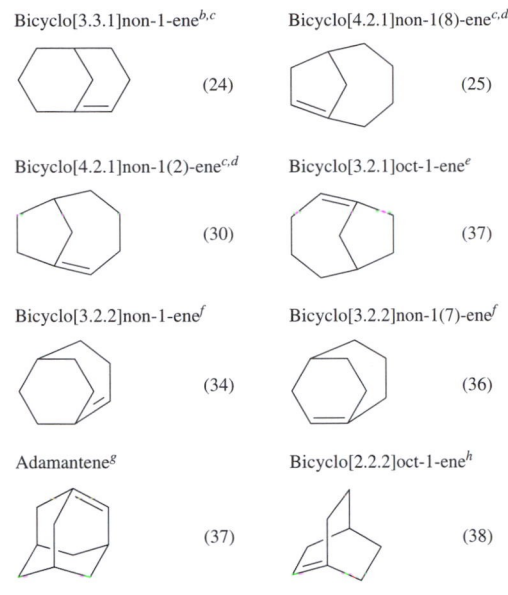

Bicyclo[3.3.1]non-1-ene[b,c]   (24)

Bicyclo[4.2.1]non-1(8)-ene[c,d]   (25)

Bicyclo[4.2.1]non-1(2)-ene[c,d]   (30)

Bicyclo[3.2.1]oct-1-ene[e]   (37)

Bicyclo[3.2.2]non-1-ene[f]   (34)

Bicyclo[3.2.2]non-1(7)-ene[f]   (36)

Adamantene[g]   (37)

Bicyclo[2.2.2]oct-1-ene[h]   (38)

a. Strain energies calculated by molecular mechanics (Ref. 111) are given in parentheses in kcal/mol.
b. J. R. Wiseman and W. A. Pletcher, *J. Am. Chem. Soc.* **92**:956 (1970); J. A. Marshall and H. Faubl, *J. Am. Chem. Soc.* **89**:5965 (1967); M. Kim and J. D. White, *J. Am. Chem. Soc.* **99**:1172 1977).
c. K. B. Becker, *Helv. Chim. Acta* **60**:81 (1977).
d. J. R. Wiseman, H.-F. Chan, and C. J. Ahola, *J. Am. Chem. Soc.* **91**:2812 (1969).
e. W. G. Dauben and J. D. Robbins, *Tetrahedron Lett.* **1975**:151.
f. Transitory existence only; J. R. Wiseman and J. A. Chong, *J. Am. Chem. Soc.* **91**:7775 (1969).
g. Transitory existence only; A. H. Alberts, J. Strating, and H. Wynberg, *Tetrahedron Lett.* **1973**:3047; J. E. Gano and L. Eizenberg, *J. Am. Chem. Soc.* **95**:972 (1973); D. J. Martella, M. Jones, Jr., and P. v. R. Schleyer, *J. Am. Chem. Soc.* **100**:2896 (1978); R. T. Conlin, R. D. Miller, and J. Michl, *J. Am. Chem. Soc.* **101**:7637 (1979).
h. A. D. Wolf and M. Jones, Jr., *J. Am. Chem. Soc.* **95**:8209 (1973); H. H. Grootveld, C. Blomberg, and F. Bickelhaupt, *J. Chem. Soc., Chem. Commun.* **1073**:542.

Although different reaction types exhibit large quantitative differences, and there are exceptions, the order $5 > 6 > 3 > 7 > 4 > 8–10$ is a rough guide of relative reactivity for many systems. Some quantitative data on typical reactions involving nucleophilic substitution or participation are shown in Scheme 3.4.

Table 3.11 gives rate data for ring closure of a series of diethyl (ω-bromoalkyl) malonate anions for ring sizes 4–13, 17, and 21. The rates range from a maximum of 6×10^2 s^{-1} for the five-membered ring to 2.9×10^{-6} s^{-1} for the 11-membered ring.[113]

$$Br-(CH_2)_nC^-(CO_2C_2H_5)_2 \longrightarrow (CH_2)_n \quad C(CO_2C_2H_5)_2$$

113. M. A. Casadei, C. Galli, and L. Mandolini, *J. Am. Chem. Soc.* **106**:1051 (1984).

Scheme 3.4. Relative Rates of Ring Closure as a Function of Ring Size

	Reaction	Ring size =	Relative rate					
			3	4	5	6	7	8
1.[a]	$Br(CH_2)_xCO_2^-$ ⟶ lactone		8.3×10^{-4}	0.31	90	1	0.0052	6×10^{-5}
2.[b]	$Br(CH_2)_xNH_2$ ⟶ cyclic amine		0.07	0.001	100	1	0.002	—
3.[c]	$PhC(CH_2)_xCl$ with $\overset{\|}{O}$ ⟶ nucleophilic participation in solvolysis		—	0.37	36	1	0.13	—
4.[d]	(aromatic ring) O^- / $O(CH_2)_xBr$ ⟶ cyclic ether formation		—	—	—	1	0.01	4×10^{-4}
5.[e]	$ArSO_2\bar{N}(CH_2)_xCl$ ⟶ cyclization		17	33	—	1	—	—

a. C. Galli, G. Illuminati, L. Mandolini, and P. Tamborra, *J. Am. Chem. Soc.* **99**:2591 (1977); L. Mandolini, *J. Am. Chem. Soc.* **100**:550 (1978).
b. D. F. DeTar and W. Brooks, Jr., *J. Org. Chem.* **43**:2245 (1978); D. F. DeTar and N. P. Luthra, *J. Am. Chem. Soc.* **102**:4505 (1980).
c. D. J. Pasto and M. P. Serve, *J. Am. Chem. Soc.* **87**:1515 (1965).
d. G. Illuminati, L. Mandolini, and B. Masci, *J. Am. Chem. Soc.* **96**:1422 (1974).
e. R. Bird, A. C. Knipe, and C. J. M. Stirling, *J. Chem. Soc., Perkin Trans. 2* **1973**:1215.

Figure 3.11 shows the relative reactivity as a function of ring size for two other intramolecular displacement reactions, namely, conversion of ω-bromoalkanecarboxylates to lactones and formation of ethers from ω-bromoalkyl monoethers of 1,2-dihydroxybenzene.

The dissection of the energy of activation of typical ring-closure reactions usually shows some consistent features. The $\Delta H^{\ddagger}$ for formation of three- and four-membered rings is normally less negative than $\Delta H^{\ddagger}$ for the corresponding five- and six-membered rings, whereas $\Delta S^{\ddagger}$ is least negative for the three-membered rings, is of comparable

Table 3.11. Relative Rates of Cyclization of Diethyl (ω-Bromoalkyl)malonate Ester Anions as a Function of Ring Size[a]

Ring size	Relative rate
4	0.58
5	833
6	1.0
7	8.7×10^{-3}
8	1.5×10^{-4}
9	1.7×10^{-5}
10	1.4×10^{-6}
11	2.9×10^{-6}
12	4.0×10^{-4}
13	7.4×10^{-4}
17	2.9×10^{-3}
21	4.3×10^{-3}

a. M. A. Casadei, G. Galli, and L. Mandolini, *J. Am. Chem. Soc.* **106**:1051 (1984).

169

SECTION 3.9.
RELATIONSHIPS
BETWEEN RING SIZE
AND RATE OF
CYCLIZATION

Fig. 3.11. Rates of ring closure of ω-bromoalkanecarboxylates (left) and ω-bromoalkyloxypheno-lates (right). [Reproduced from *Acc. Chem. Res.* **14**:95 (1981) by permission of the American Chemical Society.]

magnitude for four-, five-, and six-membered rings, and then becomes more negative as the ring size increases above seven. The $\Delta H^{\ddagger}$ term reflects the strain that develops in the closure of three-membered rings, whereas the large negative entropy associated with eight-membered and larger rings reflects the relative improbability of achieving the required molecular orientation. Because the combination of the two factors is most favorable for five- and six-membered rings, the maximum rate is observed for these ring sizes.

Superimposed on this broad relationship between enthalpy and entropy are more variable and individualized structural features, including changes in solvation and the effect of branching on the intervening chain. Most important, however, are geometric (stereoelectronic) constraints on the transition state for ring closure.[112] There is a preferred direction of approach that depends on the type of reaction that is involved. Whereas the relative rates of ring closure as a function of ring size for the reactions shown in Scheme 3.4, which are all intramolecular nucleophilic substitutions, reveal a general trend $5 > 6 > 3 > 7 > 8$, reactions with other mechanisms may exhibit a different relationship.

A systematic effort to correlate ease of ring closure with the stereoelectronic requirements of the transition state has been developed by Baldwin and co-workers. They classify ring closures with respect to three factors: (a) ring size, (b) the hybridization of the carbon at the reaction site, and (c) the relationship (endocyclic or exocyclic) of the reacting bond to the forming ring. Certain types of ring closures are found to be favorable whereas others are unfavorable for stereoelectronic reasons. The relationships are summarized in Table 3.12.

170

CHAPTER 3
CONFORMATIONAL,
STERIC, AND
STEREOELECTRONIC
EFFECTS

Table 3.12. Classification of Ring-Closure Types[a]

	Exocyclic bonds			Endocyclic bonds[b]	
Ring size	*sp* (*dig*)	*sp²* (*trig*)	*sp³* (*tet*)	*sp* (*dig*)	*sp²* (*trig*)
3	unfav	fav	fav	fav	unfav
4	unfav	fav	fav	fav	unfav
5	fav	fav	fav	fav	unfav
6	fav	fav	fav	fav	fav
7	fav	fav	fav	fav	fav

a. J. E. Baldwin, *J. Chem. Soc., Chem. Commun.* **1976**:734.
b. The category *endo-tet* also exists but is somewhat rare and is not
discussed here.

The classification can be illustrated with a few examples. All of the nucleophilic substitutions shown in Scheme 3.4 are of the *exo-tet* classification. The reacting atom is of sp^3 hybridization (tetrahedral = tet), and the reacting bond, that is, the bond to the leaving group, is exocyclic to the forming ring:

An example of an *exo-trig* process would be lactonization of a ω-hydroxycarboxylic acid:

An example of an *exo-dig* process would be the base-catalyzed cyclization of an ε-hydroxy-α,β-ynone:

$$\text{HOCH}_2\text{CH}_2\text{CH}_2\text{C} \equiv \overset{\overset{\text{O}}{\|}}{\text{C}}\text{R} \longrightarrow$$

Let us focus attention on the unfavorable ring closures. Why, for example, should formation of a five-membered ring by an *endo-trig* process be difficult? The answer is provided by a consideration of the trajectory of approach of the nucleophile.[114] If Z is an electron-attracting conjugating group of the type necessary to activate the double bond to nucleophilic attack, the reaction would involve the LUMO of the conjugated system, a π^*

114. J. E. Baldwin, *J. Chem. Soc., Chem. Commun.* **1976**:738.

171

SECTION 3.10.
TORSIONAL AND
STEREOELECTRONIC
EFFECTS ON
REACTIVITY

orbital. The nucleophile cannot approach in the nodal plane of the π system, so it must attack from above or below.

This stereoelectronic requirement would lead to a large distortion of the normal geometry of a five-membered ring and introduce strain. It is this distortion and strain that disfavor the 5-*endo-trig* cyclization. In contrast, 5-*endo-dig* cyclization is feasible because the acetylenic system provides an orbital that is available for a nearly planar mode of approach.

In agreement with these analyses, it was found that compound **5** was unreactive toward base-catalyzed cyclization to **6**, even though the double bond would be expected to be reactive toward nucleophilic conjugate addition. On the other hand the acetylene **7** is readily cyclized to **8**:[115]

The terms favored and disfavored imply just that. Other factors will determine the absolute rate of a given ring closure, but these relationships point out the need to recognize the specific stereoelectronic requirements which may be imposed on the transition state in ring-closure reactions.

3.10. Torsional and Stereoelectronic Effects on Reactivity

Torsional strain refers to the component of total molecular energy that results from nonoptimal arrangement of vicinal bonds, as in the eclipsed conformation of ethane. The origin and stereoelectronic nature of torsional strain were discussed in Section 1.1.1. The

115. J. E. Baldwin, R. C. Thomas, L. I. Kruse, and L. Silberman, *J. Org. Chem.* **42**:3846 (1977).

172

CHAPTER 3
CONFORMATIONAL,
STERIC, AND
STEREOELECTRONIC
EFFECTS

preference for staggered arrangements around single bonds is general for all alkanes, and when geometric constraints enforce an eclipsed arrangement, the molecule suffers torsional strain. Torsional strain that develops in transition states raises the activation energy of the reaction.

One case in which torsional strain plays a major role is in reactions that involve hybridization changes at a ring atom. A general relationship concerning the relative ease of conversion of carbon atoms in a ring from sp^3 to sp^2 or vice versa has been developed which is useful in comparing the reactivity of cyclohexanones with that of cyclopentanones. It is observed that reactions which convert an sp^2 carbon to an sp^3 carbon in a six-membered ring are more favorable than the corresponding reactions in a five-membered ring.

For example, cyclohexanone is reduced by sodium borohydride 23 times faster than cyclopentanone.[116] The explanation for this difference lies in the relative torsional strain in the two systems. Converting an sp^2 atom in a five-membered ring to sp^3 increases the torsional strain because of the increase in the number of eclipsing interactions in the alcohol. A similar change in a six-membered ring leads to a completely staggered (chair) arrangement and reduces torsional strain.

| staggered bonds | eclipsed bonds | eclipsed bonds | staggered bonds |

Conversely, processes which convert sp^3 carbons to sp^2 carbons are more favorable for five-membered than for six-membered rings. This can be illustrated by the data for acetolysis of cyclopentyl versus cyclohexyl tosylate. The former proceeds with an enthalpy of activation about 3 kcal/mol less than the latter.[117] A molecular mechanics analysis found that the difference was largely accounted for by the relief of torsional strain in the cyclopentyl case.[118] Notice that there is an angle-strain effect which is operating in the opposite direction, since there will be some resistance to the expansion of the bond angle at the reaction center to 120° in the cyclopentyl ring.

116. H. C. Brown and K. Ichikawa, *Tetrahedron* **1**:221 (1957).
117. H. C. Brown and G. Ham, *J. Am. Chem. Soc.* **78**:2735 (1956).
118. H.-J. Schneider and F. Thomas, *J. Am. Chem. Soc.* **102**:1424 (1980);

173

SECTION 3.10.
TORSIONAL AND
STEREOELECTRONIC
EFFECTS ON
REACTIVITY

There is another aspect to the question of the reactivity of the carbonyl group in cyclohexanone. This has to do with the preference for approach of reactants from the axial or equatorial direction. The chair conformation of cyclohexanone places the carbonyl group in an unsymmetrical environment. It is observed that small nucleophiles prefer to approach the carbonyl group of cyclohexanone from the axial direction even though this is a more sterically restricted approach than from the equatorial side.[119] How do the differences in the C−C bonds (on the axial side) as opposed to the C−H bonds (on the equatorial side) influence the reactivity of cyclohexanone?

axial attack   equatorial attack

Several possible effects have been considered. One is the interaction between the σ bonds and the $\pi*$ orbital of the carbonyl group. This interaction could distort the shape of the carbonyl LUMO. One proposal is that the (C−C σ)−(C=O $\pi*$) interaction distorts the C=O LUMO so that it has greater density on the axial (C−C) side.[120] An alternative view is that the axial C−H bonds (at C-2 and C-6) can preferentially stabilize the transition state for axial attack by electron donation into the $\sigma*$ orbital of the developing bond to the nucleophile. This proposal views C−H bonds as better electron donors than C−C bonds.[121] A third view emphasizes flattening of the carbonyl group, which makes the C-2 and C-6 axial C−H bonds more nearly antiperiplanar with the approaching nucleophile. The axial trajectory would then be favored by the nucleophile.[122] Torsional effects also play a major role in the preference for axial approach. In the initial ketone, the carbonyl group is almost eclipsed by the equatorial C-2 and C-6 C−H bonds. This torsional strain is relieved by axial attack, but equatorial approach increases it somewhat because the oxygen atom must move through a fully eclipsed arrangement.[123] Polar effects originating in bond dipoles for substituents are also an important factor.[124] It appears that for cyclohexanones the torsional effect is the most important factor, with the electrostatic and stereoelectronic effects of substituents being a secondary influence.

axial attack   equatorial attack

119. B. W. Gung, *Tetrahedron* **52**:5263 (1996).
120. J. Klein, *Tetrahedron Lett.* **1973**:4037; *Tetrahedron* **30**:3349 (1974); G. Frenking, K. F. Kohler, and M. T. Reetz, *Angew. Chem. Int. Ed. Engl.* **30**:1146 (1991).
121. A. S. Cieplak, *J. Am. Chem. Soc.* **103**:4540 (1981); A. S. Cieplak, B. D. Tait, and C. R. Johnson, *J. Am. Chem. Soc.* **111**:8447 (1989).
122. N. T. Ahn, *Top. in Curr. Chem.* **88**:195 (1980).
123. M. Cherest, H. Felkin, and N. Prudent, *Tetrahedron Lett.* **1968**:2199; M. Cherest and H. Felkin, *Tetrahedron Lett.* **1968**:2205; Y. D. Wu and K. N. Houk, *J. Am. Chem. Soc.* **109**:908 (1987); Y. D. Wu, K. N. Houk, and M. N. Paddon-Row, *Angew. Chem. Int. Ed. Engl.* **31**:1019 (1992).
124. Y.-D. Wu, J. A. Tucker, and K. N. Houk, *J. Am. Chem. Soc.* **113**:5018 (1991); Z. Shi and R. J. Boyd, *J. Am. Chem. Soc.* **115**:9614 (1993); P. Wipf and Y. Kim, *J. Am. Chem. Soc.* **116**:11678 (1994); B. Ganguly, J. Chandrasekhar, F. A. Khan, and G. Mehta, *J. Org. Chem.* **58**:1734 (1993); G. Mehta, F. A. Khan, and W. Adcock, *J. Chem. Soc., Perkin Trans 2* **1995**:2189.

174

CHAPTER 3
CONFORMATIONAL,
STERIC, AND
STEREOELECTRONIC
EFFECTS

More bulky nucleophiles usually approach the cyclohexanone carbonyl from the equatorial direction. This is called *steric approach control* and is the result of van der Waals type repulsions. Larger nucleophiles encounter the 3,5-axial hydrogens on the axial approach trajectory.[125]

Bicyclo[3.3.1]nonan-9-one is another ketone that exhibits interesting stereoselectivity. Reduction by hydride donors is preferentially *syn* to electron-attracting substituents at C-5 (X = EWG in the structure shown below) and *anti* to electron-releasing substituents (X = ERG below).[126] These effects are observed even for differentially substituted phenyl groups.[127]

These effects are attributed to differences in the σ-donor character of the C−C bonds as a result of substitution. Electron-attracting groups diminish the donor capacity and promote *syn* addition. An alternative explanation invokes a direct electrostatic effect arising from the C−X bond dipole.[128]

A very important relationship between stereochemistry and reactivity arises in the case of reaction at an sp^2 carbon adjacent to a chiral center. Using nucleophilic addition to the carbonyl group as an example, it can be seen that two diastereomeric products are possible. The stereoselectivity and predictability of such reactions are important in controlling stereochemistry in synthesis.

A number of years ago an empirical relationship, now called *Cram's rule*, was recognized. When R^1, R^2, and R^3 differ in size, and the molecule is oriented such that the largest group is *anti* to the carbonyl oxygen, the major product arises from addition of the nucleophile *syn* to the smaller substituent.[129]

R_s = smallest group
R_m = intermediate group
R_l = largest group

125. W. G. Dauben, G. Fonken, and D. S. Noyce, *J. Am. Chem. Soc.* **92**:709 (1970); H. C. Brown and W. C. Dickason, *J. Am. Chem. Soc.* **92**:709 (1970); D. C. Wigfield, *Tetrahedron* **35**:449 (1979); T. Wipke and P. Gund, *J. Am. Chem. Soc.* **98**:8107 (1976).
126. C. K. Cheung, L. T. Tseng, M.-H. Lin, S. Srivasta, and W. J. Le Noble, *J. Am. Chem. Soc.* **108**:1598 (1986); J. M. Hahn and W. J. Le Noble, *J. Am. Chem. Soc.* **114**:1916 (1992).
127. I. H. Song and W. J. Le Noble, *J. Org. Chem.* **59**:58 (1994).
128. W. Adcock, J. Cotton, and N. A. Trout, *J. Org. Chem.* **59**:1867 (1994).
129. D. J. Cram and F. A. Abd Elhafez, *J. Am. Chem. Soc.* **74**:5828 (1952).

175

SECTION 3.10.
TORSIONAL AND
STEREOELECTRONIC
EFFECTS ON
REACTIVITY

Various structural factors have been considered in interpreting this result: The most generally satisfactory approach is based on a transition-state model, advanced by Felkin and co-workers, in which the largest group is oriented perpendicularly to the carbonyl group. Nucleophilic addition to the carbonyl group occurs from the opposite side.[130]

STO-3G calculations find the corresponding transition state to be more stable than other possible conformations by several kilocalories per mole.[131] The origin of the preference for this transition-state conformation is believed to be a stabilization of the $C=O$ LUMO by the σ^* orbital of the perpendicularly oriented substituent.

The more stable the LUMO, the stronger is the interaction with the HOMO of the approaching nucleophile. The observed (Cram's rule) stereoselectivity is then a combination of stereoelectronic effects that establish a preference for a perpendicular substituent and a steric effect that establishes a preference for the nucleophile to approach from the direction occupied by the smallest substituent.

Another system in which the factors controlling the direction of reagent approach has been studied systematically is the bicyclo[2.2.1]heptane ring system. The stereochemistry of a number of reactions of the parent system and the 7,7-dimethyl derivative have been examined.[132] Some of the results are given in Table 3.13. These reactions reveal

Table 3.13. Comparison of the Stereoselectivity of Reactions with Bicyclo[2.2.1]heptene and 7,7-Dimethylbicyclo[2.2.1]heptene[a]

	Bicyclo[2.2.1]heptene		7,7-Dimethylbicyclo[2.2.1]heptene	
Reagent	exo	endo	exo	endo
B_2H_6 (hydroboration)	99.5	0.5	22	78
RCO_3H (expoxidation)	99.5	0.5	12	88
H_2, Pd (hydrogenation)	90	10	10	90

a. H. C. Brown, J. H. Kawakami, and K. T. Liu, *J. Am. Chem. Soc.* **95**:2209 (1973).

130. M. Cherest, H. Felkin, and N. Prudent, *Tetrahedron Lett.* **1968**:4199.

131. Y.-D. Wu and K. N. Houk, *J. Am. Chem. Soc.* **109**:908 (1987).

132. N. T. Ahn, *Top. Curr. Chem.* **88**:145 (1980).

176

CHAPTER 3
CONFORMATIONAL,
STERIC, AND
STEREOELECTRONIC
EFFECTS

a reversal of the preferred direction of attack with the introduction of the methyl substituents. In the parent system, the *exo* direction of attack is preferred. This is because the single CH_2 group at C-7 offers less steric resistance than the $-CH_2CH_2-$ unit on the *endo* side of the molecule. The *endo* hydrogens are in a relationship to the reaction site that is similar to the 1,3-diaxial interaction in a chair cyclohexane ring. When a *syn*-7-methyl group is present, the relative steric bulk of the two bridges is reversed. The methyl groups have a similar effect in controlling the stereochemistry of reduction of the related ketones.[133]

The preference for *endo* attack in 7,7-dimethylnorbornene is certainly steric in origin, with the 7-methyl substituent shielding the *exo* direction of approach. The origin of the preferred *exo*-attack in norbornene is more subject to discussion. A purely steric explanation views the *endo* hydrogens at C−5 and C−6 as sterically shielding the *endo* approach. There probably is also a major torsional effect. Comparison of the *exo* and *endo* modes of approach shows that greater torsional strain develops in the *endo* mode of approach.[134]

exo attack at this carbon increases angle between the two C−H bonds and decreases torsional strain in going to the transition state

endo attack at this carbon causes its bond to H to be eclipsed with other C−H bond in going to the transition state

133. H. C. Brown, J. Kawakami, and K.-T. Liu, *J. Am. Chem. Soc.* **95**:2209 (1973).
134. N. G. Rondan, M. N. Paddon-Row, P. Caramella, J. Mareda, P. Mueller, and K. N. Houk, *J. Am. Chem. Soc.* **104**:4974 (1982); M. N. Paddon-Row, N. G. Rondan, and K. N. Houk, *J. Am. Chem. Soc.* **104**:7162 (1982); K. N. Houk, N. G. Rondan, F. K. Brown, W. L. Jorgensen, J. D. Madura, and D. G. Spellmayer, *J. Am. Chem. Soc.* **105**:5980 (1983).

J. Dale, *Stereochemistry and Conformational Analysis*, Verlag Chemie, New York, 1978.

E. L. Eliel, N. L. Allinger, S. J. Angyal, and G. A. Morrison, *Conformational Analysis*, Wiley-Interscience, New York, 1965.

E. L. Eliel, S. H. Wilen, and L. Mander, *Stereochemistry of Organic Compounds*, John Wiley & Sons, New York, 1993.

A. Greenberg and J. F. Liebman, *Strained Organic Molecules*, Academic Press, New York, 1978.

L. M. Jackman and F. A. Cotton, eds., *Dynamic Nuclear Magnetic Resonance Spectroscopy*, Academic Press, New York, 1975. Chapters 3, 6, 7, and 14.

E. Juaristi and G. Cuevas, *The Anomeric Effect*, CRC Press, Boca Raton, Florida, 1995.

A. J. Kirby, *Stereoelectronic Effects*, Oxford University Press, Oxford, U.K., 1996.

A. J. Kirby, *The Anomeric Effect and Related Stereoelectronic Effects at Oxygen*, Springer-Verlag, New York, 1983.

M. S. Newman, ed., *Steric Effects in Organic Chemistry*, John Wiley & Sons, New York, 1956.

M. Oki, *Applications of Dynamic NMR Spectroscopy to Organic Chemistry*, VCH Publishers, Deerfield Beach, Florida, 1983.

Problems

(References for these problems will be found on page 793.)

1. Estimate $\Delta H°$ for each of the following conformational equilibria:

(a)

(b)

(c)

2. Draw a clear three-dimensional representation showing the preferred conformation of *cis,cis,trans*-perhydro-9b-phenalenol (**A**):

A

178

CHAPTER 3
CONFORMATIONAL,
STERIC, AND
STEREOELECTRONIC
EFFECTS

3. The *trans* : *cis* ratio at equilibrium for 4-*t*-butylcyclohexanol has been established for several solvents near 80°C:

Solvent	*trans* (%)	*cis* (%)
Cyclohexane	70.0	30.0
Benzene	72.5	27.5
1,2-Dimethoxyethane	71.0	29.0
Tetrahydrofuran	72.5	27.5
t-Butyl alcohol	77.5	22.5
i-Propyl alcohol	79.0	21.0

From these data, calculate the conformational energy of the hydroxyl group in each solvent. Do you notice any correlation between the observed conformational preference and the properties of the solvent? Explain.

4. The preferred conformations of both 1-methyl-1-phenylcyclohexane and 2-methyl-2-phenyl-1,3-dioxane have the phenyl group in the axial orientation even though the conformational free energy of the phenyl group (2.9 kcal) is greater than that for a methyl group (1.8 kcal). Explain.

5. Draw clear conformational representations of the *β*-pyranose forms of each of the following carbohydrates:

(a) D-mannose (b) D-chromose A (c) L-vancosamine

6. Explain the basis for the selective formation of the product shown over the alternative product.

(d)

(e)

(f)

(g)

7. For the following pairs of reactions, indicate which you would expect to be more favorable and explain the basis of your prediction.

(a) Which isomer will solvolyze more rapidly in acetic acid?

(b) Which will be the major reaction product?

(c) Which isomer will be converted to a quarternary salt more rapidly?

180

CHAPTER 3
CONFORMATIONAL,
STERIC, AND
STEREOELECTRONIC
EFFECTS

(d) Which lactone will be formed more rapidly?

$^-O_2C(CH_2)_2CH_2Br$ $\xrightarrow{\text{DMSO}}$ **G**

or

$^-O_2C(CH_2)_2CH_2Br$ $\xrightarrow{\text{DMSO}}$ **H**

(e) Which compound will undergo hydrolysis more rapidly?

I   or   **J**

(f) Which compound will aromatize more rapidly by loss of ethoxide ion?

K   or   **L**

(g) Which compound will be more rapidly oxidized by chromic acid?

M   or   **N**

8. Predict the most stable conformation of each of the following molecules and explain the basis of your prediction.

9. Given that the rotational barrier for a C–C bond in acetone is about 0.75 kcal, (a) sketch the relationship between energy and conformation using a Newman projection formula to define the angles of rotation, and (b) show how the energy–conformation curve will be perturbed by addition of first one (ethyl methyl ketone) and then two (isopropyl methyl ketone) methyl substituents on the methyl group undergoing rotation.

10. Using the data incorporated in Fig. 3.3 and assuming the additivity of *gauche* and eclipsing interactions of similar type, sketch the rotational energy profile you would expect for 2,3-dimethylbutane.

11. The following molecules present possibilities for stereoisomerism and/or the existence of different conformations. For each molecule, predict which stereoisomer will be the most stable and predict its preferred conformation.

(a) (b) (c) (d)

12. *trans*-3-Alkyl-2-chlorocyclohexanones (alkyl = methyl, ethyl, 2-propyl) exist in the diequatorial conformation. In contrast, the corresponding O-methyl oximes exist as diaxial conformers. Explain the preference for the diaxial conformation of the oxime ethers.

13. There are three derivatives of butadiene having one *t*-butyl substituent and nine di-*t*-butyl derivatives. Predict the preferred conformation for each of these 12 compounds.

182

CHAPTER 3
CONFORMATIONAL,
STERIC, AND
STEREOELECTRONIC
EFFECTS

14. Consider the conformations possible for 3-substituted methylenecyclohexanes. Do you expect typical substituents to exhibit larger or smaller preferences for the equatorial orientation, as compared to the same substituent on a cyclohexane ring?

15. Discuss the aspects of conformation and stereochemistry that would have to be considered for complete description of the structure of molecules having the general structure **A**. How would the size of the $(CH_2)_n$ bridge affect conformational equilibria in these molecules?

16. Predict the preferred conformation of the isomeric (Z- and E-) 3-penten-2-ones, **A**, $R = CH_3$. How would you expect the conformational picture to change as R becomes progressively larger?

$$\overset{\text{O}}{\underset{\|}{RCCH}}=CHCH_3$$
A

17. Figures 17A and 17B (p. 183) show energy as a function of rotation for a series of 2-substituted acetaldehydes, with $\theta = 0°$ in the *syn* conformation and $\theta = 180°$ in the *anti* conformation. The calculations were done using the PM3 method. Figure 17A is for a vacuum, whereas Fig. 17B is for a solvent cavity with a dielectric constant of 4.7. The table gives the calculated barriers. Discuss the following aspects: (a) rationalize the order Br > Cl > F for *anti* conformers in the gas phase; (b) rationalize the shift toward the *anti* conformation in the more polar environment.

Relative Rotamer Stabilities and Rotational Barriers for 2-Substituted Acetaldehydes (kcal/mol)

| R | $E_{syn} - E_{anti}$ | | Rotational barriers | | | |
| | | | syn→anti | | anti→anti | |
	Vacuum	$\varepsilon = 4.7$	Vacuum	$\varepsilon = 4.7$	Vacuum	$\varepsilon = 4.7$
H	0.0	0.0	0.60	0.65	0.60	0.65
F	−0.34	+0.74	1.79	2.36	0.0	0.0
Cl	−0.97	−0.02	0.68	0.87	0.0	0.11
Br	−1.36	−0.57	1.20	1.25	0.10	0.39
CN	−1.07	−0.21	0.43	0.38	0.09	0.82
NO₂	+0.46	+1.85	1.73	2.75	0.07	0.08

Fig. 17A. Potential energy of substituted acetaldehydes as a function of the OCCR angle, relative to the energy of the *syn* ($\angle$OCCR $= 0°$) rotamer; isolated molecules. [Reproduced with permission of Elsevier Science Publishing.]

Fig. 17B Potential energy of substituted acetaldehydes as a function of the OCCR angle, relative to the energy of the *syn* ($\angle$OCCR $= 0°$) rotamer; solvated molecules ($\varepsilon = 4.7$). [Reproduced with permission of Elsevier Science Publishing.]

184

CHAPTER 3
CONFORMATIONAL,
STERIC, AND
STEREOELECTRONIC
EFFECTS

18. The Baeyer–Villiger oxidation of ketones to esters (or lactones) occurs by the following mechanism.

For *endo*-substituted bicyclo[2.2.1]heptan-7-ones, the following product ratios are observed. Offer an explanation for the substituent effect.

X	A : B
CN	100 : 0
OCH_3	77 : 23
C_6H_5	51 : 49
p-$NO_2C_6H_4$	75 : 25
p-FC_6H_4	52 : 48
p-$CH_3OC_6H_4$	39 : 61

19. Amides **1–4** all adopt conformations in which a *t*-butyl group has an axial-like position and, as a result, a non-chair conformation of the six-membered ring. Comment on the origin of this structural effect.

Reproduced by permission of Verlag Helvetica Chimica Acta AG.

20. Two stereoisomers (**A** and **B**) of the structure shown below have been obtained and separated. The NMR spectrum of one isomer (**A**) shows two methyl peaks (doublets at 1.03 and 1.22 ppm) and two quartets (2.68 and 3.47 ppm) for the CH groups. The spectrum of the other isomer (**B**) shows single signals for the methyl (doublet at 1.25 ppm) and methine protons (broad quartet at 2.94 ppm). The spectra of both compounds show a change with temperature. For isomer **A** at 95°C, the pairs of methyl doublets and the methine quartet both become single signals (still a doublet and a quartet, respectively). The low-temperature spectrum (−40°C) is unchanged from the room temperature spectrum. For isomer **B** at 10°C, the methyl signals split into two doublets of *unequal intensity* (1.38 and 1.22 ppm in the ratio 9:5). The methine signal also splits into two broad signals at 3.07 and 2.89 ppm, also in the ratio 9:5. From this information, assign the stereochemistry of isomers **A** and **B** and explain the cause of the temperature dependence of the NMR spectra of each isomer.

21. Estimate the energy difference between the stable and unstable chair conformations of each of the following trimethylcyclohexanes:

22. Predict the stereochemistry of each of the following reactions:

Study and Description of Organic Reaction Mechanisms

Introduction

The chapters that follow this one will be devoted largely to the description of specific reactions. A working knowledge of organic chemistry requires understanding certain fundamental reaction types that occur in a wide variety of individual reactions. Most organic reactions occur in several steps; these steps constitute the *reaction mechanism*. Examination of the mechanisms of reactions often reveals close relationships between reactions that otherwise might appear to be unrelated. Reaction mechanisms also help us understand the effect of substituents on the rate of related reactions. Consideration of reaction mechanism often guides the development of new transformations and improvement of existing ones. In this chapter, the ways in which organic reactions are studied in order to determine the reaction mechanism will be discussed. The chapter considers the types of experimental studies that provide data about reaction mechanism and how the data are interpreted.[1]

4.1. Thermodynamic Data

Any reaction has associated with it changes in enthalpy (ΔH), entropy (ΔS), and free energy (ΔG). The principles of thermodynamics assure us that ΔH, ΔS, and ΔG are

---

1. Extensive discussions of techniques for studying reaction mechanisms are presented in: E. S. Lewis, ed., *Investigation of Rates and Mechanism of Reactions, Techniques of Chemistry*, 3rd ed., Vol. VI, Part I, John Wiley & Sons, New York, 1974; C. F. Bernasconi, ed., *Investigation of Rates and Mechanism of Reactions, Techniques of Chemistry*, 4th ed., Vol. VI, Part I, John Wiley & Sons, New York, 1986.

188

CHAPTER 4
STUDY AND
DESCRIPTION OF
ORGANIC REACTION
MECHANISMS

independent of the reaction path. They are interrelated by the fundamental equation

$$\Delta G = \Delta H - T\,\Delta S \tag{4.1}$$

Furthermore, the value of ΔG is related to the equilibrium constant K for the reaction

$$\Delta G = -RT \ln K \tag{4.2}$$

Because these various quantities are characteristics of the reactants and products but are *independent* of the reaction path, they cannot provide insight into mechanisms. Information about ΔG, ΔH, and ΔS does, however, indicate the feasibility of any specific reaction. The enthalpy change of a given reaction can be estimated from tabulated thermochemical data or from bond-energy data such as those in Table 1.3 (p. 14) The example below illustrates the use of bond-energy data for estimating the enthalpy of a reaction.

Example 4.1. Calculate the enthalpy change associated with hydrogenation of butene.

$$CH_3CH{=}CHCH_3 + H_2 \rightarrow CH_3CH_2CH_2CH_3$$

$$-\Delta H = \Sigma \text{ bond energies}_{(formed)} - \Sigma \text{ bond energies}_{(broken)}$$

Bonds formed (kcal/mol):		Bonds broken (kcal/mol):	
2	C−H 196.4	H−H	103.2
	C−C 80.5	C=C	145
	276.9		248.2

$$\Delta H = -276.9 - (-248.2) = -28.7 \text{ kcal/mol}$$

The hydrogenation is therefore calculated to be exothermic by about 29 kcal/mol.

Bond-energy calculations can provide an approximation of the enthalpy change that is associated with a given reaction. The generalized bond energies given in Table 1.3 assume that bond strengths are independent of the structure of the rest of the molecule. This is only approximately correct, as can be judged by observing the variations in C−H and C−C bond energies as a function of structure in Part B of Table 1.3. More accurate calculations of the thermodynamic parameters of a reaction can be done on the basis of tabulated thermochemical data. There are extensive compilations of ΔH_f° and ΔG_f° for many compounds. The subscript f designates these as, respectively, the enthalpies and free energies of formation of the compound from its constituent elements. The superscript $^\circ$ is used to designate data that refer to the substance in its standard state, i.e., the pure substance at 25°C and 1 atm. The compiled data can be used to calculate the enthalpy or free energy of a given reaction if the data are available for each reactant and product:

$$\Delta H^\circ = \Sigma \Delta H_{f\,products}^\circ - \Sigma \Delta H_{f\,reactants}^\circ$$

or

$$\Delta G^\circ = \Sigma\, \Delta G^\circ_{f_{products}} - \Sigma\, \Delta G^\circ_{f_{reactants}}$$

In the case of hydrogenation of 2-butene, ΔH°_f for butane (gas) is $-30.15\,kcal/mol$, ΔH°_f for E-2-butene (gas) is $-2.67\,kcal/mol$, and ΔH°_f for H_2 is 0. Thus, ΔH° of the hydrogenation reaction at standard conditions is $-27.48\,kcal/mol$.

If the thermodynamic data for a compound of interest have not been determined and tabulated, it may be possible to estimate ΔH_f or ΔG_f from tabulated data pertaining to individual structural units. Procedures have been developed for estimating thermodynamic characteristics of hydrocarbons and derivatives by summing the contributions of the constituent groups.[2] The group increments are derived from experimental thermochemical data and therefore depend on the existence of reliable data for the class of compounds of interest.

Estimation of the free-energy change associated with a reaction permits the calculation of the equilibrium position for a reaction and indicates the feasibility of a given chemical process. A positive ΔG° imposes a limit on the extent to which a reaction can occur. For example, as can be calculated using Eq. (4.2), a ΔG° of $1.0\,kcal/mol$ limits conversion to product at equilibrium to 15%. An appreciably negative ΔG° indicates that the reaction is thermodynamically favorable.

MO calculations provide another approach to obtaining estimates of thermodynamic data. The accuracy with which the various computational methods reproduce molecular energies differs. Of the semiempirical methods, only MINDO,[3] MNDO,[4] AM1,[5] and PM3[6] provide reliable estimates of molecular energies, and the range of reliability is open to some discussion.[7] With the *ab intitio* method, the 4-31G, 6-31G, G1, and G2 basis sets achieve a level of accuracy that permits comparison of energy data. Recently, good results have also been obtained with the B3LYP density functional method.[8] Users of computational thermochemical data, however, must critically assess the reliability of the method being applied in the *particular case* under study. Table 4.1 gives a comparison of some calculated ΔH_f values for some hydrocarbons with experimental values.

Calculations are frequently done on the basis of *isodesmic reactions* in order to provide for maximum cancellation of errors in the total energies (see Section 1.3). The "experimental" ΔH of the process can be obtained from the tabulated ΔH_f values of the reactants and products. Table 4.2 compares the errors in ΔH_f for some isodesmic reactions with those for the corresponding "atomization" reactions for G2 calculations on some

2. G. J. Janz, *Thermodynamic Properties of Organic Compounds*, Academic Press, New York, 1967; D. R. Stull, E. F. Westrum, Jr., and G. C. Sinke, *The Chemical Thermodynamics of Organic Compounds*, John Wiley & Sons, New York, 1969; J. D. Cox and G. Pilcher, *Thermochemistry of Organic and Organometallic Compounds*, Academic Press, New York, 1970; N. Cohen and S. W. Benson, *Chem. Rev.* **93**:2419 (1993).
3. R. C. Bingham, M. J. S. Dewar, and D. H. Lo, *J. Am. Chem. Soc.* **97**:1294 (1975).
4. M. J. S. Dewar and G. P. Ford, *J. Am. Chem. Soc.* **101**:5558 (1979).
5. M. J. S. Dewar, E. G. Zoebisch, E. F. Healy, and J. J. P.Stewart, *J. Am. Chem. Soc.* **107**:3902 (1985).
6. J. J. P. Stewart, *J. Comput. Chem.* **10**:221 (1989).
7. J. A. Pople, *J. Am. Chem. Soc.*, **97**:5307 (1975); T. A. Halgren, D. A. Kleier, J. H. Hall, Jr., L. D. Brown, and W. L. Lipscomb, *J. Am. Chem. Soc.* **100**:6595 (1978); M. J. S. Dewar and D. M. Storch, *J. Am. Chem. Soc.* **107**:3898 (1985).
8. K. Raghavachari, B. B. Stefanov, and L. A. Curtiss, *Mol. Phys.* **91**:555 (1997); B. S. Jursic, *THEOCHEM* **391**:75 (1997); B. S. Jursic, *THEOCHEM* **417**:99 (1997); J. Andzelm, J. Baker, A. Scheiner, and M. Wrinn, *Int. J. Quantum Chem.* **56**:733 (1995).

190

CHAPTER 4
STUDY AND
DESCRIPTION OF
ORGANIC REACTION
MECHANISMS

Table 4.1. Comparison of Reported Differences between Calculated Values of ΔH_f for Some Hydrocarbons and Experimental Values

	MNDO[a]	AM1[a]	PM3[b]	6-31G*[a]	G2[c]
Butane	0.7	−0.7	1.3	−0.8	−0.6
Pentane	0.7	−2.8	0.6	−0.5	
Cyclopentane	−11.9	−10.4	−5.6	4.0	
Cyclohexane	−5.3	−9.0	−1.5	3.1	3.9
Bicyclo[2.2.1]heptane	2.1	−2.0	−1.3	8.8	
Bicyclo[2.2.2]octane	−2.2	−11.9	−3.7	10.7	

a. M. J. S. Dewar and D. M. Storch, *J. Am. Chem. Soc.* **107**:3898 (1985).
b. J. J. P. Stewart, *J. Comput. Chem.* **10**:221 (1989).
c. L. A. Curtiss, K. Raghavachari, P. C. Redfern, and J. Pople, *J. Chem. Phys.* **106**:1063 (1997).

Table 4.2. Comparison of Differences from Experimental ΔH_f° for G2 Calculations Using Atomization versus Isodesmic Reactions[a]

	ΔH_f° (exp)	G2 (atomization)	G2 (isodesmic reactions)
Propane	−25.0	0.4	0.1
Cyclopropane	66.2	−2.9	−1.6
Butane	−30.0	0.4	0.2
Cyclobutane	37.4	−2.9	−1.5
Bicyclo[1.1.0]butane	51.9	−3.0	−1.5
Cyclopentane	−18.3	−1.1	−0.4
Benzene	19.7	−3.9	−0.8

a. K. Raghavachari, B. Stefanov, and L. A. Curtiss, *J. Chem. Phys.* **106**:6764 (1997). H_f data in kcal/mol.

simple hydrocarbons. For molecules of this size, calculations based the use of isodesmic reactions can usually achieve ΔH_f data within 0.5 kcal/mol.[9]

Another means of using computational energy data is based on the principle of bond additivity. This is another form of use of the concept of isodesmic reactions. The approach is to assign energies to groups and bonds from the calculations on well-characterized molecules, for which the accuracy can be checked with experimental data. Energies for more complex molecules can then be taken as the sum of those of the contributing groups, corrected for special factors.[10] Alkanes and cycloalkanes, for instance, can be formulated as the summation of primary, secondary, tertiary, and quaternary carbons and the associated hydrogens. Molecules with unsaturation or functional groups would require additional parameters. The energy of unstrained molecules can be calculated by the summation of the appropriate increments. Molecules with special structural features not incorporated into the constituent atoms, such as ring or steric strain, deviate from the calculated value, and the deviation is a measure of the extent of angle, steric, and torsional strains.[11] Density functional theory calculations can be used in place of MO calculations.[12]

9. K. Raghavachari, B. B. Stefanov, and L. A. Curtiss, *J. Chem. Phys.* **106**:6764 (1997).
10. K. B. Wiberg, *J. Org. Chem.* **50**:5285 (1985); M. R. Ibrahim and P. v. R. Schleyer, *J. Comput. Chem.* **6**:157 (1985); N. L. Allinger, L. R. Schmitz, L. R. Motoc, C. Bender, and J. K. Labanowski, *J. Phys. Org. Chem.* **3**:732 (1990); N. L. Allinger, L. R. Schmitz, I. Motoc, C. Bender, and J. K. Labanowski, *J. Am. Chem. Soc.* **114**:2880 (1992).
11. D. F. DeTar, *J. Org. Chem.* **60**:7125 (1995).
12. N. L. Allinger, K. Sakakibara, and J. Labanowski, *J. Phys. Chem.* **99**:9603 (1995); S. J. Mole, X. Zhou, and R. Liu, *J. Phys. Chem.* **100**:14665 (1996).

Table 4.3. Comparison of Computed ΔH_f° for *ab initio* Computations[a]

	3-21G	4-31G	6-31G*	6-31G**	Experimental
Ethane	−19.88	−19.92	−19.90	−19.92	−20.06
Propane	−25.17	−25.16	−25.17		−24.92
Butane[b]	−30.09	−30.09	−30.09	−30.09	−30.09
Pentane	−35.03	−34.99	−35.04		−35.06
Hexane	−39.96		−39.96		−39.95
Heptane	−44.89		−44.89		−44.87
Cyclopropane	19.76	15.48	12.93	12.19	12.71
Cyclobutane	7.71	7.04	5.60		6.58
Cyclopentane	−21.32	−16.60			−18.36
Cyclohexane	−32.20	−31.22	−30.78		−29.46
Bicyclo[1.1.1]pentane		54.38	47.68	46.85	
Bicyclo[2.1.1]hexane		13.54			
Bicyclo[2.2.1]heptane	−21.21	−18.09	−16.64		−13.12
Bicyclo[2.2.2]octane	−30.70	−27.54	−26.63		−23.66

a. In kcal/mol; D. F. DeTar, *J. Org. Chem.* **60**:7125 (1995).
b. Reference standard.

Table 4.3 gives calculated ΔH_f data for several types of *ab initio* calculations. The difference between the calculated ΔH_f and the experimental value gives an indication of the suitability of the particular method for that type of molecule.

Any of these sets of computed energies can then be used for calculation of reaction enthalpies by comparing reactants and products. Table 4.4 gives some data for hydrogenation, hydrogenolysis, and isomerization reactions at several levels of theory. The table includes some data for small-ring compounds, which represent a particularly challenging test of the accuracy of the computationally based methods.

Whether ΔH for a projected reaction is based on bond-energy data, tabulated thermochemical data, or MO computations, there remain some fundamental problems which prevent reaching a final conclusion about a reaction's feasibility. In the first place, most reactions of interest occur in solution, and the enthalpy, entropy, and free energy associated with any reaction depend strongly on the solvent medium. There is only a limited amount of tabulated thermochemical data that are directly suitable for treatment of reactions in organic solvents. Thermodynamic data usually pertain to the pure compound. MO calculations usually refer to the isolated (gas phase) molecule. Estimates of solvation effects must be made in order to apply either experimental or computational data to reactions occurring in solution.

There is an even more basic limitation to the usefulness of thermodynamic data for making predictions about reactions: Thermodynamics provides no information about the energy requirements of the pathways that a potential reaction can follow; that is, thermodynamics provides no information about the *rates of chemical reactions*. In the absence of a relatively low-energy pathway, two molecules that can potentially undergo a highly exothermic reaction will coexist without reacting. Thus, even if a reaction is thermodynamically favorable, it will not occur at a significant rate unless there is a low-energy mechanism by which it can occur. It is therefore extremely important to develop an understanding of reaction mechanisms and the energy requirements and rates of the various steps by which organic reactions proceed.

192

CHAPTER 4
STUDY AND
DESCRIPTION OF
ORGANIC REACTION
MECHANISMS

Table 4.4. Comparison of Calculated and Observed ΔH for Some Reactions[a]

Reaction	6-31G*	MP2/6-311*	B3LYP	Observed
$CH_2{=}CH_2 + H_2 \longrightarrow C_2H_6$	−36.0	−32.2	−31.7	−30.8
$CH_2{=}CHCH{=}CH_2 + 2H_2 \longrightarrow C_4H_{10}$	−62.5	−55.6	−52.4	−53.3
benzene $+ 3H_2 \longrightarrow C_6H_{12}$	−53.8	−41.1	−38.1	−44.0
$C_2H_6 + H_2 \longrightarrow 2CH_4$	−18.8	−10.4	−16.5	−15.5
cyclopropane $+ H_2 \longrightarrow C_3H_8$	−41.8	−35.6	−38.6	−35.9
cyclopropane $\longrightarrow CH_2{=}CHCH_3$	−8.9	−6.2	−10.5	−8.5
cyclobutane $+ H_2 \longrightarrow C_4H_{10}$	−40.3	−34.1	−37.3	−35.9
bicyclo $+ H_2 \longrightarrow$ cyclobutane	−53.2	−44.0	−47.0	−44.4
bicyclo $\longrightarrow CH_2{=}CHCH{=}CH_2$	−20.9	−22.4	−31.9	−26.2

a. In kcal/mol; K. B. Wiberg and J. W. Ochlerski, *J. Comput. Chem.* **18**:108 (1997).

4.2. Kinetic Data

Kinetic data can provide detailed insight into reaction mechanisms. The rate of a given reaction can be determined by following the disappearance of a reactant or the appearance of product. The extent of reaction is often measured spectroscopically, because spectroscopic techniques provide a rapid, continuous means of monitoring changes in concentration. Numerous other methods are available, however, and may be preferable in certain cases. For example, continuous pH measurement or acid–base titration can be used to follow the course of reactions that consume or generate acid or base. Conductance measurements provide a means for determining the rates of reactions that generate ionic species. Polarimetry is a convenient way of following reactions involving optically active materials. In general, any property that can be measured and quantitatively related to the concentration of a reactant or product can be used to determine a reaction rate.

The goal of a kinetic study is to establish the quantitative relationship between the concentration of reactants and catalysts and the rate of the reaction. Typically, such a study involves rate measurements at enough different concentrations of each reactant so that the *kinetic order* with respect to each reactant can be assessed. A complete investigation allows the reaction to be described by a rate law, which is an algebraic expression containing one or more *rate constants* as well as the concentrations of all reactants that are involved in the rate-determining step and steps prior to the rate-determining step. Each concentration has an exponent, which is the order of the reaction with respect to that component. The overall kinetic order of the reaction is the sum of all the exponents in the rate expression. Several examples of rate laws which illustrate the variety observed are presented in Scheme 4.1. Some are simple; others are more complex.

The relationship between a kinetic expression and a reaction mechanism can be appreciated by considering the several individual steps that constitute the overall reaction mechanism. The expression for the rate of any *single step* in a reaction mechanism will contain a concentration term for each reacting species. Thus, for the reaction sequence

$$A + B \underset{k_{-1}}{\overset{k_1}{\rightleftharpoons}} C \overset{k_2}{\rightarrow} D \overset{k_3}{\rightarrow} E + F$$

the rates for the successive steps are

step 1: $\dfrac{d[C]}{dt} = k_1[A][B] - k_{-1}[C]$

step 2: $\dfrac{d[D]}{dt} = k_2[C]$

step 3: $\dfrac{d[E]}{dt} = \dfrac{d[F]}{dt} = k_3[D]$

Scheme 4.1. Some Representative Rate Laws

1^a A   rate $= k[A]$

2^b $CH_3CHCH_2CH_2CH_2CH_3 + NaOCH_3 \longrightarrow CH_3CH=CHCH_2CH_2CH_3$
B   rate $= [B][NaOCH_3]$

3^c $CH_3Cl + GaCl_3^* \rightleftharpoons CH_3Cl^* + GaCl_2^*Cl$
rate $= k[CH_3Cl][GaCl_3^*]^2$

4^d $(CH_3)_2C=CHCH_3 + HCl \xrightarrow{CH_3NO_2} (CH_3)_2CCH_2CH_3$
C   rate $= k[C][HCl]^2$

5^e $\frac{1}{4}(CH_3Li)_4 + H_3C-$ D   rate $= k[(CH_3Li)_4]^{1/4}[D]$

6^f $F_2CHCOC_6H_5 + CH_3(CH_2)_3NH_2 \xrightarrow{dioxane} F_2CHCNH(CH_2)_3CH_3$
E   F   rate $= k_1[E][F] + k_2[E][F]^2$

a. E. N. Cain and R. K. Solly, *J. Am. Chem. Soc.* **95**:7884 (1973).
b. R. A. Bartsch and J. F. Bunnett, *J. Am. Chem. Soc.* **90**:408 (1968).
c. F. P. DeHaan, H. C. Brown, D. C. Conway, and M. G. Gibby, *J. Am. Chem. Soc.* **91**:4854 (1969).
d. Y. Pocker, K. D. Stevens, and J. J. Champoux, *J. Am. Chem. Soc.* **91**:4199 (1969).
e. S. G. Smith, L. F. Charbonneau, D. P. Novak, and T. L. Brown, *J. Am. Chem. Soc.* **94**:7059 (1972).
f. A. S. A. Shawali and S. S. Biechler, *J. Am. Chem. Soc.* **89**:3020 (1967).

194

CHAPTER 4
STUDY AND
DESCRIPTION OF
ORGANIC REACTION
MECHANISMS

Let us further specify that the first step is a very rapid but unfavorable equilibrium and that $k_2 \ll k_3$; i.e., the second step is slow relative to the third step. Under these circumstances, the overall rate of the reaction will depend on the rate of the second step, and this step is called the *rate-determining step*.

Kinetic data provide information only about the rate-determining step and steps preceding it. In the hypothetical reaction under consideration, the final step follows the rate-determining step, and because its rate will not affect the rate of the overall reaction, k_3 will not appear in the overall rate expression. The rate of the overall reaction is governed by the second step, which is the bottleneck in the process. The rate of this step is equal to k_2 multiplied by the molar concentration of intermediate C, which may not be directly measurable. It is therefore necessary to express the rate in terms of the concentrations of reactants. In the case under consideration, this can be done by recognizing that [C] is related to [A] and [B] by an equilibrium constant:

$$K = \frac{[C]}{[A][B]}$$

Furthermore, K is related to k_1 and k_{-1} by the requirement that no net change in composition occur at equilibrium:

$$k_{-1}[C] = k_1[A][B]$$

$$[C] = \frac{k_1}{k_{-1}}[A][B]$$

The rate of step 2 can therefore be written in terms of [A] and [B]:

$$\frac{d[D]}{dt} = k_2[C] = k_2\frac{k_1}{k_{-1}}[A][B] = k_{obs}[A][B]$$

Experimentally, it would be observed that the reaction rate would be proportional to both [A] and [B]. The reaction will be first-order in both reactants.

One common kind of reaction involves proton transfer occurring as a rapid equilibrium preceding the rate-determining step, for example, in the reaction of an alcohol with hydrobromic acid to give an alkyl bromide:

$$ROH + H^+ \underset{k_{-1}}{\overset{\overset{\text{fast}}{k_1}}{\rightleftharpoons}} ROH_2{}^+$$

$$ROH_2{}^+ + Br^- \overset{\overset{\text{slow}}{}}{\underset{k_2}{\longrightarrow}} RBr + H_2O$$

The overall rate being measured is that of step 2, but there may be no means of directly measuring $[ROH_2{}^+]$. The concentration of the protonated intermediate $ROH_2{}^+$ can be expressed in terms of the concentration of the starting material by taking into consideration the equilibrium constant, which relates [ROH], [Br⁻], and [H⁺]:

$$K = \frac{[ROH_2{}^+]}{[ROH][H^+]}$$

$$[ROH_2{}^+] = K[ROH][H^+]$$

$$\text{rate} = k_2K[ROH][H^+][Br^-] = k_{obs}[ROH][H^+][Br^-]$$

A useful approach that is often used in analysis and simplification of kinetic expressions is the *steady-state approximation*. It can be illustrated with a hypothetical reaction scheme:

$$A + B \underset{k_{-1}}{\overset{k_1}{\rightleftharpoons}} C$$

$$\frac{C + D \overset{k_2}{\rightarrow} E + F}{A + B + D \rightarrow E + F}$$

If C is a reactive, unstable species, its concentration will never be very large. It must then be consumed at a rate that closely approximates the rate at which it is formed. Under these conditions, it is a valid approximation to set the rate of formation of C equal to its rate of destruction:

$$k_1[A][B] = k_2[C][D] + k_{-1}[C]$$

Rearrangement of this equation provides an expression for [C]:

$$\frac{k_1[A][B]}{k_2[D] + k_{-1}} = [C]$$

By substituting this expression into the rate for the second step, the following expression is obtained:

$$\text{rate} = k_2[C][D] = k_2 \frac{k_1[A][B]}{k_2[D] + k_{-1}}[D]$$

If $k_2[D]$ is much greater than k_{-1}, the rate expression simplifies to

$$\text{rate} = \frac{k_2 k_1[A][B][D]}{k_2[D]} = k_1[A][B]$$

On the other hand, if $k_2[D]$ is much less than k_{-1}, the observed rate expression becomes

$$\text{rate} = \frac{k_1 k_2[A][B][D]}{k_{-1}}$$

The first situation corresponds to the first step being rate-determining. In the second case, it is the second step that is rate-determining, with the first step being a preequilibrium.

The normal course of a kinetic investigation involves postulating likely mechanisms and comparing the observed rate law with those expected for the various mechanisms. Those mechanisms that are incompatible with the observed kinetics can be eliminated as possibilities. Let us consider aromatic nitration by nitric acid in an inert solvent as a typical example. We will restrict the mechanisms being considered to the three shown below. In an actual case, such arbitrary restriction would not be imposed, but instead all mechanisms compatible with existing information would be considered.

(A)

$$2HONO_2 \underset{k_{-1}}{\overset{k_1}{\rightleftharpoons}} \text{(fast)} \quad H_2\overset{+}{O}NO_2 + NO_3^-$$

benzene $+ H_2\overset{+}{O}NO_2 \xrightarrow[k_2]{\text{slow}}$ [cyclohexadienyl cation with H, NO$_2$] $+ H_2O$

[cyclohexadienyl cation with H, NO$_2$] $\xrightarrow{\text{fast}}$ [nitrobenzene, NO$_2$] $+ H^+$

$$\text{rate} = k_2[H_2\overset{+}{O}NO_2][\text{benzene}] = \frac{k_2 k_1}{k_{-1}} \frac{[HONO_2]^2}{[NO_3^-]} [\text{benzene}]$$

$$= k_{\text{obs}} \frac{[HONO_2]^2}{[NO_3^-]} [\text{benzene}]$$

(B)

$$2HONO_2 \underset{k_{-1}}{\overset{k_1}{\rightleftharpoons}} \text{(fast)} \quad H_2\overset{+}{O}NO_2 + NO_3^-$$

$$H_2\overset{+}{O}NO_2 \xrightarrow[k_2]{\text{slow}} H_2O + NO_2^+$$

$NO_2^+ +$ benzene $\underset{}{\overset{\text{fast}}{\rightleftharpoons}}$ [cyclohexadienyl cation with H, NO$_2$]

[cyclohexadienyl cation with H, NO$_2$] $\xrightarrow{\text{fast}}$ [nitrobenzene, NO$_2$] $+ H^+$

$$\text{rate} = \frac{k_1 k_2}{k_{-1}} \frac{[HONO_2]^2}{[NO_3^-]} = k_{\text{obs}} \frac{[HONO_2]^2}{[NO_3^-]}$$

(C)

$$2HONO_2 \underset{k_{-1}}{\overset{k_1}{\rightleftharpoons}} \text{(fast)} \quad NO_2^+ + NO_3^- + H_2O$$

$NO_2^+ +$ benzene $\underset{k_{-2}}{\overset{k_2}{\rightleftharpoons}}$ (fast) [cyclohexadienyl cation with H, NO$_2$] **I**

[cyclohexadienyl cation with H, NO$_2$] $\xrightarrow[k_3]{\text{slow}}$ [nitrobenzene, NO$_2$] $+ H^+$

In mechanism C the third step is rate-controlling so

$$rate = k_3[\mathbf{I}]$$

[$\mathbf{I}$] can be expressed in terms of the rapid equilibria involved in its formation:

$$k_{-2}[\mathbf{I}] = k_2[NO_2{}^+][benzene]$$

$$[NO_2{}^+] = \frac{k_1[HONO_2]^2}{k_{-1}[NO_3{}^-][H_2O]}$$

$$rate = k_3 \frac{k_2[benzene]k_1[HONO_2]^2}{k_{-2}k_{-1}[NO_3^-][H_2O]}$$

$$rate = \frac{[HONO_2]^2[benzene]}{[NO_3{}^-][H_2O]}$$

$$= k_{obs}\frac{[HONO_2]^2[benzene]}{[NO_3^-]} \quad \text{if } [H_2O] \gg [benzene]$$

Mechanism B has the distinctive feature that it is zero-order in the reactant benzene, because the rate-determining step occurs prior to the involvement of benzene. Mechanism B has, in fact, been established for nitration of benzene in several organic solvents, and the absence of a benzene concentration term in the rate law is an important part of the evidence for this mechanism.[13] Mechanisms A and C, on the other hand, provide kinetic expressions that are similar in form, differing only in the inclusion of water in the expression for mechanism C. This might not be a detectable difference. If the concentration of water is several times larger than that of benzene, its overall concentration will change little during the course of the reaction. In this circumstance, the term for the concentration of water would disappear (by being a component of the observed rate constant k) so that the form of the kinetic expression alone would not distinguish between mechanisms A and C.

To illustrate the development of a kinetic expression from a postulated reaction mechanism, let us consider the base-catalyzed reaction of benzaldehyde and acetophenone.

$$PhCH=O + PhCCH_3 \longrightarrow PhCCH_2CPh \longrightarrow PhCH=CHCPh + H_2O$$

Based on a general knowledge of base-catalyzed reactions of carbonyl compounds, a reasonable sequence of steps can be written, but the relative rates of the steps is an open question. Furthermore, it is known that reactions of this type are generally reversible so that the potential reversibility of each step must be taken into account. A completely

---

13. J. H. Ridd, *Acc. Chem. Res.* **4**:248 (1971); J. H. Ridd, in *Studies on Chemical Structure and Reactivity*, J. H. Ridd, ed., John Wiley & Sons, New York, 1966, Chapter 7; J. G. Hoggett, R. B. Moodie, J. R. Penton, and K. Schofield, *Nitration and Aromatic Reactivity*, Cambridge University Press, Cambridge, 1971; K. Schofield, *Aromatic Nitration*, Cambridge University Press, Cambridge, 1980; G. A. Olah, R. Malhotra, and S. C. Narang, *Nitration, Methods and Mechanisms*, VCH Publishers, New York, 1989.

198

CHAPTER 4
STUDY AND
DESCRIPTION OF
ORGANIC REACTION
MECHANISMS

reversible mechanism is as follows:

(1) $\quad PhCCH_3 + {}^-OEt \underset{k_{-1}}{\overset{k_1}{\rightleftharpoons}} PhCCH_2^- + EtOH$

(2) $\quad PhCCH_2^- + PhCH \underset{k_{-2}}{\overset{k_2}{\rightleftharpoons}} PhCHCH_2CPh$

(3) $\quad EtOH + PhCHCH_2CPh \underset{k_{-3}}{\overset{\underset{k_3}{\text{fast}}}{\rightleftharpoons}} PhCHCH_2CPh + EtO^-$

(4) $\quad PhCHCH_2CPh + EtO^- \overset{k_4}{\longrightarrow} PhCH{=}CHCPh + {}^-OH + EtOH$

Because proton-transfer reactions between oxygen atoms are usually very fast, step 3 can be assumed to be a rapid equilibrium. With the above mechanism assumed, let us examine the rate expression which would result, depending upon which of the steps is rate-determining.

If step 1 is rate-controlling, the rate expression would be

$$\text{rate} = k_1[PhCOCH_3][{}^-OEt]$$

Under these conditions, the concentration of the second reactant, benzaldehyde, would not enter into the rate expression.

If step 1 is an equilibrium and step 2 is rate-controlling, we obtain the rate expression

$$\text{rate} = k_2[PhCOCH_2^-][PhCHO]$$

which on substituting in terms of the rapid prior equilibrium gives

$$\text{rate} = k_2K_1[PhCOCH_3][{}^-OEt][PhCHO]$$

since

$$[PhCOCH_2^-] = K_1[PhCOCH_3][{}^-OEt]$$

where K_1 is the equilibrium constant for the deprotonation in the first step. If the final step is rate-controlling, the rate is

$$\text{rate} = k_4[Ph\overset{OH}{\overset{|}{C}}HCH_2COPh][{}^-OEt]$$

The concentration of the intermediate $PhCHOHCH_2COPh$ can be expressed in terms of the three prior equilibria. Using $\mathbf{I}$ for the intermediate and $\mathbf{I}^-$ for its conjugate base, and neglecting [EtOH] because it is the solvent and will remain constant, gives the relationships:

$$K_3 = \frac{[\mathbf{I}][{}^-OEt]}{[\mathbf{I}^-]} \quad \text{and} \quad [\mathbf{I}] = K_3\frac{[\mathbf{I}^-]}{[{}^-OEt]}$$

and, since $[\mathbf{I}^-] = K_2[\text{PhCOCH}_2{}^-][\text{PhCHO}]$, substituting for $[\mathbf{I}^-]$ gives

$$[\mathbf{I}] = K_3 \frac{K_2[\text{PhCOCH}_2{}^-][\text{PhCHO}]}{[^-\text{OEt}]}$$

Substituting for $[\text{PhCOCH}_2{}^-]$ from the equilibrium expression for step 1 gives

$$[\mathbf{I}] = \frac{K_3 K_2[\text{PhCHO}]}{[^-\text{OEt}]} K_1[\text{PhCOCH}_3][^-\text{OEt}] = K'[\text{PhCHO}][\text{PhCOCH}_3]$$

and this provides the final rate expression:

$$\text{rate} = k_{\text{obs}}[^-\text{OEt}][\text{PhCHO}][\text{PhCOCH}_3]$$

The form of this third-order kinetic expression is identical to that in the case where the second step was rate-determining.

Experimental studies of this base-catalyzed condensation have revealed that it is third-order, indicating that either the second or the fourth step must be rate-determining. Studies on the intermediate $\mathbf{I}$ obtained by an alternative synthesis have shown that k_4 is about four times as large as k_{-3} so that about 80% of the intermediate goes on to product. These reactions are faster than the overall reaction under the same conditions, so the second step must be rate-controlling.[14]

These examples illustrate the relationship between kinetic results and the determination of reaction mechanism. Kinetic results can exclude from consideration all mechanisms that require a rate law different from the observed one. It is often true, however, that related mechanisms give rise to identical predicted rate expressions. In this case, the mechanisms are "kinetically equivalent," and a choice between them is not possible on the basis of kinetic data. A further limitation on the information that kinetic studies provide should also be recognized. Although the data can give the *composition* of the activated complex for the rate-determining step and preceding steps, it provides no information about the *structure* of the intermediate. Sometimes the structure can be inferred from related chemical experience, but it is never established by kinetic data alone.

The nature of the rate constants k_r can be discussed in terms of *transition-state theory*. This is a general theory for analyzing the energetic and entropic components of a reaction process. In transition-state theory, a reaction is assumed to involve the formation of an activated complex that goes on to product at an extremely rapid rate. The rate of decomposition of the activated complex has been calculated from the assumptions of the theory to be 6×10^{12} s^{-1} at room temperature and is given by the expression[15]

$$\text{rate of activated complex decomposition} = \frac{\kappa k T}{h}$$

14. E. Coombs and D. P. Evans, *J. Chem. Soc.* **1940**:1295; D. S. Noyce, W. A. Pryor, and A. H. Bottini, *J. Am. Chem. Soc.* **77**:1402 (1955).

15. For a complete development of these relationships, see M. Boudart, *Kinetics of Chemical Processes*, Prentice-Hall, Englewood Cliffs, New Jersey, 1968, pp. 35–46; I. Amdur and G. G. Hammes, *Chemical Kinetics, Principles and Selected Topics*, McGraw-Hill, New York, 1966, pp. 43–58; J. W. Moore and R. G. Pearson, *Kinetics and Mechanism*, John Wiley & Sons, New York, 1981, pp. 159–169; M. M. Kreevoy and D. G. Truhlar, in *Investigation of Rates and Mechanisms of Reaction, Techniques of Chemistry*, 4th ed., Vol. VI, Part 1, C. F. Bernasconi, ed., John Wiley & Sons, New York, 1986.

200

CHAPTER 4
STUDY AND
DESCRIPTION OF
ORGANIC REACTION
MECHANISMS

in which κ is the transmission coefficient, which is usually taken to be 1, k is Boltzmann's constant, h is Planck's constant, and T is absolute temperature. The rate of reaction is thus given by the following expression:

$$\text{rate of reaction} = \frac{\kappa kT}{h}[\text{activated complex}]$$

If the activated complex is considered to be in equilibrium with its component molecules, the attainment of the transition state (T.S.) can be treated as being analogous to a bimolecular reaction:

$$A + B \rightarrow C$$

$$A + B \rightleftharpoons T.S. \rightarrow C$$

$$K^{\ddagger} = \frac{[T.S.]}{[A][B]}$$

The position of this equilibrium is related to the free energy required for attainment of the transition state. The double-dagger superscript ($\ddagger$) is used to specify that the process under consideration involves a transition state or activated complex:

$$\Delta G^{\ddagger} = -RT \ln K^{\ddagger}$$

This free energy is referred to as the *free energy of activation*. The rate of the reaction is then given by

$$\text{rate} = \frac{\kappa kT}{h}[T.S.]$$

$$[T.S.] = K^{\ddagger}[A][B]$$

Since

$$K^{\ddagger} = e^{-\Delta G^{\ddagger}/RT}$$

$$\text{rate} = \frac{\kappa kT}{h}e^{-\Delta G^{\ddagger}/RT}[A][B] \tag{4.4}$$

Comparison with the form of the expression for the rate of any single reaction step

$$\text{rate} = k_r[A][B]$$

reveals that the magnitude of $\Delta G^{\ddagger}$ will be the factor that determines the magnitude of k_r at any given temperature.

Qualitative features of reaction mechanisms are often described in the context of transition-state theory and illustrated with potential energy diagrams. The potential energy diagrams for a hypothetical one-step bimolecular reaction and for a two-step reaction are shown in Fig. 4.1. The lower diagram depicts a two-step reaction in which an intermediate having a finite lifetime is involved. Two transition states are then involved. The higher activation energy of the first transition state implies that the first step would be slower and therefore rate-determining. These two-dimensional diagrams are useful for qualitative

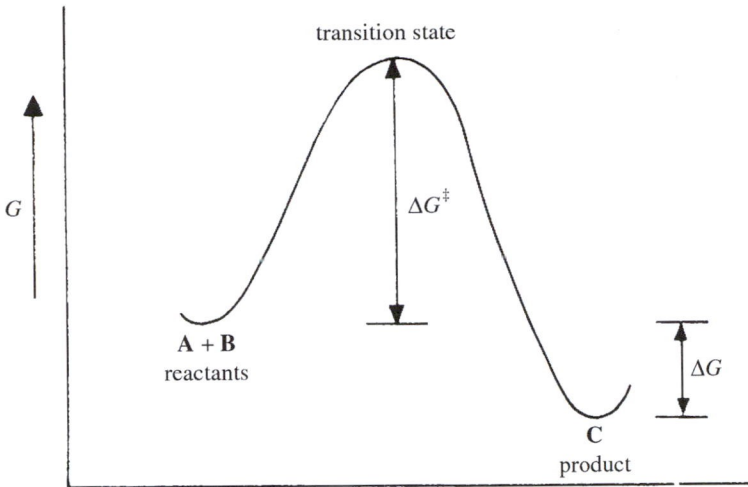

Reaction coordinate for a single-step reaction

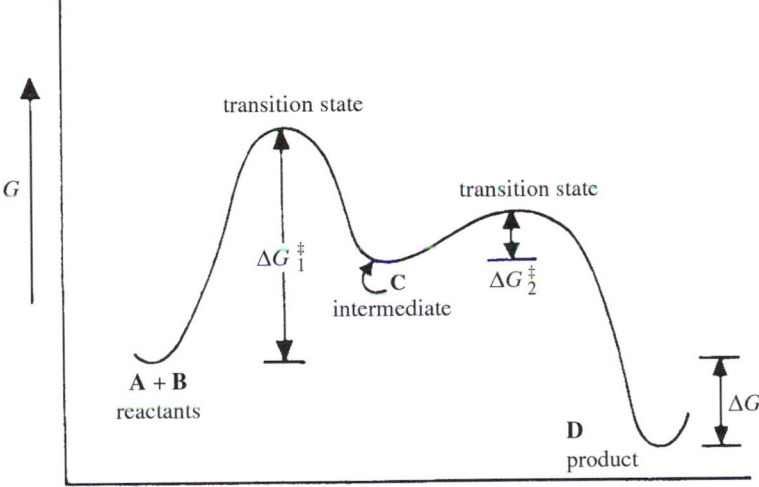

Reaction coordinate for a two-step reaction

Fig. 4.1. Potential energy diagrams for single-step and two-step reactions.

discussion of reaction mechanisms. The curve plots the free energy of the reaction complex as it progresses along the reaction coordinate from reactants to products.

Such diagrams make clear the difference between an intermediate and a transition state. An intermediate lies in a depression on the potential energy curve. Thus, it will have a finite lifetime. The actual lifetime will depend on the depth of the depression. A shallow depression implies a low activation energy for the subsequent step, and therefore a short lifetime. The deeper the depression, the longer is the lifetime of the intermediate. The situation at a transition state is quite different. It has only fleeting existence and represents an energy maximum on the reaction path.

202

CHAPTER 4
STUDY AND
DESCRIPTION OF
ORGANIC REACTION
MECHANISMS

There is one path between reactants and products that has a lower energy maximum than any other; this is the pathway that the reaction will follow. The curve in a two-dimensional potential energy plot represents this lowest-energy pathway. It represents a path across an energy surface describing energy as a function of the spatial arrangement of the atoms involved in the reaction. The *principle of microscopic reversibility* arises directly from transition-state theory. *The same pathway that is traveled in the forward direction of a reaction will be traveled in the reverse direction, since it affords the lowest energy barrier for either process.* Thus, information about the nature of a transition state or intermediate deduced by a study of a forward reaction is applicable to the discussion of the reverse process occurring under the same conditions.

Because transition states cannot be observed, there is no experimental method for establishing their structure. Theoretical descriptions of molecules have been applied to this problem. By applying one of the MO methods, structures can be calculated for successive geometries which gradually transform the reactants into products. Exploration of a range of potential geometries and calculation of the energy of the resulting ensembles can, in principle, locate and describe the minimum-energy pathway. To the extent that the calculations accurately reflect the molecular reality, this provides a structural description of the reaction path and transition state.

The temperature dependence of reaction rates permits evaluation of the enthalpy and entropy components of the free energy of activation. The terms in Eq. (4.4) corresponding to k_r can be expressed as

$$k_r = \frac{\kappa kT}{h}(e^{-\Delta H^{\ddagger}/RT})(e^{\Delta S^{\ddagger}/R}) \tag{4.5}$$

The term $(\kappa kT/h)e^{\Delta S^{\ddagger}/R}$ varies only slightly with T compared to $e^{-\Delta H^{\ddagger}/RT}$ because of the exponential nature of the latter. To a good approximation, then

$$\frac{k_r}{T} = Ce^{-\Delta H^{\ddagger}/RT} \tag{4.6}$$

$$\ln\frac{k_r}{T} = \frac{-\Delta H^{\ddagger}}{RT} + C' \tag{4.7}$$

A plot of $\ln(k_r/T)$ versus $1/T$ is then a straight line, and its slope is $-\Delta H^{\ddagger}/R$. Once $\Delta H^{\ddagger}$ is determined in this manner, $\Delta S^{\ddagger}$ is available from the relationship

$$\Delta S^{\ddagger} = \frac{\Delta H^{\ddagger}}{T} + R\ln\frac{hk_r}{\kappa kT} \tag{4.8}$$

which can be obtained by rearranging Eq. (4.5).

The temperature dependence of reactions can also be expressed in terms of the Arrhenius equation:

$$k_r = Ae^{-E_a/RT} \tag{4.9}$$

$$\ln k_r = -E_a/RT + \ln A \tag{4.10}$$

Comparing the form of Eq. (4.9) with Eq. (4.5) indicates that A in the Arrhenius equation corresponds to $(\kappa kT/h)e^{\Delta S^{\ddagger}/R}$. The Arrhenius equation shows that a plot of $\ln k_r$ versus $1/T$ will have the slope $-E_a/R$. For reactions in solution at a constant pressure, $\Delta H^{\ddagger}$ and

E_a are related by

$$E_a = \Delta H^{\ddagger} + RT \qquad (4.11)$$

The magnitudes of $\Delta H^{\ddagger}$ and $\Delta S^{\ddagger}$ reflect transition-state structure. Atomic positions in the transition state do not correspond to their positions in the ground state. In particular, the reacting bonds will be partially formed and partially broken. The energy required for bond reorganization is reflected in the higher potential energy of the activated complex and corresponds to the enthalpy of activation $\Delta H^{\ddagger}$. The entropy of activation is a measure of the degree of order produced in the formation of the activated complex. If translational, vibrational, or rotational degrees of freedom are lost in going to the transition state, there will be a decrease in the total entropy of the system. Conversely, an increase of translational, vibrational, or rotational degrees of freedom will result in a positive entropy of activation.

Wide variation in enthalpy and entropy of activation for different reaction systems is possible, as illustrated by the following two reactions.

(A) Dimerization of cyclopentadiene in the gas phase:

(Ref. 16)

$$\Delta H^{\ddagger} = 15.5 \text{ kcal/mol}$$
$$\Delta S^{\ddagger} = -34 \text{ eu}$$

(B) Decomposition of 1,1'-azobutane in the gas phase:

$$Bu-N=N-Bu \rightarrow 2Bu\cdot + N_2$$
$$\Delta H^{\ddagger} = 52 \text{ kcal/mol}$$
$$\Delta S^{\ddagger} = +19 \text{ eu}$$

(Ref. 17)

The relatively low $\Delta H^{\ddagger}$ term for the dimerization of cyclopentadiene is characteristic of concerted reactions (see Chapter 11), in which bond making accompanies bond breaking. It differs markedly from $\Delta H^{\ddagger}$ for the thermal decomposition of 1,1'-azobutane, in which the rate-determining step is a homolytic cleavage of a C—N bond, with little new bond making to compensate for the energy cost of the bond breaking. The entropy of activation, on the other hand, is more favorable in the 1,1'-azobutane decomposition because a translational degree of freedom is being gained in the transition state as the molecular fragments separate. The dimerization of cyclopentadiene is accompanied by a very negative entropy of activation because of the loss of translational and rotational degrees of freedom in formation of the transition state. The two reacting molecules must attain a

16. A. Wassermann, *Monatsh. Chem.* **83**:543 (1952).
17. A. U. Blackham and N. L. Eatough, *J. Am. Chem. Soc.* **84**:2922 (1962).

204

CHAPTER 4
STUDY AND
DESCRIPTION OF
ORGANIC REACTION
MECHANISMS

specific orientation to permit the bonding interactions that occur as the transition state is approached.

Unimolecular reactions that take place by way of cyclic transition states typically have negative entropies of activation because of the loss of rotational degrees of freedom associated with the highly ordered transition state. For example, thermal isomerization of allyl vinyl ether to 4-pentenal has $\Delta S^{\ddagger} = -8$ eu.[18]

It is important to remember that the enthalpy and entropy of activation reflect the response of the reacting system *as a whole* to formation of the activated complex. As a result, the interpretation of these parameters is more complicated for reactions taking place in solution than for gas-phase reactions. This complexity is particularly true for processes involving formation or destruction of charged species. The solvolysis of *t*-butyl chloride in 80% aqueous ethanol, for example, has as its rate-determining step unimolecular ionization of the carbon–chlorine bond to form chloride ion and the *t*-butyl cation. One might guess that this ionization should lead to a positive entropy of activation, since two independent particles are being generated. In fact, the entropy of activation is -6.6 eu. Because of its polar character, the transition state requires a greater ordering of solvent molecules than the nonpolar reactant.[19] It turns out to be generally true that reactions that generate charged species exhibit negative entropies of activation in solution. The reverse is true for reactions in which charged reactants lead to a neutral transition state.

4.3. Substituent Effects and Linear Free-Energy Relationships

In Chapter 1, Section 1.2 (p. 13), the effect of substituent groups on the acid strength of acetic acid derivatives was discussed qualitatively. It was noted in particular that the presence of groups more electronegative than hydrogen increased the acid strength relative to acetic acid. Many detailed relationships between substituent groups and chemical properties have been developed. In many cases, such relationships can be expressed quantitatively and are useful both for interpreting reaction mechanisms and for predicting reaction rates and equilibria.

The most widely applied of these relationships is the *Hammett equation*, which relates rates and equilibria of many reactions of compounds containing substituted phenyl groups. It was noted in the 1930s that there is a relationship between the acid strengths of substituted benzoic acids and the rates of many other chemical reactions, for instance, the rates of hydrolysis of substituted ethyl benzoates. The correlation is illustrated graphically in Fig. 4.2, which shows $\log k/k_0$, where $k_0 = k$ for hydrolysis of ethyl benzoate and k is the rate constant for the substituted esters, plotted against $\log K/K_0$, where K and K_0 are the corresponding acid dissociation constants.

Analogous plots for many other reactions of aromatic compounds show a similar linear correlation with the acid dissociation constants of the corresponding benzoic acids.

18. F. W. Schuler and G. W. Murphy, *J. Am. Chem. Soc.* **72**:3155 (1950).
19. E. Grunwald and S. Winstein, *J. Am. Chem. Soc.* **70**:846 (1948).

205

SECTION 4.3.
SUBSTITUENT EFFECTS
AND LINEAR
FREE-ENERGY
RELATIONSHIPS

Fig. 4.2. Correlation of acid dissociation constants of benzoic acids with rates of alkaline hydrolysis of ethyl benzoates. [From L. P. Hammett, *J. Am. Chem. Soc.* **59**:96 (1937).]

Neither the principles of thermodynamics nor theories of reaction rates require that there should be such linear relationships. There are, in fact, numerous reaction series that fail to show such correlations. Some insight into the origin of the correlation can be gained by considering the relationship between the correlation equation and the free-energy changes involved in the two processes. The line in Fig 4.2 defines an equation in which m is the slope of the line:

$$m \log \frac{K}{K_0} = \log \frac{k}{k_0} \tag{4.12}$$

Substituting K and k with the appropriate free energy of activation:

$$m(\log K - \log K_0) = \log k - \log k_0$$
$$m(-\Delta G/2.3RT + \Delta G_0/2.3RT) = -\Delta G^{\ddagger}/2.3RT + \Delta G_0^{\ddagger}/2.3RT$$
$$m(-\Delta G + \Delta G_0) = -\Delta G^{\ddagger} + \Delta G_0^{\ddagger} \tag{4.13}$$
$$m\Delta\Delta G = \Delta\Delta G^{\ddagger}$$

The linear correlation therefore indicates that the change in free energy of activation on introduction of a series of substituent groups is *directly proportional* to the change in the free energy of ionization that is caused by the same series of substituents on benzoic acid. The various correlations arising from such directly proportional changes in free energies are called *linear free-energy relatonships*.

206

CHAPTER 4
STUDY AND
DESCRIPTION OF
ORGANIC REACTION
MECHANISMS

Since ΔG and $\Delta G^{\ddagger}$ are combinations of enthalpy and entropy terms, a linear free-energy relationship between two reaction series can result from one of three circumstances: (1) ΔH is constant and the ΔS terms are proportional for the series, (2) ΔS is constant and the ΔH terms are proportional, or (3) ΔH and ΔS are linearly related. Dissection of the free-energy changes into enthalpy and entropy components has often shown the third case to be true.[20]

The Hammett free-energy relationship is expressed in the following equations for equilibria and for rate data, respectively:

$$\log \frac{K}{K_0} = \sigma\rho \tag{4.14}$$

$$\log \frac{k}{k_0} = \sigma\rho \tag{4.15}$$

The numerical values of the terms σ and ρ are defined by specifying the ionization of benzoic acids as the standard reaction to which the *reaction constant* $\rho = 1$ is assigned. The *substituent constant*, σ, can then be determined for a series of substituent groups by measurement of the acid dissociation constant of the substituted benzoic acids. The σ values so defined are used in the correlation of other reaction series, and the ρ values of the reactions are thus determined. The relationship between Eqs. (4.12) and (4.14) is evident when the Hammett equation is expressed in terms of free energy. For the standard reaction, $\log K/K_0 = \sigma\rho$. Thus,

$$-\Delta G/2.3RT + \Delta G_0/2.3RT = \sigma\rho = \sigma$$

since $\rho = 1$ for the standard reaction. Substituting into Eq. (4.12):

$$m\sigma = -\Delta G^{\ddagger}/2.3RT + \Delta G_0^{\ddagger}/2.3RT$$
$$m\sigma = \log k - \log k_0$$
$$m\sigma = \log \frac{k}{k_0} \tag{4.16}$$
$$m = \rho$$

The value of σ reflects the effect that the substituent group has on the free energy of ionization of the substituted benzoic acid. The effect of the substituent results from a combination of factors. A substituent can cause a polarization of electron density around the ring through the π system in both the reactant and the product. This will affect the position of the equilibrium. In the case of a reaction rate, the relative effect on the reactant and the transition state will determine the effect on the energy of activation. One mechanism for polarization and charge redistribution is the *resonance effect*, which is illustrated in Fig. 4.3a for several substituents. There is also an effect that originates with the bond dipoles between groups of differing electronegativity. Substituents that are more electronegative than an aromatic carbon will place a net positive charge on the substituted carbon atom, whereas atoms less electronegative than an aromatic carbon will have the opposite effect. The resulting dipoles can perturb the electronic situation at the reaction site

20. P. D. Bolton, K. A. Fleming, and F. M. Hall, *J. Am. Chem. Soc.* **94**:1033 (1972); J. E. Leffler, *J. Org. Chem.* **20**:1202 (1955).

207

SECTION 4.3.
SUBSTITUENT EFFECTS
AND LINEAR
FREE-ENERGY
RELATIONSHIPS

a. Resonance effects

b. Field effects

c. Inductive effects

Fig. 4.3. Resonance, field, and inductive components of substituent effects in substituted benzenes.

in two ways. The presence of the charge separation will influence the energy associated with development of charge elsewhere in the molecule. This is the result of through-space electrostatic interaction and is called a *field effect*. Depending on the orientation of the dipole and of the charge developing at the reaction site, a substituent can either favor or disfavor the reaction, as illustrated in Fig. 4.3b. Another possible means of interaction of a substituent and the reaction site is called the *inductive effect*. This is transmission of bond dipoles through the intervening bonds by succesive polarization of each bond. The experimental and theoretical results which are presently available indicate that the field effect outweighs the inductive effect as the primary means of transmission of the effect of bond dipoles.[21] Field effects and inductive effects can be considered together as *polar effects*, that is, originating from bond dipoles.

The Hammett equation in the form of Eq. (4.14) or Eq. (4.15) is free of complications due to steric effects, since it is applied only to *meta* and *para* substituents. The geometry of the benzene ring ensures that groups in these positions cannot interact sterically with the site of reaction. Tables of σ values for many substituents have been collected; some values are given in Table 4.5, but substituent constants are available for a much wider range of

21. M. J. S. Dewar and P. J. Grisdale, *J. Am. Chem. Soc.* **84**:3548 (1962); M. J. S. Dewar and A. P. Marchand, *J. Am. Chem. Soc.* **88**:354 (1966); H. D. Holtz and L. M. Stock, *J. Am. Chem. Soc.* **86**:5188 (1964); C. L. Liotta, W. F. Fisher, G. H. Greene, Jr., and B. L. Joyner, *J. Am. Chem. Soc.* **94**:4891 (1972); C. F. Wilcox and C. Leung, *J. Am. Chem. Soc.* **90**:336 (1968); W. F. Reynolds, *Prog. Phys. Org. Chem.* **14**:165 (1983); K. Bowden, *J. Chim. Phys.* **91**:165 (1991); K. Bowden, *J. Chim. Phys.* **89**:1647 (1992); K. Bowden and E. J. Grubbs, *Chem. Soc. Rev.* **25**:171 (1995).

208

CHAPTER 4
STUDY AND
DESCRIPTION OF
ORGANIC REACTION
MECHANISMS

Table 4.5. Substituent Constants[a]

Substituent group		σ_m	σ_P	σ^+	σ^-	σ_I	σ_R^0
Acetamido	CH$_3$CONH	**0.14**	**0.0**	−0.6	0.47		
Acetoxy	CH$_3$CO$_2$	0.39	0.31	0.18			
Acetyl	CH$_3$CO	0.36	**0.47**		0.82	0.20	0.16
Amino	NH$_2$	**−0.09**	**−0.30**	−1.3		0.12	−0.50
Bromo	Br	0.37	**0.26**	0.15		0.44	−0.16
t-Butyl	(CH$_3$)$_3$C	**−0.09**	**−0.15**	−0.26			
Carbomethoxy	CH$_3$O$_2$C	0.35	0.44		0.74	0.20	0.16
Carboxy	HO$_2$C	**0.35**	**0.44**		0.73		
Chloro	Cl	**0.37**	**0.24**	0.11		0.46	−0.18
Cyano	CN	**0.62**	**0.70**		0.99	0.56	0.08
Ethoxy	C$_2$H$_5$O	0.1	−0.14	−0.82			
Ethyl	C$_2$H$_5$	**−0.08**	**−0.13**	−0.30			
Fluoro	F	**0.34**	**0.15**	−0.07		0.50	−0.31
Hydrogen	H	**0**	**0**	0	0	0	0
Hydroxy	OH	0.13	−0.38	−0.92			
Methanesulfonyl	CH$_3$SO$_2$	**0.64**	**0.73**		1.05	0.60	0.12
Methoxy	CH$_3$O	**0.10**	**−0.12**	−0.78		0.27	−0.42
Methyl	CH$_3$	**−0.06**	**−0.14**	−0.31		−0.04	−0.13
Nitro	NO$_2$	**0.71**	0.81		1.23	0.65	0.15
Phenyl	C$_6$H$_5$	**0.05**	**0.05**	−0.18	0.08		
Trifluoromethyl	CF$_3$	**0.46**	**0.53**		0.74	0.42	0.08
Trimethylammonio	(CH$_3$)$_3$N$^+$	0.99	0.96				
Trimethylsilyl	(CH$_3$)$_3$Si	−0.04	−0.07				

a. Values of σ_m, σ_p, σ^+, and σ^- from O. Exner, in *Correlation Analysis in Chemistry*, N. B. Chapman and J. Shorter, eds., Plenum Press, New York, 1978, Chapter 10. Values of σ_I and σ_R^0 from J. Bromilow, R. T. C. Brownlee, V. O. Lopez, and R. W. Taft, *J. Org. Chem.* **44**:4766 (1979). Values of σ_m and σ_p shown in boldface type are regarded as particularly reliable.

substituents.[22] The σ value for any substituent reflects the interaction of the substituent with the reacting site by a combination of resonance and field interactions. Table 4.6 shows a number of ρ values. The ρ value reflects the sensitivity of the particular reaction to substituent effects. The examples which follow illustrate some of the ways in which the Hammett equation can be used.

Example 4.2. The pK_a of *p*-chlorobenzoic acid is 3.98; that of benzoic acid is 4.19. Calculate σ for *p*-Cl.

$$\sigma = \log \frac{K_{p-\text{Cl}}}{K_{\text{H}}} = \log K_{p-\text{Cl}} - \log K_{\text{H}}$$

$$= -\log K_{\text{H}} - (-\log K_{p-\text{Cl}})$$

$$= pK_{a_{\text{H}}} - pK_{a_{p-\text{Cl}}}$$

$$= 4.19 - 3.98 = 0.21$$

Example 4.3. The ρ value for alkaline saponification of methyl esters of substituted benzoic acids is 2.38, and the rate constant for saponification of methyl benzoate under the conditions of interest is $2 \times 10^{-4}\ M^{-1}\ s^{-1}$. Calculate the rate constant for the hydrolysis

22. C. Hansch, A. Leo, and R. W. Taft, *Chem. Rev.* **91**:165 (1991); J. Shorter, *Aust. J. Chem.* **48**:1453 (1995); J. Shorter, *Pure Appl. Chem.* **66**:2451 (1994); J. Shorter, *Aust. J. Chem.* **51**:525 (1988); J. Shorter, *Pure Appl. Chem.* **69**:2497 (1997).

209

SECTION 4.3.
SUBSTITUENT EFFECTS
AND LINEAR
FREE-ENERGY
RELATIONSHIPS

Table 4.6. Reaction Constants[a]

Reaction	ρ
$ArCO_2H \rightleftharpoons ArCO_2^- + H^+$, water	1.00
$ArCO_2H \rightleftharpoons ArCO_2^- + H^+$, EtOH	1.57
$ArCH_2CO_2H \rightleftharpoons ArCH_2CO_2^- + H^+$, water	0.56
$ArCH_2CH_2CO_2H \rightleftharpoons ArCH_2CH_2CO_2^- + H^+$, water	0.24
$ArOH \rightleftharpoons ArO^- + H^+$, water	2.26
$ArNH_3^+ \rightleftharpoons ArNH_2 + H^+$, water	3.19
$ArCH_2NH_3^+ \rightleftharpoons ArCH_2NH_2 + H^+$, water	1.05
$ArCO_2Et + {}^-OH \rightarrow ArCO_2^- + EtOH$	2.61
$ArCH_2CO_2Et + {}^-OH \rightarrow ArCH_2CO_2^- + EtOH$	1.00
$ArCH_2Cl + H_2O \rightarrow ArCH_2OH + HCl$	−1.31
$ArC(Me)_2Cl + H_2O \rightarrow ArC(Me)_2OH + HCl$	−4.48
$ArNH_2 + PhCOCl \rightarrow ArNHCOPh + HCl$	−3.21

a. From P. R. Wells, *Linear Free Energy Relationships*, Academic Press, New York, 1968, pp. 12, 13.

of methyl *m*-nitrobenzoate.

$$\log \frac{k_{m\text{-}NO_2}}{k_H} = \sigma_{m\text{-}NO_2}(\rho) = (0.71)(2.38) = 1.69$$

$$\frac{k_{m\text{-}NO_2}}{k_H} = 49$$

$$k_{m\text{-}NO_2} = 98 \times 10^{-4}\ M^{-1}\ s^{-1}$$

Example 4.4. Using data in Tables 4.5 and 4.6, calculate how much faster *p*-bromobenzyl chloride will solvolyze in water than will *p*-nitrobenzyl chloride.

$$\log \frac{k_{p\text{-}Br}}{k_H} = (-1.31)(0.26), \qquad \log \frac{k_{p\text{-}NO_2}}{k_H} = (-1.31)(0.81)$$

$$\log k_{Br} - \log k_H = -0.34, \qquad \log k_{NO_2} - \log k_H = -1.06$$

$$\log k_{Br} + 0.34 = \log k_H, \qquad \log k_{NO_2} + 1.06 = \log k_H$$

$$\log k_{Br} + 0.34 = \log k_{NO_2} + 1.06$$

$$\log k_{Br} - \log k_{NO_2} = 0.72$$

$$\log \frac{k_{Br}}{k_{NO_2}} = 0.72$$

$$\frac{k_{Br}}{k_{NO_2}} = 5.25$$

Given in Table 4.5 in addition to the σ_m and σ_p values used with the classical Hammett equation are σ^+ and σ^-. These are substituent constant sets which reflect a recognition that the extent of resonance participation can vary for different reactions. The σ^+ values are used for reactions in which there is direct resonance interaction between an electron-donor substituent and a cationic reaction center, whereas the σ^- set pertains to reactions in which there is a direct resonance interaction between the substituent and an electron-rich reaction site. These are cases in which the resonance component of the

210

CHAPTER 4
STUDY AND
DESCRIPTION OF
ORGANIC REACTION
MECHANISMS

substituent effect will be particularly important.

Direct resonance interaction with a cationic center

Direct resonance interaction with an anionic center

One underlying physical basis for the failure of Hammett σ_m and σ_p values to correlate all reaction series is that substituent interactions are some mixture of resonance, field, and inductive effects. When direct resonance interaction is possible, the extent of the resonance increases, and the substituent constants appropriate to the "normal" mix of resonance and field effects then fail. There have been many attempts to develop sets of σ values that take into account extra resonance interactions.

One approach is to correct for the added resonance interaction. This is done in a modification of the Hammett equation known as the Yukawa–Tsuno equation.[23]

$$\log \frac{K}{K_0} = \rho\sigma + \rho(r)(\sigma^+ - \sigma) \tag{4.17}$$

The additional parameter r is adjusted from reaction to reaction; it reflects the extent of the additional resonance contribution. A large r corresponds to a reaction with a large resonance component, whereas when r goes to zero, the equation is identical to the original Hammett equation. When there is direct conjugation with an electron-rich reaction center, an equation analogous to Eq. (4.17) can be employed, but σ^- is used instead of σ^+.

The Yukawa–Tsuno relationship expanded to include both the σ^+ and σ^- constants is called the LArSR equation[24]:

$$\log \frac{k}{k_0} = \rho(\sigma^0 + r^+\Delta\sigma_R^+ + r^-\Delta\sigma_R^-)$$

In this equation, the substituent parameters σ_R^+ and σ_R^- reflect the incremental resonance interaction with electron-demanding and electron-releasing reaction centers, respectively. The variables r^+ and r^- are established for a reaction series by regression analysis and are measures of the extent of the extra resonance contribution. The larger the value of r, the greater is the extra resonance contribution. Because both donor and acceptor capacity will not contribute in a single reaction process, either r^+ or r^- would be expected to be zero.

A more ambitious goal is to separate completely resonance effects from polar effects. This involves using separate substituent constants to account for resonance and polar effects. The modified equation, called a *dual-substituent-parameter equation*, takes

23. Y. Yukawa and Y. Tsuno, *Bull. Chem. Soc. Jpn.* **32**:971 (1959); J. Hine, *J. Am. Chem. Soc.* **82**:4877 (1960); B. M. Wepster, *J. Am. Chem. Soc.* **95**:102 (1973).
24. Y. Yukawa, Y. Tsuno, and M. Sawada, *Bull. Chem. Soc. Jpn.* **39**:2274 (1966); Y. Yukawa, Y. Tsuno, and M. Sawada, *Bull. Chem. Soc. Jpn* **45**:1210 (1972).

the form

211

SECTION 4.3.
SUBSTITUENT EFFECTS
AND LINEAR
FREE-ENERGY
RELATIONSHIPS

$$\log \frac{K}{K_0} \quad \text{or} \quad \log \frac{k}{k_0} = \sigma_I \rho_I + \sigma_R \rho_R$$

where σ_I and σ_R are the reaction constants which reflect the sensitivity of the system to polar and resonance effects, respectively.[25] The σ_I values have been defined from studies in aliphatic systems where no resonance component should be present. By properly scaling the σ_I values with σ values from aromatic systems, it is possible to assign values such that

$$\sigma = \sigma_I + \sigma_R$$

Statistical analysis of data from many reaction series has shown that no single σ_R is applicable to the entire range of reactions. This again reflects the fact that the resonance component is variable and responds to the nature of the particular reaction. Therefore, a series of four σ_R values was established, each of which applies to various reaction types, ranging from direct conjugation with electron-deficient reaction centers to the other extreme. We will discuss only one of these, $\sigma_R{}^0$, which applies in cases of minimal perturbation of the aromatic ring by charge development at the reaction site. The $\sigma_R{}^0$ values given in Table 4.5 are based on the use of ^{13}C-NMR chemical shifts as a measure of the sum of resonance and polar effects. The chemical shift data of substituted benzenes were analyzed to provide the best correlation with the dual-substituent-parameter equation. In nonpolar solvents, which presumably best reflect the inherent molecular properties, σ_I is 3.74 for cyclohexane and 3.38 for carbon tetrachloride. The corresponding values of σ_R are 20.59 and 20.73. The relative magnitudes of σ_I and σ_R indicate that the ^{13}C-NMR chemical shift is more responsive to the resonance effect of the substituent than to the polar effect.[26]

In general, the dissection of substituent effects need not be limited to resonance and polar components, which are of special prominence in reactions of aromatic compounds. Any type of substituent interaction with a reaction center could be characterized by a substituent constant characteristic of the particular type of interaction and a reaction parameter indicating the sensitivity of the reaction series to that particular type of interaction. For example, it has been suggested that electronegativity and polarizability can be treated as substituent effects separate from polar and resonance effects. This gives rise to the equation

$$\log \frac{k}{k_0} = \sigma_F \rho_F + \sigma_R \rho_R + \sigma_\chi \rho_\chi + \sigma_\alpha \rho_\alpha$$

where σ_F is the polar, σ_R is the resonance, σ_χ is the electronegativity, and σ_α is the polarizability substituent constant.[27] We will, in general, emphasize the resonance and field components in our discussion of substituent effects.

The existing series of substituent constants has been developed by analysis of experimental data. Separation of the various components has usually depended on correlation analysis designed to identify the contributions from various components of

25. S. Ehrenson, R. T. C. Brownlee, and R. W. Taft, *Prog. Phys. Org. Chem.* **10**:1 (1973).
26. J. Bromilow, R. T. C. Brownlee, V. O. Lopez, and R. W. Taft, *J. Org. Chem.* **44**:4766 (1979).
27. R. W. Taft and R. D. Topsom, *Prog. Phys. Org. Chem.* **16**:1 (1987).

212

CHAPTER 4
STUDY AND
DESCRIPTION OF
ORGANIC REACTION
MECHANISMS

the overall substituent effect. We might ask whether substituent effects and, in particular, the various components of empirical substituent constants, could be determined by computational approaches. There has been considerable effort in this direction.[28] One measure of intrinsic resonance effects, for example, is the transfer of electron density in ground-state molecules. The relative charge can be calculated for substituted ethylenes and benzene.[29]

Calculations have been done at the STO-3G and 4-31G levels, and the resulting substituent constants correlate well with empirical values derived from ground-state structural parameters, such as ^{13}C-NMR chemical shifts and IR absorption frequencies.

Field effects can be determined by calculating the effect of a bond dipole on a molecular probe at a specified distance. One system that has been examined is H_2 aligned with an $H-X$ molecule. The substituent effect is related to the charge which develops at $H\alpha$, relative to the case where $X = H$.

$$H-H\alpha\text{------}H-X$$
$$4\text{Å}$$

Substituent constants calculated in this way are in good agreement with empirical σ_F values.[30] The same system was used to calculate σ_R values by determining charge accumulation or depletion on the α and β carbons of substituted ethylenes using the 4-31G method.

$$CH_2=CH-X \leftrightarrow {}^-CH_2-CH=X^+$$

These computational methods provide ground-state substituent effects, but for reactivity relationships the response of substituents to developing charge at the transition state is needed. This is a challenging task because of the uncertainty of the structure at the transition state. In one approach to the problem, the stabilizing (or destabilizing) effect of substituents as a positive charge (representing an electrophile) or a negative charge (representing a nucleophile) approaches a benzene ring was calculated at the STO-3G level.[31] The effect of the substituents, as reflected in the calculated stabilization or destabilization, was parallel to that indicated by linear free-energy correlations. The amino, hydroxy, and fluoro groups, for example, were found to provide extra stabilization to the approach of an electrophile in comparison with other substituents for which a strong resonance interaction would not be expected. For the approach of a negative charge, these substituents were destabilizing, but extra stabilization was found for groups such as nitro, cyano, and sulfonyl. Although detailed calculations have been applied to only a limited number of substituents, it appears that the MO calculations give rise to the same patterns as

28. R. D. Topsom, *Acc. Chem. Res.* **16**:292 (1983); R. D. Topsom, *Prog. Phys. Org. Chem.* **16**:125 (1987).
29. S. Marriott, A. Silvestro, and R. D. Topsom, *J. Chem. Soc., Perkin Trans. 2* **1988**:457.
30. A. Exner, M. Ingr, and P. Carsky, *THEOCHEM* **397**:231 (1997).
31. E. R. Vorpagel, A. Streitwieser, Jr., and S. D. Alexandratos, *J. Am. Chem. Soc.* **103**:3777 (1981).

found on the basis of empirical correlation. The MO calculations support the idea that substituent effects in aromatic compounds are a combination of field effects and resonance effects.[32]

213

SECTION 4.3.
SUBSTITUENT EFFECTS
AND LINEAR
FREE-ENERGY
RELATIONSHIPS

Let us now consider how linear free-energy relationships can provide insight into reaction mechanisms. The choice of benzoic acid ionization as the reference reaction for the Hammett equation leads to $\sigma > 0$ for electron-withdrawing groups and $\sigma < 0$ for electron-releasing groups, since electron-withdrawing groups favor the ionization of the acid and electron-releasing groups have the opposite effect. Further inspection of the Hammett equation shows that ρ will be positive for all reactions that are favored by electron-attracting groups and negative for all reactions that are favored by electron-releasing groups. If the rates of a reaction series show a satisfactory correlation, both the sign and the magnitude of ρ provide information about the transition state for the reaction. In Example 4.3 (p. 208), the ρ value for saponification of substituted methyl benzoates is $+2.38$. This indicates that electron-withdrawing groups facilitate the reaction and that the reaction is somewhat more sensitive to substituent effects than the ionization of benzoic acids. The observation that the reaction is favored by electron-withdrawing substituents is in agreement with the accepted mechanism for ester saponification. The tetrahedral intermediate is negatively charged. Its formation should therefore be favored by electron-withdrawing substituents that can stabilize the developing charge. There is also a ground-state effect working in the same direction. Electron-withdrawing substituents will tend to make the carbonyl group more electrophilic and favor the addition of hydroxide ion.

$$\underset{\text{O}}{\overset{\text{O}}{\underset{\|}{\text{ArCOCH}_3}}} + \ ^-\text{OH} \ \overset{\text{slow}}{\rightleftharpoons} \ \underset{\overset{|}{\text{OH}}}{\overset{\overset{\text{O}^-}{|}}{\text{ArCOCH}_3}} \ \overset{\substack{\text{several} \\ \text{fast steps}}}{\longrightarrow} \ \text{ArCO}_2^- + \text{CH}_3\text{OH}$$

The solvolysis of diarylmethyl chlorides in ethanol shows a ρ value of -5.0, indicating that electron-releasing groups greatly increase the reaction rate, and supporting a mechanism involving ionization as the rate-determining step. Electron-releasing groups can facilitate the ionization by a stabilizing interaction with the electron-deficient carbocation that develops as ionization proceeds.

$$\underset{\overset{|}{\text{H}}}{\overset{\overset{\text{Ph}}{|}}{\text{ArCCl}}} \ \overset{\text{slow}}{\longrightarrow} \ \text{Ar}-\overset{\text{Ph}}{\underset{\text{H}}{\overset{+}{\text{C}}}} + \text{Cl}^- \ \overset{\text{EtOH}}{\underset{\text{fast}}{\longrightarrow}} \ \underset{\overset{|}{\text{H}}}{\overset{\overset{\text{Ph}}{|}}{\text{ArCOC}_2\text{H}_5}} + \text{HCl}$$

The relatively large ρ shows that the reaction is very sensitive to substituent effects and implies that there is a relatively large redistribution of charge in the transition state.

Not all reactions can be fitted by the Hammett equations or the multiparameter variants. There can be several reasons for this. The most common is that the mechanism of the reaction depends on the nature of the substituent. In a multistep reaction, for example, one step may be rate-determining in the case of electron-withdrawing substituents, but a different step may become rate-limiting when the substituent is electron-releasing. The rate of semicarbazone formation of benzaldehydes, for example, shows a nonlinear Hammett

32. H. Agren and P. S. Bagus, *J. Am. Chem. Soc.* **107**:134 (1985); R. D. Topsom, *Acc. Chem. Res.* **16**:292 (1983); W. F. Reynolds, P. Dais, D. W. MacIntyre, R. D. Topsom, S. Marriott, E. v. Nagy-Felsobuki, and R. W. Taft, *J. Am. Chem. Soc.* **105**:378 (1983); A. Pross and L. Radom, *Prog. Phys. Org. Chem.* **13**:1 (1980).

214

CHAPTER 4
STUDY AND
DESCRIPTION OF
ORGANIC REACTION
MECHANISMS

plot with ρ of about 3.5 for electron-releasing groups, but ρ near -0.25 for electron-withdrawing groups.[33] The change in ρ is believed to be the result of a change in the rate-limiting step.

$$\underset{\text{(rate-controlling for electron-releasing substituents)}}{ArCH{=}O + NH_2NHCNH_2 \longrightarrow ArCHNHNHCNH_2}$$

$$\underset{\text{(rate-controlling for electron-withdrawing substituents)}}{ArCHNHNHCNH_2 \longrightarrow ArCH{=}NNHCNH_2 + H_2O}$$

Any reaction which shows a major shift in transition-state structure over the substituent series would be expected to give a nonlinear Hammett plot, since a variation in the extent of resonance participation would then be expected.

By comparing σ^+, σ^-, and σ_I, individual substituents can be separated into three groups as in Table 4.7. Alkyl groups are electron-releasing by both resonance and polar effects. Substituents such as alkoxy, hydroxy, and amino, which can act as resonance donors, have negative σ_p and σ^+ values, but when polar effects are dominant, these substituents act as electron-attracting groups, as illustrated by the σ_m and σ_I values. A third group of substituents are electron-withdrawing by both resonance and polar interactions. These include carbonyl groups in aldehydes, ketones, esters, and amides, as well as cyano, nitro, and sulfonyl substituents.

The magnitude of substituent effects differs in solution and gas phase. In general, substituent effects are much stronger in the gas phase. This is because there are no "leveling effects" due to solvation. For example, in the ionization of benzoic acids, the substituent effects in terms of ΔH are about 11 times larger in the gas phase than in the aqueous phase.[34] The relative importance of direct resonance interactions also seems to be greater in aqueous solution. For example, the σ_p value of NH_2 increases in negative magnitude from -0.017 in the gas phase to -0.39 in benzene and -0.66 in water.

The development of linear free-energy relationships in aliphatic molecules is complicated because steric and conformational factors come into play along with

Table 4.7. Classification of Substituent Groups

Resonance: Electron-releasing $(+M)$	Electron-releasing $(+M)$		Electron-withdrawing $(-M)$
Field: Electron-releasing $(+I)$	Electron-withdrawing $(-I)$		Electron-withdrawing $(-I)$
Me	AcNH	Br	Ac
Et	AcO	Cl	CN
$(Me)_3C$	NH_2	F	NO_2
		OH	CF_3
		MeO	$(Me)_3N^-$
		EtO	
		Ph	

33. D. S. Noyce, A. T. Bottini, and S. G. Smith, *J. Org. Chem.* **23**:752 (1958).
34. R. W. Taft and R. D. Topsom, *Prog. Phys. Org. Chem.* **16**:1 (1987); C. Hansch, A. Leo, and R. W. Taft, *Chem. Rev.* **91**:165 (1991).

electronic effects. A number of successful treatments of aliphatic systems have been developed by separating electronic effects from steric effects. We will not discuss these methods, but there are reviews available which can be consulted for information about this area.[35]

215

SECTION 4.4.
BASIC MECHANISTIC
CONCEPTS: KINETIC
VERSUS
THERMODYNAMIC
CONTROL,
HAMMOND'S
POSTULATE, THE
CURTIN–HAMMETT
PRINCIPLE

4.4. Basic Mechanistic Concepts: Kinetic versus Thermodynamic Control, Hammond's Postulate, the Curtin–Hammett Principle

The use of two-dimensional reaction progress/energy diagrams was introduced in Section 4.2. Two-dimensional potential energy diagrams can provide insight into the important general concepts considered in this section. There are many organic reactions in which the energy requirements for competing reaction paths are rather similar. It is important to be able to analyze the factors that may permit a particular reaction path to dominate. The key issue is the relative activation energies of competing pathways because they determine the outcome of the reaction.

4.4.1. Kinetic versus Thermodynamic Control

Product composition may be governed by the equilibrium thermodynamics of the system. When this is true, the product composition is governed by *thermodynamic control*. Alternatively, product composition may be governed by competing rates of formation of products. This is called *kinetic control*.

Let us consider cases 1–3 in Fig. 4.4. In case 1, $\Delta G^{\ddagger}$'s for formation of the competing transition states A^* and B^* from the reactant R are much less than $\Delta G^{\ddagger}$'s for formation of A'^* and B'^* from A and B, respectively. If the latter two $\Delta G^{\ddagger}$'s are sufficiently large that the competitively formed products B and A do not return to R, the ratio of the products A and B at the end of the reaction will not depend on their relative stabilities, but only on their relative rates of formation. The formation of A and B is effectively irreversible in these circumstances. The reaction energy plot in case 1 corresponds to this situation and represents a case of *kinetic control*. The relative amounts of products A and B will depend on the heights of the activation barriers $\Delta G_A^{\ddagger}$ and $G_B^{\ddagger}$, not the relative stability of products A and B.

In case 2, the lowest $\Delta G^{\ddagger}$ is that for formation of A^* from R, but the $\Delta G^{\ddagger}$ for formation of B^* from A is not much larger. System 2 might be governed by either kinetic or thermodynamic factors. Conversion of R to A will be only slightly more rapid than conversion of A to B. If the reaction conditions are carefully adjusted, it will be possible for A to accumulate and not proceed to B. Under such conditions, A will be the dominant product and the reaction will be under kinetic control. Under somewhat more energetic conditions, for example, at a higher temperature, A will be transformed to B, and under these conditions the reaction will be under thermodynamic control. A and B will equilibrate, and the product ratio will depend on the equilibrium constant determined by ΔG.

In case 3, the barrier separating A and B is small relative to that for formation of A^*

35. J. Hine, *Physical Organic Chemistry*, McGraw-Hill, New York, 1962, pp. 95–98; P. R. Wells, *Linear Free Energy Relationships*, Academic Press, New York, 1968, pp. 35–44; M. Charton, *Prog. Phys. Org. Chem.* **10:** 81 (1973); S. Ehrenson, R. T. C. Brownlee, and R. W. Taft, *Prog. Phys. Org. Chem.* **10:**1 (1973).

216

CHAPTER 4
STUDY AND
DESCRIPTION OF
ORGANIC REACTION
MECHANISMS

Fig. 4.4. Kinetic versus thermodynamic control.

with product **B** (which is thermodynamically less stable) being formed because $\Delta G^{\ddagger}\mathbf{B}^* < \Delta G^{\ddagger}\mathbf{A}^*$. The dashed profile represents the reaction under a different set of conditions, such as the addition of a catalyst or change of solvent, that substantially reduces the energy of $\mathbf{A}^*$ and $\mathbf{B}^*$, such that interconversion of **A** and **B** is rapid. This will result in the formation of the more stable product **A**, even though the barrier to formation of **B** remains lower.

The idea of kinetic versus thermodynamic control can be illustrated by discussing briefly the case of formation of enolate anions from unsymmetrical ketones. This is a very important matter for synthesis and will be discussed more fully in Chapter 1 of Part B. Most ketones, highly symmetric ones being the exception, can give rise to more than one enolate. Many studies have shown that the ratio among the possible enolates that are formed depends on the reaction conditions.[36] This can be illustrated for the case of 3-methyl-2-butanone. If the base chosen is a strong, sterically hindered one and the solvent is aprotic, the major enolate formed is **3**. If a protic solvent is used or if a weaker base (one comparable in basicity to the ketone enolate) is used, the dominant enolate is **2**. Enolate **3** is the "kinetic enolate" whereas **2** is the thermodynamically favored enolate.

$$
\underset{\mathbf{2}}{\underset{H_3C}{^{-O}}\!\!\diagdown\!\!\underset{}{C}\!=\!C(CH_3)_2}
\quad\xleftarrow{\;B^-\;}\quad
\underset{\mathbf{1}}{\overset{\overset{O}{\|}}{CH_3CCH(CH_3)_2}}
\quad\xrightarrow{\;B^-\;}\quad
\underset{\mathbf{3}}{H_2C\!=\!\underset{CH(CH_3)_2}{\overset{O^-}{C}}}
$$

The structural and mechanistic basis for the relationships between kinetic versus thermodynamic control and the reaction conditions is as follows. The α hydrogens of the methyl group are sterically less hindered than the α hydrogen of the isopropyl group. As a result, removal of one of these hydrogens as a proton is faster than removal of the isopropyl hydrogen as a proton. This effect is magnified when the base is sterically hindered so that it is particularly sensitive to the difference in the steric situation of the competing hydrogens. If the base is very strong, the enolate will not be reconverted to the ketone because the enolate will be too weak a base to regain the proton. These conditions correspond to case 1 in Fig. 4.4 and represent a case of *kinetic control*. If a weaker base is used or if the solvent is protic, protons can be transferred reversibly between the isomeric enolates and the base

36. J. d'Angelo, *Tetrahedron* **32**:2979 (1976); H. O. House, *Modern Synthetic Reactions*, 2nd ed., W. A. Benjamin, Menlo Park, California, 1972.

(because the base strengths of the enolate and the base are comparable). Under these conditions, the more stable enolate will be dominant because the enolates are in equilibrium. The more substituted enolate **2** is the more stable of the pair, just as more substituted alkenes are more stable than terminal alkenes. This corresponds to case 3 in Fig. 4.4 where product (enolate) equilibration occurs at a rapid rate.

217

SECTION 4.4.
BASIC MECHANISTIC
CONCEPTS: KINETIC
VERSUS
THERMODYNAMIC
CONTROL,
HAMMOND'S
POSTULATE, THE
CURTIN–HAMMETT
PRINCIPLE

4.4.2. Hammond's Postulate

Because the rates of chemical reactions are controlled by the free energy of the transition state, information about the structure of transition states is crucial to understanding reaction mechanism. However, because transition states have only transitory existence, it is not possible to make experimental measurements that provide direct information about their structure.. Hammond has discussed the circumstances under which it is valid to relate transition-state structure to the structure of reactants, intermediates, and products.[37] His statements concerning transition-state structure are known as *Hammond's postulate*. Discussing individual steps in a reaction mechanism, Hammond's postulate states "if two states, as, for example, a transition state and an unstable intermediate, occur consecutively during a reaction process and have nearly the same energy content, their interconversion will involve only a small reorganization of molecular structure."

This statement can be discussed with reference to potential energy diagrams. Case 1 in Fig. 4.5 represents a highly exothermic step with a low activation energy. It follows from Hammond's postulate that, in this step, the transition state will structurally resemble the reactant because they are close in energy and interconverted by a small structural change. This is depicted in the potential energy diagram as a small displacement toward product along the reaction coordinate. Case 2 describes a step in which the transition state is a good deal higher in energy than either the reactant or the product. In this case, neither the reactant nor the product will be a good model of the transition state. Case 3 illustrates an endothermic step such as might occur in the formation of an unstable intermediate. In this

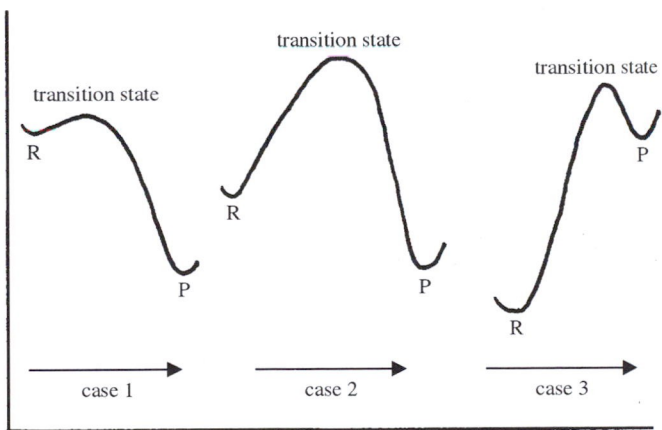

Fig. 4.5. Some typical potential energy diagrams that illustrate the application of Hammond's postulate.

37. G. S. Hammond, *J. Am. Chem. Soc.* **77**:334 (1955).

218

CHAPTER 4
STUDY AND
DESCRIPTION OF
ORGANIC REACTION
MECHANISMS

case, the energy of the transition state is similar to that of the intermediate, and the transition state should be similar in structure to the intermediate.

The significance of the concept incorporated in Hammond's postulate is that, in appropriate cases, it permits discussion of transition-state structure in terms of the reactants, intermediates, or products in a multistep reaction sequence. The postulate indicates that the cases in which such comparison is appropriate are those in which the transition state is close in energy to the reactant, intermediate, or product. Chemists sometimes speak of "early" or "late" transition states. An "early" transition state is reactant-like whereas a "late" transition state is product-like.

The case of electrophilic aromatic substitution can illustrate a situation in which it is useful to discuss transition-state structure in terms of a reaction intermediate. The *ortho–para-* and *meta*-directing effects of aromatic substituents were among the first structure–reactivity relationships to be developed in organic chemistry. Certain functional groups were found to activate aromatic rings toward substitution and to direct the entering electrophile to the *ortho* and *para* positions, whereas others were deactivating and led to substitution in the *meta* position. The bromination of methoxybenzene, benzene, and nitrobenzene can serve as examples for discussion.

It can be demonstrated that the reactions are kinetically controlled. It is therefore the $\Delta G^{\ddagger}$ value that holds the key to the connection between the rate effects and the substituent's directing effects. However, to discuss the effect of substituents on $\Delta G^{\ddagger}$, we must know something about the reaction mechanism and the nature of the competing transition states. Electrophilic aromatic substitution will be discussed in detail in Chapter 10. Evidence presented there will indicate that electrophilic aromatic substitution involves a distinct intermediate and two less well defined states. The potential energy diagram in Fig. 4.6 is believed to be a good representation of the energy changes that occur during bromination. By application of the Hammond postulate, we can conclude that the rate-determining step involves formation of a transition state that should closely resemble the intermediate σ complex. It is therefore legitimate to discuss the effect of substitutents on the transition state in terms of the structure of this intermediate.

Because the product composition is kinetically controlled, the isomer ratio will be governed by the relative magnitudes of $\Delta G_o^{\ddagger}$, $\Delta G_m^{\ddagger}$, and $\Delta G_p^{\ddagger}$, the energies of activation for the *ortho*, *meta*, and *para* transition states, respectively. In Fig. 4.7 a qualitative comparison of these $\Delta G^{\ddagger}$ values is made. At the transition state, a positive charge is present on the benzene ring, primarily at positions 2, 4, and 6 in relation to the entering bromine.

The electron-releasing methoxy group can interact directly to delocalize the charge and stabilize the intermediates leading to *o*- and *p*-bromomethoxybenzene. It cannot stabilize

219

SECTION 4.4.
BASIC MECHANISTIC
CONCEPTS: KINETIC
VERSUS
THERMODYNAMIC
CONTROL,
HAMMOND'S
POSTULATE, THE
CURTIN–HAMMETT
PRINCIPLE

Fig. 4.6. Potential energy diagram for electrophilic aromatic substitution.

the intermediate leading to *m*-bromomethoxybenzene.

The *o*- and *p*-intermediates are therefore stabilized relative to the intermediate formed from benzene but the *m*-intermediate is not, as is illustrated in Fig. 4.7. As a result, methoxybenzene reacts faster than benzene, and the products are mainly the *ortho*- and *para*-isomers.

In the case of nitrobenzene, the electron-withdrawing nitro group is not able to stabilize the positive charge in the σ-complex intermediate. In fact, it strongly destabilizes the intermediate. This destabilization is greatest in the *o*- and *p*-intermediates, which place positive charge on the nitrosubstituted carbon. The *meta* transition state is also destabilized relative to that for benzene, but not as much as the *ortho* and *para* transition states. As a result, nitrobenzene is less reactive than benzene, and the product is mainly the *meta* isomer.

The substituent effects in aromatic electrophilic substitution are dominated by resonance effects. In other systems, stereoelectronic effects or steric effects might be more important. Whatever the nature of the substituent effects, the Hammond postulate insists that structural discussion of transition states in terms of reactants, intermediates, or products is valid only when their structures and energies are similar.

220

CHAPTER 4
STUDY AND
DESCRIPTION OF
ORGANIC REACTION
MECHANISMS

Fig. 4.7. Transition-state energies in bromination of methoxybenzene, benzene, and nitrobenzene.

4.4.3. The Curtin–Hammett Principle

In Chapter 3, equilibria among conformers of organic molecules were discussed. At this point, let us consider in a general way the effect that conformational equilibria can have on a chemical reaction. Under what circumstances can the position of the conformational equilibrium for a reactant determine which of two competing reaction paths will be followed? A potential energy diagram is shown in Fig. 4.8. In most cases, the energy of activation for a chemical reaction will be greater than that for a conformational equilibration, as is illustrated in the figure. If this is the case, $\Delta G_a^{\ddagger}$ and $\Delta G_b^{\ddagger} \gg \Delta G_c$. The conformers of the reactant are in equilibrium and are interconverted at a rate much faster than that at which the competing reactions occur.

$$\mathbf{B} \rightleftharpoons \mathbf{A} \qquad K_c = \frac{[\mathbf{A}]}{[\mathbf{B}]}$$

$$\text{rate of formation of product } P_{\mathrm{A}} = \frac{dP_{\mathrm{A}}}{dt} = k_a[\mathrm{A}] = k_a K_c[\mathrm{B}]$$

$$\text{rate of formation of product } P_{\mathrm{B}} = \frac{dP_{\mathrm{B}}}{dt} = k_b[\mathrm{B}]$$

$$\text{product ratio} = \frac{dP_{\mathrm{A}}/dt}{dP_{\mathrm{B}}/dt} = \frac{k_a K_c[\mathrm{B}]}{k_b[\mathrm{B}]} = \frac{k_a K_c}{k_b}$$

221

SECTION 4.4.
BASIC MECHANISTIC
CONCEPTS: KINETIC
VERSUS
THERMODYNAMIC
CONTROL,
HAMMOND'S
POSTULATE, THE
CURTIN–HAMMETT
PRINCIPLE

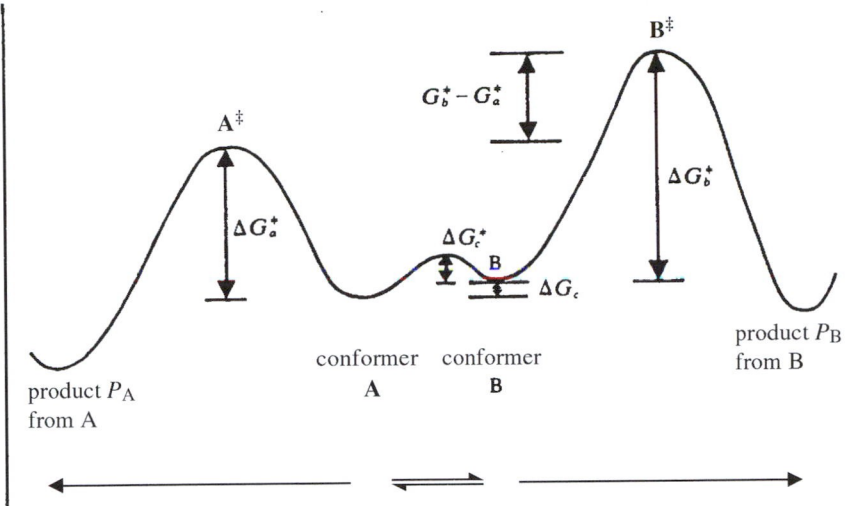

Fig. 4.8. Effect of conformation on product distribution.

According to transition-state theory,

$$k_r = \frac{\kappa kT}{h} e^{-\Delta G^{\ddagger}/RT} \quad \text{and} \quad K_c = e^{-(-\Delta G_c)/RT}$$

$$\text{product ratio} = \frac{(\kappa kt/h) e^{-\Delta G_a^{\ddagger}/RT} e^{+\Delta G_c/RT}}{(\kappa kT/h) e^{-\Delta G_b^{\ddagger}/RT}}$$

$$= e^{(-\Delta G_a^{\ddagger} = \Delta G_b^{\ddagger} + \Delta G_c)/RT}$$

But from Fig. 4.8,

$$\Delta G_b^{\ddagger} - \Delta G_a^{\ddagger} + \Delta G_c = G_b^{\ddagger} - G_a^{\ddagger}$$

The product ratio is therefore determined not by ΔG_c but by the relative energy of the two transition states **A*** and **B***. Although the rate of the formation of the products is dependent upon the relative concentration of the two conformers, since $\Delta G_b^{\ddagger}$ is decreased relative to $\Delta G_a^{\ddagger}$ to the extent of the difference in the two conformational energies, the conformational preequilibrium is established rapidly, relative to the two competing product-forming steps.[38] The position of the conformational equilibrium cannot control the product ratio. The reaction may proceed through a minor conformation if it is the one that provides access to the lowest-energy transition state. The conclusion that the ratio of products formed from conformational isomers is not determined by the conformation population ratio is known as the *Curtin–Hammett principle*.[39]

38. For a more complete discussion of the relationship between conformational equilibria and reactivity, see J. I. Seeman, *Chem. Rev.* **83**:83 (1983).

39. D. Y. Curtin, *Rec. Chem. Prog.* **15**:111 (1954); E. L. Eliel, *Stereochemistry of Carbon Compounds*, McGraw-Hill, New York, 1962, pp. 151–152, 237–238.

222

CHAPTER 4
STUDY AND
DESCRIPTION OF
ORGANIC REACTION
MECHANISMS

The same arguments can be applied to other energetically facile interconversions of two potential reactants. For example, many organic molecules undergo rapid proton shifts (tautomerism), and the chemical reactivity of the two isomers may be quite different. It is not valid, however, to deduce the ratio of two tautomers on the basis of subsequent reactions that have activation energies greater than that of the tautomerism. Just as in the case of conformational isomerism, the ratio of products formed in subsequent reactions will not be controlled by the position of the facile equilibrium.

4.5. Isotope Effects

A special type of substituent effect which has proved very valuable in the study of reaction mechanisms is the replacement of an atom by one of its isotopes. Isotopic substitution most often involves replacing protium by deuterium (or tritium) but is applicable to nuclei other than hydrogen. The quantitative differences are largest, however, for hydrogen, because its isotopes have the largest relative mass differences. Isotopic substitution usually has no effect on the qualitative chemical reactivity of the substrate, but often has an easily measured effect on the rate at which reaction occurs. Let us consider how this modification of the rate arises. Initially, the discussion will concern *primary kinetic isotope effects*, those in which a bond to the isotopically substituted atom is broken in the rate-determining step. We will use C−H bonds as the specific topic of discussion, but the same concepts apply for other elements.

Any C−H bond has characteristic vibrations which impart some energy to the molecule in its normal state. This energy is called the *zero-point* energy. The energy associated with these vibrations is related to the mass of the vibrating atoms. Because of the greater mass of deuterium, the vibrations associated with a C−D bond contribute less to the zero-point energy than those associated with the corresponding C−H bond. For this reason, substitution of protium by deuterium lowers the zero-point energy of a molecule. For a reaction involving cleavage of a bond to hydrogen (or deuterium), a vibrational degree of freedom in the normal molecule is converted to a translational degree of freedom as the bond is broken. The energy difference due to this vibration disappears at the transition state. The transition state has the same energy for the protonated and deuterated species. Because the deuterated molecule has the lower zero-point energy, it has a higher activation energy to reach the transition state, as illustrated in Fig. 4.9.

Just how large the rate difference is depends on the nature of the transition state. The maximum effect occurs when the hydrogen being transferred is bound about equally to two other atoms at the transition state. The calculated maximum for the isotope effect k_H/k_D involving C−H bonds is about 7 at room temperature.[40] When bond breaking is more or less than half complete at the transition state, the isotope effect is smaller and can be close to 1 if the transition state is very reactant-like or very product-like. Primary isotope effects can provide two very useful pieces of information about a reaction mechanism. First, the existence of a substantial isotope effect—i.e., if k_H/k_D is 2 or more—is strong evidence that the bond to the substituted hydrogen atom is being broken in the rate-determining step. Second, the magnitude of the isotope effect provides a qualitative indication of where the transition state lies with regard to product and reactant. A relatively low primary isotope effect implies that the bond to hydrogen is either only slightly or nearly completely

40. K. B. Wiberg, *Chem. Rev.* **55**:713 (1955); F. H. Westheimer, *Chem. Rev.* **61**:265 (1961).

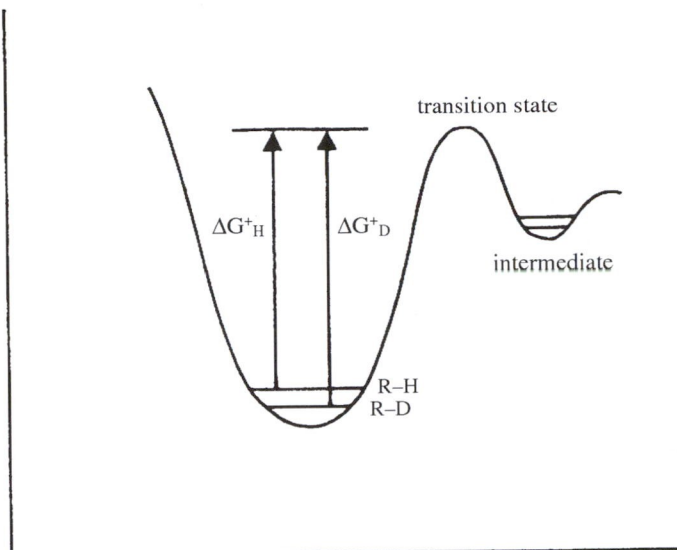

Fig. 4.9. Differing zero-point energies of protium- and deuterium-substituted molecules as the cause of primary kinetic isotope effects.

broken at the transition state. That is, the transition state must occur quite close to reactant or to product. An isotope effect near the theoretical maximum is good evidence that the transition state involves strong bonding of the hydrogen to both its new and its old bonding partners.

Isotope effects may also be observed when the substituted hydrogen atom is not directly involved in the reaction. Such effects are called *secondary kinetic isotope effects*. Secondary isotope effects are smaller than primary ones and are usually in the range of $k_H/k_D = 0.7-1.5$. Secondary isotope effects may be normal ($k_H/k_D > 1$) or inverse ($k_H/k_D < 1$). They are also classified as α or β, etc., depending on the location of the isotopic substitution with respect to the reacting carbon. Secondary isotope effects result from a tightening or loosening of a C—H bond at the transition state. The strength of the bond may change because of a hybridization change or a change in the extent of hyperconjugation, for example. If sp^3-hybridized carbon is converted to sp^2 as reaction occurs, a hydrogen bound to the carbon will experience decreased resistance to C—H bending. The freeing of the vibration for a C—H bond is greater that that for a C—D bond because the C—H bond is slightly longer and the vibration therefore has a larger amplitude. This will result in a normal isotope effect. Entry 5 in Scheme 4.2 is an example of such a reaction since it proceeds through a carbocation intermediate.

An inverse isotope effect will occur if coordination at the reaction center increases in the transition state. The bending vibration will become more restricted. Entry 4 in Scheme 4.2 exemplifies such a case involving conversion of a tricoordinate carbonyl group to a tetravalent cyanohydrin. In this case the secondary isotope effect is 0.73.

Secondary isotope effects at the β position have been especially thoroughly studied in nucleophilic substitution reactions. When carbocations are involved as intermediates, substantial β-isotope effects are observed. This is because the hyperconjugative stabiliza-

224

CHAPTER 4
STUDY AND
DESCRIPTION OF
ORGANIC REACTION
MECHANISMS

Scheme 4.2. Some Representative Kinetic Isotope Effects

Reaction	$k_H/k_D(°C)^a$
A. Primary kinetic isotope effects	
1^b $PhCH_2-H^* + Br· \longrightarrow Ph-CH_2· + H^*-Br$	4.6 (77)
2^c $(CH_3)_2C-\overset{O}{\overset{\|}{C}}-C(CH_3)_2 + OH^- \longrightarrow (CH_3)_2C-\overset{O^-}{\overset{\|}{C}}=C(CH_3)_2$ (with H^*, H^*, H^*)	6.1 (25)
3^d	4.0 (191)
B. Secondary kinetic isotope effects	
4^e	0.73 (25)
5^f	1.30 (25)
6^g	1.37 (50)

a. Temperature of measurement is indicated in parentheses.
b. K. W. Wiberg and L. H. Slaugh, *J. Am. Chem. Soc.* **80**:3033 (1958).
c. R. A. Lynch, S. P. Vincenti, Y. T. Lin, L. D. Smucker, and S. C. Subba Rao, *J. Am. Chem. Soc.* **94**:8351 (1972).
d. W. H. Saunders, Jr., and T. A. Ashe, *J. Am. Chem. Soc.* **91**:473 (1969).
e. L. do Amaral, H. G. Bull, and E. H. Cordes, *J. Am. Chem. Soc.* **94**:7579 (1972).
f. V. J. Shiner, Jr., M. W. Rapp, and H. R. Pinnick, Jr., *J. Am. Chem. Soc.* **92**:232 (1970).
g. M. Taagepera and E. R. Thornton, *J. Am. Chem. Soc.* **94**:1168 1972).

tion by the β hydrogens weakens the C—H bond.[41] The observed secondary isotope effects are normal, as would be predicted since the bond is weakened.

Detailed analysis of isotope effects reveals that there are many other factors that can contribute to the overall effect in addition to the dominant change in bond vibrations. For that reason, it is not possible to quantitatively predict the magnitude of either primary or seconday isotope effects for a given reaction. Furthermore, there is not a sharp numerical division between primary and secondary effects, especially in the range between 1 and 2.

41. V. J. Shiner, W. E. Buddenbaum, B. L. Murr, and G. Lamaty, *J. Am. Chem. Soc.* **90**:809 (1968); A. J. Kresge and R. J. Preto, *J. Am. Chem. Soc.* **89**:5510 (1967); G. J. Karabatsos, G. C. Sonnichsen, C. G. Papaioannou, S. E. Scheppele, and R. L. Shone, *J. Am. Chem. Soc.* **89**:463 (1967); D. D. Sunko and W. J. Hehre, *Prog. Phys. Org. Chem.* **14**:205 (1983).

For these reasons, isotope effects are usually used in conjunction with other criteria in the description of reaction mechanisms.[42]

4.6. Isotopes in Labeling Experiments

A quite different use of isotopes in mechanistic studies is their use as labels for ascertaining the location of a given atom involved in a reaction. As in kinetic experiments, the substitution of an isotope will not qualitatively affect the course of the reaction. The nuclei most commonly used for isotopic tracer experiments in organic chemistry are deuterium, tritium, and the ^{13}C and ^{14}C isotopes of carbon. There are several means of locating isotopic labels. Deuterium can frequently be located by analysis of NMR spectra. In contrast to the normal ^1H isotope, deuterium does not show an NMR signal under the usual operating circumstances. The absence of a specific signal can therefore be used to locate deuterium. Both mass spectrometry and IR spectroscopy also can be used to locate deuterium. Tritium and ^{14}C and other radioactive isotopes can be detected on the basis of the radioactivity. This is a very sensitive method. In most experiments in which radioactive labels are used, only a small fraction of the atoms at the site of substitution are the radioactive nuclide. The location of ^{14}C requires a degradative process to separate the atoms that might conceivably be labeled. Carbon-13 has become an important isotope for tracer experiments relatively recently. Unlike ^{12}C, ^{13}C has a nuclear magnetic moment and can be detected in NMR spectrometers. The appearance of strongly enhanced ^{13}C resonances permits assignment of the labeled postion(s). This method avoids the necessity of developing a degradative scheme to separate specific carbon atoms.

There are many excellent examples of experiments using isotopic labeling in both organic chemistry and biochemistry.[43] An interesting example is the case of hydroxylation of the amino acid phenylalanine which is carried out by the enzyme phenylalanine hydroxylase.

This reaction was studied by use of tritium. The phenylalanine was labeled with tritium at the 4-position of the phenyl ring. When the product, tyrosine, was isolated, it retained much of the original radioactivity, even though the 4-position was now substituted by a

---

42. For more complete discussion of isotope effects, see W. H. Saunders, in *Investigation of Rates and Mechanisms of Reactions*, E. S. Lewis, ed., *Techniques of Chemistry*, 3rd ed., Vol. VI, Part 1, John Wiley & Sons, New York, 1974, pp. 211–255; L. Melander and W. H. Saunders, Jr., *Reaction Rates of Isotopic Molecules*, John Wiley & Sons, New York, 1980; W. H. Saunders, in *Investigation of Rates and Mechanisms of Reactions*, C. F. Bernasconi, ed., *Techniques of Chemistry*, 4th ed., Vol. VI, Part 1, John Wiley & Sons, New York, 1986, Chapter VIII.

43. For other examples of use of isotopic labels in mechanistic studies, see V. F. Raaen, in *Investigation of Rates and Mechanisms of Reactions*, E. S. Lewis, ed., *Techniques of Chemistry*, 3rd ed., Vol. VI, Part 1, John Wiley & Sons, New York, 1974, pp. 257–284; E. Buncel and C. C. Lee, eds., *Isotopes in Organic Chemistry*, Vols. 1–4, Elsevier, New York, 1975–1978; C. Wentrup, in *Investigation of Rates and Mechanisms of Reactions*, C. F. Bernasconi, ed., *Techniques of Chemistry*, 4th ed., Vol. VI, Part 1, John Wiley & Sons, New York, 1986, Chapter IX.

226

CHAPTER 4
STUDY AND
DESCRIPTION OF
ORGANIC REACTION
MECHANISMS

hydroxyl group. When this result was studied in detail, it was found that the ^{3}H originally at the 4-position had rearranged to the 3-position in the course of oxidation. This hydrogen shift, called the *NIH shift*,[44] has subsequently been found to occur in many biological oxidations of aromatic compounds. The oxidation occurs via an epoxide intermediate.

4.7. Characterization of Reaction Intermediates

Identification of the intermediates in a multistep reaction is a major objective of studies of reaction mechanisms. When the nature of each intermediate is fairly well understood, a great deal is known about the reaction mechanism. The amount of an intermediate present in a reacting system at any instant of time will depend on the rates of the steps by which it is formed and the rate of its subsequent reaction. A qualitative indication of the relationship between intermediate concentration and the kinetics of the reaction can be gained by considering a simple two-step reaction mechanism:

$$\text{reactants} \xrightarrow{k_1} \text{intermediate} \xrightarrow{k_2} \text{products}$$

In some reactions, the situation $k_1 > k_2$ exists. Under these conditions, the concentration of the intermediate will build up as it goes on more slowly to product. The possibility of isolating, or at least observing, the intermediate then exists. If both k_1 and k_2 are large, the reaction may proceed too rapidly to permit isolation of the intermediate but spectroscopic studies, for example, should reveal the existence of two distinct phases for the overall reaction. It should be possible to analyze such a system and determine the two rate constants.

If the two steps are of about equal rates, only a small concentration of the intermediate will exist at any time. It is sometimes possible to interrupt such a reaction by lowering the temperature rapidly or adding a reagent that stops the reaction and isolate the intermediate. Intermediates can also be "trapped." A compound which is expected to react specifically with the intermediate is added to the reaction system. If trapping occurs, the intermediate is diverted from its normal course, and evidence for the existence of the intermediate is obtained if the structure of the trapped product is consistent with expectation.

Often, it is more practical to study intermediates present in low concentration by spectroscopic methods. The most common methods in organic chemistry include

44. From its discovery at the National Institutes of Health (NIH); for an account of this discovery, see G. Guroff, J. W. Daly, D. M. Jerina, J. Renson, B. Witkop, and S. Udenfriend, *Science* **157**:1524 (1967).

ultraviolet-visible (UV-VIS), infrared (IR), nuclear magnetic resonance (NMR), and electron spin resonance (ESR) spectroscopy. UV-VIS spectrometers can scan the electronic region of the spectrum and provide evidence of the development of characteristic *chromophores*. The major limitation imposed is that the compound to be detected must have a characteristic absorbance in the range 220–700 nm. The region corresponds to energy levels associated with promotion of electrons to higher-energy states. In organic molecules, absorbance in this region usually requires the presence of multiple bonds, especially two or more such bonds in conjugation. Saturated molecules normally do not absorb significantly in this region. The amount of an intermediate that can be detected depends on how strongly it absorbs relative to other components of the reaction systems. In favorable cases, concentrations as low as $10^{-6}\ M$ can be detected.

IR spectrometers measure absorption of energy by excitation of molecular vibrations, including stretching, bending, and twisting of various functional groups. Most organic molecules have a large number of bands in the IR spectrum. Although it is usually not possible to assign all bands to specific vibrations, individual bands can be highly characteristic of a specific molecule. Nearly all of the organic functional groups also have one or more regions of characteristic absorption. If it is suspected that a particular functional group is present in an intermediate, examination of the changes of the spectrum in the characteristic region may permit detection. An example of IR detection of intermediates can be drawn from a study of the photochemical conversion of **1** to **3**. It was suspected that the ketene **2** might be an intermediate.[45]

Ketenes absorb near 2100–2130 cm^{-1}. When the photolysis was carried out and the IR spectrum of the solution monitored, it was found that a band appeared at 2118 cm^{-1}, grew, and then decreased as photolysis proceeded. The observation of this characteristic absorption constitutes good evidence for a ketene intermediate. As with UV-VIS spectroscopy, the amount of intermediate that can be detected depends both on the intensity of the absorption band and the presence of interfering bands. In general, IR spectroscopy requires somewhat higher concentration for detection than does UV-VIS spectroscopy.

Either UV-VIS or IR spectroscopy can be combined with the technique of *matrix isolation* to detect and identify highly unstable intermediates. In this method, the intermediate is trapped in a solid inert matrix, usually one of the inert gases, at very low temperatures. Because each molecule is surrounded by inert gas atoms, there is no possiblity for intermolecular reactions and the rates of intramolecular reactions are slowed by the low temperature. Matrix isolation is a very useful method for characterizing intermediates in photochemical reactions. The method can also be used for gas-phase reactions which can be conducted in such a way that the intermediates can be rapidly condensed into the matrix.

NMR spectroscopy is very widely used for detection of intermediates in organic reactions. Proton magnetic resonance is most useful because it provides the greatest

45. O. L. Chapman and J. D. Lassila, *J. Am. Chem. Soc.* **90**:2449 (1968).

228

CHAPTER 4
STUDY AND
DESCRIPTION OF
ORGANIC REACTION
MECHANISMS

sensitivity of the nuclei of interest in organic chemistry. Fluorine-19 and phosphorus-31 are other nuclei that provide high sensitivity. Carbon-13, oxygen-17, and nitrogen-15 have relatively lower sensitivity, but the development of high-field instruments and the use of Fourier transform methods have greatly increased sensitivity so that NMR can be used to detect characteristic signals of reaction intermediates.

Free radicals and other intermediates with unpaired electrons can be detected in extremely low concentration by electron paramagnetic resonance (EPR). This technique measures the energy absorbed to reorient an electronic spin in a magnetic field. It provides structural information on the basis of splitting of the signal by adjacent nuclei, much as in NMR interpretation. EPR is not only extremely sensitive but also very specific. Diamagnetic molecules present in solution give no signals, and the possibility for interference is therefore greatly decreased. The method can only be applied to reactions involving paramagnetic intermediates. However, because the method is so sensitive, it is important to demonstrate that any paramagnetic species that are detected are true intermediates, rather than being involved only in minor pathways.

All other spectroscopic methods are applicable, in principle, to the detection of reaction intermediates so long as the method provides sufficient structural information to assist in the identification of the transient species. In the use of all methods, including those discussed above, it must be remembered that simple detection of a species does not prove that it is an intermediate. It also must be shown that the species is converted to product. In favorable cases, this may be done by isolation or trapping experiments. More often, it may be necessary to determine the kinetic behavior of the appearance and disappearance of the intermediate and demonstrate that this behavior is consistent with the species being an intermediate.

4.8. Catalysis by Acids and Bases

A detailed understanding of reaction mechanisms requires knowledge of the role that catalysts play in the reaction. Catalysts do not affect the position of equilibrium of a reaction. They function by increasing the rate of one or more steps in a reaction mechanism by providing a reaction path having a lower activation energy. The most general family of catalytic processes is those that involve transfer of a proton. Many reactions are strongly catalyzed by proton donors (Brønsted acids) or proton acceptors (Brønsted bases). Catalysis occurs when the conjugate base or conjugate acid of the substrate is more reactive than the neutral species. For example, reactions involving addition of neutral nucleophiles at carbonyl groups are often accelerated by acids. This type of catalysis occurs because the conjugate acid of the carbonyl compound is much more electrophilic than the neutral molecule.

$$
\underset{\substack{\| \\ RCR}}{\overset{O}{}} + H^+ \; \rightleftharpoons \; \underset{\substack{\| \\ RCR}}{\overset{+OH}{}}
$$

$$
\underset{\substack{\| \\ RCR}}{\overset{+OH}{}} + \mathbf{Nu:} \longrightarrow R-\underset{\substack{| \\ +\mathbf{Nu}}}{\overset{OH}{\underset{}{C}}}-R \longrightarrow product
$$

Many important organic reactions involve nucleophilic carbon species (carbanions). The properties of carbanions will be discussed in detail in Chapter 7 and in Part B,

Chapters 1 and 2. Most C–H bonds are very weakly acidic and have no tendency to ionize spontaneously to form carbanions. Reactions that involve carbanion intermediates are therefore usually carried out in the presence of a base which can generate the reactive carbanion intermediate. Base-catalyzed condensation reactions of carbonyl compounds provide many examples of this type of reaction. The reaction between acetophenone and benzaldehyde, which was considered in Section 4.2, for example, requires a basic catalyst to proceed, and the kinetics of the reaction show that the rate is proportional to the catalyst concentration. This is because the neutral acetophenone molecule is not nucleophilic and does not react with benzaldehyde. The much more nucleophilic enolate (carbanion) formed by deprotonation is the reactive nucleophile.

$$\text{PhCCH}_3 + \text{B}^- \ \rightleftharpoons\ \text{PhC}{=}\text{CH}_2 + \text{BH}$$

$$\text{PhC}{=}\text{CH}_2 + \text{PhCH}{=}\text{O}\ \longrightarrow\ \text{PhCCH}_2\text{CHPh}\ \longrightarrow\ \text{product}$$

The role that acid and base catalysts play can be quantitatively studied by kinetic techniques. It is possible to recognize several distinct types of catalysis by acids and bases. The term *specific acid catalysis* is used when the reaction rate is dependent on the *equilibrium* for protonation of the reactant. This type of catalysis is independent of the concentration and specific structure of the various proton donors present in solution. Specific acid catalysis is governed by the *hydrogen-ion concentration* (pH) of the solution. For example, for a series of reactions in an aqueous buffer system, the rate of the reaction would be a function of the pH, but not of the concentration or identity of the acidic and basic components of the buffer. The kinetic expression for any such reaction will include a term for hydrogen-ion concentration, $[\text{H}^+]$. The term *general acid catalysis* is used when the nature and concentration of proton donors present in solution affect the reaction rate. The kinetic expression for such a reaction will include a term for each of the potential proton donors that acts as a catalyst. The terms *specific base catalysis* and *general base catalysis* apply in the same way to base-catalyzed reactions.

Specific acid catalysis:

$$\text{rate} = k[\text{H}^+][\text{X}][\text{Y}], \quad \text{where [X] and [Y] are the concentration of the reactants}$$

General acid catalysis:

$$\text{rate} = k_1[\text{H}^+][\text{X}][\text{Y}] + k_2[\text{HA}^1][\text{X}][\text{Y}] + k_3[\text{HA}^2][\text{X}][\text{Y}] + \cdots + k_n[\text{HA}^n][\text{X}][\text{Y}],$$

where $\text{HA}^1, \text{HA}^2, \ldots, \text{HA}^n$ are all kinetically significant proton donors.

The experimental detection of general acid catalysis is done by rate measurements at constant pH but differing buffer concentration. Because under these circumstances $[\text{H}^+]$ is constant but the weak acid component(s) of the buffer (HA^1, HA^2, etc.) changes, the observation of a change in rate is evidence of general acid catalysis. If the rate remains constant, the reaction exhibits specific acid catalysis. Similarly, general base-catalyzed reactions show a dependence of the rate on the concentration and identity of the basic constituents of the buffer system.

230

CHAPTER 4
STUDY AND
DESCRIPTION OF
ORGANIC REACTION
MECHANISMS

Specific acid catalysis is observed when a reaction proceeds through a protonated intermediate that is in equilibrium with its conjugate base. Because the position of this equilibrium is a function of the concentration of solvated protons, only a single acid-dependent term appears in the kinetic expression. For example, in a two-step reaction involving rate-determining reaction of one reagent with the conjugate acid of a second, the kinetic expression will be as follows:

$$B + H^+ \underset{\text{fast}}{\rightleftharpoons} BH^+ \qquad BH^+ + C \xrightarrow{\text{slow}} B - C + H^+$$

$$\text{rate} = k_2[BH^+][C] = k_2[C]K[B][H^+]$$
$$= k_{\text{obs}}[H^+][B][C]$$

Several situations can lead to the observation of general acid catalysis. General acid catalysis can occur as a result of hydrogen bonding between the reactant $\mathbf{R}$ and a proton donor $\mathbf{D-H}$ to form a reactive complex $\{\mathbf{D-H-R}\}$ which then reacts with a substance $\mathbf{Z}$:

$$\mathbf{D-H} + \mathbf{R} \rightleftharpoons \mathbf{D-H} \cdots \mathbf{R}$$

$$\mathbf{D-H} \cdots \mathbf{R} + \mathbf{Z} \xrightarrow{\text{slow}} \mathbf{D-H} + \mathbf{R-Z}$$

Under these circumstances, a distinct contribution to the overall rate will be seen for each potential hydrogen-bond donor $\mathbf{D-H}$. General acid catalysis is also observed when a rate-determining proton transfer occurs from acids other than the solvated proton:

$$\mathbf{R} + HA \xrightarrow{\text{slow}} {}^+\mathbf{RH} + A^-$$

$${}^+\mathbf{RH} \rightarrow \text{product}$$

Each acid HA^1, HA^2, etc., will then make a contribution to the overall rate of the reaction.

A kinetic expression which is equivalent to that for general acid catalysis also occurs if a prior equilibrium between reactant and the acids is followed by rate-controlling proton transfer. Each individual conjugate base will appear in the overall rate expression:

$$\mathbf{R} + HA \rightleftharpoons {}^+\mathbf{RH} + A^-$$

$${}^+\mathbf{RH} + A^- \xrightarrow{\text{slow}} \text{product} + HA$$

Notice that specific acid catalysis describes a situation in which the reactant is in equilibrium with regard to proton transfer, and proton transfer is not rate-determining. On the other hand, each case that leads to general acid catalysis involves proton transfer in the rate-determining step. Because of these differences, the study of rates as a function of pH and buffer concentrations can permit conclusions about the nature of proton-transfer processes and their relationship to the rate-determining step in a reaction.

As might be expected intuitively, there is a relationship between the effectiveness of general acid catalysts and the acid strength of a proton donor as measured by its acid dissociation constant K_a. This relationship is expressed by the following equation, which is known as the *Brønsted catalysis law*:

$$\log k_{\text{cat}} = \alpha \log K_a + b \qquad (4.18)$$

An analogous equation holds for catalysis by bases. This equation requires that the free energies of activation for the catalytic step for a series of acids be directly proportional to the free energy of dissociation for the same series of acids. The proportionality constant α is an indication of the sensitivity of the catalytic step to structural changes, relative to the effect of the same structural changes on acid dissociation. It is often found that a single proportionality constant α is restricted to only structurally related types of acids and that α values of different magnitudes are revealed by each type.

Figure 4.10 is plot of the Brønsted relationship for hydrolysis of an enol ether. The plot shows that the effectiveness of the various carboxylic acids as catalysts is related to their dissociation constants. In this particular case, the constant α is 0.79:[46]

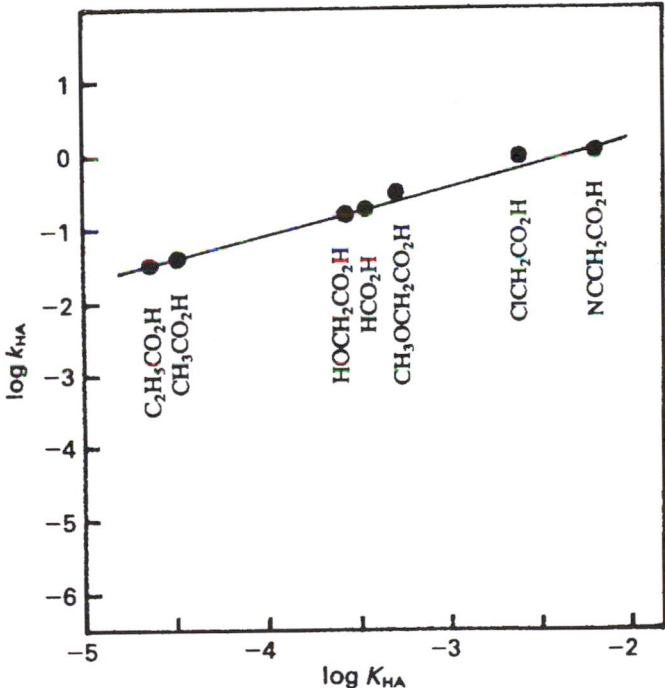

Fig. 4.10. Brønsted relation for the hydrolysis of cyclohexenyl methyl ether. [Adapted from Ref. 46 by permission of the American Chemical Society.]

46. A. J. Kresge, H. L. Chen, Y. Chiang, E. Murrill, M. A. Payne, and D. S. Sagatys, *J. Am. Chem. Soc.* **93**:413 (1971).

232

CHAPTER 4
STUDY AND
DESCRIPTION OF
ORGANIC REACTION
MECHANISMS

Because α relates the sensitivity to structural changes that the proton-transfer process exhibits to that exhibited by dissociation of the acid, it is frequently assumed that the value of α can be used as an indicator of transition-state structure. The closer α approaches unity, the greater is the degree of proton transfer in the transition state. There are limits to the generality of this interpretaton, however.[47]

The details of proton-transfer processes can also be probed by examination of *solvent isotope effects*, for example, by comparing the rates of a reaction in H_2O versus D_2O. The solvent isotope effect can be either normal or inverse, depending on the nature of the proton-transfer process in the reaction mechanism. D_3O^+ is a stronger acid than H_3O^+. As a result, reactants in D_2O solution are somewhat more extensively protonated than in H_2O at identical acid concentration. A reaction that involves a rapid equilibrium protonation will proceed faster in D_2O than in H_2O because of the higher concentration of the protonated reactant. On the other hand, if proton transfer is part of the rate-determining step, the reaction will be faster in H_2O than in D_2O because of the normal primary kinetic isotope effect of the type considered in Section 4.5.

Many organic reactions involve acid concentrations considerably higher than can be accurately measured on the pH scale, which applies to relatively dilute aqueous solutions. It is not difficult to prepare solutions in which the formal proton concentration is $10\,M$ or more, but these formal concentrations are not a suitable measure of the *activity* of protons in such solutions. For this reason, it has been necessary to develop *acidity functions* to measure the proton-donating strength of concentrated acidic solutions. The activity of the hydrogen ion (solvated proton) can be related to the extent of protonation of a series of bases by the equilibrium expression for the protonation reaction,

$$B + H^+ \rightleftharpoons {}^+BH$$

$$K = \frac{(a_{BH^+})}{(a_{H^+})(a_B)} = \frac{[{}^+BH]\gamma_{+BH}}{a_{H^+} + [B]\gamma_B}$$

where γ is the activity coefficient for the base and its conjugate acid. A common measure of acidity is referred to as h_0 and is defined by measuring the extent of protonation of a series of bases for which K has been measured. The relative concentrations of the base and its conjugate acid then define h_0 for any particular acidic solution.

$$h_0 = \frac{[{}^+BH]\gamma_{+BH}}{K[B]\gamma_B}$$

The quantity H_0, defined as $-\log h_0$, is commonly tabulated and corresponds to the "pH" of very concentrated acidic solutions.

The problem of determining K independently of measurement of H_0 is the principal issue to be faced in establishing the H_0 scale for a series of acidic solutions. What is done is to measure K for some base in aqueous solution where $H_0 \approx$ pH. This base can than be used to find the H_0 of a somewhat more acidic solution. The K of a second, somewhat weaker base is then determined in the more acidic solution. This second base can then be used to extend H_0 into a still more acidic solution. The process is continued by using a

47. A. J. Kresge, *J. Am. Chem. Soc.* **92**:3210 (1970); R. A. Marcus, *J. Am. Chem. Soc.* **91**:7224 (1969); F. G. Bordwell and W. J. Boyle, Jr., *J. Am. Chem. Soc.* **94**:3907 (1972); D. A. Jencks and W. P. Jencks, *J. Am. Chem. Soc.* **99**:7948 (1977); A. Pross, *J. Org. Chem.* **49**:1811 (1984).

Table 4.8. H_0 as a Function of Composition of Aqueous Sulfuric Acid^a

%H_2SO_4	H_0	%H_2SO_4	H_0
5	0.24	55	−3.91
10	−0.31	60	−4.46
15	−0.66	65	−5.04
20	−1.01	70	−5.80
25	−1.37	75	−6.56
30	−1.72	80	−7.34
35	−2.06	85	−8.14
40	−2.41	90	−8.92
45	−2.85	95	−9.85
50	−3.38	98	−10.41

a. From J. J. Jorgenson and D. R. Hartter, *J. Am. Chem. Soc.* **85**:878 (1963).

series of bases to establish H_0 for successively more acidic solutions. The H_0 is thereby referenced to the original aqueous measurement.[48] The assumption involved in this procedure is that the ratio of the activity coefficients for the series of bases and the series of cations does not change from solvent to solvent; that is,

$$\frac{\gamma_{+B_1H}}{\gamma_{B_1}} = \frac{\gamma_{+B_2H}}{\gamma_{B_2}} = \frac{\gamma_{+B_3H}}{\gamma_{B_3}} = \cdots$$

Not unexpectedly, this procedure reveals some dependence on the particular type of base used, so no universal H_0 scale can be established. Nevertheless, this technique provides a very useful measure of the relative hydrogen-ion activity of concentrated acid solutions which can be used in the study of reactions that proceed only at high acid concentration. Table 4.8 gives H_0 values for some water–sulfuric acid mixtures.

4.9. Lewis Acid Catalysis

Lewis acids are defined as molecules that act as electron-pair acceptors. The proton is an important special case, but many other species can play an important role in the catalysis of organic reactions. The most important in organic reactions are metal cations and covalent compounds of metals. Metal cations that play prominent roles as catalysts include the alkali-metal monocations Li^+, Na^+, K^+, Cs^+, and Rb^+, divalent ions such as Mg^{2+}, Ca^{2+}, and Zn^2, many of the transition-metal cations, and certain lanthanides. The most commonly employed of the covalent compounds include boron trifluoride, aluminum chloride, titanium tetrachloride, and tin tetrachloride. Various other derivatives of boron, aluminum, and titanium also are employed as Lewis acid catalysts.

48. For reviews and discussion of acidity functions, see E. M. Arnett, *Prog. Phys. Org. Chem.* **1**:223 (1963); C. H. Rochester, *Acidity Functions*, Academic Press, New York, 1970; R. A. Cox and K. Yates, *Can. J. Chem.* **61**:225 (1983); C. D. Johnson and B. Stratton, *J. Org. Chem.* **51**:4100 (1986).

234

CHAPTER 4
STUDY AND
DESCRIPTION OF
ORGANIC REACTION
MECHANISMS

The catalytic activity of metal ions originates in the formation of a donor–acceptor complex between the cation and the reactant, which must act as a Lewis base. The result of the complexation is that the donor atom becomes effectively more electronegative. All functional groups that have unshared electron pairs are potential electron donors, but especially prominent in reaction chemistry are carbonyl (sp^2) oxygens, hydroxyl and ether (sp^3) oxygens, and nitrogen-, and sulfur-containing functional groups. Halogen substituents can act as donors to very strong Lewis acids. The presence of two potential donor atoms in a favorable geometric relationship permits formation of bidentate "chelate" structures and may lead to particularly strong complexes.

If the complexation results in a full covalent bond between the donor and the Lewis base, there is a net transfer of one unit of formal charge to the metal ion from the donor atom. This enhances the effective electronegativity of the donor atom. The complexes of carbonyl groups, for example, are more reactive to nucleophilic attack. Hydroxyl groups complexed to metal cations are stronger acids and better leaving groups than the uncomplexed hydroxyl. Ether or sulfide groups complexed with metal ions are better leaving groups.

The strength of the complexation is a function of both the donor atom and the metal ion. The solvent medium is also an important factor because solvent molecules that are potential electron donors can compete for the Lewis acid. Qualitative predictions about the strength of donor–acceptor complexation can be made on the basis of the hard–soft-acid–base concept (see Section 1.2.3). The better matched the donor and acceptor, the stronger is the complexation. Scheme 4.3 gives an ordering of hardness and softness for some neutral and ionic Lewis acids and bases.

Neutral compounds such as boron trifluoride and aluminum chloride form Lewis acid–base complexes by accepting an electron pair from the donor molecule. The same functional groups that act as lone-pair donors to metal cations can form complexes with boron trifluoride, aluminum chloride, and related compounds.

Because in this case the complex is formed between two neutral species, it too is neutral, but a formal positive charge develops on the donor atom and a formal negative charge develops on the acceptor atom. The result is to increase the effective electronegativity of the donor atom and increase the electrophilicity of the complexed functional group.

Scheme 4.3. Relative Hardness and Softness

| | Lewis acids | | Lewis bases | |
	Cationic	Neutral	Neutral	Anionic
Hard	H^+	BF_3, $AlCl_3$	H_2O	F^-, SO_4^{2-}
	Li^+, Na^+, Ca^{2+}	R_3B	Alcohols,	Cl^-
			ketones,	Br^-
			ethers	
	Zn^{2+}, Cu^{2+}		Amines	N_3^-
			(aliphatic)	
			Amines	
			(aromatic)	
	Pd^{2+}, Hg^{2+}, Ag^+			CN^-
	RS^+, RSe^+			I^-
Soft	I^+		Sulfides	S^{2-}

Titanium tetrachloride and tin tetrachloride can form complexes that are related in character to both those formed by metal ions and those formed by neutral Lewis acids. Complexation can occur with an increase in the coordination number at the Lewis acid or with displacement of a chloride from the metal coordination sphere.

The crystal structure of the adduct of titanium tetrachloride and the ester formed from ethyl 2-hydroxypropanoate (ethyl lactate) and acrylic acid has been solved.[49] It is a chelated structure with the oxygen donor atoms being incorporated into the titanium coordination sphere along with the four chloride anions.

49. T. Poll, J. O. Melter, and G. Helmchen, *Angew. Chem. Int. Ed. Engl.* **24**:112 (1985).

236

CHAPTER 4
STUDY AND
DESCRIPTION OF
ORGANIC REACTION
MECHANISMS

Diels–Alder reactions in the presence of Lewis acids represent a case in which the Lewis acid is often used in catalytic quantities. The complexed ester (ethyl acrylate in the example given below) is substantially more reactive than the uncomplexed molecule, and the reaction proceeds through the complex. The reactive complex is regenerated by exchange of the Lewis acid from the adduct.

There are more variables to consider in catalysis by Lewis acids than in the case of catalysis by protons. In addition to the hard/soft relationship, steric, geometric, and stereoelectronic factors can come into play. This makes the development of an absolute scale of "Lewis acid strength" difficult, because it depends on the specific characteristics of the base. There are also variations in the strength of donor–acceptor bonds. Bond strengths calculated for complexes such as $H_3N^+-BF_3^-$ (22.0 kcal/mol) and $(CH_3)_3N^+-BH_3^-$ (41.1 kcal/mol) are substantially less than for covalent bonds between similar elements. Some Lewis acid–base complexes have long weak bonds that are primarily electrostatic in nature (e.g. $CH_3CN^+-BF_3^-$, 9.1 kcal/mol).[50] The Lewis acid strengths of a number of compounds used commonly in synthesis are given in Table 4.9. The relative acidity values given are derived from the LUMO energy of the π^* orbital of

Table 4.9. Relative Lewis Acidity[a]

Acid	ΔH^b	NMR $\Delta\delta^c$	Relative acidity
BCl_3	−6.6	1.35	1.00
$AlCl_3$	−25.6	1.23	0.91
$EtAlCl_2$	−20.0	1.15	0.80
BF_3	+4.1	1.17	0.76
Et_2AlCl	−5.6	0.91	0.71
Et_3Al	−10.1	0.63	0.63
$SnCl_4$	+10.0	0.87	0.61

a. From P. Laszlo and M. Teston, J. Am. Chem. Soc., 112:8750 (1990).
b. Enthalpy of interaction in kcal/mol by MNDO.
c. Change in chemical shift of H-3 in 2-butenal.

50. V. Jonas, G. Frenking, and M. T. Reetz, J. Am. Chem. Soc. 116:8741 (1994).

the compound with the BCl_3 complex defined as 1.00 and the uncomplexed 2-butenal LUMO energy taken as 0.[51]

Stereoelectronic factors are also important in determining the structure and reactivity of complexes. Complexes of carbonyl groups with trivalent boron and aluminum compounds tend to adopt a geometry consistent with directional interaction with one of the oxygen lone pairs. Thus the C—O—M bond angle tends to be in the trigonal (120–140°) range, and the boron or aluminum is usually close to the carbonyl plane.[52]

4.10. Solvent Effects

Most organic reactions are done in solution, and it is therefore important to recognize some of the ways in which solvent can affect the course and rates of reactions. Some of the more common solvents can be roughly classified as in Table 4.10 on the basis of their structure and dielectric constant. There are important differences between *protic* solvents—solvents that contain relatively mobile protons such as those bonded to oxygen, nitrogen, or sulfur—and *aprotic* solvents, in which all hydrogens are bound to carbon. Similarly, *polar* solvents, those that have high dielectric constants, have effects on reaction rates that are different from those of nonpolar solvent media.

When discussing solvent effects, it is important to distinguish between the macroscopic effects and those which depend upon details of structure. Macroscopic properties refer to properties of the bulk solvent. An important example is the dielectric constant, which is a measure of the ability of the bulk material to increase the capacitance of a condenser. In terms of structure, the dielectric constant is a function of both the permanent dipole of the molecule and its *polarizability*. Polarizability refers to the ease of distortion of the molecule's electron distribution. Dielectric constants increase with dipole moment and with polarizability because of the ability of both the permanent and the induced molecular dipole to align with an external electric field. An important property of solvent molecules is the response of the solvent to changes in charge distribution as reaction occurs. The dielectric constant of a solvent is a good indicator of the ability of the solvent to accommodate separation of charge. It is not the only factor, however, because, being a

Table 4.10. Dielectric Constants of Some Common Solvents[a]

Aprotic solvents				Protic solvents	
Nonpolar		Polar			
Hexane	1.9	Pyridine	12	Acetic acid	6.1
Carbon tetrachloride	2.2	Acetone	21	Trifluoroacetic acid	8.6
Dioxane	2.2	Hexamethylphosphoramide	30	*t*-Butyl alcohol	12.5
Benzene	2.3	Nitromethane	36	Ammonia	(22)
Diethyl ether	4.3	Dimethylformamide	37	Ethanol	24.5
Chloroform	4.8	Acetonitrile	38	Methanol	32.7
Tetrahydrofuran	7.6	Dimethyl sulfoxide	47	Water	78

a. Dielectric constant data are abstracted from the compilation of solvent properties in J. A. Riddick and W. B. Bunger, eds., *Organic Solvents*, Vol. II of *Techniques of Organic Chemistry*, 3rd ed., Wiley–Interscience, New York, 1970.

51. R. F. Childs, D. L. Mulholland, and A. Nixon, *Can. J. Chem.* **60**:801 (1982); P. Laszlo and M. Teston, *J. Am. Chem. Soc.* **112**:8750 (1990).
52. S. Shambayati, W. E. Crowe, and S. L. Schreiber, *Angew. Chem. Int. Ed. Engl.* **29**:256 (1990).

238

CHAPTER 4
STUDY AND
DESCRIPTION OF
ORGANIC REACTION
MECHANISMS

Fig. 4.11. Solvation changes during ionization of *t*-butyl chloride.

macroscopic property, it conveys little information about the ability of the solvent molecules to interact with the solute molecules at close range. These direct solute–solvent interactions will depend on the specific structures of the molecules.

Let us consider how the nature of the solvent might affect the solvolysis of *t*-butyl chloride. Much evidence, which will be discussed in detail in Chapter 5, indicates that the rate-determining step of the reaction is ionization of the carbon–chlorine bond to give a carbocation and the chloride ion. The transition state must reflect some of the charge separation that occurs in the ionization. Figure 4.11 gives a schematic interpretation of the solvation changes that would take place during the ionization of *t*-butyl chloride, with S representing surrounding solvent molecules. The bulk dielectric constant may be a poor indicator of the ability of solvent molecules to facilitate the charge separation in the transition state. The fact that the carbon and chlorine remain partially bonded at the transition state prevents the solvent molecules from actually intervening between the developing centers of charge. Instead, the solvent molecules must stabilize the charge development by acting around the periphery of the activated complex. This interaction will depend upon the detailed structure of the activated complex and solvent. The ability of a number of solvents to stabilize the transition state of *t*-butyl chloride ionization has been measured by comparing the rate of the reaction in the various solvents. The reference solvent was taken as 80 : 20 ethanol–water. The Y value of other solvents is defined by the equation

$$\log \frac{k_{\text{solvent}}}{k_{80\% \text{ ethanol}}} = Y$$

Table 4.11 lists the Y values for some alcohol–water mixtures and for some other solvents. The Y value reflects primarily the ionization power of the solvent. It is largest for polar

Table 4.11. Y Values for Some Solvent Systems[a]

Ethanol–water		Methanol–water			
Percent ethanol	Y	Percent methanol	Y	Other solvents	Y
100	−2.03	100	−1.09	Acetic acid	−1.64
80	0.0	80	0.38	Formic acid	2.05
50	1.65	50	1.97	*t*-Butyl alcohol	−3.2
20	3.05	10	3.28	90% Acetone–water	−1.85
0	3.49			90% Dioxane–water	−2.03

a. From A. H. Fainberg and S. Winstein, *J. Am. Chem. Soc.* **78**:2770 (1956).

solvents such as water and formic acid and becomes progressively smaller and eventually negative as the solvent becomes less polar and contains more (or larger) nonpolar alkyl groups. Notice that, among the solvents listed, the Y values range from -3.2 for t-butyl alcohol to $+3.49$ for water, corresponding to a spread of more than 10^6 in the measured rate of reaction. This large range of reaction rates demonstrates how important solvent effects can be.

Solvents that fall in the nonpolar aprotic class are much less effective at stabilizing the development of charge separation. These molecules have small dipole moments and do not have hydrogens capable of forming hydrogen bonds. Reactions that involve charge separation in the transition state therefore usually proceed much more slowly in this class of solvents than in protic or polar aprotic solvents. The reverse is true for reactions in which species having opposite charges come together in the transition state. Because in this case the transition state is less highly charged than the individual reactants, it is favored by weaker solvation, which leaves the oppositely charged reactants in a more reactive state. On the basis of arguments along these lines, the broad relationships between reactivity and solvent type shown in Scheme 4.4 can be deduced.

Many other measures of solvent polarity have been developed.[53] One of the most useful is based on shifts in the absorption spectrum of a reference dye. The positions of absorption bands are, in general, sensitive to solvent polarity because the electronic distribution, and therefore the polarity, of the excited state is different from that of the ground state. The shift in the absorption maximum reflects the effect of solvent on the energy gap between the ground-state and excited-state molecules. An empirical solvent polarity measure called $E_T(30)$ is based on this concept.[54] Some values of this measure for common solvents are given in Table 4.12 along with the dielectric constants for the solvents. It can be seen that there is a rather different order of polarity given by these two quantities.

The electrostatic solvent effects discussed in the preceding paragraphs are not the only possible modes of interaction of solvent with reactants and transition states. Specific structural effects may cause either the reactants or the transition state to be particularly strongly solvated. Figure 4.12 shows how such solvation can affect the relative energies of the ground state and transition state and cause rate variations from solvent to solvent.

Scheme 4.4. Effect of Solvent Polarity on Reactions of Various Charge Types

Reaction	Solvent effect
$A^- + B^+ \longrightarrow \overset{\delta-}{A} \text{---} \overset{\delta+}{B} \longrightarrow A-B$	Favored by nonpolar solvent
$A-B \longrightarrow \overset{\delta-}{A} \text{---} \overset{\delta+}{B} \longrightarrow A^- + B^+$	Favored by polar solvent
$A + B \longrightarrow A\text{---}B \longrightarrow A-B$	Relatively intensitive to solvent polarity
$[A-B]^+ \longrightarrow \overset{\delta+}{A} \text{---} \overset{\delta+}{B} \longrightarrow A + B^+$	Slightly favored by polar solvent
$A^+ + B \longrightarrow \overset{\delta+}{A} \text{---} \overset{\delta+}{B} \longrightarrow [A-B]^+$	Slightly favored by nonpolar solvent

53. C. Reichardt, *Angew. Chem. Int. Ed. Engl.* **18**:98 (1979); C. Reichardt, *Solvent Effects in Organic Chemistry*, Verlag Chemie, Weinheim, 1979; J. Catalan, V. Lopez, P. Perez, R. Martin-Villamil, and J. G. Rodriquez, *Liebigs Ann.* **1995**:241; C. Laurance, P. Nicolet, M. T. Dalati, J. L. M. Abboud, and R. Notario, *J. Phys. Chem.* **98**:5807 (1994).
54. C. Reichardt and K. Dimroth, *Fortschr. Chem. Forsch.* **11**:1 (1968); C. Reichardt, *Liebigs Ann. Chem.* **752**:64 (1971).

240

CHAPTER 4
STUDY AND
DESCRIPTION OF
ORGANIC REACTION
MECHANISMS

Table 4.12. $E_T(30)$, an Empirical Measure of Polarity, Compared with Dielectric Constant

	$E_T(30)$	ε		$E_T(30)$	ε
Water	63.1	78	Dimethylformamide	43.8	37
Trifluoroethanol	59.5		Acetone	42.2	21
Methanol	55.5	32.7	Dichloromethane	41.1	8.9
80 : 20 Ethanol–water	53.7		Chloroform	39.1	4.8
Ethanol	51.9	24.5	Ethyl acetate	38.1	6.0
Acetic acid	51.2	6.1	Tetrahydrofuran	37.4	7.6
Isopropanol	48.6	19.9	Diethyl ether	34.6	4.3
Acetonitrile	46.7	38	Benzene	34.5	2.3
Nitromethane	46.3	36	Hexane	30.9	1.0
Dimethyl sulfoxide	45.0	47			

a. Data from C. Reichardt, *Angew. Chem. Int. Ed. Engl.* **18**:98 (1979).

Unfortunately, no general theory for quantitatively predicting such specific effects has been developed to date.

Because a solvent may affect the rates of two competing reactions to different extents, a change in solvent may strongly modify the composition of a product mixture arising from competing reaction paths. Many such instances have been encountered in synthetic chemistry. An important example of solvent effects is the enhanced nucleophilicity of many anions in polar aprotic solvents as compared with protic solvents.[55] In protic solvents, anions are strongly solvated by hydrogen bonding. This is particularly true for anions that have a high concentration of charge on oxygen or nitrogen. Hydrogen bonding decreases the availability of the electrons of the anion to participate in reactions as a nucleophile. Stated another way, the energy required to disrupt hydrogen bonding adds to the activation energy of the reaction. In aprotic solvents, no hydrogens suitable for

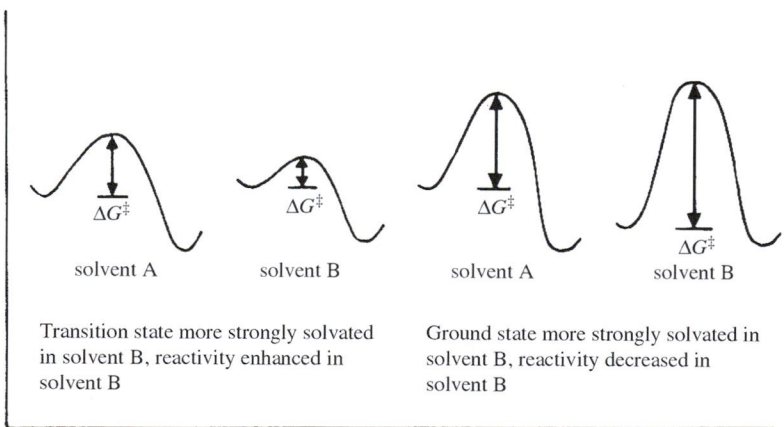

Fig. 4.12. Potential energy diagrams showing effect of preferential solvation of transition state (a) and ground state (b) on the activation energy.

55. A. J. Parker, *Q. Rev. Chem. Soc.* **16**:163 (1962); C. D. Ritchie, in *Solute–Solvent Interactions*, J. F. Coetzee and C. D. Ritchie, eds., Marcel Dekker, New York, 1969, Chapter 4; E. Buncel and H. Wilson, *Adv. Phys. Org. Chem.* **14**:133 (1977).

hydrogen bonding are present. As a result, the electrons of the anion are more easily available for reaction. The anion is at a higher energy level because of the absence of solvent stabilization. The polarity of the aprotic solvents is important because nonpolar solvents have very low solubility for ionic compounds. Dissolved ionic compounds are likely to be present as ion pairs or larger aggregates in which the reactivity of the anion is diminished by the electrostatic interaction with the cation. Energy must be expended against this electrostatic attraction to permit the anion to react as a nucleophile. Metal cations such as K^+ and Na^+ are strongly solvated by polar aprotic solvents such as dimethyl sulfoxide (DMSO) and dimethylformamide. The oxygen atoms in these molecules act as electron donors toward the cations. The dissolved salts are dissociated, and, as a result, the anions are highly reactive because they are poorly solvated and not associated with cations.

$$CH_3\overset{+}{S}CH_3 \qquad (CH_3)NCH \longleftrightarrow (CH_3)_2\overset{+}{N}{=}CH$$
$$\underset{O^-}{|} \qquad\qquad \underset{O}{\|} \qquad\qquad \underset{O^-}{|}$$

The realization that the nucleophilicity of anions is strongly enhanced in polar aprotic solvents has led to important improvements of several types of synthetic processes that involve nucleophilic substitutions or additions.

Particularly striking examples of the effect of specific solvation can be cited from among the *crown ethers*. These are macrocyclic polyethers that have the property of specifically solvating cations such as Na^+ and K^+.

18-crown-6   dibenzo-18-crown-6

When added to nonpolar solvents, the crown ethers increase the solubility of ionic materials. For example, in the presence of 18-crown-6, potassium fluoride is soluble in benzene and acts as a reactive nucleophile:

$$CH_3(CH_2)_7Br + KF \xrightarrow[\text{benzene}]{\text{18-crown-6C}} H_3(CH_2)_7F \qquad\qquad \text{Ref. 56}$$

In the absence of the polyether, potassium fluoride is insoluble in benzene and unreactive toward alkyl halides. Similar enhancement of solubility and reactivity of other salts is observed in the presence of crown ethers The solubility and reactivity enhancement result because the ionic compound is dissociated to a tightly complexed cation and a "naked" anion. Figure 4.13 shows the tight coordination that can be achieved with a typical crown ether. The complexed cation, because it is surrounded by the nonpolar crown ether, has high solubility in the nonpolar media. To maintain electroneutrality, the anion is also transported into the solvent. The cation is shielded from interaction with the anion as a

56. C. Liotta and H. P. Harris, *J. Am. Chem. Soc.* **96**:2250 (1974).

242

CHAPTER 4
STUDY AND
DESCRIPTION OF
ORGANIC REACTION
MECHANISMS

Fig. 4.13. Space-filling molecular model depicting a metal cation complexed by 18-crown-6.

result of the surrounding crown ether molecule. As a result, the anion is unsolvated and at a relatively high energy and therefore highly reactive.

A closely related solvation phenomenon is the basis for *phase-transfer catalysis*.[57] Phase-transfer catalysts are salts in which one of the ions (usually the cation) has large nonpolar substituent groups that confer good solubility in organic solvents. The most common examples are tetraalkylammonium and tetraalkylphosphonium ions. In two-phase systems consisting of water and a nonpolar organic solvent, these cations are extracted into the organic phase, and, as a result, anions are also present in the organic phase. The anions are weakly solvated and display high reactivity. Reactions are carried out between a salt containing the desired nucleophilic anion and an organic reactant, typically, an alkyl halide. The addition of the phase-transfer catalysts causes migration of the anion into the organic phase, and, because of the high nucleophilicity of the anion, reaction occurs under exceptionally mild conditions. Section 3.2.1 of Part B gives some specific examples of the use of phase-transfer catalysis in nucleophilic displacement reactions.

It should always be borne in mind that solvent effects can modify the energy of both the reactants and the transition state. It is the *difference* in the two solvation effects that is the basis for changes in activation energies and reaction rates. Thus, although it is common to express solvent effects solely in terms of reactant solvation or transition-state solvation,

57. C. M. Starks, C. L. Liotta, and M. Halpern, *Phase Transfer Catalysis: Fundamentals, Applications and Industrial Perspectives*, Chapman and Hall, New York, 1994.

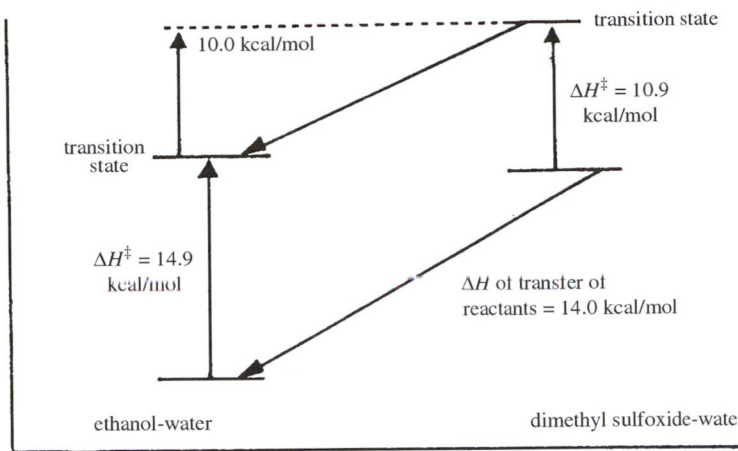

Fig. 4.14. Reactant and transition-state solvation in the reaction of ethyl acetate with hydroxide ion. [From P. Haberfield, J. Friedman, and M. F. Pinkson, *J. Am. Chem. Soc.* **94**:71 (1972).]

this is usually an oversimplification. One case that illustrates this point is the hydrolysis of esters by hydroxide ion.

$$^{-}OH + CH_3\overset{\displaystyle O}{\overset{\|}{C}}OC_2H_5 \;\rightleftharpoons\; H_3C-\overset{\displaystyle O^-}{\underset{\displaystyle HO^-}{\overset{\|}{C}}}-OC_2H_5 \;\longrightarrow\; CH_3CO_2^- + HOC_2H_5$$

The reaction is found to be much more rapid in DMSO–water than in ethanol–water. Reactant solvation can be separated from transition-state solvation by calorimetric measurement of the heat of solution of the reactants in each solvent system. The data in Fig. 4.14 compare the energies of the reactants and transition state for ethyl acetate and hydroxide ion reacting in aqueous ethanol versus aqueous DMSO. It can be seen that both the reactants and the transition state are more strongly solvated in the ethanol–water medium. The enhancement in reaction rate comes from the fact that the difference is greater for the small hydroxide ion than for the larger anionic species present at the transition state. It is generally true that solvation forces are strongest for the small, hard anions and decrease with increasing size and softness.

4.11. Substituent Effects in the Gas Phase

Having considered how solvents can affect the reactivities of molecules in solution, let us consider some of the special features that arise in the gas phase, where solvation effects are totally eliminated. Although the majority of organic preparative reactions and mechanistic studies have been conducted in solution, some important reactions are carried out in the gas phase. Also, because most theoretical calculations do not treat solvent effects, experimental data from the gas phase are the most appropriate basis for comparison with theoretical results. Frequently, quite different trends in substituent effects are seen when systems in the gas phase are compared to similar systems in solution.

244

CHAPTER 4
STUDY AND
DESCRIPTION OF
ORGANIC REACTION
MECHANISMS

It is possible to measure equilibrium constants and heats of reaction in the gas phase by using mass spectrometers of special configuration.[58] With proton-transfer reactions, for example, the equilibrium constant can be determined by measuring the ratio of two reactant species competing for protons. Table 4.13 compares ΔH_{gas} with ΔH_{aq} for a series of phenol ionizations.

A key point to recognize is that the relative magnitude of the substituent effects is much larger in the gas phase. In general terms, this can be explained on the basis of the absence of solvation in the gas phase. Whereas a phenolate anion in aqueous solution is stabilized by hydrogen bonding, there is no such stabilization in the gas phase. Because the solvent stabilization of the phenolate anions will be rather similar, on an absolute scale, for all the substituted phenols, the substituent effects are "leveled" to some extent. In contrast, in the gas phase the importance of the internal substituent effect on the stability of the anion is undiminished. The importance of the solvation can also be judged by noting that entropy makes a larger contribution to ΔG than enthalpy for the reaction series in solution. This reflects the extensive solvent organization that accompanies solvation.[59]

A comparison of phenol acidity in DMSO versus the gas phase also shows an attenuation of substituent effects, but not nearly as much as in water. Whereas the effect of substituents on ΔG for deprotonation in aqueous solution is about one-sixth that in the gas phase, the ratio for DMSO is about one-third. This result points to hydrogen bonding of the phenolate anion by water as the major difference in the solvating properties of water and DMSO.[60]

Table 4.13. Comparison of Substituent Contributions to Phenol Ionization in the Gas Phase and Solution[a]

| | Substituent increment in kcal/mol[b] | | | |
X	ΔG_{gas}	ΔH_{gas}	ΔG_{H_2O}	ΔH_{H_2O}
m-CH$_3$	+0.4	+0.4	+0.18	+0.02
p-CH$_3$	+1.3	+1.3	+0.42	+0.02
m-Cl	−7.9	−7.9	−1.2	−0.3
p-Cl	−6.6	−6.6	−0.7	−0.2
p-NO$_2$	−25.8	−25.8	−3.8	−0.8

a. Data are from T. B. McMahon and P. Kebarle, *J. Am. Chem. Soc.* **99**:2222 (1977).
b. The tabulated increments give the change in ΔG and ΔH resulting from replacement of hydrogen by the substituent specified.

58. Discussion of the techniques for gas-phase equilibrium measurements can be found in T. A. Lehman and M. M. Bursey, *Ion Cyclotron Resonance Spectrometry*, Wiley-Interscience, New York, 1976; M. T. Bowers, ed., *Gas Phase Ion Chemistry*, Vols. 1 and 2, Academic Press, New York, 1979.
59. L. P. Fernandez and L. G. Hepler, *J. Am. Chem. Soc.* **81**:1783 (1959); C. L. Liotta, H. P. Hopkins, Jr., and P. T. Kasudia, *J. Am. Chem. Soc.* **96**:7153 (1974).
60. M. Mashima, R. R. McIver, Jr., R. W. Taft, F. G. Bordwell, and W. N. Olmstead, *J. Am. Chem. Soc.* **106**:2717 (1984); M. Fujio, R. T. McIver, and R. W. Taft, *J. Am. Chem. Soc.* **103**:4017 (1981).

Another example of enhanced sensitivity to substituent effects in the gas phase can be seen in a comparison of the gas-phase basicity for a series of substituted acetophenones and methyl benzoates. It was found that sensitivtiy of the free energy to substituent changes was about four times that in solution, as measured by the comparison of ΔG for each substituent.[61] The gas-phase data for both series were correlated by the Yukawa–Tsuno equation. For both series, the ρ value was about 12. However, the parameter r^+, which reflects the contribution of extra resonance effects, was greater in the acetophenone series than in the methyl benzoate series. This can be attributed to the substantial resonance stabilization provided by the methoxy group in the esters, which diminishes the extent of conjugation with the substituents.

$\rho = 12.2, r^+ = 0.76$

$\rho = 11.9, r^+ = 0.45$

Another area of gas-phase substituent effects that has attracted interest is the acidity of simple alcohols. In the gas phase, the order is $t\text{-BuOH} > \text{EtOH} > \text{MeOH} \gg \text{H}_2\text{O}$.[62] This is opposite from the order in solution as revealed by the pK_a data in water and DMSO shown in Table 4.14. These changes in relative acidity can again be traced to solvation effects. In the gas phase, any substituent effect can be analyzed directly in terms of its stabilizing or destabilizing effect on the anion. Replacement of hydrogen by alkyl substituents normally increases electron density at the site of substitution, but this effect cannot be the dominant one, because it would lead to an ordering of gas-phase acidity opposite to that observed. The dominant effect is believed to be polarizability. The methyl

Table 4.14. Acidities of Simple Alcohols in Solution[a]

	pK_a	
	H_2O	DMSO
H_2O	15.7	31.4
CH_3OH	15.5	29.0
$\text{CH}_3\text{CH}_2\text{OH}$	15.9	29.8
$(\text{CH}_3)_2\text{CHOH}$		30.2
$(\text{CH}_3)_3\text{COH}$		32.2

a. Data are from W. N. Olmstead, Z. Margolin, and F. G. Bordwell, *J. Org. Chem.* **45**:3295 (1980).

61. M. Mishima, M. Fujo, and Y. Tsuno, *Tetrahedron Lett.* **27**:939, 951 (1986).

62. J. I. Brauman and L. K. Blair, *J. Am. Chem. Soc.* **92**:5986 (1970); J. E. Bartmess and R. T. McIver, Jr., *Gas Phase Ion Chemistry*, Vol. 2, M. T. Bowers, ed., Academic Press, New York, 1979.

246

CHAPTER 4
STUDY AND
DESCRIPTION OF
ORGANIC REACTION
MECHANISMS

substituents, each consisting of four atoms, are better able to undergo local electronic distortion (induced polarizability) to accommodate the negative charge than is a single hydrogen atom. Thus, each methyl-for-hydrogen substitution increases gas-phase acidity.[63] An additional factor that accounts for the greater acidity of alcohols in comparison to water is the fact that the water O—H bond is somewhat stronger. In solution, these factors are swamped by solvation effects, and the observed order of acidity is MeOH > EtOH > i-PrOH > t-BuOH. This order results from the fact that the small anions can be better solvated than the more substituted ones.[64] These trends are also evident in MP4/6-31G*-level MO calculations.[65]

Within an isomeric series of alcohols, the order *sec* > *tert* > *pri* holds for C_4 and C_5 alcohols, although the differences between the isomeric *sec* and *tert* alcohols are small.[66] Analysis of a large number of alcohols showed the effect of adding an additional methyl substituent at the α, β, and γ positions. The effect at the α carbon was the greatest, but methyl groups at the β and γ positions were also stabilizing.[67] Figure 4.15 gives gas-phase data illustrating the trend of increasing acidity with size. This is again indicative of the importance of the entire molecule in stabilizing the charge in the gas phase.[68]

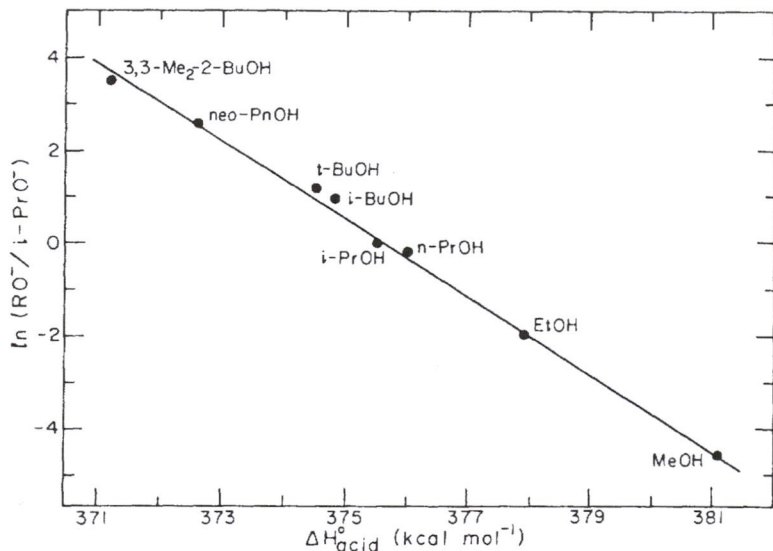

Fig. 4.15. Plot of $\ln(RO^-/i\text{-}PrO^-)$ vs. ΔH°_{acid} for several alcohols under N_2 collision induced dissociation at 50 eV. [Reproduced from *Int. J. Mass Spectrom. Ion Processes* by permission of Elsevier Science Publishers.]

63. R. W. Taft, M. Taagepera, J. L. M. Abboud, J. F. Wolf, D. J. DeFrees, W. J. Hehre, J. E. Bartmess, and R. T. McIver, *J. Am. Chem. Soc.* **100**:7765 (1978).
64. W. N. Olmstead, Z. Margolin, and F. G. Bordwell, *J. Org. Chem.* **45**:3295 (1980).
65. I. Tunon, E. Silla, and J.-L. Pascual-Ahuir, *J. Am. Chem. Soc.* **115**:2226 (1993).
66. G. Boaud, R. Houriet, and T. Gäuman, *J. Am. Chem. Soc.* **105**:2203 (1983).
67. M. J. Haas and A. G. Harrison, *Int. J. Mass Spectrom. Ion Processes* **124**:115 (1993).
68. For a broad review of substituent and solvent effects on acidity and basicity, see R. W. Taft, *Prog. Phys. Org. Chem.* **14**:247 (1983).

4.12. Stereochemistry

The study of the stereochemical course of organic reactions often leads to detailed insight into reaction mechanisms. Mechanistic postulates frequently make distinctive predictions about the stereochemical outcome of the reaction. Throughout the chapters dealing with specific types of reactions, consideration will be given to the stereochemistry of a reaction and its relationship to the reaction mechanism. As an example, the bromination of alkenes can be cited. A very simple mechanism for bromination is given below:

According to this mechanism, a molecule of bromine becomes complexed to the double bond of the alkene, and reorganization of the bonding electrons gives the product. This mechanism can be shown to be incorrect for most alkenes on the basis of stereochemistry. Most alkenes give bromination products in which the two added bromines are on opposite sides of the former carbon–carbon double bond. The above mechanism does not account for this and therefore must be incorrect.

Another example of a reaction in which the stereochemistry of the process provides some valuable information about the mechanism is the thermal rearrangement of 1,5-dienes and substituted analogs:

$X = CH_2, O$
$Y = H$, alkyl, OR, etc.

These reactions will be discussed in more detail under the topic of 3,3-sigmatropic rearrangements in Chapter 11. For the present, we simply want to focus on the fact that the reaction is *stereospecific*; the *E*-isomer gives one diastereomeric product whereas the related *Z*-isomer gives a different one. The stereochemical relationship between reactants and products can be explained if the reaction occurs through a chairlike transition state in

248

CHAPTER 4
STUDY AND
DESCRIPTION OF
ORGANIC REACTION
MECHANISMS

which the configurational relationships in the original double bonds are maintained.

A large number of stereochemical results on the 3,3-sigmatropic family of reactions have been correlated by this type of analysis.[69] The basic assumption is that the transition state is similar to a cyclohexane ring in its response to steric effects. The success of the model in interpreting the stereochemical results supports its correctness. It can be concluded that the transition state must be a rather tightly organized arrangement of the atoms, with strong bonding both between atoms 3 and 4 and between atoms 1 and 6.

4.13. Conclusion

To conclude this chapter, it is important to emphasize a logical point about the determination of reaction mechanisms. *A proposed mechanism can never really be proven; rather, it is a case of alternative mechanisms being eliminated.* Having in mind a mechanism that explains all the facts does not constitute proof that the mechanism is correct. That conclusion is possible only when all alternatives have been excluded. A key stage in a mechanistic investigation then is the enumeration of the various possible mechanisms and the design of experiments that distinguish between them. The principal basis for enumerating mechanistic possibilities is accumulated chemical experience with related systems and the inherent structural features of the system. A chemist approaching a mechanistic study must cast as broad as possible vision on the problem so as not to exclude possibilities.

General References

General

C. F. Bernasconi, ed., *Investigation of Rates and Mechanisms of Reactions, Techniques of Chemistry*, 4th ed., Vol. VI, Part 1, John Wiley & Sons, New York, 1986.

B. K. Carpenter, *Determination of Organic Reaction Mechanisms*, Wiley-Interscience, New York, 1984.

J. Hine, *Structural Effects on Equilibria in Organic Chemistry*, Wiley-Interscience, New York, 1984.

J. A. Hirsch, *Concepts in Theoretical Organic Chemistry*, Allyn and Bacon, Boston, 1974.

E. S. Lewis, ed., *Investigation of Rates and Mechanisms of Reactions, Techniques of Chemistry*, 3rd ed., Vol. VI, Part 1, John Wiley & Sons, New York, 1974.

69. R. K. Hill, *Asymmetric Syntheses*, Vol. 3, J. D. Morrison, ed., Academic Press, New York, 1984, Chapter 8.

C. D. Ritchie, *Physical Organic Chemistry*, Marcel Dekker, New York, 1975.
K. B. Wiberg, *Physical Organic Chemistry*, Wiley, New York, 1964.

Thermodynamics

J. D. Cox and G.Pilcher, *Thermochemistry of Organic and Organometallic Compounds*, Academic Press, New York, 1970.
G. Janz, *Thermodynamic Properties of Organic Compounds*, Academic Press, New York, 1967.
D. R. Stull, E. F. Westrum, Jr., and G. C. Sinke, *The Chemical Thermodynamics of Organic Compounds*, John Wiley & Sons, New York, 1969.

Kinetics

K. A. Connors, *Chemical Kinetics*, VCH Publishers, New York, 1990.
G. G. Hammes, *Principles of Chemical Kinetics*, Academic Press, New York, 1978.
J. E. House, *Principles of Chemical Kinetics*, W. C. Brown Publishers, Dubuque, Iowa, 1997.
K. J. Laidler, *Chemical Kinetics*, McGraw-Hill, New York, 1965.
K. J. Laidler, *Theory of Chemical Reaction Rates*, McGraw-Hill, New York, 1969.
M. J. Pilling and P. Seakins, *Reaction Kinetics*, 2nd ed., Oxford University Press, Oxford, U.K., 1995.

Linear Free-Energy Relationships and Substituent Effects

C. Hansch, A. Leo, and R. W. Taft, *Chem. Rev.* **91:**165 (1991).
C. D. Johnson, *The Hammett Equation*, Cambridge University Press, Cambridge, 1973.
P. R. Wells, *Linear Free Energy Relationships*, Academic Press, New York, 1968.

Isotope Effects

C. J. Collins and N. S. Borman, eds., *Isotope Effects on Chemical Reactions*, Van Nostrand Reinhold, New York, 1970.
L. Melander, *Isotope Effects on Reaction Rates*, Ronald Press, New York, 1960.
L. Melander and W. H. Saunders, Jr., *Reaction Rates of Isotopic Molecules*, New York, 1980.

Solvent Effects

J. F. Coetzee and C. D. Ritchie, *Solute–Solvent Interactions*, Marcel Dekker, New York, 1969.
C. M. Starks, C. L. Liotta, and M. Halpern, *Phase Transfer Catalysis: Fundamentals, Applications and Perspectives*, Chapman and Hall, New York, 1994.

Catalysis

R. P. Bell, *The Proton in Chemistry*, Chapman and Hall, London, 1971.
W. P. Jencks, *Catalysis in Chemistry and Enzymology*, McGraw-Hill, New York, 1969.
C. H. Rochester, *Acidity Functions*, Academic Press, New York, 1970.

250

CHAPTER 4
STUDY AND
DESCRIPTION OF
ORGANIC REACTION
MECHANISMS

Problems

(References for these problems will be found on page 794.)

1. Measurement of the equilibrium constant for the interconversion of the dithiete **A** and the dithione **B** yielded the data given below. Calculate $\Delta G°$, $\Delta H°$, and $\Delta S°$.

Temp. (°C)	K
−2.9	16.9
11.8	11.0
18.1	8.4
21.9	7.9
29.3	6.5
32.0	6.1
34.9	5.7
37.8	5.3
42.5	4.6

2. (a) Calculate the enthalpy and entropy of activation ($\Delta H^{\ddagger}$ and $\Delta S^{\ddagger}$) at 40°C for the acetolysis of *m*-chlorobenzyl *p*-toluenesulfonate from the data given:

Temp. (°C)	$k \times 10^5$ (s^{-1})
25.0	0.0136
40.0	0.085
50.1	0.272
58.8	0.726

(b) Calculate the activation parameters ΔE_a, $\Delta H^{\ddagger}$, and $\Delta S^{\ddagger}$ at 100°C from the data given for the reaction shown below:

Temp. (°C)	$k \times 10^4$ (s^{-1})
60.0	0.30
70.0	0.97
75.0	1.79
80.0	3.09
90.0	8.92
95.0	15.90

3. 2-Vinylmethylenecyclopropane rearranges in the gas phase to 3-methylenecyclopentene. Two possible reaction mechanisms are:

(a) Sketch a reaction energy diagram for each process.

(b) How might an isotopic labeling experiment distinguish between these mechanisms?

4. In Table 4.5 the phenyl group is assigned both a σ^+ and a σ^- value. Furthermore, the signs of σ^+ and σ^- are different. Discuss the reasons that the phenyl group has both a σ^+ and a σ^- value and explain why they are of different signs.

5. Match the ρ values with the appropriate reactions. Explain your reasoning. Reaction constants: $+2.45$, $+0.75$, -2.39, -7.29

   Reactions:
   (a) nitration of substituted benzenes   cation
   (b) ionization of substituted benzenethiols anion
   (c) ionization of substituted benzenephosphonic acids anion
   (d) reaction of substituted N,N-dimethylanilines with methyl iodide. cation

6. (a) Determine the value of ρ for the reaction shown from the data given:

Y$-$⟨benzene⟩$-$SO$_2$$-$O (ring with CH$_2$) $+$ OH$^-$ $\longrightarrow$ Y$-$⟨benzene⟩$-$CH$_2$SO$_3^-$, OH

Y	$k\ (M^{-1}\,s^{-1})$
H	37.4
CH$_3$O	21.3
CH$_3$	24.0
Br	95.1
NO$_2$	1430

(b) The pseudo-first-order rate constants for the acid-catalyzed hydration of substituted styrenes in 3.5 M HClO$_4$ at 25°C are given. Plot the data against σ and σ^+ and determine ρ and ρ^+. Interpret the significance of the results.

Substituent	$k \times 10^8\ (s^{-1})$
p-CH$_3$O	488,000
p-CH$_3$	16,400
H	811
p-Cl	318
p-NO$_2$	1.44

252

CHAPTER 4
STUDY AND
DESCRIPTION OF
ORGANIC REACTION
MECHANISMS

(c) The acidity of a series of substituted benzyldimethylammonium ions has been measured. Determine whether these data are correlated by the Hammett equation. What is the value of ρ? What interpretation do you put on its sign?

$$X-\text{C}_6\text{H}_4-\text{CH}_2\overset{+}{\text{N}}\text{H}(\text{CH}_3)_2 \overset{K_a}{\rightleftharpoons} X-\text{C}_6\text{H}_4-\text{CH}_2\text{N}(\text{CH}_3)_2 + \text{H}^+$$

X	pK_a
p-CH$_3$O	9.32
p-CH$_3$	9.22
p-F	8.94
H	9.03
m-NO$_2$	8.19
p-NO$_2$	8.14
p-Cl	8.83
m-Cl	8.67

7. Write the rate law that would describe the rate of product formation for each of the following systems:

(a)

$$\begin{array}{c}\text{H}\\ \text{Br}\end{array}\text{C=C}\begin{array}{c}\text{H}\\ \text{Br}\end{array} + \text{R}_3\text{N} \underset{k_{-1}}{\overset{k_1}{\rightleftharpoons}} \begin{array}{c}\text{H}\\ \text{Br}\end{array}\text{C=C}\begin{array}{c}\\ \text{Br}\end{array}^- + \text{R}_3\text{NH}^+$$

$$\begin{array}{c}\text{H}\\ \text{Br}\end{array}\text{C=C}\begin{array}{c}\\ \text{Br}\end{array}^- \overset{k_2}{\longrightarrow} \text{H-C}\equiv\text{C-Br} + \text{Br}^-$$

$$\text{H-C}\equiv\text{C-Br} + \text{R}_3\text{N} \overset{k_3}{\longrightarrow} \text{H-C}\equiv\text{C-}\overset{+}{\text{N}}\text{R}_3 + \text{Br}^-$$

if the second step is rate-controlling and the first step is a preequilibrium.

(b)

$$(\text{CH}_3)_3\text{CO}_2\text{C-C}_6\text{H}_4-\text{NO}_2 \xrightarrow[\text{acetone}]{\underset{\text{H}_2\text{O}}{k_1}} (\text{CH}_3)_3\text{C}^+ + {}^-\text{O}_2\text{C-C}_6\text{H}_4-\text{NO}_2$$

$$(\text{CH}_3)_3\text{C}^+ + \text{H}_2\text{O} \overset{k_2}{\longrightarrow} (\text{CH}_3)_3\text{COH} + \text{H}^+$$

and

$$(\text{CH}_3)_3\text{C}^+ \overset{k_3}{\longrightarrow} (\text{CH}_3)_2\text{C=CH}_2 + \text{H}^+$$

if the competing product-forming steps are faster than the first step.

(c)

assuming that the σ complex is a steady-state intermediate. The final step is a rapid equilibrium that converts some of the initial Br_2 to unreactive Br_3^-. What is the rate expression if the intermediate goes to product much faster than it reverts to starting material and if the equilibrium constant for tribromide ion formation is large?

(d)

where no assumption is made as to the relative magnitude of k_1, k_{-1}, and k_2.

8. The rates of brominolysis of a series of 1,2-diarylcyclopropanes under conditions where the rate is determined by Br· attack and leads to 1,3-dibromo-1,3-diarylpropanes are given below.

Set up an equation which you would expect to correlate the observed rate of reaction with both the Ar^1 and Ar^2 substituents. Check the performance of your equation by comparing the correlation with the data given below. Discuss the results of the correlation.

254

CHAPTER 4
STUDY AND
DESCRIPTION OF
ORGANIC REACTION
MECHANISMS

Ar^1	Ar^2	Relative rate	Ar^1	Ar^2	Relative rate
p-MeO	p-MeO	1.6×10^4	p-Cl	p-Ph	48
p-Ph	p-MeO	4.4×10^3	p-CN	p-Ph	45
H	p-MeO	2.5×10^3	p-Br	p-Cl	18
p-Cl	p-MeO	2.3×10^3	p-Br	p-Br	15
m-Br	p-MeO	2.1×10^3	m-Br	H	10
p-CN	p-MeO	1.6×10^3	m-Br	p-Cl	4.8
p-NO$_2$	p-MeO	1.3×10^3	p-CN	H	1.8
H	p-Ph	6.8×10^2	p-NO$_2$	H	0.88
p-Cl	p-Ph	4.3×10^2	p-Cl	p-CN	0.55
m-Br	p-Ph	1.8×10^2	p-Cl	p-NO$_2$	0.37
H	H	1.4×10^2	p-CN	p-CN	0.046

9. Predict whether normal or inverse isotope effects will be observed for each reaction below. Explain. Indicate any reactions in which you would expect $k_H k_D > 2$. The isotopically substituted hydrogens are marked with asterisks.

(a)

$$CH_3CH_2\text{*}CHCH_2\text{*}CH_3 \xrightarrow[H_2O]{EtOH} CH_3CH_2\text{*}CHCH_2\text{*}CH_3 \\ \quad \overset{|}{OSO_2Ar} \qquad\qquad\qquad\qquad +$$

$$\longrightarrow CH_3CH_2\text{*}CHCH_2\text{*}CH_3 + CH_3CH_2\text{*}CHCH_2\text{*}CH_3 \\ \qquad\qquad\qquad \overset{|}{OH} \qquad\qquad\qquad\qquad \overset{|}{OCH_2CH_3}$$

(b)

$$CH_3\text{*}CHCH_3\text{*} + {}^-OH \longrightarrow CH_3\text{*}CCH_3\text{*} + H_2O \\ \quad\; \overset{|}{NO_2} \qquad\qquad\qquad\qquad \overset{\|}{NO_2^-}$$

(c)

$$Ph_2C=C=O + PhCH=CH_2\text{*} \longrightarrow$$

(d)

$$CH_3CH_2C\overset{+}{=}OC_2H_5 + H_2O \xrightarrow{fast} CH_3CH_2CH\text{*}=O + C_2H_5OH \\ \quad\;\; \overset{|}{H\text{*}}$$

(e)

(f)

(g)

$$H_2{}^*C{=}CHCH_2O\overset{\overset{\text{S}}{\|}}{C}Ph \xrightarrow{100°C} Ph\overset{\overset{\text{O}}{\|}}{C}SCH_2{}^*CH{=}CH_2$$

(h)

$$H_2{}^*C{=}CH_2{}^* + Ag^+ \rightleftharpoons \left[H_2{}^*C\overset{\overset{\text{Ag}}{|}}{\overset{}{=}}CH_2{}^* \right]^+$$

10. Reactions of dialkylaluminum hydrides with acetylenes give addition products:

$$R_2AlH + R'C{\equiv}CR'' \longrightarrow \underset{\substack{R' \qquad\ \ R''}}{R_2Al}\overset{\qquad\ H}{\underset{}{\diagup}C{=}C\diagdown}$$

$$A$$

The rate expression for the reaction is

$$-\frac{d[A]}{dt} = k[A][(R_2AlH)_3]^{1/3}$$

Propose a mechanism that could account for the overall four-thirds-order kinetics and the appearance of the dialkylaluminum hydride concentration to the one-third power.

11. The Cannizzaro reaction is a disproportionation that takes place in strongly basic solution and converts benzaldehyde to benzyl alcohol and sodium benzoate.

$$2PhCH{=}O + NaOH \xrightarrow{H_2O} PhCO_2{}^-Na^+ + PhCH_2OH$$

Several mechanisms have been postulated, all of which propose a hydride ion transfer as a key step. On the basis of the following results, postulate one or more mechanisms that are consistent with all the data provided. Indicate the significance of each observation with respect to the mechanism(s) you postulate.

(1) When the reaction is carried out in D_2O, the benzyl alcohol contains no deuterium in the methylene group.
(2) When the reaction is carried out in $H_2{}^{18}O$, both the benzyl alcohol and sodium benzoate contain ^{18}O.
(3) The reaction rate is given by the expression

$$\text{rate} = k_{obs}[PhCH{=}O]^2[^-OH]$$

(4) The rates of substituted benzaldehydes are correlated by the Hammett equation with $\rho = +3.76$.
(5) The solvent isotope effect k_{D_2O}/k_{H_2O} is 1.90.

12. A mechanism for alkene arylation by palladium(II) is given below. The isotope effect k_H/k_D was found to be 5 when benzene-d_6 was used. When styrene-β,β-d_2 was used,

256

CHAPTER 4
STUDY AND
DESCRIPTION OF
ORGANIC REACTION
MECHANISMS

no isotope effect was observed. Which step is rate-determining?

13. A scale for solvent ionizing power, Y^+, applicable in solvolysis reactions of cationic substrates, has been developed. For example,

$$S + (C_2H_5)_3O^+PF_6^- \rightarrow (C_2H_5)_2O + C_2H_5S^+ + PF_6^-$$

$$S = \text{solvent}$$

The numerical values of Y^+ are found to be related to Y, the measure of solvent ionizing power for neutral substrates, by the equation

$$Y^+ = -0.09Y$$

Explain, in qualitative terms, (a) why Y^+ is negative with respect to Y and (b) why Y^+ is smaller in magnitude than Y (as is indicated by the coefficient 0.09).

14. Two mechanisms are among those that have been postulated for decomposition of aryl diazonium salts in aqueous solution containing nucleophilic anions, A^-:

Mechanism A

and

Mechanism B

and

Indicate how each of the following techniques might be applied to distinguishing between these mechanisms:

(a) kinetic studies
(b) rate and product composition as a function of [A⁻]
(c) solvent isotope effect studies
(d) isotope effect resulting from substitution of D for H at *ortho* positions
(e) substituent effect studies

15. Cycloheptatrienes are in many cases in rapid equilibrium with an isomeric bicyclo[4.1.0]heptadiene. The thermodynamics of the valence isomerism has been studied in a number of instances, and some of the data are given below. Calculate the equilibrium constant for each case at 25°C. Calculate the temperature at which $K = 1$ for each system. Are the signs of the enthalpy and entropy as you would expect them to be? Can you discern any pattern of substituent effects from the data?

Ar	$\Delta H°$ (kcal/mol)	$\Delta S°$ (eu)
Phenyl	−5.4	−16.8
p-Nitrophenyl	−3.5	−11.0
p-Methoxyphenyl	−2.3	−7.4

16. Bicyclo[2.2.1]heptadiene rearranges at elevated temperatures to cycloheptatriene and toluene. The reaction is facilitated by substituents at C-7 such as phenyl and alkoxy, in which case cycloheptatrienes are the dominant products.

For R = *t*-butoxy, the rate data are given for several temperatures in decane.

Temp. (°C)	k (s⁻¹)
139.8	7.28×10^{-6}
154.8	3.37×10^{-5}
170.3	1.43×10^{-4}

The reaction is about 50% faster in ethoxyethanol than in decane. Calculate the activation parameters at 150°C. Although precisely comparable data are not available, E_a for the gas-phase isomerization of norbornadiene is ~50 kcal/mol. Draw a sketch

258

CHAPTER 4
STUDY AND
DESCRIPTION OF
ORGANIC REACTION
MECHANISMS

showing the degree of transition-state stabilization or destabilization caused by the alkoxy substituent. Is there any basis for regarding the bond cleavage in the rate-determining step to be heterolytic or homolytic? How do you propose that the effect of the substituent group R operates? Can you propose an experiment that might support your proposal?

17. A study of the aromatic nitration reaction in aqueous nitric acid revealed that when no aromatic substrate was present, an incorporation of ^{18}O from labeled water into nitric acid occurred.

$$HNO_3 + H_2{}^{18}O \rightarrow HN^{18}O_3 + H_2O$$

The rate of this exchange process was identical with the rate of nitration of several reactive aromatic hydrocarbons. Discuss how this result is consistent with mechanism B on p. 196, but not with mechanisms A or C.

18. Comparison of several series of solvolysis reactions that proceed via carbocation intermediates reveals that an α-cyano group retards the reactions by a factor of about 10^3. A β-cyano group is even more strongly rate-retarding, with the factor being as high as 10^7. Why are both α- and β-cyano groups rate-retarding? What might cause the β-cyano group to be more rate-retarding than the α-cyano group?

19. Comparison of the gas-phase acidity of benzoic acids with pK_a values of the same compounds in aqueous solution provides some interesting relationships.

(1) The trend in acidity as a function of substituent is the same but the magnitude of the substituent effects is much larger in the gas phase. (The $\Delta\Delta G°$ for any given substituent is about 10 times larger in the gas phase.)

(2) Whereas acetic acid and benzoic acid are of comparable acidity in water, benzoic acid is much more acidic in the gas phase.

(3) While the substituent effect in the gas phase is assumed to be nearly entirely an enthalpy effect, it can be shown that in solution the substituent effect is largely the result of changes in ΔS.

Discuss how the change from gas phase to water solution can cause each of these effects.

20. It has been suggested that the chemical shift of aromatic ring carbons might provide a good indication of the intrinsic electron-releasing or electron-attracting capacity of substituents in circumstances where there is no perturbation by an approaching reagent. Such a perturbation is always present in substituent effects determined on the basis of reactivity. The measured chemical shifts from benzene for the carbon *para* to the substituent are given below. Plot these against σ, σ^+, and σ^-. What conclusions do you draw from these plots?

R	$\Delta\delta^a$	R	$\Delta\delta^a$
NH_2	-9.86	CF_3	3.19
OCH_3	-7.75	CN	3.80
F	-4.49	CH_3CO	4.18
Cl	-2.05	CH_3O_2C	4.12
Br	-1.62	CH_3SO_2	4.64
CH_3	-2.89	NO_2	5.53

a. $\Delta\delta$ is the change in chemical shift in CCl_4 from benzene in ppm; a negative sign indicates increased shielding.

21. (a) The σ_I, σ_R, and σ_R^- substituent constants for the nitroso and nitro groups are given below. How do you account for their implications for the electronic character of these groups, i.e., that while NO_2 has the stronger field/inductive effect, the N=O group has the stronger resonance effect?

	σ_I	σ_R	σ_R^-
N=O	0.34	0.33	1.46
NO_2	0.56	0.19	1.27

(b) Dual-substituent-parameter correlations have been found for the ^{15}N and [17]O chemical shifts in a series of substituted nitro compounds, as shown below. Give a structural interpretation of these results.

meta	para
$^{15}N = -7.5\sigma_I - 2.5\sigma_R^0$	$^{15}N = -6.0\sigma_I - 1.0\sigma_R^0$
$^{17}O = -6.5\sigma_I - 0.6\sigma_R^-$	$^{17}O = 13.5\sigma_I + 15.6\sigma_R^+$

22. The ionization of a series of 4-substituted pyridines has been studied, and both equilibrium acidities (pK_a) and enthalpies of ionization have been recorded at 25°C:

$$R-\underset{}{\bigcirc}N^+\!-H \;\rightleftharpoons\; R-\underset{}{\bigcirc}N + H^+$$

R	pK_a	$\Delta H°$ (kcal/mol)
H	5.21	4.8
NH_2	9.12	11.3
OCH_3	6.58	6.8
CH_3	6.03	6.1
Cl	3.83	3.6
Br	3.75	3.5
CN	1.86	1.3

260

CHAPTER 4
STUDY AND
DESCRIPTION OF
ORGANIC REACTION
MECHANISMS

Calculate $\Delta S°$ for ionization of each compound. Comment on the contribution of $\Delta H°$ and $\Delta S°$ to the free energy of ionization. Test the data for linear free-energy correlations. Are the linear free-energy correlations dominated by entropy or enthalpy terms?

23. Allyl esters undergo rearrangement reactions at 300°C and above. Two examples are shown, one of which is "degenerate," since the product and reactant are identical:

$$CH_3CH{=}CHCH_2OCCH_3 \longrightarrow CH_3CHCH{=}CH_2$$
(with C=O on the acetyl group; product has OCCH$_3$ / O substituent)

$$H_2C{=}CHCH_2OCCH_3 \longrightarrow CH_3COCH_2CH{=}CH_2$$

At least three distinct mechanisms can be written for these reactions. Write down some possible mechanisms, and suggest isotopic labeling studies that could distinguish among the possibilities you have proposed.

24. Estimates of the heat of solvation of various species in DMSO as compared to water have been made and can be expressed as enthalpies of transfer. Some data for some common ions are given below. Discuss their significance.

$$X^{\ddagger}_{H_2O} \rightarrow X^{\ddagger}_{DMSO}$$

X	$\Delta H_{transfer}$ (kcal/mol)
K$^+$	−8.8
Na$^+$	−7.1
Cl$^-$	+4.9

25. Norbornene, norbornadiene, nortricyclane, and quadicyclane can all be hydrogenated to norbornane. The measured heats of hydrogenation are given in the scheme below. The table gives the $\Delta H_f°$ values obtained for each of these compounds by MM and three semiempirical MO methods. Compare the accuracy of the semiempirical methods in predicting the experimental heats of hydrogenation.

Enthalpies of Formation of the Norbornadiene Cycle[a]

Compound	MM[b]	MNDO	AM1	PM3	Exp.
Norbornadiene	55.5 (30.9)	62.7	67.7	58.8	57.4
Norbornene	19.5 (22.7)	25.3	26.0	22.0	21.4
Norbornane	12.8 (18.1)	−10.4	−14.4	−13.7	−12.4
Nortricyclane	19.5 (52.6)	27.1	33.8	26.0	20.2
Quadricyclane	79.4 (108.1)	79.1	104.4	86.3	79.5

a. Units are kcal/mol.
b. Calculated strain energies are given in parentheses.

26. The second-order rate constants for the reaction of a number of amines with benzyl chloride are tabulated below. Calculate $\Delta H^{\ddagger}$ and $\Delta S^{\ddagger}$ from the data. Offer an explanation for the relative reactivity order for the amines. What trends do you observe in $\Delta H^{\ddagger}$ with reactivity?

Rate Constants for the Reaction of Tertiary Amines with Benzyl Chloride

	Rate constants $10^5 k_2$ $(M^{-1}\ s^{-1})$								
	20°C	24°C	30°C	40°C	50°C	60°C	70°C	80°C	90°C
$(CH_3)_3N$	38.2	50.2	72.8						
$(C_2H_5)_3N$				1.67	3.07	4.54	7.49		
$(C_3H_7)_3N$				0.354	0.633	1.05	1.76	3.04	
$(C_4H_9)_3N$				0.471	0.844	1.24	1.94	3.22	
$(C_6H_{13})_3N$				0.290	0.566	0.860	1.54	2.62	
$(C_8H_{17})_3N$				0.336	0.570	0.912	1.60	2.73	
$PhN(CH_3)_2$						0.135	0.233	0.384	0.698
Pyridine				0.168	0.337	0.910	1.55	2.63	
Quinoline					0.051	0.105	0.226	0.457	0.820

Nucleophilic Substitution

Introduction

Nucleophilic substitution at carbon is of broad synthetic utility and has received exceptionally detailed mechanistic study by organic chemists. The goal of developing a coherent mechanistic interpretation was first undertaken by C. K. Ingold and E. D. Hughes in England in the 1930s. Their studies laid the basis for current understanding.[1] Since those initial investigations, organic chemists have continued to study substitution reactions, and much detailed information about this type of reaction is available. From these accumulated data, a broad conceptual interpretation of nucleophilic substitution has developed. We can provide only a small selection of these details to illustrate the general concepts. The area of nucleophilic substitution also illustrates the fact that while a broad conceptual framework can outline the general features to be expected for a given system, precise details will reveal aspects that are characteristic of specific systems. As the chapter unfolds, the reader should come to appreciate both the breadth of the general concepts and the depth of knowledge about special characteristics of some of the individual systems.

Nucleophilic substitution reactions may involve several different combinations of charged and uncharged species as reactants. The equations in Scheme 5.1 illustrate the four most common charge types. These reactions illustrate the relationship of reactants and products in nucleophilic substitution reactions but say nothing about mechanism. In order to approach an understanding of the mechanisms of such reactions, let us begin with a review of the limiting cases as defined by Hughes and Ingold. These limiting cases are the *ionization mechanism* (S_N1, substitution–nucleophilic–unimolecular) and the *direct displacement mechanism* (S_N2, substitution–nucleophilic–bimolecular).

1. C. K. Ingold, *Structure and Mechanism in Organic Chemistry*, 2nd ed., Cornell University Press, Ithaca, New York, 1969.

Scheme 5.1. Representative Nucleophilic Substitution Reactions

A. Neutral substrate + neutral nucleophile

$RX + Y: \longrightarrow RY^+ + X:^-$

1^a $CH_3CH_2I + (CH_3CH_2CH_2CH_2)_3P: \xrightarrow[\text{acetone}]{} (CH_3CH_2CH_2CH_2)_3\overset{+}{P}CH_2CH_3\ I^-$

   quantitative

2^b $PhC(CH_3)_2Cl + CH_3CH_2OH \longrightarrow PhC(CH_3)_2\overset{+}{\underset{\underset{H}{|}}{O}}CH_2CH_3 \xrightarrow{-H^+} PhC(CH_3)_2OCH_2CH_3$

   87%

3^c $CH_3CHCH_2CH_3 + H_2O \xrightarrow[\text{acetone}]{} CH_3CHCH_2CH_3 \xrightarrow{-H^+} CH_3CHCH_2CH_3$

   with substituent $O=S=O$ with p-tolyl / $\overset{+}{O}H_2$ / OH

   77%

(The p-toluenesulfonate group is commonly referred to as *tosylate* and abbreviated –OTs.)

B. Neutral substrate + anionic nucleophile

$RX + Y:^- \longrightarrow RY + X:^-$

4^d $CH_3CHCH_2C\equiv N \xrightarrow[\text{acetone}]{NaI} CH_3CHCH_2C\equiv N$

   $\underset{Br}{|}$ $\underset{I}{|}$ 96%

5^e (bicyclic)$CH_2OTs \xrightarrow[\text{acetone}]{LiBr}$ (bicyclic)CH_2Br

   94%

6^f $CH_3CH(CH_2)_5CH_3 + PhS^- \xrightarrow[\text{ethanol}]{} CH_3CH(CH_2)_5CH_3$

   $\underset{OTs}{|}$ $\underset{SPh}{|}$

5.1. The Limiting Cases—Substitution by the Ionization (S_N1) Mechanism

The ionization mechanism for nucleophilic substitution proceeds by rate-determining heterolytic dissociation of the reactant to a tricoordinate carbocation (also sometimes referred to as a *carbonium ion* or *carbenium ion*)[2] and the leaving group. This dissociation is followed by rapid combination of the highly electrophilic carbocation with a Lewis base (nucleophile) present in the medium. A two-dimensional potential energy diagram representing this process for a neutral reactant and anionic nucleophile is shown in Fig. 5.1.

This mechanism has several characteristic features. Because the ionization step is rate-determining, the reaction will exhibit first-order kinetics, with the rate of decom-

2. Tricoordinate carbocations are frequently called *carbonium ions*. The terms methyl cation, butyl cation, etc., are used to describe the corresponding tricoordinate cations. *Chemical Abstracts* uses as specific names *methylium*, *ethylium*, *propylium*. We will use *carbocation* as a generic term for trivalent carbon cations.

265

SECTION 5.1.
THE LIMITING CASES—
SUBSTITUTION BY THE
IONIZATION (S_N1)
MECHANISM

C. Cationic substrate + neutral nucleophile

$$RX^+ + Y: \longrightarrow RY^+ + X:$$

7^g $PhC\overset{+}{H}\overset{|}{\underset{CH_3}{S}}(CH_3)_2 + (H_2N)_2C{=}S \xrightarrow[\text{acetonitrile}]{60°C} PhCH{-}S{-}\overset{\overset{+}{N}H_2}{\underset{NH_2}{C}}$
$\underset{CH_3}{\overset{|}{}}$

8^h $Ph_2C{=}\overset{+}{N}{=}\overset{..}{\underset{..}{N}}: \xrightarrow{TsOH} Ph_2CH{-}\overset{+}{N}{\equiv}N \xrightarrow{CH_3CH_2OH} Ph_2CH\overset{+}{\underset{H}{O}}CH_2CH_3 \xrightarrow{-H^+} Ph_2CHOCH_2CH_3$

D. Cationic substrate + anionic nucleophile

$$RX^+ + Y:^- \longrightarrow RY + X:$$

9^i $(CH_3CH_2)_3O^+ + (CH_3)_3CCO_2^- \longrightarrow \underset{90\%}{(CH_3)_3CCO_2CH_2CH_3} + (CH_3CH_2)_2O$

10^j $CH_2{=}CHCH_2CH_2\overset{+}{\underset{CH_3}{S}}C_6H_5 \xrightarrow[\text{dimethyformamide}]{NaI} \underset{52\%}{CH_2{=}CHCH_2CH_2I}$

11^k

a. S. A. Buckler and W. A. Henderson, *J. Am. Chem. Soc.* **82**:5795 (1960).
b. R. L. Buckson and S. G. Smith, *J. Org. Chem.* **32**:634 (1967).
c. J. D. Roberts, W. Bennett, R. E. McMahon, and E. W. Holroyd, *J. Am. Chem. Soc.* **74**:4283 (1952).
d. M. S. Newman and R. D. Closson, *J. Am. Chem. Soc.* **66**:1553 (1944).
e. K. B. Wiberg and B. R. Lowry, *J. Am. Chem. Soc.* **85**:3188 (1963).
f. H. L. Goering, D. L. Towns, and B. Dittmar, *J. Org. Chem.* **27**:736 (1962).
g. H. M. R. Hoffmann and E. D. Hughes, *J. Chem. Soc.* **1964**:1259.
h. J. D. Roberts and W. Watanabe, *J. Am. Chem. Soc.* **72**:4869 (1950).
i. D. J. Raber and P. Gariano, *Tetrahedron Lett.* **1971**:4741.
j. E. J. Corey and M. Jautelat, *Tetrahedron Lett.* **1968**:5787.
k. H. Hellman, I. Loschmann, and F. Lingens, *Chem. Ber.* **87**:1690 (1954).

position of the substrate being independent of the concentration or nature of the nucleophile.

$$RX \xrightarrow[\text{slow}]{k_1} R^+ + X:^-$$

$$R^+ + Y:^- \xrightarrow[\text{fast}]{k_2} RY$$

$$\text{rate} = -\frac{d[RX]}{dt} = -\frac{d[Y:^-]}{dt} = k_1[RX]$$

Application of Hammond's postulate indicates that the transition state should resemble the product of the first step, the carbocation intermediate. Ionization is facilitated by factors that either lower the energy of the carbocation or raise the energy of the reactant. The rate of ionization depends primarily on how reactant structure and solvent ionizing power affect these energies.

Ionization reaction rates are subject to both electronic and steric effects. The most important electronic effects are stabilization of the carbocation by electron-releasing

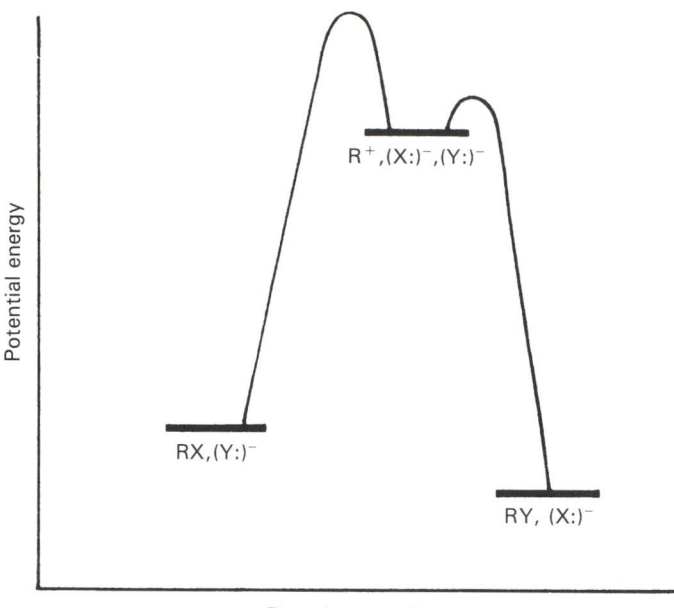

Fig. 5.1. Potential energy diagram for nucleophilic substitution by the ionization (S_N1) mechanism.

substituents and the ability of the leaving group to accept the electron pair from the covalent bond that is broken. Steric effects are also significant because of the change in coordination that occurs on ionization. The three remaining substituents are spread apart as ionization occurs so that steric compression by bulky groups favors the ionization. Geometrical constraints that preclude planarity of the carbocation are unfavorable and increase the energy of activation.

The ionization process is very sensitive to the medium. The effects of the medium depend on the charge type of the reactants. These relationships follow the general pattern for solvent effects discussed in Section 4.10. Ionization of a neutral substrate results in charge separation at the transition state, and solvent polarity will have a greater effect at the transition state than for the reactants. Solvents of higher dielectric constant will lower the energy of the transition state more than will solvents of lower polarity. Ionization of cationic substrates such as alkyl diazonium ions or trialkylsulfonium ions leads to dispersal of charge in the transition state and is moderately enhanced by less polar solvents, because the reactants are more strongly solvated than the transition state.

Stereochemical analysis can add detail to the mechanistic picture of the S_N1 substitution reaction. The ionization mechanism results in formation of a carbocation intermediate which is planar because of its sp^2 hybridization. If the carbocation is sufficiently long-lived under the reaction conditions to diffuse away from the leaving group, it becomes symmetrically solvated and gives racemic product. If this condition is not met, the solvation is dissymmetric, and product with net retention or inversion of configuration may be obtained, even though an achiral carbocation is formed. The extent of inversion or retention depends upon the details of the system. Examples of this effect will be discussed in later sections of the chapter.

A further consequence of the ionization mechanism is that if the same carbocation can be generated from more than one precursor, its subsequent reactions should be independent of its origin. However, as in the case of stereochemistry, certainty about this must be tempered by the fact that the ionization initially produces an ion pair. If the subsequent reaction takes place from this ion pair, rather than from the completely dissociated carbocation, the leaving group may influence the course of the reaction.

267

SECTION 5.2.
THE LIMITING CASES—
SUBSTITUTION BY THE
DIRECT DISPLACEMENT
(S_N2) MECHANISM

5.2. The Limiting Cases—Substitution by the Direct Displacement (S_N2) Mechanism

The direct displacement mechanism is concerted, without an intermediate, and proceeds through a single rate-determining transition state. According to this mechanism, the reactant is attacked by a nucleophile from the side opposite the leaving group, with bond making occurring simultaneously with bond breaking between the carbon atom and the leaving group. The transition state involves trigonal bipyramidal geometry with a pentacoordinate carbon. The nucleophile and the leaving group are both coordinated to the central carbon in the transition state. A potential energy diagram for direct displacement is given in Fig. 5.2.

The frontier molecular orbital approach provides a description of the bonding interactions that occur in the S_N2 process. The orbitals involved are depicted in Fig. 5.3. The frontier orbitals are a filled lone-pair orbital on the approaching nucleophile Y: and the σ^* antibonding orbital associated with the carbon undergoing substitution and the leaving group X. This antibonding orbital has a large lobe on carbon directed away from the C−X bond.[3] Back-side approach by the nucleophile is favored because the strongest

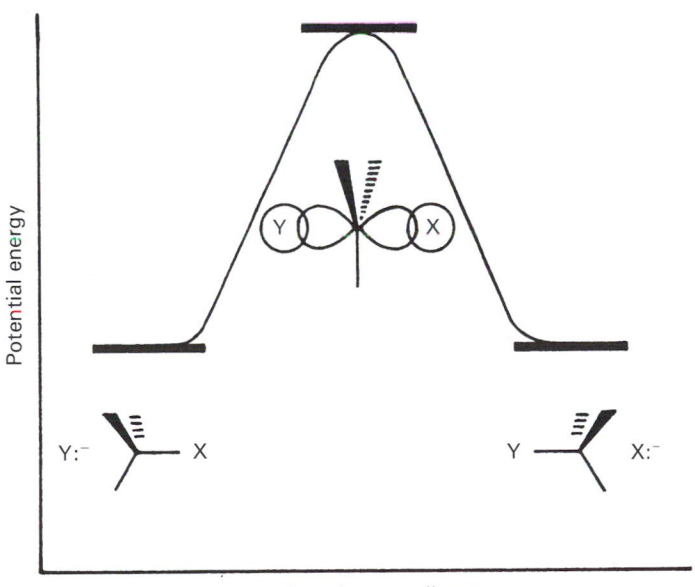

Fig. 5.2. Potential energy diagram for nucleophilic substitution by the direct displacement (S_N2) mechanism.

3. L. Salem, *Chem. Brit.* **5**:449 (1969); L. Salem, *Electrons in Chemical Reactions: First Principles*, John Wiley & Sons, New York, 1982, pp. 164–165.

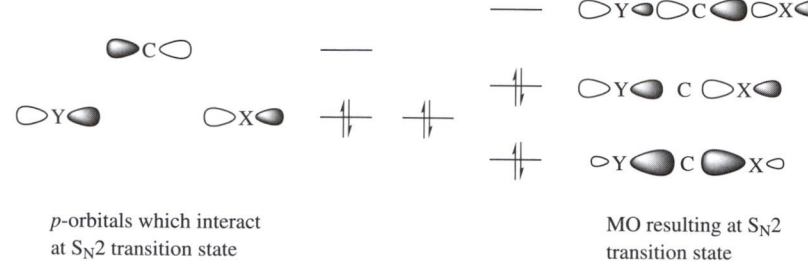

p-orbitals which interact
at S$_N$2 transition state

MO resulting at S$_N$2
transition state

Fig. 5.3. MO description of the transition state for an S$_N$2 displacement at carbon.

initial interaction is between the filled orbital on the nucleophile and the antibonding σ^* orbital.

σ^* orbital

Front-side approach is disfavored both because the density of the σ^* orbital is less in the region between the carbon and the leaving group and because front-side approach would involve both a bonding and an antibonding interaction with the σ^* orbital since it has a nodal surface between the atoms.

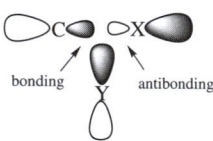

bonding   antibonding

The MO picture predicts that the reaction will proceed with inversion of configuration because the development of the transition state is accompanied by rehybridization of the carbon to the trigonal bipyramidal geometry. As the reaction proceeds on to product and sp^3 hybridization is reestablished, the product is formed with inversion of configuration.

The MO description of the S$_N$2 transition state presented in Fig. 5.3 shows a carbon p orbital interacting with two equivalent occupied orbitals, one from the leaving group and one from the nucleophile. These interacting orbitals give rise to three MOs describing the reacting bonds of the transition state. The nucleophile and leaving-group orbitals are shown as p orbitals, but the qualitative picture would not change if they were sp^3 or some other hybrid orbital. The HOMO at the transition state is π in character at the reacting carbon. The energy of this orbital is lowered by conjugation with adjacent substituents. The S$_N$2 transition state should therefore be stabilized by substituents that have an adjacent π orbital. The vinyl, phenyl, and carbonyl groups can all provide such stabilization, and, as we shall see later, each of these groups does enhance S$_N$2 reactivity.

The concerted displacement mechanism implies both kinetic and stereochemical consequences. The reaction will exhibit second-order kinetics, first-order in both reactant and nucleophile.

269

SECTION 5.3.
DETAILED
MECHANISTIC
DESCRIPTION AND
BORDERLINE
MECHANISMS

$$RX + Y:^- \xrightarrow{k} RY + X:^-$$

$$\text{rate} = -\frac{d[RX]}{dt} = -\frac{d[Y:^-]}{dt} = k[RX][Y:^-]$$

Because the nucleophile is intimately involved in the rate-determining step, not only will the rate depend on its concentration, but the nature of the nucleophile will be very important in determining the rate of the reaction. This is in marked contrast to the ionization mechanism, in which the identity and concentration of the nucleophile do not affect the rate of the reaction.

Because the degree of coordination increases at the reacting carbon atom, the rate of direct displacement is expected to be sensitive to the steric bulk of the other substituents. The optimum substrate from a steric point of view is CH_3-X since it provides minimum steric resistance to approach of the nucleophile. Each replacement of hydrogen by an alkyl group decreases the rate of reaction. As in the case of the ionization mechanism, the better the leaving group is able to accommodate an electron pair, the more facile is the reaction. That is, the better the leaving group, the faster is the reaction. However, because the nucleophile assists in the departure of the leaving group in the displacement mechanism, the effect of the leaving group on rate is less pronounced than in the ionization mechanism.

The points that we have emphasized in this brief overview of the S_N1 and S_N2 mechanisms are kinetics and stereochemistry. These features of a reaction provide important evidence for ascertaining whether a particular nucleophilic substitution follows an ionization or a direct displacement pathway. There are limitations to the generalization that reactions exhibiting first-order kinetics react by the S_N1 mechanism and those exhibiting second-order kinetics react by the S_N2 mechanism. Many nucleophilic substitutions are carried out under conditions in which the nucleophile is present in large excess. When this is the case, the concentration of the nucleophile is essentially constant during the reaction and the observed kinetics become *pseudo-first-order*. This is true, for example, when the solvent is the nucleophile (solvolysis). In this case, the kinetics of the reaction provide no evidence as to whether the S_N1 or S_N2 mechanism operates.

Stereochemistry also sometimes fails to provide a clear-cut distinction between the two limiting mechanisms. Many reactions proceed with partial inversion of configuration rather than complete racemization or inversion. Some reactions exhibit inversion of configuration but other features of the reaction suggest that an ionization mechanism must operate. Many systems exhibit "borderline" behavior such that it is difficult to distinguish between the ionization and direct displacement mechanisms. The types of reactants most likely to exhibit borderline behavior are secondary alkyl and primary and secondary benzylic systems. In the next section, we will examine in more detail the characteristics of these borderline systems.

5.3. Detailed Mechanistic Description and Borderline Mechanisms

The ionization and direct displacement mechanisms can be viewed as the extremes of a mechanistic continuum. At the S_N1 extreme, there is *no covalent interaction* between the reactant and the nucleophile in the transition state for cleavage of the bond to the leaving group. At the S_N2 extreme, the bond formation to the nucleophile is *concerted* with the bond-breaking step. In between these two limiting cases lies the borderline area, in which the degree of covalent interaction between the nucleophile and the reactant is intermediate between the two limiting cases. The concept of *ion pairs* is important in the consideration of the borderline area. The concept of ion pairs was introduced by Saul Winstein, who proposed that there were two distinct types of ion pairs involved in substitution reactions.[4] The ion-pair concept has been elaborated in detailed interpretations of substitution mechanisms.[5]

Winstein suggested that two intermediates preceding the dissociated carbocation were required to reconcile data on kinetics, salt effects, and stereochemistry of solvolysis reactions. The process of ionization initially generates a carbocation and counterion in proximity to each other. This species is called an *intimate ion pair* (or contact ion pair). This species can proceed to a *solvent-separated ion pair*, in which one or more solvent molecules have inserted between the carbocation and the leaving group but in which the ions have not diffused apart. The "free carbocation" is formed by diffusion away from the anion, which is called *dissociation*.

$$R—X \underset{\text{ionization}}{\rightleftharpoons} R^+X^- \rightleftharpoons R^+\|X^- \underset{\text{dissociation}}{\rightleftharpoons} R^+ + X^-$$

intimate ion pair   solvent-separated ion pair   dissociated ions

Attack by a nucleophile or the solvent can occur at either of the ion pairs. Nucleophilic attack on the intimate ion pair would be expected to occur with inversion of configuration, since the leaving group would still shield the front side of the carbocation. At the solvent-separated ion pair stage, the nucleophile might approach from either face, particularly in the case where solvent is the nucleophile. Reactions through dissociated carbocations should occur with complete racemization. According to this interpretation, the identity and stereochemistry of the reaction products will be determined by the extent to which reaction occurs on the un-ionized reactant, the intimate ion pair, the solvent-separated ion pair, or the dissociated carbocation.

Various specific experiments support this general scheme. For example, in 80% aqueous acetone, the rate constant for racemization of *p*-chlorobenzhydryl *p*-nitrobenzoate (k_{rac}) and that for exchange of ^{18}O label in the carbonyl group (k_{ex}) can both be measured.[6] At 100°C, $k_{ex}/k_{rac} = 2.3$.

optically active   racemic

4. S. Winstein, E. Clippinger, A. H. Fainberg, R. Heck, and G. C. Robinson, *J. Am. Chem. Soc.* **78**:328 (1956); S. Winstein, B. Appel, R. Baker, and A. Diaz, *Chem. Soc. Spec. Publ.* No. 19, 109 (1965).
5. J. M. Harris, *Prog. Phys. Org. Chem.* **11**: 89 (1974); D. J. Raber, J. M. Harris, and P. v. R. Schleyer, in *Ion Pairs*, M. Szwarc, ed., John Wiley & Sons, New York, 1974, Chapter 3; T. W. Bentley and P. v. R. Schleyer, *Adv. Phys. Org. Chem.* **14**:1 (1977); J. P. Richard, *Adv. Carbocation Chem.* **1**:121 (1989); P. E. Dietze, *Adv. Carbocation Chem.* **2**:179 (1995).
6. H. L. Goering and J. F. Levy, *J. Am. Chem. Soc.* **86**:120 (1964).

If it is assumed that ionization would result in complete randomization of the ^{18}O label in the carboxylate ion, k_{ex} is a measure of the rate of ionization with ion-pair return, and k_{rac} is a measure of the extent of racemization associated with ionization. The fact that the rate of isotope exchange exceeds that of racemization indicates that ion-pair collapse occurs with predominant retention of configuration. When a nucleophile is added to the system ($0.14\,M$ NaN$_3$), k_{ex} is found to be unchanged, but no racemization of reactant is observed. Instead, the intermediate that would return with racemization is captured by azide ion and converted to substitution product with inversion of configuration. This must mean that the intimate ion pair returns to reactant more rapidly than it is captured by azide ion, whereas the solvent-separated ion pair is captured by azide ion faster than it returns to racemic reactant.

Several other cases have been studied in which isotopic labeling reveals that the bond between the leaving group and carbon is able to break without net substitution occurring. A particularly significant case, since it applies to secondary sulfonates, which frequently exhibit borderline behavior, is isopropyl benzenesulfonate. During solvolysis of this compound in trifluoroacetic acid, it is found that exchange among the sulfonate oxygens occurs at about one-fifth the rate of solvolysis.[7] This implies that about one-fifth of the ion pairs recombine rather than react with the nucleophile.

271

SECTION 5.3.
DETAILED
MECHANISTIC
DESCRIPTION AND
BORDERLINE
MECHANISMS

$$(CH_3)_2CH-O-\overset{\overset{O^*}{\|}}{\underset{\underset{O^*}{\|}}{S}}-C_6H_5 \quad \xrightarrow[\substack{CF_3CO_2Na \\ k=36\times10^{-4}\,s^{-1}}]{CF_3CO_2H} \quad (CH_3)_2CHO_2CCF_3$$

$$k=8\times10^{-4}\,s^{-1}\,\updownarrow$$

$$(CH_3)_2CH-O^*-\overset{\overset{O}{\|}}{\underset{\underset{O^*}{\|}}{S}}-C_6H_5 \qquad O^* = {}^{18}O\ \text{label}$$

A study of the exchange reaction of benzyl tosylates during solvolysis in several solvents showed that for those with electron-donating substituents, e.g., p-methylbenzyl tosylate, the degree of exchange was quite high, implying reversible formation of a carbocation by an S_N1 mechanism. For benzyl tosylates with electron-attracting substituents, such as m-Cl, the amount of exchange was negligible, indicating that reaction occurred only by displacement by solvent. In this study, it was also demonstrated that there was no exchange with added "external" tosylate anion, proving that exchange occurred only at the ion-pair stage.[8]

The ion-pair return phenomenon can also be demonstrated by comparing the rate of loss of enantiomeric purity of reactant with the rate of product formation. For a number of systems, including 1-arylethyl tosylates,[9] the rate of decrease of optical rotation is greater than the rate of product formation. This indicates the existence of an intermediate that can re-form racemic reactant. The solvent-separated ion pair is the most likely intermediate in the Winstein scheme to play this role.

$$\underset{\underset{ArSO_2O}{|}}{ArCHCH_3} \;\rightleftharpoons\; \overset{+}{Ar}CHCH_3\|^-O_3SAr \;\xrightarrow{Nu^-}\; \underset{\underset{Nu}{|}}{ArCHCH_3}$$

7. C. Paradisi and J. F. Bunnett, *J. Am. Chem. Soc.* **107**:8223 (1985).
8. Y. Tsuji, S. H. Kim, Y. Saek, K. Yatsugi, M. Fuji, and Y. Tsuno, *Tetrahedron Lett.* **36**:1465 (1995).
9. A. D. Allen, V. M. Kanagasabapathy, and T. T. Tidwell, *J. Am. Chem. Soc.* **107**:4513 (1985).

Racemization, however, does not always accompany isotopic scrambling. In the case of *sec*-butyl 4-bromobenzenesulfonate, isotopic scrambling occurs in trifluoroethanol solution without any racemization. Two mechanisms are possible. Scrambling may involve an "intimate" ion pair in which the sulfonate can rotate with respect to the carbocation without allowing migration to the other face of the carbocation. The alternative is a concerted mechanism, which avoids a carbocation intermediate but violates the prohibition of front-side displacement.[10]

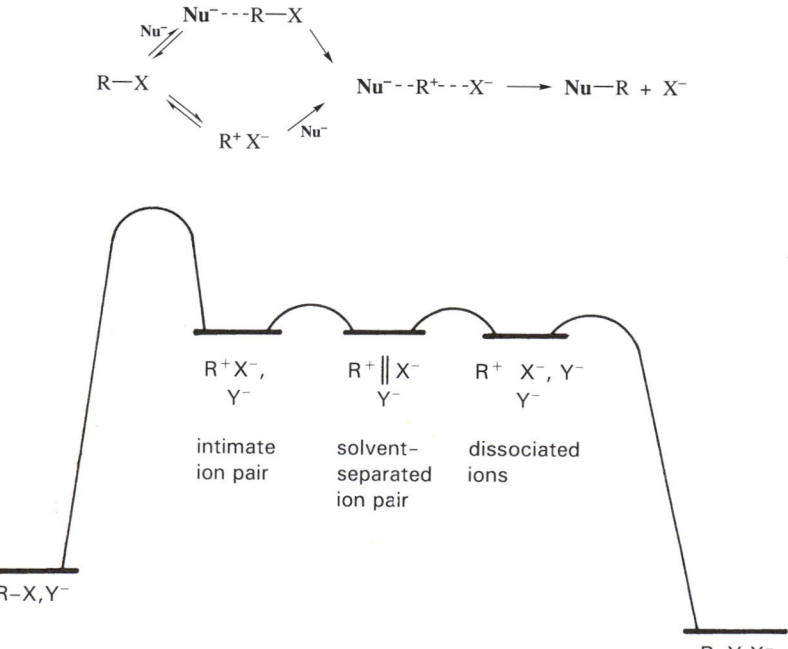

The concept of ion pairs in nucleophilic substitution is now generally accepted. Presumably, the barriers separating the intimate, solvent-separated, and dissociated ion pairs are quite small. The potential energy diagram in Fig. 5.4 depicts the three ion-pair species as being roughly equivalent in energy and separated by small barriers.

An elaboration of the ion-pair concept includes an "ion sandwich" in which a preassociation occurs between a potential nucleophile and a reactant. Such an ion sandwich might be a kinetic intermediate which accelerates dissociation. Alternatively, if a carbocation were quite unstable, it might always return to reactant unless a nucleophile was properly positioned to capture the carbocation.

Fig. 5.4. Schematic relationship between reactants, intermediate species, and products in substitution proceeding through ion pairs.

10. P. E. Dietze and M. Wojciechowski, *J. Am. Chem. Soc.* **112**:5240 (1990).

For many secondary sulfonates, nucleophilic substitution seems to be best explained by a concerted mechanism with a high degree of carbocation character at the transition state. This has been described as an "exploded transition state." Both the breaking and forming bonds are relatively weak so that the carbon has a substantial positive charge. However, the carbocation *per se* has no lifetime because bond breaking and formation occur concurrently.[11]

273

SECTION 5.3.
DETAILED
MECHANISTIC
DESCRIPTION AND
BORDERLINE
MECHANISMS

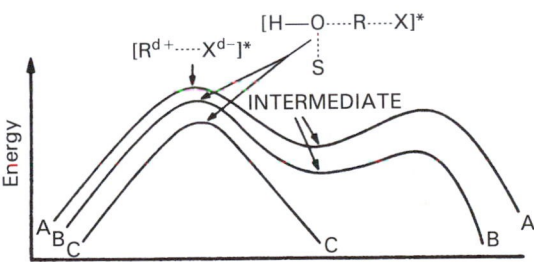

The gradation from S_N1 to S_N2 mechanisms can be described in terms of the shape of the potential energy diagrams for the reactions as illustrated in Fig. 5.5. Curves A and C represent the S_N1 and S_N2 limiting mechanisms, as described in the earlier sections. The transition from the S_N1 mechanism to the S_N2 mechanism involves greater and greater nucleophilic participation of the solvent or nucleophile in the transition state.[12] An ion pair with strong nucleophilic participation represents a mechanistic variation between the S_N1 and S_N2 processes. This mechanism is designated S_N2(intermediate) and pictures a carbocation-like transition state requiring back-side nucleophilic participation and therefore exhibiting second-order kinetics.

Jencks[13] has discussed how the gradation from the S_N1 to the S_N2 mechanism is related to the stability and lifetime of the carbocation intermediate, as illustrated in Fig. 5.6. In the S_N1 mechanism, the carbocation intermediate has a relatively long lifetime and is equilibrated with solvent prior to capture by a nucleophile. The reaction is clearly a stepwise one, and the energy minimum in which the carbocation intermediate resides is significant. As the stability of the carbocation decreases, its lifetime becomes shorter. The barrier to capture by a nucleophile becomes less and eventually disappears. This is described as the "uncoupled" mechanism. Ionization proceeds without nucleophilic

Fig. 5.5. Potential energy diagrams for substitution mechanisms. A is the S_N1 mechanism. B is the S_N2 mechanism with intermediate ion-pair or pentacoordinate species. C is the classical S_N2 mechanism. [Reproduced from T. W. Bentley and P. v. R. Schleyer, *Adv. Phys. Org. Chem.* **14**:1 (1977) by permission of Academic Press.]

11. B. L. Knier and W. P. Jencks, *J. Am. Chem. Soc.* **102**:6789 (1980); M. T. Skoog and W. P. Jencks, *J. Am. Chem. Soc.* **106**:7597 (1984).

12. T. W. Bentley and P. v. R. Schleyer, *Adv. Phys. Org. Chem.* **14**:1 (1977).

13. W. P. Jencks, *Acc. Chem. Res.* **13**:161 (1980).

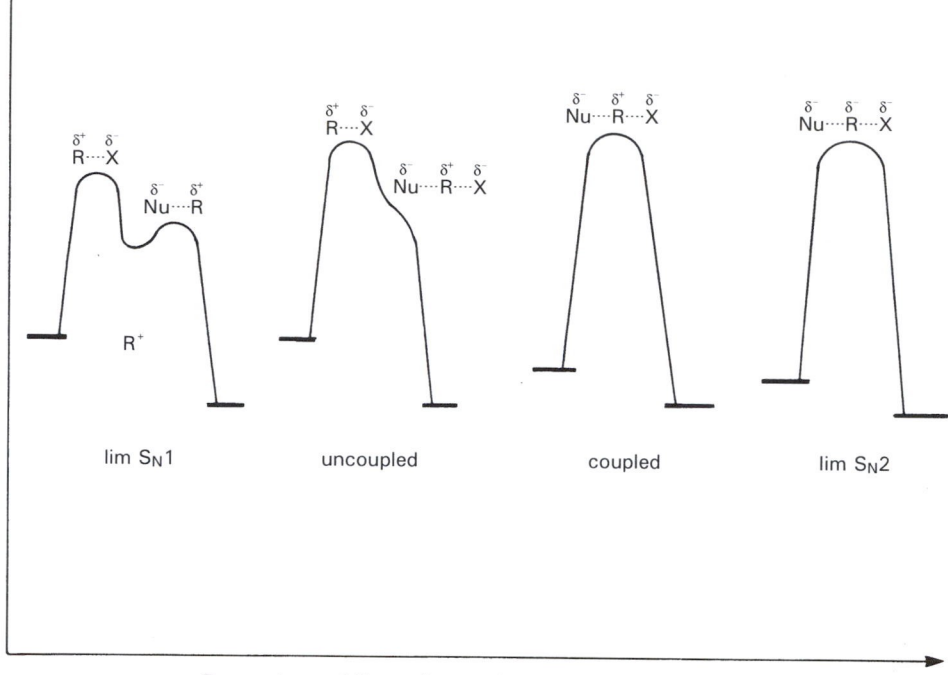

Fig. 5.6. Relationship between stability and potential lifetime of carbocation intermediate and mechanism for substitution.

participation, but the carbocation does not exist as a free intermediate. Such a reaction would still exhibit S_N1 kinetics, since there is no nucleophilic participation in the ionization. At still lesser carbocation stability, the lifetime of the ion pair would be so short that it always returns to reactant unless a nucleophile is present to capture it as it is formed. At this stage, the reaction would exhibit second-order kinetics, since the nucleophile must be present for reaction to occur. Jencks describes this as the "coupled" substitution process. Finally, as the stability of the (potential) carbocation becomes so low that it cannot form without direct participation of the nucleophile, the limiting S_N2 mechanism is reached. The continuum in Fig. 5.6 corresponds to decreasing carbocation character at the transition state proceeding from lim S_N1 to lim S_N2 mechanisms.

Figure 5.7 summarizes these ideas in a More O'Ferrall diagram.[14] The lim S_N2 mechanism corresponds to the concerted pathway through the middle of the diagram. It is imposed by high-energy carbocation intermediates that require nucleophilic participation. The lim S_N1 mechanism is the path along the edge of the diagram corresponding to separate bond-breaking and bond-forming steps. An "ion sandwich" mechanism implies a true intermediate in which the nucleophile is present in the transition state, but at which bond formation has not progressed. The "exploded transition state" mechanism describes a very similar structure, but one which is a transition state, not an intermediate.[15]

14. R. A. More O'Ferrall, *J. Chem. Soc. B.* **1970**:274.
15. For discussion of the borderline mechanisms, see J. P. Richard, *Adv. Carbocation Chem.* **1**:121 (1989); P. E. Dietze, *Adv. Carbocation Chem.* **2**:179 (1995).

275

SECTION 5.3.
DETAILED
MECHANISTIC
DESCRIPTION AND
BORDERLINE
MECHANISMS

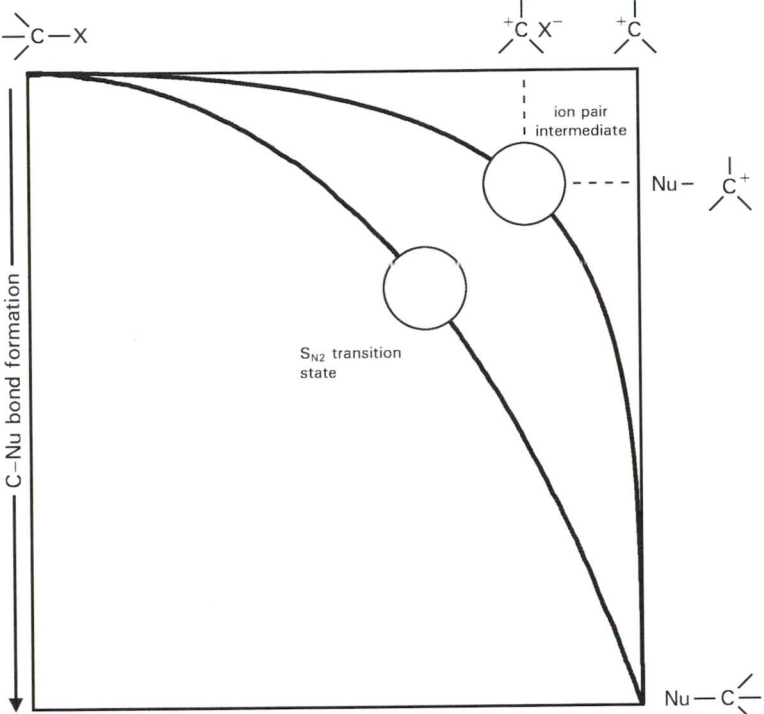

Fig. 5.7. More O'Ferrall–Jencks diagram showing concerted, ion-pair and stepwise mechanisms for nucleophilic substitution.

An example with the characteristics of the coupled displacement is the reaction of azide ion with substituted 1-phenylethyl chlorides. Although the reaction exhibits second-order kinetics, it has a substantially negative ρ value, indicative of an electron deficiency at the transition state.[16] The physical description of this type of activated complex is the "exploded" S_N2 transition state.

The importance of solvent participation in the borderline mechanisms should be noted. Nucleophilic participation is minimized by high electronegativity, which reduces the Lewis basicity and polarizability of the solvent molecules. Trifluoroacetic acid and perfluoro alcohols are among the least nucleophilic of the solvents used in solvolysis studies.[17] These solvents are used to define the characteristics of reactions proceeding without nucleophilic solvent participation. Solvent nucleophilicity increases with the electron-donating capacity of the molecule. The order trifluoroacetic acid < trifluoroethanol <acetic acid < water < ethanol gives a qualitative indication of the trend in solvent nucleophilicity. More will be said about solvent nucleophilicity in Section 5.5.

Reactant structure will also influence the degree of nucleophilic solvent participation. Solvation is minimized by steric hindrance. The 2-adamantyl system is regarded as being a

16. J. P. Richard and W. P. Jencks, *J. Am. Chem. Soc.* **106**:1383 (1984).
17. T. W. Bentley, C. T. Bowen, D. H. Morten, and P. v. R. Schleyer, *J. Am. Chem. Soc.* **103**:5466 (1981).

secondary substrate that cannot accommodate significant back-side nucleophilic participation.

The 2-adamantyl system has been used as a model reactant for defining the characteristics of ionization without nucleophilic participation. The degree of nucleophilic participation in other reactions can then be estimated by comparison with the 2-adamantyl system.[18]

5.4. Carbocations

Because carbocations are key intermediates in many nucleophilic substitution reactions, it is important to develop a grasp of their structural properties and the effect substituents have on stability. The critical step in the ionization mechanism of nucleophilic substitution is the generation of the tricoordinate carbocation intermediate. For this mechanism to operate, it is essential that this species not be prohibitively high in energy. Carbocations are inherently high-energy species. The ionization of t-butyl chloride is endothermic by 153 kcal/mol in the gas phase.[19]

$$(CH_3)_3CCl \rightarrow (CH_3)_3C^+ + Cl^-$$

An activation energy of this magnitude would lead to an unobservably slow reaction at normal temperature. Carbocation formation in solution is feasible because of the solvation of the ions that are produced.

One of the earliest pieces of evidence for the existence of carbocation intermediates was the observation that triphenylmethyl chloride (trityl chloride) gave conducting solutions when dissolved in liquid sulfur dioxide, a polar nonnucleophilic solvent. Trityl chloride also reacted with Lewis acids, such as aluminum chloride, to give colored saltlike solids.[20] In contrast to triphenylmethyl chloride, which has the properties of a covalent compound, triphenylmethyl perchlorate behaves as an ionic compound. The presence of triphenylmethyl cations in this solid has been confirmed by an X-ray crystal structure determination.[21] The central carbon is planar but the three phenyl rings are at an angle of 54° to the plane of the trigonal carbon so that the overall cation has a propeller-like shape. The temperature-dependent NMR spectrum of the carbocation indicates that it also has this structure in solution.[22] The twisting of the aromatic rings with respect to each other is evidently the result of van der Waals repulsions between the *ortho* hydrogens. The

18. F. L. Schadt, T. W. Bentley, and P. v. R. Schleyer, *J. Am. Chem. Soc.* **98**:7667 (1976).
19. D. W. Berman, V. Anicich, and J. L. Beauchamp, *J. Am. Chem. Soc.* **101**:1239 (1979).
20. For reviews of the arylmethyl cations, see C. D. Nenitzescu, in *Carbonium Ions*, Vol. I, G. A. Olah and P. v. R. Schleyer, eds., Wiley-Interscience, New York, 1968, Chapter 1; H. H. Freeman, in *Carbonium Ions*, Vol. IV, G. A. Olah and P. v. R. Schleyer, eds., Wiley-Interscience, New York, 1973, Chapter 28.
21. A. H. Gomes de Mesquita, C. H. MacGillavry, and K. Eriks, *Acta Crystallogr.* **18**:437 (1965).
22. I. I. Schuster, A. K. Colter, and R. J. Kurland, *J. Am. Chem. Soc.* **90**:4679 (1968).

difference in structure between the chloride and the perchlorate reflects the fact that the perchlorate is more stable as an anion.

The triarylmethyl cations are particularly stable because of the conjugation with the aryl groups, which delocalizes the positive charge. Because of their stability and ease of generation, the triarylmethyl cations have been the subject of studies aimed at determining the effect of substituents on carbocation stability. Many of these studies used the characteristic UV absorption spectra of the cations to determine their concentration. In acidic solution, equilibrium is established between triarylcarbinols and the corresponding carbocations.

$$ROH + H^+ \rightleftharpoons R^+ + H_2O$$

The relative stability of the carbocation can be expressed in terms of its pk_{r^+} which is defined as

$$pK_{R+} = \log\frac{[R^+]}{[ROH]} + H_R$$

where H_R is a measure of acidity defined for the medium.[23] In dilute aqueous solution, H_R is equivalent to pH, and pK_{R+} is equal to the pH at which the carbocation and alcohol are present in equal concentrations. By measuring the extent of carbocation formation at several acidities and applying the definition of pK_{R+}, the values shown in Table 5.1 were determined.

The pK_{R+} values allow for a comparison of the stability of carbocations. The carbocations that can be studied in this way are all relatively stable carbocations. The data in Table 5.1 reveal that electron-releasing substituents on the aryl rings stabilize the carbocation (more positive pK_{R+}) whereas electron-withdrawing groups such as nitro are destabilizing. This is what would be expected from the electron-deficient nature of the carbocation.

The diarylmethyl cations listed in Table 5.1 are 6–7 pK_{R+} units less stable than the corresponding triarylmethyl cations. This indicates that the additional aryl group has a cumulative, although not necessarily additive, effect on stability of the carbocation. Primary benzylic cations (monoarylmethyl cations) are generally not sufficiently stable for determination of pK_{R+} values. A particularly stable benzylic ion, the 2,4,6-trimethyl-phenylmethyl cation, has a pK_{R+} of -17.4.

One of the most important and general trends in organic chemistry is the increase in carbocation stability with additional alkyl substitution. This stability relationship is fundamental to understanding many aspects of reactivity, especially of nucleophilic

23. N. C. Deno, J. J. Jaruzelski, and A. Schriesheim, *J. Am. Chem. Soc.* **77**:3044 (1955).

Table 5.1. Values of pK_{R+} for Some Carbocations^a

Carbocation	pK_{R+}
Triarylmethyl cations	
Triphenylmethyl	−6.63
4,4′, 4″-Trimethyltriphenylmethyl	−3.56
4-Methoxytriphenylmethyl	−3.40
4,4′-Dimethoxytriphenylmethyl	−1.24
4,4′, 4,′ -Trimethoxytriphenylmethyl	+0.82
4,4′, 4″-Trichlorotriphenylmethyl	−7.74
4-Nitrotriphenylmethyl	−9.15
4,4′, 4″-Trinitrotriphenylmethyl	−16.27
4,4′, 4″-Tri(dimethylamino)triphenylmethyl	+9.36
Sesquixanthydrul^b	+9.05
Diarylmethyl cations	
Diphenylmethyl	−13.3
4,4′-Dimethyldiphenylmethyl	−10.4
4,4′-Dimethoxydiphenylmethyl	−5.71
2,2′, 4, 4′, 6, 6′-Hexamethyldiphenylmethyl	−6.6
4,4′-Dichlorodiphenylmethyl	−13.96
Miscellaneous carbocations	
Benzyl^c	−20
2-Phenyl-2-propyl^c	−12.3
Tricyclopropylmethyl^d	−2.3
Tropylium (cycloheptatrienyl)^a	+4.7
Triphenylcyclopropenyl^e	+3.1
Trimethylcyclopropenyl^f	+7.8
Tricyclopropylcyclopropenyl^g	+9.7

a. Unless otherwise indicated, the pK_{R+} values are taken from N. C. Deno, J. J. Jaruzelski, and A. Schriesheim, *J. Am. Chem. Soc.* **77**:3044 (1995); for an extensive compilation of similar data, see H. H. Freedman, in *Carbonium Ions*, Vol. IV, G. A. Olah and P. v. R. Schleyer, eds., Wiley-Interscience, New York, 1973, Chapter 28.
b. J. C. Martin and R. G. Smith, *J. Am. Chem. Soc.* **86**:2252 (1964).
c. T. L. Amyes, J. R. Richard, and M. Novak, *J. Am. Chem. Soc.* **114**:8032 (1992).
d. N. C. Deno, H. G. Richey, Jr., J. S. Liu, D. N. Lincoln, and J. O. Turner, *J. Am. Chem. Soc.* **87**:4533 (1965).
e. R. Breslow, H. Höver, and H. W. Chang, *J. Am. Chem. Soc.* **83**:2375 (1961).
f. J. Ciabattoni and E. C. Nathan III, *Tetrahedron Lett.* **1969**:4997.
g. K. Komatsu, I. Tomioka, and K. Okamoto, *Tetrahedron Lett.* **1980**:947; R. A. Moss and R. C. Munjal, *Tetrahedron Lett.* **1980**:1221.

substitution. In recent years, it has become possible to put the stabilization effect on a quantitative basis. One approach has been gas-phase measurements that determine the proton affinity of alkenes leading to carbocation formation. From these data, the hydride affinity of the carbocation can be obtained.

$$R^+ + H^- \rightarrow R - H \quad -\Delta G^\circ = \text{hydride affinity}$$

These data provide a thermodynamic basis for comparison of the relative stability of nonisomeric carbocations. Some representative results are shown in Table 5.2. The

Hydride affinity (kcal/mol)[a]			
CH_3^+	314[b]	$CH_2=CHCH_2^+$	256
$CH_3CH_2^+$	274[b]	$CH_2=CH\overset{+}{C}HCH_3$	237
$(CH_3)_2CH^+$	247[b]	$CH_2=CH\overset{+}{C}(CH_3)_2$	225
$(CH_3)_3C^+$	230[b]	$CH_3CH=CH\overset{+}{C}HCH_3$	225
$CH_2=CH^+$	287[c]		298[c]
	223		233
	258		226
	200		220

a. Except where noted, data are from D. H. Aue and M. T. Bowers, in *Gas Phase Ion Chemistry*, M. T. Bowers, ed. Academic Press, New York, 1979.
b. F. A. Houle and J. L. Beauchamp, *J. Am. Chem. Soc.* **101**:4067 (1979).
c. D. W. Berman, V. Anicich, and J. L. Beauchamp, *J. Am. Chem. Soc.* **101**:1239 (1979).

stability order *tert* > *sec* > *prim* > methyl is the same order as established on the basis of solvolysis rate in solution.

There is also a less dramatic but consistent trend which reveals that, within each structural class (primary, secondary, tertiary) larger ions are more stable than smaller ones, e.g., $t\text{-}C_4H_9 < t\text{-}C_5H_{11} < t\text{-}C_6H_{13}$.[24] The same trend is observed for primary cations.[25] The greater stability of the larger ions in the gas phase reflects their ability to disperse the positive charge over a larger number of atoms.

Because these stability measurements pertain to the gas phase, it is important to consider the effects that solvation might have on the structure–stability relationships. Hydride affinity values based on solution measurements can be derived from thermodynamic cycles that relate hydrocarbon pK, bond dissociation energy and electrochemical potentials. The hydride affinity, ΔG, for the reaction

$$R^+ + H^- \rightleftharpoons R\text{-}H$$

is a measure of carbocation stability. This quantity can be related to an electrochemical potential by summation with the energy for hydrogen atom removal, the bond dissociation energy.

$$
\begin{array}{ll}
R\text{-}H \rightleftharpoons R^+ + H^- & -\Delta G_{H^-} \\
R^{\cdot} + H^{\cdot} \rightleftharpoons R\text{-}H & -\Delta G_{hom} \\
\hline
R^{\cdot} + H^- \rightleftharpoons R^+ + H^{\cdot} & -\mathscr{F}\Delta\mathscr{E}^{\circ}\,[(H^{\cdot}/H^-) - (R^+/R^{\cdot})]
\end{array}
$$

24. F. P. Lossing and J. J. Holmes, *J. Am. Chem. Soc.* **106**:6917 (1984).
25. J. C. Schultz, F. A. Houle, and J. L. Beauchamp, *J. Am. Chem. Soc.* **106**:3917 (1984).

$$-\Delta G_{H^-} = \Delta G_{hom} - \mathscr{F}\Delta\mathscr{E}^\circ[(H^\cdot/H^-) - (R^+/R^\cdot)]$$

where (H/H^-) and $(R^+/R^\cdot)$ are one electron oxidation potentials for H^- and $R^\cdot$.[26] The former potential is about $-0.55\,V$ in DMSO.

Measurement of $(R^+/R^\cdot)$ can be accomplished by cyclic voltammetry for relatively stable species and by other methods for less stable cations. The values obtained for ΔG_{H^-} range from 83 kcal/mol for the aromatic tropylium ion to 130 kcal/mol for destabilized benzylic cations. For stable carbocations, the results obtained by this method correlate with cation stability as measured by pK_{R^+}. Some of these data are presented in Table 5.3.

It has been possible to obtain thermodynamic data for the ionization of alkyl chlorides by reaction with SbF_5, a Lewis acid, in the nonnucleophilic solvent SO_2ClF.[27] It has been found that the solvation energies of the carbocations in this medium are small and do not differ much from one another, making comparison of the nonisomeric systems possible. As long as subsequent reactions of the carbocation can be avoided, the thermodynamic characteristics of this reaction provide a measure of the relative ease of carbocation formation in solution.

$$RCl + SbF_5 \xrightarrow{SO_2ClF} R^+ + [SbF_5Cl]^-$$

There is an excellent correlation between these data and the gas-phase data, in terms both of the stability order and the energy differences between carbocations. A plot of the gas-phase hydride affinity versus the ionization enthalpy gives a line of slope 1.63 with a correlation coefficient of 0.973. This result is in agreement with the expectation that the gas-phase stability would be somewhat more sensitive to structure than the solution-phase stability. The energy gap between tertiary and secondary ions is about 17 kcal/mol in the gas phase and about 9.5 kcal/mole in the SO_2ClF solution.

An independent measurement of the energy difference between secondary and tertiary cations in solution is available from calorimetric measurement of the enthalpy

Table 5.3. Solution Hydride Affinity of Some Carbocations

	Hydride affinity (kcal/mol)[a]
Tropylium ion	83
Ph_3C^+	96
Ph_2CH^+	105
$PhCH_2^+$	118
p-MeOPhCH$_2^+$	106
p-NCPhCH$_2^+$	122

a. J.-P. Cheng, K. L. Handoo, and V. D. Parker, *J. Am. Chem. Soc.* **115**:2655 (1993).

26. J. P. Cheng, K. L. Handoo, and V. D. Parker, *J. Am. Chem. Soc.* **115**:2655 (1993).
27. E. M. Arnett and N. J. Pienta, *J. Am. Chem. Soc.* **102**:3329 (1980).

of isomerization of the *s*-butyl cation to the *t*-butyl cation. This value has been found to be -14.5 kcal/mol in SO_2ClF solution.[28]

$$CH_3CH_2\overset{+}{C}HCH_3 \xrightarrow{SO_2ClF} CH_3\overset{CH_3}{\underset{+}{\overset{|}{C}}}CH_3, \quad \Delta H = -14.5 \text{ kcal/mol}$$

A wide range of carbocation stability data has been obtained by measuring the heat of ionization of a series of chlorides and carbinols in nonnucleophilic solvents in the presence of Lewis acids.[29] Some representative data are given in Table 5.4 These data include the diarylmethyl and triarylmethyl systems for which pK_{R^+} data are available (Table 5.1) and give some basis for comparison of the stabilities of secondary and tertiary alkyl carbocations with those of the more stable aryl-substituted ions.

Any structural effect which reduces the electron deficiency at the tricoordinate carbon will have the effect of stabilizing the carbocation. Allyl cations are stabilized by delocalization involving the adjacent double bond.

$$H_2C\overset{\overset{H}{\underset{|}{C}}}{=}\overset{+}{C}H_2 \longleftrightarrow \overset{+}{H_2C}\overset{\overset{H}{\underset{|}{C}}}{}CH_2 \qquad H_2C\overset{\overset{H}{\underset{|}{C}}}{+}CH_2$$

The π-electron delocalization requires proper orbital alignment. As a result, there is a significant barrier to rotation about the carbon–carbon bonds in the allyl cation. The results of 6-31G/MP2 calculations show the perpendicular allyl cation to be 37.8 kcal/mol higher than the planar ion.[30] Related calculations indicate that rotation does not occur but that

Table 5.4. ΔH for Ionization of Chlorides and Alcohols in SO_2ClF over a Wide Structural Range[a]

	ΔH (kcal/mol)	
Reactant	X = Cl	X = OH
$(CH_3)_2CH-X$	-15	
$(Ph)_2\overset{|}{\underset{CH_3}{C}}-X$	-16	
$(CH_3)_3C-X$	-25	-35
$(CH_3)_2\overset{|}{\underset{Ph}{C}}-X$	-30	-40
$(Ph)_2\overset{|}{\underset{CH_3}{C}}-X$		-37.5
$(Ph)_3C-X$		-49
$(\triangleright\!\!-)_3C-X$		-59

a. Data from E. M. Arnett and T. C. Hofelich, *J. Am. Chem. Soc.* **105**:2889 (1983).

28. E. W. Bittner, E. M. Arnett, and M. Saunders, *J. Am. Chem. Soc.* **98**:3734 (1976).
29. E. M. Arnett and T. C. Hofelich, *J. Am. Chem. Soc.* **105**:2889 (1983).
30. A. Gobbi and G. Frenking, *J. Am. Chem. Soc.* **116**:9275 (1994).

Scheme 5.2. Rotational Energy Barriers for Allyl Cations
(kcal/mol)^a

a. From J. M. Bollinger, J. M. Brinich, and G. A. Olah, *J. Am. Chem. Soc.* **92**:4025 (1970).

instead a hydrogen migration intervenes.[31] For substituted allylic ions, the height of the barrier depends upon the substituent groups a, b, and R. Alkyl substituents lower the barrier. For the 1,1,3,3-tetramethyl ion, a barrier of 19.4 kcal/mol has been calculated. Some other values which have been measured are shown in Scheme 5.2.

Benzylic cations are well known to be stabilized by resonance interaction with the aromatic ring. The crystal structure of the 2-phenyl-2-propyl (cumyl) cation has been determined.[32] The cation is nearly planar and the structural parameters are consistent with delocalization. Benzyl cation stabilization is strongly affected by the substituents on the benzene ring. Substituent effects can be correlated by the Yukawa–Tsuno equation.[33] For example, gas-phase chloride-ion affinities correlate with the Yukawa–Tsuno equation with $\rho = -14.0$ and $r^+ = 1.29$, indicating a strong resonance interaction.[34] An MO calculation estimating the stabilization was done using STO-3G-level basis functions. The electron-donating *p*-amino and *p*-methoxy groups were found to stabilize a benzyl cation by 26 and 14 kcal/mol, respectively. On the other hand, the electron-attracting *p*-cyano and *p*-nitro groups are destabilizing by 12 and 20 kcal/mol, respectively.[35]

Adjacent atoms with one or more lone pairs of electrons strongly stabilize a carbocation. Table 1.13 (p. 30) indicates the stabilization of the methyl cation by such

31. J. B. Foresman, M. W. Wong, K. B. Wiberg, and M. J. Frisch, *J. Am. Chem. Soc.* **115**:2220 (1993).
32. T. Laube, G. A. Olah, and R. Bau, *J. Am. Chem. Soc.* **119**:3087 (1997).
33. Y. Tsuno and M. Fujio, *Chem. Soc. Rev.* **25**:129 (1996).
34. M. Mishima, K. Arima, H. Inoue, S. Usui, M. Fujio, and Y. Tsuno, *Bull. Chem. Soc. Jpn.* **68**:3199 (1995).
35. W. J. Hehre, M. Taagepera, R. W. Taft, and R. D. Topsom, *J. Am. Chem. Soc.* **103**:1344 (1981).

substituents. Alkoxy and dialkylamino groups are important examples of substituents that exert this stabilizing effect.

$$CH_3\overset{..}{\underset{..}{O}}-\overset{+}{C}H_2 \longleftrightarrow CH_3\overset{+}{\underset{..}{O}}=CH_2 \qquad (CH_3)_2\overset{..}{N}-\overset{+}{C}H_2 \longleftrightarrow (CH_3)_2\overset{+}{N}=CH_2$$

Although these structures have a positive charge on a more electronegative atom, they benefit from an additional bond which satisfies the octet requirement of the tricoordinate carbon. These "carbocations" are well represented by the doubly bonded resonance structures. One indication of the participation of adjacent oxygen substituents is the existence of a barrier to rotation about the C—O bonds in this type of carbocation.

The barrier in **A** is about 14 kcal/mol ($\Delta G^{\ddagger}$) as measured by NMR coalescence of the signals of the nonidentical vinyl protons.[36] The gas-phase barrier is calculated by MO methods to be 26 kcal/mol. The observed barrier for **B** is 19 kcal/mol.[37,38] Even halogen substituents can stabilize carbocations as a result of resonance donation from the halogen lone pairs. A fluorine or chlorine substituent is nearly as stabilizing as a methyl group in the gas phase.[39]

$$CH_3\overset{+}{C}HF \longleftrightarrow CH_3CH=F^+ \qquad CH_3\overset{+}{C}HCl \longleftrightarrow CH_3CH=Cl^+$$

Electron-withdrawing groups that are substituted directly on the cationic site are destabilizing. Table 5.5 gives an indication of the relative retardation of the rate of ionization and the calculated destabilization for several substituents. The trifluoromethyl group, which exerts a powerful polar effect, is strongly destabilizing on the basis of both the kinetic data and the MO calculations. The cyano and formyl groups are less so. In fact, the destabilizing effect of these groups is considerably less than would be predicted on the basis of their polar substituent constants. Both the cyano and formyl groups can act as π donors, even though the effect is to place partial positive charge on nitrogen and oxygen atoms, respectively. The relevant resonance structures are depicted below.

36. D. Cremer, J. Gauss, R. F. Childs, and C. Blackburn, *J. Am. Chem. Soc.* **107**:2435 (1985).
37. R. F. Childs and M. E. Hagar *Can. J. Chem.* **58**:1788 (1980).
38. There is another mechanism for equilibration of the cation pairs $A_1 \rightleftharpoons A_2$ and $B_1 \rightleftharpoons B_2$, namely, inversion at oxygen. However, the observed barrier represents at least the *minimum* for the C=O rotational barrier and therefore demonstrates that the C-O bond has double-bond character.
39. C. H. Reynolds, *J. Am. Chem. Soc.* **114**:8676 (1992).

Table 5.5. Destabilization of 2-Substituted 2-Propyl Cation by Electron-Withdrawing Substituents

Z	Solvolysis rate relative to Z = H	Destabilization (kcal/mol) energy relative to Z = H
CN	$\sim 10^{-3}$ [a]	9.9[b]
CF$_3$	$\sim 10^{-3}$ [c]	37.3[b]
CH=O	–	6.1[b]

a. P. G. Gassman and J. J. Talley, *J. Am. Chem. Soc.* **102**:1214 (1980).
b. M. N. Paddon-Row, C. Santiago, and K. N. Houk, *J. Am. Chem. Soc.* **102**:6561 (1980).
c. K. M. Koshy and T. T. Tidwell, *J. Am. Chem. Soc.* **102**:1216 (1980).

These interactions are reflected in MO energies, bond lengths, and charge distributions calculated for such cations.[40] The resonance structures are the nitrogen and oxygen analogs of those for the allyl cation. The effect of this π delocalization is to attenuate the dipolar destabilization by these substituents.[41] Calculations of substituent effects using the PM3 semiempirical method have been reported for a range of substituents,[42] and *ab initio* 6-31G**/MP2 calculations are available for some of the same substituents.[43] Some results of these calculations are shown in Table 5.6.

Several very stable carbocations are among the "Miscellaneous carbocation" listed in Table 5.1. These ions are remarkably stable, considering that they do not bear electron-releasing heteroatom substituents such as oxygen or nitrogen. The tricyclopropylmethyl cation, for example, is more stable than the triphenylmethyl cation.[44] The stabilization of carbocations by cyclopropyl substituents results from the interaction of the electrons in the cyclopropyl C−C bonds with the positive carbon. The electrons in these orbitals are at relatively higher energy than normal σ electrons and are therefore particularly effective in interacting with the vacant p orbital of the carbocation. This stabilization involves interaction of the cyclopropyl bonding orbitals with the carbon p orbital. This interaction imposes a preference for the bisected conformation of the cyclopropylmethyl cation in comparison to the perpendicular conformation.

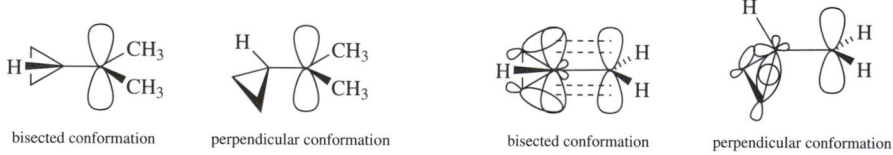

bisected conformation　　perpendicular conformation　　bisected conformation　　perpendicular conformation

40. D. A. Dixon, P. A. Charlier, and P. G. Gassman, *J. Am. Chem. Soc.* **102**:3957 (1980); M. N. Paddon-Row, C. Santiago, and K. N. Houk, *J. Am. Chem. Soc.* **102**:6561 (1980); D. A. Dixon, R. A. Eades, R. Frey, P. G. Gassman, M. L. Hendewerk, M. N. Paddon-Row, and K. N. Houk, *J. Am. Chem. Soc.* **106**:3885 (1984); X. Creary, Y.-X. Wang, and Z. Jiang, *J. Am. Chem. Soc.* **117**:3044 (1995).
41. T. T. Tidwell, *Angew. Chem. Int. Ed. Engl.* **23**:20 (1984); P. G. Gassman and T. T. Tidwell, *Acc. Chem. Res.* **16**:279 (1983); J. L. Holmes and P. M. Mayer, *J. Phys. Chem.* **99**:1366 (1995); J. L. Holmes, F. P. Lossing, and P. M. Mayer, *Chem. Phys. Lett.* **212**:134 (1993).
42. A. M. El-Nahas and T. Clark, *J. Org. Chem.* **60**:8023 (1995).
43. X. Creary, Y.-X. Wang and Z. Jiang, *J. Am. Chem. Soc.* **117**:3044 (1995).
44. For reviews of the cyclopropylmethyl cation, see H. G. Richey, Jr., in *Carbonium Ions*, Vol. III, G. A. Olah and P. v. R. Schleyer, eds., Wiley-Interscience, New York, 1972, Chapter 25; G. A. Olah, V. P. Reddy, and G. K. S. Prakash, *Chem. Rev.* **92**:69 (1992); G. A. Olah, V. Reddy, and G. K. S. Prakash, *Chemistry of the Cyclopropyl Group*, Part 2, Z. Rappoport, ed., John Wiley & Sons, Chichester, U.K., 1995, pp. 813–859.

**Table 5.6. Calculated Substituent Effects on Carboca-
tion Stabilization (kcal/mol)**

Substituent	PM3[a]	6-31G**/MP2[b]
NH$_2$	−80.2	
CH$_3$O	−57.6	
OH	−51.4	
Ph	−56.3	
CH$_2$=CH	−43.3	−66.1
CH$_3$	−29.0	−41.5
F	−5.5	
CN	−5.0	−4.3
CH=O	−1.7	−0.2
NO$_2$	+30.8	+22.3

a. A. M. El-Nahas and T. Clark, *J. Org. Chem.* **60**:8023 (1995).
b. X. Creary, Y.-X. Wang, and Z. Jiang, *J. Am. Chem. Soc.* **117**:3044 (1995).

Only the bisected conformation aligns the cyclopropyl C−C orbitals for effective overlap. Crystal structure determinations on two cyclopropylmethyl cations with additional stabilizing substituents, **C** and **D**, have confirmed the preference for the bisected geometry. The crystal structures of **C** and **D** are shown in Fig. 5.8.

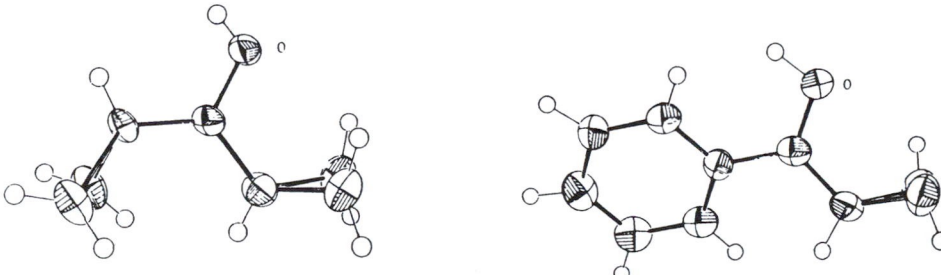

In ion **D**, in which the phenyl group would be expected to be coplanar with the cationic center to maximize delocalization, the observed angle is 25–30°. This should permit effective benzylic stabilization. The planes of the cyclopropyl groups in both structures are at ∼85° to the plane of the trigonal carbon, in agreement with expectation for the bisected ion.[45]

Fig. 5.8. Crystal structures of bis(cyclopropyl)hydroxymethyl cation and 1-cyclopropyl-1-phenylhydroxymethyl cation. (Structural diagrams are reproduced from Ref. 45 with permission of the American Chemical Society.)

45. R. F. Childs, R. Faggiani, C. J. Lock, M. Mahendran, and S. D. Zweep, *J. Am. Chem. Soc.* **108**:1692 (1986).

Solvolysis rate studies also indicate that there is greater stabilization by a cyclopropyl group in a bisected geometry. In tosylate **1**, the cyclopropane ring is locked into an orientation which affords a perpendicular arrangement. It reacts 300 times more slowly than the model compound **2**. Tosylate **3**, which corresponds to the bisected geometry, undergoes acetolysis at least 10^5 times faster than the model 2-adamantyl tosylate **4**.[46]

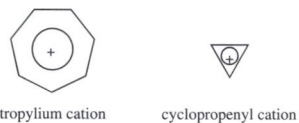

The tropylium and the cyclopropenyl cations are stabilized aromatic systems. These ions are aromatic according to Hückel's rule, with the cyclopropenium ion having two π electrons and the tropylium ion six (see Section 9.3). Both ring systems are planar and possess cyclic conjugation, as is required for aromaticity.

tropylium cation   cyclopropenyl cation

A major advance in the direct study of carbocations occurred during the 1960s when methods for observation of the NMR spectra of the cations in "superacid" media were developed. The term superacid refers to media of very high proton-donating capacity, for example, more acidic than 100% sulfuric acid. The solution is essentially nonnucleophilic, so carbocations of only moderate stability can be generated and observed.[47] A convenient medium for these NMR measurements is FSO_3H–SbF_5–SO_2. The fluorosulfonic acid acts as a proton donor, and antimony pentafluoride is a powerful Lewis acid that can assist ionization. This particular combination has been dubbed "magic acid" because of its powerful protonating ability.

Alkyl halides and alcohols, depending on the structure of the alkyl group, react with magic acid to give rise to carbocations. Primary and secondary alcohols are protonated at $-60°C$, but do not ionize. Tertiary alcohols ionize, giving rise to the cation. As the temperature is increased, carbocation formation also occurs from secondary alcohols. *s*-Butyl alcohol ionizes with rearrangement to the *t*-butyl cation. At $-30°C$ the protonated primary alcohol isobutanol ionizes, also forming the *t*-butyl cation. Protonated *n*-butanol is stable to $0°C$, at which point it too gives rise to the *t*-butyl cation. It is typically observed that ionizations in superacids give rise to the most stable of the isomeric carbocations which could be derived from the alkyl group. The *t*-butyl cation is generated from C_4 systems, whereas C_5 and C_6 alcohols give rise to the *t*-pentyl and *t*-hexyl ions, respectively. These and related observations are illustrated in Scheme 5.3. Entries 6–9 and 10–12 further illustrate the tendency for rearrangement to the most stable cation to occur. The tertiary 1-methylcyclopentyl cation is the only ion observed from a variety of five- and six-membered ring derivatives. The tertiary bicyclo[3.3.0]octyl cation is formed from all

46. J. E. Baldwin and W. D. Fogelsong, *J. Am. Chem. Soc.* **90**:4303 (1968).
47. A review of the extensive studies of carbocations in superacid media is available in G. A. Olah, G. K. S. Prakash, and J. Sommer, *Super Acids*, John Wiley & Sons, New York, 1985.

Scheme 5.3. Protonation and Ionization of Organic Substrates in Superacid Media

Aliphatic alcohols in FSO_3H–SbF_5–SO_2

1^a $ROH \xrightarrow{-60°C} R\overset{+}{O}H_2$

      R = methyl, ethyl, *n*-propyl, isopropyl, *n*-butyl, *s*-butyl,
      n-amyl, isoamyl, neopentyl, *n*-hexyl, neohexyl

2^b $(CH_3)_2CHCH_2OH \xrightarrow{-60°C} (CH_3)_2CHCH_2\overset{+}{O}H_2 \xrightarrow{-30°C} (CH_3)_3C^+$

3^c $(CH_3)_3COH \xrightarrow{-60°C} (CH_3)_3C^+$

Alkyl halides in antimony pentafluoride

4^b $\underset{\underset{F}{|}}{CH_3CHCH_2CH_3} \xrightarrow{-110°C} \underset{+}{CH_3CHCH_2CH_3} \xrightarrow{-40°C} (CH_3)_3C^+$

5^c $CH_3OCH_2Cl \xrightarrow{-60°C} CH_3OCH_2^+$

Cyclopentylmethyl and cyclohexyl systems

(continued)

bicyclooctyl precursors. The tendency to rearrange to the thermodynamically stable ions by multiple migrations is a consequence of the very low nucleophilicity of the solvent system. In the absence of nucleophilic capture by solvent, the carbocations have a long lifetime and undergo extensive skeletal rearrangement and accumulate as the most stable isomer.

Up to this point in our discussion, we have considered only carbocations in which the cationic carbon can be sp^2-hybridized and planar. When this hybridization cannot be achieved, the carbocations are of higher energy. In a classic experiment, Bartlett and Knox demonstrated that the tertiary chloride 1-chloroapocamphane was inert to nucleophilic substitution.[48] Starting material was recovered unchanged even after refluxing for 48 h in ethanolic silver nitrate. The unreactivity of this compound is attributed to the structure of

48. P. D. Bartlett and L. H. Knox, *J. Am. Chem. Soc.* **61**:3184 (1939).

Scheme 5.3. (*continued*)

Bicyclooctyl systems in SbF$_5$–SO$_2$ClF, –78°C

10^e

11^e

12^e

Benzylic and cyclopropylcarbinyl systems

13^f OCH$_3$ SbF$_5$–SO$_2$ClF, –78°C   OCH$_3$

CH$_2$OH   CH$_2$

14^g CH$_3$ FSO$_3$H–SbF$_5$–SO$_2$, –75°C   CH$_3$

C—OH   C

CH$_3$ CH$_3$

a. G. A. Olah, J. Sommer, and E. Namanworth, *J. Am. Chem. Soc.* **89**:3576 (1967).
b. M. Saunders, E. L. Hagen, and J. Rosenfeld, *J. Am. Chem. Soc.* **90**:6882 (1968).
c. G. A. Olah and J. M. Bollinger, *J. Am. Chem. Soc.* **89**:2993 (1967).
d. G. A. Olah, J. M. Bollinger, C. A. Cupas, and J. Lukas, *J. Am. Chem. Soc.* **89**:2692 (1967).
e. G. A. Olah and G. Liang, *J. Am. Chem. Soc.* **93**:6873 (1971).
f. G. A. Olah, R. D. Porter, C. L. Juell, and A. M. White, *J. Am. Chem. Soc.* **94**:2044 (1972).
g. C. U. Pittman, Jr., and G. A. Olah, *J. Am. Chem. Soc.* **87**:2998 (1965).

the bicyclic system, which prevents rehybridization to a planar sp^2 carbon. Direct displacement by back-side attack is also precluded, because of the bridgehead location of the C—Cl bond.

The apocamphyl structure is particularly rigid, and bridgehead carbocations become accessible in more flexible structures. The relative solvolysis rates of the bridgehead bromides 1-bromoadamantane, 1-bromobicyclo[2.2.2]octane, and 1-bromobicyclo[2.2.1]-

heptane illustrate this trend. The relative rates for solvolysis in 80% ethanol at 25°C are shown below.[49]

The relative reactivity of tertiary bridgehead systems toward solvolysis is well correlated with the strain, calculated by molecular mechanics, resulting from conversion of the ring structure to a carbocation.[50] This result implies that the increased energy associated with a nonplanar carbocation is proportional to the strain energy present in the ground-state reactant. The solvolysis rates also correlate with bridgehead cation stability measured by gas-phase hydride affinity and 6-311G**/MP2 MO calculations.[51]

Carbocations in which the cationic carbon is sp-hybridized are of higher energy than those in which the cationic center is sp^2-hybridized.[52] This is because of the higher electronegativity of the orbital with greater s character. The vinyl cation, $CH_2=CH^+$, lies between the ethyl cation and the methyl cation in stability (see Table 5.2). The intermediacy of substituted vinyl cations in solvolysis reactions has been demonstrated, but direct observation has not been possible for simple vinyl cations.[53] Most examples of solvo-lytic generation of vinyl cations involve very reactive leaving groups, especially trifluoromethylsulfonates (triflates). Typical products include allenes, acetylenes, and vinyl esters.[54]

The phenyl cation is an extremely unstable cation, as is reflected by the high hydride affinity shown in Table 5.2. In this case, the ring geometry opposes rehybridization so the vacant orbital retains sp^2 character. Because the empty orbital is in the nodal plane of the ring, it receives no stabilization from the π electrons.

Phenyl cations are formed by thermal decomposition of aryl diazonium ions.[56] The cation is so extremely reactive that under some circumstances it can recapture the nitrogen

49. For a review of bridgehead carbocations, see R. C. Fort, Jr., in *Carbonium Ions*, Vol. IV, G. A. Olah and P. v. R. Schleyer, eds., Wiley-Interscience, New York, 1973, Chapter 32.
50. T. W. Bentley and K. Roberts, *J. Org. Chem.* **50**:5852 (1985); R. C. Bingham and P. v. R. Schleyer, *J. Am. Chem. Soc.* **93**:3189 (1971); P. Müller and J. Mareda, *Helv. Chim. Acta* **70**:1017 (1987); P. Müller, J. Mareda, and D. Milin, *J. Phys. Org. Chem.* **8**:507 (1995).
51. E. W. Della and W. K. Janowski, *J. Org. Chem.* **60**:7756 (1995); J. L. M. Abboud, O. Castano, E. W. Della, M. Herreros, P. Müller, R. Notario, and J.-C. Rossier, *J. Am. Chem. Soc.* **119**:2262 (1997).
52. V. D. Nefedov, E. N. Sinotova, and V. P. Lebedev, *Russ. Chem. Rev.* **61**:283 (1992).
53. H.-U. Siehl and M. Hanack, *J. Am. Chem. Soc.* **102**:2686 (1980).
54. For reviews of vinyl cations, see Z. Rappoport, in *Reactive Intermediates*, Vol. 3, R. A. Abramovitch, ed., Plenum Press, New York, 1983; P. J. Stang, *Prog. Phys. Org. Chem.* **10**:205 (1973); G. Modena and U. Tonellato, *Adv. Phys. Org. Chem.* **9**:185 (1971).
55. R. H. Summerville, C. A. Senkler, P. v. R. Schleyer, T. F. Dueber, and P. J. Stang, *J. Am. Chem. Soc.* **96**:1100 (1974).
56. C. G. Swain, J. E. Sheats, and K. G. Harbison, *J. Am. Chem. Soc.* **97**:783 (1975).

generated in the decomposition.[57] Attempts to observe formation of phenyl cations by ionization of aryl triflates have only succeeded when especially stabilizing groups, such as trimethylsilyl groups, are present at the 2- and 6-positions of the aromatic ring.[58]

5.5. Nucleophilicity and Solvent Effects

The term *nucleophilicity* refers to the effect of a Lewis base on the *rate* of a nucleophilic substitution reaction and may be contrasted with *basicity*, which is defined in terms of the position of an equilibrium reaction with a proton or some other acid. Nucleophilicity is used to describe trends in the *kinetic* aspects of substitution reactions. The relative nucleophilicity of a given species may be different toward various reactants, and it has not been possible to devise an absolute scale of nucleophilicity. We need to gain some impression of the structural features that govern nucleophilicity and to understand the relationship between nucleophilicity and basicity.[59]

The factors that influence nucleophilicity are best assessed in the context of the limiting S_N2 mechanism, since it is here that the properties of the nucleophile are most important. The rate of an S_N2 reaction is directly related to the effectiveness of the nucleophile in displacing the leaving group. In contrast, the relative nucleophilicity has no effect on the rate of an S_N1 reaction but does affect the product distribution resulting from partitioning of the carbocation intermediate among the competing nucleophiles.

Many properties have an influence on nucleophilicity. Those considered to be most significant are (1) the solvation energy of the nucleophile; (2) the strength of the bond being formed to carbon; (3) the size of the nucleophile; (4) the electronegativity of the attacking atom; and (5) the polarizability of the attacking atom.[60] Let us consider how each of these factors affects nucleophilicity:

1. A high solvation energy lowers the ground-state energy relative to the transition state, in which the charge is more diffuse. This results in an increased activation energy. Viewed from another perspective, the solvation energy affects nucleophilicity because the solvation shell must be disrupted to arrive at the transition state, and this desolvation energy contributes to the activation energy.

2. A stronger bond between the nucleophilic atom and carbon is reflected in a more stable transition state and therefore a reduced activation energy. Since the S_N2 process is concerted, the strength of the partially formed new bond is reflected in the energy of the transition state.

3. A sterically restricted nucleophile is less reactive than a more accessible one because of nonbonded repulsions which develop in the transition state. The trigonal bipyramidal geometry of the S_N2 transition state is sterically more demanding than the tetrahedral reactant, so steric congestion increases as the transition state is approached.

57. R. G. Bergstrom, R. G. M. Landells, G. W. Wahl, Jr., and H. Zollinger, *J. Am. Chem. Soc.* **98**:3301 (1976).
58. Y. Apeloig and D. Arad, *J. Am. Chem. Soc.* **107**:5285 (1985); Y. Himeshima, H. Kobayashi, and T. Sonoda, *J. Am. Chem. Soc.* **107**:5286 (1985).
59. For general reviews of nucleophilicity, see R. F. Hudson, in *Chemical Reactivity and Reaction Paths*, G. Klopman, ed., John Wiley & Sons, New York, 1974, Chapter 5; J. M. Harris and S. P. McManus, eds., *Nucleophilicity, Advances in Chemistry Series*, No. 215, American Chemical Society, Washington, D.C., 1987.
60. A. Streitwieser, Jr., *Solvolytic Displacement Reactions*, McGraw-Hill, New York, 1962; J. F. Bunnett, *Annu. Rev. Phys. Chem.* **14**:271 (1963).

4. A more electronegative atom binds its electrons more tightly than a less electro-negative one. Since the S_N2 process requires donation of electron density to an antibonding orbital of the reactant, high electronegativity is unfavorable.

5. Polarizability describes the ease of distortion of the electron cloud of the attacking atom of the nucleophile. Again, since the S_N2 process requires bond formation by an electron pair from the nucleophile, the more easily distorted the electric field of the atom, the higher is its nucleophilicity. Polarizability increases going down and to the left in the periodic table.

Empirical measures of nucleophilicity may be obtained by comparing relative rates of reaction of a standard reactant with various nucleophiles. One measure of nucleophilicity is the *nucleophilic constant* (n), defined originally by Swain and Scott.[61] Taking methanolysis of methyl iodide as the standard reaction, n was defined as

$$n_{CH_3I} = \log(k_{nucleophile}/k_{CH_3OH}) \quad \text{in } CH_3OH, \ 25°C$$

Table 5.7 lists the nucleophilic constants for a number of species according to this definition. It is apparent from Table 5.7 that nucleophilicity toward methyl iodide does not correlate directly with basicity. Azide ion, phenoxide ion, and bromide are all equivalent in nucleophilicity but differ greatly in basicity. Conversely, azide ion and acetate ion are

Table 5.7. Nucleophilic Constants of Various Nucleophiles[a]

Nucleophile	n_{CH_3I}	pK_a of conjugate acid
CH_3OH	0.0	−1.7
NO_3^-	1.5	−1.3
F^-	2.7	3.45
$CH_3CO_2^-$	4.3	4.8
Cl^-	4.4	−5.7
$(CH_3)_2S$	5.3	
NH_3	5.5	9.25
N_3^-	5.8	4.74
$C_6H_5O^-$	5.8	9.89
Br^-	5.8	−7.7
CH_3O^-	6.3	15.7
HO^-	6.5	15.7
NH_2OH	6.6	5.8
NH_2NH_2	6.6	7.9
$(CH_3CH_2)_3N$	6.7	10.70
CN^-	6.7	9.3
$(CH_3CH_2)_3As$	7.1	
I^-	7.4	−10.7
HO_2^-	7.8	
$(CH_3CH_2)_3P$	8.7	8.69
$C_6H_5S^-$	9.9	6.5
$C_6H_5Se^-$	10.7	
$(C_6H_5)_3Sn^-$	11.5	

a. Data from R. G. Pearson and J. Songstad, *J. Am. Chem. Soc.* **89**:1827 (1967); R. G. Pearson, H. Sobel, and J. Songstad, *J. Am. Chem. Soc.* **90**:319 (1968); P. L. Bock and G. M. Whitesides, *J. Am. Chem. Soc.* **96**:2826 (1974).

61. C. G. Swain and C. B. Scott, *J. Am. Chem. Soc.* **75**:141 (1953).

nearly identical in basicity, but azide ion is 30 times (1.5 log units) more nucleophilic. Among neutral nucleophiles, whereas triethylamine is *more basic* than triethylphosphine (pK_a of the conjugate acid is 10.70 versus 8.69), the phosphine is more nucleophilic ($n = 8.7$ versus 6.7, a factor of 100). Correlation of nucleophilicity with basicity is better if the attacking atom is the same. Thus, for the series of oxygen nucleophiles $CH_3O^- > C_6H_5O^- > CH_3CO_2^- > NO_3^-$, nucleophilicity parallels basicity.

Nucleophilicity usually decreases across a row in the periodic table. For example, $HO^- > F^-$ or $C_6H_5S^- > Cl^-$. This order is primarily determined by electronegativity. Nucleophilicity also increases going down the periodic table; for example, nucleophilicity decreases in the order $I^- > Br^- > Cl^- > F^-$ and $C_6H_5Se^- > C_6H_5S^- > C_6H_5O^-$. Three factors work together to determine this order. Electronegativity decreases going down the periodic table. Probably more important are the greater polarizability and weaker solvation of the heavier atoms, which have more diffuse electron distributions.

There is clearly a conceptual relationship between the properties called nucleophilicity and basicity. Both describe a process involving formation of a new bond to an electrophile by donation of an electron pair. The pK_a values in Table 5.7 refer to basicity toward a proton. There are many reactions in which a given chemical species might act either as a nucleophile or as a base. It is therefore of great interest to be able to predict whether a chemical species $Y:^-$ will act as a nucleophile or as a base under a given set of circumstances. Scheme 5.4 lists some examples.

The definition of basicity is based on the ability of a substance to remove protons and refers to an *equilibrium*:

$$B: + H_2O \rightleftharpoons \overset{+}{B}H + {}^-OH \qquad K_b = \frac{[\overset{+}{B}H][{}^-OH]}{[B:]}$$

Scales for bases that are too weak to study in aqueous solution employ other solvents but are related to the equilibrium in aqueous solution. These equilibrium constants provide a measure of *thermodynamic basicity*, but we also need to have some concept of *kinetic basicity*. For the reactions in Scheme 5.4, for example, it is important to be able to make generalizations about the rates of competing reactions.

Scheme 5.4. Competition Between Nucleophilicity and Basicity

S_N1 Substitution	$Y:^-$ acts as a nucleophile	$Y:^- + R_2\overset{+}{C}CHR_2' \longrightarrow R_2CCHR_2' \ (\text{Y})$
versus		
E_1 Elimination	$Y:^-$ acts as a base	$Y:^- + R_2\overset{+}{C}CHR_3' \rightarrow R_2C = CR_2' + HY$
S_N2 Substitution	$Y:^-$ acts as a nucleophile	$Y:^- + R_2CHCH_2Br \rightarrow R_2CHCH_2Y + Br^-$
versus		
E_2 Elimination	$Y:^-$ acts as a base	$Y:^- + R_2CHCH_2Br \rightarrow R_2C = CH_2 + HY + Br^-$
Nucleophilic addition at a carbonyl carbon	$Y:^-$ acts as a nucleophile	$Y:^- + R_2CHCR' \ (\overset{O}{\|}) \longrightarrow R_2CHCR' \ (\overset{O^-}{\|}, \text{Y})$
versus		
Enolate formation	$Y:^-$ acts as a base	$Y:^- + R_2CHCR' \ (\overset{O}{\|}) \longrightarrow R_2C = CR' \ (\overset{O^-}{\|}) + HY$

Table 5.8. Hardness and Softness of Some Common Ions and Molecules

	Bases (nucleophiles)	Acids (electrophiles)
Soft:	RSH, RS^-, I^-, R_3P $^-C{\equiv}N$, $^-C{\equiv}O^+$ $RCH{=}CHR$ benzene	I_2, Br_2, $RS-X$, $RSe-X$, RCH_2-X $Cu(I)$, $Ag(I)$, $Pd(II)$, $Pt(II)$, $Hg(II)$ zerovalent metal complexes
Borderline:	Br^-, N_3^-, $ArNH_2$ pyridine	$Cu(II)$, $Zn(II)$, $Sn(II)$ R_3C^+, R_3B
Hard:	H_2O, HO^-, ROH, RO^-, RCO_2^- F^-, Cl^-, NO_3^-, NH_3, RNH_2	$H-X$, H^+, Li^+, Na^+, K^+ Mg^{2+}, Ca^{2+}, $Al(III)$, $Sn(IV)$, $Ti(IV)$ R_3Si-X

The most useful qualitative approach for making predictions of this type is the hard–soft-acid–base (HSAB} concept.[62] This concept proposes that reactions will occur most readily between species that are matched in hardness and softness. Hard nucleophiles prefer hard electrophiles, whereas soft nucleophiles prefer soft electrophiles. This concept can be applied to the problem of competition between nucleophilic substitution and elimination in the reaction of anions with alkyl halides, for example. The sp^3 carbon is a soft electrophile whereas the proton is a hard electrophile. Thus, according to the HSAB theory, a soft anion should act primarily as a nucleophile, giving the substitution product, whereas a hard anion is more prone to abstract a proton, giving the elimination product. The property of softness correlates with high polarizability and low electronegativity. Species in Table 5.7 that exhibit high nucleophilicity toward methyl iodide include CN^-, I^-, and $C_6H_5S^-$. Hardness reflects a high charge density and is associated with small, highly electronegative species. Examples from Table 5.7 include F^- and CH_3O^-. Table 5.8 classifies some representative chemical species with respect to softness and hardness. Numerical values of hardness were presented in Table 1.9.

The soft-nucleophile–soft-electrophile combination is also associated with a late transition state, in which the strength of the newly forming bond contributes significantly to the stability of the transition state. The hard-nucleophile–hard-electrophile combination implies an early transition state with electrostatic attraction being more important than bond formation. The reaction pathway is chosen early on the reaction coordinate and primarily on the basis of charge distribution.

Nucleophilicity is also correlated with oxidation potential for comparisons between nucleophiles involving the same element.[63] Good nucleophilicity correlates with ease of oxidation, as would be expected from the electron-donating function of the nucleophile in S_N2 reactions. Hard–soft considerations would also suggest that better nucleophilicity would be associated with species having relatively high-energy electrons. Remember (Section 1.2.3) that soft species have relatively high-lying HOMOs.

Another significant structural effect that imparts high nucleophilicity is the *alpha effect*. It is observed that atoms which are directly bonded to an atom with one or more unshared pairs of electrons tend to be stronger nucleophiles than would otherwise be expected. Examples in Table 5.7 include HO_2^-, which is more nucleophilic than HO^-, and

62. R. G. Pearson and J. Songstad, *J. Am. Chem. Soc.* **89:**1827 (1967); R. G. Pearson, *J. Chem. Educ.* **45:**581, 643 (1968); T. L. Ho, *Chem. Rev.* **75:**1 (1975).

63. M. E. Niyazymbetov and D. H. Evans *J. Chem. Soc., Perkin Trans. 2* **1993:**1333; M. E. Niyazymbetov, Z. Rongfeng, and D. H. Evans, *J. Chem. Soc., Perkin Trans., 2* **1996:**1957.

64. G. Klopman, K. Tsuda, J. B. Louis, and R. E. Davis, *Tetrahedron* **26:**4549 (1970); W. B. England, P. Kovacic, S. M. Hanrah, and M. B. Jones, *J. Org. Chem.* **45:**2057 (1980).

NH_2NH_2 (hydrazine) and NH_2OH (hydroxylamine), both of with are more nucleophilic than ammonia. Various explanations for the alpha effect have been put forward.[64] One view is that the ground state of the nucleophile is destabilized by lone pair–lone pair repulsions which are decreased as bond formation occurs in the transition state. In MO terms, this would imply a relatively high energy of the nucleophile HOMO that participates in bond formation.[65] Another view is that the adjacent electron pair can act to stabilize charge deficiency at the transition state. As discussed in Section 5.3, there are many S_N2 reactions in which the transition state is electron-poor.

The nucleophilicity of anions is highly dependent on the degree of solvation. Much of the data which forms the basis for quantitative measurement of nucleophilicity is for reactions in hydroxylic solvents. In protic hydrogen-bonding solvents, anions are subject to strong interactions with solvent. Hard nucleophiles are more strongly solvated by protic solvents than soft nucleophiles, and this difference contributes to the greater nucleophilicity of soft anions in such solvents. Nucleophilic substitution reactions often occur more rapidly in polar aprotic solvents than they do in protic solvents. This is because anions are weakly solvated in such solvents (see Section 4.10). Nucleophilicity is also affected by the solvation of the cations in solution. The cations associated with nucleophilic anions are strongly solvated in solvents such as N,N-dimethylformamide (DMF), dimethyl sulfoxide (DMSO), hexamethylphosphoric triamide (HMPA), N-methylpyrrolidone, and sulfolane.[66] As a result, the anions are dissociated from the cations, which further enhances their nucleophilicity.

$$\begin{array}{ccccc} & O & & O & \\ & \parallel & & \parallel & \\ & HCN(CH_3)_2 & & CH_3SCH_3 & & O{=}P[N(CH_3)_2]_3 \end{array}$$

In the absence of the solvation typical of protic solvents, the relative nucleophilicity of anions changes. Hard nucleophiles increase in reactivity more than do soft nucleophiles. As a result, the relative reactivity order changes. In methanol, for example, the relative reactivity order is $N_3^- > I^- > CN^- > Br^- > Cl^-$, whereas in DMSO the order becomes $CN^- > N_3^- > Cl^- > Br^- > I^-$.[67] In methanol, the reactivity order is dominated by solvent effects, and the more weakly solvated N_3^- and I^- ions are the most reactive nucleophiles. The iodide ion is large and very polarizable. The anionic charge on the azide ion is dispersed by delocalization. When the effect of solvation is diminished in DMSO, other factors become more important. These include the strength of the bond being formed, which would account for the reversed order of the halides in the two series. There is also evidence that S_N2 transition states are better solvated in aprotic dipolar solvents than in protic solvents.

In interpreting many aspects of displacement reactions, particularly solvolysis, it is important to be able to characterize the nucleophilicity of the solvent. Assessment of

65. M. M. Heaton, J. Am. Chem. Soc. **100**:2004 (1978).

66. T. F. Magnera, G. Caldwell, J. Sunner, S. Ikuta, and P. Kebarle, J. Am. Chem. Soc. **106**:6140 (1984); T. Mitsuhashi, G. Yamamoto, and H. Hirota, Bull Chem. Soc. Jpn. **67**:831 (1994); K. Okamoto, Adv. Carbocation Chem. **1**:181 (1989).

67. R. L. Fuchs and L. L. Cole, J. Am. Chem. Soc. **95**:3194 (1973); R. Alexander, E. C. F. Ko, A. J. Parker, and T. J. Broxton, J. Am. Chem. Soc. **90**:5049 (1968); D. Landini, A. Maia, and F. Montanari, J. Am. Chem. Soc. **100**:2796 (1978).

Table 5.9. Solvent Nucleophilicity (N_{Tos}) and Ionization (Y_{Tos}) Parameters[a]

Solvent	N_{Tos}	Y_{Tos}
Ethanol	+0.09	−1.75
Methanol	+0.01	−0.92
50 % Aqueous ethanol	−0.20	1.29
Water	−0.26	
Acetic acid	−2.05	−0.61
Formic acid	−2.05	3.04
Trifluoroethanol	−2.78	1.80
97% $(CF_3)_2CHOH-H_2O$	−3.93	1.83
Trifluoroacetic acid	−4.74	4.57

a. From F. L. Schadt, T. W. Bentley, and P. v. R. Schleyer, *J. Am. Chem. Soc.* **98**:7667 (1976).

solvent nucleophilicity can be done by comparing rates of a standard substitution process in various solvents. One such procedure is based on the Winstein–Grunwald equation:

$$\log(k/k_0) = lN + mY$$

where N and Y are measures of the solvent nucleophilicity and ionizing power, respectively. The variable parameters l and m are characteristic of specific reactions.[68] The value of N, the indicator of solvent nucleophilicity, can be determined by specifying a standard substrate for which l is assigned the value 1.00 and a standard solvent for which N is assigned the value 0.00. 2-Adamantyl tosylate has been taken as a standard for which nucleophilic participation of the solvent is considered to be negligible, and 80 : 20 ethanol–water is taken as the standard solvent. The resulting solvent characteristics are called N_{Tos} and Y_{Tos}. Some representative values for solvents that are frequently used in solvolysis studies are given in Table 5.9.

5.6. Leaving-Group Effects

The identity of the leaving group influences the rate of nucleophilic substitution proceeding by either the direct displacement or the ionization mechanism. Because the leaving group departs with the pair of electrons from the covalent bond to the reacting carbon atom, a correlation with electronegativity is expected. Provided the reaction series consists of structurally similar leaving groups, such relationships are observed. For example, a linear relationship has been demonstrated between the ionization of substituted benzoic acids and the rate of reaction of substituted arenesulfonates with ethoxide ion in ethanol (Hammett-type equation).[69] A qualitative trend of increasing reactivity with increasing acidity of the conjugate acid of the leaving group also holds for less similar

68. S. Winstein, E. Grunwald, and H. W. Jones, *J. Am. Chem. Soc.* **73**:2700 (1951); F. L. Schadt, T. W. Bentley, and P. v. R. Schleyer, *J. Am. Chem. Soc.* **98**:7667 (1976).
69. M. S. Morgan and L. H. Cretcher, *J. Am. Chem. Soc.* **70**:375 (1948).

Table 5.10. Relative Solvolysis Rates of 1-Phenylethyl Esters and Halides[a,b]

Leaving group	k_{rel}
$CF_3SO_3^-$ (triflate)	1.4×10^8
p-Nitrobenzenesulfonate	4.4×10^5
p-Toluenesulfonate	3.7×10^4
$CH_3SO_3^-$ (mesylate)	3.0×10^4
I^-	91
Br^-	14
$CF_3CO_2^-$	2.1
Cl^-	1.0
F^-	9×10^{-6}
p-Nitrobenzoate	5.5×10^{-6}
$CH_3CO_2^-$	1.4×10^{-6}

a. Data from D. S. Noyce and J. A. Virgilio, *J. Org. Chem.* **37**:2643 (1972).
b. In 80% aqueous ethanol at 75°C.

systems, although no generally applicable quantitative system for specifying leaving-group ability has been established.

Table 5.10 lists estimated relative rates of solvolysis of 1-phenylethyl esters and halides in 80% aqueous ethanol at 75°C.[70] The reactivity of the leaving groups generally parallels their electron-attracting capacity. Trifluoroacetate, for example, is about 10^6 times more reactive than acetate, and p-nitrobenzenesulfonate is about 10 times more reactive than p-toluenesulfonate. The order of the halide leaving groups is $I^- > Br^- > Cl^- \gg F^-$. This order is opposite to that of electronegativity and is dominated by the strength of the bond to carbon, which ranges from ~ 50 kcal/mol for the C–I bond to ~ 100 kcal/mol for the C–F bond.

Sulfonate esters are especially useful substrates in nucleophilic substitution reactions used in synthesis. They have a high level of reactivity, and, unlike alkyl halides, they can be prepared from alcohols by reactions that do not directly involve bonds to the carbon atom undergoing substitution. The latter aspect is particularly important in cases in which the stereochemical and structural integrity of the reactant must be maintained. Sulfonate esters are usually prepared by reaction of an alcohol with a sulfonyl halide in the presence of pyridine:

$$ROH + R'SO_2Cl \xrightarrow{\text{pyridine}} ROSO_2R'$$

Tertiary alcohols are more difficult to convert to sulfonate esters, and because of their high reactivity they are often difficult to isolate.[71]

Trifluoromethanesulfonate (triflate) ion is an exceptionally good leaving group. It can be used for nucleophilic substitution reactions on unreactive substrates. Acetolysis of cyclopropyl triflate, for example, occurs 10^5 times faster than acetolysis of cyclopropyl tosylate.[72] Table 5.11 gives a comparison of the triflate group with some other common leaving groups.

70. D. S. Noyce and J. A. Virgilio, *J. Org. Chem.* **37**:2643 (1972).
71. H. M. R. Hoffmann, *J. Chem. Soc.* **1965**:6748.
72. T. M. Su, W. F. Sliwinski, and P. v. R. Schleyer, *J. Am. Chem. Soc.* **91**:5386 (1969).

Table 5.11. Relative Solvolysis Rates of Ethyl Sulfonates and Halides[a]

Derivatives compared	k_{rel}	Solvent 25°C
Triflate/tosylate	3×10^4	Acetic acid
Triflate/brosylate	5×10^3	Acetic acid
Triflate/iodide	4.5×10^5	Ethanol
Triflate/bromide	1.5×10^5	80 % Ethanol

a From A. Streitwieser, Jr., C. L. Wilkins, and E. Kiehlmann, *J. Am. Chem. Soc.* **90**:1598 (1968).

It would be anticipated that the limiting S_N1 and S_N2 mechanisms would differ in their sensitivity to the nature of the leaving group. The ionization mechanism should exhibit a greater dependence on leaving-group ability because it requires cleavage of the bond to the leaving group without assistance by the nucleophile. Table 5.12 presents data on the variation of the relative leaving-group abilities of tosylate and bromide as a function of substrate structure. The dependence is as expected, with smaller differences in reactivity between tosylate and bromide being observed for systems that react by the S_N2 mechanism.

A poor leaving group can be made more reactive by coordination to an electrophilic species. Hydroxide is a very poor leaving group. Normally, alcohols therefore do not undergo direct nucleophilic substitution. It has been estimated that the reaction

$$CH_3OH + Br^- \rightarrow CH_3Br + HO^-$$

is endothermic by 16 kcal/mol.[73] Because the activation energy for the reverse process is about 21 kcal/mol, the reaction would have an activation energy of 37 kcal/mol. As would be predicted on the basis of this activation energy, the reaction is too slow to detect at normal temperature. The reaction is, however, greatly accelerated in acidic solution. Protonation of the hydroxyl group provides the much better leaving group water, which is about as good a leaving group as bromide ion. The practical result is that primary alcohols can be converted to alkyl bromides by heating with sodium bromide and sulfuric acid or with concentrated hydrobromic acid.

Table 5.12. Tosylate/Bromide Rate Ratios for Solvolysis of RX in 80% Ethanol[*]

R	k_{OTs}/k_{Br}
Methyl	11
Ethyl	10
Isopropyl	40
t-Butyl	4000
1-Adamantyl	9750

a. From J. L. Fry, C. J. Lancelot, L. K. M. Lam, J. M. Harris, R. C. Bingham, D. J. Raber, R. E. Hall, and P. v. R. Schleyer, *J. Am. Chem. Soc.* **92**:2539 (1970).

73. R. A. Ogg, Jr., *Trans. Faraday Soc.* **31**:1385 (1935).

One of the best leaving groups is molecular nitrogen in alkyl diazonium ions. Diazonium ions are generated by nitrosation of primary amines. The diazonium ions generated from alkyl amines are very unstable and immediately decompose with loss of nitrogen.

$$RNH_2 + HONO \longrightarrow R-\underset{H}{N}-N=O + H_2O$$

$$R-\underset{H}{N}-N=O \longrightarrow RN=NOH \overset{H^+}{\longrightarrow} R-\overset{+}{N}\equiv N + H_2O$$

$$R-\overset{+}{N}\equiv N \longrightarrow R^+ + N_2$$

$$R^+ + \mathbf{Nu:} \longrightarrow R-\mathbf{Nu}$$

Because a neutral molecule is eliminated, rather than an anion, there is no electrostatic attraction (ion pairing) between the products of the dissociation step. As a result, the carbocations generated by diazonium-ion decomposition frequently exhibit somewhat different behavior from those generated from halides or sulfonates under solvolytic conditions.[74]

5.7. Steric and Strain Effects on Substitution and Ionization Rates

Examples of effects of reactant structure on the rate of nucleophilic substitution reactions have appeared in the preceding sections of this chapter. The general trends of reactivity of primary, secondary, and tertiary systems and the special reactivity of allylic and benzylic systems have been discussed in other contexts. This section will emphasize the role that steric effects can play in nucleophilic substitution reactions.

Reactions with good nucleophiles in solvents of low ionizing power are sensitive to the degree of substitution at the carbon atom undergoing reaction. Reactions which proceed by the direct displacement mechanism are retarded by increased steric repulsions at the transition state. This is the principal cause for the relative reactivities of methyl, ethyl, and i-propyl chloride, which are, for example, in the ratio 93 : 1:0.0076 toward iodide ion in acetone.[75] A statistical analysis of rate data for 18 sets of nucleophilic substitution reactions of substrates of the type RCH_2Y, where Y is a leaving group and R is H or alkyl, indicated that steric effects of R were the dominant factor in determining rates.[76] Table 5.13 records some of the data. Notice that the fourth entry, involving solvolysis in acetic acid, shows a diminished sensitivity to steric effects. Because acetic acid is a much weaker nucleophile than the solvents involved in the other examples, the transition state is more ionic in character with less nucleophilic participation than in the other examples. The relative rates of formolysis of alkyl bromides at 100°C are: methyl, 0.58; ethyl, 1.000; i-propyl, 26.1; and t-butyl 10^8.[77] This order is clearly dominated by carbocation stability. The effect of substituting a methyl group for hydrogen can be seen

74. C. J. Collins, *Acc. Chem. Res.* **4**:315 (1971); A. Streitwieser, Jr., *J. Org. Chem.* **22**:861 (1957); E. H. White, K. W. Field, W. H. Hendrickson, P. Dzadzic, D. F. Roswell, S. Paik, and R. W. Mullen, *J. Am. Chem. Soc.* **114**:8023 (1992).
75. J. B. Conant and R. E. Hussey, *J. Am. Chem. Soc.* **47**:476 (1925).
76. M. Charton, *J. Am. Chem. Soc.* **97**:3694 (1975).
77. L. C. Bateman and E. D. Hughes, *J. Chem. Soc.* **1937**:1187; *J. Chem. Soc.* **1940**:945.

299

SECTION 5.7.
STERIC AND STRAIN
EFFECTS ON
SUBSTITUTION AND
IONIZATION RATES

Table 5.13. Rate Constants for Nucleophilic Substitution in Primary Alkyl Substrates[a]

Reaction	$10^5 k$ for RCH_2X				
	R=H−	$CH_3−$	$CH_3CH_2−$	$(CH_3)_2CH−$	$(CH_3)_3C−$
$RCH_2Br + LiCl$, acetone	600	9.9	6.4	1.5	0.00026
$RCH_2Br + Bu_3P$, acetone	26,000	154	64	4.9	
$RCH_2Br + NaOCH_3$, methanol	8140	906	335	67	
RCH_2OTs, acetic acid	0.052	0.044		0.018	0.0042

a. From M. Charton, *J. Am. Chem. Soc.* **97**:3694 (1975).

from this type of data to depend on the extent of nucleophilic participation in the transition state. A large $CH_3 : H$ rate ratio is expected if nucleophilic participation is weak and stabilization of the cationic nature of the transition state is important. A low ratio is expected when nucleophilic participation is strong.

The rate of acetolysis of *t*-butyl bromide relative to that of *i*-propyl bromide at 25°C is $10^{3.7}$, whereas the rate of acetolysis of 2-methyl-2-adamantyl bromide relative to that of 2-adamantyl bromide is $10^{8.1}$.[78]

The reason the adamantyl system is much more sensitive to the substitutions of CH_3 for H is that its cage structure prevents solvent participation whereas the *i*-propyl system has much stronger solvent participation. The internal stabilizing effect of the methyl substituent is therefore more important in the adamantyl system.

Steric effects of another kind become important in highly branched substrates, in which ionization is facilitated by relief of steric crowding in going from the tetrahedral ground state to the transition state for ionization.[79] The ratio of the hydrolysis rates in 80% aqueous acetone of *t*-butyl *p*-nitrobenzoate and 2,3,3-trimethyl-2-butyl *p*-nitrobenzoate is 1 : 4.4.

The cause of this effect has been called *B-strain* (back-strain), and in this example only a modest rate enhancement is observed. As the size of the groups is increased, the effect on rate becomes larger. When all three of the groups in the above example are *t*-butyl, the solvolysis occurs 13,500 times faster than in the case of *t*-butyl *p*-nitrobenzoate.[80]

78. J. L. Fry, J. M. Harris, R. C. Bingham, and P. v. R. Schleyer, *J. Am. Chem. Soc.* **92**:2540 (1970).
79. H. C. Brown, *Science* **103**:385 (1946); E. N. Peters and H. C. Brown, *J. Am. Chem. Soc.* **97**:2892 (1975).
80. P. D. Bartlett and T. T. Tidwell, *J. Am. Chem. Soc.* **90**:4421 (1968).

Table 5.14. Relative Hydrolysis Rates of 2-Alkyl-2-adamantyl *p*-Nitrobenzoates[a]

R	k_{rel}, 25°C[b]
CH_3-	2.0
CH_3CH_2-	15.4
$(CH_3)_3CCH_2-$	20.0
$(CH_3)_2CH-$	67.0
$(CH_3)_3C-$	4.5×10^5

a. From J. L. Fry, E. M. Engler, and P. v. R. Schleyer, *J. Am. Chem. Soc.* **94**:4628 (1972).
b. Relative to *t*-butyl *p*-nitrobenzoate = 1

Large B-strain effects are observed in rigid systems such as the 2-alkyl-2-adamantyl *p*-nitrobenzoates. Table 5.14 shows some pertinent data. The repulsive van der Waals interaction between the substituent and the *syn*-axial hydrogens is relieved as the hybridization at C-2 goes from sp^3 to sp^2. As the alkyl group becomes more sterically demanding, the ground-state energy is increased more than the transition-state energy, and reactivity is enhanced.

Another feature of systems that are subject to B-strain is their reluctance to form strained substitution products. The cationic intermediates usually escape to elimination products in preference to capture by a nucleophile. Rearrangements are also common. 2-Methyl-2-adamantyl *p*-nitrobenzoate gives 82% methyleneadamantane by elimination and 18% 2-methyl-2-adamantanol by substitution in aqueous acetone. Elimination accounts for 95% of the product from 2-neopentyl-2-adamantyl *p*-nitrobenzoate. The major product (83%) from 2-*t*-butyl-2-adamantyl *p*-nitrobenzoate is the rearranged alkene **5**.

5.8. Effects of Conjugation on Reactivity

In addition to steric effects, there are other important substituent effects which determine both the rate and mechanism of nucleophilic substitution reactions. It was

81. J. L. Fry, E. M. Engler, and P. v. R. Schleyer, *J. Am. Chem. Soc.* **94**:4628 (1972).

mentioned on p. 281 that benzylic and allylic cations are stabilized by electron delocalization. It is therefore easy to understand why substitution reactions of the ionization type proceed more rapidly in such systems than in alkyl systems. It has also been observed that direct displacement reactions also take place particularly rapidly in benzylic and allylic systems. Allyl chloride is 33 times more reactive than ethyl chloride toward iodide ion

in acetone, for example.[82] These enhanced rates reflect stabilization of the S_N2 transition state through overlap of the π-type orbital which develops at the α carbon in the transition state.[83] The π-systems of the allylic and benzylic groups provide extended conjugation.

 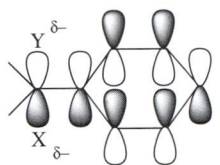

Substitution reactions by the ionization mechanism proceed very slowly on α-halo derivatives of ketones, aldehydes, acids, esters, nitriles, and related compounds. As discussed on p. 284, such substituents destabilize a carbocation intermediate. Substitution by the direct displacement mechanism, however, proceeds especially readily in these systems. Table 5.15 indicates some representative relative rate accelerations. Steric effects may be responsible for part of the observed acceleration, since an sp^2 carbon, such as in a carbonyl group, will provide less steric resistance to the incoming nucleophile than an alkyl group. The major effect is believed to be electronic. The adjacent π-LUMO of the carbonyl group can interact with the electron density that is built up at the pentacoordinate carbon. This can be described in resonance terminology as a contribution from an enolate-like structure to the transition state. In MO terminology, the low-lying LUMO has a

Table 5.15 α-Substituent Effects[a]

$$X-CH_2Cl + I^- \rightarrow X-CH_2I + Cl^-$$

X	Relative rate	X	Relative rate
$CH_3CH_2CH_2-$	1	$\underset{\text{PhC}-}{\overset{O}{\underset{\|}{}}}$ PhC–	3.2×10^4
$PhSO_2-$	0.25	$N{\equiv}C-$	3×10^3
$\underset{CH_3C-}{\overset{O}{\underset{\|}{}}}$ CH₃C–	3.5×10^4	$\underset{C_2H_5OC-}{\overset{O}{\underset{\|}{}}}$ C₂H₅OC–	1.7×10^3

a. Data from F. G. Bordwell and W. T. Branner, Jr., *J. Am. Chem. Soc.* **86**:4545 (1964).

82. see Ref. 75.
83. A. Streitwieser, Jr., *Solvolytic Displacement Reactions*, McGraw-Hill, New York, 1962, p. 13; F. Carrion and M. J. S. Dewar, *J. Am. Chem. Soc.* **106**:3531 (1984).

stabilizing interaction with the developing π orbital of the transition state.[84]

resonance representation of electronic
interaction with carbonyl group at the
transition state for substitution which
delocalizes negative charge

MO representation of stabilization
by interaction with π^* orbital

It should be noted that not all electron-attracting groups enhance reactivity. The sulfonyl and trifluoro groups, which cannot participate in this type of π conjugation, retard the rate of S_N2 substitution at an adjacent carbon.[85]

The extent of the rate enhancement due to adjacent substituents is dependent on the nature of the transition state. The most important factor is the nature of the π-type orbital that develops at the trigonal bipyramidal carbon in the transition state. If this carbon is cationic in character, electron donation from adjacent substituents becomes stabilizing. If bond formation at the transition state is advanced, resulting in charge buildup at carbon, electron withdrawal should be more stabilizing. Substituents such as carbonyl groups therefore have their greatest effect on reactions with strong nucleophiles. Adjacent alkoxy substituents can stabilize S_N2 transition states that are cationic in character. Because the vinyl and phenyl groups can stabilize either type of transition state, the allyl and benzyl systems show enhanced reactivity toward both strong and weak nucleophiles.[86]

5.9. Stereochemistry of Nucleophilic Substitution

Studies of the stereochemical course of nucleophilic substitution reactions are a powerful tool for investigation of the mechanisms of these reactions. Bimolecular direct displacement reactions by the $\lim S_N2$ mechanism are expected to result in 100% inversion of configuration. The stereochemical outcome of the $\lim S_N1$ ionization mechanism is less predictable because it depends on whether reaction occurs via one of the ion-pair intermediates or through a completely dissociated ion. Borderline mechanisms may also show variable stereochemistry, depending upon the lifetime of the intermediates and the extent of internal return. It is important to dissect the overall stereochemical outcome into the various steps of such reactions.

84. R. D. Bach, B. A. Coddens, and G. J. Wolber, *J. Org. Chem.* **51**:1030 (1986); F. Carrion and M. J. S. Dewar, *J. Am. Chem. Soc.* **106**:3531 (1984); S. S. Shaik, *J. Am. Chem. Soc.* **105**:4389 (1983); D. McLennon and A. Pross, *J. Chem. Soc., Perkin Trans. 2* **1984**:981; T. I. Yousaf and E. S. Lewis, *J. Am. Chem. Soc.* **109**:6137 (1987).

85. F. G. Bordwell and W. T. Brannen, *J. Am. Chem. Soc.* **86**:4645 (1964).

86. D. N. Kost and K. Aviram, *J. Am. Chem. Soc.* **108**:2006 (1986); S. S. Shaik, *J. Am. Chem. Soc.* **105**:4359 (1983).

Table 5.16 presents data on some representative nucleophilic substitution processes. The first entry illustrates the use of 1-butyl-1-d p-bromobenzenesulfonate to demonstrate that primary systems react with inversion, even under solvolysis conditions in formic acid. The observation of inversion indicates a concerted mechanism in this weakly nucleophilic solvent.

Neopentyl (2,2-dimethylpropyl) systems are resistant to nucleophilic substitution reactions. They are primary and do not form carbocation intermediates, but the t-butyl substituent effectively hinders back-side attack. The rate of reaction of neopentyl bromide with iodide ion is 470 times slower than that of n-butyl bromide.[87] Usually, the neopentyl system reacts with rearrangement to the t-pentyl system, although use of good nucleophiles in polar aprotic solvents permits direct displacement to occur. Entry 2 shows that such a reaction with azide ion as the nucleophile proceeds with complete inversion of configuration. The primary benzyl system in entry 3 exhibits high, but not complete, inversion. This is attributed to racemization of the reactant by ionization and internal return.

Entry 4 shows that reaction of a secondary 2-octyl system with the moderately good nucleophile acetate ion occurs with complete inversion. The results cited in entry 5 serve to illustrate the importance of solvation of ion-pair intermediates in reactions of secondary substrates. The data show that partial racemization occurs in aqueous dioxane but that an added nucleophile (azide ion) results in complete inversion, both in the product resulting from reaction with azide ion and in the alcohol resulting from reaction with water. The alcohol of retained configuration is attributed to an intermediate oxonium ion resulting from reaction of the ion pair with the dioxane solvent. This would react with water to give product of retained configuration. When azide ion is present, dioxane does not effectively compete for the ion-pair intermediate, and all of the alcohol arises from the inversion mechanism.[88]

Nucleophilic substitution in cyclohexyl systems is quite slow and is often accompanied by extensive elimination. The stereochemistry of substitution has been determined with the use of a deuterium-labeled substrate (entry 6). In the example shown, the substitution process occurs with complete inversion of configuration. By NMR analysis, it can be determined that there is about 15% of rearrangement by hydride shift accompanying solvolysis in acetic acid. This increases to 35% in formic acid and 75% in trifluoroacetic acid. The extent of rearrangement increases with decreasing solvent

87. P. D. Bartlett and L. J. Rosen, J. Am. Chem. Soc. **64**:543 (1942).
88. H. Weiner and R. A. Sneen, J. Am. Chem. Soc. **87**:292 (1965).

Table 5.16 Stereochemical Course of Nucleophilic Substitution Reactions

	Substrate [a]	Reaction conditions	Product[a]	Stereochemistry	Reference
1	$CH_3CH_2CH_2CHDOBs$	Acetic acid, 99°C	$CH_3CH_2CH_2CHDOAc$	$96 \pm 8\%$ inversion	b
		Formic acid, 99°C	$CH_3CH_2CH_2CHDO_2CH$	$99 \pm 6\%$ inversion	b
2	$(CH_3)_3CCHDOTs$	Sodium azide in HMPA, 90°C	$(CH_3)_3CCHDN_3$	$98 \pm 2\%$ inversion	c
3	$C_6H_5CHDOTs$	Acetic acid, 25°C	$C_6H_5CHDOAc$	$82 \pm 1\%$ inversion	d

Secondary

	Substrate [a]	Reaction conditions	Product[a]	Stereochemistry	Reference
4	$CH_3CH(CH_2)_5CH_3$ $\|$ OTs	Tetraethylammonium acetate in acetone, reflux	$CH_3CH(CH_2)_5CH_3$ $\|$ OAc	100% inversion	d
5	$CH_3CH(CH_2)_5CH_3$ $\|$ OBs	75% aqueous dioxane, 65°C	$CH_3CH(CH_2)_5CH_3$ $\|$ OH	77% inversion	e
		75% aqueous dioxane containing $0.06\,M$ sodium azide, 65°C	$CH_3CH(CH_2)_5CH_3$ $\|$ OH 22% yield	100% inversion	e
			$CH_3CH(CH_2)_5CH_3$ $\|$ N_3 78% yield	100% inversion	e
6		Acetic acid		100% inversion	f
7		80% ethanol-water		> 97% inversion	g
8	$C_6H_5CHCH_3$ $\|$ Cl	Potassium acetate in acetic acid, 50°C	$C_6H_5CHCH_3$ $\|$ OAc	15% inversion	h
		Tetraethylammonium acetate in acetone, 50%	$C_6H_5CHCH_3$ $\|$ OAc	65% inversion	h
		60% aqueous ethanol	$C_6H_5CHCH_3$ $\|$ OH	33% inversion	i

Tertiary

9	<chemical structure: phenyl with CH₃CH₂CCH₃ / OPNB>	Potassium acetate in acetic acid, 23°C	<chemical structure: phenyl with CH₃CH₂CCH₃ / OAc>	$5 \pm 2\%$ inversion	j
		Sodium azide in methanol, 65°C	<chemical structure: phenyl with CH₃CH₂CCH₃ / N₃>	$56 \pm 1\%$ inversion	j
			<chemical structure: phenyl with CH₃CH₂CCH₃ / OCH₃>	14% inversion	j
		90% aqueous acetone	<chemical structure: phenyl with CH₃CH₂CCH₃ / OH>	38% retention	k

a. Abbreviations used: OBs=p-bromobenzenesulfonate; OTs = p-toluenesulfonate; OAc = acetate; OPNB = p-nitrobenzoate.

b. A. Streitwieser, Jr., *J. Am. Chem. Soc.* **77**:1117 (1995).

c. B. Stephenson, G. Solladie, H. S. Mosher, *J. Am. Chem. Soc.* **94**:4184 (1972).

d. A. Streitwieser, Jr., T. D. Walsh, and J. R. Wolfe, Jr., *J. Am. Chem. Soc.* **87**:3682 (1965).

e. H. Weiner and R. A. Sneen, *J. Am. Chem. Soc.* **87**:287 (1965).

f. J. B. Lambert, G. J. Putz, and C. E. Mixan, *J. Am. Chem. Soc.* **94**:5132 (1972); see also J. E. Nordlander and T. J. McCrary, *J. Am. Chem. Soc.* **94**:5133 (1972).

g. K. Humski, V. Sendijarevic, and V. J. Shiner, *J. Am. Chem. Soc.* **98**:2865 (1976); K. Humski, V. Sendijarevic, and V. J. Shiner, *J. Am. Chem. Soc.* **95**:7722 (1973).

h. J. Steigman and L. P. Hammett, *J. Am. Chem. Soc.* **59**:2536 (1937).

i. V. J. Shiner, Jr., S. R. Hartshorn, and P. C. Vogel, *J. Org. Chem.* **38**:3604 (1973).

j. L. H. Sommer and F. A. Carey, *J. Org. Chem.* **32**:800 (1967).

k. H. L. Goering and S. Chang, *Tetrahedron Lett.* **1965**:3607.

306

CHAPTER 5
NUCLEOPHILIC
SUBSTITUTION

nucleophilicity, as would be expected. Rearrangement implies involvement of the hydride-bridged species as an intermediate or transition state.

Stabilization of a carbocation intermediate by benzylic conjugation, as in the 1-phenylethyl system shown in entry 8, leads to substitution with diminished stereospecificity. A thorough analysis of stereochemical, kinetic, and isotope effect data on solvolysis reactions of 1-phenylethyl chloride has been carried out.[89] The system has been analyzed in terms of the fate of the intimate ion-pair and solvent-separated ion-pair intermediates. From this analysis, it has been estimated that for every 100 molecules of 1-phenylethyl chloride that undergo ionization to an intimate ion pair (in trifluoroethanol), 80 return to starting material of retained configuration, 7 return to inverted starting material, and 13 go on to the solvent-separated ion pair.

$$RX \underset{80}{\rightleftharpoons} R^+X^- \overset{13}{\rightleftharpoons} R^+\|X^- \overset{0}{\rightleftharpoons} R^+ + X^-$$
$$\big\updownarrow 7$$
$$XR \overset{6}{\rightleftharpoons} X^-R^+ \overset{1}{\rightleftharpoons} X^-\|R^+$$

As is evident from the result shown for the tertiary benzylic substrate 2-phenyl-2-butyl p-nitrobenzoate in entry 9, the simple expectation of complete racemization is not realized. In moderately nucleophilic media such as potassium acetate in acetic acid, this ideal is almost achieved, with just a slight excess of inversion. Use of a better nucleophile like azide ion, however, leads to product with a significant (56%) degree of inversion. This can be attributed to nucleophilic attack on an ion-pair intermediate prior to symmetrical solvation or dissociation. More surprising is the observation of net retention of configuration in the hydrolysis of 2-phenyl-2-butyl p-nitrobenzoate in aqueous acetone. It is possible that this is the result of preferential solvent collapse from the front side at the solvent-separated ion-pair stage. The bulky tertiary system may hinder solvation from the rear side. It is also possible that hydrogen bonding between a water molecule and the anion of the ion pair facilitates capture of a water molecule from the front side of the ion pair.

Nucleophilic substitution reactions that occur under conditions of amine diazotization often have significantly different stereochemistry, as compared with that in halide or sulfonate solvolysis. Diazotization generates an alkyl diazonium ion, which rapidly decomposes to a carbocation, molecular nitrogen, and water:

$$R-NH_2 \rightarrow R-\underset{H}{N}-N{=}O \rightarrow R-N{=}N-OH \overset{H^+}{\longrightarrow} R-\overset{+}{N}{\equiv}N + H_2O \rightarrow R^+ + N_2$$

Thus, in contrast to an ionization process from a neutral substrate, which initially generates an intimate ion pair, deamination reactions generate a cation which does not have an anion closely associated with it. Furthermore, the leaving group, molecular nitrogen, is very stable so that little, if any, nucleophilic participation is needed for bond cleavage. The

89. V. J. Shiner, Jr., S. R. Hartshorn, and P. C. Vogel, *J. Org. Chem.* **38**:3604 (1973).

Table 5.17. Stereochemical Course of Deamination Reactions in Acetic Acid

$$RNH_2 \xrightarrow{NaNO_2, CH_3CO_2H} ROAc$$

	Amine	Stereochemistry of acetate ester formation
1[a]	$CH_3CH_2CH_2CHDNH_2$	69% inversion
2[b]	$CH_3CHCH_2CH_3$ $\ \ \ \ \ \ \ \ \|$ $\ \ \ \ \ \ NH_2$	28% inversion
3[c]	—$CHCH_3$ $\ \ \ \ \ \|$ $\ \ \ NH_2$	10% retention
4[d]	CH_3 $\ \ \|$ —CCH_2CH_3 $\ \ \|$ $\ \ NH_2$	24% retention

a. A. Streitwieser, Jr., and W. D. Schaeffer, *J. Am. Chem. Soc.* **79**:2888 (1957).
b. K. B. Wiberg, Dissertation, Columbia University, 1950.
c. R. Huisgen and C. Ruchardt, *Justus Liebigs Ann. Chem.* **601**:21 (1956).
d. E. H. White and J. E. Stuber, *J. Am. Chem. Soc.* **85**:2168 (1963).

stereochemistry of substitution is shown for four representative amines in Table 5.17. Displacement in the primary 1-butyl system is much less stereospecific than the 100% inversion observed on acetolysis of the corresponding brosylate (Table 5.16, entry 1). Similarly, the 2-butyl diazonium ion affords 2-butyl acetate with only 28% net inversion of configuration. Small net retention is seen in the deamination of 1-phenylethylamine. The tertiary benzylic amine 2-phenyl-2-butylamine reacts with 24% net retention. These results indicate that the lifetime of the carbocation is so short that a symmetrically solvated state is not reached. Instead, the composition of the product is determined by a nonselective collapse of the solvent shell.

An analysis of the stereochemistry of deamination has also been done using the conformationally rigid 2-decalylamines:

In solvents containing low concentrations of water in acetic acid, dioxane, or sulfolane, most of the alcohol is formed by capture of water with retention of configuration. This result has been explained as involving a solvent-separated ion pair which would arise as a result of concerted protonation and nitrogen elimination.[90]

$$
\begin{array}{ccc}
CH_3CO_2H \ \ HO_2CCH_3 & CH_3CO_2H \ \ HO_2CCH_3 & CH_3CO_2H \ \ HO_2CCH_3 \\
R{-}N{\equiv}N{-}OH & \longrightarrow \ \ R^+ \ \ N{\equiv}N \ \ OH_2 & \longrightarrow \ \ ROH + RO_2CCH_3 \\
CH_3CO_2H \ \ \ \ H{-}O_2CCH_3 & CH_3CO_2H \ \ \ {}^-O_2CCH_3 & CH_3CO_2H \ \ HO_2CCH_3
\end{array}
$$

90. H. Maskill and M. C. Whiting, *J. Chem. Soc., Perkin Trans. 2* **1976**:1462; T. Cohen, A. D. Botelhjo, and E. Jankowksi, *J. Org. Chem.* **45**:2839 (1980).

**Table 5.18. Product Composition from Deamination of
Stereoisomeric Amines**

| | Product composition[a] | | | |
| | Alcohol | | Ester | |
	Ret	Inv	Ret	Inv
cis-4-*t*-Butylcyclohexylamine (ax)[b]	33	8	25	33
trans-4-*t*-Butylcyclohexylamine (eq)[b]	43	2	43	12
trans,trans-2-Decalylamine (ax)[c]	26	2	32	40
trans,cis-2-Decalylamine (eq)[c]	18	1	55	26

a. Composition of total of alcohol and acetate ester. Considerable, and variable,
   amounts of alkene are also formed.
b. H. Maskill and M. C. Whiting, *J. Chem. Soc., Perkin Trans. 2.* **1976**:1462.
c. T. Cohen, A. D. Botelhjo, and E. Jankowski, *J. Org. Chem.* **45**:2839 (1980).

In such a process, the water molecule formed in the elimination step is captured primarily
from the front side, leading to net retention of configuration for the alcohol. For the ester,
the extent of retention and inversion is more balanced, although it varies among individual
systems. It is clear from the data in Table 5.18 that the two pairs of stereoisomeric amines
do not form the same intermediate, even though a simple mechanistic interpretation would
suggest that both would form the 2-decalyl cation. The collapse of the ions to product is
evidently so rapid that there is not time for relaxation of the initially formed intermediates
to reach a common structure.

A few nucleophilic substitution reactions have been observed to proceed with a high
degree of retention of configuration. One example is reaction of alcohols with thionyl
chloride, which under some conditions gives predominantly product of retained configuration. This reaction is believed to proceed by formation of a chlorosulfite ester. This can
then react with chloride to give inverted product.

When the reaction is performed in dioxane solution, an oxonium ion is formed from the
solvent and the chlorosulfite ester. The oxonium ion then undergoes substitution by
chloride. Two inversions are involved so that the result is overall retention.[91]

91. E. S. Lewis and C. E. Boozer, *J. Am. Chem. Soc.* **74**:308 (1952).

When a molecule that is a substrate for nucleophilic substitution also contains a group that can act as a nucleophile, it is often observed that the kinetics and stereochemistry of nucleophilic substitution are strongly affected. The involvement of nearby nucleophilic substituents in a substitution process is called *neighboring-group participation*.[92]

A classic example of neighboring-group participation involves the solvolysis of compounds in which an acetoxy substituent is present next to a carbon that is undergoing nucleophilic substitution. For example, the rates of solvolysis of the *cis* and *trans* isomers of 2-acetoxycyclohexyl *p*-toluenesulfonate differ by a factor of about 670, the *trans* compound being the more reactive one:[93]

$k = 1.9 \times 10^{-4} \, s^{-1} \, (100°C)$ $k = 2.9 \times 10^{-7} \, s^{-1} \, (100°C)$

Besides the pronounced difference in rate, the products obtained from the isomeric compounds reveal a marked difference in stereochemistry. The diacetate obtained from the *cis* isomer is the *trans* compound (inverted stereochemistry), whereas retention of configuration is observed for the *trans* isomer.

These results can be explained by the *participation* of the *trans* acetoxy group in the ionization process. The assistance provided by the acetoxy carbonyl group facilitates the ionization of the tosylate group, accounting for the rate enhancement. This kind of backside participation by the adjacent acetoxy group is both sterically and energetically favorable. The cation which is formed by participation is stabilized by two oxygen atoms and is far more stable than a secondary carbocation. The acetoxonium-ion intermediate is subsequently opened by nucleophilic attack with inversion at one of the two equivalent carbons, leading to the observed *trans* product.[94]

92. B. Capon, *Q. Rev. Chem. Soc.* **18**:45 (1964); B. Capon and S. P. McManus, *Neighboring Group Participation*, Plenum Press, New York, 1976.
93. S. Winstein, E. Grunwald, R. E. Buckles, and C. Hanson, *J. Am. Chem. Soc.* **70**:816 (1948).
94. S. Winstein, C. Hanson, and F. Grunwald, *J. Am. Chem. Soc.* **70**:812 (1948).

When enantiomerically pure *trans*-2-acetoxycyclohexyl tosylate is solvolyzed, the product is racemic *trans*-diacetate. This is consistent with the proposed mechanism, since the acetoxonium intermediate is achiral and can only give rise to racemic material.[95] Additional evidence for this interpretation comes from the isolation of a cyclic *ortho* ester when the solvolysis is carried out in ethanol. In this solvent the acetoxonium ion is captured by the solvent.

(Ref. 96)

The hydroxy group can act as an intramolecular nucleophile. Solvolysis of 4-chlorobutanol in water gives as the product the cyclic ether tetrahydrofuran.[97] The reaction is much faster than solvolysis of 3-chloropropanol under similar conditions.

In basic solution, the alkoxide ions formed by deprotonation are even more effective nucleophiles. In ethanol containing sodium ethoxide, 2-chloroethanol reacts about 5000 times faster than ethyl chloride. The product is ethylene oxide, confirming the involvement of the oxygen atom as a nucleophile.

$$HOCH_2CH_2Cl \rightleftharpoons {}^-OCH_2CH_2Cl \longrightarrow H_2C\overset{O}{\underset{}{\diagup\diagdown}}CH_2 + Cl^-$$

As would be expected, the effectiveness of neighboring-group participation depends on the molecular geometry required for participation. The rate of cyclization of ω-hydroxyalkyl halides, for example, shows a strong dependence on the length of the chain separating the hydroxy and halo substituents. Some data are given in Table 5.19. The maximum rate occurs for the 4-hydroxybutyl system involving a five-membered ring. As

Table 5.19 Solvolysis Rates of ω-Chloro Alcohols[a]

ω-Chloro alcohol	Approximate relative rate
$Cl(CH_2)_2OH$	2000
$Cl(CH_2)_3OH$	1
$Cl(CH_2)_4OH$	5700
$Cl(CH_2)_5OH$	20

a. B. Capon, *Q. Rev. Chem. Soc.* **18**:45 (1964); W. H. Richardson, C. M. Golino, R. H. Wachs, and M. B. Yelvington, *J. Org. Chem.* **36**:943 (1971).

95. S. Winstein, H. V. Hess, and R. E. Buckles, *J. Am. Chem. Soc.* **64**:2796 (1942).
96. S. Winstein and R. E. Buckles, *J. Am. Chem. Soc.* **65**:613 (1943).
97. H. W. Heine, A. D. Miller, W. H. Barton, and R. W. Greiner, *J. Am. Chem. Soc.* **75**:4778 (1953).

discussed in Section 3.9, intramolecular processes involving five-membered ring formation are often more rapid than those forming either four- or six-membered rings.

Like the un-ionized hydroxyl group, an alkoxy group is a weak nucleophile. Nevertheless, it can operate as a neighboring nucleophile. For example, solvolysis of the isomeric p-bromobenzenesulfonate esters **6** and **7** leads to identical product mixtures, suggesting the involvement of a common intermediate. This can be explained by involvement of the cyclic oxonium ion which would result from intramolecular participation.[98]

The occurrence of nucleophilic participation is also indicated by a rate enhancement relative to the rate of solvolysis of n-butyl p-bromobenzenesulfonate. The solvolysis rates of a series of ω-methoxyalkyl p-bromobenzenesulfontes have been determined. A maximum rate is again observed where participation of a methoxy group via a five-membered ring is possible (see Table 5.20).

Transannular participation of ether oxygen has also been identified by kinetic studies of a series of cyclic ethers. The relative rates for compounds **10–13** show that there is a large acceleration in the case of replacement of the 5-CH_2 group by an ether oxygen.[99]

| relative rate: | 1.0 | 0.014 | 0.14 | 4.85×10^4 |

**Table 5.20. Relative Solvolysis Rates of Some
ω-Methoxyalkyl p-Bromobenzenesulfonates
in Acetic Acid[a]**

$CH_3(CH_2)_2OSO_2Ar$	1.00
$CH_3O(CH_2)_2OSO_2Ar$	0.28
$CH_3O(CH_2)_3OSO_2Ar$	0.67
$CH_3O(CH_2)_4OSO_2Ar$	657
$CH_3O(CH_2)_5OSO_2Ar$	123
$CH_3O(CH_2)_6OSO_2Ar$	1.16

a. From S. Winstein, E. Allred, R. Heck, and R. Glick, *Tetrahedron* **3**:1 (1958).

98. E. L. Allred and S. Winstein, *J. Am. Chem. Soc.* **89**:3991 (1967).
99. L. A. Paquette and M. K. Scott, *J. Am. Chem. Soc.* **94**:6760 (1972).

The huge difference in rate that results from the alternative placement of oxygen in the eight-membered rings reflects the relative stability of the various oxonium ions that result from participation. The ion **16** is much more favorable than **14** or **15**.

The rate retardation evident for **11** and **12** can be attributed to an unfavorable polar effect of the C−O bond.

In general, any system that has a nucleophilic substituent group situated properly for back-side displacement of a leaving group at another carbon atom of the molecule can be expected to display neighboring-group participation. The extent of the rate enhancement will depend on how effectively the group acts as an internal nucleophile. The existence of participation may be immediately obvious from the structure of the product if some derivative of the cyclic intermediate is stable. In other cases, demonstration of kinetic acceleration or stereochemical consequences may provide the basis for identifying neighboring-group participation.

The π electrons of carbon–carbon double bonds can also become involved in nucleophilic substitution. This can facilitate the ionization step and may lead to a carbocation having special stability. Solvolysis reactions of the *syn* and *anti* isomers of 7-substituted norbornenes provide some dramatic examples of the influence of participating double bonds on reaction rates and stereochemistry. The *anti* tosylate is more reactive toward acetolysis than the saturated analog by a factor of about 10^{11}. The acetolysis product, *anti*-7-acetoxynorbornene, is the product of retention of configuration. These results can be explained by participation of the π electrons of the double bond to give the ion **17**, which would be stabilized by delocalization of the positive charge.[100]

In contrast, the *syn* isomer, in which the double bond is not in a position to participate in the ionization step, reacts 10^7 times slower than the *anti* isomer. The reaction product is derived from a rearranged carbocation ion that is stabilized by virtue of being allylic.[101]

100. S. Winstein, M. Shavatsky, C. Norton, and R. B. Woodward, *J. Am. Chem. Soc.* **77**:4183 (1955); S. Winstein and M. Shavatsky, *J. Am. Chem. Soc.* **78**:592 (1956); S. Winstein, A. H. Lewin, and K. C. Pande, *J. Am. Chem. Soc.* **85**:2324 (1963).
101. S. Winstein and E. T. Stafford, *J. Am. Chem. Soc.* **79**:505 (1957).

The extent of participation of the carbon–carbon double bond in the ionization of *anti*-7-norbornenyl systems is a function of the substitution at C-7. Placement of an aryl substituent at C-7 diminishes the relative rate acceleration due to participation by the double bond. Evidently, the extent of participation is a function of the stability of the potential carbocation. When an aryl group is present at C-7, the resulting benzyl-type stabilization decreases the relative importance of participation by the double bond. The degree of stabilization is sensitive to substituents on the phenyl ring. For *p*-methoxyphenyl, phenyl, and *p*-trifluoromethylphenyl, the rate enhancement factors for the unsaturated relative to the saturated system are 3, 40, and 3.5×10^4, respectively.[102] The double bond clearly has a much larger effect on the poorly stabilized *p*-trifluoromethylphenyl intermediate. This dependence of the extent of participation on other stabilizing features is a general trend and has been observed with other types of carbocations.[103]

Participation of π electrons from an adjacent double bond controls the stereochemistry of substitution in the case of cyclopent-3-enyl tosylates, even though no strong rate enhancement is observed. The stereochemistry has been demonstrated by solvolysis of a stereospecifically labeled analog.[104] The formolysis product is formed with complete retention of configuration, in contrast to the saturated system, which reacts with complete inversion under similar conditions.[105] The retention of configuration is explained by a structure similar to that shown in the case of *anti*-7-norbornenyl cation.

Evidently, since there is no appreciable rate acceleration, this participation is not very strong at the transition state. Nevertheless, the participation is strong enough to control stereochemistry. When more nucleophilic solvents are used (e.g., acetic acid), participation is not observed, and the product is 100% of inverted configuration.

Participation of carbon–carbon double bonds in solvolysis reactions is revealed in some cases by isolation of products with new carbon–carbon σ bonds. A particularly

102. P. G. Gassman and A. F. Fentiman, Jr., *J. Am. Chem. Soc.* **91**:1545 (1969); *J. Am. Chem. Soc.* **92**:2549 (1970).
103. H. C. Brown, *The Nonclassical Ion Problem*, Plenum Press, New York, 1977, pp. 163–175.
104. J. B. Lambert and R. B. Finzel, *J. Am. Chem. Soc.* **105**:1954 (1983).
105. K. Huniski, V. Sendijarevič, and V. J. Shiner, Jr., *J. Am. Chem. Soc.* **95**:7722 (1973).

significant case is the formation of the bicyclo[2.2.1]heptane system during solvolysis of 2-cyclopent-3-enylethyl tosylate[106]:

In this case, the participation leads to the formation of the norbornyl cation, which is captured as the acetate. More will be said later about this important cation in Section 5.12.

A system in which the details of aromatic π-electron participation have been thoroughly probed is the case of the "phenonium" ions, the species resulting from participation by a β-phenyl group.

"phenonium" ion

Such participation leads to a bridged ion with the positive charge delocalized into the aromatic ring. Evidence for this type of participation was first obtained by a study of the stereochemistry of solvolysis of 3-phenyl-2-butyl tosylates. The *erythro* isomer gave largely retention of configuration, a result that can be explained via the bridged-ion intermediate. The *threo* isomer, where participation leads to an achiral intermediate, gave racemic *threo* product.[107]

106. R. G. Lawton, *J. Am. Chem. Soc.* **83**:2399 (1961).
107. D. J. Cram, *J. Am. Chem. Soc.* **71**:3863 (1949); *J. Am. Chem. Soc.* **74**:2129 (1952).

**Table 5.21. Extent of Aryl Rearrangement in
2-Phenylethyl Tosylate Solvolysis[a]**

Solvent	Rearrangement (%)
C_2H_5OH	0.3
CH_3CO_2H	5.5
$H_2O:HCO_2H(10:90)$	40
HCO_2H	45
CF_3CO_2H	100

a. C. C. Lee, G. P. Slater, and J. W. T. Spinks, *Can. J. Chem.* **35**:1417 (1957);
   J. E. Nordlander and W. G. Deadman, *Tetrahedron Lett.* **1967**:4409.

Both primary and secondary carbocations with β-phenyl substituents usually give evidence of aryl participation. For example, isotopically labeled carbons are scrambled to some extent during solvolysis of 2-phenylethyl tosylates. Either a bridged-ion intermediate or rapidly reversible rearrangement of a primary carbocation could account for the randomization of the label. The extent of label scrambling increases as solvent nucleophilicity decreases. The data are shown in Table 5.21. This trend can be attributed to competition between S_N2 displacement by solvent and ionization with participation of the aryl group. Whereas reaction in more nucleophilic solvents such as ethanol proceeds almost exclusively by direct displacement, the nonnucleophilic solvent trifluoroacetic acid leads to complete randomization of the label.

The relative importance of aryl participation is a function of the substituents on the aryl ring. The extent of participation can be quantitatively measured by comparing the rate of direct displacement, k_s, with the rate of aryl-assisted solvolysis, designated k_Δ.[108] The relative contributions to individual solvolyses can be dissected by taking advantage of the assisted mechanism's higher sensitivity to aryl substituent effects. In systems with electron-withdrawing substituents, the aryl ring does not participate effectively, and only the process described by k_s contributes to the rate. Such compounds give a Hammett correlation with ρ values (-0.7 to -0.8) characteristic of a weak substituent effect. Compounds with electron-releasing substituents deviate from the correlation line because of the aryl participation. The extent of reaction proceeding through the k_s process can be estimated from the correlation line for electron-withdrawing substituents. Table 5.22 gives data indicating the extent of aryl participation under a variety of conditions. This method of analysis also confirms that the relative extent of participation of the β-phenyl groups is highly dependent on the solvent.[109] In solvents of good nucleophilicity (e.g., ethanol), the

108. A. Diaz, I. Lazdins, and S. Winstein, *J. Am. Chem. Soc.* **90**:6546 (1968).
109. F. L. Schadt, III, C. J. Lancelot, and P. v. R. Schleyer, *J. Am. Chem. Soc.* **100**:228 (1978).

**Table 5.22. Extent of Solvolysis with Aryl Participation
as a Function of Substituent and Solvent for
1-Aryl-2-propyl Tosylates**

X	Solvent		
	80% EtOH[a]	CH_3CO_2H[b]	HCO_2H[b]
NO_2	0	–	–
CF_3	0	–	–
Cl	7	–	–
H	21	38	78
CH_3	63	71	94
OCH_3	93	94	99

a. D. J. Raber, J. M. Harris, and P. v. R. Schleyer, *J. Am. Chem. Soc.* **93**:4829
   (1971).
b. C. C. Lancelot and P. v. R. Schleyer, *J. Am. Chem. Soc.* **91**:4296 (1969).

normal solvent displacement mechanism makes a larger contribution. As solvent nucleophilicity decreases, the relative extent of aryl participation increases.

The bridged form of the β-phenylethyl cation can be observed in superacid media and
can be characterized by carbon and proton NMR spectroscopy.[110] The bridged ion
subsequently rearranges to the more stable α-methylbenzyl cation, with E_a for rearrangement being 13 kcal/mol. High-level MO calculations have been performed. The phenonium ion has a good deal of delocalization of the electron density, resulting in positive
charge in the cyclopropane ring.[111]

5.11. Mechanism of Rearrangements of Carbocations

The discussion of the behavior of carbocation intermediates in superacid media and
of neighboring-group participation has already provided examples of carbocation rearrangements. Such rearrangements are a characteristic feature of the chemistry of carbocations.
Rearrangements can occur by shift of an alkyl, vinyl, or aryl group or hydrogen.
Rearrangement creates a new cation with the positive charge located on the atom from
which the migration occurred. 1,2-Shifts are the most common type of rearrangement.[112]

110. G. A. Olah, R. J. Spear, and D. A. Forsyth, *J. Am. Chem. Soc.* **98**:6284 (1976).
111. S. Sieber and P. v. R. Schleyer, *J. Am. Chem. Soc.* **115**:6987 (1993).
112. For a review, see V. G. Shubin, *Top Curr. Chem.* **116–117**:267 (1984).

A thermodynamic driving force exists for rearrangement of the carbon skeleton in the direction of forming a more stable carbocation. Activation energies for skeletal migrations are not large, and it is not uncommon to observe overall rearrangements that must have involved individual steps that proceed from a more stable to a less stable species. Thus, although rearrangement of a tertiary cation to a secondary cation is endothermic by about 10 kcal/mol, this barrier is not prohibitive if the rearrangement can lead to a more stable cation. Formation of primary cations by rearrangement is less likely to occur, because the primary ions are ~ 20 and ~ 35 kcal/mol higher in energy than secondary and tertiary cations, respectively[113] (See Section 5.4.) The barriers for conversion of ions to more stable ions are apparently very low, and such rearrangements occur very rapidly. For example, in superacid media at $-160°C$, the equilibration of the five methyl groups of the 2,3,3-trimethylbutyl cation is so rapid that the barrier must be less than 5 kcal/mol.[114]

$$(CH_3)_3C-\overset{+}{C}(CH_3)_2 \rightleftharpoons (CH_3)_2\overset{+}{C}-C(CH_3)_3$$

The extent to which rearrangement occurs depends on the structure of the cation and the nature of the reaction medium. Capture of carbocations by nucleophiles is a process with a very low activation energy, so that only very fast rearrangements can occur in the presence of nucleophiles. Neopentyl systems, for example, often react to give t-pentyl products. This is very likely to occur under solvolytic conditions but can be avoided by adjusting reaction conditions to favor direct substitution, for example, by use of an aprotic dipolar solvent to enhance the reactivity of the nucleophile.[115] In contrast, in nonnucleophilic media, in which the carbocations have a longer lifetime, several successive rearrangement steps may occur. This accounts for the fact that the most stable possible ion is usually the one observed in superacid systems.

Although many overall rearrangements can be formulated as a series of 1,2-shifts, both isotopic tracer studies and computational work have demonstrated the involvement of other species. These are bridged ions in which hydride or alkyl groups are partially bound to two other carbons. Such structures can be transition states for hydride and alkyl-group shifts, but some evidence indicates that these structures can also be intermediates.

alkyl-bridged ion

corner-protonated
cyclopropane

edge-protonated
cyclopropane

hydride-bridged ion

The alkyl-bridged structures can also be described as "corner-protonated" cyclopropanes, since if the bridging C−C bonds are considered to be fully formed, there is an "extra" proton on the bridging carbon. In another possible type of structure, called "edge-protonated" cyclopropanes, the carbon–carbon bonds are depicted as fully formed, with the "extra" proton associated with one of the "bent" bonds. MO calculations, structural studies under stable-ion conditions, and product and mechanistic studies of reactions in solution have all been applied to understanding the nature of the intermediates involved in carbocation rearrangements.

The energy surface for $C_3H_7^+$ has been calculated at the 6-311G**/MP4 level. The 1- and 2-propyl cations and corner- and edge-protonated cyclopropane structures were

113. G. J. Karabatsos and F. M. Vane. *J. Am. Chem. Soc.* **85**:729 (1963).
114. G. A. Olah and A. M. White, *J. Am. Chem. Soc.* **91**:5801 (1969).
115. B. Stephenson, G. Solladie, and H. S. Mosher, *J. Am. Chem. Soc.* **94**:4184 (1972).

compared. The secondary carbocation was found to be the most stable structure.[116] Hydrogen migration was found to occur through a process which involves the bridged cyclopropyl species. Similar conclusions were drawn at the G2 and B3LYP levels of calculation.[117] The relative energies are shown below.

	$CH_3CH_2CH_2^+$	$CH_3\overset{+}{C}HCH_3$	$\triangle\text{--}H^+$	$H\overset{+}{\triangle}H$ H
6-311G**/MP4	+19.3 kcal	0	8.6 kcal	7.3 kcal
G2		0	8.2 kcal	7.2 kcal
B3LYP		0	16.0 kcal	12.2 kcal

The 2-butyl cation has been extensively investigated both computationally and experimentally. C-2 and C-3 are rapidly interconverted by a hydride shift. Scrambling of C-3 and C-4 (or C-1 and C-2) occurs with an E_a of about 7–8 kcal/mol. The rearrangement of 2-butyl cation to the t-butyl ion is rather slow, occurring with an activation energy of 18 kcal/mol.[118] The C-3/C-4 scrambling can occur via an edge-protonated intermediate.

The 2-butyl cation can be observed under stable-ion conditions. The NMR spectrum corresponds to a symmetrical species, which implies either very rapid hydride shift or a symmetrical H-bridged structure.

A maximum barrier of 2.5 kcal/mol can be assigned from the NMR data.[119] There have been two extensive MO calculations of the $C_4H_9^+$ species. At the 6-311G**/MP4 level of theory, the H-bridged structure was the most stable found and was about 2 kcal/mol more

116. W. Koch, B. Liu, and P. v. R. Schleyer, *J. Am. Chem. Soc.* **111**:3479 (1989).
117. M. V. Frash, V. B. Kazansky, A. M. Rigby, and P. A. van Santen, *J. Phys. Chem.* **101**:5346 (1997).
118. D. M. Brouwer, *Recl. Trav. Chim. Pays-Bas* **87**:1435 (1968); D. M. Brouwer and H. Hogeveen, *Prog. Phys. Org. Chem.* **9**:179 (1972); M. Boronat, P. Viruela, and A. Corma, *J. Phys. Chem.* **100**:633 (1996).
119. M. Saunders and M. R. Kates, *J. Am. Chem. Soc.* **100**:7082 (1978).

stable that the open 2-butyl cation.[120] The methyl-bridged ion was only slightly less stable. The structure has also been calculated at the 6-31G**/MP4 level of theory.[121] The energies relative to the *t*-butyl cation are similar, although the methyl-bridged ion was found to be slightly more stable than the hydride-bridged ion.

6-311G*/MP4	13.4	13.8	21.9	15.7	33.0	0.0
6-31**/MP4	12.8	12.0	20.5		32.6	0.0

Along with the minimal barrier for H shift, the 2-butyl to *t*-butyl rearrangement gives the energy surface shown in Fig. 5.9. This diagram indicates that the mechanism for C-3/C-4 scrambling in the 2-butyl cation involves the edge-protonated cyclopropane intermediate.

Fig. 5.9. Energy profile for the scrambling and rearrangement of $C_4H_9^+$ cation. **A**: H-bridged; **B**: methyl-bridged; **C**: Edge protonated methycyclopropane; **D**: classical secondary; **E**: classical primary; **F**: tertiary. Adapted from refs 120 and 121.

120. J. W. de M. Carneiro, P. v. R. Schleyer, W. Koch, and K. Raghavachari, *J. Am. Chem. Soc.* **112**:4064 (1990). S. Sieber, P. Buzek, P. v. R. Schleyer, W. Koch, and J. W. de M. Carneiro, *J. Am. Chem. Soc.* **115**:259 (1993).
121. M. Boronat, P. Viruela, and A. Corma, *J. Phys. Chem.* **100**:633 (1996).

The primary cation is an intermediate in the isomerization to the *t*-butyl ion, which accounts for the relatively slow rate of this process.

The occurrence and extent of rearrangement of the 2-butyl cation have also been investigated by solvolysis studies using isotopic labeling. When 2-butyl tosylate is solvolyzed in acetic acid, C-2/C-3 rearrangement occurs only to the extent of 9% in the 2-butyl acetate which is isolated.[122] Thus, under these conditions, most of the reaction proceeds by direct participation of the solvent.

When 2-butyl tosylate is solvolyzed in the less nucleophilic solvent trifluoroacetic acid, a different result emerges. The extent of migration approaches the 50% that would result from equilibration of the two possible secondary cations.[123]

Two hydride shifts resulting in interchange of the C(2) and C(3) hydrogens account for the two minor products.

122. J. D. Roberts, W. Bennett, R. F. McMahon, and E. W. Holroyd, Jr., *J. Am. Chem. Soc.* **74**:4283 (1952).
123. J. J. Dannenberg, B. J. Goldberg, J. K. Barton, K. Dill, D. H. Weinwurzel, and M. O. Longas, *J. Am. Chem. Soc.* **103**:7764 (1981); J. J. Dannenberg, J. K. Barton, B. Bunch, B. J. Goldberg, and T. Kowalski, *J. Org. Chem.* **48**:4524 (1983); A. D. Allen, I. C. Ambidge, and T. Tidwell, *J. Org. Chem.* **48**:4527 (1983).

Similar studies have been carried out to characterize $C_5H_{11}{}^+$ species. The barrier to the hydride and methyl shifts which interconvert the methyl groups in the *t*-pentyl cation is 10–15 kcal/mol.[124]

These rearrangements must pass through a secondary ion or related bridged species. The solvolysis product of 3-methyl-2-butyl tosylate in trifluoroacetic acid consists of 98.5% of the ester derived from the rearranged 2-methyl-2-butyl cation and 1.5% of the 3-methyl-2-butyl ester. Even the 1.5% of product having the 3-methyl-2-butyl structure has undergone some rearrangement.[125] The energies of possible intermediates have been calculated at several levels of theory.[126] The energies assigned some of the ions are shown in the chart below.

							CH₃⁺	C₂H₅⁺
6-31G*/MP2	11.7	7.2	0	34.1	17.8	14.4		
B3P86	12.3	10.5	0	35.4	20.4	14.7		
6-31G*/MP4		13.6	0		18.5		9.6	12.4

These results indicate an energy profile for the 3-methyl-2-butyl cation to 2-methyl-2-butyl cation rearrangement in which the open secondary cations are transition states, rather than intermediates, with the secondary cations represented as methyl-bridged species (corner-protonated cyclopropanes) (Fig. 5.10).

The 2-methyl-2-butyl cation provides the opportunity to explore the effect of C−C hyperconjugation. At the 6-31G**/MP4 level of calculation, little energy difference is found between structures **C** and **D** which differ in alignment of CH₃ or H with the empty *p* orbital.[127]

124. M. Saunders and E. L. Hagen, *J. Am. Chem. Soc.* **90**:2436 (1968).

125. D. Farcasiu, G. Marino, J. M. Harris, B. A. Hovanes, and C. S. Hsu, *J. Org. Chem.* **59**:154 (1994); D. Farcasiu, G. Marino, and S. Hsu, *J. Org. Chem.* **59**:163 (1994).

126. M. Boronat, P. Viruela, and A. Corma, *J. Phys. Chem.* **100**:16514 (1996); D. Farcasiu and S. H. Norton, *J. Org. Chem.* **62**:5374 (1997).

127. P. v. R. Schleyer, J. W. de Carneiro, W. Koch, and D. A. Forsyth, *J. Am. Chem. Soc.* **113**:3990 (1991).

Fig. 5.10. Energy surface for the rearrangement of the 3-methyl-2-butyl cation to the 2-methyl-2-butyl cation. [Adapted from D. Farcasiu and S. H. Norton, *J. Org. Chem.* **62**:5374 (1997).]

Structure **C**, however, gives a much closer approximation to the observed ^{13}C chemical shift and thus may represent the preferred structure. The calculations also indicate a lengthened C(3)−C(4) (1.58Å) and a contraction of the C(2)−C(3)−C(4) bond angle to 101.5°, both of which would be consistent with C−C hyperconjugation.

The question of relative preference for migration of different groups, which is sometimes referred to as "migratory aptitude," is a complex one, and there is no absolute order. In general, aryl groups and branched alkyl groups migrate in preference to unbranched alkyl groups, but, because the barriers involved are low, selectivity is not high. Very often, the preferred migration involves that group which is best positioned from a stereoelectronic point of view. The course of migration is also influenced by strain. In general, a shift that will reduce strain is favored.

The preferred alignment of orbitals for a 1,2-hydride or 1,2-alkyl shift involves coplanarity of the p orbital at the carbocation ion center and the σ orbital of the migrating group.

The transition state involves a three-center, two-electron bond. This corresponds to a symmetrically bridged structure, which, in some cases, may actually be an intermediate. The migration process can be concerted with the formation of the carbocation; that is, the migration can begin before the bond to the leaving group at the adjacent carbon atom is completely broken. The phenonium ion case discussed in the previous section is one example. When migration is concerted, the group that is aligned *anti* to the leaving group will migrate preferentially. In some cases, this alignment will be controlled by conformational equilibria. Conformation can also be a factor in migrations under nonconcerted conditions when the barrier to migration is of the same magnitude as the conformational barrier. Rearrangements following deamination reactions seem to be particularly likely to be governed by the conformation of the reactant, and this may reflect the high energy of the cations generated by deamination.[128]

The 3-cyclohexenyl ion provides another example of the dependence of the extent of rearrangement on reaction conditions. Under solvolysis conditions, the major product is the result of direct solvent displacement. There is also some product resulting from hydride shift to the more stable allylic ion as well as a trace of the bicyclo[3.1.0]hexane product arising from participation of the double bond.[129]

Deamination of the corresponding amine gives the allylic alcohol resulting from hydride shift as the main product and an increased amount of the cyclization product.

Formation of the 3-cyclohexenyl cation from the alcohol in superacid media is followed by more extensive rearrangement to give the methylcyclopentenyl ion, which is tertiary and allylic.[130]

128. J. A. Berson, J. W. Foley, J. M. McKenna, H. Junge, D. S. Donald, R. T. Luibrand, N. G. Kundu, W. J. Libbey, M. S. Poonian, J. J. Gajewski, and J. B. E. Allen, *J. Am. Chem. Soc.* **93**:1299 (1971).
129. M. Hanack and W. Keverie, *Chem. Ber.* **96**:2937 (1963).
130. G. A. Olah, G. Liang, and Y. K. Mo, *J. Am. Chem. Soc.* **94**:3544 (1972).

This trend of increasing amount and extent of rearrangement can be readily interpreted. In the acetolysis, a large part of the reaction must be occurring via direct nucleophilic participation by the solvent or ion-pair capture so that only a relatively small amount of hydride shift occurs. As is characteristic of deamination, the carbocations formed in the deamination reaction are more reactive cations, and, as a result, the extent of rearrangement is greater. Finally, in the nonnucleophilic superacid media, the cations are relatively long-lived and undergo several rearrangements, eventually leading to the most stable accessible ion.

The ring contraction of a cyclohexyl cation to a methylcyclopentyl cation is thermodynamically favorable but would require a substantial energy of activation if the rearrangement proceeded through a primary cyclopentylmethyl cation.

It is believed that this process involves migration through a pentacoordinate *protonated cyclopropane* in which an alkyl group acts as a bridge in an electron-deficient carbocation structure. The cyclohexyl→methylcyclopentyl rearrangement is postulated to occur by rearrangement between two such structures.

Shifts of hydride between carbon atoms separated by several atoms are possible if the molecular geometry is favorable. Particularly clear-cut examples have been found in medium-sized rings. For example, solvolysis of cyclononyl-1-^{14}C tosylate can be shown by degradation of the product cyclononene to occur with about 20% of the ^{14}C becoming located at the 5-, 6-, and 7-positions.

$* = {}^{14}C$ label

This result can be explained by a "transannular" 1,5-hydride shift. Many similar processes have been documented.[131]

131. V. Prelog and J. G. Trayham, in *Molecular Rearrangements*, P. de Mayo, ed., Interscience, New York, 1963, p. 593.

Reaction of cyclooctene with trifluoroacetic acid occurs by a hydride-shift process.

325

SECTION 5.11.
MECHANISM OF
REARRANGEMENTS OF
CARBOCATIONS

The main reaction path is *stereospecific*, with the trifluoroacetate being added *syn* to the proton. This implies that the reaction proceeds through a discrete hydride-bridged intermediate rather than a conformationally mobile cyclooctyl cation.[132]

Hydride-bridged ions of this type are evidently quite stable in favorable cases and can be observed under stable-ion conditions. The hydride-bridged cyclooctyl and cyclononyl cations can be observed at −150°C but rearrange, even at that temperature, to methylcycloheptyl and methylcyclooctyl ions, respectively.[133] However, a hydride-bridged ion in which the bridging hydride is located in a bicyclic cage is stable in trifluoroacetic acid at room temperature.[134]

In some cases, NMR studies in superacid media have permitted the observation of successive intermediates in a series of rearrangements. An example is the series of cations originating with the bridgehead ion **F**, generated by ionization of the corresponding chloride. Rearrangement eventually proceeds to the tertiary ion **K**. The bridgehead ion is

132. J. E. Nordlander, K. D. Kotian, D. E. Raff II, F. G. Njoroge, and J. J. Winemiller, *J. Am. Chem. Soc.* **106**:1427 (1984).
133. R. P. Kirchen and T. S. Sorensen, *J. Am. Chem. Soc.* **101**:3240 (1979).
134. J. E. McMurry and C. N. Hodge, *J. Am. Chem. Soc.* **106**:6450 (1984).

stable below $-75°C$. The unrearranged methyl ether is obtained if the solution is quenched with sodium methoxide in methanol at $-90°C$. At about $-65°C$, ion **F** rearranges to ion **J**. This is believed to involve the methyl-bridged ion **G** as an intermediate. Ion **J** is stable below $-30°C$ but above $-30°C$ **K** is formed. This latter rearrangement involves a sequence of steps, again including a methyl-bridged species.[135] This multistep sequence terminating in the most stable $C_9H_{15}^+$ ion is quite typical of carbocations in superacid media. In the presence of nucleophilic anions or solvent, rearrangement usually does not proceed all the way to the most stable ion, because nucleophilic trapping captures one or more of the rearranged species.

Carbocation rearrangement is particularly facile if a functional group can stabilize the rearrangement product. A good example of this is the rearrangement of carbocations having a hydroxyl group on an adjacent carbon atom. Migration in this case generates a protonated carbonyl compound and is very favorable. These rearrangements are referred to as "pinacol rearrangements."[136]

The pinacol rearrangement is frequently observed when vicinal diols react with acid. The structure of the products from unsymmetrical diols can be predicted on the basis of ease of carbocation formation. For example, 1,1-diphenyl-2-methyl-1,2-propanediol rearranges to 3,3-diphenyl-2-butanone because the diarylcarbinol is most readily ionized.[137] Synthetically useful examples of this type of rearrangement are discussed in Section 10.1 of Part B.

135. G. A. Olah, G. Liang, J. R. Wiseman, and J. A. Chong, *J. Am. Chem. Soc.* **94**:4927 (1972).
136. Y. Pocker, in *Molecular Rearrangements*, P. de Mayo, ed., Interscience, New York, 1963, pp. 15–25.
137. W. M. Schubert and P. H. LeFevre, *J. Am. Chem. Soc.* **94**:1639 (1972).

327

SECTION 5.12.
THE NORBORNYL
CATION AND OTHER
NONCLASSICAL
CARBOCATIONS

5.12. The Norbornyl Cation and Other Nonclassical Carbocations

Throughout the discussion of carbocation rearrangements, we have encountered examples of bridged species which require expansion of bonding concepts beyond the two-center, two-electron bonds that suffice for most organic molecules. These bridged structures, which involve delocalization of σ electrons and formation of three-center, two-electron bonds, are called *nonclassical ions*. The case for the importance of such bridged structures largely originated with a specific carbocation, the norbornyl cation, and the issue of whether it has a classical or nonclassical (bridged) structure.[138] The special properties of this intermediate were first recognized on the basis of studies by Saul Winstein and his collaborators. The behavior of norbornyl systems in solvolytic displacement reactions was suggestive of neighboring-group participation by a carbon–carbon single bond. Evidence for both enhanced rate and abnormal stereochemistry was developed by study of the acetolysis of *exo*-2-norbornyl sulfonates.

The acetolyses of both *exo*-2-norbornyl brosylate and *endo*-2-norbornyl brosylate produce exclusively *exo*-2-norbornyl acetate. The *exo*-brosylate is more reactive than the *endo* isomer by a factor of 350.[139] Furthermore, enantiomerically enriched *exo*-brosylate gave completely racemic *exo*-acetate, and the *endo*-brosylate gave acetate that was at least 93% racemic.

Both acetolyses were considered to proceed by way of a rate-determining formation of a carbocation. The rate of ionization of the *endo*-brosylate was considered normal, because its reactivity was comparable to that of cyclohexyl brosylate. Elaborating on a suggestion made earlier concerning rearrangement of camphene hydrochloride,[140] Winstein proposed that ionization of the *exo*-brosylate was assisted by the C(1)−C(6) bonding electrons and led directly to the formation of a nonclassical ion as an *intermediate*.

This intermediate serves to explain the formation of racemic product, since it is achiral. The cation has a plane of symmetry passing through C-4, C-5, C-6, and the midpoint of

138. H. C. Brown, *The Nonclassical Ion Problem*, Plenum Press, New York, 1977; H. C. Brown, *Tetrahedron* **32**:179 (1976); P. D. Bartlett, *Nonclassical Ions*, W. A. Benjamin, New York, 1965; S. Winstein, in *Carbonium Ions*, Vol. III, G. A. Olah and P. v. R. Schleyer, eds., Wiley-Interscience, New York, 1972, Chapter 22; G. D. Sargent, *ibid.*, Chapter 24; C. A. Grob, *Angew. Chem. Int. Ed. Engl.* **21**:87 (1982); G. M. Kramer and C. G. Scouten, *Adv. Carbocation Chem.* **1**:93 (1989).
139. S. Winstein and D. S. Trifan, *J. Am. Chem. Soc.* **71**:2953 (1949); *J. Am. Chem. Soc.* **74**:1147, 1154 (1952); S. Winstein, E. Clippinger, R. Howe, and E. Vogelfanger, *J. Am. Chem. Soc.* **87**:376 (1965).
140. T. P. Nevell, F. deSalas, and C. L. Wilson, *J. Chem. Soc.* **1939**:1188.

the C(1)–C(2) bond. The plane of symmetry is seen more easily in an alternative, but equivalent, representation.

Carbon-6, which bears two hydrogens, is pentacoordinate and serves as the bridging atom in the cation.

Attack by acetate at C-1 or C-2 would be equally likely and would result in equal amounts of the enantiomeric acetates. The acetate ester would be *exo* because reaction must occur from the direction opposite the bridging interaction. The nonclassical ion can be formed directly only from the *exo*-brosylate because it has the proper *anti* relationship between the C(1)–C(6) bond and the leaving group. The bridged ion can be formed from the *endo*-brosylate only after an unassisted ionization. This would explain the rate difference between the *exo* and *endo* isomers.

The nonclassical-ion concept proved to be an intriguing one, and many tests for the intermediacy of nonclassical ions in the norbornyl system were employed.[141] Other bicyclic systems were also investigated to explore the generality of the concept of nonclassical ions. Whereas the classical ion in the norbornyl system is chiral and the nonclassical ion is achiral, the situation is reversed in the bicyclo[2.2.2]octane system.

When bicyclo[2.2.2]octyl brosylate was solvolyzed in acetic acid containing sodium acetate, the products were a mixture of bicyclo[2.2.2]octyl acetate and bicyclo[3.2.1]octyl acetate, each of which was optically active. The formation of bicyclo[2.2.2]octyl acetate was found to proceed with $82 \pm 15\%$ *retention* of configuration, a result which is in

141. Much of the extensive work on the norbornyl cation is discussed in a series of reviews in *Accounts of Chemical Research*: C. A. Grob, *Acc. Chem. Res.* **16**:426 (1983); H. C. Brown, *Acc. Chem. Res.* **16**:432 (1983); G. A. Olah, G. K. S. Prakash, and M. Saunders, *Acc. Chem. Res.* **16**:440 (1983). C. Walling, *Acc. Chem. Res.* **16**:448 (1983).

accord with involvement of a bridged-ion intermediate.[142] The achiral classical cation could not have been the major intermediate.

329

SECTION 5.12.
THE NORBORNYL
CATION AND OTHER
NONCLASSICAL
CARBOCATIONS

achiral   chiral

The description of the nonclassical norbornyl cation developed by Winstein implies that the nonclassical ion is stabilized, relative to a secondary ion, by C–C σ bond delocalization. H. C. Brown of Purdue University put forward an alternative interpretation.[143] He argued that all the available data were consistent with describing the intermediate as a rapidly equilibrating classical ion. The 1,2-shift that interconverts the two ions was presumed to be rapid relative to capture of the nucleophile. Such a rapid rearrangement would account for the isolation of racemic product, and Brown proposed that the rapid migration would lead to preferential approach of the nucleophile from the *exo* direction.

The two competing descriptions of the norbornyl cation have been very extensively tested. In essence, the question that is raised has to do with the relative energy of the bridged structure. Is it lower in energy than the classical ion and therefore an *intermediate* to which the classical ion would collapse, or is it a transition state (or higher-energy intermediate) in the rapid isomerization of two classical structures? Figure 5.11 illustrates the potential energy diagrams corresponding to the various possibilities.

Many studies of substituent effects on rate and product stereochemistry were undertaken to resolve this issue. When techniques for direct observation of carbocations became available, the norbornyl cation was subjected to intense study from this perspective. The norbornyl cation was generated in SbF_5–SO_2–SOF_2, and the temperature dependence of the proton NMR spectrum was examined.[144] Subsequently, the ^{13}C-NMR spectrum was studied, and the proton spectrum was determined at higher field strength. These studies excluded rapidly equilibrating classical ions as a description of the norbornyl cation under stable-ion conditions.[145] It was also determined that 3,2- and 6,2-hydride shifts were occurring under stable-ion conditions. Activation energies of 10.8 and 5.9 kcal/mol were measured for these processes, respectively.

The resonances observed in the ^{13}C-NMR spectrum have been assigned. None of the signals appears at a position near that where the signal of the C-2 carbon of the classical

142. H. M. Walborsky, M. F. Baum, and A. A. Youssef, *J. Am. Chem. Soc.* **83**:988 (1961).
143. H. C. Brown, *The Transition State, Chem. Soc. Spec. Publ.*, No. 16, 140 (1962); *Chem. Brit.* **1966**:199; *Tetrahedron* **32**:179 (1976).
144. P. v. R. Schleyer, W. E. Watts, R. C. Fort, Jr., M. B. Comisarow, and G. A. Olah, *J. Am. Chem. Soc.* **86**:5679 (1964); M. Saunders, P. v. R. Schleyer, and G. A. Olah, *J. Am. Chem. Soc.* **86**:5680 (1964).
145. G. A. Olah, G. K. S. Prakash, M. Arvanaghi, and F. A. L. Anet, *J. Am. Chem. Soc.* **104**:7105 (1982).

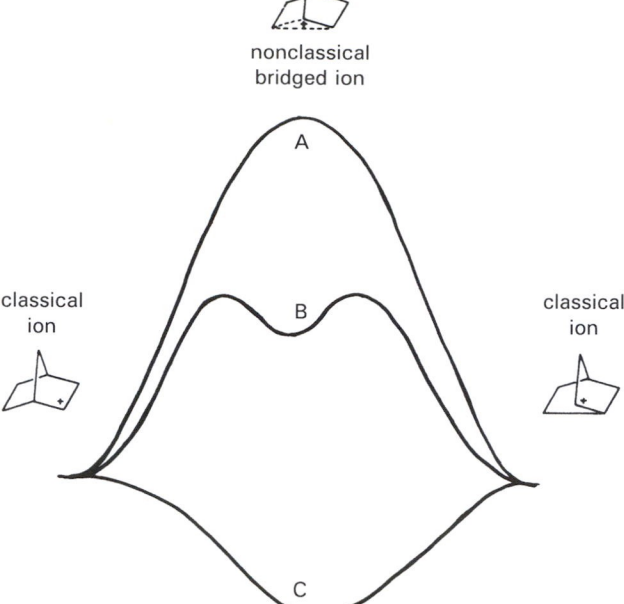

Fig. 5.11. Contrasting potential energy diagrams for stable and unstable bridged norbornyl cation. (A) Bridged ion is a transition state for rearrangement between classical structures. (B) Bridged ion is an intermediate in rearrangement of one classical structure to the other. (C) Bridged nonclassical ion is the only stable structure.

secondary 2-propyl cation is found. Instead, the resonances for the norbornyl cation appear at relatively high field and are consistent with the pentacoordinate nature of the bridged-ion structure.[146] Other NMR techniques have also been applied to the problem and confirm the conclusion that the spectra observed under stable-ion conditions cannot be the result of averaging of the spectra of two rapidly equilibrating ions.[147]

These results, which pertain to *stable-ion conditions*, provide strong evidence that the most stable structure for the norbornyl cation is the symmetrically bridged nonclassical ion. How much stabilization does the σ bridging provide? An estimate based on molecular mechanics calculations and a thermodynamic cycle suggests a stabilization of about 6 ± 1 kcal/mol.[148] An experimental value based on mass-spectrometric measurements is 11 kcal/mol.[149] Gas-phase hydride affinity and chloride affinity data also show the norbornyl cation to be especially stable.[150]

High-level MO calculations (6-31G*/MP2) give a nonclassical structure that is 13.6 kcal more stable than the classical structure and predict ^{13}C chemical shifts and

146. G. A. Olah, G. Liang, G. D. Mateescu, and J. L. Riemenschneider, *J. Am. Chem. Soc.* **95**:8698 (1973); G. A. Olah, *Acc. Chem. Res.* **9**:41 (1976).
147. C. S. Yannoni, V. Macho, and P. C. Myhre, *J. Am. Chem. Soc.* **104**:7105 (1982); M. Saunders and M. R. Kates, *J. Am. Chem. Soc.* **102**:6867 (1980); M. Saunders and M. R. Kates, *J. Am. Chem. Soc.* **105**:3571 (1983).
148. P. v. R. Schleyer and J. Chandrasekhar, *J. Org. Chem.* **46**:225 (1981).
149. M. C. Blanchette, J. L. Holmes, and F. P. Lossing, *J. Am. Chem. Soc.* **109**:1392 (1987).
150. R. B. Sharma, D. K. S. Sharma, K. Hiraoka, and P. Kebarle, *J. Am. Chem. Soc.* **107**:3747 (1985).

331

SECTION 5.12.
THE NORBORNYL
CATION AND OTHER
NONCLASSICAL
CARBOCATIONS

coupling in excellent agreement with the experimental results.[151] The difference in energy between the two structures is reduced only slightly when calculations include the effect of solvation, indicating that the bridged ion should be more stable than the classical ion even in solution.[152] There is also good agreement between calculated and observed IR spectra.[153]

X-ray crystal structure determinations have been completed on two salts containing bicyclo[2.2.1]heptyl cations (Fig. 5.12). Both are more stable than the 2-norbornyl cation itself; **18** is tertiary whereas **19** contains a stabilizing methoxy group. The crystal structure of **18** shows an extremely long (1.74 Å) C—C bond between C-1 and C-6. The C(1)—C(2) bond is shortened to 1.44 Å. The distance between C-2 and C-6 is shortened from 2.5 Å in norbornane to 2.09 Å.[154] These structural changes can be depicted as a partially bridged structure.

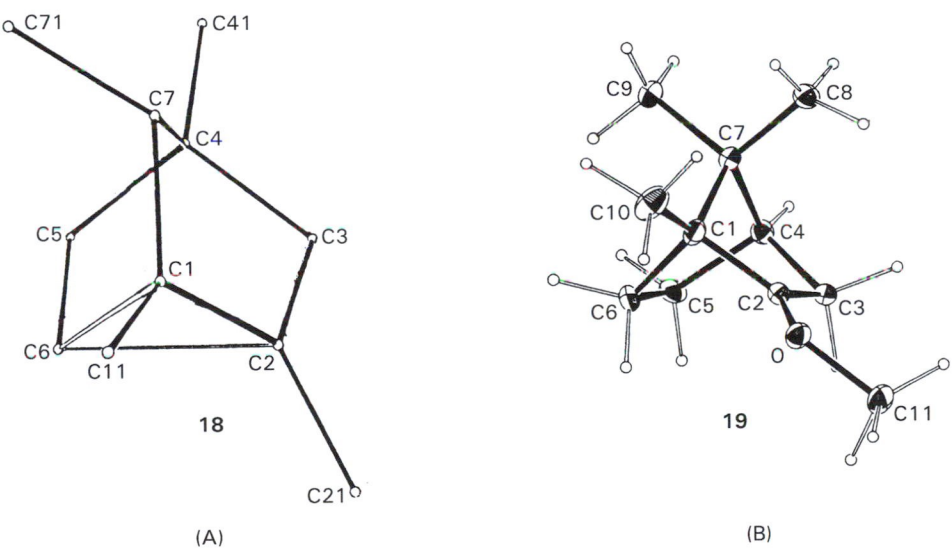

Fig. 5.12. Crystal structures of substituted norbornyl cations. (A) 1,2,4,7-Tetramethylnorbornyl cation (reproduced from Ref. 154 by permission of Wiley-VCH). (B) 2-Methoxy-1,7,7-trimethyl-norbornyl cation (reproduced from Ref. 155 by permission of the American Chemical Society).

151. P. v. R. Schleyer and S. Sieber, *Angew. Chem. Int. Ed. Engl.* **32**:1606 (1993); S. A. Perera and R. J. Bartlett, *J. Am. Chem. Soc.* **118**:7849 (1996).
152. P. R. Schreiner, D. L. Severance, W. L. Jorgensen, P. v. R. Schleyer III, *J. Am. Chem. Soc.* **117**:2663 (1995).
153. W. Koch, B. Liu, D. J. DeFrees, D. E. Sunko, and H. Vancik, *Angew. Chem. Int. Ed. Engl.* **29**:183 (1990).
154. T. Laube, *Angew. Chem. Int. Ed. Engl.* **26**:560 (1987).
155. L. K. Montgomery, M. P. Grendez, and J. C. Huffman, *J. Am. Chem. Soc.* **109**:4749 (1987).

The evidence for this tendency toward bridging in a tertiary cation is supportive of the bridged structure for the less stable secondary cation. The ion **19** shows some expected features, such as a shortened C(2)−O bond, corresponding to resonance interaction with the methoxy group. The C(1)−C(6) bond is abnormally long (1.60 Å), although not as long as in **18**.[155] This and other features of the structure are consistent with some σ delocalization even in this cation, which should be strongly stabilized by the methoxy group.

Let us now return to the question of solvolysis and how it relates to the structure under stable-ion conditions. To relate the structural data to solvolysis conditions, the primary issues that must be considered are the extent of solvent participation in the transition state and the nature of solvation of the cationic intermediate. The extent of solvent participation has been probed by comparison of solvolysis characteristics in trifluoroacetic acid with the solvolysis in acetic acid. The *exo* : *endo* reactivity ratio in trifluoroacetic acid is 1120 : 1, compared to 280 : 1 in acetic acid. Whereas the *endo* isomer shows solvent sensitivity typical of normal secondary tosylates, the *exo* isomer reveals a reduced sensitivity. This indicates that the transition state for solvolysis of the *exo* isomer possesses a greater degree of charge dispersal, which would be consistent with a bridged structure. This fact, along with the rate enhancement of the *exo* isomer, indicates that the σ participation commences prior to the transition state being attained, so that it can be concluded that bridging is a characteristic of the solvolysis intermediate, as well as of the stable-ion structure.[156]

Another line of evidence that bridging is important in the transition state for solvolysis has to do with substituent effects for groups placed at C-4, C-5, C-6, and C-7 on the norbornyl system. The solvolysis rate is most strongly affected by C-6 substituents, and the *exo* isomer is more sensitive to these substituents than is the *endo* isomer. This implies that the transition state for solvolysis is especially sensitive to C-6 substituents, as would be expected if the C(1)−C(6) bond participates in solvolysis.[157]

Computation of the transition state in the ionization of the protonated *exo* and *endo* alcohols has been done using B3LYP calculations.[158] The calculations confirm that participation occurs in the ionization process and is greater for the *exo* than the *endo* system. However, the stabilization resulting from the participation is considerably less than the full stabilization energy of the nonclassical carbocation. A difference of 3.7 kcal/mol is calculated between the *exo* and *endo* transition states. Figure 5.13 gives the relative energy relationships.

Many other cations besides the norbornyl cation have nonclassical structures.[159] Scheme 5.5 shows some examples which have been characterized by structural studies or by evidence derived from solvolysis reactions. To assist in interpretation of the nonclassical structures, the bond representing the bridging electron pair is darkened in a corresponding classical structure. Not surprisingly, the borderline between classical structures and nonclassical structures is blurred. There are two fundamental factors

156. J. E. Nordlander, R. R. Gruetzmacher, W. J. Kelly, and S. P. Jindal, *J. Am. Chem. Soc.* **96**:181 (1984).
157. F. Fuso, C. A. Grob, P. Sawlewicz, and G. W. Yao, *Helv. Chim. Acta* **69**:2098 (1986); P. Flury and C. A. Grob, *Helv. Chim. Acta* **66**:1971 (1983).
158. P. R. Schreiner, P. v. R. Schleyer, and H. F. Schaefer III, *J. Org. Chem.* **62**:4216 (1997).
159. V. A. Barkhash, *Top. Curr. Chem.* **116–117**:1 (1984); G. A. Olah and G. K. S. Prakash, *Chem. Brit.* **19**:916 (1983).

333

SECTION 5.12.
THE NORBORNYL
CATION AND OTHER
NONCLASSICAL
CARBOCATIONS

3.7 kcal mol^{-1}
including ground state
energy difference

TS$_{exo}$

TS$_{endo}$

reverse
association:
$\Delta H^{\ddagger} = 5.6$
kcal mol^{-1}

forward
reaction:
$\Delta H^{\ddagger} = 5.3$
kcal mol^{-1}

forward
reaction:
$\Delta H^{\ddagger} = 2.8$
kcal mol^{-1}

reverse
association:
$\Delta H^{\ddagger} = 1.9$
kcal mol^{-1}

1 (X = OH$_2^+$)
(0.0 kcal mol^{-1})

6
(0.9 (−2.3)
kcal mol^{-1})

2 (X = OH$_2^+$)
(1.2(1.6)
kcal mol^{-1})

Fig. 5.13. Computational energy diagram (B3LYP/6-311 + G*//B3LYP/6-31G*) for intermediates in ionization and rearrangement of protonated norbornanol. (Adapted from Ref. 158.)

Scheme 5.5. Other Examples of Nonclassical Carbocations

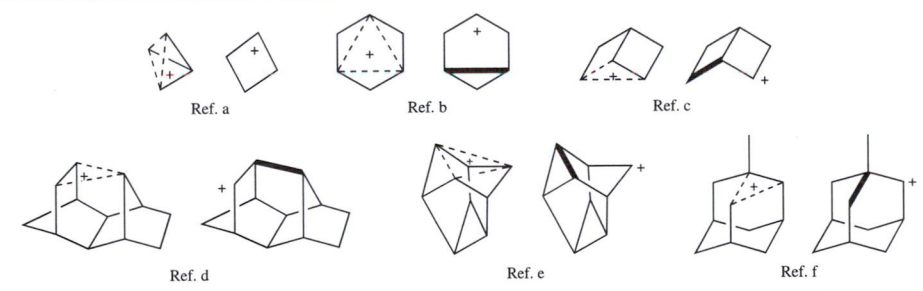

Ref. a

Ref. b

Ref. c

Ref. d

Ref. e

Ref. f

a. M. Saunders and H.-U. Siehl, *J. Am. Chem. Soc.* **102**:6868 (1980); J. S. Starat, J. D. Roberts, G. K. S. Prakash, D. J. Donovan, and G. A. Olah, *J. Am. Chem. Soc.* **100**:8016, 8018 (1978); W. Koch, B. Liu, D. J. Frees, *J. Am. Chem. Soc.* **110**: 7325 (1988); M. Saunders, K. E. Laidig, K. B. Wiberg and P. V. R. Schleyer *J. Am. Chem. Soc.* **110**: 7652 (1988); P. C. Myhre, G. C. Webb and C. S. Yannoni, *J. Am. Chem. Soc.* **112**: 8992 (1990).
b. G. A. Olah, G. K. S. Prakash, T. N. Rawdah, D. Whittaker, and J. C. Rees, *J. Am. Chem. Soc.* **101**: 3935 (1979).
c. R. N. McDonald and C. A. Curi, *J. Am. Chem. Soc.* **101**:7116, 7118 (1979).
d. S. Winstein and R. L. Hansen, *Tetrahedron Lett.* **25**:4 (1960).
e. R. M. Coates and E. R. Fretz, *J. Am. Chem. Soc.* **99**:297 (1977); H. C. Brown and M. Ravindranathan, *J. Am. Chem. Soc.* **99**:299 (1977).
f. J. E. Nordlander and J. E. Haky, *J. Am. Chem. Soc.* **103**:1518 (1981).

which prevent an absolute division. (1) The energies of the two (or more) possible structures may be so close as to prevent a clear distinction as to stability. (2) The molecule may adopt a geometry that is intermediate between a classical geometry and a symmetrical bridged structure.

The $C_4H_7^+$ cation shown as the first entry in Scheme 5.5 is a particularly interesting example. It is a bridged ion which can be reached from isomeric cyclopropylmethyl and cyclobutyl ions.

NMR studies show that all three methylene groups are equivalent, and that the *exo* and *endo* sets of hydrogens do not exchange. The barrier for exchange among the three CH_2 groups is <2 kcal. MO calculations at the 6-31G*/MP4SDTQ level indicate that both the cyclopropylmethyl and the nonclassical (cyclobutonium) cations correspond to energy minima, differing by only 0.26 kcal. The cyclobutyl cation is about 12 kcal higher in energy.[160] The nonclassical structure represents a tetracyclic cation which incorporates three pentacoordinate carbons.[161]

To summarize, it now appears that nonclassical or bridged structures are readily attainable intermediates or transition states for many cations and are intimately involved in rearrangement processes. In some cases, such as the norbornyl cation, the bridged structure is the most stable structure. As a broad generalization, tertiary cations are nearly always more stable than related bridged ions and therefore have classical structures. Primary carbocations can be expected to undergo rearrangement to more stable secondary or tertiary ions, with bridged ions being likely transition states (or intermediates) on the rearrangement path. The energy balance between classical secondary structure and bridged structures is close and depends on the individual system. Bridged structures are most likely to be stable in cases in which a strained bond can participate in bridging or in which solvation of the positive charge is difficult. Because of poor solvation, bridged structures are particularly likely to be favored in superacid media and in the gas phase.

General References

D. Bethell and V. Gold, *Carbonium Ions, An Introduction*, Academic Press, London, 1967.
C. A. Bunton, *Nucleophilic Substitution at a Saturated Carbon Atom*, Elsevier, New York, 1963.

160. M. Saunders, K. E. Laidig, K. B. Wiberg, and P. v. R. Schleyer, *J. Am. Chem. Soc.* **110**:7652 (1988).
161. For further discussion of this structure, see, R. F. W. Bader and K. E. Laidig, *THEOCHEM* **261**:1 (1992).

T. L. Ho, *Hard and Soft Acids and Bases Principle in Organic Chemsitry*, Academic Press, New York, 1977.

S. P. McManus and C. U. Pittman, Jr., in *Organic Reactive Intermediates*, S. P. McManus, ed., Academic Press, New York, 1973, Chapter 4.

G. A. Olah and P. v. R. Schleyer, eds., *Carbonium Ions*, Vols. I–IV, Wiley-Interscience, New York, 1968–1973.

M. Saunders, J. Chandrasekhar, and P. v. R. Schleyer, in *Rearrangements in Ground and Excited States*, P. de Mayo, ed., Academic Press, New York, 1980, Chapter 1.

A. Streitwieser, Jr., *Solvolytic Displacement Reactions*, McGraw-Hill, New York, 1962.

E. R. Thornton, *Solvolysis Mechanisms*, Ronald Press, New York, 1964.

Problems

(References for these problems will be found on page 795.)

1. Provide an explanation for the relative reactivity or stability relationships revealed by the following sets of data.

   (a) For solvolysis of

   $$R-\underset{\underset{CH_3}{|}}{\overset{\overset{CH_3}{|}}{C}}-O-\overset{\overset{O}{\|}}{C}-\underset{}{\boxed{}}-NO_2$$

   in aqueous acetone, the relative rates are: R = CH_3, 1; *i*-Pr, 2.9; *t*-Bu, 4.4; Ph, 10^3; cyclopropyl, 5×10^5.

   (b) For solvolysis of

   $$Ar-\underset{\underset{X}{|}}{\overset{\overset{CH_3}{|}}{C}}-OSO_2R$$

   the Hammett ρ^+ reaction constant varies with the substituent X as shown below:

X	ρ^+
CH_3	−4.5
CF_3	−6.9
CH_3SO_2	−8.0

   (c) The hydride affinities measured for the methyl cation with increasing fluorine substitution are in the order CH_3^+(312) > CH_2F^+(290) > CHF_2^+(284) < CF_3^+(299).

   (d) The rates of solvolysis of a series of 2-alkyl-2-adamantyl *p*-nitrobenzoates are: R = CH_3, 1.4×10^{-10} s^{-1}; C_2H_5, 1.1×10^{-9} s^{-1}; *i*-C_3H_7, 5.0×10^{-9} s^{-1}; *t*-C_4H_9, 3.4×10^{-5} s^{-1}; $(CH_3)_3CCH_2$, 1.5×10^{-9} s^{-1}.

   (e) The relative rates of methanolysis of a series of alkyl chlorides having ω-phenylthio substituents as a function of chain length are: $n = 1$, 3.3×10^4; $n = 2$, 1.5×10^2; $n = 3$, 1.0; $n = 4$, 1.3×10^2; $n = 5$, 4.3.

2. Which reaction in each pair would be expected to be faster? Explain.

(a)

or

solvolysis in
80% ethanol

(b)

solvolysis in
100% ethanol

(c)

or

solvolysis in
acetic acid

(d)

or

solvolysis in
aqueous acetone

OPNB = *p*-nitrobenzoate

(e)

or

solvolysis in
acetic acid

(f) $C_6H_5SO_2CH_2CH_2Cl$ *or* $C_6H_5SO_2CH_2Cl$ reaction with KI
in acetone

(g) $(CH_3)_3CCH_2OTs$ *or*

solvolysis in
aqueous dioxane

(h)

or

solvolysis in
aqueous dioxane

(i) $H_2C{=}CHCH_2CH_2OTs$ *or* $CH_3CH{=}CHCH_2OTs$ solvolysis in
98% formic acid

(j)

or

solvolysis in
acetic acid

(k) $PhS(CH_2)_3Cl$ *or* $PhS(CH_2)_4Cl$ solvolysis in
aqueous dioxane

(l)

or

solvolysis in
acetic acid

3. Suggest reasonable mechanisms for each of the following reactions. The starting materials were the racemic substance, except where noted otherwise.

(a) $PhCH_2Cl + P(OCH_3)_3 \longrightarrow PhCH_2P(O)(OCH_3)_2 + CH_3Cl$

(b)

no ^{18}O label in product

(c)

(d)

but

(e) $HC{\equiv}CCH_2CH_2CH_2Cl \xrightarrow{CF_3CO_2H}$

(f)

(g)

(h)

(i)

(j)

optically
active

racemic

(j)

$$(PhC)_2NCH_2CH_2Cl \xrightarrow[CH_3C\equiv N,\ heat]{H_2O} PhCNHCH_2CH_2OCPh$$

(l)

(m)

4. The solvolysis of 2R, 3S-3-(4-methoxyphenyl)but-2-yl p-toluensulfonate in acetic acid can be followed by several kinetic measurements: (a) rate of decrease of observed rotation (k_α); rate of release of the leaving group (k_t); and, when ^{18}O-labeled sulfonate is used, the rate of equilibration of the sulfonate oxygens (k_{ex}). At 25°C, the rate constants are

$$k_\alpha = 25.5 \times 10^{-6}\,s^{-1};\quad k_t = 5.5 \times 10^{-6}\,s^{-1};\quad k_{ex} = 17.2 \times 10^{-6}\,s^{-1}.$$

Describe the nature of the process measured by each of these rate constants, and devise a mechanism which includes each of these processes. Rationalize the order of the rates $k_\alpha > k_{ex} > k_t$.

5. Both *endo*- and *exo*-norbornyl p-bromobenzenesulfonates react with $R_4P^+N_3^-$ (R = long-chain alkyl) in toluene to give azides of inverted configuration. The yield from the *endo* reactant is 95%, and from the *exo* reactant, the yield is 80%. In the case of the *exo* reactant, the remaining material is converted to nortricyclane (tricyclo[2.2.1.0$^{2.6}$]heptane). The measured rates of azide formation are first-order in both reactant and azide ion. No rearrangement of deuterium is observed in the azides when deuterium-labeled reactants are used. What conclusion about the mechanism of the substitution process do you draw from these results? How do the reaction conditions relate to the mechanism you have suggested? How is the nortricyclane formed?

6. The following observations have been made concerning the reaction of Z-1-phenyl-1,3-butadiene (A) and Z-4-phenyl-3-buten-2-ol (B) in 3–7 M H_2SO_4 and 0.5–3 M $HClO_4$:

   (a) Both compounds are converted to the corresponding E-isomer by acid-catalyzed processes with rates given by

$$\text{rate} = k[\text{reactant}][H^+]$$

   when $[H^+]$ is measured by the H_0 acidity function.

   (b) The rate of isomerization of A is slower in deuterated (D_2SO_4–D_2O) media by a factor of 2–3. For B, the rate of isomerization is faster in D_2SO_4–D_2O by a factor of 2.5.

   (c) When ^{18}O-labeled B is used, the rate of loss of ^{18}O to the solvent is equal to the rate of isomerization.

   (d) The activation energies are 19.5 ± 1 kcal/mol for A and 22.9 ± 0.7 kcal/mol for B.

   Write a mechanism for each isomerization which is consistent with the information given.

7. Treatment of 2-(p-hydroxyphenyl)ethyl bromide with basic alumina produces a white solid: mp, 40–43°C; IR, 1640 cm^{-1}; UV, 282 nm in H_2O, 261 mm in ether; NMR, two singlets of equal intensity at 1.69 and 6.44 ppm from TMS. *Anal:* C, 79.97; H, 6.71. Suggest a reasonable structure for this product and a rationalization for its formation.

8. The solvolysis of the tosylate of 3-cyclohexenol has been studied in several solvents. The rate of solvolysis is not very solvent-sensitive, being within a factor of 5 for all solvents. The product distribution is solvent-sensitive, however, as shown below.

Solvent (ROH)	1	2	3	Cyclohexadienes
acetic acid	20%	a	10%	70%
formic acid	58%	a	a	42%
CF$_3$CHCF$_3$ OH	10%	65%	a	25%

a. Minor product, less than 3% yield.

Furthermore, the stereochemistry of the product 1 changes as the solvent is changed. In aqueous dioxane, the reaction proceeds with complete inversion, but in 1,1,1,3,3,3 hexafluoro-2-propanol with 100% retention. In acetic acid, the reaction occurs mainly with inversion (83%), but in formic acid the amount of retention (40%) is comparable to the amount of inversion (60%). Discuss these results, particularly with respect to the change of product composition and stereochemistry as a function of solvent.

9. Each of the following carbocations can rearrange to a cation with special stabilization. Indicate likely routes for the rearrangement to a more stable species for each ion.

10. In the discussion of the *syn*- and *anti*-norbornenyl tosylates (p. 312), it was pointed out that, relative to 7-norbornyl tosylate, the reactivities of the *syn* and *anti* isomers were 10^4 and 10^{11}, respectively. The high reactivity of the *anti* isomer was attributed to participation of the carbon–carbon double bond. What is the source of the 10^4 factor of acceleration in the *syn* isomer relative to the saturated model?

11. Indicate the structure of the ion you expect to be formed as the stable species when each of the following compounds is dissolved in superacid media at $-30°C$:

12. The behavior of compounds **A** and **B** on solvolysis in acetic acid containing acetate ion has been studied. The solvolysis of **A** is about 13 times faster than that of **B**. Kinetic studies in the case of **A** show that **A** is racemized competitively with solvolysis. A single product is formed from **A**, but **B** gives a mixture. Explain these results.

13. A series of ^{18}O-labeled sulfonate esters was prepared, and the extent of ^{18}O scrambling which accompanies solvolysis was measured. The rate of ^{18}O exchange was compared with the rate of solvolysis.

$$R-O-\overset{\overset{^{18}O}{\|}}{\underset{\underset{^{18}O}{\|}}{S}}-Ph \xrightarrow[\text{NaO}_2\text{CCF}_3]{\text{CF}_3\text{CO}_2\text{H}} R-^{18}O-\overset{\overset{^{18}O}{\|}}{\underset{\underset{O}{\|}}{S}}-Ph + R-O_2CCF_3$$

R	k_{sol}	k_{ex}
$(CH_3)_2CH$	3.6×10^{-5}	7.9×10^{-6}
cyclopentyl	3.8×10^{-3}	8.5×10^{-4}
2-adamantyl	1.5×10^{-3}	1.8×10^{-3}
$(CH_3)_3CCHCH_3$[a]	7.3×10^{-3}	negligible

a. Solvolysis product is $(CH_3)_2CCH(CH_3)_2 \cdot \underset{O_2CCF_3}{|}$

Discuss the variation in the ratio $k_{sol} : k_{ex}$. Offer an explanation for the absence of exchange in the 3,3-dimethyl-2-butyl case.

14. The relative stabilities of 1-phenylvinyl cations can be measured by determining the gas-phase basicity of the corresponding alkynes. The table below gives some data on free energy of protonation for substituted phenylethynes and 1-phenylpropynes. These give rise to the corresponding Yukawa-Tsuno relationships.

for $ArC\equiv CH$: $\qquad \delta\Delta G° = -14.1(\sigma° + 1.21\bar{\sigma}_R^+)$

for $ArC\equiv CCH_3$: $\qquad \delta\Delta G° = -13.3(\sigma° + 1.12\bar{\sigma}_R^+)$

How do you interpret the values of p and r in these equations? Which system is more sensitive to the aryl substituent? How would you explain this difference in sensitivity? Sketch the resonance, field and hyperconjugative interactions which you believe would contribute to these substituent effects. What, if any, geometric constraints would these interactions place on the ions?

**Relative Gas-Phase Basicities of 1-Phenylpropynes (1)
and Phenylethynes (2)**

	ΔG (kcal/mol)[a]	
Substituent	1	2
p-OMe	11.8	13.0
p-OMe-m-Cl	7.9	9.1
p-Me	4.7	5.5
m-Me	1.9	2.2
p-Cl	−0.5	0.1
m-F	−5.1	−5.6
m-Cl	−4.5	−5.1
m-CF$_3$	−6.5	−6.6
3,5-F$_2$	−8.4	
H	0.0	0.0

a. ΔG = change in free energy of protonation relative to unsubstituted compound.

15. It is observed that solvolyses of the tertiary benzylic p-nitrobenzoates **A**, **B** and **C** are correlated by Yukawa–Tsuno equation with reduced resonance components as indicated by a lower coefficient of the resonance parameter r. Offer an explanation.

$(CH_3)_2CO_2C_6H_4NO_2$

$CH_2C(CH_3)_3$
$(CH_3)_3CCO_2C_6H_4NO_2$

$CH(CH_3)_2$
$(CH_3)_3CCO_2C_6H_4NO_2$

A **B** **C**

$$\log k/k_0 = -3.96(\sigma^\circ + 1.0\,\sigma^+) \quad \log k/k_0 = -3.37(\sigma^\circ + 0.78\,\sigma^-) \quad \log /k_0 = -3.09(\sigma^\circ + 0.68\sigma^-)$$

16. Studies of the solvolysis of 1-phenylethyl chloride and its p-substituted derivatives in aqueous trifluorethanol containing azide anion as a potential nucleophile provide details relative to the mechanism of nucleophilic substitution in this system.

   (a) The reaction is independent of the azide ion concentration for *para* substituents that have σ^+ values more negative than -0.3 but is first-order in $[N_3^-]$ for substituents with σ^+ more positive than -0.08.
   (b) When other good nucleophiles are present that can compete with azide ion, e.g., $CH_3CH_2CH_2SH$, substrates undergoing solvolysis at rates that are zero-order in $[N_3^-]$ show little selectivity between the nucleophiles.
   (c) For substrates that solvolyze at rates independent of $[N_3^-]$, the ratio of 1-arylethyl azide to 1-arylethanol in the product increases as the σ^+ of the substituent becomes more negative.
   (d) The major product in reactions in which the solvolysis is first-order in $[N_3^-]$ is the 1-arylethyl azide.

   Consider these results with respect to the mechanisms outlined in Fig. 5.6 (p. 274). Delineate the types of substituted 1-arylethyl halides which react with azide ion according to each of these mechanisms on the basis of the data given above.

17. Offer a mechanistic interpretation of each of the following phenomena:

   (a) Although there is a substantial difference in the rate at which **A** and **B** solvolyze (**A** reacts 4.4×10^4 times faster in acetic acid), both compounds give products of completely retained configuration.

A **B**

(b) The solvyolysis of **C** is much more sensitive to substituent effects than that of **D**.

C   ρ = –5.27   **D**   ρ = –3.27

(c) Although the stereoisomers **E** and **F** solvolyze in acetone at comparable rates, the products of the solvolysis reactions are very different.

(d) Solvolysis of *endo*-2-chloro-7-thiabicyclo[2.2.1]heptane occurs 4.7×10^9 times faster than that of the *exo* isomer. The product from either isomer is the presence of sodium acetate is the *endo* acetate.

endo   *exo*   *endo*

(e) Solvolysis of 2-octyl *p*-bromobenzenesulfonate in 80% methanol: 20% acetone gives, in addition to the expected methyl 2-octyl ether, a 15% yield of 2-octanol. The 2-octanol could be shown not to result from the presence of adventitious water in the medium.

(f) Addition of CF_3CHN_2 to fluorosulfonic acid at $-78°C$ gives a solution the ^1H-NMR spectrum of which shows a quartet ($J_{HF} = 6.1$ Hz) at δ 6.3 ppm from external TMS. On warming to $-20°C$, this quartet disappears and is replaced by another one ($J_{HF} = 7.5$ Hz) at δ 5.50 ppm.

(g) 2-*t*-Butyl-*exo*-norbornyl *p*-nitrobenzoate is an extremely reactive compound, undergoing solvolysis 2.8×10^6 times faster than *t*-butyl *p*-nitrobenzoate. The *endo* isomer is about 500 times less reactive. In contrast to the unsubstituted norbornyl system, which gives almost exclusively *exo* product, both *t*-butyl isomers give about 5% of the *endo* product.

X = *p*-nitrobenzoate
endo

exo
C(CH₃)₃

95%
exo

5%
endo

(h) Solvolyisis of 2,4,6-trimethylbenzyl chloride in 80% aqueous ethanol is characterized by $\Delta S^{\ddagger} = -11.0$ eu. Solvolysis of 2,4,6-tri-*t*-butylbenzyl chloride, however, has $\Delta S^{\ddagger} = +0.3$ eu. Suggest an explanation for the difference in the entropy of activation for solvolysis of these two systems.

(i) Solvolysis of the *p*-nitrobenzoates of both the *syn* and *anti* isomers of 2-hydroxybicyclo[6.1.0]nonane gives as the major solvolysis product the corresponding alcohol of retained stereochemistry when carried out in buffered aqueous acetone.

(j) Solvolysis of compounds **G** and **H** gives a product mixture which is quite similar for both compounds.

G, Ar = 3,5-dinitrophenyl

H

83%–85%

8%–12%

On the other hand, compound **I** gives a completely different mixture.

I

39%

51%

(k) The isomeric tosylates **J** and **K** give an identical product mixture consisting of the alcohol **L** and ether **M** when solvolyzed in aqueous ethanol.

J

L, R = H
M, R = C₂H₅

K

(l) The solvolysis of isomeric 5-fluoro and 5-trimethylstannyl 2-adamantyl tosylates has been examined. The relative rates depend on the substituents and the stereochemistry of the reactants as shown.

X	anti-isomer		syn-isomer	
F	2.5×10^{-6}	4% retention	5×10^{-4}	100% retention
$(CH_3)_3Sn$	10	100% retention	15	63% inversion

(m) In the synthesis of nonracemic 1-phenylethyl fluorides, the use of CsF in DMF, formamide, or N-methylformamide was found to be successful for $X = NO_2$, CN, and $CO_2C_2H_5$, but only racemic product was obtained for $X = Br$.

18. A detailed study of the solvolysis of **L** has suggested the following mechanism, with the reactivity of the intermediate **M** being comparable to that of **L**. Evidence for the existence of steps k_2 and k_{-2} was obtained from isotopic scrambling in the sulfonate **M** when it was separately solvolyzed and by detailed kinetic analysis. Derive a rate expression which correctly describes the non-first-order kinetics for the solvolysis of **L**.

19. Both experimental studies on gas-phase ion stability and MO calculations indicate that the two vinyl cations shown below benefit from special stabilization. Indicate what structural features present in these cations can provide this stabilization.

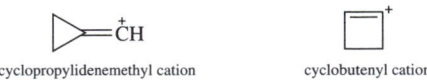

cyclopropylidenemethyl cation   cyclobutenyl cation

20. The rats of solvolysis of four isomeric tricyclooctane derivatives have been determined. After correction for leaving-group and temperature effects, the relative
reactivities are as shown.

| exo–anti (1) | exo–syn (10⁴) | endo–anti (10¹²) | endo–syn (10) |

In aqueous dioxane, the *endo-anti* isomer gave a product mixture consistent of alcohol
N and the corresponding ester (derived from capture of the leaving group *p*-
nitrobenzoate). The other isomers gave much more complex product mixtures
which were not completely characterized. Explain the trend in rates and discuss the
structural reason for the stereochemical course of the reaction in the case of the *endo-
anti* isomer.

N

21. The ^{13}C-NMR chemical shift of the trivalent carbon is a sensitive indicator of
carbocation structure. Given below are the data for three carbocations with varying
aryl substituents. Generally, the larger the chemical shift, the lower is the electron
density at the carbon atom.

O   **P**   **Q**

Aryl substituents	Chemical shift		
3,5-di-CF₃	287	283	73
4 CF₃	284	278	81
H	272	264	109
4 CH₃	262	252	165
4 OCH₃	235	230	220

How do you explain the close similarity in the substituent group trends for ions **O** and
P as contrasted to the opposing trend in **Q**?

22. A variety of kinetic data permit the assignment of relative reactivities toward solvolysis
of a series of systems related to the norbornane skeleton. Offer a general discussion of

the structural effects that are responsible for the observed relative rates.

347

PROBLEMS

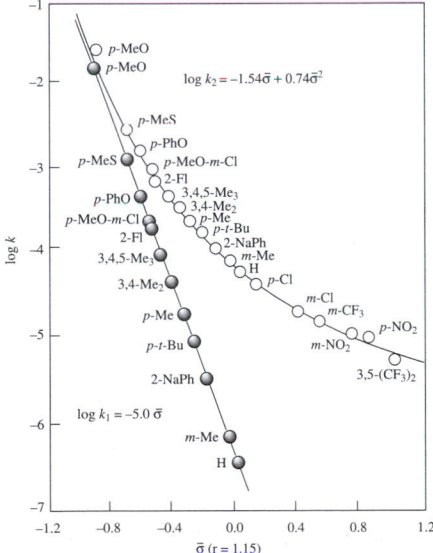

23. Fujio and co-workers studied the reactions of pyridine with a wide range of 1-arylethyl bromides in acetonitrile. By analysis of the kinetic data, they were able to dissect each reaction into a first-order and a second-order component, as shown in the table below. The first-order rate components were correlated by a Yukawa-Tsuno equation: $\log k/k_0 = 5.0(\sigma^\circ + 1.15\sigma^-)$. The second-order component gave a curved plot, as shown in Fig. 5P23. Working from the assumption that there are distinctive and separate S_N2 and S_N1 reactions under these conditions, analyze the differing responses of the reaction to substituents in terms of transition-state structures.

Substituent	$10^5 k_1 (s^{-1})$	$10^5 k_2 (M^{-1}s^{-1})$	Substituent	$10^5 k_1 (s^{-1})$	$10^5 k_2 (M^{-1}s^{-1})$
p-MeO	1660	2820	2-Naph	0.28	11.6
p-MeS	103	215	*m*-Me	0.055	7.29
p-PhO	41.5	119	H	0.032	5.54
p-MeO-*m*-Cl	21.2	79.0	*P*-Cl		4.37
2-Fluorenyl (2-FI)	18.3	59.5	*m*-Cl		2.085
3,4,5-Me$_3$	8.56	41.1	*m*-CF$_3$		1.77
3,4-Me$_2$	3.67	28.3	*m*-NO$_2$		1.21
p-Me	1.46	19.2	*p*-NO$_2$		1.19
p-t-Bu	0.82	15.2	3,5-(CF$_3$)$_2$		0.651

Fig. 5.P23. The substituent effect in the Menschutkin reaction of 1-arylethyl bromides with pyridine in acetonitrile at 35°C. Circles represent k_2 for the bimolecular process and squares k_1 (for the unimolecular process.

24. The relative solvolysis rates in 50% ethanol–water of four isomeric p-bromobenzenesulfonates are given below. **R** and **T** give an identical product mixture comprised of **V** and **W**, whereas **S** gives **X** and **Y**. Analyze these data in terms of possible participation of the oxygen atom in nucleophilic substitution.

R 1.1×10^5 **S** 1.0 **T** 3.9×10^5 **U** $\ll 1.0$ OSO_2Ar

V X = OH   **X** X = OH
W X = OC_2H_5 **Y** X = OC_2H_5

25. The Yukawa–Tsuno r values have been measured for the solvolysis reactions forming benzyl cations and several α-substituted derivatives, 6-31G* charges and bond orders have been calculated for the presumed cationic intermediates. Analyze the data for relationships between r and the structural parameters. (*Hint*: Plot r versus the bond orders and the charges at C-1, C-2, C-3, and C-4.)

X, Y:	H, H	CH$_3$, H	CH$_3$, CH$_3$	H, CF$_3$
r	1.28	1.15	1.00	1.51
Mulliken chape				
C-1	−0.0024	−0.050	−0.068	−0.211
C-2	0.189	0.164	0.140	0.053
C-3	0.051	0.046	0.043	0.053
C-4	0.213	0.190	0.171	0.233
Bond order				
C(1)−C(7)	1.584	1.465	1.363	1.622
C(1)−C(2)	1.158	1.193	1.225	1.134
C(2)−C(3)	1.167	1.543	1.524	1.585
C(3)−C(4)	1.343	1.361	1.375	1.329

26. Reactions of substituted cumyl benzoates in $50:50$ trifluoroethanol–water show no effect of [NaN$_3$] on the rate between 0 and 0.5 M. The product ratio, however, is highly dependent on the cumyl substituent. Electron-releasing substituents favor azide formation whereas electron-withdrawing substituents result in solvent capture. Formu-

late a detailed mechanism which is consistent with these observations.

27. The comparison of activation parameters for reactions in two different solvents requires consideration of differences in solvation of both the reactants and the transition states. This can be done using a potential energy diagram such as that illustrated below, where A and B refer to two different solvents. By thermodynamic methods, it is possible to establish $\Delta H_{transfer}$ values which correspond to the enthalpy change associated with transfer of a solute from one solvent to another.

$\Delta H_{transfer}$ data for n-hexyl tosylate and several ions are given in Table 5.P27A. In Table 5.P27B, activation parameters are given for the reaction of these ions with n-hexyl tosylate by the S_N2 mechanism. Use these data to construct a chart such as that above for each of the nucleophiles. (Show the enthalpy in kcal/mol on each chart.) Use these charts to interpret the relative reactivity data given in Table 5.P27C. Discuss each of the following aspects of the data.

(a) Why is Cl^- more reactive in DMSO than Br^-, while the reverse is true in MeOH?

Table 5.P27A. Enthalpies of Transfer of Ions and of n-Hexyl Tosylate from Methanol to Dimethyl Sulfoxide at 25°C

Species	$\Delta\Delta H_s$ (kcal/mol)
n-$C_6H_{13}OTs$	−0.4
Cl^-	6.6
N_3^-	3.6
Br^-	2.3
SCN^-	1.0
I^-	−1.1

(b) Why does the rate of thiocyanate (SCN^-) ion change the least of the five nucleophiles in going from MeOH to DMSO?

Table 5.P27B. Activation Parameters for Nucleophile–Hexyl Tosylate Reactions

Nu^-	Solvent	$\Delta H^{\ddagger}$ (kcal/mol)	$\Delta S^{\ddagger}$ (eu)	$\Delta G^{\ddagger}$ (kcal/mol)
Cl^-	MeOH	24.3 ± 0.2	-4.2	25.5
	DMSO	20.2 ± 0.1	-4.4	21.6
N_3^-	MeOH	21.2 ± 0.1	-8.2	23.5
	DMSO	18.6 ± 0.1	-7.8	21.0
Br^-	MeOH	22.9 ± 0.2	-6.4	24.9
	DMSO	20.5 ± 0.1	-5.6	22.0
SCN^-	MeOH	19.8 ± 0.2	-15.8	24.2
	DMSO	20.0 ± 0.2	-12.4	23.7
I^-	MeOH	22.4 ± 0.1	-6.0	23.9
	DMSO	20.9 ± 0.2	-5.8	22.8

(c) Why does thiocyanate ion have the most negative entropy of activation?

Table 5.P27C. Rates^a of Displacement on Hexyl Tosylate by Nucleophiles

Nu^-	Solvent	k (40°C)	k (30°C)	k (20°C)
Cl^-	DMSO	50.4 ± 0.4	16.7 ± 0.1	5.06 ± 0.04
	MeOH	0.0852 ± 0.0004	0.0226 ± 0.0003	0.00550 ± 0.00007
N_3^-	DMSO	135 ± 0.7	48.3 ± 0.2	16.1 ± 0.1
	MeOH	1.66 ± 0.01	0.514 ± 0.002	0.152 ± 0.001
Br^-	DMSO	17.8 ± 0.1	5.60 ± 0.04	1.75 ± 0.01
	MeOH	0.250 ± 0.002	0.0721 ± 0.0007	0.0191 ± 0.0002
SCN^-	DMSO	1.11 ± 0.01	0.365 ± 0.003	0.115 ± 0.002
	MeOH	0.481 ± 0.005	0.165 ± 0.002	0.0512 ± 0.0005
I^-	DMSO	5.50 ± 0.07	1.75 ± 0.02	16.0 ± 0.1 (50°)
	MeOH	0.956 ± 0.007	0.275 ± 0.001	0.0767 ± 0.0006

a. $k_2 \times 10^4 \, M^{-1} \, s^{-1}$. Error limits are standard deviations.

(d) Do you see any correlation between softness (which is in the order $I^- >$ $^-SCN > N_3^- > Br^- > Cl^-$) and the effect of solvent on rate?

Polar Addition and Elimination Reactions

Introduction

Addition and elimination processes are the reverse of one another in a formal sense. There is also a close mechanistic relationship between the two reactions, and in many systems reaction can occur in either direction. For example, hydration of alkenes and dehydration of alcohols are both familiar reactions that are related as an addition–elimination pair.

$$RCH{=}CHR' + H_2O \xrightarrow{\ H^+\ } \underset{\underset{OH}{|}}{RCHCH_2R'}$$

$$\underset{\underset{OH}{|}}{RCHCH_2R'} \xrightarrow{\ H^+\ } RCH{=}CHR' + H_2O$$

Another familiar pair of reactions is hydrohalogenation and dehydrohalogenation.

$$RCH{=}CHR + HX \longrightarrow \underset{\underset{X}{|}}{RCHCH_2R}$$

$$\underset{\underset{X}{|}}{RCHCH_2R} \xrightarrow{\ base\ } RCH{=}CHR$$

When the addition and elimination reactions are mechanistically reversible, they proceed by identical mechanistic paths but in opposite directions. In these circumstances, mechanistic conclusions about the addition reaction are applicable to the elimination reaction and vice versa. The principle of microscopic reversibility states that the mechanism (pathway) traversed in a reversible reaction is the same in the reverse as in the forward direction. Thus, if an addition–elimination system proceeds by a reversible mechanism, the intermediates and transition states involved in the addition process are the same as those involved in the elimination reaction. The reversible acid-catalyzed reaction of alkenes with water is an example.

The initial discussion in this chapter will focus on addition reactions. The discussion is restricted to reactions that involve polar or ionic mechanisms. There are other important classes of addition reactions which are discussed elsewhere; these include concerted addition reactions proceeding through nonpolar transition states (Chapter 11), radical additions (Chapter 12), photochemical additions (Chapter 13), and nucleophilic addition to electrophilic alkenes (Part B, Chapter 1, Section 1.10).

Several limiting generalized mechanisms can be described for polar additions:

Mechanism A indicates prior dissociation of the electrophile and implies that a carbocation is generated which is free of the counterion Y^- at its formation. Mechanism B also involves a carbocation intermediate, but it is generated in the presence of an anion and exists initially as an ion pair. Depending on the mutual reactivity of the two ions, they might or might not become free of one another before combining to give product. Both mechanisms A and B would be referred to as Ad_E2 reactions; that is, they are *bimolecular electrophilic additions*. Mechanism C is a process that has been observed for several electrophilic additions. It implies concerted transfer of the electrophilic and nucleophilic components of the reagent from two separate molecules. It would be described as a *termolecular electrophilic addition*, Ad_E3. Examples of each of these mechanistic types will be encountered as specific reactions are discussed in the sections that follow. The discussion will focus on a few reactions that have received the most detailed mechanistic study. Other synthetically important polar additions are described in Chapter 4 of Part B.

6.1. Addition of Hydrogen Halides to Alkenes

The addition of hydrogen halides to alkenes has been studied from a mechanistic point of view over a period of many years. One of the first aspects of the mechanism to be established was its regioselectivity, that is, the direction of addition. A reaction is described as *regioselective* if an unsymmetrical alkene gives a predominance of one of the two possible addition products; the term *regiospecific* is used if one product is formed exclusively.[1]

$$RCH{=}CH_2 + E{-}Y \longrightarrow \underset{\underset{\text{(major)}}{E}}{RCHCH_2Y} + \underset{\underset{\text{(minor)}}{Y}}{RCHCH_2E}$$

regioselective reaction

$$RCH{=}CH_2 + E{-}Y \longrightarrow \underset{Y}{RCHCH_2E}$$

regiospecific reaction

In the addition of hydrogen halides to alkenes, it is generally found that the halogen atom becomes attached to the more substituted carbon atom. The statement of this general observation is called *Markownikoff's rule*. The basis for this regioselectivity lies in the relative abilities of the carbon atoms to accept positive charge. The addition of hydrogen halide is initiated by an electrophilic protonation of the alkene. The new C−H bond is formed from the π electrons of the carbon–carbon double bond. It is easy to see that if a carbocation is formed as an intermediate, the halide would be added to the more substituted carbon, because addition of the proton at the less substituted carbon atom provides the more stable carbocation intermediate.

$$R_2C{=}CHR \xrightarrow{HX} \underset{+}{R_2CCH_2R} \xrightarrow{X^-} \underset{X}{R_2CCH_2R}$$

more favorable

$$R_2C{=}CHR \xrightarrow{HX} \underset{+}{R_2CHCHR} \xrightarrow{X^-} \underset{X}{R_2CHCHR}$$

less favorable

As will be indicated when the mechanism is discussed in more detail, discrete carbocations may not be formed in all cases. An unsymmetrical alkene will nevertheless follow Markownikoff's rule, because the partial positive charge that develops will be located preferentially at the carbon that is better able to accommodate the electron deficiency, that is, the more substituted one.

The regioselectivity of addition of hydrogen bromide to alkenes can be complicated if a free-radical chain addition occurs in competition with the ionic addition. The free-radical reaction is readily initiated by peroxidic impurities or by light and leads to the *anti-Markownikoff* addition product. The mechanism of this reaction will be considered more fully in Chapter 12. Conditions that minimize the competing radical addition include use of high-purity alkene and solvent, exclusion of light, and addition of free-radical inhibitors.[2]

The order of reactivity of the hydrogen halides is HI > HBr > HCl, and reactions of simple alkenes with HCl are quite slow.[3] The studies that have been applied to determining mechanistic details of hydrogen halide addition to alkenes have focused on the kinetics and stereochemistry of the reaction and on the effect of added nucleophiles. The kinetic studies often reveal complex rate expressions which demonstrate that more than one process contributes to the overall reaction rate. For addition of hydrogen bromide or hydrogen

1. A. Hassner, *J. Org. Chem.* **33**:2684 (1968).
2. D. J. Pasto, G. R. Meyer, and B. Lepeska, *J. Am. Chem. Soc.* **96**:1858 (1974).
3. P. J. Kropp, K. A. Daus, M. W. Tubergen, K. D. Kepler, V. P. Wilson, S. L. Craig, M. M. Baillargeon, and G. W. Breton, *J. Am. Chem. Soc.* **115**:3071 (1993).

chloride to alkenes, an important contribution to the overall rate is often made by a third-order process:

$$\text{rate} = k[\text{alkene}]\,[\text{HX}]^2$$

Among the cases in which this type of kinetics have been observed are the addition of hydrogen chloride to 2-methyl-1-butene, 2-methyl-2-butene, 1-methylcyclopentene,[4] and cyclohexene.[5] The addition of hydrogen bromide to cyclopentene also follows a third-order rate expression.[2] The transition state associated with the third-order rate expression involves proton transfer to the alkene from one hydrogen halide molecule and capture of the halide ion from the second:

The third-order process presumably involves reaction of a complex formed between the alkene and hydrogen halide with the second hydrogen halide molecule, since there is little likelihood of productive termolecular collisions.

The stereochemistry of addition of hydrogen halides to unconjugated alkenes is predominantly *anti*. This is true for addition of hydrogen bromide to 1,2-dimethylcyclohexene,[6] cyclohexene,[7] 1,2-dimethylcyclopentene,[8] cyclopentene,[2] *Z*- and *E*-2-butene,[2] and 3-hexene,[2] among others. *Anti* stereochemistry is also dominant for addition of hydrogen chloride to 1,2-dimethylcyclohexene[9] and 1-methylcyclopentene.[3] Temperature and solvent can modify the stereochemistry, however. For example, although the addition of hydrogen chloride to 1,2-dimethylcyclohexene in ether is *anti* near room temperature, *syn* addition dominates in CH_2Cl_2 at $-78°C$.[10]

Anti stereochemistry can be explained by a mechanism in which the alkene interacts simultaneously with the proton-donating hydrogen halide and with a source of halide ion, either a second molecule of hydrogen halide or a free halide ion. The *anti* stereochemistry is consistent with the expectation that the attack of halide ion would be from the opposite

4. Y. Pocker, K. D. Stevens, and J. J. Champoux, *J. Am. Chem. Soc.* **91**:4199 (1969); Y. Pocker and K. D. Stevens, *J. Am. Chem. Soc.* **91**:4205 (1969).
5. R. C. Fahey, M. W. Monahan, and C. A. McPherson, *J. Am. Chem. Soc.* **92**:2810 (1970).
6. G. S. Hammond and T. D. Nevitt, *J. Am. Chem. Soc.* **76**:4121 (1954).
7. R. C. Fahey and R. A. Smith, *J. Am. Chem. Soc.* **86**:5035 (1964); R. C. Fahey, C. A. McPherson, and R. A. Smith, *J. Am. Chem. Soc.* **96**:4534 (1974).
8. G. S. Hammond and C. H. Collins, *J. Am. Chem. Soc.* **82**:4323 (1960).
9. R. C. Fahey and C. A. McPherson, *J. Am. Chem. Soc.* **93**:1445 (1971).
10. K. B. Becker and C. A. Grob, *Synthesis* **1973**:789.

side of the π bond from which proton delivery occurs.

A significant modification in the stereochemistry is observed when the double bond is conjugated with a group that can stabilize a carbocation intermediate. Most of the specific cases involve an aryl substituent. Examples of alkenes that give primarily *syn* addition are *Z*- and *E*-1-phenylpropene,[11] *Z*- and *E*-β-*t*-butylstyrene,[12] 1-phenyl-4-*t*-butylcyclohexene,[13] and indene.[14] The mechanism proposed for these additions features an ion pair as the key intermediate. Because of the greater stability of the carbocations in these molecules, concerted attack by halide ion is not required for complete carbon–hydrogen bond formation. If the ion pair formed by alkene protonation collapses to product faster than reorientation takes place, the result will be *syn* addition, since the proton and halide ion are initially on the same side of the molecule.

Kinetic studies of the addition of hydrogen chloride to styrene support the conclusion that an ion-pair mechanism operates because aromatic conjugation is involved. The reaction is first-order in hydrogen chloride, indicating that only one molecule of hydrogen chloride participates in the rate-determining step.[15]

There is usually a competing reaction with solvent when hydrogen halide additions to alkenes are carried out in nucleophilic solvents:

(Ref. 15)

(Ref. 5)

It is not difficult to incorporate this result into the general mechanism for hydrogen halide additions. These products are formed as the result of solvent competing with halide ion as the nucleophilic component in the addition. Solvent addition can occur via a concerted mechanism or by capture of a carbocation intermediate. Addition of a halide salt increases the likelihood of capture of a carbocation intermediate by halide ion. The effect of added halide salt can be detected kinetically. For example, the presence of tetramethylammonium

11. M. J. S. Dewar and R. C. Fahey, *J. Am. Chem. Soc.* **85**:3645 (1963).
12. R. J. Abraham and J. R. Monasterios, *J. Chem. Soc., Perkin Trans. 2* **1975**:574.
13. K. D. Berlin, R. O. Lyerla, D. E. Gibbs, and J. P. Devlin, *J. Chem. Soc., Chem. Commun.* **1970**:1246.
14. M. J. S. Dewar and R. C. Fahey, *J. Am. Chem. Soc.* **85**:2248 (1963).
15. R. C. Fahey and C. A. McPherson, *J. Am. Chem. Soc.* **91**:3865 (1969).

chloride increases the rate of addition of hydrogen chloride to cyclohexene.[9] Similarly, lithium bromide increases the rate of addition of hydrogen bromide to cyclopentene.[6]

Skeletal rearrangements have been observed in hydrogen halide additions when hydrogen or carbon migration can lead to a more stable carbocation.

$$(CH_3)_2CHCH=CH_2 \xrightarrow[CH_3NO_2]{HCl} (CH_3)_2CHCHCH_3 + (CH_3)_2CCH_2CH_3 \qquad \text{(Ref. 4)}$$

40%   Cl   60%   Cl

$$(CH_3)_3CCH=CH_2 \xrightarrow[CH_3NO_2]{HCl} (CH_3)_2CCH(CH_3)_2 + (CH_3)_3CCHCH_3 \qquad \text{(Ref. 4)}$$

83%   Cl   17%   Cl

Even though the rearrangements suggest that discrete carbocation intermediates are involved, these reactions frequently show kinetics consistent with the presence of at least two hydrogen chloride molecules in the rate-determining transition state. A termolecular mechanism in which the second hydrogen chloride molecule assists in the ionization of the electrophile has been suggested.[4]

$$\text{H}\overset{\frown}{-}\text{Cl}^{--}\text{H}-\text{Cl}$$
$$(CH_3)_2CHCH=CH_2 \longrightarrow (CH_3)_2CH\overset{+}{C}HCH_3 + [Cl-H-Cl]^-$$
$$(CH_3)_2CH\overset{+}{C}HCH_3 \longrightarrow (CH_3)_2\overset{+}{C}CH_2CH_3$$
$$(CH_3)_2CH\overset{+}{C}HCH_3 + [Cl-H-Cl]^- \longrightarrow (CH_3)_2CHCHCH_3 + HCl$$
Cl

and

$$(CH_3)_2\overset{+}{C}CH_2CH_3 + [Cl-H-Cl]^- \longrightarrow (CH_3)_2CCH_2CH_3 + HCl$$
Cl

An alternative view of these addition reactions is that the rate-determining step is halide-assisted proton transfer, followed by capture of the carbocation, with or without rearrangement. Bromide ion accelerates addition of HBr to 1-, 2-, and 4-octene in 20% trifluoroacetic acid in CH_2Cl_2. In the same system, 3,3-dimethyl-1-butene shows substantial rearrangement. Even 1- and 2-octene show some evidence of rearrangement, as detected by hydride shifts. These results can all be accounted for by a halide-assisted protonation.[16] The key intermediate in this mechanism is an ion sandwich. An estimation of the fate of the 2-octyl cation under these conditions has been made:

$$RCH=CH_2 \longrightarrow \underset{+}{RCHCH_3}$$

RCHCH$_3$ / O$_2$CCF$_3$	3%
hydride shift	3%
deprotonation	32%
RCHCH$_3$ / Br	62%

(−O$_2$CCF$_3$, Br$^-$)

The addition of hydrogen halides to dienes can result in either 1,2- or 1,4-addition. The extra stability of the allylic cation formed by proton transfer to a diene makes the ion-

16. H. M. Weiss and K. M. Touchette, *J. Chem. Soc., Perkin Trans. 2* **1998**:1517.

pair mechanism favorable. 1,3-Pentadiene, for example, gives a mixture of products favoring the 1,2-addition product by a ratio of from 1.5:1 to 3.4:1, depending on the temperature and solvent.[17]

With 1-phenyl-1,3-butadiene, the addition is exclusively at the 3,4-double bond. This reflects the greater stability of this product, which retains styrene-type conjugation. Initial protonation at C-4 is favored by the fact that the resulting carbocation benefits from both allylic and benzylic stabilization.

The kinetics of this reaction are second-order, as would be expected for the formation of a stable carbocation by an Ad_E2 mechanism.[18]

The addition of hydrogen chloride or hydrogen bromide to norbornene is an interesting case because such factors as the stability and facile rearrangement of the norbornyl cation come into consideration. Addition of deuterium bromide to norbornene gives *exo*-norbornyl bromide. Degradation of the product to locate the deuterium atom shows that about half of the product is formed via the bridged norbornyl cation, which leads to deuterium at both the 3- and 7-positions. The *exo* orientation of the bromine atom and the redistribution of the deuterium indicate the involvement of the bridged ion.

Similar studies have been carried out on the addition of hydrogen chloride to norbornene.[19]

17. J. E. Nordlander, P. O. Owuor, and J. E. Haky, *J. Am. Chem. Soc.* **101**:1288 (1979).
18. K. Izawa, T. Okuyama, T. Sakagami, and T. Fueno, *J. Am. Chem. Soc.* **95**:6752 (1973).
19. H. Kwart and J. L. Nyce, *J. Am. Chem. Soc.* **86**:2601 (1964); J. K. Stille, F. M. Sonnenberg, and T. H. Kinstle, *J. Am. Chem. Soc.* **88**:4922 (1966).

Again, the chloride is almost exclusively the *exo* isomer. The distribution of deuterium in the product was determined by NMR spectroscopy. The fact that **1** and **2** are formed in unequal amounts excludes the symmetrical bridged ion as the only intermediate.[20] The excess of **1** over **2** indicates that some *syn* addition occurs by ion-pair collapse before the bridged ion achieves symmetry with respect to the chloride ion. If the amount of **2** is taken as an indication of the extent of bridged-ion involvement, one would conclude that 82% of the reaction proceeds through this intermediate, which must give equal amounts of **1** and **2**.

6.2. Acid-Catalyzed Hydration and Related Addition Reactions

The formation of alcohols by acid-catalyzed addition of water to alkenes is a fundamental organic reaction. At the most rudimentary mechanistic level, it can be viewed as involving a carbocation intermediate. The alkene is protonated, and the carbocation is then captured by water.

$$RCH{=}CH_2 \xrightarrow{H^+} R\overset{+}{C}HCH_3 \xrightarrow{H_2O} \underset{\underset{OH}{|}}{RCHCH_3} + H^+$$

This mechanism explains the observed formation of the more highly substituted alcohol from unsymmetrical alkenes (Markownikoff's rule). A number of other points must be considered in order to provide a more complete picture of the mechanism. Is the protonation step reversible? Is there a discrete carbocation intermediate, or does the nucleophile become involved before proton transfer is complete? Can other reactions of the carbocation, such as rearrangement, compete with capture by water?

Much of the mechanistic work on hydration reactions has been done with conjugated alkenes, particularly styrenes. With styrenes, the rate of hydration is increased by electron-releasing substituents, and there is an excellent correlation with σ^+.[21] A substantial solvent isotope effect, $k_{H_2O}/k_{D_2O} = 2\text{--}4$, is observed. Both of these observations are in accord with a rate-determining protonation to give a carbocation intermediate. Capture of the resulting cation by water is usually fast relative to deprotonation. This has been demonstrated by showing that in the early stages of hydration of styrene deuterated at C-2, there is no loss of deuterium from the unreacted alkene that is recovered by quenching the reaction.

$$PhCH{=}CD_2 \underset{H^+}{\rightleftharpoons} Ph\overset{+}{C}HCD_2H \underset{\underset{slow}{-H^+}}{\rightleftharpoons} PhCH{=}CHD$$

$$\Big\updownarrow \begin{array}{c} H_2O \\ fast \end{array}$$

$$\underset{\underset{OH}{|}}{PhCHCD_2H}$$

The overall process is reversible, however, and some styrene remains in equilibrium with the alcohol, so exchange eventually occurs.

20. H. C. Brown and K.-T. Liu, *J. Am. Chem. Soc.* **97**:600 (1975).
21. W. M. Schubert and J. R. Keefe, *J. Am. Chem. Soc.* **94**:559 (1972); W. M. Schubert and B. Lamm, *J. Am. Chem. Soc.* **88**:120 (1966); W. K. Chwang, P. Knittel, K. M. Koshy, and T. T. Tidwell, *J. Am. Chem. Soc.* **99**:3395 (1977).

359

SECTION 6.2.
ACID-CATALYZED
HYDRATION AND
RELATED ADDITION
REACTIONS

Table 6.1. Rates of Hydration of Some Alkenes in Aqueous Sulfuric Acid[a]

Alkene	k_2 (M^{-1} s^{-1})	k_{rel}
$H_2C{=}CH_2$	1.46×10^{-15}	1
$CH_3CH{=}CH_2$	2.38×10^{-8}	1.6×10^7
$CH_3(CH_2)_3CH{=}CH_2$	4.32×10^{-8}	3.0×10^7
$(CH_3)_2C{=}CHCH_3$	2.14×10^{-3}	1.5×10^{12}
$(CH_3)_2C{=}CH_2$	3.71×10^{-3}	2.5×10^{12}
$PhCH{=}CH_2$	2.4×10^{-6}	1.6×10^9

a. W. K. Chwang, V. J. Nowlan, and T. T. Tidwell, *J. Am. Chem. Soc.* **99**:7233 (1977).

Alkenes lacking phenyl substituents appear to react by a similar mechanism. Both the observation of general acid catalysis[22] and the kinetic evidence of a solvent isotope effect[23] are consistent with rate-limiting protonation with simple alkenes such as 2-methylpropene and 2,3-dimethyl-2-butene.

$$R_2C{=}CHR + H^+ \xrightarrow{slow} \underset{+}{R_2CCH_2R} \xrightarrow[H_2O]{fast} \underset{OH}{\overset{|}{R_2CCH_2R}} + H^+$$

Relative rate data in aqueous sulfuric acid for a series of simple alkenes reveal that the reaction is strongly accelerated by alkyl substituents. This is expected because alkyl groups both increase the electron density of the double bond and stabilize the carbocation intermediate. Table 6.1 gives some representative data. These same reactions show solvent isotope effects consistent with the reaction proceeding through a rate-determining protonation.[24] Strained alkenes show enhanced reactivity toward acid-catalyzed hydration. *E*-Cyclooctene is about 2500 times as reactive as the *Z* isomer, for example.[25] This reflects the higher ground-state energy of the strained alkene.

Other nucleophilic solvents can add to alkenes under the influence of strong acid catalysts. The mechanism is presumably analogous to that for hydration, with the solvent replacing water as the nucleophile. The strongest acid catalysts are likely to react via discrete carbocation intermediates, whereas addition catalyzed by weaker acids may involve reaction of the solvent with an alkene–acid complex. In the addition of acetic acid to *Z*- or *E*-2-butene, the use of DBr as the catalyst results in stereospecific *anti* addition, whereas use of a stronger acid, trifluoromethanesulfonic acid, leads to loss of stereospecificity. This difference in stereochemistry can be explained by a stereospecific Ad_E3 mechanism in the case of hydrogen bromide and an Ad_E2 mechanism in the case of trifluoromethanesulfonic acid.[26] The dependence on acid strength reflects the degree to

22. A. J. Kresge, Y. Chiang, P. H. Fitzgerald, R. S. McDonald, and G. H. Schmid, *J. Am. Chem. Soc.* **93**:4907 (1971); H. Slebocka-Tilk and R. S. Brown, *J. Org. Chem.* **61**:8079 (1998).
23. V. Gold and M. A. Kessick, *J. Chem. Soc.* **1965**:6718.
24. V. J. Nowlan and T. T. Tidwell, *Acc. Chem. Res.* **10**:252 (1977).
25. Y. Chiang and A. J. Kresge, *J. Am. Chem. Soc.* **107**:3636 (1985).
26. D. J. Pasto and J. F. Gadberry, *J. Am. Chem. Soc.* **100**:1469 (1978).

which nucleophilic solvent participation is required to complete proton transfer.

$$CH_3CH{=}CHCH_3 \;\underset{DBr}{\rightleftharpoons}\; CH_3CH{=}CHCH_3 \quad (D{-}Br)$$

$$CH_3CH{=}CHCH_3 \;(D{-}Br, CH_3CO_2H) \longrightarrow CH_3CH{-}CHCH_3 \;(D, CH_3CO_2) \qquad \textit{anti}\text{-addition}$$

$$CH_3CH{=}CHCH_3 + CF_3SO_3D \longrightarrow CH_3CHCHCH_3 \;(D, +)$$

$$CH_3CHCHCH_3 \;(D, +) + CH_3CO_2H \longrightarrow CH_3CH{-}CHCH_3 \;(D, CH_3CO_2) \qquad \text{nonstereospecific addition}$$

Strong acids also catalyze the addition of alcohols to alkenes to give ethers, and the mechanistic studies which have been done indicate that the reaction closely parallels the hydration process.[27]

Trifluoroacetic acid adds to alkenes without the necessity for a stronger acid catalyst.[28] The mechanistic features of this reaction are similar to those of addition of water catalyzed by strong acids. For example, there is a substantial isotope effect when CF_3CO_2D is used,[29] and the reaction rates of substituted styrenes are correlated with σ^+.[30]

The reactivity of carbon–carbon double bonds toward acid-catalyzed addition of water is greatly increased by electron-releasing substituents. The reaction of vinyl ethers with water in acidic solution is an example that has been extensively studied. With these substrates, the initial addition products are unstable hemiacetals which decompose to a ketone and an alcohol. Nevertheless, the hydration step is rate-determining, and the kinetic results pertain to this step. The mechanistic features that have been examined are similar to those for hydration of simple alkenes. Proton transfer is the rate-determining step, as is demonstrated by general acid catalysis and solvent isotope effect data.[31]

$$R''CH{=}COR\;(R') \xrightarrow[\text{slow}]{H^+} R''CH_2C{=}\overset{+}{O}R\;(R') \xrightarrow{H_2O} R''CH_2\overset{OH}{\underset{R'}{C}}OR + H^+ \longrightarrow R''CH_2CR'\;(\overset{\|}{O}) + ROH$$

27. N. C. Deno, F. A. Kish, and H. J. Peterson, *J. Am. Chem. Soc.* **87**:2157 (1965).

28. P. E. Peterson and G. A. Allen, *J. Am. Chem. Soc.* **85**:3608 (1963); A. D. Allen and T. T. Tidwell, *J. Am. Chem. Soc.* **104**:3145 (1982).

29. J. J. Dannenberg, B. J. Goldberg, J. K. Barton, K. Dill, D. M. Weinwurzel, and M. O. Longas, *J. Am. Chem. Soc.* **103**:7764 (1981).

30. A. D. Allen, M. Rosenbaum, N. O. L. Seto, and T. T. Tidwell, *J. Org. Chem.* **47**:4234 (1982).

31. A. J. Kresge and H. J. Chen, *J. Am. Chem. Soc.* **94**:2818 (1972); A. J. Kresge, D. S. Sagatys, and H. L. Chen, *J. Am. Chem. Soc.* **99**:7228 (1977).

6.3. Addition of Halogens

Alkene chlorinations and brominations are very general organic reactions, and mechanistic study of these reactions has provided much detailed insight into the electrophilic addition reactions of alkenes.[32] Two of the principal points at issue in the description of the mechanism for a given reaction are: (1) Is there a discrete positively charged intermediate, or is the addition concerted? (2) If there is a positively charged intermediate, is it a carbocation or a bridged halonium ion? Stereochemical studies have provided much of the data pertaining to these points. The results of numerous stereo chemical studies can be generalized as follows. For brominations, *anti* addition is preferred for alkenes that do not have substituent groups that would strongly stabilize a carbocation intermediate. When the alkene is conjugated with an aryl group, the extent of *syn* addition becomes greater, and *syn* addition can become the dominant pathway. Chlorination is not as stereospecific as bromination, but it tends to follow the same pattern. Some specific cases are given in Table 6.2.

Table 6.2. Stereochemistry of Halogenation

Alkene	Solvent	Ratio *anti* : *syn*	Reference
Bromination			
Z-2-Butene	CH_3CO_2H	>100 : 1	a
E-2-Butene	CH_3CO_2H	>100 : 1	a
Cyclohexene	CCl_4	very large	b
Z-1-Phenylpropene	CCl_4	83 : 17	c
E-1-Phenylpropene	CCl_4	88 : 12	c
E-2-Phenylbutene	CH_3CO_2H	68 : 32	a
Z-2-Phenylbutene	CH_3CO_2H	63 : 37	a
cis-Stilbene	CCl_4	>10 : 1	d
	CH_3NO_2	1 : 9	d
Chlorination			
Z-2-Butene	none	>100 : 1	e
	CH_3CO_2H	>100 : 1	f
E-2-Butene	none	>100 : 1	e
	CH_3CO_2H	>100 : 1	f
Cyclohexene	none	>100 : 1	g
E-1-Phenylpropene	CCl_4	45 : 55	f
	CH_3CO_2H	41 : 59	f
Z-1-Phenylpropene	CCl_4	32 : 68	f
	CH_3CO_2H	22 : 78	f
cis-Stilbene	$ClCH_2CH_2Cl$	92 : 8	h
trans-Stilbene	$ClCH_2CH_2Cl$	65 : 35	h

a J. H. Rolston and K. Yates, *J. Am. Chem. Soc.* **91**:1469, 1477 (1969).
b. S. Winstein, *J. Am. Chem. Soc.* **64**:2792 (1942).
c R. C. Fahey and H.-J. Schneider, *J. Am. Chem. Soc.* **90**:4429 (1968).
d. R. E. Buckles, J. M. Bader, and R. L. Thurmaier, *J. Org. Chem.* **27**:4523 (1962).
e. M. L. Poutsma, *J. Am. Chem. Soc.* **87**:2172 (1965).
f. R. C. Fahey and C. Schubert, *J. Am. Chem. Soc.* **87**:5172 (1965).
g M. L. Poutsma, *J. Am. Chem. Soc.* **87**:2161 (1965).
h. R. E. Buckles and D. F. Knaack, *J. Org. Chem.* **25**:20 (1980).

32. For reviews, see D. P. de la Mare and R. Bolton, *Electrophilic Additions to Unsaturated Systems*, 2nd ed., Elsevier, New York, 1982, pp. 136–197; G. H. Schmidt and D. G. Garratt, in *The Chemistry of Double-Bonded Functional Groups*, Supplement A, Part 2, S. Patai, ed., John Wiley & Sons, New York, 1977, Chapter 9; M.-F. Ruasse, *Ind. Chem. Libr.* **7**:100 (1995); R. S. Brown, *Ind. Chem. Libr.* **7**:113 (1995); G. Bellucci and R. Bianchini, *Ind. Chem. Libr.* **7**:128 (1995); R. S. Brown, *Acc. Chem. Res.* **30**:131 (1997).

Interpretation of reaction stereochemistry has focused attention on the role played by bridged halonium ions.

$$RCH{=}CH_2 \xrightarrow{Br_2} \underset{\substack{\text{bridged} \\ \text{bromonium} \\ \text{ion}}}{R-\overset{\overset{+}{Br}}{\underset{H}{\overset{|}{C}}}\diagdown CH_2} \; \rightleftharpoons \; \underset{\substack{\text{open} \\ \text{carbocation}}}{R\overset{+}{C}H-CH_2Br}$$

If the addition of Br^+ to the alkene results in a bromonium ion, the *anti* stereochemistry can be readily explained. Nucleophilic ring opening by bromide ion would occur by backside attack at carbon, with rupture of one of the C$-$Br bonds, giving overall *anti* addition.

On the other hand, a freely rotating open carbocation would be expected to give both the *syn* and *anti* addition products. If the principal intermediate were an ion pair that collapsed faster than rotation about the C$-$C bond, *syn* addition could predominate.

Whether a bridged intermediate or a carbocation is involved in bromination depends primarily on the stability of the potential cation. Aliphatic systems normally go through the bridged intermediate, but styrenes are borderline cases. When the phenyl ring has electron-releasing substituents, there is sufficient stabilization to permit carbocation formation, whereas electron-attracting groups favor the bridged intermediate.[33] As a result, styrenes with electron-attracting substituents give a higher proportion of the *anti* addition product.

Substituent effects on addition reactions of stilbenes also give insight into the role of bridged ions versus nonbridged carbocation intermediates. The compounds react with second-order kinetics in protic solvents. In aprotic solvents, stilbene gives clean *anti* addition, but 4,4′-dimethoxystilbene gives a mixture of the *syn* and *anti* addition products, indicating a carbocation intermediate.[34] In nucleophilic solvents, solvent competes with bromide, but *anti* stereoselectivity is still observed, except in the case of stilbenes with donor substituents. It is possible that *anti* stereoselectivity can result not only from bridged-ion intermediates, but also from very fast capture of a carbocation intermediate.

The stereochemistry of chlorination can be explained in similar terms. Chlorine would be expected to be a somewhat poorer bridging group than bromine because it is less polarizable and more resistant to becoming positively charged. Comparison of the data for bromination and chlorination of *E*- and *Z*-1-phenylpropene confirms this trend (see Table 6.2). Although *anti* addition is dominant in bromination, *syn* addition is slightly preferred

33. M. F. Ruasse, A. Argile, and J. E. Dubois, *J. Am. Chem. Soc.* **100**:7645 (1978).
34. G. Bellucci, C. Chiappe, and G. Lo Moro, *J. Org. Chem.* **62**:3176 (1997).

in chlorination. Styrenes generally appear to react with chlorine via ion-pair intermediates.[35] For nonconjugated alkenes, stereospecific *anti* addition is usually observed for both halogens.

Interpretation of the ratio of capture of competing nucleophiles has led to the estimate that bromonium ions have lifetimes on the order of 10^{-10} s in methanol. This lifetime is about 100 times longer than that for secondary carbocations.[36] There is also direct evidence for the existence of bromonium ions. The bromonium ion related to propene can be observed by NMR spectroscopy when 1-bromo-2-fluoropropane is subjected to superacid conditions. The terminal bromine adopts a bridging position in the resulting cation.

$$CH_3CHCH_2Br \xrightarrow[SO_2, -60°C]{SbF_5} CH_3CH{-}CH_2 + SbF_6^- \qquad (Ref.\ 37)$$

Bromonium ions can be also produced by an electrophilic attack by a species that should generate a positive bromine:

$$(CH_3)_2C{=}C(CH_3)_2 + Br{-}C{\equiv}N{-}\bar{S}bF_5 \longrightarrow (CH_3)_2C{-}C(CH_3)_2 + CNSbF_5 \qquad (Ref.\ 38)$$

The highly hindered alkene adamantylideneadamantane forms a bromonium ion which crystallizes as a tribromide salt. An X-ray crystal structure (Fig. 6.1) has confirmed the cyclic nature of the bromonium ion species.[39] This particular bromonium ion does not react further because of extreme steric hindrance to back-side approach by bromide ion.

An interpretation of activation parameters has led to the conclusion that the bromination transition state resembles a three-membered ring, even in the case of alkenes that eventually react via open carbocation intermediates. It was found that for *cis–trans* pairs of alkenes the difference in enthalpy at the transition state for bromination was *greater than* the enthalpy difference for the isomeric alkenes, as shown in Fig. 6.2. This

Fig. 6.1. Crystal structure of bromonium ion from adamantylideneadamantane. (Reproduced from Ref. 39 by permission of the American Chemical Society.)

35. K. Yates and H. W. Leung, *J. Org. Chem.* **45**:1401 (1980).
36. P. W. Nagorski and R. S. Brown, *J. Am. Chem. Soc.* **114**:7773 (1992).
37. G. A. Olah, J. M. Bollinger, and J. Brinich, *J. Am. Chem. Soc.* **90**:2587 (1968).
38. G. A. Olah, P. Schilling, P. W. Westerman, and H. C. Lin, *J. Am. Chem. Soc.* **96**:3581 (1974).
39. H. Slebocka-Tilk, R. G. Ball, and R. S. Brown, *J. Am. Chem. Soc.* **107**:4504 (1985).

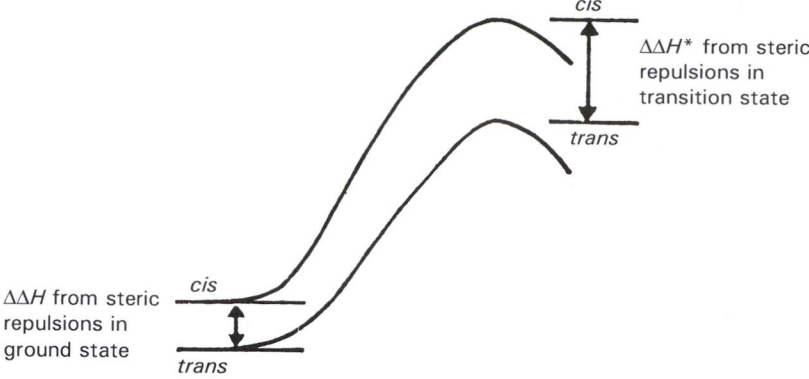

Fig. 6.2. Enthalpy differences of starting alkenes and transition states in bromination.

finding indicates that the steric repulsions between *cis* groups *increase* on going from reactant to transition state. This would be consistent with a cyclic transition state in which the *cis* substituents are still eclipsed and somewhat closer together than in the alkene.[40]

The kinetics of brominations are often complex, with at least three terms making contributions under given conditions:

$$\text{rate} = k_1[\text{alkene}][\text{Br}_2] + k_2[\text{alkene}][\text{Br}_2]^2 + k_3[\text{alkene}][\text{Br}_2][\text{Br}^-]$$

In methanol, pseudo-second-order kinetics are observed when a high concentration of Br^- is present.[41] Under these conditions, the dominant contribution to the overall rate comes from the third term of the general expression. The occurrence of third-order terms suggests the possibility of a mechanism similar to the Ad_E3 mechanism for addition of hydrogen halides to alkenes, namely, attack of halide ion on an alkene–halogen complex. There is good evidence that the initial complex is a charge-transfer complex. The formation of the charge-transfer complex and its subsequent disappearance, with kinetics corresponding to the formation of bromination product can be observed spectroscopically under appropriate conditions.[42,43]

As in the case of hydrogen halide additions, this mode of attack should lead to *anti* addition.

In nonpolar solvents, the observed rate of bromination is frequently found to be described as a sum of the first two terms in the general expression. The second-order term

40. K. Yates and R. S. McDonald, *J. Org. Chem.* **38**:2465 (1973).
41. J.-E. Dubois and G. Mouvier, *Tetrahedron Lett.* **1963**:1325; *Bull. Soc. Chim. Fr.* **1968**:1426.
42. S. Fukuzumi and J. K. Kochi, *J. Am. Chem. Soc.* **104**:7599 (1982).
43. G. Bellucci, R. Bianchini, and R. Ambrosetti, *J. Am. Chem. Soc.* **107**:2464 (1985).

is interpreted as corresponding to the collapse of an alkene–halogen complex to an ion pair which then goes on to product. The cationic intermediate can have the bromonium-ion structure, which would lead to *anti* addition.

Several mechanisms have been considered for the term that is overall third-order and second-order in bromine.[44–46]

The first possibility envisages essentially the same mechanism as for the second-order process, but with Br_2 replacing solvent in the rate-determining conversion to an ion pair. The second mechanism pictures Br_2 attack on a reversibly formed ion-pair intermediate. The third mechanism postulates collapse of a ternary complex that is structurally similar to the initial charge-transfer complex but has $2:1$ bromine : alkene stoichiometry. There are very striking similarities between the second-order and third-order processes in terms of magnitude of ρ values and product distribution.[45] In fact, there is a quantitative correlation between the rates of the two processes over a broad series of alkenes, which can be expressed as

$$\Delta G_3^{\ddagger} = \Delta G_2^{\ddagger} + \text{constant}$$

where $\Delta G_3^{\ddagger}$ and $\Delta G_2^{\ddagger}$ are the free energies of activation for the third-order and second-order processes, respectively.[46] These correlations suggest that the two mechanisms must be very similar.

Another aspect of the mechanism is the reversibility of formation of the bromonium ion. Reversibility has been demonstrated for highly hindered alkenes.[47] This can be

44. G. Bellucci, R. Bianchini, R. A. Ambrosetti, and G. Ingrosso, *J. Chem. Soc.* **50**:3313 (1985); G. Bellucci, G. Berti, R. Bianchini, G. Ingrosso, and R. Ambrosetti, *J. Am. Chem. Soc.* **102**:7480 (1980).
45. K. Yates, R. S. McDonald, and S. Shapiro, *J. Org. Chem.* **38**:2460 (1973); K. Yates and R. S. McDonald, *J. Org. Chem.* **38**:2465 (1973).
46. S. Fukuzumi and J. K. Kochi, *Int. J. Chem. Kinet.* **15**:249 (1983).
47. R. S. Brown, H. Slebocka-Tilk, A. J. Bennet, G. Bellucci, R. Bianchini, and R. Ambrosetti, *J. Am. Chem. Soc.* **112**:6310 (1990); G. Bellucci, R. Bianchini, C. Chiappe, F. Marioni, R. Ambrosetti, R. S. Brown, and H. Slebocka-Tilk, *J. Am. Chem. Soc.* **111**:2640 (1989).

attributed to a relatively slow rate of nucleophilic capture. However, even the bromonium ion from cyclohexene appears to be able to release Br_2 on reaction with Br^-. When the bromonium ion is generated by neighboring-group participation in the solvolysis of *trans-*2-bromocyclohexyl trifluoromethanesulfonate, the product mixture is identical to that formed by bromination under the same conditions. If cyclopentene, which is more reactive than cyclohexene, is included, bromination products from cyclopentene are formed. This indicates that free Br_2 is generated by reversal of bromonium ion formation.[48] Other examples of reversible bromonium ion formation have been found.[49]

SOH = solvent

In summary, it appears that bromination usually involves a charge-transfer complex which collapses to an ion-pair intermediate. The cation can be a carbocation, as in the case of styrenes, or a bromonium ion. The complex can evidently also be captured by bromide ion when it is present in sufficiently high concentration.

Chlorination generally exhibits second-order kinetics, first-order in both alkene and chlorine.[50] The reaction rate also increases with alkyl substitution, as would be expected for an electrophilic process. The magnitude of the rate increase is quite large, as shown in Table 6.3.

48. C. Y. Zheng, H. Slebocka-Tilk, R. W. Nagorski, L. Alvarado, and R. S. Brown, *J. Org. Chem.* **58**:2122 (1993).
49. R. Rodebaugh and B. Fraser-Reid, *Tetrahedron* **52**:7663 (1996).
50. G. H. Schmid, A. Modro, and K. Yates, *J. Org. Chem.* **42**:871 (1977).

Table 6.3. Relative Reactivity of Alkenes toward Halogenation

Alkene	Relative reactivity		
	Chlorination[a]	Bromination[b]	Bromination[c]
Ethylene		0.01	0.0045
1-Butene	1.00	1.00	1.00
3,3-Dimethyl-1-butene	1.15	0.27	1.81
Z-2-Butene	63	27	173
E-2-Butene	50	17.5	159
2-Methylpropene	58	57	109
2-Methyl-2-butene	11,000	1,380	
2,3-Dimethyl-2-butene	430,000	19,000	

a. M. L. Poutsma, *J. Am. Chem. Soc.* **87**:4285 (1965); solvent is excess alkene.
b. J. E. Dubois and G. Mouvier, *Bull. Soc. Chim. Fr.* **1968**:1426; Solvent is methanol.
c. A. Modro, G. H. Schmid, and K. Yates, *J. Org. Chem.* **42**:3637 (1977); solvent is carbon tetrachloride.

In chlorination, loss of a proton can be a competitive reaction of the cationic intermediate. This process leads to formation of products resulting from net substitution with double-bond migration:

Isobutylene and tetramethylethylene give products of this type.

$$(CH_3)_2C=CH_2 + Cl_2 \longrightarrow H_2C=\overset{\overset{\displaystyle CH_3}{|}}{C}CH_2Cl$$
87%

(Ref. 51)

$$(CH_3)_2C=C(CH_3)_2 + Cl_2 \longrightarrow H_2C=\overset{\overset{\displaystyle CH_3}{|}}{\underset{\underset{\displaystyle Cl}{|}}{C}}C(CH_3)_2$$
100%

Alkyl migrations can also occur.

$$(CH_3)_3CCH=CH_2 + Cl_2 \longrightarrow H_2C=\overset{\overset{\displaystyle CH_3}{|}}{\underset{\underset{\displaystyle CH_3}{|}}{C}}CHCH_2Cl$$
~10%

(Ref. 51)

$$(CH_3)_3CCH=CHC(CH_3)_3 + Cl_2 \longrightarrow CH_2=\overset{\overset{\displaystyle CH_3}{|}}{\underset{\underset{\displaystyle CH_3\ \ Cl}{|\ \ \ |}}{C}}CHCHC(CH_3)_3$$
46%

(Ref. 52)

Both proton loss and rearrangement reflect the greater positive charge at carbon in a chloronium ion than in a bromonium ion because of the weaker bridging by chlorine.

51. M. L. Poutsma, *J. Am. Chem. Soc.* **87**:4285 (1965).
52. R. C. Fahey, *J. Am. Chem. Soc.* **88**:4681 (1966).

The relative reactivities of some alkenes toward chlorination and bromination are given in Table 6.3. The relative reactivities are solvent-dependent.[53] The reaction is faster in more polar solvents, and, in all media, reactivity increases with additional substitution of hydrogen by electron-releasing alkyl groups at the double bond.[54] Quantitative estimation of the solvent effect using the Winstein–Grunwald Y values indicates that the transition state has a high degree of ionic character. The Hammett correlation for bromination of styrenes is best with σ^+ substituent constants and gives $\rho = -4.8$.[55] All these features are in accord with an electrophilic mechanism.

Much less detail is available about the mechanism of fluorination and iodination of alkenes. Elemental fluorine reacts violently with alkenes, giving mixtures that include products resulting from degradation of the carbon chain. Electrophilic addition of fluorine to alkenes can be achieved with xenon difluoride[56] or electrophilic derivatives of fluorine,[57] or by use of highly dilute elemental fluorine at low temperature.[58] Under the latter conditions, *syn* stereochemistry is observed. The reaction is believed to proceed by rapid formation and then collapse of β-fluorocarbocation : fluoride ion pair. From both the stereochemical results and theoretical calculations,[59] it appears unlikely that a bridged fluoronium species is involved. Acetyl hypofluorite, which is prepared by reaction of fluorine with sodium acetate at $-75°C$ in halogenated solvents,[60] reacts with alkenes to give β-acetoxyalkyl fluorides.[61] The reaction gives predominantly *syn* addition, which is also consistent with rapid collapse of a β-fluorocarbocation:acetate ion pair.

There have also been relatively few mechanistic studies of the addition of iodine. One significant feature of iodination is that it is easily reversible, even in the presence of excess alkene.[62] The addition is stereospecifically *anti*, but it is not entirely clear whether a polar or a radical mechanism is involved.[63]

As with addition of other electrophiles, halogenation of conjugated dienes can give 1,2- or 1,4-addition products. When molecular bromine is used as the brominating agent in chlorinated hydrocarbon solvent, the 1,4-addition product dominates by $\sim 7:1$ in the case of butadiene.[64]

$$CH_2{=}CHCH{=}CH_2 \xrightarrow[25°C]{Br_2} \underset{\substack{| \\ Br \\ 12\%}}{BrCH_2CHCH{=}CH_2} + \underset{88\%}{BrCH_2CH{=}CHCH_2Br}$$

53. F. Garnier and J.-E. Dubois, *Bull. Soc. Chim. Fr.* **1968**:3797; A. Modro, G. H. Schmid, and K. Yates, *J. Org. Chem.* **42**:3673 (1977).

54. F. Garnier, R. H. Donnay, and J.-E. Dubois, *J. Chem. Soc., Chem. Commun.* **1971**:829; M.-F. Ruasse and J.-E. Dubois, *J. Am. Chem. Soc.* **97**:1977 (1975).

55. K. Yates, R. S. McDonald, and S. A. Shapiro, *J. Org. Chem.* **38**:2460 (1973).

56. M. Zupan and A. Pollak, *J. Chem. Soc., Chem. Commun.* **1973**:845; M. Zupan and A. Pollak, *Tetrahedron Lett.* **1974**:1015.

57. For reviews of fluorinating agents, see A. Haas and M. Lieb, *Chimia* **39**:134 (1985); W. Dmowski, *J. Fluorine Chem.* **32**:255 (1986); H. Vyplel, *Chimia* **39**:305 (1985).

58. S. Rozen and M. Brand, *J. Org. Chem.* **51**:3607 (1986); S. Rozen, *Acc. Chem. Res.* **29**:243 (1996).

59. W. J. Hehre and P. C. Hiberty, *J. Am. Chem. Soc.* **96**:2665 (1974); T. Iwaoka, C. Kaneko, A. Shigihara, and H. Ichikawa, *J. Phys. Org. Chem.* **6**:195 (1993).

60. O. Lerman, Y. Tov, D. Hebel, and S. Rozen, *J. Org. Chem.* **49**:806 (1984).

61. S. Rozen, O. Lerman, M. Kol, and D. Hebel, *J. Org. Chem.* **50**:4753 (1985).

62. P. W. Robertson, J. B. Butchers, R. A. Durham, W. B. Healy, J. K. Heyes, J. K. Johannesson, and D. A. Tait, *J. Chem. Soc.* **1950**:2191.

63. M. Zanger and J. L. Rabinowitz, *J. Org. Chem.* **40**:248 (1975); R. L. Ayres, C. J. Michejda, and E. P. Rack, *J. Am. Chem. Soc.* **93**:1389 (1971); P. S. Skell and R. R. Pavlis, *J. Am. Chem. Soc.* **86**:2956 (1964).

64. G. Bellucci, G. Berti, R. Bianchini, G. Ingrosso, and K. Yates, *J. Org. Chem.* **46**:2315 (1981).

The product distribution can be shifted to favor the 1,2-product by use of such milder brominating agents as the pyridine–bromine complex or the tribromide ion, $Br_3{}^-$. It is believed that molecular bromine reacts through a cationic intermediate, whereas the less reactive brominating agents involve a process more like the Ad_E3 *anti*-addition mechanism.

The stereochemistry of both chlorination and bromination of several cyclic and acyclic dienes has been determined. The results show that bromination is often stereospecifically *anti* for the 1,2-addition process, whereas *syn* addition is preferred for 1,4-addition. Comparable results for chlorination show much less stereospecificity.[65] It appears that chlorination proceeds primarily through ion-pair intermediates, whereas in bromination a stereospecific *anti*-1,2-addition may compete with a process involving a carbocation intermediate. The latter can presumably give *syn* or *anti* product.

6.4. Electrophilic Additions Involving Metal Ions

Certain metal cations are capable of electrophilic attack on alkenes. Addition is completed when a nucleophile adds to the alkene–cation complex. The nucleophile may be the solvent or a ligand from the metal ion's coordination sphere.

The best characterized of these reactions involve the mercuric ion, Hg^{2+}, as the cation.[66] The same process occurs for other transition-metal cations, especially Pd^{2+}, but the products often go on to react further. Synthetically important reactions involving Pd^{2+} will be discussed in Section 8.2 of Part B. The mercuration products are stable, and this allows a relatively uncomplicated study of the addition reaction itself. The usual nucleophile is the solvent, either water or an alcohol. The term *oxymercuration* is used to refer to reactions in

65. G. E. Heasley, D.C. Hayes, G. R. McClung, D. K. Strickland, V. L. Heasley, P. D. Davis, D. M. Ingle, K. D. Rold, and T. L. Ungermann, *J. Org. Chem.* **41**:334 (1976).
66. W. Kitching, *Organomet. Chem. Rev.* **3**:61 (1968); R. C. Larock, *Solvomercuration/Demercuration Reactions in Organic Synthesis*, Springer-Verlag, New York, 1986.

which water or an alcohol acts as the nucleophile.

$$ROH + RCH{=}CHR \xrightarrow{Hg^{2+}} \underset{\underset{RO}{|}}{RCH}{-}\overset{\overset{Hg^+}{|}}{CHR} + H^+$$

In interesting contrast to protonation and halogenation reactions, the mercuration reaction is not accelerated by alkyl substituents on the alkene. For example, 1-pentene is about 10 times more reactive than Z-2-pentene and 40 times more reactive than E-2-pentene.[67] This reversal of reactivity has been attributed to steric effects which evidently outweigh the normal electron-releasing effect of alkyl substituents. When steric factors are taken into account, the reactivity trends are similar to those for other electrophilic additions.[68] As expected for an electrophilic reaction, the ρ value is negative.[69] A bridged mercurinium ion is considered to be formed in the rate-determining step. The addition of the nucleophile follows Markownikoff's rule, and the regioselectivity of oxymercuration is ordinarily very high.

$$RCH{=}CH_2 \xrightarrow{Hg^{2+}} \underset{\underset{H\ \ H}{|\ \ \ |}}{RC}{-}CH \xrightarrow{NaH} \underset{Nu}{R}\overset{H}{C}{-}\overset{Hg^+}{C}{\cdots}H + H^+$$

A mercurinium ion has both similarities and differences as compared with the intermediates that have been described for other electrophilic additions. The proton that initiates acid-catalyzed addition processes is a hard acid and has no unshared electrons. It can form either a carbocation or a hydrogen-bridged cation. Either species is *electron-deficient* and highly reactive.

$$H^+ + \underset{/}{\overset{\backslash}{C}}{=}\underset{\backslash}{\overset{/}{C}} \longrightarrow \underset{/}{\overset{\backslash}{\overset{+}{C}}}{-}\underset{|}{\overset{\overset{H}{|}}{C}}{-} \quad or \quad {-}\underset{/}{\overset{\backslash}{\overset{\pm}{C}}}\overset{H}{\cdots}\underset{\backslash}{\overset{/}{C}}{-}$$

The positive bromine which leads to bromonium ion intermediates is softer and also has unshared electron pairs which can permit a total of *four* electrons to participate in the bridged bromonium ion intermediate. This would be expected to lead to a more strongly bridged and more stable species than is possible in the case of the proton. The bromonium ion can be represented as having two covalent bonds to bromine and is electrophilic but not electron-deficient.

$$-\underset{/}{\overset{\backslash}{C}}\overset{\overset{+}{Br}}{\underset{\diagdown}{\diagup}}\underset{\backslash}{\overset{/}{C}}-$$

The electrophile in oxymercuration reactions, ^+HgX or Hg^{2+}, is a soft acid and strongly polarizing. It polarizes the π electrons of an alkene to the extent that a three-center, two-

67. H. C. Brown and P. J. Geoghegan, Jr., *J. Org. Chem.* **37**:1937 (1972).
68. S. Fukuzuini and J. K. Kochi. *J. Am. Chem. Soc.* **103**:2783 (1981).
69. A. Lewis and J. Arozo, *J. Org. Chem.* **46**:1764 (1981); A. Lewis, *J. Org. Chem.* **49**:4682 (1984).

electron bond is formed between mercury and the two carbons of the double bond. A three-center, two-electron bond implies weaker bridging in the mercurinium ion than in the three-center, four-electron bonding of the bromonium ion. Oxymercuration of simple alkenes is usually a stereospecific *anti* addition. This result is in agreement with the involvement of a mercurinium intermediate that is opened by nucleophilic attack. The formation of the mercurinium ion is normally fast and reversible, with nucleophilic capture being rate-determining.[70]

$$RCH{=}CHR + Hg^{2+} \rightleftharpoons RHC{=}CHR \xrightarrow{Nu^-} RHC{-}CHR$$

The reactivity of mercury salts is a function of both the solvent and the counterion in the mercury salt.[71] Mercuric chloride, for example, is unreactive, and mercuric acetate is usually used. When higher reactivity is required, salts of electronegatively substituted carboxylic acids such as mercuric trifluoroacetate can be used. Mercuric nitrate and mercuric perchlorate are also highly reactive. Soft anions reduce the reactivity of the Hg^{2+} ion by coordination, which reduces the electrophilicity of the cation. The harder oxygen anions leave the mercuric ion in a more reactive state. Organomercury compounds have a number of valuable synthetic applications, and these will be discussed in Chapter 8 of Part B.

6.5. Additions to Alkynes and Allenes

Reactions of alkynes with electrophiles are generally similar to those of alkenes. Because the HOMO of alkynes (acetylenes) is also of π type, it is not surprising that there is a good deal of similarity between alkenes and alkynes in their reactivity toward electrophilic reagents.[72] The fundamental questions about additions to alkynes include the following. How reactive are alkynes in comparison with alkenes? What is the stereochemistry of additions to alkynes? And what is the regiochemistry of additions to alkynes? The important role of halonium ions and mercurinium ions in addition reactions of alkenes raises the question of whether similar species can be involved with alkynes, where the ring would have to include a double bond:

The three basic mechanisms that have been considered to be involved in electrophilic additions to alkynes are shown below. The first involves a discrete vinyl cation. In general, it can lead to either of the two stereoisomeric addition products. The second mechanism is a termolecular process which would be expected to lead to stereospecific *anti* addition. The

70. H. B. Vardhan and R. D. Bach, *J. Org. Chem.* **57**:4948 (1992).
71. H. C. Brown, J. T. Kurek, M.-H. Rei, and K. L. Thompson, *J. Org. Chem.* **49**:2551 (1984).
72. G. H. Schmid, *The Chemistry of the Carbon–Carbon Triple Bond*, Part 1, S. Patai, ed., John Wiley & Sons, New York, 1978, Chapter 3.

third mechanism postulates a bridged-ion intermediate. Mechanisms A and C are of the Ad_E2 type whereas mechanism B would be classified as Ad_E3.

Further details must be added for a complete description, but these outlines encompass most reactions of alkynes with simple electrophiles.

Hydrogen chloride adds to aryl acetylenes in acetic acid to give mixtures of α-chlorostyrenes and the corresponding vinyl acetate.[73] A vinyl cation, which would be stabilized by the aryl substituent, is believed to be an intermediate. The ion pair formed by protonation can either collapse to give the vinyl halide or capture solvent to give the acetate. Aryl-substituted alkynes give mainly the *syn* addition product.

Alkyl-substituted alkynes can react by either the Ad_E3 or the Ad_E2 mechanism. The Ad_E3 mechanism leads to *anti* addition. The preference for one or the other mechanism depends on the individual structure and the reaction conditions.[74] Added Cl^- promotes the Ad_E3 mechanism and increases the overall rate of reaction.

Reaction of 4-octyne with trifluoroacetic acid in CH_2Cl_2 containing $0.1-1.0\,M\,Br^-$ leads mainly to Z-4-bromo-4-octene by an *anti* addition. The presence of Br^- greatly accelerates the reaction as compared to reaction with trifluoroacetic acid alone, indicating the involvement of the Br^- in the rate-determining step.[75]

73. R. C. Fahey and D.-J. Lee, *J. Am. Chem. Soc.* **90**:2124 (1968).
74. R. C. Fahey, M. T. Payne, and D.-J. Lee, *J. Org. Chem.* **39**:1124 (1974).
75. H. M. Weiss and K. M. Touchette, *J. Chem. Soc., Perkin Trans. 2* **1998**:1523.

1-Octyne and 2-octyne also give >95% *anti* addition under these conditions. The reactions are formulated as concerted Ad_E3 processes.

$$RC{\equiv}CR \xrightarrow{\substack{HO_2CCF_3 \\ Br^-}} \underset{Br}{\overset{R}{\diagdown}}C{=}C\underset{R}{\overset{H}{\diagup}}$$

Compared to alkene additions carried out under similar conditions, there is much less involvement of a cationic intermediate, which is consistent with the higher energy of the vinyl cation.

Alkynes can be hydrated in concentrated aqueous acid solutions. The initial product is an enol, which isomerizes to the more stable ketone.

$$CH_3C{\equiv}CH \xrightarrow{H^+} CH_3\overset{+}{C}{=}CH_2 \xrightarrow{H_2O} \underset{OH}{CH_3C{=}CH_2} \longrightarrow \underset{O}{CH_3CCH_3}$$

Alkyne reactivity increases with addition of electron-donating substituents. The reactivity of alkynes is somewhat more sensitive to substituent effects than is the case for alkenes.[76] Solvent isotope effects are indicative of a rate-determining protonation.[77] These reactions are believed to proceed by rate-determining proton transfer to give a vinyl cation. A hydrogen-bridge structure is not regarded as energetically feasible. Various MO calculations place the bridged ion 30–45 kcal/mol above the vinyl cation in energy.[78] Reactions proceeding through a vinyl cation would not be expected to be stereospecific, since the cation would adopt *sp* hybridization.

$$R{-}\overset{\overset{+}{H}}{C}{=}C{-}H \quad \text{less stable than} \quad R{-}\overset{+}{C}{=}CH_2$$

Alkynes react when heated with trifluoroacetic acid to give addition products. Mixtures of *syn* and *anti* addition products are obtained.[79] Similar addition reactions occur with trifluoromethanesulfonic acid.[80] These reactions are analogous to acid-catalyzed hydration and proceed through a vinyl cation intermediate.

$$R{-}C{\equiv}C{-}R + CF_3CO_2H \longrightarrow \underset{CF_3CO_2}{\overset{R}{\diagdown}}C{=}C\underset{H}{\overset{R}{\diagup}} + \underset{CF_3CO_2}{\overset{R}{\diagdown}}C{=}C\underset{R}{\overset{H}{\diagup}}$$

Alkynes undergo addition reactions with halogens. The reaction has been thoroughly examined from a mechanistic point of view. In the presence of excess halogen, tetrahaloalkanes are formed, but mechanistic studies can be carried out with a limited

76. A. D. Allen, Y. Chiang, A. J. Kresge, and T. T. Tidwell, *J. Org. Chem.* **47**:775 (1982).

77. P. Cramer and T. T. Tidwell, *J. Org. Chem.* **46**:2683 (1981).

78. H.-J. Kohler and H. Lischka, *J. Am. Chem. Soc.* **101**:3479 (1979).

79. P. E. Peterson and J. E. Dudley, *J. Am. Chem. Soc.* **88**:4990 (1966); R. H. Summerville and P. v. R. Schleyer, *J. Am. Chem. Soc.* **96**:1110 (1974).

80. P. J. Stang and R. H. Summerville, *J. Am. Chem. Soc.* **91**:4600 (1969); R. H. Summerville, C. A. Senkler, P. v. R. Schleyer, T. E. Dueber, and P. J. Stang, *J. Am. Chem. Soc.* **96**:1100 (1974); G. I. Crisp and A. G. Meyer, *Synthesis* **1994**:667.

amount of halogen so that the initial addition step can be characterized. In general, halogenation of alkynes is slower than halogenation of the corresponding alkenes. We will discuss the reason for this shortly. The reaction shows typical characteristics of an electrophilic reaction. For example, the rates of chlorination of substituted phenylacetylenes are correlated by σ^+ with $\rho = -4.2$. In acetic acid, the reaction is overall second-order, first-order in both reactants. The addition is not very stereoselective, and a considerable amount of solvent capture product is formed. All of these features are consistent with reaction proceeding through a vinyl cation intermediate.[81]

For alkyl-substituted alkynes, there is a difference in stereochemistry between mono- and disubstituted derivatives. The former give *syn* addition whereas the latter react by *anti* addition. The disubstituted (internal) compounds are considerably ($\sim$100 times) more reactive than the monosubstituted (terminal) ones. This result suggests that the transition state of the rate-determining step is stabilized by *both* of the alkyl substituents and points to a bridged intermediate. This would be consistent with the overall stereochemistry of the reaction for internal alkynes.

The monosubstituted intermediate does not seem to be effectively bridged, since *syn* addition predominates. A very short-lived vinyl cation appears to be the best description of the intermediate in this case.[82]

Chlorination of 1-hexyne in acetic acid leads mainly to 1,1-dichlorohexan-2-one via chlorination and deacetylation of the initial product, 2-acetoxy-1-chlorohexene.

The corresponding intermediate, *E*-3-acetoxy-4-chlorohexene, can be isolated from 3-hexyne. In dichloromethane, both 1-hexyne and 3-hexyne give mixtures of the expected dichlorohexenes, with the *E*-isomer predominating.[83]

81. K. Yates and T. A. Go, *J. Org. Chem.* **45**:2377 (1980).
82. K. Yates and T. A. Go, *J. Org. Chem.* **45**:2385 (1980).
83. G. E. Heasley, C. Codding, J. Sheehy, K. Gering, V. L. Heasley, D. F. Shellhamer, and T. Rempel, *J. Org. Chem.* **50**:1773 (1985).

The rates of bromination of a number of alkynes have been measured under conditions that permit comparison with the corresponding alkenes. The rate of bromination of styrene exceeds that of phenylacetylene by about 10^3.[84] For internal alkyne–disubstituted alkene comparisons, the ratios range form 10^3 to 10^7, being greatest in the least nucleophilic solvents.[85] Bromination of alkyl-substituted alkynes shows rate enhancement by both alkyl substituents, and this indicates that the transition state has bridged character.[86]

The stereochemistry of addition is usually *anti* for alkyl-substituted alkynes, whereas the addition to aryl-substituted compounds is not stereospecific. This suggests a termolecular mechanism in the alkyl case, as opposed to an aryl-stabilized vinyl cation intermediate in the aryl case.[87] Aryl-substituted alkynes can be shifted toward *anti* addition by including bromide salts in the reaction medium. Under these conditions, a species preceding the vinyl cation must be intercepted by bromide ion. This species can be represented as a complex of molecular bromine with the alkyne. An overall mechanistic summary is shown in the following scheme.

This scheme represents an alkyne–bromine complex as an intermediate in all alkyne brominations. This is analogous to the case of alkenes. The complex may dissociate to a vinyl cation when the cation is sufficiently stable, as is the case when there is an aryl substituent. It may collapse to a bridged bromonium ion or undergo reaction with a nucleophile. The latter is the dominant reaction for alkyl-substituted alkynes and leads to stereospecific *anti* addition. Reactions proceeding through vinyl cations are expected to be nonstereospecific.

Alkynes react with mercuric acetate in acetic acid to give addition products. In the case of 3-hexyne, the product has *E*-stereochemistry, but the *Z*-isomer is isolated from diphenylacetylene.[88] The kinetics of the addition reaction are first-order in both alkyne and mercuric acetate.[89]

84. M.-F. Ruasse and J.-E. Dubois, *J. Org. Chem.* **42**:2689 (1977).
85. K. Yates, G. H. Schmid, T. W. Regulski, D. G. Garratt, H.-W. Leung, and R. McDonald, *J. Am. Chem. Soc.* **95**:160 (1973); J. M. Kornprobst and J.-E. Dubois, *Tetrahedron Lett.* **1974**:2203; G. Modena, F. Rivetti, and U. Tonellato, *J. Org. Chem.* **43**:1521 (1978).
86. G. H. Schmid, A. Modro, and K. Yates, *J. Org. Chem.* **45**:665 (1980).
87. J. A. Pincock and K. Yates, *Can. J. Chem.* **48**:3332 (1970).
88. R. D. Bach, R. A. Woodard, T. J. Anderson, and M. D. Glick, *J. Org. Chem.* **47**:3707 (1982).
89. M. Bassetti and B. Floris, *J. Org. Chem.* **51**:4140 (1986).

The most common synthetic application of mercury-catalyzed addition to alkynes is the conversion of alkynes to ketones. This reaction is carried out under aqueous acidic conditions, where the addition intermediate undergoes protonation to regenerate Hg^{2+}.

$$RC{\equiv}CH \longrightarrow \underset{\underset{OH}{|}}{RC}{=}CH{-}Hg \longrightarrow \underset{\underset{OH}{|}}{RC}{=}CH{-}Hg \overset{H^+}{\longrightarrow} \underset{\underset{OH}{|}}{RC}{=}CH_2 \longrightarrow \underset{\underset{O}{\|}}{RC}CH_3$$

Several examples are given in Section 4.8 of Part B.

We can understand many of the general characteristics of electrophilic additions to alkynes by recognizing the possibility for both bridged ions and vinyl cations as intermediates. Reactions proceeding through vinyl cations can be expected to be nonstereospecific, with the precise stereochemistry depending upon the lifetime of the vinyl cation and the identity and concentration of the potential nucleophiles. Stereospecific *anti* addition can be expected from processes involving nucleophilic attack on either a bridged-ion intermediate or an alkyne–electrophile complex. These general mechanisms can also explain the relative reactivity of alkenes and alkynes in comparable addition processes. In general, reactions that proceed through vinyl cations, such as those involving rate-determining protonation, are only moderately slower for alkynes as compared to similar alkenes. This can be attributed to the relatively higher energy of vinyl cations compared to cations with sp^2 hybridization. It has been estimated that this difference is around 10–15 kcal/mol, a significant but not enormous difference.[90] This difference is also partially compensated by the higher ground-state energy of alkynes. Reactions that proceed through transition states leading to bridged intermediates typically show much larger rate retardation for the alkyne addition. Bromination is the best studied example of this type. This presumably reflects the greater strain of bridged species in the case of alkynes. Bridged intermediates derived from alkynes must incorporate a double bond in the three-membered ring.[91] The activation energies for additions to alkynes through bridged intermediates are thus substantially greater than for alkenes.

Electrophilic additions to allenes represent an interesting reaction type which is related to additions to both alkenes and alkynes.[92] An allene could, for example, conceivably be protonated at either a terminal sp^2 carbon or the central sp carbon.

$$RCH{=}C{=}CHR \overset{H^+}{\longrightarrow} \underset{\underset{H}{|}}{R\overset{+}{C}}{=}CH{=}CHR \quad \text{versus} \quad RCH{=}C{=}CHR \overset{H^+}{\longrightarrow} RCH_2{-}\overset{+}{C}{=}CHR$$

The allylic carbocation resulting from protonation of the center carbon might seem the obvious choice, but, in fact, the kinetically favored protonation leads to the vinyl cation

90. K. Yates, G. H. Schmid, T. W. Regulski, D. G. Garratt, H.-W. Leung, and R. McDonald, *J. Am. Chem. Soc.* **95**:160 (1973); Z. Rappoport, in *Reactive Intermediates*, Vol. 3, R. A. Abramovitch, ed., Plenum Press, New York, 1985, Chapter 7.
91. G. Melloni, G. Modena, and U. Tonellato, *Acc. Chem. Res.* **8**:227 (1981).
92. For a review of electrophilic additions to allenes, see W. Smadja, *Chem. Rev.* **83**:263 (1983).

intermediate. The reason for this is stereoelectronic. The allene structure is nonplanar, so that an initial protonation of the center carbon leads to a twisted structure that is devoid of allylic conjugation. This twisted cation is about 36–38 kcal/mol higher in energy than that formed by protonation at a terminal carbon.[93]

Addition of hydrogen halides to simple allenes initially gives the vinyl halide, and if the second double bond reacts, a geminal dihalide is formed.[94]

$$RCH{=}C{=}CH_2 \xrightarrow{\ HX\ } RCH_2{-}\underset{X}{C}{=}CH_2 \xrightarrow{\ HX\ } RCH_2{-}\underset{X}{\overset{X}{C}}{-}CH_3$$

The regioselectivity of the second step is consistent with Markownikoff's rule because a halogen atom can stabilize a carbocation by resonance.

$$RCH_2\overset{+}{\underset{:\ddot{X}:}{C}}CH_3 \longleftrightarrow RCH_2\underset{:\ddot{X}^+}{\overset{\|}{C}}CH_3$$

Strong acids in aqueous solution convert allenes to ketones via an enol intermediate. This process also involves protonation at a terminal carbon.

$$CH_2{=}C{=}CH_2 \xrightarrow[H_2O]{H^+} CH_3{-}\underset{OH}{C}{=}CH_2 \longrightarrow CH_3\underset{O}{\overset{\|}{C}}CH_3$$

The kinetic features of this reaction, including the solvent isotope effect, are consistent with a rate-determining protonation to form a vinyl cation.[95]

Allenes react with other typical electrophiles such as the halogens and mercuric ion. In systems where bridged-ion intermediates would be expected, nucleophilic capture generally occurs at the allylic position. This pattern is revealed, for example, in the products of solvent capture in halogen additions[96] and by the structures of mercuration products.[97]

93. K. B. Wiberg, C. M. Breneman, and T. J. Le Page, *J. Am. Chem. Soc.* **112**:61 (1990); A. Gobbi and G. Frenking, *J. Am. Chem. Soc.* **116**:9275 (1994).
94. T. L. Jacobs and R. N. Johnson, *J. Am. Chem. Soc.* **82**:6397 (1960); R. S. Charleston, C. K. Dalton, and S. R. Schraeder, *Tetrahedron Lett.* **1969**:5147; K. Griesbaum, W. Naegle, and G. G. Wanless, *J. Am. Chem. Soc.* **87**:3151 (1965).
95. P. Cramer and T. T. Tidwell, *J. Org. Chem.* **46**:2683 (1981).
96. H. G. Peer, *Recl. Trav. Chim. Pays-Bas.* **81**:113 (1962); W. R. Dolbier, Jr., and B. H. Al-Sader, *Tetrahedron Lett.* **1975**:2159.
97. W. Waters and E. F. Kieter, *J. Am. Chem. Soc.* **89**:6261 (1967).

6.6. The E2, E1, and E1cb Mechanisms

An elimination reaction—the removal of another molecule from a reactant—can be classified according to the relative placement of the carbon atoms from which elimination occurs.

$$
\begin{array}{ccc}
\overset{\displaystyle H}{\underset{\displaystyle R}{R-\overset{|}{\underset{|}{C}}-X}} & \longrightarrow & \overset{\displaystyle}{\underset{\displaystyle R}{R-\overset{}{\underset{|}{C}}:}} \; + \; HX \quad \alpha \text{ elimination}
\end{array}
$$

$$
\overset{\displaystyle H \;\; X}{\underset{\displaystyle H \;\; H}{R-\overset{|}{\underset{|}{C}}-\overset{|}{\underset{|}{C}}-R}} \longrightarrow RCH{=}CHR \; + \; HX \quad \beta \text{ elimination}
$$

$$
\overset{\displaystyle H \;\; H \;\; X}{\underset{\displaystyle H \;\; H \;\; H}{R-\overset{|}{\underset{|}{C}}-\overset{|}{\underset{|}{C}}-\overset{|}{\underset{|}{C}}-R}} \longrightarrow \overset{R \qquad R}{\triangle} \; + \; HX \quad \gamma \text{ elimination}
$$

The products of α eliminations are unstable divalent carbon species called carbenes. They will be discussed in Chapter 10 of Part B. In this chapter, attention will be focused on β-elimination reactions.[98] Some representative examples of β-elimination reactions are given in Scheme 6.1.

The β eliminations can be further subdivided by closer examination of the mechanisms involved. Three distinct limiting mechanisms are outlined below.

E2 Mechanism

$$
\begin{array}{c}
RCH_2CHR' + B^- \\
\underset{X}{\big|}
\end{array}
\longrightarrow
\;\; [\text{transition state}] \;\;
\longrightarrow RCH{=}CHR' + BH + X^-
$$

E1 Mechanism

$$
\underset{X}{RCH_2CHR'} \longrightarrow \underset{X^-}{RCH_2\overset{+}{C}HR'} \overset{B^-}{\longrightarrow} RCH{=}CHR' + BH
$$

E1cb Mechanism

$$
\underset{X}{RCH_2CHR'} + B^- \longrightarrow \underset{X}{R\overset{-}{C}HCHR'} + BH \longrightarrow RCH{=}CHR' + X^-
$$

As depicted, the E2 mechanism involves a bimolecular transition state in which removal of a proton β to the leaving group is concerted with departure of the leaving group. In contrast, the rate-determining step in the E1 mechanism is the unimolecular ionization of

98. For reviews, see E. Baciocchi in *Chemistry of Halides, Pseudo-Halides and Azides*, Part 2, S. Patai and Z. Rappoport, eds., John Wiley & Sons, New York, 1983, Chapter 23; W. H. Saunders, Jr., and A. F. Cockerill, *Mechanisms of Elimination Reactions*, John Wiley & Sons, New York, 1973; D. J. McLennan, *Tetrahedron* **31**:2999 (1975).

Scheme 6.1. Some Examples of β-Elimination Reactions

Dehydrohalogenations

1^a $CH_3(CH_2)_5CH_2CH_2Br \xrightarrow{K^+t\text{-}BuO^-} CH_3(CH_2)_5CH=CH_2$ 85%

2^b $\xrightarrow{K^+i\text{-}PrO^-}$ 34–40%

3^c $C_4H_9CH_2CH-\underset{Cl}{|}\cdots \xrightarrow{KF} C_4H_9CH=CH-\cdots$

4^d 75% + 25%

5^e $\xrightarrow[DMF]{LiCl}$

6^f $\underset{Br\ Br}{PhCHCHCCH_3} \xrightarrow{NaOAc} \underset{Br}{PhCH=CCH_3}$ 64%–73%

Dehydrohalogenations to acetylenes

7^g $\underset{Br}{PhCHCH_2Br} \xrightarrow[NH_3]{NaNH_2} PhC≡CH$ 45%–52%

8^h $\underset{Cl}{CH_3C=CHCH_2OH} \xrightarrow[NH_3]{NaNH_2} \xrightarrow{NH_4Cl} CH_3C≡CCH_2OH$ 75%–85%

the reactant. This is the same process as the rate-determining step in the S_N1 mechanism. Elimination is completed by removal of a β proton. The E1cb mechanism, like the E1 mechanism, involves two steps, but the order is reversed. Proton abstraction precedes expulsion of the leaving group. The correlation of many features of β-elimination reactions is greatly aided by recognition that these three mechanisms represent *variants of a continuum of mechanistic possibilities*. Many β-elimination reactions occur via mechanisms that are intermediate between the limiting mechanistic types. This idea, called the *variable E2 transition state theory*, is outlined in Fig. 6.3.

We will discuss shortly the most important structure–reactivity features of the E2, E1, and E1cb mechanisms. The variable transition state theory allows discussion of reactions proceeding through transition states of intermediate character in terms of the limiting mechanistic types. The most important structural features to be considered in such a discussion are (1) the nature of the leaving group, (2) the nature of the base, (3) electronic and steric effects of substituents in the reactant molecule, and (4) solvent effects.

Scheme 6.1 (continued)

Eliminations using sulfonates

9^i

$\left\langle\ \right\rangle\!-\!OSO_2C_7H_7 \xrightarrow{K^+t\text{-BuO}^-}$ 76%

10^i $Ph_2CHCH_2OSO_2C_7H_7 \xrightarrow{NaOMe} Ph_2C{=}CH_2$ 92%

11^k $HC{\equiv}CCH_2CHCH_3 \xrightarrow[H_2O]{KOH} HC{\equiv}CCH{=}CHCH_3$
$\qquad\qquad\quad |$
$\qquad\quad OSO_2C_7H_7$ 91%

Eliminations involving quaternary ammonium hydroxides

12^l $(CH_3)_3CCH_2CH_2\overset{+}{N}(CH_3)_3\,{}^-OH \xrightarrow{\Delta} (CH_3)_3CCH{=}CH_2$ 81%

13^m $(CH_3)_2\overset{+}{C}N(CH_3)_3$ $\xrightarrow[H_2O]{\Delta}$ 98%

a. P. Veeravagu, R. T. Arnold, and E. W. Eigemann, *J. Am. Chem. Soc.* **86**:3072 (1964).
b. J. P. Schaeffer and L. Endres, *Org. Synth.* **47**:31 (1967).
c. E. Elkik, *Bull. Soc. Chim. Fr.* **1968**:283.
d. S. A. Acharya and H. C. Brown, *J. Chem. Soc., Chem. Commun.* **1968**:305.
e. E. W. Warnhoff, D. G. Martin, and W. S. Johnson, *Org. Synth.* **IV**:162 (1963).
f. N. H. Cromwell, D. J. Cram, and C. E. Harris, *Org. Synth.* **III**:125 (1953).
g. K. N. Campbell and B. K. Campbell, *Org. Synth.* **IV**:763 (1963).
h. P. J. Ashworth, G. H. Mansfield, and M. C. Whiting, *Org. Synth.* **IV**:128 (1963).
i. C. H. Snyder and A. R. Soto, *J. Org. Chem.* **29**:742 (1964).
j. P. J. Hamrick, Jr., and C. R. Hauser, *J. Org. Chem.* **26**:4199 (1961).
k. G. Eglinton and M. C. Whiting, *J. Chem. Soc.* **950**:3650.
l. A. C. Cope and D. L. Ross, *J. Am. Chem. Soc.* **83**:3854 (1961).
m. L. C. King, L. A. Subluskey, and E. W. Stern, *J. Org. Chem.* **21**:1232 (1956).

Fig. 6.3. Variable transition state theory of elimination reactions. J. F. Bunnett, *Angew. Chem. Int. Ed. Engl.* **1**, 225 (1962); J. F. Bunnett, *Surv. Prog. Chem.* **5**, 53 (1969); W. H. Saunders, Jr., and A. F. Cockerill, *Mechanisms of Elimination Reactions*, Wiley, New York, 1973, pp. 48–55; D. J. McLennan, *Tetrahedron* **31**, 2999 (1975); W. H. Saunders, Jr., *Acc. Chem. Res.* **9**, 19 (1976).

There is another useful way of depicting the ideas embodied in the variable transition state theory of elimination reactions. This is to construct a *three-dimensional* potential energy diagram.[99] Suppose that we consider the case of an ethyl halide. The two stepwise reaction paths both require the formation of high-energy intermediates. The E1 mechanism requires formation of a carbocation whereas the E1cb mechanism proceeds via a carbanion intermediate.

$$CH_3CH_2X \longrightarrow CH_3CH_2^+ + X^- \quad \text{E1 mechanism}$$

$$CH_3CH_2X \longrightarrow {}^-CH_2CH_2X + BH \quad \text{E1cb mechanism}$$

In the absence of other stabilizing substituent groups, both a primary carbocation and a primary carbanion are highly unstable intermediates. If we construct a three-dimensional diagram in which progress of $C-H$ bond breaking is one dimension, progress of $C-X$ bond breaking is the second, and the energy of the reacting system is the third, we obtain a diagram such as that in Fig. 6.4. In Fig. 6.4A, only the two horizontal (bond-breaking) dimensions are shown. We see that the E1 mechanism corresponds to complete $C-X$ cleavage before $C-H$ cleavage starts. The E1cb mechanism corresponds to complete $C-X$ cleavage before $C-H$ cleavage begins. In Fig. 6.4B, the energy dimension is added. The front right and back left corners correspond to the E1 and E1cb intermediates, respectively.

Because of the high energy of both the E1 and E1cb intermediates, the lowest-energy path will be the concerted E2 path, more or less diagonally across the energy surface. This pathway is of lower energy because the partially formed double bond provides some compensation for the energy required to break the $C-H$ and $C-X$ bonds and the high-energy intermediates are avoided.

If a substituent is added to the ethyl group which would stabilize the carbocation intermediate, this would cause a lowering of the right front corner of the diagram, which indicates the energy of the carbocation intermediate. Similarly, if a substituent is added which would stabilize a carbanion intermediate, the back left corner of the diagram would be lowered in energy. For this reason, a substituent that would stabilize carbocation character will move the E2 transition state to a point where it more closely resembles the E1 transition state. A structural change that effects stabilization of carbanion character

Fig. 6.4. Three-dimensional (More O'Ferrall) diagrams depicting transition-state locations for E1, E1cb, and E2 mechanisms.

99. R. A. More O'Ferrall, *J. Chem. Soc.* **1970**:274.

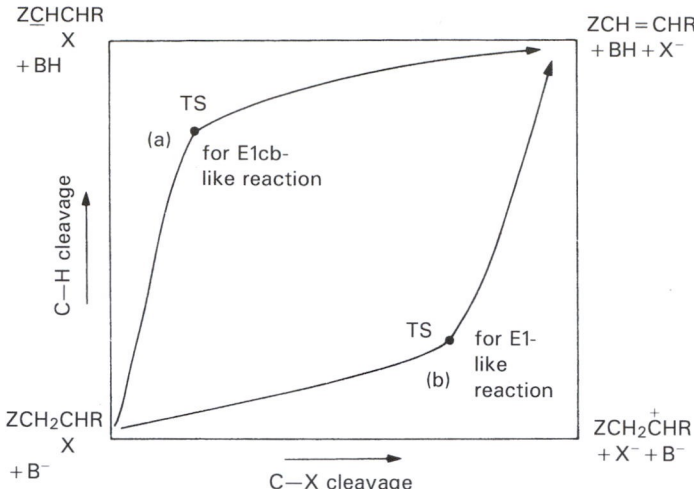

Fig. 6.5. Representation of changes in transition-state character in the variable transition state E2 elimination reaction, showing displacement of transition-state location as a result of substituent effects: (a) substituent Z stabilizes carbanion character of E1cb-like transition state; (b) substituent R stabilizes carbocation character of E1-like transitions state.

would cause the E2 transition state to become more similar to the E1cb transition state. In the E1-like transition state, C—X bond cleavage will be more advanced than C—H cleavage, whereas in the E1cb-like transition state, the C—H bond breaking will be more advanced. Figure 6.5 illustrates how these changes can be depicted with this type of energy diagram.

We will now use these general ideas to discuss specific structural effects that favor the various possible mechanisms for elimination reactions. We have a background which is pertinent to the structure–reactivity effects in E1 reactions from the discussion of S_N1 reactions in Chapter 5. Ionization is favored by (1) electron-releasing groups that stabilize the positive charge in the carbocation intermediate, (2) readily ionized, i.e., "good," leaving groups, and (3) solvents of high ionizing strength. The base plays no role in the rate-determining step in the E1 mechanism, but its identity cannot be ignored. Once ionization has occurred, the cationic intermediate is subject to two competing reactions: nucleophilic capture (S_N1) or proton removal (E1). Stronger bases favor the E1 path over the S_N1 path.

E2 reactions are distinguished from E1 reactions in that the base is present in the transition state for the rate-determining step. The reactions therefore exhibit overall second-order kinetics. The precise nature of the transition state is a function of variables such as the strength of the base, the identity of the leaving group, and the solvent. For example, an elimination reaction proceeding by an E2 transition state will be moved in the E1cb direction by an increase in base strength or by a change to a poorer leaving group. On the other hand, a good leaving group in a highly ionizing solvent will result in an E2 transition state that resembles an E1 process, with extensive weakening of the bond to the leaving group. Reactions that proceed by the limiting E1cb mechanism require substituent groups that can effectively stabilize the intermediate carbanion. This mechanism is not observed with simple alkyl halides or sulfonates. It is more likely to be involved when the leaving group is β to a carbonyl, nitro, cyano, sulfonyl, or other carbanion-stabilizing group.

The nature of the transition state in elimination reactions is of great importance, since it controls the regiochemistry of β elimination in compounds in which the double bond can be introduced in one of several positions. These effects are discussed in the next section.

6.7. Regiochemistry of Elimination Reactions

The most useful generalizations and predictions regarding regioselectivity in elimination reactions are drawn from the variable transition state theory. As shown in Fig. 6.3, this theory proposes that the transition states in E2 reactions may vary over a mechanistic range spanning the gap between the E1 and E1cb extremes. As long as the base is present at the transition state, the reaction will exhibit second-order kinetics. In all such cases, the cleavage of the C—H bond and the C—X bond must be concerted. The relative extent of breaking of the two bonds at the transition state may differ a great deal, however, depending on the nature of the leaving group X and the ease of removal of the hydrogen as a proton. If there are several different β hydrogens, these factors will determine which one is removed. If one compares E1 and E1cb eliminations, it is seen that quite different structural features govern the direction of elimination for these two mechanisms. The variable transition state theory suggests that the regiochemistry of E2 eliminations proceeding through "E1-like" transition states will resemble that of E1 eliminations, whereas E2 eliminations proceeding through "E1cb-like" transition states will show regioselectivity similar to that found for E1cb reactions. It is therefore instructive to consider these limiting mechanisms before discussing the E2 case.

In the E1 mechanism, the leaving group has completely ionized before C—H bond breaking occurs. The direction of the elimination therefore depends on the structure of the carbocation and the identity of the base involved in the proton transfer that follows C—X heterolysis. Because of the relatively high energy of the carbocation intermediate, quite weak bases can effect proton removal. The solvent may often serve this function. The counterion formed in the ionization step may also act as the proton acceptor:

The product composition of the alkenes formed in E1 elimination reactions usually favors the more substituted, and therefore more stable, alkene. One factor is that the energies of the transition states parallel those of the isomeric alkenes. However, because the activation energy for proton removal from a carbocation is low, the transition state should resemble the carbocation intermediate more than the alkene product. In the carbocation, there will be hyperconjugation involving each β hydrogen.[100] Because the hyperconjugation structures possess some double-bond character, the interaction with hydrogen will be greatest at more highly substituted carbon atoms. That is, there will be greater weakening of C—H bonds

100. P. B. D. de la Mare, *Pure Appl. Chem.* **56**:1755 (1984).

Fig. 6.6. Product-determining step for E1 elimination.

and more double-bond character at more highly substituted carbon atoms. This structural effect in the carbocation intermediate will then govern the direction of elimination as indicated in Figure 6.6.

In the E1cb mechanism, the direction of elimination is governed by the kinetic acidity of the individual β protons, which, in turn, is determined by the polar and resonance effects of nearby substituents and by the degree of steric hindrance to approach of base to the proton. Alkyl substituents will tend to retard proton abstraction both electronically and sterically. Preferential proton abstraction from less substituted positions leads to the formation of the less substituted alkene. This regiochemistry is opposite to that of the E1 reaction.

The preferred direction of elimination via the E2 mechanism depends on the precise nature of the transition state. The two extreme transition states for the E2 elimination will resemble the E1 and E1cb mechanisms in their orientational effects. At the "E1cb-like" end of the E2 range, a highly developed bond is present between the proton and the base. The leaving group remains tightly bound to carbon, and there is relatively little development of the carbon–carbon double bond. When the transition state of an E2 reaction has extensive E1cb character, the direction of the elimination is governed by the ease of proton removal. In this case, the less-substituted alkene usually dominates. At the "E1-like" end of the spectrum, the transition state is characterized by well-advanced cleavage of the C−X bond and a largely intact C−H bond. "E1-like" transition states for E2 reactions lead to formation of the more highly substituted of the possible alkenes. In a more synchronous E2 reaction, the new double bond is substantially formed at the transition state at the expense of partial rupture of both the C−H and C−X bonds. E2 eliminations usually give the more substituted alkene. This is because the transition states leading to the possible alkenes will reflect the partial double-bond character, and the greater stability of the more substituted double bond will favor the corresponding transition state. Concerted E2 reactions are also subject to the stereoelectronic requirement that the reacting C−H and

C—X bonds be antiperiplanar.

Prior to development of the mechanistic ideas outlined above, it was recognized by experience that some types of elimination reactions gave the more substituted alkene as the major product. Such eliminations were said to follow the "Saytzeff rule." This behavior is characteristic of E1 reactions and E2 reactions involving relatively good leaving groups, such as halides and sulfonates. These are now recognized as reactions which proceed with C—X cleavage being well advanced in the transition state. E2 reactions involving poor leaving groups, particularly those involving quaternary ammonium salts, were said to follow the "Hofmann rule" and gave mainly the less substituted alkene. We now recognize that such reactions would proceed through transition states with E1cb character.

The data recorded in Table 6.4 for the 2-hexyl system illustrate two general trends that have been recognized in other systems as well. First, poorer leaving groups favor elimination according to the "Hofmann rule," as shown, for example, by the increasing amount of terminal olefin in the halogen series as the leaving group is changed from iodide to fluoride. Poorer leaving groups move the transition state in the E1cb direction. A higher negative charge must build up on the β carbon to induce loss of the leaving group. This charge buildup is accomplished by more complete proton abstraction.

Comparison of the data for methoxide with those for t-butoxide in Table 6.4 illustrates a second general trend: Stronger bases favor formation of the less substituted alkene.[101–103] A stronger base leads to an increase in the carbanion character at the transition state and thus shifts the transition state in the E1cb direction. A linear correlation between the strength of the base and the difference in $\Delta G^{\ddagger}$ for the formation of 1-butene versus 2-butene has been established.[102] Some of the data are given in Table 6.5.

The direction of elimination is also affected by steric effects, and if both the base and the reactant are highly branched, steric factors may lead to preferential removal of the less hindered hydrogen.[104] Thus, when 4-methyl-2-pentyl iodide reacts with very hindered bases such as potassium tricyclohexylmethoxide, there is preferential formation of the

Table 6.4. Product Ratios for Some E2 Eliminations[a]

Substrate: $CH_3CH_2CH_2CH_2CHCH_3$ X	Base, solvent	Percent composition of alkene		
		1-Hexene	2-Hexene	
			trans	cis
$X = I$	MeO^-, MeOH	19	63	18
Cl	MeO^-, MeOH	33	50	17
F	MeO^-, MeOH	69	21	9
$OSO_2C_7H_7$	MeO^-, MeOH	33	44	23
I	t-BuO$^-$, t-BuOH	78	15	7
Cl	t-BuO$^-$, t-BuOH	91	5	4
F	t-BuO$^-$, t-BuOH	97	1	1
$OSO_2C_7H_7$	t-BuO$^-$, t-BuOH	83	4	14

a. Data from R. A. Bartsch and J. F. Bunnett, *J. Am. Chem. Soc.* **91**:1376 (1967).

101. D. H. Froemsdorf and M. D. Robbins, *J. Am. Chem. Soc.* **89**:1737 (1967); I. N. Feit and W. H. Saunders, Jr., *J. Am. Chem. Soc.* **92**:5615 (1970).

102. R. A. Bartsch, G. M. Pruss, B. A. Bushaw, and K. E. Wiegers, *J. Am. Chem. Soc.* **95**:3405 (1973).

103. R. A. Bartsch, K. E. Wiegers, and D. M. Guritz, *J. Am. Chem. Soc.* **96**:430 (1974).

104. R. A. Bartsch, R. A. Read, D. T. Larsen, D. K. Roberts, K. J. Scott, and B. R. Cho, *J. Am. Chem. Soc.* **101**:1176 (1979).

Table 6.5. Orientation in E2 Elimination as a Function of Base Strength

Base (potassium salt)	pK	% 1-Butene from 2-iodobutane[a]	% 1-Butene from 2-butyl tosylate[b]
p-Nitrobenzoate	8.9	5.8	c
Benzoate	11.0	7.2	c
Acetate	11.6	7.4	c
Phenolate	16.4	11.4	30.6
2,2,2-Trifluoroethoxide	21.6	14.3	46.0
Methoxide	29.0	17.0	c
Ethoxide	29.8	17.1	56.0
t-Butoxide	32.2	20.7	58.5

a. From R. A. Bartsch, G. M. Pruss, B. A. Bushaw, and K. E. Wiegers, *J. Am. Chem. Soc.* **95**:3405 (1973). The pK values refer to DMSO solution.
b. R. A. Bartsch, R. A. Read, D. T. Larsen, D. K. Roberts, K. J. Scott, and B. R. Cho, *J. Am. Chem. Soc.* **101**:1176 (1979).
c. Not reported.

terminal alkene. In this case, potassium *t*-butoxide favors the internal alkene, although by a smaller ratio than for less branched alkoxides.

$$(CH_3)_2CHCH_2CHCH_3 \longrightarrow (CH_3)_2CHCH{=}CHCH_3 + (CH_3)_2CHCH_2CH{=}CH_2$$
$$\overset{|}{I}$$

base		
$K^+ {}^-OC(C_6H_{11})_3$	42%	58%
$K^+ {}^-OC(CH_3)_3$	61%	39%
$K^+ {}^-OCH_2CH_2CH_3$	75%	25%

The leaving group also affects the amount of internal versus terminal alkene that is formed. The poorer the leaving group, the more E1cb-like is the transition state. This trend is illustrated for the case of the 2-butyl system by the data in Table 6.6. Positively charged leaving groups, such as in dimethylsulfonium and trimethylammonium salts, may favor a more E1cb-like transition state because their inductive and field effects increase the acidity of the β protons.

6.8. Stereochemistry of E2 Elimination Reactions

In principle, elimination may proceed in *syn* or *anti* fashion:

In most cases, E2 elimination proceeds via a transition state involving the *anti* arrangement. Nevertheless, *syn* elimination is possible, and, when special structural features retard *anti* elimination, *syn* elimination becomes the dominant mode.

Cyclohexyl systems have a very strong preference for *anti* elimination via conforma-

Table 6.6. Orientation of Elimination in the 2-Butyl System under Various E2 Conditions

		1-Butene (%)	2-Butene (%)	Reference
CH$_3$CHCH$_2$CH$_3$ | I	$\xrightarrow{\text{PhCO}_2^-}$ DMSO	7	93	a
CH$_3$CHCH$_2$CH$_3$ | I	$\xrightarrow{\text{C}_2\text{H}_5\text{O}^-}$ DMSO	17	83	a
CH$_3$CHCH$_2$CH$_3$ | I	$\xrightarrow{(\text{CH}_3)_3\text{CO}^-}$ DMSO	21	79	b
CH$_3$CHCH$_2$CH$_3$ | Br	$\xrightarrow{(\text{CH}_3)_3\text{CO}^-}$ DMSO	33	67	b
CH$_3$CHCH$_2$CH$_3$ | Cl	$\xrightarrow{(\text{CH}_3)_3\text{CO}^-}$ DMSO	43	57	b
CH$_3$CHCH$_2$CH$_3$ | Br	$\xrightarrow{\text{C}_2\text{H}_5\text{O}^-}$ C$_2$H$_5$OH	19	81	c
CH$_3$CHCH$_2$CH$_3$ | OSO$_2$C$_7$H$_7$	$\xrightarrow{\text{C}_2\text{H}_5\text{O}^-}$ C$_2$H$_5$OH	35	65	d
CH$_3$CHCH$_2$CH$_3$ | OSO$_2$C$_7$H$_7$	$\xrightarrow{(\text{CH}_3)_3\text{CO}^-}$ DMSO	61	39	d
CH$_3$CHCH$_2$CH$_3$ | $^+$S(CH$_3$)$_2$	$\xrightarrow{\text{C}_2\text{H}_5\text{O}^-}$ C$_2$H$_5$OH	74	26	e
CH$_3$CHCH$_2$CH$_3$ | $^+$N(CH$_3$)$_3$	$\xrightarrow{^-\text{OH}}$	95	5	f

a. R. A. Bartsch, B. M. Pruss, B. A. Bushaw, and K. E. Wiegers, *J. Am. Chem. Soc.* **95**:3405 (1973).
b. D. L. Griffith, D. L. Meges, and H, C, Brown, *J. Chem. Soc., Chem. Commun.* **1968**:90.
c. M. L. Dhar, E. D. Hughes, and C. K. Ingold, *J. Chem. Soc.* **1948**:2058.
d. D. H. Froemsdorf and M. D. Robbins, *J. Am. Chem. Soc.* **89**:1737 (1967).
e. E. D. Hughes, C. K. Ingold, G. A. Maw, and L. I. Woolf, *J. Chem. Soc.* **1948**:2077.
f. A. C. Cope, N. A. LeBel, H.-H. Lee, and W. R. Moore, *J. Am. Chem. Soc.* **79**:4720 (1957).

tions in which both the departing proton and the leaving group occupy axial positions. The orientation permits the alignment of the involved orbitals so that concerted *anti* elimination can occur.

For example, *cis*-4-t-butylcyclohexyl bromide undergoes E2 elimination at a rate about 500 times greater than the *trans* isomer because only the *cis* isomer permits *anti* elimination from the favored chair conformation.[105]

E2 rate constant
with K$^+$ $^-$OC(CH$_3$)$_3$ 4.1×10^{-3} 8.0×10^{-6}

Other cyclic systems are not so selective. In the decomposition of *N,N,N*-trimethylcyclobutylammonium hydroxide, elimination is 90% *syn*.[106] The cyclobutyl ring resists the conformation required for *anti* elimination. The more flexible five-membered ring analog undergoes about 50% *syn* elimination. Elimination form the *N,N,N*-trimethylnorbornylammonium ion is exclusively *syn*[107]. This is another case in which the rigid ring prohibits attainment of an *anti*-elimination process. There is also a steric effect operating against removal of an *endo* proton, which is required for *anti* elimination.

Although there is usually a preference for *anti* elimination in acyclic systems, *syn* elimination is competitive in some cases. In acyclic systems, the extent of *anti* versus *syn* elimination can be determined by use of stereospecifically deuterated substrates or by use of diastereomeric reactants which will give different products by *syn* and *anti* elimination. The latter approach showed that elimination from 3-phenyl-2-butyl tosylate is a stereospecific *anti* process.[108]

The occurrence of *syn* elimination in 5-decyl systems has been demonstrated with the use of diastereomeric deuterium-labeled substrates. Stereospecifically labeled 5-substituted decane derivatives were prepared and subjected to appropriate elimination conditions. By comparison of the amount of deuterium in the *E* and *Z* isomers of the product, it was

105. J. Zavada, J. Krupicka, and J. Sicher, *Collect. Czech. Chem. Commun.* **33**:1393 (1968).
106. M. P. Cooke, Jr., and J. L. Coke, *J. Am. Chem. Soc.* **90**:5556 (1968).
107. J. P. Coke and M. P. Cooke, *J. Am. Chem. Soc.* **89**:6701 (1967).
108. W.-B. Chiao and W. H. Saunders, *J. Org. Chem.* **45**:1319 (1980).

possible to determine the extent of *syn* and *anti* elimination.[109]

389

SECTION 6.8.
STEREOCHEMISTRY OF
E2 ELIMINATION
REACTIONS

Data obtained for three different leaving groups are presented in Table 6.7. The results show that *syn* elimination is extensive for quaternary ammonium salts. With better leaving groups, the extent of *syn* elimination is small in the polar solvent DMSO but quite significant in benzene. The factors which promote *syn* elimination will be discussed shortly.

Table 6.8 summarizes some data on *syn* versus *anti* elimination in acyclic systems. The general trend revealed by these and other data is that *anti* stereochemistry is normally preferred for reactions involving good leaving groups such as bromide and tosylate. With poorer leaving groups (e.g., fluoride, trimethylamine), *syn* elimination becomes important. The amount of *syn* elimination is small in the 2-butyl system, but it becomes a major pathway with 3-hexyl compounds and longer chains. *Syn* elimination is especially prevalent in the medium-sized alicyclic compounds.[110]

Table 6.7. Extent of *Syn* Elimination as a Function of the Leaving Group in the 5-Decyl System[a]

	Percent *syn* elimination			
	E product		Z product	
Leaving group	DMSO	Benzene	DMSO	Benzene
Cl	6	62	7	39
OTs	4	27	4	16
$^+N(CH_3)_3$	93	92	76	84

a. Data from M. Pankova, M. Svoboda, and J. Zavada, *Tetrahedron Lett.* **1972**:2465. The base used was potassium *t*-butoxide.

109. M. Pankova, M. Svoboda, and J. Zavada, *Tetrahedron Lett.* **1972**:2465. The analysis of the data also requires that account be taken of (a) isotope effects and (b) formation of 4-decene. The method of analysis is described in detail by J. Sicher, J. Zavada, and M. Pankova, *Collect. Czech. Chem. Commun.* **36**:314 (1971).
110. J. Sicher, *Angew. Chem. Int. Ed. Engl.* **11**:200 (1972).

Table 6.8. Stereochemistry of E2 Eliminations for Some Acyclic Substrates

Substrate	Base, solvent	% *anti*	% *syn*	Reference
$CH_3CHCHCH_3$ 		 D Br	$K^{+-}OC(CH_3)_3, (CH_3)_3COH$	100	0	a
$CH_3CHCHCH_3$ 		 D $OSO_2C_7H_7$	$K^{+-}OC(CH_3)_3, (CH_3)_3COH$	>98	<2	b
$CH_3CHCHCH_3$ 		 D $^+N(CH_3)_3$	$K^{+-}OC(CH_3)_3$, DMSO	100	0	c
$CH_3CH_2CHCHCH_2CH_3$ 		 D F	$K^{+-}OC(CH_3)_3, (CH_3)_3COH$	32	68	d
$CH_3(CH_2)_3CHCH(CH_2)_3CH_3$ 		 D $^+N(CH_3)_3$	$K^{+-}OC(CH_3)_3$, DMSO	24	76	e
$CH_3(CH_2)_3CHCH(CH_2)_3CH_3$ 		 D $OSO_2C_7H_7$	$K^{+-}OC(CH_3)_3, (CH_3)_3COH$	93	7	f
$CH_3(CH_2)_3CHCH(CH_2)_3CH_3$ 		 D Cl	$K^{+-}OC(CH_3)_3$, benzene	62	38	g
$CH_3(CH_2)_3CHCH(CH_2)_3CH_3$ 		 D F	$K^{+-}OC(CH_3)_3$, benzene	<20	>80	g
$CH_3(CH_2)_3CHCH(CH_2)_3CH_3$ 		 D Cl	$K^{+-}OC(CH_3)_3$, DMSO	93	7	g
$CH_3(CH_2)_3CHCH(CH_2)_3CH_3$ 		 D F	$K^{+-}OC(CH_3)_3$, DMSO	80	20	g

a. R. A. Bartsch, *J. Am. Chem. Soc.* **93**:3683 (1971).
b. D. H. Froemsdorf, W. Dowd, W. A. Gifford, and S. Meyerson, *J. Chem. Soc., Chem. Commun.* **1968**:449.
c. D. H. Froemsdorf, H. R. Pinnick, Jr., and S. Meyerson, *J. Chem. Soc., Chem. Commun.* **1968**:1600.
d. J. K. Borchardt, J. C. Swanson, and W. H. Saunders, Jr., *J. Am. Chem. Soc.* **96**:3918 (1974).
e. J. Sicher, J. Závada, and M. Pánková, *Collect. Czech. Chem. Commun.* **36**:3140 (1971).
f. J. Závada, M. Pánková, and J. Sicher, *J. Chem. Soc., Chem. Commun.* **1968**:1145.
g. M. Pánková, M. Svoboda, and J. Závada, *Tetrahedron Lett.* **1972**:2465.

The factors that determine whether *syn* or *anti* elimination predominates are still subject to investigation. One factor that is believed to be important is whether the base is free or present in an ion pair.[111] The evidence is that an ion pair promotes *syn* elimination of anionic leaving groups. This effect can be explained by proposing a transition state in which the anion functions as a base and the cation assists in the departure of the leaving group.

111. R. A. Bartsch, G. M. Pruss, R. L. Buswell, and B. A. Bushaw, *Tetrahedron Lett.* **1972**:2621; (1972); J. K. Borchardt, J. C. Swanson, and W. H. Saunders, Jr., *J. Am. Chem. Soc.* **96**:3918 (1974).

This interpretation is in agreement with the solvent effect that is evident in the 5-decyl system, as revealed in Table 6.7. The extent of *syn* elimination is much higher in the nondissociating solvent benzene than in DMSO. The ion-pair interpretation is also favored by the fact that addition of specific metal-ion-complexing agents (crown ethers) that would favor dissociation of the ion pair leads to diminished amounts of *syn* elimination.[112] Another factor that affects the *syn* : *anti* ratio is the strength of the base. Strong bases are more likely to exhibit a high proportion of *syn* elimination.[113]

Steric effects also play a significant role. With N-(β,β-disubstituted-ethyl)-N,N,N-trimethylammonium ions, *syn* elimination is important when the β substituents are aryl or branched. As the β groups become less bulky, the amount of *syn* elimination diminishes. This effect is demonstrated by the data below.[114]

$$R^1-\underset{\underset{H}{|}}{\overset{\overset{R^2}{|}}{C}}-CHD\overset{+}{N}(CH_3)_3 \longrightarrow \underset{R^1}{\overset{R^2}{\diagdown}}C=C\underset{D}{\overset{H}{\diagup}}$$

R^1	R^2	% *syn*	% *anti*
Ph	MeO−Ph	62	38
Ph	$(CH_3)_2CH$	68	32
Ph	Me	26	74
n-Bu	H	<5	>95

The dependence on steric bulk is attributed to the steric requirements imposed by the bulky trimethylamine leaving group. In the transition state for *anti* elimination, steric repulsion is increased as R^1 and R^2 increase in size. When the repulsion is sufficiently large, the transition state for *syn* elimination is preferred.

The proportion of *cis* and *trans* isomers of internal alkenes formed during elimination reactions depends on the identity of the leaving group. Halides usually give predominantly the *trans* alkenes.[115] Bulkier groups, particularly arenesulfonates, give higher proportions of the *cis* alkene. Sometimes, more *cis* isomer is formed than *trans*. The normal preference for *trans* alkene probably reflects the greater stability of the *trans* alkene; i.e., the unfavorable steric repulsions present in the *cis* alkene are also present in the E2 transition state leading to *cis* alkene. High *cis* : *trans* ratios are attributed to a second steric effect that

112. R. A. Bartsch, E. A. Mintz, and R. M. Parlman, *J. Am. Chem. Soc.* **96**:3918 (1974).
113. K. C. Brown and W. H. Saunders, Jr. *J. Am. Chem. Soc.* **92**:4292 (1970); D. S. Bailey and W. H. Saunders, Jr., *J. Am. Chem. Soc.* **92**:6904 (1970).
114. Y.-T. Tao and W. H. Saunders, Jr., *J. Am. Chem. Soc.* **105**:3183 (1983).
115. H. C. Brown and R. L. Kliminsch, *J. Am. Chem. Soc.* **87**:5517 (1965); I. N. Feit and W. H. Saunders, Jr., *J. Am. Chem. Soc.* **92**:1630 (1970).

becomes important only when the leaving group is large. The conformations leading to *cis* and *trans* alkene by *anti* elimination are depicted below.

When the leaving group and base are both large, conformation **2** is favored because it permits the leaving group to occupy a position removed from the β-alkyl substituents. *Anti* elimination through a transition state arising from conformation **2** gives *cis* alkene.

6.9. Dehydration of Alcohols

The dehydration of alcohols is an important elimination reaction that takes place under acidic rather than basic conditions. It involves an E1 mechanism.[116] The function of the acidic reagent is to convert the hydroxyl group to a better leaving group by protonation:

This elimination reaction is the reverse of acid-catalyzed hydration, which was discussed in Section 6.2. Because a carbocation or closely related species is the intermediate, the elimination step would be expected to favor the more substituted alkene as discussed on p. 384. The E1 mechanism also explains the general trends in relative reactivity. Tertiary alcohols are the most reactive, and reactivity decreases going to secondary and primary alcohols. Also in accord with the E1 mechanism is the fact that rearranged products are found in cases where a carbocation intermediate would be expected to rearrange:

For many alcohols, exchange of the hydroxyl group with solvent competes with dehydration.[117] This exchange indicates that the carbocation can undergo S_N1 capture in competion with elimination. Under conditions where proton removal is rate-determining, it

116. D. V. Banthorpe, *Elimination Reactions*, Elsevier, New York, 1963, pp. 145–156.
117. C. A. Bunton and D. R. Llewellyn, *J. Chem. Soc.* **1957**:3402; J. Manassen and F. S. Klein, *J. Chem. Soc.* **1960**:4203.

would be expected that a significant isotope effect would be seen. This is, in fact, observed.

$$\text{PhCHCHPh} \xrightarrow[\text{H}_2\text{O}]{\text{H}_2\text{SO}_4} \text{PhCH=CHPh} \qquad \text{(Ref. 118)}$$
$$\overset{|}{\text{HO}}\ \overset{|}{\text{H*}} \qquad\qquad k_H/k_D = 1.8$$

6.10. Eliminations Not Involving C–H Bonds

The discussion of β-elimination processes thus far has focused on reactions that involve removal of a proton bound to carbon. It is the electrons in the C–H σ bond, however, that are essential to the elimination process. Compounds bearing other substituents that can release electrons can undergo similar β eliminations. Many such processes are known, and frequently the reactions are stereospecific.

Vicinal dibromides may be debrominated by treating them with certain reducing agents, including iodide ion and zinc. The stereochemical course in the case of 1,1,2-tribromocyclohexane was determined using a ^{82}Br-labeled sample prepared by *anti* addition of ^{82}Br$_2$ to bromocyclohexene. Exclusive *anti* elimination would give unlabeled bromocyclohexene whereas ^{82}Br-labeled product would result from *syn* elimination. Debromination with sodium iodide was found to be cleanly an *anti* elimination, whereas debromination with zinc gave mainly, but not entirely, *anti* elimination.[119]

Reagent	anti	syn
Sodium iodide/MeOH	100%	0%
Zinc/EtOH	89%	11%

The iodide-induced reduction is essentially the reverse of a halogenation. Application of the principle of microscopic reversibility would suggest that the reaction would proceed through a bridged intermediate as shown below.[120]

The rate-determining expulsion of bromide ion through a bridged intermediate requires an *anti* orientation of the two bromides. The nucleophilic attack of iodide at one bromide enhances its nucleophilicity and permits formation of the bridged ion. The stereochemical preference in noncyclic systems is also *anti*, as indicated by the fact that *meso*-stilbene

118. D. S. Noyce, D. R. Hartter, and R. M. Pollack, *J. Am. Chem. Soc.* **90**:3791 (1968).
119. C. L. Stevens and J. A. Valicenti, *J. Am. Chem. Soc.* **87**:838 (1965).
120. C. S. T. Lee, I. M. Mathai, and S. I. Miller, *J. Am. Chem. Soc.* **92**:4602 (1970).

dibromide yields E-stilbene, while d,l-stilbene dibromide gives mainly Z-stilbene under these conditions.[120]

The zinc-induced debromination may proceed by formation of an organozinc intermediate, with the loss of stereospecificity occurring during the formation of this intermediate. Similar nonstereospecific debrominations occur with one-electron donors, such as chromium(II) salts, and have been interpreted as resulting from a free-radical intermediate.[121] The organozinc and organochromium species that are postulated as intermediates in these reductive eliminations are representatives of a general structural type M−C−C−X, in which M is a metal and X is a leaving group. These structures are, in general, very prone to elimination with formation of a double bond.

One example of elimination reactions of this type is acid-catalyzed deoxymercuration.[122] The β-oxyorganomercurials are much more stable than similar reagents derived from more electropositive metals but are much more reactive than simple alcohols. For example, $CH_3CH(OH)CH_2HgI$ is converted to propene under acid-catalyzed conditions at a rate 10^{11} times greater than dehydration of 2-propanol under the same conditions. These reactions are pictured as proceeding through a bridged mercurinium ion by a mechanism that is the reverse of oxymercuration (see Section 6.4).

One of the pieces of evidence in favor of this mechanism is the fact that the $\Delta H^{\ddagger}$ for deoxymercuration of *trans*-2-methoxycyclohexylmercuric iodide is about 8 kcal/mol less than for the *cis* isomer. Only the *trans* isomer can undergo elimination by an *anti* process

121. K. Kochi and D. M. Singleton, *J. Am. Chem. Soc.* **90**:1582 (1968).
122. M. M. Kreevoy and F. R. Kowitt, *J. Am. Chem. Soc.* **82**:739 (1960).

through a chair conformation.

Comparing the rates of acid-catalyzed β elimination of compounds of the type MCH_2CH_2OH yields the following reactivity order for β substituents: $IHg \sim Ph_3Pb \sim Ph_3Sn > Ph_3Si > H$. The relative rates are within a factor of 10 for the first three, but these are 10^6 greater than for Ph_3Si and 10^{11} greater than for a proton. There are two factors involved in these very large rate accelerations. One is bond energies. The relevant values are $Hg\text{-}C = 27 < Pb\text{-}C = 31 < Sn\text{-}C = 54 < Si\text{-}C = 60 < H\text{-}C = 96$ kcal/mol.[123] The metal substituents also have a very strong stabilizing effect for carbocation character at the β carbon. This stabilization can be pictured as an orbital–orbital interaction in which the electron-rich carbon–metal bond donates electron density to the adjacent p orbital or as formation of a bridged species.

There are a number of synthetically important β-elimination processes involving organosilicon[124] and organotin[125] compounds. Treatment of β-hydroxyalkylsilanes or β-hydroxyalkylstannanes with acid results in stereospecific *anti* eliminations which are much more rapid than for compounds lacking the group IV substituent.

(Ref. 126)

(Ref. 127)

123. D. D. Davis and H. M. Jacocks III, *J. Organomet. Chem.* **206**:33 (1981).
124. A. W. P. Jarvie, *Organomet. Chem. Rev., Sect. A* **6**:153 (1970); W. P. Weber, *Silicon Reagents for Organic Synthesis*, Springer-Verlag, Berlin, 1983; E. W. Colvin, *Silicon in Organic Synthesis*, Butterworths, London, 1981.
125. M. Pereyre, J.-P. Quintard, and A. Rahm, *Tin in Organic Synthesis*, Butterworths, London, 1987.
126. P. F. Hudrlick and D. Peterson, *J. Am. Chem. Soc.* **97**:1464 (1975).
127. D. D. Davis and C. E. Gray, *J. Org. Chem.* **35**:1303 (1970).

β-Halosilanes also undergo facile elimination when treated with methoxide ion.

$$\underset{\underset{Br}{|}}{\overset{\overset{Br}{|}}{C_4H_9CHCHSi(CH_3)_3}} \xrightarrow{NaOCH_3} C_4H_9CH=CHBr \qquad \text{(Ref. 128)}$$

Fluoride-induced β-elimination reactions of silanes having leaving groups in the β position are important processes in synthetic chemistry, as, for, example in the removal of β-trimethylsilylethoxy groups.

$$RCO_2CH_2CH_2Si(CH_3)_3 + R_4N^+F^- \rightarrow RCO_2^- + CH_2 = CH_2 + FSi(CH_3)_3 \qquad \text{(Ref.129)}$$

These reactions proceed by alkoxide or fluoride attack at silicon which results in C−Si bond cleavage and elimination of the leaving group from the β carbon. These reactions are stereospecific *anti* eliminations.

$$\underset{\underset{X}{|}}{\overset{(CH_3)_3Si \quad Nu^-}{RCH{-}CHR}} \longrightarrow RCH=CHR + X^- + (CH_3)_3SiNu$$

β-Elimination reactions of this type can also be effected by converting a β-hydroxy group to a good leaving group. For example, conversion of β-hydroxyalkylsilanes to the corresponding methanesulfonates leads to rapid elimination.[130]

$$\underset{\underset{OH}{|}}{(CH_3)_3SiCH_2CR_2} \xrightarrow{CH_3SO_2Cl} H_2C=CR_2$$

β-Trimethylsilylalkyl trifluoroacetates also undergo facile *anti* elimination.[131]

The ability to promote β elimination and the electron-donor capacity of the β-metalloid substituents can be exploited in a very useful way in synthetic chemistry.[132] Vinylstannanes and vinylsilanes react readily with electrophiles. The resulting intermediates then undergo elimination of the stannyl or silyl substituent, so that the net effect is replacement of the stannyl or silyl group by the electrophile. An example is the replacement of a trimethylsilyl substituent by an acetyl group by reaction with acetyl chloride.

(Ref. 133)

128. A. W. P. Jarvie, A. Holt, and J. Thompson, *J. Chem. Soc. B* **1969**:852; B. Miller and G. J. McGarvey, *J. Org. Chem.* **43**:4424 (1978).
129. P. Sieber, *Helv. Chim. Acta* **60**:2711 (1977).
130. F. A. Carey and J. R. Toler, *J. Org. Chem.* **41**:1966 (1976).
131. M. F. Connil, B. Jousseaume, N. Noiret, and A. Saux, *J. Org. Chem.* **59**:1925 (1994).
132. T. H. Chan and I. Fleming, *Synthesis* **1979**:761; I. Fleming, *Chem. Soc. Rev.* **10**:83 (1981).
133. I. Fleming and A. Pearce, *J. Chem. Soc., Chem. Commun.* **1975**:633.

The silyl and stannyl substituents are crucial to these reactions in two ways. In the electrophilic addition step, they act as electron-releasing groups promoting addition and also control the regiochemistry. A silyl or stannyl substituent strongly stabilizes carbocation character at the β-carbon atom and thus directs the electrophile to the α-carbon. The reaction is then completed by the elimination step, in which the carbon–silicon or carbon–tin bond is broken.

$$E^+ + RCH{=}CHMR_3 \longrightarrow R\overset{+}{C}H{-}\underset{\underset{E}{|}}{C}HMR_3 \longrightarrow RCH{=}CHE$$

Computational investigations of vinylsilanes indicate that there is a ground-state interaction between the alkene π orbital and the carbon–silicon bond which raises the energy of the π HOMO and enhances reactivity.[134] Furthermore, this stereoelectronic interaction favors attack of the electrophile *anti* to the silyl substituent.

MO calculations indicate a stabilization of 38 kcal/mol, which is about the same as the value calculated for an α-methyl group.[135]

Allylsilanes and allylstannanes are also reactive toward electrophiles and usually undergo a concerted elimination of the silyl substituent. Several examples are shown below.

$$(CH_3)_3SiCH_2CH{=}CH_2 + I_2 \longrightarrow CH_2{=}CHCH_2I \qquad \text{(Ref. 136)}$$

$$(CH_3)_3SiCH_2CH{=}CHC_6H_{13} + (CH_3)_3CCl \xrightarrow{\;TiCl_4\;} CH_2{=}CH\overset{\overset{\displaystyle C(CH_3)_3}{\displaystyle |}}{C}HC_6H_{13} \qquad \text{(Ref. 137)}$$

$$(CH_3)_3Sn\underset{\underset{CH_2{=}CH}{|}}{C}H_2C{=}CH_2 + BrCH_2CH{=}C(CH_3)_2 \longrightarrow CH_2{=}\underset{\underset{CH_2{=}CH}{|}}{C}CH_2CH_2CH{=}C(CH_3)_2 \qquad \text{(Ref. 138)}$$

$$(CH_3)_3SnCH_2CH{=}CH_2 + (MeO)_2CHCH_2CH_2Ph \xrightarrow{\;(Et_2Al)_2SO_4\;} CH_2{=}CHCH_2\underset{\underset{OMe}{|}}{C}HCH_2CH_2Ph \qquad \text{(Ref. 139)}$$

Further examples of these synthetically useful reactions can be found in Chapter 9 of Part B.

134. S. D. Kahn, C. F. Pau, A. R. Chamberlin, and W. J. Hehre, *J. Am. Chem. Soc.* **109**:650 (1987).
135. S. E. Wierschke, J. Sandrasekhar, and W. L. Jorgensen, *J. Am. Chem. Soc.* **107**:1496 (1985).
136. D. Grafstein, *J. Am. Chem. Soc.* **77**:6650 (1955).
137. I. Fleming and I. Paterson, *Synthesis* **1979**:445.
138. J. P. Godschalx and J. K. Stille, *Tetrahedron Lett.* **24**:1905 (1983).
139. A. Hosomi, H. Iguchi, M. Endo, and H. Sakurai, *Chem. Lett.* **1979**:977.

General References

A. F. Cockerill and R. G. Harrison, *The Chemistry of Double-Bonded Functional Groups*, Part 1, S. Patai, ed., John Wiley & Sons, New York, 1977, Chapter 4.

G. V. Boyd, in *The Chemistry of Triple-Bonded Functional Groups*, Supplement 2, S. Patai, ed., John Wiley & Sons, New York, 1994, Chapter 6.

P. B. D. de la Mare and R. Bolton, *Electrophilic Additions to Unsaturated Systems*, 2nd ed., Elsevier, New York, 1982.

R. C. Fahey in *Topics in Stereochemistry*, Vol. 3, E. L. Eliel and N. L. Allinger, eds., Wiley-Interscience, New York, 1968, pp 237–342.

J. G. Gandler, in *The Chemistry of Double-Bonded Functional Groups*, Supplement A, Vol. 2, S. Patai, ed., John Wiley & Sons, New York, 1989, Chapter 12.

G. H. Schmid, in *The Chemistry of the Carbon–Carbon Triple Bond*, Part 1, S. Patai, ed., John Wiley & Sons, New York, 1978, Chapter 8.

G. H. Schmid in *The Chemistry of Double-Bonded Functional Groups*, Supplement A, Vol. 2, S. Patai, ed., John Wiley & Sons, New York, 1989, Chapter 11.

G. H. Schmid and D. G. Garratt, in *The Chemistry of Double-Bonded Functional Groups*, Part 2, S. Patai, ed., John Wiley & Sons, New York, 1977, Chapter 9.

P. J. Stang and F. Diederich, eds., *Modern Acetylene Chemistry*, VCH Publishers, Weinheim, 1995.

W. H. Saunders, Jr., and A. F. Cockerill, *Mechanisms of Elimination Reactions*, John Wiley & Sons, New York, 1973.

Problems

(References for these problems will be found on page 796.)

1. Which compounds in each of the following pairs will react faster with the indicated reagent?

   (a) 1-hexene or *E*-3-hexene with bromine in acetic acid

   (b) *cis*- or *trans*-$(CH_3)_3C$—⟨ ⟩—CH_2Br with potassium *t*-butoxide in *t*-butyl alcohol

   (c) 2-phenylpropene or 4-(2-propenyl)benzoic acid with sulfuric acid in water

   (d)
   toward acid-catalyzed hydration

   (e) $CH_3CH(CH_2)_3CH_3$ or $CH_3CH(CH_2)_3CH_3$ with potassium *t*-butoxide in
       $\overset{|}{SO_2C_6H_5}$ $\overset{|}{OSO_2C_6H_5}$ *t*-butyl alcohol

   (f) Br—⟨ ⟩—$C\equiv CH$ or CH_3—⟨ ⟩—$C\equiv CH$ with chlorine in acetic acid

   (g) ⟨ ⟩O or ⟨ ⟩—OC_2H_5 toward acid-catalyzed hydration

2. Predict the structure, including stereochemistry, of the product(s) of the following reactions. If more than one product is expected, indicate which will be the major

product and which the minor product.

(a)

$$\underset{\underset{CH_2OH}{|}}{\overset{\overset{CH_2OH}{|}}{PhCCH_2CH=CH_2}} \xrightarrow[CCl_4]{Br_2} C_{12}H_{15}O_2Br$$

(b)

$\xrightarrow{DCl} C_6H_8DCl$

(c) *erythro*

$\xrightarrow[\text{ethanol}]{C_2H_5O^-Na^+} C_{15}H_9Cl_3O$

(d)

$\xrightarrow{H_2O, \Delta} C_{10}H_{18}$

(e)

$\xrightarrow[CCl_4]{Cl_2} C_{11}H_{17}Cl$

(f)

$\xrightarrow[\text{xylene}]{K^+ \,^-OC(C_2H_5)_3} C_7H_{12}$

(g)

$\xrightarrow[\text{(2) NaCl}]{\overset{\text{(1) Hg(O_2CCH_3)_2}}{CH_3OH}} C_{10}H_{13}ClHgO$

(h) $PhC\equiv CCH_2CH_3 \xrightarrow[CH_3CO_2H,\ 25°C]{Cl_2} C_{10}H_{10}Cl_2 + C_{12}H_{13}ClO_2$

3. The reactions of the *cis* and *trans* isomers of 4-*t*-butylcyclohexyltrimethylammonium chloride with potassium *t*-butoxide in *t*-butanol have been compared. The *cis* isomer gives 90% 4-*t*-butylcyclohexane and 10% *N,N*-dimethyl-4-*t*-butylcyclohexylamine, while the *trans* isomer gives only the latter product in quantitative yield. Explain the different behavior of the two isomers.

4. For E2 eliminations in 2-phenylethyl systems with several different leaving groups, both the primary isotope effect and Hammett ρ values for the reactions are known. Deduce from these data the relationship between the location on the E2 transition state spectrum and the nature of the leaving group; i.e., deduce which system has the most E1-like transition state and which has the most E1cb-like. Explain your reasoning.

$$PhCH_2CH_2X \xrightarrow{C_2H_5O^-} PhCH=CH_2$$

X	k_H/k_D	ρ
Br	7.11	2.1
$OSO_2C_7H_7$	5.66	2.3
$^+S(CH_3)_2$	5.07	2.7
$^+N(CH_3)_3$	2.98	3.7

5. When 2-bromo-2-methylpentane is dissolved in DMF, the formation of 2-methyl-1-pentene (A) and 2-methyl-2-pentene (B) occurs. The ratio of alkenes formed is not

constant throughout the course of the reaction, however. Initially, the A : B ratio is about 1 : 1, but this drops to about 1 : 4 by the time the reaction is 25% complete and then remains fairly constant. In a similar reaction, but with NaBr present in excess, the A : B ratio is constant at about 1 : 5 throughout the reaction. Suggest an explanation for this observation.

6. For the reactions given below, predict the effect on the rate of the isotope substitution which is described. Explain the basis of your prediction.

   (a) The effect on the rate of dehydration of 1,2-diphenylethanol of introduction of deuterium at C-2.

   (b) The effect on the rate of dehydration of 1,2-diphenylethanol of using $D_2O–D_2SO_4$ in place of $H_2O–H_2SO_4$ as the reaction medium.

   (c) The effect on the rate of bromination of styrene when deuterium is introduced on the α carbon.

7. Predict the effect on the 1-butene: Z-2-butene: E-2-butene product ratio when the E2 elimination of *erythro*-3-*deuterio*-2-bromobutane is compared with that of 2-bromobutane. Which alkene(s) will increase in relative amount and which will decrease in relative amount? Explain the basis of your prediction.

8. Arrange the following compounds in order of increasing rate of acid-catalyzed hydration: ethylene, 2-cyclopropylpropene, 2-methylpropene, propene, 1-cyclopropyl-1-methoxyethene. Explain the basis of your prediction.

9. Discuss the factors that are responsible for the stereochemistry observed for the following reactions.

(a)

(b)

(c)

(d)

10. Explain the mechanistic basis of the following observations and discuss how the observation provides information about the reaction mechanism.

(a) When substituted 1-aryl-2-methyl-2-propyl chlorides react with sodium methoxide, a mixture of terminal and internal alkene is formed:

By using the product ratio, the overall rate can be dissected into the individual rates for formation of **2** and **3**. These rates are found to be substituent/dependent for formation of **2** ($\rho = +1.4$) but substituent independent for formation of **3** ($\rho = -0.1 \pm 0.1$). The reactions are both second-order, first-order in base and first-order in substrate.

(b) When 1,3-pentadiene reacts with DCl it forms more *E*-4-chloro-5-*deuterio*-2-pentene than *E*-4-chloro-1-*deuterio*-2-pentene

(c) When indene (**4**) is brominated in carbon tetrachloride, it gives some *syn* addition (~15%), but indenone (**5**) gives only *anti* addition under the same conditions.

(d) The acid-catalyzed hydration of allene gives acetone, not allyl alcohol or propionaldehyde.

(e) In the addition of hydrogen chloride to cyclohexene in acetic acid, the ratio of cyclohexyl acetate to cyclohexyl chloride drops significantly when tetramethylammonium chloride is added in increasing concentration. This effect is not observed with styrene.

(f) the ρ values for base-catalyzed elimination of HF from a series of 1-aryl-2-fluorethanes increase from the mono- to the di- and trifluoro compounds, as shown by the data below:

$$ArCH_2CF_3 \quad ArCH_2CHF_2 \quad ArCH_2CH_2F$$
$$\rho = +4.04 \quad \rho = +3.56 \quad \rho = +3.24$$

11. Suggest reasonable mechanisms for each of the following reactions:

(a)

(b)

(c) *syn*-5-trimethylsilyl-4-octanol $\xrightarrow[\text{CH}_3\text{CO}_2\text{H}]{\text{NaO}_2\text{CCH}_3}$ Z-4-octene

(d)

(e)

12. The rates of bromination of dialkylacetylenes are roughly 100 times greater than for the corresponding monosubstituted alkynes. For hydration, however, the rates of reaction are less than 10 times greater for disubstituted derivatives. Account for this observation by comparison of the mechanisms for bromination and hydration.

13. The bromination of 3-aroyloxycyclohexanes gives rise to a mixture of stereoisomeric and positionally isomeric addition products. The product composition for Ar = phenyl is shown. Account for the formation of each of the products and describe the factors which will affect the product ratio.

14. The reaction of substituted 1-arylethyl chlorides with $K^{+-}OC(CH_3)_3$ in DMSO does not follow a Hammett correlation. Instead, the reactivity order is p-NO$_2$ > p-MeO > p-CF$_3$ > p-CH$_3$ > H > p-Cl. What explanation can you offer for the failure to observe a Hammett relationship?

15. The ratio of terminal to internal alkene from decomposition of some sulfonium salts under alkaline conditions is as indicated:

n	X	term : inter
6	H	93 : 7
2	OH	100 : 0
3	OH	89 : 11
2	OPh	50 : 50
3	OPh	25 : 75

What explanation can you offer for the change in product ratio?

16. The Hammett correlation of the acid-catalyzed dehydration of 1,2-diarylethanols has been studied.

$$ArCHCH_2Ar' \xrightarrow{H^+} ArCH{=}CHAr'$$
$$\overset{|}{OH}$$

The equation that correlates the data resulting from substitution in the Ar and Ar′ rings is

$$\log k - -3.78(\sigma_{Ar}^+ + 0.23\sigma_{Ar'}) - 3.18$$

Give a rationalization for the form of this correlation equation. What information does it give regarding involvement of the Ar′ ring in the rate-determining step?

17. The addition of hydrogen chloride to olefins in nitromethane follows the rate expression

$$\text{rate} = k[HCl]^2[\text{alkene}]$$

Two other features of the reaction that have been established are the following: (1) When DCl is used instead of HCl, unreacted olefin recovered by stopping the reaction at 50% completion contains no deuterium; (2) added chloride salts ($R_4N^+Cl^-$) decrease the reaction rate, but other salts ($R_4N^+ClO_4^-$) do not. Write a mechanism for this reaction that is in accord with the data given.

18. In the bromination of styrene, a $\rho\sigma^+$ plot is noticeably curved. If the extremes of the curves are taken to represent straight lines, the curve can be resolved into two Hammett relationships with $\rho = -2.8$ for electron-attracting substituents and $\rho = -4.4$ for electron-releasing substituents. When the corresponding β-methylstyrenes are examined, a similarly curved $\sigma\rho$ plot is obtained. Furthermore, the stereospecificity of the reaction in the case of the β-methylstyrenes varies with the aryl substituents. The reaction is a stereospecific *anti* addition for strongly electron-attracting substituents but becomes only weakly stereoselective for electron-releasing substituents, e.g., 63% *anti*, 37% *syn*, for *p*-methoxy. Discuss the possible mechanistic basis for the Hammett plot curvature and its relationship to the stereochemical results.

19. The second-order rate constants for hydration and the kinetic solvent isotope effect for hydration of several 2-substituted 1,3-butadienes are given below. Discuss the information these data provide about the hydration mechanism.

$$\underset{\underset{R}{|}}{CH_2}{=}\overset{\overset{R}{|}}{C}{-}CH{=}CH_2 \xrightarrow[H_2O]{H^+} CH_3\overset{\overset{R}{|}}{\underset{\underset{OH}{|}}{C}}CH{=}CH_2 + CH_3\overset{\overset{R}{|}}{C}{=}CHCH_2OH$$

R	k_2 (M^{-1} s^{-1}) (25°C)	k_{H^+/D^+}
▷—	1.22×10^{-2}	1.2
CH_3	3.19×10^{-5}	1.8
Cl	2.01×10^{-8}	1.4
H	3.96×10^{-8}	1.8
C_2H_5O	6×10^{1}	—

20. The reaction of both E- and Z-2-butene with acetic acid to give 2-butyl acetate can be catalyzed by various strong acids. Using DBr, DCl, and CH_3SO_3D, in CH_3CO_2D, it was possible to demonstrate that the reaction proceeded largely with *anti* addition ($84\% \pm 2\%$). If the reaction was stopped short of completion, there was no interconversion of Z-2-butene with either E-2-butene or 1-butene. When CF_3SO_3D was used as the catalyst, several features of the reaction changed.

   (1) The recovered butene showed small amounts of conversion to 1-butene and partial isomerization to the stereoisomeric 2-butene.
   (2) The recovered 2-butene contained small amounts of deuterium.
   (3) The stereoselectivity was somewhat reduced (60%–70% *anti* addition).

   How do you account for the changes which occur when CF_3SO_3D is used as a catalyst, as compared with the other acids?

21. A comparison of rate and product composition of the products from reaction of t-butyl chloride with NaOMe in methanol and methanol–DMSO mixtures containing NaOMe has been done. Interpret the effect of the change of solvent composition and NaOMe concentration.

	100% MeOH			36.8% DMSO			64.2% DMSO		
		Product composition (%)			Product composition (%)			Product composition (%)	
[NaOMe] (M)	$k \times 10^2\,\mathrm{s}^{-1}$	Ether	Alkene	$k \times 10^4\,\mathrm{s}^{-1}$	Ether	Alkene	$k \times 10^4\,\mathrm{s}^{-1}$	Ether	Alkene
0.0	2.15	73.8	26.2	0.81	50	50	0.24	24	76
0.20	2.40			1.52			5.3	1	99
0.25	2.30	62.9	32.1						
0.30	2.26			1.90	10.5	89.5	10.3	0	100
0.40	2.36			2.65			17.5	0	100
0.50	2.56	58.6	41.4				24.1	0	100
0.70				4.11	1.1	98.9			
0.75	2.58	51.7	48.3						
0.80				4.59					
0.90	2.64			6.16	4.1	95.9			
1.00	2.74	52.2	47.8	6.81	3.8	96.2			

Carbanions and Other Nucleophilic Carbon Species

Introduction

This chapter is concerned with carbanions, which are the conjugate bases (in the Brønsted sense) formed by deprotonation of a carbon atom. Carbanions vary widely in stability, depending on the hybridization of the carbon atom and the ability of substituent groups to stabilize negative charge. In the absence of substituents that are effective at stabilizing the charge, proton removal from a C−H bond is difficult. Carbanions are very useful in synthesis, since formation of new carbon–carbon bonds often requires a nucleophilic carbon species. There has therefore been much study of the methods of generating carbanions and of substituent effects on stability and reactivity.

7.1. Acidity of Hydrocarbons

In the discussion of the relative acidity of carboxylic acids in Chapter 1, the thermodynamic acidity, expressed as the acid dissociation constant, was taken as the measure of acidity. It is straightforward to determine dissociation constants of such acids in aqueous solution by measurement of the titration curve with a pH-sensitive electrode (pH meter). Determination of the acidity of carbon acids is more difficult. Because most are very weak acids, very strong bases are required to cause deprotonation. Water and alcohols are far more acidic than most hydrocarbons and are unsuitable solvents for generation of hydrocarbon anions. Any strong base will deprotonate the solvent rather than the hydrocarbon. For synthetic purposes, aprotic solvents such as ether, tetrahydrofuran (THF), and dimethoxyethane (DME) are used, but for equilibrium measurements solvents that promote dissociation of ion pairs and ion clusters are preferred. Weakly acidic solvents such as DMSO and cyclohexylamine are used in the preparation of moderately basic carbanions. The high polarity and cation-solvating ability of DMSO facilitate

406

CHAPTER 7
CARBANIONS AND
OTHER
NUCLEOPHILIC
CARBON SPECIES

dissociation of ion pairs so that the equilibrium data obtained refer to the free ions rather than to ion aggregates.

The basicity of a solvent–base system may be specified by a basicity constant H_-. The value of H_- corresponds essentially to the pH of strongly basic nonaqueous solutions. The larger the value of H_-, the greater is the proton-removing ability of the medium. Use of a series of overlapping indicators permits assignment of H_- values to solvent–base systems and allows pK's to be determined over a range of 0–30 pK units. The indicators employed include substituted anilines and arylmethanes, which have significantly different electronic (UV-VIS) spectra in their neutral and anionic forms. Table 7.1 presents H_- values for some representative solvent–base systems.

The acidity of a hydrocarbon can be determined in an analogous way.[1] If the electronic spectra of the neutral and anionic forms are sufficiently different, the concentrations of each can be determined directly, and the equilibrium constant for

$$RH + B^- \rightleftharpoons R^- + BH$$

is related to pK by the equation

$$pK_{RH} = H_- + \log\frac{[RH]}{[R^-]}$$

A measurement of the ratio [RH]:[R$^-$] at a known H_- yields the pK. If, as is frequently the case, the electronic spectrum of the hydrocarbon and its anion are not sufficiently different, one of the indicators is used and its spectrum is monitored. The equilibrium established between the indicator and hydrocarbon in the basic medium

$$RH + In^- \rightleftharpoons R^- + HIn$$

then provides a way to relate the concentrations that are not directly measured, [RH] and [R$^-$], to quantities that are, [HIn] and [In$^-$].

When the acidities of hydrocarbons are discussed in terms of the relative stabilities of neutral and anionic forms, particularly with respect to the extent of electron delocalization in the anion, the appropriate data are equilibrium acidity measurements. We have just seen

Table 7.1. Values of H_- for Some Representative Solvent–Base Systems[a]

Solution	H_-[b]
5 M KOH	15.5
10 M KOH	17.0
15 M KOH	18.5
0.01 M NaOMe in 1:1 DMSO–MeOH	15.0
0.01 M NaOMe in 10:1 DMSO–MeOH	18.0
0.01 M NaOEt in 20:1 DMSO–EtOH	21.0

a. Values are rounded to the nearer 0.5 pH unit; this is typical of the range of disagreement using different indicator series.
b. Selected values from J. R. Jones, *The Ionization of Carbon Acids*, Academic Press, New York, 1973, Chapter 6.

1. D. Dolman and R. Stewart, *Can. J. Chem.* **45**:911 (1967); E. C. Steiner and J. M. Gilbert, *J. Am. Chem. Soc.* **87**:382 (1965); K. Bowden and R. Stewart, *Tetrahedron* **21**:261 (1965).

how such data may be obtained, but in many instances it is not possible to obtain equilibrium data. In such cases, it may be possible to compare the rates of deprotonation, that is, the *kinetic acidity*. Such comparison can be made between different protons in the same compound or between two different compounds. This can be done by following an isotopic exchange. In the presence of a source of deuterons, such as deuterated solvent (S−D), the rate of incorporation of deuterium into the organic molecule is a measure of the rate of carbanion formation.[2]

$$R-II + D^- \rightleftharpoons R^- + DII$$
$$R^- + S-D \rightleftharpoons R-D + S^-$$
$$S^- + BH \rightleftharpoons SH + B^-$$

It has been found that there is often a correlation between the rate of deprotonation (kinetic acidity) and the thermodynamic stability of the carbanion (thermodynamic acidity). Because of this relationship, kinetic measurements can be used to construct orders of hydrocarbon acidities. These kinetic measurements have the advantage of not requiring the presence of a measurable concentration of the carbanion at any time; instead, the relative ease of carbanion formation is judged from the rate at which exchange occurs. This method is therefore applicable to very weak acids, for which no suitable base will generate a measurable carbanion concentration.

The kinetic method of determining relative acidity suffers from one serious complication, however. This complication has to do with the fate of the ion pair that is formed immediately on removal of the proton.[3] If the ion pair separates and diffuses into the solution rapidly, so that each deprotonation results in exchange, the exchange rate is an accurate measure of the rate of deprotonation. Under many conditions of solvent and base, however, an ion pair may return to reactants at a rate exceeding protonation of the carbanion by the solvent. This phenomenon is called *internal return*:

$$R_3C-H + M^+B^- \underset{\substack{\text{internal} \\ \text{return}}}{\overset{\text{ionization}}{\rightleftharpoons}} [R_3C^-M^+ + BH] \xrightarrow{\text{separation}} R_3C^- + M^+ + BH$$

$$\xrightarrow[\text{exchange}]{S-D} R_3CD + S^-$$

When internal return occurs, a deprotonation event escapes detection because exchange does not result. One experimental test for the occurrence of internal return is racemization at chiral carbanionic sites that occurs without exchange. Even racemization cannot be regarded as an absolute measure of the deprotonation rate because, under some conditions, hydrogen–deuterium exchange has been shown to occur with retention of configuration. Because of these uncertainties about the fate of ion pairs, it is important that a linear relationship between exchange rates and equilibrium acidity be established for representative examples of the compounds under study. A satisfactory correlation provides a basis for using kinetic acidity data for compounds of that structural type.

The pK values determined are influenced by the solvent and other conditions of the measurement. The nature of the solvent in which the extent or rate of deprotonation is determined has a significant effect on the apparent acidity of the hydrocarbon. In general,

2. A. I. Shatenshtein, *Adv. Phys. Org. Chem.* **1**: 155 (1963).
3. W. T. Ford, E. W. Graham, and D. J. Cram, *J. Am. Chem. Soc.* **89**:155 (1963); D. J. Cram, C. A. Kingsbury, and B. Rickborn, *J. Am. Chem. Soc.* **83**:3688 (1961).

the extent of ion pairing is primarily a function of the ability of the solvent to solvate the ionic species. Ion pairing is greatest in nonpolar solvents such as ethers. In dipolar aprotic solvents, especially DMSO, ion pairing is much less likely to be significant.[4] The identity of the cation present also can have a significant effect if ion pairs are present. Because of these factors, the numerical pK values are not absolute and are specific to the solvent. Nevertheless, they provide a useful measure of relative acidity. The two solvents that have been used for most quantitative measurements on weak carbon acids are cyclohexylamine and DMSO.

An extensive series of hydrocarbons has been studied in cyclohexylamine, with the use of cesium cyclohexylamide as base. For many of the compounds studied, spectroscopic measurements were used to determine the relative extent of deprotonation of two hydrocarbons and thus establish relative acidity.[5] For other hydrocarbons, the acidity was derived by kinetic measurements. It was shown that the rate of tritium exchange for a series of related hydrocarbons is linearly related to the equilibrium acidities of these hydrocarbons in the solvent system. This method was used to extend the scale to hydrocarbons such as toluene for which the exchange rate, but not equilibrium data, can be obtained.[6] Representative values of some hydrocarbons with pK values ranging from 16 to above 40 are given in Table 7.2.

The pK values of a wide variety of organic compounds have been determined in DMSO.[7] Some of these values are included in Table 7.2. It is not expected that the values will be numerically identical with aqueous pK_a values, but for most compounds the same relative order of acidity is observed for hydrocarbons of similar structural type.

For synthetic purposes, carbanions are usually generated in ether solvents, often THF or DME. There are relatively few quantitative data available on hydrocarbon acidity in such solvents. Table 7.2 contains a few entries for Cs^+ salts. The numerical values are scaled with reference to the pK of 9-phenylfluorene.[8]

Some of the relative acidities in Table 7.2 can be easily explained. The order of decreasing acidity $Ph_3CH > Ph_2CH_2 > PhCH_3$, for example, reflects the ability of each successive phenyl group to delocalize the negative charge on carbon and thereby stabilize the carbanion. The much greater acidity of fluorene relative to dibenzocycloheptatriene (entries 5 and 6) reflects the aromatic stabilization of the cyclopentadienide ring in the fluorene anion. Cyclopentadiene is an exceptionally acidic hydrocarbon, comparable in acidity to simple alcohols, because of the aromatic stabilization of the anion.

Allylic conjugation stabilizes carbanions, and pK values of 43 (in cyclohexylamine)[9] and 47–48 (in THF–HMPA)[10] have been determined for propene. On the basis of exchange rates with cesium cyclohexylamide, cyclohexene and cycloheptene have been found to have pK values of about 45 in cyclohexylamine.[11] The hydrogens on the sp^2

4. E. M. Arnett, T. C. Moriarity, L. E. Small, J. P. Rudolph, and R. P. Quirk, *J. Am. Chem. Soc.* **95**:1492 (1973); T. E. Hogen-Esch and J. Smid, *J. Am. Chem. Soc.* **88**:307 (1966).

5. A. Streitwieser, Jr., J. R. Murdoch, G. Hafelinger, and C. J. Chang, *J. Am. Chem. Soc.* **95**:4248 (1973); A. Streitwieser, Jr., E. Ciuffarin, and J. H. Hammons, *J. Am. Chem. Soc.* **89**:63 (1967); A. Streitwieser, Jr., E. Juaristi, and L. L. Nebenzahl, in *Comprehensive Carbanion Chemistry*, Part A, E. Buncel, and T. Durst, eds., Elsevier, New York, 1980, Chapter 7.

6. A. Streitwieser, Jr., M. R. Granger, F. Mares, and R. A. Wolf, *J. Am. Chem. Soc.* **95**:4257 (1973).

7. F. G. Bordwell, *Acc. Chem. Res.* **21**:456 (1988).

8. D. A. Bors, M. J. Kaufman, and A. Streitwieser, Jr., *J. Am. Chem. Soc.* **107**:6975 (1985).

9. D. W. Boerth and A. Streitwieser, Jr., *J. Am. Chem. Soc.* **103**:6443 (1981).

10. B. Jaun, J. Schwarz, and R. Breslow, *J. Am. Chem. Soc.* **102**:5741 (1980).

11. A. Streitwieser, Jr., and D. W. Boerth *J. Am. Chem. Soc.* **100**:755 (1978).

Table 7.2. Acidities of Some Hydrocarbons

	Hydrocarbon	pK^a		
		Cs^+ (cyclohexylamine)[b]	Cs^+ (THF)[c]	K^+ (DMSO)[d]
1	PhCH$_2$$-$H	41.2	40.9	43
2	(H$_3$C$-$⟨⟩$-$)$_2$ CH$-$H	35.1	33.1	
3	(Ph)$_2$CH$-$H	33.4	33.3	32.3
4	(Ph)$_3$C$-$H	31.4	31.3	30.6
5		31.2		
6		22.7	22.9	22.6
7		19.9		20.1
8		18.5	18.2	17.9
9		16.6[e]		18.1

a. Values refer to indicated solvent medium containing salt of cation specified.
b. A. Streitwieser, Jr., J. R. Murdoch, G. Häfelinger and C. J. Chang, *J. Am. Chem. Soc.* **95**:4248 (1973); A. Streitwieser, Jr., E. Ciuffarin, and J. H. Hammons, *J. Am. Chem. Soc.* **89**:63 (1967); A. Streitwieser, Jr., and F. Guibe, *J. Am. Chem. Soc.* **100**:4532 (1978).
c. M. J. Kaufman, S. Gronert and A. Streitwieser, *J. Am. Chem. Soc.* **110**:2829 (1988); A. Streitwieser, J. C. Ciula, J. A. Krom, and G. Thiele, *J. Org. Chem.* **56**:1074 (1991).
d. C. D. Ritchie and R. E. Uschold, *J. Am. Chem. Soc.* **90**:2821 (1968); F. G. Bordwell, J. E. Bartmess, G. E. Drucker, Z. Margolin, and W. S. Matthews, *J. Am. Chem. Soc.* **97**:3226 (1975); W. S. Matthews, J. E. Bares, J. E. Bartmess, F. G. Bordwell, F. J. Cornforth, G. E. Drucker, Z. Margolin, R. J. McCallum, G. J. McCollum, and N. R. Vanier, *J. Am. Chem. Soc.* **97**:7006 (1975); F. G. Bordwell, G. E. Drucker, and H. E. Fried, *J. Org. Chem.* **46**:632 (1981); F. G. Bordwell, *Acc. Chem. Res.* **21**:456, 463 (1988).
e. A. Streitwieser, Jr., and L. L. Nebenzahl, *J. Am. Chem. Soc.* **98**:2188 (1976); in water, the pK_a of cyclopentadiene is 16.0.

410

CHAPTER 7
CARBANIONS AND
OTHER
NUCLEOPHILIC
CARBON SPECIES

carbons in benzene and ethylene would be expected to be more acidic than the hydrogens in saturated hydrocarbons. A pK of 43 has been estimated for benzene on the basis of extrapolation from a series of fluorobenzenes.[12] Electrochemical measurements have been used to establish a lower limit of about 46 for the pK of ethylene.[10]

For saturated hydrocarbons, exchange is too slow and reference points are so uncertain that direct determination of pK values by exchange measurements is not feasible. The most useful approach to obtain pK data for such hydrocarbons involves making a measurement of the electrochemical potential for the reaction

$$R^\cdot + e^- \rightleftharpoons R^-$$

From this value and known C$-$H bond dissociation energies, pK values can be calculated. Early application of these methods gave estimates of the pK's of toluene and propene of about 45 and 48, respectively. Methane was estimated to have a pK in the range of 52–62.[10] Electrochemical measurements in DMF have given the results shown in Table 7.3.[13] These measurements put the pK of methane at about 48, with benzylic and allylic stabilization leading to values of 39 and 38 for toluene and propene, respectively. The electrochemical values overlap with the pK_{DMSO} scale for compounds such as diphenylmethane and triphenylmethane.

Terminal alkynes are among the most acidic of the hydrocarbons. For example, in DMSO, phenylacetylene is found to have a pK near 26.5.[14] In cyclohexylamine, the value is 23.2.[15] An estimate of the pK in aqueous solution of 20 is based on a Brønsted relationship.[16] The relatively high acidity of acetylenes is associated with the greater s character of the C$-$H bond. The s character is 50%, as opposed to 25% in sp^3 bonds. The electrons in orbitals with high s character experience decreased shielding from the nuclear charge. The carbon is therefore effectively more electronegative, as viewed from

Table 7.3. pK Values for Less Acidic Hydrocarbons

Hydrocarbon	pK (DMF)^a
CH$_3$$-$H	48
CH$_3$CH$_2$$-$H	51
⬠$-$H	49
⬡$-$H	49
CH$_2$=CHCH$_2$$-$H	38
PhCH$_2$$-$H	39
(Ph)$_2$CH$-$H	31
(Ph)$_3$C$-$H	29

a. K. Daasbjerg, *Acta Chem. Scand.* **49**:878 (1995).

12. A. Streitwieser, Jr., P. J. Scannon, and H. M. Niemeyer, *J. Am. Chem. Soc.* **94**:7936 (1972).
13. K. Daasbjerg, *Acta Chem. Scand.* **49**:878 (1995).
14. F. G. Bordwell and W. S. Matthews, *J. Am. Chem. Soc.* **96**:1214 (1974).
15. A. Streitwieser, Jr., and D. M. E. Reuben, *J. Am. Chem. Soc.* **93**:1794 (1971).
16. D. B. Dahlberg, M. A. Kuzemko, Y. Chiang, A. J. Kresge, and M. F. Powell, *J. Am. Chem. Soc.* **105**:5387 (1983).

the proton sharing an *sp* hybrid orbital, and hydrogens on *sp* carbons exhibit greater acidity. This same effect accounts for the relatively high acidity of the hydrogens on cyclopropane rings, which have increased *s* character in the C−H bonds.[17]

Knowledge of the structure of carbanions is important to understanding the stereo-chemistry of their reactions. Theoretical calculations at the 4-31G level indicate a pyramidal geometry at carbon in the methyl and ethyl anions. The optimum H−C−H angle in these two carbanions is calculated to be 97–100°. An interesting effect is observed in that the proton affinity (basicity) of methyl anion decreases in a regular manner as the H−C−H angle is decreased.[18] This increase in acidity with decreasing internuclear angle has a parallel in small-ring compounds, in which the acidity of hydrogens is substantially greater than in compounds having tetrahedral geometry at carbon. Pyramidal geometry at carbanions can also be predicted on the basis of qualitative considerations of the orbital occupied by the unshared electron pair. In a planar carbanion, the lone pair would occupy a *p* orbital. In a pyramidal geometry, the orbital has substantial *s* character. Because the electron pair is of lower energy in an orbital with some *s* character, it is predicted that a pyramidal geometry will be favored.

As was discussed in Section 5.4 for carbocations, measurements in the gas phase, which eliminate the effect of solvation, show structural trends that parallel those followed in solution measurements but with larger absolute energy differences. Table 7.4 gives some data for the ΔH of proton dissociation for key hydrocarbons. These data show a correspondence with hybridization and delocalization effects observed in solution. The very large heterolytic dissociation energies reflect both the inherent instability of the carbanions and also the electrostatic attraction between the oppositely charged carbanion and proton. For comparison, enthalpy measurements in DMSO using KO-*t*-Bu or KCH_2SOCH_3 as base give values of -15.4 and -18.2 kcal/mol, respectively, for fluorene, a hydrocarbon with a pK of about 20.[19]

The stereochemistry observed in hydrogen-exchange reactions of carbanions is very dependent on the conditions under which the anion is formed and trapped by proton

Table 7.4. Enthalpy of Proton Dissociation for Some Hydrocarbons (Gas Phase)

	ΔH (kcal/mol)[a]
CH_4	418.8
$CH_2{=}CH_2$	407.5
	411.5
	400.8
—CH_3	381

a. S. T. Graul and R. R. Squires, *J. Am. Chem. Soc.* **112**:2517 (1990).

17. A. Streitwieser, Jr., R. A. Caldwell, and W. R. Young, *J. Am. Chem. Soc.* **9**:529 (1969).

18. A. Streitwieser, Jr., and P. H. Owens, *Tetrahedron Lett.* **1973**:5221; A. Streitwieser, Jr., P. H. Owens, R. A. Wolf, and J. E. Williams, Jr., *J. Am. Chem. Soc.* **96**:5448 (1974); E. D. Jemmis, V. Buss, P. v. R. Schleyer, and L. C. Allen, *J. Am. Chem. Soc.* **98**:6483 (1976).

19. E. M. Arnett and K. G. Venkatasubramaniam, *J. Org. Chem.* **48**:1569 (1983).

412

CHAPTER 7
CARBANIONS AND
OTHER
NUCLEOPHILIC
CARBON SPECIES

transfer. The dependence on solvent, counterion, and base is the result of ion-pairing effects. The base-catalyzed cleavage of **1** is noteworthy. The anion of **1** is cleaved at elevated temperature to 2-butanone and 2-phenyl-2-butyl anion, which under the conditions of the reaction is protonated by the solvent.

Use of resolved **1** allows the stereochemical features of the anion to be probed by measuring the enantiomeric purity of the 2-phenylbutane product. Retention of configuration was observed in solvents of low dielectric constant, whereas increasing amounts of inversion occurred as the proton-donating ability and dielectric constant of the solvent increased. Cleavage of **1** with potassium *t*-butoxide in benzene gave 2-phenylbutane with 93% net retention of configuration. The stereochemical course changed to 48% net inversion of configuration when potassium hydroxide in ethylene glycol was used. In DMSO with potassium *t*-butoxide as base, completely racemic 2-phenylbutane was formed.[20] The retention of configuration in benzene presumably reflects a short lifetime for the carbanion in a tight ion pair. Under these conditions, the carbanion does not become symmetrically solvated before proton transfer from either the protonated base or the ketone occurs. The solvent benzene would not be an effective proton donor. In ethylene glycol, the solvent provides a good proton source, and, since net inversion is observed, the protonaton must be occurring on an unsymmetrically solvated species that favors backside protonation. The racemization that is observed in DMSO indicates that the carbanion has a sufficient lifetime to become symmetrically solvated. The stereochemistry observed in the three solvents is in good accord with their solvating properties. In benzene, which is nonpolar, reaction occurs primarily through ion pairs. Ethylene glycol provides a ready source of protons, and fast proton transfer accounts for the observed inversion. DMSO promotes ion-pair dissociation and equilibration, as indicated by the observed racemization.

The stereochemistry of hydrogen–deuterium exchange at the chiral carbon in 2-phenylbutane shows a similar trend. When potassium *t*-butoxide is used as the base, the exchange occurs with retention of configuration in *t*-butanol, but racemization occurs in DMSO.[21] The retention of configuration is visualized as occurring through an ion pair in which a solvent molecule coordinated to the metal ion acts as the proton donor:

20. D. J. Cram, A. Langemann, J. Allinger, and K. R. Kopecky, *J. Am. Chem. Soc.* **81**:5740 (1959).
21. D. J. Cram, C. A. Kingsbury, and B. Rickborn, *J. Am. Chem. Soc.* **83**:3688 (1961).

In DMSO, symmetrical solvation is achieved prior to protonation, and complete racemization occurs.

The organometallic derivatives of lithium, magnesium, and other strongly electropositive metals have some of the properties expected for salts of carbanions but are significantly covalent in character. Because of the very weak acidity of most hydrocarbons, the simple organolithium compounds usually cannot be prepared by proton-transfer reactions. Instead, the most general preparative methods start with the corresponding halogen compound:

$$CH_3I + 2\,Li \longrightarrow CH_3Li + LiI$$

$$CH_3(CH_2)_3Br + 2\,Li \longrightarrow CH_3(CH_2)_3Li + LiBr$$

Other preparative methods will be considered in Chapter 7 of Part B.

Organolithium compounds are strong bases, as would be expected for the carbanions derived from simple hydrocarbons. Organolithium compounds react rapidly with any molecule having an acidic $-OH$, $-NH$, or $-SH$ group to form the hydrocarbon. Organolithium compounds derived from saturated hydrocarbons are extremely strong bases. Accurate pK values are not known but would range upward from the estimate of ~50 for methane. The order of basicity $CH_3Li < CH_3(CH_2)_3Li < (CH_3)_3CLi$ is expected on the basis of the electron-releasing effect of alkyl substituents and is consistent with increasing reactivity as bases in proton-abstraction reactions in the order $CH_3Li < CH_3(CH_2)_3Li < (CH_3)_3CLi$. Phenyl-, methyl, n-butyl-, and t-butyllithium are certainly all stronger bases than the anions of the hydrocarbons listed in Table 7.2. Unlike proton transfers involving oxygen, nitrogen, or sulfur atoms, proton transfer between carbon atoms is usually not a fast reaction. Thus, even though t-butyllithium is thermodynamically capable of deprotonating toluene, for example, the reaction is quite slow in a hydrocarbon solvent medium. In part, the reason is that the organolithium compounds exist as tetramers, hexamers, and higher aggregates in hydrocarbon and ether solvents.[22]

The gas-phase structure of monomeric methyllithium has been determined to be tetrahedral with an $H-C-H$ bond angle of $106°$.[23] These structural parameters are close to those calculated at the 6-311G*/MP2 level of theory.[24] In solution, organolithium compounds exist as aggregates, with the degree of aggregation depending on the structure of the organic group and the solvent. The nature of the species present in solution can be studied by low-temperature NMR spectroscopy. n-Butyllithium, for example, in THF is present as a tetramer–dimer mixture.[25] The tetrameric species is dominant.

$$[(BuLi)_4 \cdot (THF)_4] + 4\,THF \rightleftharpoons 2[(BuLi)_2 \cdot (THF)_4]$$

In solutions of n-propyllithium in cyclopropane at $0°C$, the hexamer is the main species, but higher aggregates are present at lower temperatures.[22] The reactivity of the organo-

22. G. Fraenkel, M. Henrichs, J. M. Hewitt, B. M. Su, and M. J. Geckle, *J. Am. Chem. Soc.* **102**:3345 (1980); G. Fraenkel, M. Henrichs, M. Hewitt, and B. M. Su, *J. Am. Chem. Soc.* **106**:255 (1984).
23. D. B. Grotjahn, T. C. Pesch, J. Xin, and L. M. Ziurys, *J. Am. Chem. Soc.* **119**:12368 (1997).
24. E. Kaufmann, K. Raghavachari, A. E. Reed, and P. v. R. Schleyer, *Organometallics* **7**:1597 (1988).
25. D. Seebach, R. Hassig, and J. Gabriel, *Helv. Chim. Acta* **66**:308 (1983); J. P. McGarrity and C. A. Ogle, *J. Am. Chem. Soc.* **107**:1805 (1984).

414

CHAPTER 7
CARBANIONS AND
OTHER
NUCLEOPHILIC
CARBON SPECIES

lithium compounds is increased by adding molecules capable of solvating the organometallic species. Tetramethylethylenediamine (TMEDA) has been commonly used for organolithium reagents. This tertiary amine can chelate lithium. The resulting complexes generally are able to effect deprotonation at accelerated rates.[26]

In the case of phenyllithium, it has been possible to demonstrate by NMR studies that the compound is tetrameric in 1:2 ether–cyclohexane but dimeric in 1:9 TMEDA–cyclohexane.[27] X-ray crystal structure determinations have been done on both dimeric and tetrameric structures. A dimeric structure crystallizes from hexane containing TMEDA.[28] This structure is shown in Fig. 7.1A. A tetrameric structure incorporating four ether molecules forms from ether–hexane solution.[29] This structure is shown in Fig. 7.1B. There is a good correspondence between the structures that crystallize and those indicated by the NMR studies.

Tetrameric structures based on distorted cubic structures are also found for $(CH_3Li)_4$ and $(C_2H_5Li)_4$.[30] These tetrameric structures can also be represented as being based on a

(A) [(PhLi)₂(TMEDA)₂]

(B) [(PhLi)₄(Et₂O)₄]

Fig. 7.1. Crystal structures of phenyllithium: (A) dimeric structure incorporating tetramethylethylenediamine; (B) tetrameric structure incorporating diethyl ether. (Reproduced from Refs. 28 and 29 with permission of Wiley-VCH and the American Chemical Society.)

26. G. G. Eberhardt and W. A. Butte, *J. Org. Chem.* **29:**2928 (1964); R. West and P. C. Jones, *J. Am. Chem. Soc.* **90:**2656 (1968).

27. L. M. Jackman and L. M. Scarmoutzos, *J. Am. Chem. Soc.* **106:**4627 (1984).

28. D. Thoennes and E. Weiss, *Chem. Ber.* **111:**3157 (1978).

29. H. Hope and P. P. Power, *J. Am. Chem. Soc.* **105:**5320 (1983).

30. E. Weiss and E. A. C. Lucken, *J. Organomet. Chem.* **2:**197 (1964); E. Weiss and G. Hencken, *J. Organomet. Chem.* **21:**265 (1970); H. Köster, D. Thoennes, and E. Weiss, *J. Organomet. Chem.* **160:**1 (1978); H. Dietrich, *Acta Crystallogr.* **16:**681 (1963); H. Dietrich, *J. Organomet. Chem.* **205:**291 (1981).

Fig. 7.2. Crystal structures of *n*-butyllithium. (A) (*n*-BuLi · TMEDA)₂; (B) [(*n*-BuLi · THF)₄ · hexane]; (C) [(*n*-BuLi · DME)₄]∞; (D) [(*n*-BuLi)₄ · TMEDA]∞. Hydrogen atoms have been omitted. (Reproduced from *J. Am. Chem. Soc.*, **115**, 1568, 1573 (1993).

416

CHAPTER 7
CARBANIONS AND
OTHER
NUCLEOPHILIC
CARBON SPECIES

tetrahedron of lithium ions with each face occupied by a carbanion.

The THF solvate of lithium *t*-butylacetylide is another example of a tetrameric structure.[31]

Crystal structure determination has also been done with *n*-butyllithium. A 4:1 *n*-BuLi:TMEDA complex is a tetramer accommodating two TMEDA molecules, which, rather than chelating a lithium, link the tetrameric units. The 2:2 *n*-BuLi:TMEDA complex has a structure similar to that of [PhLi]₂:[TMEDA]₂. Both 1:1 *n*-BuLi:THF and 1:1 *n*-BuLi:DME complexes are tetrameric with ether molecules coordinated at each lithium (Fig. 7.2).[32] These and many other organolithium structures have been compared in a review of this topic.[33]

The relative slowness of removal of protons from carbon acids by organolithium reagents is probably due to the compact character of the carbon–lithium clusters. Because the electrons of the carbanion are tightly associated with the cluster of lithium cations, some activation energy is required to break the bond before the carbanion can act as a base. This relative sluggishness of organometallic compounds as bases permits important reactions in which the organometallic species acts as a nucleophile in preference to functioning as a strong base. Reactions involving addition of organolithium and organo-magnesium compounds to carbonyl groups in aldehydes, ketones, and esters are important examples. As will be seen in the next section, carbonyl compounds are much more acidic than hydrocarbons. Nevertheless, in most cases, the proton-transfer reaction is slower than nucleophilic attack at the carbonyl group. It is this feature of the reactivity of organome-tallics that permits the very extensive use of organometallic compounds in organic synthesis. The reactions of organolithium and organomagnesium compounds with carbonyl compounds will be discussed in a synthetic context in Chapter 7 of Part B.

7.2. Carbanions Stabilized by Functional Groups

Functional groups that permit the negative charge of a carbanion to be delocalized to a more electronegative atom such as oxygen cause very large increases in the acidity of C−H bonds. Among the functional groups that exert a strong stabilizing effect on carbanions are the carbonyl, nitro, sulfonyl, and cyano groups. Both polar and resonance effects are involved in the ability of these functional groups to stabilize the negative charge. Perhaps the best basis for comparing these groups is the data on the various substituted methanes.[34] Bordwell and co-workers determined the relative acidities of the substituted methanes with reference to aromatic hydrocarbon indicators in DMSO. The data are given in Table 7.5. The ordering $NO_2 > C=O > CO_2R \sim SO_2 \sim CN > CONR_2$

31. W. Neuberger, E. Weiss, and P. v. R. Schleyer, quoted in Ref. 33.
32. M. A. Nichols and P. G. Williard, *J. Am. Chem. Soc.* **115**:1568 (1993); N. D. R. Barnett, R. E. Mulvey, W. Clegg, and P. A. O'Neil, *J. Am. Chem. Soc.* **115**:1573 (1993).
33. W. N. Setzer and P. v. R. Schleyer, *Adv. Organomet. Chem.* **24**:353 (1985).
34. F. G. Bordwell and W. S. Matthews, *J. Am. Chem. Soc.* **96**:2116 (1974); W. S. Matthews, J. E. Bares, J. E. Bartmess, F. G. Bordwell, F. J. Cornforth, G. E. Drucker, Z. Margolin, R. J. McCallum, G. J. McCollum, and N. R. Vanier, *J. Am. Chem. Soc.* **97**:7006 (1975).

Table 7.5. Equilibrium Acidities of Substituted Methanes in Dimethyl Sulfoxide[a]

Compound	pK
CH_3NO_2	17.2
CH_3COPh	24.7
CH_3COCH_3	26.5
CH_3SO_2Ph	29.0
$CH_3CO_2C_2H_5$	30.5[b]
$CH_3SO_2CH_3$	31.1
CH_3CN	31.3
$CH_3CON(C_2H_5)_2$	34.5[b]

a. Except where noted otherwise, data are from W. S. Matthews, J. E. Bares, J. E. Bartmess, F. G. Bordwell, F. J. Cornforth, G. E. Drucker, Z. Margolin, R. J. McCallum, G. J. McCollum, and N. R. Vanier, *J. Am. Chem. Soc.* **97**:7006 (1975).
b. Accurate to ±0.5; F. G. Bordwell and H. E. Fried, *J. Org. Chem.* **46**:4327 (1981).

for anion stabilization emerges from these data.

Carbanion-stabilizing effects have been calculated at several levels of theory. Table 7.6 gives some gas-phase data. The AM1 and PM3 semiempirical calculations have also been done in water. The order $NO_2 > CH=O > CN > Ph > CH_2=CH$ is in accord with the experimental trends and reflects charge delocalization. The electronegative substituents F, OH, and NH_2 are stabilizing by virtue of polar effects. The small stabilization provided by CH_3 is presumably a polarization effect.

Table 7.7 gives some additional pK data. The presence of two electron-withdrawing groups further stabilizes the negative charge. Pentane-2,4-dione, for example, has a pK around 13 in DMSO. β-Diketones are even more acidic in hydroxylic solvents and the carbanions can be generated using the conjugate bases of hydroxylic solvents such as water or alcohols, which have pK values of 15–20. Stronger bases are required for compounds that have a single stabilizing functional group. Alkali-metal salts of ammonia or amines or sodium hydride are sufficiently strong bases to form carbanions from most ketones, aldehydes, and esters. The anion of di-*i*-propylamine is a popular strong base for use in synthetic procedures. It is prepared by reaction of *n*-BuLi with di-*i*-propylamine. The generation of carbanions stabilized by electron-attracting groups is very important from a synthetic point of view, and the synthetic aspects of the chemistry of these carbanions will be discussed in Chapters 1 and 2 of Part B.

Carbanions derived from carbonyl compounds are often referred to as *enolates*. This name is derived from the enol tautomer of carbonyl compounds. The resonance-stabilized enolate anion is the conjugate base of both the keto and enol forms of carbonyl

418

CHAPTER 7
CARBANIONS AND
OTHER
NUCLEOPHILIC
CARBON SPECIES

Table 7.6. Carbanion Stabilization by Substituent Groups (kcal/mol)

Substituent	Method						
	AM1[a]	AM1[a] (H$_2$O)	PM[a]	PM[a] (H$_2$O)	6-31G*/MP2[b]	QCI50[c]	Exp[d]
F	19.4	14.4	24.6	21.0	16.4	10.4	
OH	15.9	12.0	19.4	16.4	10.9		
NH$_2$	20.1	13.3	20.9	15.1	5.6	−0.5	
CH$_3$	14.5	4.6	14.7	5.7	3.3	−3.2	
Ph	60.1	33.7	60.1	29.5			37.6
CH=CH$_2$	45.5	29.4	33.0	31.4	37.1		25.8
CH=O	62.7		60.3				50.2
CN	55.0	41.2	59.4	49.3	57.9		44.4
NO$_2$	85.8	69.7	91.9	72.5			57.9

a. A. M. El-Nahas, *J. Chem. Res. (Synop).* **1996**:310.
b. A. Pross, D. E. Defrees, B. A. Levi, S. K. Pollock, L. Radom, and W. J. Hehre, *J. Org. Chem.* **46**:1693 (1981).
c. A. M. El-Nahas and P. v. R. Schleyer, *J. Comput. Chem.* **15**:596 (1994).
d. J. E. Bartmess, J. A. Scott, and R. T. McIver, *J. Am. Chem. Soc.* **101**:604 (1979).

Table 7.7. pK Values of Other Compounds[a]

Ketones

O
‖
CH$_3$CCH$_3$ 26.5

O
‖
PhCCH$_3$ 24.7

O
‖
PhCH$_2$CCH$_3$ 19.9

O
‖
PhCH$_2$CCH$_2$Ph 18.7

=O 25.8

=O 26.4

=O 16.95

Diketones

O   O
‖   ‖
CH$_3$CCH$_2$CCH$_3$ 13.3

O   O
‖   ‖
PhCH$_2$CCH$_2$CCH$_2$Ph 13.35

11.2

Ketoesters

O
‖
CH$_3$CCH$_2$CO$_2$CH$_2$CH$_3$ 14.2

Esters

PhCH$_2$CO$_2$C$_2$H$_5$ 22.6

PhSCH$_2$CO$_2$C$_2$H$_5$ 21.4

25.2

Diesters

CH$_2$(CO$_2$C$_2$H$_5$)$_2$ 16.4

7.3

Nitroalkanes

CH$_3$NO$_2$ 17.2

PhCH$_2$NO$_2$ 12.3

—NO$_2$ 16.0

—NO$_2$ 17.9

a. F. G. Bordwell, *Acc. Chem. Res.* **21**:456 (1988).

compounds:

There have been numerous studies of the rates of deprotonation of carbonyl compounds. These data are of interest not only because they define the relationship between thermodynamic and kinetic acidity for these compounds, but also because they are necessary for understanding mechanisms of reactions in which enolates are involved as intermediates. Rates of enolate formation can be measured conveniently by following isotopic exchange using either deuterium or tritium:

Another technique is to measure the rate of halogenation of the carbonyl compound. Ketones and aldehydes in their carbonyl forms do not react rapidly with the halogens, but the enolate is rapidly attacked. The rate of halogenation is therefore a measure of the rate of deprotonation.

Table 7.8 gives data on the rates of deuteration of some simple alkyl ketones. From these data, the order of reactivity toward deprotonation is $CH_3 > RCH_2 > R_2CH$. Steric hindrance to the approach of the base is the major factor in establishing this order. The importance of steric effects can be seen by comparing the CH_2 group in 2-butanone with the more hindered CH_2 group in 4,4-dimethyl-2-pentanone. The two added methyl groups on the adjacent carbon decrease the rate of proton removal by a factor of about 100. The rather slow rate of exchange at the CH_3 group of 4,4-dimethyl-2-pentanone must also reflect a steric factor arising from the bulky nature of the neopentyl group. If bulky groups interfere with effective solvation of the developing negative charge on oxygen, the rate of

420

CHAPTER 7
CARBANIONS AND
OTHER
NUCLEOPHILIC
CARBON SPECIES

Table 7.8. Relative Rates and E_a of Base-Catalyzed Deuteration of Some Ketones[a]

Ketones	Relative rate	E_a[b]
$\underset{\substack{\|\| \\ O}}{CH_3C}CH_2\underline{H}$	100	11.9
$CH_3\underset{\substack{\| \\ \underline{H}}}{C}CHCH_3$ ($\underset{O}{\|\|}$)	41.5	12.1
$\underline{H}-CH_2\underset{O}{\overset{\|\|}{C}}CH_2CH_3$	45	
$CH_3\underset{\substack{\| \\ \underline{H}}}{C}C(CH_3)_2$ ($\underset{O}{\|\|}$)	<0.1	
$\underline{H}-CH_2\underset{O}{\overset{\|\|}{C}}CH(CH_3)_2$	45	12.3
$CH_3\underset{\substack{\| \\ \underline{H}}}{C}CHC(CH_3)_3$ ($\underset{O}{\|\|}$)	0.45	
$\underline{H}-CH_2\underset{O}{\overset{\|\|}{C}}CH_2C(CH_3)_3$	5.1	

a. In aqueous dioxane with sodium carbonate as base. The data of C. Rappe and W. H. Sachs, *J. Org. Chem.* **32**:4127 (1967), given on a per-group basis, have been converted to a per-hydrogen basis.
b. MeO-catalyzed exchange in MeOD. T. Niiya, M. Yukawa, H. Morishita, H. Ikeda, and Y. Goto, *Chem. Pharm. Bull.* **39**:2475 (1991).

proton abstraction will be reduced. The observed activation energies parallel the rates.[35]

Structural effects on the rates of deprotonation of ketones have also been studied using very strong bases under conditions where complete conversion to the enolate occurs. In solvents such as THF or DME, bases such as lithium di-*i*-propylamide (LDA) and potassium hexamethyldisilylamide (KHMDS) give solutions of the enolates in relative proportions that reflect the relative rates of removal of the different protons in the carbonyl compound (kinetic control). The least hindered proton is removed most rapidly under these

35. T. Niiya, M. Yukawa, H. Morishita, H. Ikeda, and Y. Goto, *Chem. Pharm. Bull.* **39**:2475 (1991).

conditions, so that for unsymmetrical ketones the major enolate is the less substituted one. Scheme 7.1 shows some representative data.

Equilibrium between the various enolates of a ketone can be established by the presence of an excess of the ketone, which permits proton transfer. Equilibration is also favored by the presence of dissociating solvents such as HMPA. The composition of the equilibrium enolate mixture is usually more closely balanced than for kinetically

Scheme 7.1. Composition of Ketone–Enolate Mixtures Formed under Kinetic and Thermodynamic Conditions[a]

Ketone	Conditions	Enolate mixture		
$CH_3CH_2CCH_3$	Kinetic (LDA, 0°C)	71%	13%	16%
$(CH_3)_2CHCCH_3$	Kinetic (K HMDS, –78°C) Thermodynamic (KH)	99% 88%	1% 12%	
$CH_3(CH_2)_3CCH_3$	Kinetic (LDA, –78°C) Thermodynamic (KH, 20°C)	100% 42%	0% 46%	0% 12%
$PhCH_2CCH_3$	Kinetic (LDA, 0°C) Thermodynamic (NaH)	14% 2%	86% (combined) 98% (combined)	

(*continued*)

422

CHAPTER 7
CARBANIONS AND
OTHER
NUCLEOPHILIC
CARBON SPECIES

Scheme 7.1. (*continued*)

Ketone	Conditions	Enolate mixture	

Kinetic (LDA, 0°C)
Thermodynamic (NaH)

99% / 1%
26% / 74%

Kinetic (LiCPh₃)
Thermodynamic (KCPh₃)

100% / 0%
35% / 65%

Kinetic (LiCPh₃)
Thermodynamic (KCPh₃)

82% / 18%
52% / 48%

Kinetic (LDA)
Thermodynamic (NaH)

98% / 2%
50% / 50%

a. Selected from a more complete compilation by D. Caine, *Carbon–Carbon Bond Formation*, R. L. Augustine, ed., Marcel Dekker, New York, 1979.

controlled conditions. In general, the more highly substituted enolate is the preferred isomer, but if the alkyl groups are sufficiently branched as to interfere with solvation of the enolate, there can be exceptions. The equilibrium ratios of enolates for several ketone–enolate systems are shown in Scheme 7.1.

Nitroalkanes show an inverse relationship between kinetic acidity and thermodynamic acidity. Additional alkyl substituents on nitromethane retard the rate of proton removal although the equilibrium is more favorable for the more highly substituted derivatives.[36] The alkyl groups have a strong stabilizing effect on the nitronate ion, but unfavorable steric effects are dominant at the transition state for proton removal. As a result, kinetic and thermodynamic acidity show opposite responses to alkyl substitution.

	Kinetic acidity k (M^{-1} min^{-1})	Thermodynamic acidity pK
CH_3NO_2	238	10.2
$CH_3CH_2NO_2$	39.1	8.5
$(CH_3)_2CHNO_2$	2.08	7.7

36. F. G. Bordwell, W. J. Boyle, Jr., and K. C. Yee, *J. Am. Chem. Soc.* **92**:5926 (1970).

The cyano group is also effective at stabilizing negative charge on carbon. It has been possible to synthesize a number of hydrocarbon derivatives that are very highly substituted with cyano groups. Table 7.9 gives pK values for some of these compounds. As can be seen, the highly substituted derivatives are quite strong acids.

Second-row elements, particularly phosphorus and sulfur, stabilize adjacent carbanions. The pK's of some pertinent compounds are given in Table 7.10.

The conjugate base of 1,3-dithiane has proven valuable in synthetic applications as a nucleophile (Part B, Chapter 13). The anion is generated by deprotonation using *n*-butyllithium:

The pK of 1,3-dithiane is 36.5 (Cs^+ ion pair in THF).[37] The value for 2-phenyl-1,3-dithiane is 30.5. There are several factors which can contribute to the anion-stabilizing effect of sulfur substituents. Bond dipole effects contribute but cannot be the dominant factor because oxygen substituents do not have a comparable stabilizing effect. Polarizability of sulfur can also stabilize the carbanion. Delocalization can be described as involving $3d$ orbitals on sulfur or hyperconjugation with the σ^* orbital of the C—S bond.[38] MO calculations favor the latter interpretation. An experimental study of the rates of deprotonation of phenylthionitromethane indicates that sulfur polarizability is a major factor.[39] Whatever the structural basis is, there is no question that thio substituents enhance

Table 7.9. Acidities of Some Cyano Compounds[a]

Compound	pK
CH_3CN	>25.0
$NCCH_2CN$	11.2
$(NC)_3CH$	−5.0
	< − 8.5
	< − 11.0

a. Selected from Tables 5.1 and 5.2 in J. R. Jones, *The Ionization of Carbon Acids*, Academic Press, New York, 1973, pp. 64, 65.

37. L. Xie, D. A. Bors, and A. Streitwieser, *J. Org. Chem.* **57**:4986 (1992).
38. W. T. Borden, E. R. Davidson, N. H. Andersen, A. D. Denniston, and N. D. Epiotis, *J. Am. Chem. Soc.* **100**:1604 (1978); A. Streitwieser, Jr., and S. P. Ewing, *J. Am. Chem. Soc.* **97**:190 (1975); A. Streitwieser, Jr., and J. E. Williams, Jr., *J. Am. Chem. Soc.* **97**:191 (1975); N. D. Epiotis, R. L. Yates, F. Bernardi, and S. Wolfe, *J. Am. Chem. Soc.* **98**:5435 (1976); J.-M. Lehn and G. Wipff, *J. Am. Chem. Soc.* **98**:7498 (1976); D. A. Bors and A. Streitwieser, Jr., *J. Am. Chem. Soc.* **108**:1397 (1986).
39. C. F. Bernasconi and K. W. Kittredge, *J. Org. Chem.* **63**:1944 (1998).

424

CHAPTER 7
CARBANIONS AND
OTHER
NUCLEOPHILIC
CARBON SPECIES

**Table 7.10. Acidities of Some Compounds with
Sulfur and Phosphorus Substituents**

Compound	pK (DMSO)	Reference
$PhCH_2SPh$	30.8	a
$PhSO_2CH_3$	29.0	b
$PhSO_2CH_2Ph$	23.4	b
$PhCH(SPh)_2$	23.0	a
$(PhS)_2CH_2$	38.0	a
$PhSO_2CH_2SPh$	20.3	b
$PhSO_2CH_2PPh_2$	20.2	b
$H_5C_2O_2CCH_2P^+Ph_3$	9.2[c]	d
$PhCOCH_2P^+Ph_3$	6.0[c]	d

a. F. G. Bordwell, J. E. Bares, J. E. Bartmess, G. E. Drucker, J.
   Gerhold, G. J. McCollum, M. Van Der Puy, N. R. Vanier, and W.
   S. Matthews, *J. Org. Chem.* **42**:326 (1977).
b. F. G. Bordwell, W. S. Matthews, and N. R. Vanier, *J. Am. Chem.
   Soc.* **97**:442 (1975).
c. In methanol.
d. A. J. Speziale and K. W. Ratts, *J. Am. Chem. Soc.* **85**:2790 (1963).

the acidity of hydrogens on the adjacent carbons. On hydrocarbons, the phenylthio group increases the acidity by at least 15 pK units. The magnitude of the effect is from 5 to 10 pK units in carbanions stabilized by another electron-accepting group.[40]

Trialkyl- and triarylsilyl substituents have a modest carbanion-stabilizing effect. This is attributed mainly to polarizability and is somewhat greater for the triarylsilyl substituents. This stabilization results in a reduction of the pK by 1 (trimethylsilyl) to 4 (triphenylsilyl) log units in fluorenes and 3 to 7.5 log units respectively, in sulfones.[41]

Two other important groups of nucleophilic carbon species are the phosphorus and sulfur ylides. These species are of great synthetic importance, and their reactivity will be considered in some detail in Chapters 1 and 2 of Part B. Here, we will discuss the structures of some of the best known ylides. *Ylide* is the name given to molecules for which one of the contributing resonance structures has opposite charges on adjacent atoms when both atoms have octets of electrons. Since we are dealing with nucleophilic carbon species, our interest is in ylides with negative charge on carbon. The three groups of primary synthetic importance are phosphonium ylides, sulfoxonium ylides, and sulfonium ylides. Ylides of ammonium ions also have some significance.

The question of which resonance structure is the principal contributor has been a point of discussion. Because the nonpolar *ylene* resonance structures have 10 electrons at the

40. F. G. Bordwell, J. E. Bares, J. E. Bartmess, G. E. Drucker, J. Gerhold, G. J. McCollum, M. Van Der Puy, N. R. Vanier, and W. S. Matthews, *J. Org. Chem.* **42**:326 (1977); F. G. Bordwell, M. Van Der Puy, and N. R. Vanier, *J. Org. Chem.* **41**:1885 (1976).
41. S. Zhang, X.-M. Zhan, and F. G. Bordwell. *J. Am. Chem. Soc.* **117**:602 (1995); A. Streitwieser, L. Xie, P. Wang, and S. M. Bachrach, *J. Org. Chem.* **58**:1778 (1993).

phosphorus or sulfur atom, these structures imply participation of d orbitals on the heteroatoms. Such resonance structures are not possible for ammonium ylides. Structural studies indicate that the dipolar ylide structure is probably the main contributor.[42] The stabilizing effect of phosphonium substituents is primarily the result of polar (field) and polarization effects.[43]

The ylides are formed by deprotonation of the corresponding "onium salts."

$$R_2CH-\overset{+}{P}R_3' + B^- \rightleftharpoons R_2\overline{C}-\overset{+}{P}R_3' + BH$$

$$R_2CH-\overset{+}{S}R_3' + B^- \rightleftharpoons R_2\overline{C}-\overset{+}{S}R_2' + BH$$

The stability of the resulting neutral species is increased by substituent groups that can help to stabilize the electron-rich carbon. Phosphonium ions with acylmethyl substituents, for example, are quite acidic. A series of aroylmethyl phosphonium ions have pK values of 4–7, with the precise value depending on the aryl substituents[44]:

In the absence of the carbonyl or similar stabilizing group, the onium salts are much less acidic. The pK_{DMSO} of methyltriphenylphosphonium ion is estimated to be 22. Strong bases such as amide ion or the anion of DMSO are required to deprotonate alkylphosphonium salts:

$$RCH_2\overset{+}{P}R_3' \xrightarrow[\text{base}]{\text{strong}} R\overline{C}H-\overset{+}{P}R_3$$

Similar considerations apply to the sulfonium and sulfoxonium ylides. These ylides are formed by deprotonation of the corresponding positively charged sulfur-containing cations.

The additional electronegative oxygen atom in the sulfoxonium salts stabilizes these ylides considerably, relative to the sulfonium ylides.[45]

7.3. Enols and Enamines

The study of the chemistry of carbonyl compounds has shown that they can act as carbon nucleophiles in the presence of acid catalysts as well as bases. The nucleophilic reactivity of carbonyl compounds in acidic solution is due to the presence of the enol tautomer. Enolization in acidic solution is catalyzed by O-protonation. Subsequent deprotonation at carbon gives the enol:

42. H. Schmidbaur, W. Buchner, and D. Scheutzow, *Chem. Ber.* **106**:1251 (1973).
43. X.-M. Zhang and F. G. Bordwell, *J. Am. Chem. Soc.* **116**:968 (1994).
44. S. Fliszar, R. F. Hudson, and G. Salvadori, *Helv. Chim. Acta* **46**:1580 (1963).
45. E. J. Corey and M. Chaykovsky, *J. Am. Chem. Soc.* **87**:1353 (1965).

426

CHAPTER 7
CARBANIONS AND
OTHER
NUCLEOPHILIC
CARBON SPECIES

Like simple alkenes, enols are nucleophilic by virtue of their π electrons. Enols are much more reactive than simple alkenes, however, because the hydroxyl group can participate as an electron donor during the reaction process. The strong C=O bond is re-formed, providing a favorable energy contribution.

$$HO \atop RC=CHR + E^+ \longrightarrow HO^{\delta+} \atop RC=CHR \atop E \longrightarrow HO^+ \atop RC-CHR \atop E \longrightarrow O \atop RCCHR + H^+ \atop E$$

Enols are not as reactive as enolate anions, however. This lower reactivity simply reflects the presence of the additional proton in the enol, which decreases the electron density of the enol relative to the enolate. In MO terminology, the $-OH$ and $-O^-$ donor substituents raise the energy of the π-HOMO.

enol   enolate

   A number of studies of the acid-catalyzed mechanism of enolization have been done. The case of cyclohexanone is illustrative.[46] The reaction is catalyzed by various carboxylic acids and substituted ammonium ions. The effectiveness of these proton donors as catalysts correlates with their pK_a values. When plotted according to the Brønsted catalysis law (Section 4.8), the value of the slope α is 0.74. When deuterium or tritium is introduced in the α position, there is a marked decrease in the rate of acid-catalyzed enolization: $k_H/k_D \approx 5$. This kinetic isotope effect indicates that the C–H bond cleavage is part of the rate-determining step. The generally accepted mechanism for acid-catalyzed enolization pictures the rate-determining step as deprotonation of the protonated ketone:

   Rates of enolization can be measured in several ways. One method involves determining the rate of halogenation of the ketone. In the presence of a sufficient concentration of bromine or iodine, halogenation is much faster than enolization or its reverse and can therefore serve to measure the rate of enolization:

$$O \atop RCCH_2R \underset{k_{-1}}{\overset{k_1 \atop HA}{\rightleftharpoons}} OH \atop RC=CHR \overset{k_2 \atop X_2}{\underset{fast}{\longrightarrow}} O \atop RCCHR + HX \atop X$$

$$k_2 \gg k_1, k_{-1}$$

It is also possible to measure the rate of enolization by isotopic exchange. NMR spectroscopy provides a very convenient method for following hydrogen–deuterium exchange, and this is now the preferred method. Data for several ketones are given in

46. G. E. Lienhard and T.-C. Wang, *J. Am. Chem. Soc.* **91**:1146 (1969).

Table 7.11.

427

SECTION 7.3.
ENOLS AND ENAMINES

A point of contrast with the data for base-catalyzed removal of a proton (Table 7.8, p. 420) is the tendency for acid-catalyzed enolization to result in preferential formation of the more substituted enol. For 2-butanone, the ratio of exchange at CH_2 to that at CH_3 is $4.2:1$, after making the statistical correction for the number of hydrogens. The preference for acid-catalyzed enolization to give the more substituted enol is explained in terms of the stabilizing effect that alkyl groups have on carbon–carbon double bonds. To the extent that the transition state resembles product,[47] alkyl groups would be expected to stabilize the more branched transition state. There is an opposing steric effect that appears to be significant for 4,4-dimethyl-2-pentanone, in which the methylene group that is flanked by a t-butyl group is less reactive than the methyl group. The overall range of reactivity differences in acid-catalyzed exchange is much less than for base-catalyzed exchange, however (compare Tables 7.8 and 7.11). This is consistent with the C-deprotonation of the

**Table 7.11. Relative Rates of Acid-Catalyzed
Enolization for Some Ketones**[a]

Ketone	Relative rate
$CH_3CCH_2{-}H$ (O)	100
CH_3CCHCH_3 with H (O)	220
$H{-}CH_2CCH_2CH_3$ (O)	76
$CH_3CCHCH_2CH_3$ with H (O)	171
$CH_3CC(CH_3)_2$ with H (O)	195
$H{-}CH_2CCH(CH_3)_2$ (O)	80
$CH_3CCHC(CH_3)_3$ with H (O)	46
$H{-}CH_2CCH_2C(CH_3)_3$ (O)	105

a. In D_2O-dioxane with DC1 catalyst. The data of C. Rappe and W. H. Sachs, *J. Org. Chem.* **32**:3700 (1967), given on a per-group basis, have been converted to a per-hydrogen basis.

47. C. G. Swain, E. C. Stivers, J. F. Reuwer, Jr., and L. J. Schaad, *J. Am. Chem. Soc.* **80**:5885 (1958).

428

CHAPTER 7
CARBANIONS AND
OTHER
NUCLEOPHILIC
CARBON SPECIES

O-protonated compound having a much earlier transition state.

The amount of enol present in equilibrium with a carbonyl group is influenced by other substituent groups. In the case of compounds containing a single ketone, aldehyde, or ester function, there is very little of the enol present at equilibrium. When two such groups are close to one another in a molecule, however, particularly if they are separated by a single carbon atom, a major amount of the enol may be present. The enol forms of β-diketones and β-ketoesters are stabilized by intramolecular hydrogen bonds and by conjugation of the carbon–carbon double bond with the carbonyl group.

The simplest compound with this type of enolic structure is malonaldehyde. The structure determined by microwave spectroscopy on a deuterated analog has provided the bond length data shown below.[48] The barrier for shift of the enolic hydrogen (or deuterium) between the two oxygen atoms is about 4–5 kcal.[49] The structural data shown above for the enol form of 2,4-pentanedione were obtained by an electron diffraction study.[50] In this case, the data pertain to the time-averaged structure resulting from proton transfer between the two hydrogen-bonded oxygens.

Table 7.12 gives data on the amount of enol present at equilibrium for some representative compounds. The precise percent of enol present at equilibrium is solvent-dependent. For example, for ethyl acetoacetate, the amount of enol is higher in nonpolar solvents (15–30%) such as carbon tetrachloride and benzene than in more polar solvents such as water and acetone (5% enol in acetone, 1% enol in water).[51] The strong intramolecular hydrogen bond in the enol form minimizes the molecular dipole by reducing the negative charge on the oxygen of the carbonyl group. In more polar solvents, this stabilization is less important, and in protic solvents such as water, hydrogen bonding by the solvent is dominant.

α-Dicarbonyl compounds also have an enhanced tendency toward enolization,

48. S. L. Baughcum, R. W. Duerst, W. F. Rowe, Z. Smith, and E. B. Wilson, *J. Am. Chem. Soc.* **103**:6296 (1981).
49. S. L. Baughcum, Z. Smith, E. B. Wilson, and R. W. Duerst, *J. Am. Chem. Soc.* **106**:2260 (1984).
50. A. H. Lowrey, C. George, P. D'Antonio, and J. Karle, *J. Am. Chem. Soc.* **93**:6399 (1971).
51. K. D. Grande and S. M. Rosenfeld, *J. Org. Chem.* **45**:1626 (1980); S. G. Mills and P. Beak, *J. Org. Chem.* **50**:1216 (1985).

Table 7.12. Equilibrium Constants for Enolization of Some Carbonyl Compounds

Compound	$k = \text{enol/keto}$	Reference
$CH_3CH{=}O \rightleftharpoons CH_2{=}CHOH$	10^{-5}	a
$(CH_3)_2CHCH{=}O \rightleftharpoons (CH_3)_2C{=}CHOH$	1.4×10^{-4}	b
$\underset{\text{O}}{CH_3\overset{\|}{C}CH_3} \rightleftharpoons \underset{\text{OH}}{H_2C{=}\overset{\|}{C}CH_3}$	8×10^{-8} / 6×10^{-9}	a / c
$\underset{\text{O}}{CH_3CH_2\overset{\|}{C}CH_2CH_3} \rightleftharpoons \underset{\text{OH}}{CH_3CH{=}\overset{\|}{C}CH_2CH_3}$	2×10^{-8} / 3.6×10^{-8}	a / c
cyclopentanone ⇌ cyclopentenol	1×10^{-7} / 1.7×10^{-8}	a / c
cyclohexanone ⇌ cyclohexenol	5×10^{-6} / 3.9×10^{-7}	a / c
$\underset{\text{O}}{PhC\overset{\|}{C}H_3} \rightleftharpoons \underset{\text{OH}}{PhC{=}CH_2}$	2×10^{-7} / 1.2×10^{-8}	a / c
$CH_3\overset{O}{C}CH_2\overset{O}{C}OC_2H_5 \rightleftharpoons CH_3\overset{OH}{C}{=}CH\overset{O}{C}OC_2H_5$	$7 \times 10^{-2}(H_2O)$ / $3 \times 10^{-1}(CCl_4)$	d / d
$CH_3\overset{O}{C}CH_2\overset{O}{C}CH_3 \rightleftharpoons CH_3\overset{OH}{C}{=}CH\overset{O}{C}CH_3$	$2.3 \times 10^{-1}(H_2O)$ / $29\ (CCl_4)$	d / d
dimedone (5,5-dimethyl-1,3-cyclohexanedione) ⇌ enol	$20\ (H_2O)$	d
2-formylcyclohexanone ⇌ enol (76%) ⇌ enol (24%)	>50	e

a. In water; J. P. Guthrie and P. A. Cullimore, *Can. J. Chem.* **57**:240 (1979).
b. In water; Y. Chiang, A. J. Kresge, and P. A. Walsh, *J. Am. Chem. Soc.* **108**:6314 (1986).
c. In water; J. E. Dubois, M. El-Alauoi, and J. Toullec, *J. Am. Chem. Soc.* **103**:5393 (1981); J. Toullec, *Tetrahedron Lett.* **25**:4401 (1984).
d. S. G. Mills and P. Beak, *J. Org. Chem.* **50**:1216 (1985).
e. E. W. Garbisch, *J. Am. Chem. Soc.* **85**:1696 (1963).

430

CHAPTER 7
CARBANIONS AND
OTHER
NUCLEOPHILIC
CARBON SPECIES

although it is not so pronounced as for β-dicarbonyl compounds. The K_{enol} for pyruvic acid is about 10^{-3}.[52] There is resonance stabilization between the enol double bond and the ester carbonyl as well as a contribution from hydrogen bonding.

Enols of simple ketones can be generated in high concentration as metastable species by special techniques.[53] Vinyl alcohol, the enol of acetaldehyde, can be generated by very careful hydrolysis of any of several ortho ester derivatives in which the group RCO_2^- is acetate acid or a chlorinated acetate acid.[54]

The enol can be observed by NMR spectroscopy and at $-20°C$ has a half-life of several hours. At $+20°C$ the half-life is only 10 minutes. The presence of bases causes very rapid isomerization to acetaldehyde via the enolate. Solvents have a significant effect on the lifetime of such unstable enols. Solvents such as DMF and DMSO, which are known to slow rates of proton exchange by hydrogen bonding, increase the lifetime of unstable enols.[55]

Solutions of unstable enols of simple ketones and aldehydes can also be generated in water by addition of a solution of the enolate to water.[56] The initial protonation takes place on oxygen, generating the enol, which is then ketonized at a rate that depends on the solution pH. The ketonization exhibits both acid and base catalysis.[57] Acid catalysis involves C-protonation with concerted O-deprotonation.

In agreement with expectation for a rate-determining proton transfer, the reaction shows general acid catalysis. Base-catalyzed ketonization occurs by C-protonation of the enolate.

52. Y. Chiang, A. J. Kresge, and P. Pruszynski, *J. Am. Chem. Soc.* **114**:3103 (1992); J. Damitio, G. Smith, J. E. Meany, and Y. Pocker, *J. Am. Chem. Soc.* **114**:3081 (1992).

53. B. Capon, B. Z. Guo, F. C. Kwok, A. F. Siddhanta, and C. Zucco, *Acc. Chem. Res.* **21**:135 (1988).

54. B. Capon, D. S. Rycroft, T. W. Watson, and C. Zucco, *J. Am. Chem. Soc.* **103**:1761 (1981).

55. E. A. Schmidt and H. M. R. Hoffmann, *J. Am. Chem. Soc.* **94**:7832 (1972).

56. Y. Chiang, A. J. Kresge, and P. A. Walsh, *J. Am. Chem. Soc.* **104**:6122 (1982); Y. Chiang, A. J. Kresge, and P. A. Walsh, *J. Am. Chem. Soc.* **108**:6314 (1986).

57. B. Capon and C. Zucco, *J. Am. Chem. Soc.* **104**:7567 (1982).

As would be expected on the basis of electronegativity arguments, enols are much more acidic than the corresponding keto forms. It has been possible to determine the pK of the enol form of acetophenone as being 10.5. The pK of the keto form is 18.4.[58] Because the enolate is the same for both equilibria, the pK values are related to the enol $\rightleftharpoons$ keto equilibrium.

$$
\begin{array}{c}
O \\
\parallel \\
PhCCH_3
\end{array}
$$

$K = 10^{-1.9}$ ⟋   ⟍ $K = 10^{-18.4}$

$$
\underset{CH_2=CPh}{\overset{OH}{|}} \quad \xrightleftharpoons{K = 10^{-10.5}} \quad \underset{CH_2=CPh}{\overset{O^-}{|}}
$$

Similar measurements have been made for the equilibria involving acetone and its enol, 2-hydroxypropene.[59]

$$
\begin{array}{c}
O \\
\parallel \\
CH_3CCH_3
\end{array}
$$

$K = 10^{-8.2}$ ⟋   ⟍ $K = 10^{-19.2}$

$$
\underset{CH_2=CCH_3}{\overset{OH}{|}} \quad \xrightleftharpoons{K = 10^{-11}} \quad \underset{CH_2=CCH_3}{\overset{O^-}{|}} + H^+
$$

The accessibility of enols and enolates, respectively, in acidic and basic solutions of carbonyl compounds makes possible a wide range of reactions that depend on the nucleophilicity of these species. The reactions will be discussed in Chapter 8 and in Chapters 1 and 2 of Part B.

Amino substituents on a carbon–carbon double bond enhance the nucleophilicity of the β carbon to an even greater extent than the hydroxyl group in enols. This is because of the greater electron-donating power of nitrogen. Such compounds are called *enamines*.[60]

$$
\underset{R}{\overset{R_2\ddot{N}}{\diagdown}}C=C\underset{R}{\overset{R}{\diagup}} \quad \longleftrightarrow \quad \underset{R}{\overset{R_2\overset{+}{N}}{\diagdown}}C-C\underset{R}{\overset{R}{\diagup}}
$$

An interesting and useful property of enamines of 2-alkylcyclohexanones is the fact that there is a substantial preference for the less substituted isomer to be formed. This tendency is especially pronounced for enamines derived from cyclic secondary amines such as pyrrolidine. This preference can be traced to a strain effect called A^{1,3} or allylic strain (see Section 3.3). In order to accommodate conjugation between the nitrogen lone pair and the carbon–carbon double bond, the nitrogen substituent must be coplanar with the double bond. This creates a steric repulsion when the enamine bears a β substituent and leads to a

58. Y. Chiang, A. J. Kresge, and J. Wirz, *J. Am. Chem. Soc.* **106**:6392 (1984).
59. Y. Chiang, A. J. Kresge, Y. S. Tang, and J. Wirz, *J. Am. Chem. Soc.* **106**:460 (1984).
60. A. G. Cook, *Enamines*, 2nd ed., Marcell Dekker, New York, 1988, Chapter 1; Z. Rappoport, ed., *Chemistry of Enamines*, John Wiley & Sons, Chichester, U.K., 1994.

432

CHAPTER 7
CARBANIONS AND
OTHER
NUCLEOPHILIC
CARBON SPECIES

preference for the unsubstituted enamine.

strongly favored   steric repulsion

Because of the same preference for coplanarity in the enamine system, α alkyl substituents adopt an axial conformation to minimize steric interaction with the amino group.

favored   disfavored

These steric factors are also indicated by the relative basicity of enamines derived from five-, six-, and seven-membered ketones.[61] The five- and seven-membered enamines are considerably stronger bases, indicating better conjugation between the amine lone pair and the double bond. The reduced basicity of the cyclohexanone enamines is related to the preference for *exo* and *endo* double bonds in six-membered rings (see Section 3.10).

$$n = 2, pK_a\ 11.3$$
$$n = 3, pK_a\ \ 9.5$$
$$n = 4, pK_a\ 12.3$$

The preparation of enamines will be discussed in Chapter 8, and their application as carbon nucleophiles in synthesis is discussed in Chapter 1 of Part B.

7.4. Carbanions as Nucleophiles in S_N2 Reactions

Carbanions are very useful intermediates in the formation of carbon–carbon bonds. This is true both for unstabilized structures found in organometallic reagents and stabilized structures such as enolates. Carbanions can participate as nucleophiles both in addition and in substitution reactions. At this point, we will discuss aspects of the reactions of carbanions as nucleophiles in reactions that proceed by the S_N2 mechanism. Other synthetic applications of carbanions will be discussed more completely in Part B.

Carbanions are classified as soft nucleophiles. It would be expected that they would be good nucleophiles in S_N2 reactions, and this is generally true. The reactions of aryl-,

61. A. G. Cook, M. L. Absi, and V. F. Bowden, *J. Org. Chem.* **60**:3169 (1995).

alkenyl-, and alkyllithium reagents with primary alkyl halides and tosylates appear to proceed by S_N2 mechanisms. Similar reactions occur between arylmagnesium halides (Grignard reagents) and alkyl sulfates and sulfonates. Some examples of these reactions are given in Scheme 7.2.

Scheme 7.2. Alkylation of Some Organometallic Reagents

A. Organolithium reagents

B. Organomagnesium compounds

a. H. Neumann and D. Seebach, *Chem. Ber.* **111**:2785 (1978).
b. J. Millon, R. Lorne, and G. Linstrumelle, *Synthesis* **1975**:434.
c. T. L. Shih, M. J. Wyvratt, and H. Mrozik, *J. Org. Chem.* **52**:2029 (1987).
d. A. J. Quillinan and F. Scheinman, *Org. Synth.* **58**:1 (1978).
e. H. Gilman and W. E. Catlin, *Org. Synth. Coll. Vol. I*, 471 (1941).
f. L. I. Smith, *Org. Synth. Col. Vol. II*, 360 (1943).
g. J. Eustache, J. M. Bernardon, and B. Shroot, *Tetrahedron Lett.* **28**:4681 (1987).

434

CHAPTER 7
CARBANIONS AND
OTHER
NUCLEOPHILIC
CARBON SPECIES

Evidence for an S_N2-type mechanism in the reaction of allyl and benzyl lithium reagents has been obtained from stereochemical studies. With 2-bromobutane, both of these reagents react with complete inversion of configuration.[62] n-Butyllithium, however, gives largely racemic product, indicating that some competing process must also occur.[63] A general description of the mechanism for reaction of organolithium compounds with alkyl halides must take account of the structure of the organometallic compound. It is known that halide anions are accommodated into typical organolithium cluster structures and can replace solvent molecules as ligands. A similar process in which the alkyl halide became complexed at lithium would provide an intermediate structure that could account for the subsequent alkylation. This process is represented below for a tetrameric structure, with the organic group simply represented by C.

In general terms, the reactions of organolithium reagents with alkylating agents could occur at any of the aggregation stages present in solution, and there could be reactivity differences among them. There has been little detailed mechanistic study which would distinguish among these possibilities.

The reaction of phenyllithium and allyl chloride labeled with ^{14}C reveals that allylic rearrangement occurs. About three-fourths of the product results from bond formation at C-1 rather than C-1. This can be accounted for by a cyclic transition state.[64]

Substituted allylic halides give mixtures of products resulting from bond formation at both C-1 and C-3 of the allylic system, with the product ratio favoring the product formed by reaction at the less substituted site. The portion of the product formed by reaction at C-1 in allylic systems may result from direct substitution, but it has also been suggested that a

62. L. H. Sommer and W. D. Korte, J. Org. Chem. **35**:22 (1970).

63. D. Zook and R. N. Goldey, J. Am. Chem. Soc. **75**:3975 (1953).

64. R. M. Magid and J. G. Welch, J. Am. Chem. Soc. **90**:5211 (1968); R. M. Magid, E. C. Nieh, and R. D. Gandour, J. Org. Chem. **36**:2099 (1971).

cyclic transition state involving an aryllithium dimer might be involved.

These mechanisms ascribe importance to the Lewis acid–Lewis base interaction between the allyl halide and the organolithium reagent. When substitution is complete, the halide ion is incorporated into the lithium cluster in place of one of the carbon ligands.

From a synthetic point of view, direct alkylation of lithium and magnesium organometallic compounds has largely been supplanted by transition-metal-catalyzed processes. We will discuss these reactions in Chapter 8 of Part B.

The alkylation reactions of enolate anions of both ketones and esters have been extensively utilized in synthesis. Both very stable enolates, such as those derived from β-ketoesters, β-diketones, and malonate esters, as well as less stable enolates of monofunctional ketones, esters, nitriles, etc., are reactive. Many aspects of the relationships between reactivity, stereochemistry, and mechanism have been clarified. A starting point for the discussion of these reactions is the structure of the enolates. Because of the delocalized nature of enolates, an electrophile can attack either at oxygen or at carbon.

Soft electrophiles will prefer carbon, and it is found experimentally that most alkyl halides react to give C-alkylation. Because of the π character of the HOMO of the anion, there is a stereoelectronic preference for attack of the electrophile approximately perpendicular to the plane of the enolate. The frontier orbital is ψ_2, with electron density mainly at O and C-2. The ψ_1 orbital is transformed into the C=O bond. The transition state for an S_N2 alkylation of an enolate can be represented as below.

A more detailed representation of the reaction requires more intimate knowledge of the enolate structure. Studies of ketone enolates in solution indicate that both tetrameric and dimeric clusters can exist. Tetrahydrofuran, a solvent in which many synthetic reactions are performed, favors tetrameric structures for the lithium enolate of isobutyrophenone, for example.[65]

---

65. L. M. Jackman and N. Szeverenyi, *J. Am. Chem. Soc.* **99**:4954 (1977); L. M. Jackman and B. C. Lange, *Tetrahedron* **33**:2737 (1977).

436

CHAPTER 7
CARBANIONS AND
OTHER
NUCLEOPHILIC
CARBON SPECIES

The structures of several lithium enolates of ketones have been determined by X-ray crystallography. Figure 7.3 illustrates some of the observed structures. Figure 7.3A shows an unsolvated enolate of methyl *t*-butyl ketone (pinacolone).[66] The structures in Fig. 7.3B and 7.3C are the THF solvates of the enolates of methyl *t*-butyl ketone and cyclopentanone, respectively.[67] Each of these structures consists of clusters of four enolate anions and four lithium cations arranged with lithium and oxygen at alternating corners of a distorted cube. The structure in Fig. 7.3D includes only two enolate anions. Four lithium ions are present along with two di-*i*-propylamide anions. An interesting feature of this structure is the coordination of the remote siloxy oxygen atom to one of the lithium cations.[68] This is an example of the Lewis acid–Lewis base interaction that is frequently involved in organizing transition-state structure in the reactions of lithium enolates. A common feature of all four of the structures is the involvement of the enolate oxygen in multiple contacts with lithium cations in the cluster. An approaching electrophile will clearly be somewhat hindered from direct contact with oxygen in such structures.

Several ester enolates have also been examined by X-ray crystallography.[69] The enolates of *t*-butyl propionate and *t*-butyl 3-methylpropionate were obtained as TMEDA solvates of enolate dimers. The enolate of methyl 3,3-dimethylbutanoate was obtained as a THF-solvated tetramer.

Computational methods can also be used to describe enolate structure. Most of the structural features of enolates are correctly modeled by B3LYP computations with dimethyl ether as the solvent molecule.[70] Although semiempirical PM3 calculations give adequate representations of the geometries of the aggregates, the energy values are not accurate.

Computational methods also indicate the stability of aggregated structures. Both *ab initio* and semiempirical calculations of the structure of the lithium enolate of methyl

66. P. G. Williard and G. B. Carpenter, *J. Am. Chem. Soc.* **107**:3345 (1985).
67. R. Amstutz, W. B. Schweizer, D. Seebach, and J. D. Dunitz, *Helv. Chim. Acta* **64**:2617 (1981).
68. P. G. Williard and M. J. Hintze, *J. Am. Chem. Soc.* **109**:5539 (1987).
69. D. Seebach, R. Amstutz, T. Laube, W. B. Schweizer, and J. D. Dunitz, *J. Am. Chem. Soc.* **107**:5403 (1985).
70. A. Abbotto, A. Streitwieser, and P. v. R. Schleyer, *J. Am. Chem. Soc.* **119**:11255 (1997).

Fig. 7.3. Crystal structures of some lithium enolates of ketones. (A) Unsolvated hexameric enolate of methyl *t*-butyl ketone; (B) tetrahydrofuran solvate of tetramer of enolate of methyl *t*-butyl ketone; (C) tetrahydrofuran solvate of tetramer of enolate of cyclopentanone; (D) dimeric enolate of 3,3-dimethyl-4-(*t*-butyldimethylsiloxy)-2-pentanone. (Structural diagrams are reproduced from Refs. 66–69.) by permission of the American Chemical Society and Verlag Helvetica Chimica Acta AG.

isobutyrate have been reported. Dimeric and tetrameric structures give calculated ^{13}C chemical shifts in agreement with the experimental values.[71]

   One of the general features of the reactivity of enolate anions is the sensitivity of both the reaction rate and the ratio of C- versus O-alkylation to the degree of aggregation of the enolate. For example, addition of HMPA frequently increases the rate of enolate alkylation reactions.[72] Use of dipolar aprotic solvents such as DMF and DMSO in place of THF also leads to rate acceleration.[73] These effects can be attributed, at least in part, to dissociation of the lithium enolate aggregates. Similar effects are observed when crown ethers or similar cation-complexing agents are added to reaction mixtures.[74]

71. H. Weiss, A. V. Yakimansky, and A. H. E. Muller, *J. Am. Chem. Soc.* **118:**8897 (1996).
72. L. M. Jackman and B. C. Lange, *J. Am. Chem. Soc.* **103:**4494 (1981); C. L. Liotta and T. C. Caruso, *Tetrahedron Lett.* **26:**1599 (1985).
73. H. D. Zook and J. A. Miller, *J. Org. Chem.* **36:**1112 (1971); H. E. Zaugg, J. F. Ratajczyk, J. E. Leonard, and A. D. Schaeffer, *J. Org Chem.* **37:**2249 (1972); H. E. Zaugg, *J. Am. Chem. Soc.* **83:**837 (1961).
74. A. L. Kurts, S. M. Sakembaeva, J. P. Beletskaya, and O. A. Reutov, *Z. Org. Khim. SSSR* (Engl. Transl.) **10:**1588 (1974).

438

CHAPTER 7
CARBANIONS AND
OTHER
NUCLEOPHILIC
CARBON SPECIES

The order of enolate reactivity also depends on the metal cation that is present. The general order is $BrMg < Li < Na < K$. This order, too, is in the order of greater dissociation of the enolate–cation ion pairs and ion aggregates. Carbon-13 chemical shift data provide an indication of electron density at the nucleophilic carbon in enolates. These shifts have been found to be both cation-dependent and solvent-dependent. Apparent electron density increases in the order $K^+ > Na^+ > Li^+$ and THF/HMPA > DME > THF >ether.[75] There is a good correlation with observed reactivity under the corresponding conditions.

The C- versus O-alkylation ratio has been studied for the potassium salt of ethyl acetoacetate as a function of both solvent and leaving group.[76]

Leaving group, X	Solvent	C:O ratio
$OSO_2C_2H_5$	t-BuOH	100:00
$OSO_2C_2H_5$	THF	100:00
$OSO_2C_2H_5$	HMPA	17:83
$OSO_2C_5H_7$	HMPA	12:88
Cl	HMPA	40:60
Br	HMPA	61:39
I	HMPA	87:13

These data show that a change from a hard leaving group (sulfonate, sulfate) to a softer leaving group (bromide, iodide) favors C-alkylation. One major influence on the C:O ratios is presumably the degree of aggregation. The reactivity at oxygen should be enhanced by dissociation because the electron density will be less tightly associated with the cation. It also appears that the nature of the aggregate, that is, the anions incorporated into it, may also be a major influence on reactivity. The leaving group in the alkylating reagent has a major effect on whether C- or O-alkylation occurs. In the case of the lithium enolate of acetophenone, for example, C-alkylation is predominant with methyl iodide, but C- and O-alkylation occur to approximately equal extents with dimethyl sulfate. The C:O ratio is shifted more to O-alkylation by addition of HMPA or other cation-complexing agents. Thus, with four equivalents of HMPA the C:O ratio for methyl iodide drops from >200:1 to 10:1 whereas with dimethyl sulfate the C:O ratio changes from 1.2:1 to 0.2:1 when HMPA is added.[77]

Steric and stereoelectronic effects control the direction of approach of an electrophile to the enolate. Electrophiles approach from the least hindered side of the enolate. Numerous examples of such effects have been observed.[78] In ketone and ester enolates that are exocyclic to a conformationally biased cyclohexane ring there is a slight

75. H. O. House, A. V. Prabhu, and W. V. Phillips, *J. Org. Chem.* **41**:1209 (1976).
76. A. L. Kurts, A. Masias, N. K. Gerkina, I. P. Beletskaya, and O. A. Reutov *Dokl. Akad. Nauk. SSSR* (Engl. Transl.) **187**:595 (1969); A. L. Kurts, N. K. Gerkina, A. Masias, I. P. Beletskaya, and O. A. Reutov, *Tetrahedron* **27**:4777 (1971).
77. L. M. Jackman and B. C. Lange, *J. Am. Chem. Soc.* **103**:4494 (1981).

If the axial face is further hindered by addition of a substituent, the selectivity is increased.

Endocyclic cyclohexanone enolates with 2-alkyl groups show a small preference $(1:1–5:1)$ for approach of the electrophile from the direction that permits the chair conformation to be maintained.[80]

The 1(9)-enolate of 1-decalone exhibits a preference for alkylation to form a *cis* ring juncture.

This is the result of a steric differentiation on the basis of the approach of electrophile from the side of the enolate where the 10-position is occupied by the smaller hydrogen rather than the ring.

In general, the stereoselectivity of enolate alkylation can be predicted and interpreted on the basis of the stereoelectronic requirement for approximately perpendicular approach to the enolate in combination with selection between the two faces on the basis of steric factors.

General References

E. Buncel, *Carbanions: Mechanistic and Isotopic Aspects*, Elsevier, Amsterdam, 1975.

E. Buncel and T. Durst, eds., *Comprehensive Carbanion Chemistry*, Elsevier, New York, 1981.

D. J. Cram, *Fundamentals of Carbanion Chemistry*, Academic Press, New York, 1965.

78. For reviews, see D. A. Evans, in *Asymmetric Synthesis*, Vol. 3, J. D. Morrison, ed., Academic Press, New York, 1984, Chapter 1; D. Caine, in *Carbon–Carbon Bond Formation*, R. L. Augustine, ed., Marcel Dekker, New York, 1979.

79. A. P. Krapcho and E. A. Dundulis, *J. Org. Chem.* **45**:3236 (1980); H. O. House and T. M. Bare, *J. Org. Chem.* **33**:943 (1968).

80. H. O. House, B. A. Tefertiller, and H. D. Olmstead, *J. Org. Chem.* **33**:935 (1996); H. O. House and M. J. Umen, *J. Org. Chem.* **38**:1000 (1973); J. M. Conia and P. Briet, *Bull. Soc. Chim. Fr.* **1966**:3881, 3886; C. Djerassi, J. Burakevich, J. W. Chamberlin, D. Elad, T. Toda, and G. Stork, *J. Am. Chem. Soc.* **86**:465 (1964); C. Agami, J. Levisalles, and B. Lo Cicero, *Tetrahedron* **35**:961 (1979).

440

CHAPTER 7
CARBANIONS AND
OTHER
NUCLEOPHILIC
CARBON SPECIES

J. R. Jones, *The Ionization of Carbon Acids*, Academic Press, New York, 1973.

E. M. Kaiser and D. W. Slocum, in *Organic Reactive Intermediates*, S. P. McManus, ed., Academic Press, New York, 1973, Chapter 5.

Z. Rappoport, ed., *The Chemistry of Enols*, John Wiley & Sons, New York, 1990.

M. Szwarc, *Ions and Ion Pairs in Organic Reactions*, John Wiley & Sons, New York, 1972.

J. Toullec, "Enolization of Simple Carbonyl Compounds," *Adv. Phys. Org. Chem.* **18**:1 (1982).

Problems

(References for these problems will be found on page 797).

1. Predict the order of increasing thermodynamic acidity in each series of compounds:

   (a) benzene, 1,4-cyclohexadiene, cyclopentadiene, cyclohexane

   (b) CH_3CN, CH_3NO_2, CH_3CCH_3, CH_3SCH_3, CH_3SCH_3

   (c) $PhCH_3$, $PhCH_2Si(CH_3)_3$, $PhSCH_3$, $PhCH_2SCH_3$

   (d) $CH_3CCH_2CCH_3$, $CH_3CH_2CCH_2CCH_2CH_3$, $PhCCH_2CCH_3$, $PhCCH_2CCF_3$

   (e) 9-(*m*-chlorophenyl)fluorene, 9-(*p*-methoxyphenyl)fluorene, 9-phenylfluorene, 9-(*m*-methoxyphenyl)fluorene, 9-(*p*-methylphenyl)fluorene.

2. Indicate which portion is the most acidic in each of the following molecules. Explain your reasoning.

   (a) $H_3CC \equiv CH$

   (b)

   (c)

   (d)

   (e)

   (f)

   (g) $CH_3SCH_2SCH_3$

   (h)

   (i) $(CH_3)_2CHN \overset{+}{=} NOCH_3$

   (j)

3. Offer an explanation for the following observations.

(a) Exchange rates indicate the hydrocarbon cubane to be much more acidic than cyclobutane, and even more acidic than cyclopropane.

cubane

(b) Phenyl cyclopropyl ketone (pK = 28.2) is less acidic than acetophenone (pK = 24.7) and undergoes C—H exchange more slowly than phenyl isopropyl ketone.

(c) The order of acidity for cyclopentadiene, indene, and fluorene is as indicated by the pK data given:

pK   18.0   20.1   22.6

(d) The rotational barrier for the allyl anion in THF, as measured by NMR, is a function of the cation that is present:

M$^+$	$\Delta G^{\ddagger}$
Li	10.7
K	16.7
Cs	18.0

4. (a) The relative rates of hydroxide ion-catalyzed deuterium exchange at C-3 (the CH$_2\alpha$ to the C=O) have been measured for the bicyclic ketones shown below. Analyze the factors that would be involved in the relative ease of exchange in these compounds.

Rate constant for
exchange at 25°C $k = 5.6 \times 10^{-2}$ M^{-1} s^{-1} $k = 6.04 \times 10^{-6}$ M^{-1} s^{-1}

(b) Treatment of (+)-camphenilone with potassium t-butoxide in t-butyl alcohol-O-d at 185°C results in H—D exchange accompanied by racemization at an equal rate. Prolonged reaction periods result in the introduction of three deuterium atoms. Suggest a mechanism to account for these observations.

5. Using data from Tables 7.1 (p. 406) and 7.2 (p. 409), estimate the extent of

442

CHAPTER 7
CARBANIONS AND
OTHER
NUCLEOPHILIC
CARBON SPECIES

deprontonation for each hydrocarbon-solvent-base combination. Discuss the uncertainties involved in your calculations.

(a) indene by 0.01 M $NaOCH_3$ in 1:1 DMSO–C_2H_5OH
(b) fluorene by 0.01 M $NaOCH_2H_5$ in 20:1 DMSO–C_2H_5OH
(c) triphenylmethane by 5 M $KOCH_3$ in CH_3OH

6. The rates of removal of axial and equatorial protons from 4-t-butylcyclohexane in NaOD/dioxane have been measured by an NMR technique. The rate of removal of an axial proton is 5.5 times faster than for an equatorial proton. What explanation can you offer for this difference?

7. The following table gives exchange rates in methanolic sodium methoxide for a number of hydrocarbons and equilibrium acidities for some. Determine whether there is a correlation between kinetic and thermodynamic acidity in this series of compounds. If so, predict the thermodynamic acidity of the hydrocarbons for which no values are listed.

Compound	k (exchange) $(M^{-1}\,s^{-1})$	pK
9-Phenylfluorene	173×10^{-4}	18.5
Indene	50×10^{-4}	19.9
3,4-Benzfluorene	90.3×10^{-4}	
1,2-Benzfluorene	31.9×10^{-4}	20.3
2,3-Benzfluorene	2.15×10^{-4}	
Fluorene	3.95×10^{-4}	22.7

8. The acidity of various substituted acetophenones has been measured in DMSO. Would you expect the ρ value for a Hammett correlation to be positive or negative? Would you expect the best correlation with σ, σ^+, or σ^-? Justify your prediction, considering each of the σ values explicitly. The data are given below. Check your prediction by plotting the pK versus σ, σ^+, and σ^-.

X	pK (DMSO)	X	pK (DMSO)	X	pK (DMSO)
p-$(CH_3)_2N$	27.48	H	24.70	m-Cl	23.18
p-CH_3O	25.70	p-F	24.45	m-Br	23.19
m-$(CH_3)_2N$	25.32	m-CH_3O	24.52	m-CF_3	22.76
p-CH_3	25.19	p-Br	23.81	p-CF_3	22.69
m-CH_3	24.95	p-Cl	23.78	p-CN	22.04
p-Ph	24.51	m-F	23.45		

9. Suggest mechanisms for each of the following reactions:

(a) Ph_3CCPh (with =O) $\xrightarrow[\text{2) }H_3O^+]{\text{1) LiAlH}_4\text{, pyridine}}$ $Ph_3CH + PhCH_2OH$

(b)

$\xrightarrow[\text{heat}]{\text{pyridine}}$ $PhCCH_2OH + PhCHCH{=}O$ (with carbonyl on first; OH on second)

(c) Treatment of either of the diastereomers shown below with $0.025\,M$ $Na^{+-}CH_2SOCH_3$ in DMSO produces the same equilibrium mixture of 72% *trans* and 28% *cis*.

(d) The hemiacetal of aflatoxin B_1 racemizes readily in basic solution.

(e)

$\xrightarrow[\text{H}_2\text{O}]{\text{0.5\% KOH}}$

10. Meldrum's acid, pK 7.4, is exceptionally acidic in comparison to an acyclic analog such as dimethyl malonate, pK 15.9. For comparison, 5,5-dimethyl-1,3-cyclohexane-dione is only moderately more acidic than 2,4-pentanedione (11.2 versus 13.43). The pK values are those for DMSO solution. It is also found that the enhanced acidity of Meldrum's acid derivatives decreases as the ring size is increased. Analyze factors that could contribute to the enhanced acidity of Meldrum's acid.

15.9   7.3   n = 5, 13.0 / n = 8, 15.2 / n = 12, 15.1

444

CHAPTER 7
CARBANIONS AND
OTHER
NUCLEOPHILIC
CARBON SPECIES

11. In some solvents, it can be shown that the equilibrium k_1/k_{-1} is fast relative to the process governed by k_2:

$$R_3C-H + B^- \underset{k_{-1}}{\overset{k_1}{\rightleftharpoons}} [R_3C^- + H-B]$$

$$[R_3C^- + H-B] \xrightarrow{k_2} [R_3C^- + D-B] + H-B$$

$$[R_3C^- + D-B] \xrightarrow{k_3} R_3CD + B^-$$

This process is referred to as *internal return*, i.e., the base returns the proton to the carbanion faster than exchange of the protonated base with other solvent molecules occurs. If internal return is important under a given set of conditions, how would the correlation between kinetics of exchange and equilibrium acidity be affected? How could the occurrence of internal return be detected experimentally?

12. The pK_a values of the conjugate acids of several enamines derived from 2-methyl-propanal have been reported. Rationalize the observed variation with the structure of the amino constituent.

pK_a: 5.5   8.5   8.7

13. Identify factors which especially stabilize the enol from the following compounds.

(a)

$K_{enol} = 10^{-2}$

(b)

$K_{enol} = 8.6 \times 10^{-1}$

14. Certain cyclic compounds exhibit enhanced acidity relative to acyclic models. Offer an explanation for the examples given.

(a)

$pK = 12.2$ versus $pK = 15.9$

(b)

$pK = 18.5$ versus $pK = 24.6$

(c)

$pK = 7.42$ versus $CH_3O_2CCH_2CO_2CH_3$

$pK = 15.87$

15. The stereoselectivity of alkylation of 3-acetylbutyrolactone is influenced by additional alkyl substituents on the ring at C-4 and C-5. Analyze possible conformations of the enolate and develop an explanation of the stereoselectivity.

R^4	$R^{5\alpha}$	$R^{5\beta}$	A:B
CH$_3$	H	H	29:71
CH$_3$	H	Pr	<3:>97
CH$_3$	Pr	H	80:20
H	H	Et	25:75

16. Metal ions, in particular Zn^{2+}, Ni^{2+}, and Cu^{2+}, enhance the rate of general base-catalyzed enolization of 2-acetylpyridine by several orders of magnitude. Account for this effect.

17. The C-2 equatorial proton is selectively removed when 1,3-dithianes are deprotonated. Furthermore, if the resulting carbanion is protonated, there is a strong preference for equatorial protonation, even if this leads to a less stable axial orientation for the 2-substituent.

Discuss the relevance of these observations to the structure of sulfur-stabilized carbanions and rationalize your conclusion about the structure of the carbanions in MO terms.

18. (a) It is found that when 2-methyl-2-butene is converted to a dianion, it first gives the 2-methylbutadiene dianion **A** but this is converted to the more stable anion **B**,

446

CHAPTER 7
CARBANIONS AND
OTHER
NUCLEOPHILIC
CARBON SPECIES

which can be referred to as "methyltrimethylene–methane dianion."

Does simple Hückel MO theory offer an explanation for this result?

(b) The Hückel MO diagrams for several conceivable dianions which might be formed by double deprotonation of 2-methyl-1,5-hexadiene are given. On the basis of these diagrams, which of the dianions would be expected to be the most stable species?

19. Which of the two plausible structures given for methylketene dimer is more consistent with its observed pK_a of 2.8? Why?

20. Predict the products of each of the following reactions:

(a)

+ Br$_2$ $\xrightarrow{\text{ether}}$

(b)

+ Br$_2$ $\xrightarrow[\text{H}_2\text{O}]{\text{NaO}_2\text{CCH}_3}$

(c)

+ Cl$_2$ $\xrightarrow[\text{CH}_3\text{CO}_2\text{H}]{\text{NaO}_2\text{CCH}_3}$

(d)

+ Br$_2$ $\xrightarrow[\text{CH}_3\text{CO}_2\text{H}]{\text{NaO}_2\text{CCH}_3}$

(e)

$$\underset{\text{CH}_3\overset{\text{O}}{\overset{\|}{\text{C}}}\text{CH}_2\text{CH}_3}{} \xrightarrow[\text{H}_2\text{SO}_4,\ \text{H}_2\text{O}]{\text{Cl}_2}$$

21. Predict the structure and stereochemistry of the products that would be obtained under the specified reaction conditions. Explain the basis of your prediction.

(a)

$\xrightarrow[\text{2) ICH}_2\text{CH}=\text{CHCO}_2\text{C(CH}_3)_3]{\text{1) LDA, }-20°\text{C}}$

(b)

$\xrightarrow[\text{2) CH}_3\text{I}]{\text{1) LDA, }-40°\text{C}}$

(c)

$\xrightarrow{\text{CH}_3\text{I}}$

(d)

$\xrightarrow{\text{C}_2\text{H}_5\text{I}}$

(e)

$\xrightarrow{\text{Na}^+\,^-\text{CH}_2\overset{\text{O}}{\overset{\|}{\text{S}}}\text{CH}_3}$

(f)

$\xrightarrow{\text{Na}^+\,^-\text{CH}_2\overset{\text{O}}{\overset{\|}{\text{S}}}\text{CH}_3}$

448

CHAPTER 7
CARBANIONS AND
OTHER
NUCLEOPHILIC
CARBON SPECIES

22. The alkylation of 3-methyl-2-cyclohexenone with several dibromides led to the products shown below. Discuss the course of each reaction and suggest an explanation for the dependence of the product structure on the identity of the dihalide.

(31%) (25%) (42%)

$n = 3$ (55%) $n = 4$ (42%)

23. The stereochemistry of base-catalyzed deuterium exchange has been examined for **A** where X=CN and X=CPh:

$$\overset{O}{\underset{\|}{C}}$$

A   **B**

When X=CN, the isotopic exchange occurs with 99% retention of configuration, but when X=CPh, only about 30% net retention is observed. Explain this result.

$$\overset{O}{\underset{\|}{C}}$$

24. The distribution of α-bromoketones formed in the reaction of acetylcyclopentane with bromine was studied as a function of deuterium substitution. On the basis of the data given below, calculate the primary kinetic isotope effect (k_H/k_D) for enolization of acetylcyclopentane.

R = H 94%
R = D 80%

6%
20%

Reactions of Carbonyl Compounds

Introduction

The carbonyl group is one of the most prevalent of the functional groups and is involved in many synthetically important reactions. Reactions involving carbonyl groups are also exceptionally important in biological processes. Most of the reactions of aldehydes, ketones, esters, amides, and the other carboxylic acid derivatives directly involve the carbonyl group. In Chapter 7, the role of the carbonyl group in stabilizing carbanions was discussed. In this chapter, the primary topic will be the mechanistic patterns of addition and substitution reactions at carbonyl centers. The first two chapters of Part B deal with the use of carbonyl compounds in synthesis to form carbon–carbon bonds.

In many reactions at carbonyl groups, a key step is addition of a nucleophile, generating a tetracoordinate carbon atom. The overall course of the reaction is then determined by the fate of this *tetrahedral intermediate*.

$$
\underset{\text{RCR'}}{\overset{\text{O}}{\|}} + \text{Nu:}^- \longrightarrow \underset{\underset{\text{Nu}}{|}}{\overset{\text{O}^-}{\underset{|}{\text{RCR'}}}} \quad \text{or} \quad \underset{\text{RCR'}}{\overset{\text{O}}{\|}} + \text{Nu:} \longrightarrow \underset{\underset{\text{Nu}^+}{|}}{\overset{\text{O}^-}{\underset{|}{\text{RCR'}}}}
$$

The reactions of the specific classes of carbonyl compounds are related by the decisive importance of tetrahedral intermediates, and differences in reactivity can often be traced to structural features present in the tetrahedral intermediates.

8.1. Hydration and Addition of Alcohols to Aldehydes and Ketones

For most simple carbonyl compounds, the equilibrium constant for addition of water to the carbonyl group is unfavorable:

$$\underset{\substack{\|\\ O}}{RCR'} + H_2O \; \rightleftharpoons \; \underset{\substack{|\\ OH}}{\overset{OH}{RCR'}} \qquad K < 1$$

The exceptions are formaldehyde, which is nearly completely hydrated in aqueous solution, and aldehydes and ketones with highly electronegative substituents, such as trichloroacetaldehyde and hexafluoroacetone. The data given in Table 8.1 illustrate that the equilibrium constant for hydration decreases with increasing alkyl substitution.

Although the equilibrium constant for hydration is unfavorable, the equilibrium between an aldehyde or ketone and its hydrate is established rapidly and can be detected by isotopic exchange, using water labeled with ^{18}O, for example:

$$R_2C{=}O + H_2O^* \; \rightleftharpoons \; \underset{\substack{|\\ OH}}{\overset{*OH}{R_2COH}} \; \rightleftharpoons \; R_2C{=}O^* + H_2O$$

The hydration reaction has been extensively studied because it is the mechanistic prototype for many reactions at carbonyl centers that involve more complex molecules.[1] For acetaldehyde, the half-life of the exchange reaction is on the order of one minute under neutral conditions but is considerably faster in acidic or basic media. The second-order rate constant for acid-catalyzed hydration of acetaldehyde is on the order of 500 M^{-1} s^{-1}.[2] Acid catalysis involves either protonation or hydrogen bonding at the carbonyl oxygen.

Table 8.1. Hydration of Carbonyl Compounds

$$K = \frac{[\text{hydrate}]}{[\text{carbonyl}]}^{\,a}$$

Carbonyl compound	K (in water, 25°C)
CH_2O	2.28×10^{3b}
CH_3CHO	1.06^b
CH_3CH_2CHO	0.85^b
$(CH_3)_2CHCHO$	0.61^b
$(CH_3)_3CCHO$	0.23^b
CF_3CHO	2.9×10^{4b}
C_6H_5CHO	8×10^{-3c}
CH_3COCH_3	1.4×10^{-3b}
FCH_2COCH_3	0.11^c
$ClCH_2COCH_3$	0.11^b
CF_3COCH_3	35^b
CF_3COCF_3	1.2×10^{6b}
$C_6H_5COCH_3$	9.3×10^{-6c}
$C_6H_5COCF_3$	78^b
$CH_3COCOCH_3$	0.6^d
$CH_3COCO_2CH_3$	0.8^d

a. $K = K_{eq}[H_2O] = 55.5 K_{eq}$.
b. J. P. Guthrie, *Can. J. Chem.* **53**:898 (1975).
c. J. P. Guthrie, *Can. J. Chem.* **56**:962 (1978).
d. T. J. Burkey and R. C. Fahey, *J. Am. Chem. Soc.* **105**:868 (1983).

1. R. P. Bell, *Adv. Phys. Org. Chem.* **4**:1 (1966); W. P. Jencks, *Chem. Rev.* **72**:705 (1972).
2. P. Greenzaid, Z. Luz, and D. Samuel, *J. Am. Chem. Soc.* **89**:756 (1967).

Both specific acid catalysis and general acid catalysis can be observed.[3] (Review Section 4.8 for the discussion of specific and general acid catalysis.)

451

SECTION 8.1.
HYDRATION AND
ADDITION OF
ALCOHOLS TO
ALDEHYDES AND
KETONES

Specific acid-catalyzed hydration

$$RCH{=}O + H^+ \rightleftharpoons RCH{=}\overset{+}{O}H$$

$$RCH{=}\overset{+}{O}H + H_2O \xrightarrow{slow} \underset{\underset{{}^+OH_2}{|}}{RCHOH} \rightleftharpoons RCH(OH)_2 + H^+$$

General acid-catalyzed hydration

$$RCH{=}O + HA \rightleftharpoons RCH{=}O\text{-}\text{-}HA$$

$$RCH{=}O\text{-}\text{-}HA + H_2O \xrightarrow{slow} \underset{\underset{{}^+OH_2}{|}}{RCHOH} + A^- \rightleftharpoons RCH(OH)_2 + HA$$

Basic catalysts function by deprotonating water to give the more nucleophilic hydroxide ion.

Base-catalyzed hydration

$$B^- + H_2O + RCH{=}O \rightleftharpoons \underset{\underset{OH}{|}}{RCHO^-} + BH$$

$$\underset{\underset{OH}{|}}{RCHO^-} + H_2O \rightleftharpoons RCH(OH)_2 + {}^-OH$$

$$BH + {}^-OH \rightleftharpoons B^- + H_2O$$

MO (STO-3G) calculations on the gas-phase hydration reaction of formaldehyde suggest a concerted process involving two water molecules as a low-energy mechanism for hydration.

The calculated activation energy for this process in water is 16 kcal/mol, which is in good agreement with the experimentally observed value.[4]

Aldehydes and ketones undergo reversible addition reactions with alcohols. The product of addition of one mole of alcohol to an aldehyde or ketone is referred to as a *hemiacetal* or *hemiketal*, respectively. Dehydration followed by addition of a second molecule of alcohol gives an *acetal* or *ketal*. This second phase of the process can be catalyzed only by acids, since a necessary step is elimination of hydroxide (as water) from the tetrahedral intermediate. There is no low-energy mechanism for base assistance of this

3. L. H. Funderburk, L. Aldwin, and W. P. Jencks, *J. Am. Chem. Soc.* **100**:5444 (1978); R. A. McClelland and M. Coe, *J. Am. Chem. Soc.* **105**:2718 (1983).
4. I. H. Williams, D. Spangler, D. A. Femec, G. M. Maggiora, and R. L. Schowen, *J. Am. Chem. Soc.* **105**:31 (1983).

elimination step. For this reason, acetals and ketals are stable toward hydrolysis in alkaline aqueous solution but are hydrolyzed rapidly in acidic solution.

$$R_2C=O + R'OH \rightleftharpoons \overset{OH}{\underset{|}{R_2COR'}} \quad \textit{hemiketal}$$

$$\overset{OH}{\underset{|}{R_2COR'}} + H^+ \rightleftharpoons \overset{^+OH_2}{\underset{|}{R_2COR'}}$$

$$\overset{^+OH_2}{\underset{|}{R_2COR'}} \rightleftharpoons R_2C=\overset{+}{O}R' + H_2O$$

$$R_2C=\overset{+}{O}R' + R'OH \rightleftharpoons \overset{OH}{\underset{|}{R_2COR'}} \rightleftharpoons R_2C(OR')_2 + H^+$$
$$\underset{\underset{+}{HOR'}}{} \quad \textit{ketal}$$

The equilibrium constants for addition of alcohols to carbonyl compounds to give hemiacetals or hemiketals show the same response to structural features as the hydration reaction. Equilibrium constants for addition of methanol to acetaldehyde in both water and chloroform solution are near $0.8\ M^{-1}$. The comparable value for addition of water is about $0.02\ M^{-1}$.[5] The overall equilibrium constant for formation of the dimethyl acetal of acetaldehyde is $1.58\ M^{-1}$. Because the position of the equilibrium does not strongly favor product, the synthesis of acetals and ketals must be carried out in such a way as to drive the reaction to completion. One approach is to use a dehydrating reagent or azeotropic distillation so that the water that is formed is irreversibly removed from the system.

Because of the unfavorable equilibrium constant in aqueous solution and the relative facility of the hydrolysis, acetals and ketals are rapidly converted back to aldehydes and ketones in acidic aqueous solution.

$$R_2C(OR')_2 + H_2O \overset{H^+}{\longrightarrow} R_2C=O + 2R'OH$$

The mechanism of this hydrolysis reaction has been studied in great detail.[6] The mechanism is the reverse of that for acetal or ketal formation.

$$R_2C(OR')_2 + H^+ \rightleftharpoons R_2C\overset{\diagup OR'}{\underset{\diagdown OR'}{}} \rightleftharpoons R_2C=\overset{+}{O}R' + R'OH$$
$$\underset{^+H}{}$$

$$R_2C=\overset{+}{O}R' + H_2O \rightleftharpoons R_2C\overset{\diagup OR'}{\underset{\diagdown \underset{+}{OH_2}}{}} \rightleftharpoons R_2C\overset{\diagup OR}{\underset{\diagdown OH}{}} + H^+$$

$$R_2C\overset{\diagup OR'}{\underset{\diagdown OH}{}} + H^+ \rightleftharpoons R_2C=\overset{+}{O}H + R'OH \qquad R_2C=\overset{+}{O}H \rightleftharpoons R_2C=O + H^+$$

5. R. Bone, P. Cullis, and R. Wolfenden, *J. Am. Chem. Soc.* **105**:1339 (1983).
6. E. H. Cordes and H. G. Bull, *Chem. Rev.* **74**:581 (1974).

Some of the evidence which has helped to establish the general mechanism is as follows:

453

SECTION 8.1.
HYDRATION AND
ADDITION OF
ALCOHOLS TO
ALDEHYDES AND
KETONES

1. Isotopic labeling experiments have established that C−O bond rupture occurs between the carbonyl carbon and oxygen; substitution at the alcohol C−O bond is not involved.
2. For most acetals and ketals, the reaction is *specific-acid-catalyzed*. This is consistent with the existence of a preequilibrium in which the ketal is protonated. The role of the proton is to assist the departure of the alkoxy group by converting it to a better leaving group. In essence, this cleavage step is an S_N1 reaction with the remaining alkoxy group stabilizing the carbocation formed by ionization.
3. Hammett treatments show good correlations with large negative ρ values for the hydrolysis of acetals of aromatic aldehydes. This is consistent with the development of a positive charge at the carbonyl center in the rate-determining step.
4. Solvent isotope effects are usually in the range $k_{D_2O^+}/k_{H_2O^+} = 2–3$. These values reflect the greater equilibrium acidity of deuterated acids (Section 4.5) and indicate that the initial protonation is a fast preequilibrium.

Acetal and ketal hydrolyses usually exhibit specific acid catalysis, in agreement with a mechanism involving rate-determining cleavage of the conjugate acid of the reactant. General acid catalysis is observed, however, in certain acetals and ketals in which special structural features reduce the energy required for C−O bond cleavage.[7] Thus, hydrolysis of each of the acetals shown in Scheme 8.1 exhibits general acid catalysis, and each acetal has a special structural feature that facilitates C−O bond heterolysis. Easing the energy

Scheme 8.1. Acetals and Ketals That Exhibit General Acid Catalysis in Hydrolysis

	Acetal or ketal	Special structural feature
1[a]		Very stable carbocation (stabilized by both alkoxy function and aromaticity)
2[b]		Good leaving group
3[c]		Good leaving group
4[d]		Ring strain relieved in cleavage step
5[e]	$(Ar)_2C(OC_2H_5)_2$	Aryl substituents stabilize carbocation
6[f]	$PhCH[OC(CH_3)_3]_2$	Aryl stabilization and relief of steric strain

a. E. Anderson and T. H. Fife. *J. Am. Chem. Soc.* **91**:7163 (1969).
b. T. H. Fife and L. H. Brod, *J. Am. Chem. Soc.* **92**:1681 (1970).
c. J. L. Jensen and W. B. Wuhrman, *J. Org. Chem.* **48**:4686 (1983).
d. R. F. Atkinson and T. C. Bruice, *J. Am. Chem. Soc.* **96**:819 (1974).
e. R. H. DeWolfe, K. M. Ivanetich, and N. F. Perry, *J. Org. Chem.* **34**:848 (1969).
f. E. Anderson and T. H. Fife, *J. Am. Chem. Soc.* **93**:1701 (1971).

7. T. H. Fife, *Acc. Chem. Res.* **5**:264 (1972).

requirement for C—O bond cleavage permits the proton-transfer step to become partially rate-determining, which results in the observation of general acid catalysis.

Three-dimensional potential energy diagrams of the type discussed in connection with the variable E2 transition state theory for elimination reactions can be used to consider structural effects on the reactivity of carbonyl compounds and the tetrahedral intermediates involved in carbonyl-group reactions. Many of these reactions involve the formation or breaking of two separate bonds. This is the case in the first stage of acetal hydrolysis, which involves both a proton transfer and breaking of a C—O bond. The overall reaction might take place in several ways. There are two mechanistic extremes:

1. The proton could be completely transferred and then the departing alcohol molecule would leave to form a carbocation in a distinct second step. This is the specific acid-catalyzed mechansism.
2. The acetal might undergo ionization with formation of an alkoxide ion and a carbocation. In a second step, the alkoxide would be protonated. This mechanism is extremely rare, if not impossible, because an alkoxide ion is a poor leaving group.

There is an intermediate mechanism between these extremes. This is a general acid catalysis in which the proton transfer and the C—O bond rupture occur as a *concerted* process. The concerted process need not be perfectly synchronous; that is, proton transfer might be more complete at the transition state than C—O rupture, or vice versa. These ideas are represented in a three-dimensional energy diagram in Fig. 8.1.

The two paths around the edge of the diagram represent the stepwise processes described as the mechanistic extremes 1 and 2. We know that process 2, represented by path a, is a high-energy process, so the upper-left corner of the diagram would have a high energy. The paths designated b and c indicate concerted but nonsynchronous mechanisms in which there is both partial proton transfer and partial C—O bond rupture at the transition state. In path b, C—O cleavage is more complete than proton transfer at the transition state, whereas the reverse is true for path c. Both these paths represent concerted, general acid-catalyzed processes. Path d represents the specific acid-catalyzed process in which proton transfer precedes C—O cleavage.

If it is possible to estimate or calculate the energy of the reacting system at various stages, the energy dimension can be added, as in Fig. 8.2. The actual mechanism will then

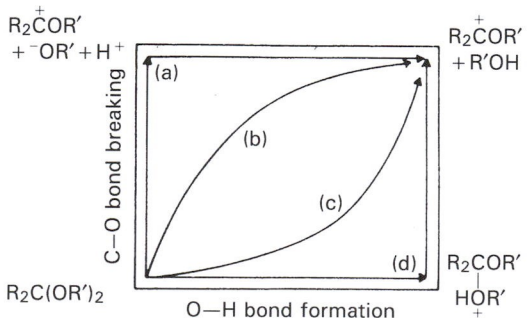

Fig. 8.1. Representation of transition states for the first stage of acetal hydrolysis. (a) Initial C—O bond breaking; (b) concerted mechanism with C—O bond breaking leading O—H bond formation; (c) concerted mechanism with proton transfer leading C—O bond breaking; (d) initial proton transfer.

455

SECTION 8.1.
HYDRATION AND
ADDITION OF
ALCOHOLS TO
ALDEHYDES AND
KETONES

Fig. 8.2. Contour plot showing a favored concerted mechanism for the first step in acetal hydrolysis, in which proton transfer is more complete in the transition state than C—O bond breaking.

be the process which proceeds over the lowest energy barrier. The energy dimension can be shown as contours. The diagram in Fig. 8.2 shows the initial ionization to an alkoxide and carbocation as being very high in energy. The stepwise path of protonation followed by ionization is shown with small barriers, with the protonated ketal as an intermediate. The lowest-energy path is shown as a concerted process represented by the dashed line. The transition state, which lies at the highest-energy point on this line, would exhibit more complete proton transfer than C—O cleavage.

Structural effects can be discussed by asking how they will affect the position of the transition state on the potential energy surface. The stepwise path via the protonated acetal should be followed in the case of alcohols that are poor leaving groups. If the alcohol is more acidic, its conjugate base is a better leaving group, and the transition state will be shifted to a point where C—O bond breaking begins before proton transfer is complete. This would mean that the mechanism is concerted, although the transition state would still have much of the character of a carbocation. Three-dimensional reaction energy diagrams can be used to describe how structural changes can affect the nature of the transition state. Just as the two-dimensional diagrams can give meaning to such phrases as an "early" or a "late" transition state, the three-dimensional diagrams are illustrative of statements such as "C—O cleavage is more advanced than proton transfer."

Consideration of the types of acetals shown in Scheme 8.1, the hydrolysis reactions of which exhibit general acid catalysis, indicates why the concerted mechanism operates in these molecules. The developing aromatic character of the cation formed in the case of entry 1 will lower the energy requirement for C—O bond rupture. The bond can begin to break before protonation is complete. Entries 2 and 3 represent cases in which good leaving groups (a stable phenolate anion) reduce the energy requirement for C—O bond cleavage. In entry 4, a four-membered ring is broken in the reaction. Cleavage in this case is facilitated by release of strain energy. Entries 5 and 6 are similar to entry 1 because the aryl groups provide stabilization for developing cationic character.

The second step in acetal and ketal hydrolysis is conversion of the hemiacetal or hemiketal to the carbonyl compound. The mechanism of this step is similar to that of the first step. Usually, the second step is faster than the initial one.[8] Hammett σ–ρ plots and solvent isotope effects both indicate that the transition state has less cationic character than

8. Y. Chiang and A. J. Kresge, *J. Org. Chem.* **50**:5038 (1985); R. A. McClelland, K. M. Engell, T. S. Larsen, and P. O. Sorensen, *J. Chem. Soc., Perkin Trans. 2* **1994**:2199.

in the case of the first step. These features of the mechanism suggest that a concerted removal of the proton at the hydroxyl group occurs as the alcohol is eliminated.

$$\begin{array}{c} \overset{\delta+}{O\text{-}\text{-}H\text{-}\overset{\delta+}{\underset{\smile}{OH_2}}} \\ Ph\text{—}\overset{\delta+}{\underset{\underset{\smile}{|}}{C}}\text{—}H \\ (\overset{|}{OC_2H_5} \\ \overset{\delta+}{H\text{-}\underset{\smile}{OH_2}} \end{array} \longrightarrow PhCH{=}O \qquad\qquad \begin{array}{l} H_3O^+ \\[4pt] HOC_2H_5 \\[4pt] H_2O \end{array}$$

This would disperse the positive charge over several atoms and diminish the sensitivity of the reaction to substituent effects. The ρ values that are observed are consistent with this interpretation. Whereas ρ is -3.25 for acetal hydrolysis, it is only -1.9 for hemiacetal hydrolysis.[9]

In contrast to acetals, which are base-stable, hemiacetals undergo base-catalyzed hydrolysis. In the alkaline pH range, the mechanism shifts toward a base-catalyzed elimination.

$$Ph\text{—}\overset{\overset{\displaystyle OH}{|}}{\underset{\underset{\displaystyle OC_2H_5}{|}}{C}}\text{—}H + {}^-OH \rightleftharpoons Ph\text{—}\overset{\overset{\displaystyle O^-}{|}}{\underset{\underset{\displaystyle OC_2H_5}{(}}{C}}\text{—}H \longrightarrow PhCH{=}O + C_2H_5O^-$$

There are two opposing substituent effects on this reaction. Electron-attracting aryl substituents favor the deprotonation but disfavor the elimination step. The observed substituent effects are small, and under some conditions the Hammett plot is nonlinear.[10]

8.2. Addition–Elimination Reactions of Ketones and Aldehydes

The mechanistic pattern established by study of hydration and alcohol addition reactions of ketones and aldehydes is followed in a number of other reactions of carbonyl compounds. Reactions at carbonyl centers usually involve a series of addition and elimination steps proceeding through tetrahedral intermediates. These steps can be either acid-catalyzed or base-catalyzed. The rate and products of the reaction are determined by the reactivity of these tetrahedral intermediates.

In general terms, there are three possible mechanisms for addition of a nucleophile and a proton to give a tetrahedral intermediate in a carbonyl addition reaction.

(A) Protonation followed by nucleophilic attack on the protonated carbonyl group:

$$H^+ + \overset{\backslash}{\underset{/}{C}}{=}O \longrightarrow \overset{\backslash}{\underset{/}{C}}{=}\overset{+}{O}\text{—}H$$

$$Nu{:}^- + \overset{\backslash}{\underset{/}{C}}{=}\overset{+}{O}\text{—}H \longrightarrow Nu\text{—}\overset{\overset{\displaystyle |}{}}{\underset{\underset{\displaystyle |}{}}{C}}\text{—}O\text{—}H$$

9. T. J. Przystas and T. H. Fife, *J. Am. Chem. Soc.* **103**:4884 (1981).
10. R. L. Finley, D. G. Kubler, and R. A. McClelland, *J. Org. Chem.* **45**:644 (1980).

(B) Nucleophilic addition at the carbonyl group followed by protonation:

457

SECTION 8.2.
ADDITION–
ELIMINATION
REACTIONS OF
KETONES AND
ALDEHYDES

$$Nu:^- + \underset{/}{\overset{\backslash}{C}}=O \longrightarrow Nu-\underset{|}{\overset{|}{C}}-O^-$$

$$Nu-\underset{|}{\overset{|}{C}}-O^- + H^+ \longrightarrow Nu-\underset{|}{\overset{|}{C}}-O-H$$

(C) Concerted proton transfer and nucleophilic attack:

$$Nu:^- \quad \underset{/}{\overset{\backslash}{C}}=O \quad H^+ \longrightarrow Nu-\underset{|}{\overset{|}{C}}-O-H$$

There are examples of each of these mechanisms, and a three-dimensional potential energy diagram can provide a useful general framework within which to consider specific addition reactions. The breakdown of a tetrahedral intermediate involves the same processes but operates in the opposite direction, so the principles that are developed will apply equally well to the reactions of the tetrahedral intermediates. Let us examine the three general mechanistic cases in relation to the energy diagram in Fig. 8.3.

Case A should be favored for weak nucleophiles. The protonated carbonyl compound will be much more highly reactive toward such a nucleophile. This mechanism is most likely to operate in relatively acidic conditions. Case B should be favored for strong nucleophiles. Such nucleophiles will generally be more basic than carbonyl groups and would be protonated in preference to the carbonyl group. In such systems, proton donors would diminish the overall reaction rate by decreasing the amount of anionic nucleophile that is available for reaction. The addition reactions of primary amines, for example, are generally not catalyzed by acids because the protonated amines are not nucleophilic toward the carbonyl group and the carbonyl group does not compete with the amine as a site for protonation. Similarly, carbanions generally cannot exist under acidic conditions so carbanion additions are entirely nucleophilic in character. The concerted mechanism C is observed for less basic nucleophiles. The simultaneous transfer of the proton at the carbonyl oxygen assists in achieving addition by species that are not sufficiently nucleophilic to react by mechanism B. The general trend then is that the weaker and

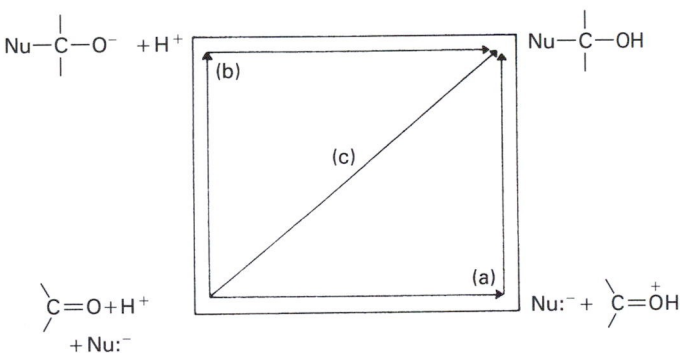

Fig. 8.3. Three-dimensional potential energy diagram for addition of a proton and nucleophile to a carbonyl group. (a) Proton transfer complete before nucleophilic addition begins; (b) nucleophilic addition complete before proton transfer begins; (c) concerted proton transfer and nucleophilic addition.

less basic the nucleophile, the more important is partial or complete protonation of the carbonyl group. If we consider the reverse process, the same general relationships will hold. The good leaving groups (which are poor nucleophiles) can be expected to follow path A, poor leaving groups will follow path B, and intermediate cases are likely to react by the concerted mechanism.

Certain nucleophilic species add to carbonyl groups to give tetrahedral intermediates that are unstable and break down to form a new double bond. An important group of such reactions are those with compounds containing primary amino groups. Scheme 8.2 lists some of the more familiar classes of such reactions. In general, these reactions are reversible, and mechanistic information can be obtained by study of either the forward or the reverse process.

The hydrolysis of simple imines occurs readily in aqueous acid and has been studied in great detail by kinetic methods. The precise mechanism is a function of the reactant structure and the pH of the solution. The overall mechanism consists of an addition of water to the $C{=}N$ bond, followed by expulsion of the amine from a tetrahedral intermediate.[11]

$$RCH{=}NR' + H^+ \rightleftharpoons RCH{=}\overset{+}{N}HR'$$

$$RCH{=}\overset{+}{N}HR' + H_2O \rightleftharpoons \underset{\overset{|}{{}^+OH_2}}{RCHNHR'} \rightleftharpoons \underset{\overset{|}{OH}}{RCH\overset{+}{N}H_2R'}$$

or

$$RCH{=}\overset{+}{N}HR' + {}^-OH \rightleftharpoons \underset{\overset{|}{OH}}{RCHNHR'}$$

$$\underset{\overset{|}{OH}}{RCHNHR'} + H^+ \rightleftharpoons \underset{\overset{|}{OH}}{RCH\overset{+}{N}H_2R'}$$

$${}^-OH + \underset{\overset{|}{OH}}{RCH\overset{+}{N}H_2R'} \rightleftharpoons H_2O + \underset{\overset{|}{\_O}}{RCH\overset{+}{N}H_2R'} \rightleftharpoons \underset{\overset{||}{O}}{RCH} + H_2NR'$$

**Scheme 8.2. Addition–Elimination Reactions
of Aldehydes and Ketones**

$$R_2C{=}O + R'NH_2 \rightleftharpoons R_2C{=}NR'$$

imine (also often called Schiff base, especially when the amine is an aniline derivative)

$$R_2C{=}O + HONH_2 \rightleftharpoons R_2C{=}NOH$$

oxime

$$R_2C{=}O + NH_2NHR' \rightleftharpoons R_2C{=}NNHR'$$

hydrazone

$$R_2C{=}O + NH_2NH\overset{\overset{O}{||}}{C}NH_2 \rightleftharpoons R_2C{=}NNHCNH_2$$

semicarbazone

11. J. Hine, J. C. Craig, Jr., J. G. Underwood II, and F. A. Via, *J. Am. Chem. Soc.* **92:**5194 (1970); E. H. Cordes and W. P. Jencks, *J. Am. Chem. Soc.* **85:**2843 (1963).

459

SECTION 8.2.
ADDITION–
ELIMINATION
REACTIONS OF
KETONES AND
ALDEHYDES

The relative rates of the various steps are a function of the pH of the solution and the basicity of the imine. In the alkaline range, the rate-determining step is usually nucleophilic attack by hydroxide ion on the protonated C=N bond. At intermediate pH values, water replaces hydroxide as the dominant nucleophile. In acidic solution, the rate-determining step becomes the breakdown of the tetrahedral intermediate. A mechanism of this sort, in which the overall rate is sensitive to pH, can be usefully studied by constructing a pH–rate profile, which is a plot of the observed rate constants versus pH. Figure 8.4 is an example of the pH–rate profile for hydrolysis of a series of imines derived from substituted aromatic aldehydes and *t*-butylamine. The form of pH–rate profiles can be predicted on the basis of the detailed mechanism of the reaction. The value of the observed rate can be calculated as a function of pH if a sufficient number of the individual rate constants and the acid dissociation constants of the participating species are known. Agreement between the calculated and observed pH–rate profile can serve as a sensitive test of the adequacy of the postulated mechanism. Alternatively, one may begin with the experimental pH–rate profile and deduce details of the mechanism from it.

Complete understanding of the shapes of the curves in Fig. 8.4 requires a kinetic expression somewhat more complicated than we wish to deal with here. The nature of the extremities of the curves can be understood, however, on the basis of qualitative arguments. The rate decreases with a decrease in pH in the acidic region because formation of the zwitterionic tetrahedral intermediate is required for expulsion of the

Fig. 8.4. Logarithm of the first-order rate constants for the hydrolysis of substituted benzylidene-1,1-dimethyl-ethylamines as a function of pH. [Reproduced from *J. Am. Chem. Soc.* **85:** 2843 (1963) by permission of the American Chemical Society.]

amine. The concentration of the zwitterionic species decreases with decreasing acidity, because it is governed by an acid–base equilibrium:

$$K = \frac{[H^+][RCH\overset{+}{N}H_2R]}{[RC\overset{|}{H}NH_2R]}$$

with O^- above the first bracket group and OH below the second bracket group.

As the hydrogen-ion concentration increases, the concentration of the reactive form of the intermediate decreases. Note also that, in the acidic region, electron-withdrawing substituents accelerate the reaction. This is due to a more favorable equilibrium for the hydration step. In the alkaline region, the rate is pH-independent. In this region, the rate-controlling step is attack of the hydroxide ion on the protonated imine. The concentration of both of these species is pH-dependent, but in opposite, compensating ways. The overall rate is therefore pH-independent in the alkaline range. (Work Problem 6 at the end of this chapter to establish that this is so.)

The formation of imines takes place by a mechanism that is the reverse of the hydrolysis. Preparative procedures often ensure completion of the reaction by removing water as it is formed by azeotropic distillation or by the use of an irreversible dehydrating agent.

The other C=N systems included in Scheme 8.2 are more stable to aqueous hydrolysis than are the imines. For many of these compounds, the equilibrium constants for formation are high, even in aqueous solution. The additional stability can be attributed to the participation of the atom adjacent to the nitrogen in delocalized bonding. This resonance interaction tends to increase electron density at the sp^2 carbon and reduces its reactivity toward nucleophiles.

$$R_2C{=}N{-}\ddot{O}H \longleftrightarrow R_2\bar{C}{-}N{=}\overset{+}{O}H$$

$$R_2C{=}N{-}\ddot{N}H_2 \longleftrightarrow R_2\bar{C}{-}N{=}\overset{+}{N}H_2$$

The formation of oximes, hydrazones, and related imine derivatives is usually catalyzed by both general acids and general bases. General base catalysis of dehydration of the tetrahedral intermediate involves nitrogen deprotonation concerted with elimination of hydroxide ion.[12]

$$R_2C{=}O + NH_2R' \underset{}{\overset{fast}{\rightleftharpoons}} R_2\overset{OH}{\underset{}{CH}}NR'$$

$$\overset{\frown}{B} \quad H{-}\overset{R}{\underset{R'}{N}}{-}\overset{R}{\underset{}{C}}{-}OH \xrightarrow{slow} \overset{+}{B}H + R'N{=}CR_2 + \bar{O}H$$

12. W. P. Jencks, *Prog. Phys. Org. Chem.* **2**:63 (1964); J. M. Sayer, M. Peskin, and W. P. Jencks, *J. Am. Chem. Soc.* **95**:4277 (1973).

General acid catalysis of the breakdown of the carbinolamine intermediate occurs by assistance of the expulsion of water.

461

SECTION 8.2.
ADDITION–
ELIMINATION
REACTIONS OF
KETONES AND
ALDEHYDES

$$R_2C{-}\overset{H}{\underset{\underset{R'}{\overset{|}{NH}}}{\overset{|}{O}}} \; H{-}A \longrightarrow R_2C{=}\overset{+}{\underset{H}{N}}R' + H_2O + A^-$$

As with simple imines, the identity of the rate-limiting step changes with solution pH. As the pH decreases, the rate of the addition decreases because protonation of the amino compound reduces the concentration of the nucleophilic unprotonated form. Thus, whereas the dehydration step is normally rate-determining in neutral and basic solution, addition becomes rate-determining in acidic solutions.

Secondary amines cannot form imines, and dehydration proceeds to give carbon–carbon double bonds bearing amino substituents (enamines). Enamines were mentioned in Chapter 7 as examples of nucleophilic carbon species, and their synthetic utility is discussed in Chapter 1 of Part B. The equilibrium for the reaction between secondary amines and carbonyl compounds ordinarily lies far to the left in aqueous solution, but the reaction can be driven forward by dehydration methods.

$$R_2''NH + RCH_2\overset{\overset{O}{\|}}{C}R' \rightleftharpoons RCH_2\overset{\overset{OH}{|}}{\underset{NR_2''}{C}}R' \rightleftharpoons RCH{=}\overset{}{\underset{NR_2''}{C}}R' + H_2O$$

The mechanism of hydrolysis of enamines has been studied kinetically over a range of pH. In alkaline solution, rate-determining C-protonation is followed by attack of hydroxide ion on the resulting iminium ion. The carbinolamine intermediate then breaks down as in imine hydrolysis. In the neutral and weakly acidic pH range, water attack on the C-protonated enamine becomes rate-limiting. As in imine hydrolysis, decomposition of the tetrahedral intermediate becomes rate-limiting in strongly acidic solution.[13]

$$RCH{=}\overset{}{\underset{R'}{C}}NR_2'' + H^+ \longrightarrow RCH_2\overset{}{\underset{R'}{C}}{=}\overset{+}{N}R_2''$$

$$RCH_2\overset{}{\underset{R'}{C}}{=}\overset{+}{N}R_2'' + {}^-OH \longrightarrow RCH_2\overset{\overset{OH}{|}}{\underset{R'}{C}}NR_2'' \longrightarrow RCH_2\overset{\overset{O}{\|}}{C}R' + HNR_2''$$

Certain reactions between carbonyl compounds and nucleophiles are catalyzed by amines. Some of these reactions are of importance for forming carbon–carbon bonds, and these are discussed in Chapter 2 of Part B. The mechanistic principle can be illustrated by considering the catalysis of the reaction between aldehydes and hydroxylamine by aniline derivatives.

$$ArCH{=}O + NH_2OH \xrightarrow{Ar'NH_2} ArCH{=}NOH + H_2O$$

13. P. Y. Sollenberger and R. B. Martin, *J. Am. Chem. Soc.* **92**:4261 (1970); W. Maas, M. J. Janssen, E. J. Stamhuis, and H. Wynberg, *J. Org. Chem.* **32**:1111 (1967); E. J. Stamhuis and W. Maas, *J. Org. Chem.* **30**:2156 (1965).

Analysis of the kinetics of this catalysis points to the protonated imine as the key intermediate.

$$ArCH{=}O + Ar'NH_2 + H^+ \longrightarrow ArCH{=}\overset{+}{N}HAr' + H_2O$$

$$ArCH{=}\overset{+}{N}HAr' + NH_2OH \xrightarrow{\text{fast}} ArCH{=}NOH + Ar'NH_3$$

Because the imine is much more basic than the carbonyl compound, it is more extensively protonated at any given pH than is the aldehyde. The protonated imine is also more reactive as an electrophile than the neutral aldehyde. There are four possible electrophiles in the system:

$$ArCH{=}\overset{+}{O}H > ArCH{=}\overset{+}{N}HAr' > ArCH{=}O > ArCH{=}NAr'$$
$$\underrightarrow{\text{decreasing activity}}$$

The protonated imine is the dominant reactive form. Although the protonated aldehyde is more reactive, its concentration is very low because it is much less basic than the imine or the reactant hydroxylamine. On the other hand, even though the aldehyde may be present in a greater concentration than the protonated imine, its reactivity is sufficiently less that the iminium ion is the major reactant.[14]

8.3. Addition of Carbon Nucleophiles to Carbonyl Groups

The addition of carbon nucleophiles, including organometallic compounds, enolates, or enols, and ylides to carbonyl groups is an important method of formation of carbon–carbon bonds. Such reactions are extremely important in synthesis and will be discussed extensively in Part B. Here, we will examine some of the fundamental mechanistic aspects of addition of carbon nucleophiles to carbonyl groups.

Organolithium and organomagnesium reagents are highly reactive toward most carbonyl compounds. With aldehydes and ketones, the tetrahedral adduct is stable, and alcohols are isolated after protonation of the adduct, which is an alkoxide ion.

$$R'M + R_2C{=}O \longrightarrow R'{-}\overset{\overset{\displaystyle R}{|}}{\underset{\underset{\displaystyle R}{|}}{C}}{-}O^-M^+ \xrightarrow{H^+} R'{-}\overset{\overset{\displaystyle R}{|}}{\underset{\underset{\displaystyle R}{|}}{C}}{-}OH$$

$$M = \text{Li or MgX}$$

In the case of esters, carboxylate anions, amides, and acid chlorides, the tetrahedral adduct may undergo elimination. The elimination forms a ketone, permitting a second addition step to occur. The rate at which breakdown of the tetrahedral adduct occurs is a function of the reactivity of the heteroatom substituent as a leaving group. The order of stability of the

14. E. H. Cordes and W. P. Jencks, *J. Am. Chem. Soc.* **84**:826 (1962); J. Hine, R. C. Dempsey, R. A. Evangelista, E. T. Jarvi, and J. M. Wilson, *J. Org. Chem.* **42**:1593 (1977).

tetrahedral adducts is shown below.

— decreasing stability of tetrahedral adduct —→

In most cases, the product ratio can be controlled by choice of reaction conditions. Ketones are isolated under conditions where the tetrahedral intermediate is stable until hydrolyzed, whereas tertiary alcohols are formed when the tetrahedral intermediate decomposes while unreacted organometallic reagent remains. Examples of synthetic application of these reactions will be discussed in Chapter 7 of Part B.

The reaction of organolithium reagents with simple carbonyl compounds is very fast, and there is relatively little direct kinetic evidence concerning the details of the reaction. It would be expected that one important factor in determining reactivity would be the degree of aggregation of the organolithium reagent. It has been possible to follow the reaction of benzaldehyde with n-butyllithium at $-85°C$, using NMR techniques which are capable of monitoring fast reactions. The reaction occurs over a period of 50–300 milliseconds. It has been concluded that the dimer of n-butyllithium is more reactive than the tetramer by a factor of about 10. As the reaction proceeds, the product alkoxide ion is incorporated into butyllithium aggregates. This gives rise to additional species with different reactivity.[15]

The rates of the reactions of several aromatic ketones with alkyllithium reagents have been examined. The reaction of 2,4-dimethyl-4'-(methylthio)benzophenone with methyl-lithium in ether exhibits the rate expression:

$$\text{rate} = k[\text{MeLi}]^{1/4}[\text{ketone}]$$

This is consistent with a mechanism in which monomeric methyllithium in equilibrium

15. J. F. McGarrity, C. A. Ogle, Z. Brich, and H.-R. Loosli, *J. Am. Chem. Soc.* **107**:1810 (1985).

with the tetramer is the reactive nucleophile.[16]

$$[MeLi]_4 \rightleftharpoons 4\ MeLi$$

$$MeLi + ketone \xrightarrow{slow} product$$

Most other studies have indicated considerably more complex behavior. The rate data for reaction of 3-methyl-1-phenylbutanone with s-butyllithium or n-butyllithium in cyclohexane can be fit to a mechanism involving product formation both through a complex of the ketone with alkyllithium aggregate and by reaction with dissociated alkyllithium.[17] Evidence for the initial formation of a complex can be observed in the form of a shift in the carbonyl absorption band in the IR spectrum. Complex formation presumably involves a Lewis acid–Lewis base interaction between the carbonyl oxygen and lithium ions in the alkyllithium cluster.

In general terms, it appears likely that alkyllithium reagents have the possibility of reacting through any of several aggregated and dissociated forms.

MO modeling (3-21G) of the reaction of organolithium compounds with carbonyl groups has examined the interaction of formaldehyde with the dimer of methyllithium. The reaction is predicted to proceed by initial complexation of the carbonyl group at lithium, followed by a rate-determining step involving formation of the new carbon–carbon bond. The cluster then reorganizes to incorporate the newly formed alkoxide ion.[18] The transition state is reached very early in the second step, with only slight formation of the C—C bond.

The kinetics of addition of alkyllithium reagents to esters have been studied using a series of ethyl benzoates. The rates show a rather complex dependence on both alkyllithium concentration and the nature of aryl substituents in the ester. The rapid formation of an initial ester–alkyllithium complex can be demonstrated. It is believed that

16. S. G. Smith, L. F. Charbonneau, D. P. Novak, and T. L. Brown, *J. Am. Chem. Soc.* **94**:7059 (1972).
17. M. A. Al-Aseer and S. G. Smith, *J. Org. Chem.* **49**:2608 (1984).
18. E. Kaufman, P. v. R. Schleyer, K. N. Houk, and Y.-D. Wu, *J. Am. Chem. Soc.* **107**:5560 (1985).

product can be formed by reaction with both aggregated and monomeric alkyllithium reagent. TMEDA greatly accelerates the reaction, presumably by dissociating the organometallic aggregate (see Section 7.1).

$$(RLi)_n + {}_nTMEDA \rightleftharpoons \frac{n}{2}[(RLi)_2 \cdot 2TMEDA]$$

$$[(RLi)_2 \cdot 2TMEDA] + R'O_2CPh \longrightarrow \left[\begin{array}{c} O-R' \\ | \\ R-C-O^-Li^+ \cdot RLi \cdot 2TMEDA \\ | \\ Ph \end{array} \right]$$

The kinetics of reaction of Grignard reagents with ketones are also subject to a number of complications. The purity of the magnesium metal used in the preparation of the Grignard reagent is crucial because trace transition-metal impurities have a major effect on the observed reaction rates. One of the most thorough studies involves the reaction of methylmagnesium bromide with 2-methylbenzophenone in diethyl ether.[19] The results suggest that the reaction mechanism is similar to that discussed for alkyllithium reactions. There is initial complexation between the ketone and Grignard reagent. The main Grignard species CH_3MgBr is in equilibrium with $(CH_3)_2Mg$, and the latter species can contribute to the overall rate. Finally, the product alkoxide complexes with the Grignard reagent to give another reactive species. The general mechanistic scheme is outlined below.

$$2RMgX \rightleftharpoons MgX_2 + R_2Mg$$

$$RMgX + R_2'C{=}O \longrightarrow [R_2'C{=}O{\cdots}Mg{-}R] \longrightarrow R{-}\underset{\underset{R'}{|}}{\overset{\overset{R'}{|}}{C}}{-}O^{-+}MgX$$

$$R{-}\underset{\underset{R'}{|}}{\overset{\overset{R'}{|}}{C}}{-}O^{-+}MgX + RMgX \longrightarrow \left[R{-}\underset{\underset{R'}{|}}{\overset{\overset{R'}{|}}{C}}{-}O^-Mg^+\overset{X}{\underset{X}{\cdots}}Mg{-}R \right]$$

$$\left[R{-}\underset{\underset{R'}{|}}{\overset{\overset{R'}{|}}{C}}{-}O^-Mg^+\overset{X}{\underset{X}{\cdots}}Mg{-}R \right] + R_2'C{=}O \longrightarrow R{-}\underset{\underset{R'}{|}}{\overset{\overset{R'}{|}}{C}}{-}O^-Mg^+\overset{X}{\underset{X}{\diamondsuit}}Mg^+{-}O^-\underset{\underset{R'}{|}}{\overset{\overset{R'}{|}}{C}}{-}R$$

There is another possible mechanism for addition of organometallic reagents to carbonyl compounds. This involves a discrete electron-transfer step.[20]

$$(R{-}M)_n + O{=}CR_2' \longrightarrow [(R{-}M)_n \, O{=}CR_2'] \longrightarrow [(R{-}\overset{+}{M})_n \, {}^-O{-}\!\!\!\!\cdot CR_2'] \longrightarrow (R{-}\overset{+}{M})_{n-1} + M^+ {}^-O{-}\underset{\underset{R'}{|}}{\overset{\overset{R'}{|}}{C}}{-}R$$

The distinguishing feature of this mechanism is the second step, in which an electron is transferred from the organometallic reagent to the carbonyl compound to give the radical

19. E. C. Ashby, J. Laemmle, and H. M. Neumann, *J. Am. Chem. Soc.* **94**:5421 (1972).
20. E. C. Ashby, *Pure Appl. Chem.* **52**:545 (1980); E. C. Ashby, J. Laemmle, and H. M. Neumann, *Acc. Chem. Res.* **7**:272 (1974).

anion of the carbonyl compound. Subsequent collapse of the ion pair with transfer of an alkyl group to carbon gives the same product as is formed in the normal mechanism. The electron-transfer mechanism would be expected to be favored by structural features which stabilize the radical-anion intermediate. Aryl ketones and diones fulfill this requirement, and much evidence for the electron-transfer mechanism has been accumulated for such ketones. In several cases, it is possible to observe a radical-anion intermediate by EPR spectroscopy.[21] (See Section 4.7 for a discussion of some of the limitations of this methodology.)

The stereoselectivity of organometallic additions with carbonyl compounds fits into the general pattern for nucleophilic attack discussed in Chapter 3. With 4-t-butylcyclohexanone, there is a preference for equatorial approach but the selectivity is low. Enhanced steric factors promote stereoselective addition.

(Ref. 22)

Stereoselectivity in approach of methyllithium to cyclic ketones

Addition of Grignard reagents to ketones and aldehydes was one of the reactions which led to the formulation of Cram's rule.[23] Many ketones and aldehydes have subsequently been examined to determine the degree of stereoselectivity. Cram's rule is obeyed when no special complexing functional groups are present near the reaction site. One series of studies is summarized in Table 8.2.

Enolates can also serve as carbon nucleophiles in carbonyl addition reactions. The addition reaction of enolates with carbonyl compounds is of very broad scope and is of great synthetic importance. Essentially all of the enolates considered in Chapter 7 are capable of adding to carbonyl groups. The reaction is known as the *generalized aldol addition*.

Enolates of aldehydes, ketones, and esters and the carbanions of nitriles and nitro compounds, as well as phosphorus- and sulfur-stabilized carbanions and ylides, undergo the reaction. The synthetic applications of this group of reactions will be discussed in detail in Chapter 2 of Part B. In this section, we will discuss the fundamental mechanistic aspects of the reaction of ketone enolates with aldehydes and ketones.

The aldol addition can be carried out under either of two broad sets of conditions, with the product being determined by kinetic factors under one set of conditions and by thermodynamic factors under the other. To achieve *kinetic control*, the enolate that is to

21. K. Maruyama and T. Katagiri, *J. Am. Chem. Soc.* **108**:6263 (1986); E. C. Ashby and A. B. Goel, *J. Am. Chem. Soc.* **103**:4983 (1981); T. Lund, M. L. Pedersen, and L. A. Frandsen, *Tetrahedron Lett.* **35**:9225 (1994).
22. E. C. Ashby and S. A. Noding, *J. Org. Chem.* **44**:4371 (1979).
23. D. J. Cram and F. A. Abd Elhafez, *J. Am. Chem. Soc.* **74**:5828 (1952); D. J. Cram and J. D. Knight, *J. Am. Chem. Soc.* **74**:5835 (1952); F. A. Abd Elhafez and D. J. Cram, *J. Am. Chem. Soc.* **74**:5846 (1952).

Table 8.2. Stereoselectivity in Addition of Organometallic Reagents to Some Chiral Aldehydes and Ketones[a]

R	L	M	S	R'M	Percent of favored product
H	Ph	CH_3	H	CH_3MgBr	71
H	Ph	CH_3	H	$PhMgBr$	78
H	t-C_4H_9	CH_3	H	$PhMgBr$	98
CH_3	Ph	CH_3	H	C_2H_5Li	93
CH_3	Ph	CH_3	H	C_2H_5MgBr	88
CH_3	Ph	CH_3	H	t-C_4H_9MgBr	96
C_2H_5	Ph	CH_3	H	CH_3MgBr	86
C_2H_5	Ph	CH_3	H	CH_3Li	94
C_2H_5	Ph	CH_3	H	$PhLi$	85
i-C_3H_7	Ph	CH_3	H	CH_3MgBr	90
i-C_3H_7	Ph	CH_3	H	CH_3Li	96
i-C_3H_7	Ph	CH_3	H	$PhLi$	96
t-C_4H_9	Ph	CH_3	H	CH_3MgBr	96
t-C_4H_9	Ph	CH_3	H	CH_3Li	97
t-C_4H_9	Ph	CH_3	H	$PhLi$	98
Ph	Ph	CH_3	H	CH_3MgBr	87
Ph	Ph	CH_3	H	CH_3Li	97
Ph	Ph	CH_3	H	t-C_4H_9MgBr	96

a. Data from O. Arjona, R. Perez-Ossorio, A. Perez-Rubalcaba, and M. L. Quiroga, *J. Chem. Soc., Perkin Trans. 2* **1981**:597; C. Alvarez-Ibarra, P. Perez-Ossorio, A. Perez-Rubalcaba, M. L. Quiroga, and M. J. Santesmases, *J. Chem. Soc., Perkin Trans. 2* **1983**:1645.

serve as the nucleophile is generated stoichiometrically, usually with lithium as the counterion in an aprotic solvent. Under these conditions, enolates are both structurally and stereochemically stable. They do not equilibrate with the other isomeric or stereoisomeric enolates that can be formed from the same ketone. The electrophilic carbonyl compound is then added. Under these conditions, the reaction product is determined primarily by two factors: (1) the structure of the initial enolate and (2) the stereoselectivity of the addition to the electrophilic carbonyl group.

For the other broad category of reaction conditions, the reaction proceeds under conditions of *thermodynamic control*. This can result from several factors. Aldol condensations can be effected for many compounds using less than a stoichiometric amount of base. Under these conditions, the aldol reaction is reversible, and the product ratio will be determined by the relative stability of the various possible products. Conditions of thermodynamic control also permit equilibration among all the enolates of the nucleophile. The conditions that permit equilibration include higher reaction temperatures, protic solvents, and the use of less tightly coordinating cations.

The fundamental mechanistic concept by which the stereochemical course of the aldol addition under conditions of kinetic control has been analyzed involves a cyclic transition state in which both the carbonyl and enolate oxygens are coordinated to a Lewis acid. We will use the Li^+ cation in our discussion, but another metal cation or electron-deficient atom may play the same role.

According to this concept, the aldol condensation normally occurs through a chairlike transition state. It is further assumed that the structure of this transition state is sufficiently similar to that of chair cyclohexane to allow the conformational concepts developed for cyclohexane derivatives to be applied. Thus, in the example above, the reacting aldehyde is shown with R rather than H in the equatorial-like position. The differences in stability of the various transition states, and therefore the product ratios, are governed by the steric interactions between substituents.

A consequence of this mechanism is that the reaction is *stereospecific* with respect to the *E*- or *Z*-configuration of the enolate. The *E*-enolate will give the *anti* aldol product whereas the *Z*-enolate will give the *syn* aldol.

E-enolate ||| *Z*-enolate |||

anti-product   *syn*-product

Numerous observations on the reactions of enolates of both ketones and esters are consistent with this general concept.[24] However, the *specific ratio* of *syn* and *anti* product from any given reaction process depends on several variables, and the prediction or interpretation of this ratio requires that the following assessments be made. (1) What is the stereochemical composition of the enolate which is reacting? (2) How strong is the selectivity between the two faces of the carbonyl group? This is a function of the steric interactions in the diastereomeric transition states. (3) Does the Lewis acid character of the cation promote a tight coordination with both the carbonyl and enolate oxygen atoms and thereby favor a cyclic transition state? (4) Are there any special structural features, such as additional Lewis base coordination sites, which could override steric effects? (5) Are the reaction conditions such as to promote kinetic control? Chapter 2 of Part B will give a more complete discussion of the ways in which these factors can be controlled to provide specific reaction products.

24. C. H. Heathcock, in *Asymmetric Syntheses*, Vol. 3, J. D. Morrison, ed., Academic Press, New York, 1984, Chapter 2; C. H. Heathcock, in *Comprehensive Carbanion Chemistry*, Part B, E. Buncel and T. Durst, eds., Elsevier, Amsterdam, 1984, Chapter 4; T. Mukaiyama, *Org. React.* **28**:203 (1982); D. A. Evans, J. V. Nelson, and T. R. Taber, *Top. Stereochem.* **13**:1 (1982); A. T. Nielsen and W. J. Houlihan, *Org. React.* **16**:1 (1968).

When the aldol reaction is carried out under thermodynamic conditions, the product selectivity is often not as high as under kinetic conditions. All the regioisomeric and stereoisomeric enolates may participate as nucleophiles. The adducts can return to reactants, and so the difference in stability of the stereoisomeric *anti* and *syn* products will determine the product composition.

It is also possible to carry out the aldol condensation under acidic conditions. The reactive nucleophile is then the enol. The mechanism, as established in detail for acetaldehyde,[25] involves nucleophilic attack of the enol on the protonated aldehyde.

There has been little study of the stereoselectivity of the reaction under acidic conditions. In the absence of a coordinating Lewis acid, there is no preference for a cyclic transition state. When regioisomeric enols are possible, acid-catalyzed reactions tend to proceed through the more substituted of the enols. This reflects the predominance of this enol. (See Section 7.2.)

25. L. M. Baigrie, R. A. Cox, H. Slebocka-Tilk, M. Tencer, and T. T. Tidwell, *J. Am. Chem. Soc.* **107**:3640 (1985).

Under both basic and acidic conditions, the aldol adduct can proceed to dehydrated product.

$$\text{base-catalyzed dehydration} \qquad \text{acid-catalyzed dehydration}$$

The dehydration reactions have somewhat higher activation energies than the addition step and are not usually observed under strictly controlled kinetic conditions. Detailed kinetic studies have provided rate and equilibrium constants for the individual steps in some cases. The results for the acetone–benzaldehyde system in the presence of hydroxide ion are given below. Note that K_2 is sufficiently large to drive the first equilibrium forward.

$$CH_3CCH_3 + PhCH{=}O \underset{k_{-1}}{\overset{K_1, k_1}{\rightleftharpoons}} CH_3CCH_2CHPh \underset{k_{-2}}{\overset{K_2, k_2}{\rightleftharpoons}} CH_3CCH{=}CHPh$$

(Ref. 26)

$$K_1 = 0.83\ M^{-1} \qquad\qquad K_2 = 24$$
$$k_1 = 0.17\ M^{-2}\,s^{-1} \qquad\quad k_2 = 7.5 \times 10^{-3}\ M^{-1}\,s^{-1}$$
$$k_{-1} = 1.4 \times 10^{-2}\ M^{-1}\,s^{-1} \qquad k_{-2} = 3.2 \times 10^{-4}\ M^{-1}\,s^{-1}$$

8.4. Reactivity of Carbonyl Compounds toward Addition

We would like at this point to consider some general relationships concerning the reactivity of carbonyl compounds toward addition of nucleophiles. The discussion in Sections 8.2 and 8.3 indicates that many factors influence the overall rate of a reaction under typical conditions. Among the crucial factors are (1) the role of protons or other Lewis acids in activating the carbonyl group toward nucleophilic attack, (2) the reactivity of the nucleophilic species and its influence on the mechanism which is followed, and (3) the stability of the tetrahedral intermediate and the extent to which it proceeds to product rather than reverting to starting material. Because consideration of all of these factors complicates the interpretation of the inherent reactivity of the carbonyl compound itself, examination of an irreversible process where the addition product is stable is the most direct means of comparing the reactivity of carbonyl compounds. Under these conditions, the relative rates of reaction of different carbonyl compounds can be directly compared. One such reaction is hydride reduction. In particular, reduction by sodium borohydride in alcohol solvents is a fast, irreversible reaction that has provided a convenient basis for comparing the reactivity of different carbonyl compounds.[27]

$$RCR' + BH_4^- \longrightarrow \underset{H}{RCR'}\,(OBH_3^-) \longrightarrow RCHR'\,(OH)$$

26. J. P. Guthrie, J. Cossar, and K. F. Taylor, *Can. J. Chem.* **62**:1958 (1984).

27. H. C. Brown, O. H. Wheeler, and K. Ichikawa, *Tetrahedron* **1**:214 (1957); H. C. Brown and K. Ichikawa, *Tetrahedron* **1**:221 (1957).

471

SECTION 8.4.
REACTIVITY OF
CARBONYL
COMPOUNDS TOWARD
ADDITION

**Table 8.3. Rates of Reduction of Aldehydes and Ketones
by Sodium Borohydride**

Carbonyl compound	$k_2 \times 10^4 \ (M^{-1} \ s^{-1})^a$
Benzaldehyde	$12,400^b$
Benzophenone	1.9
Acetophenone	2.0
Acetone	15.1
Cyclobutanone	264
Cyclopentanone	7
Cyclohexanone	161

a. In isopropyl alcohol at $0°C$.
b. Extrapolated from data at lower temperatures.

The reaction is second-order overall, with the rate given by $k[R_2C=O][NaBH_4]$. The interpretation of the rate data is complicated slightly by the fact that the alkoxyborohydrides produced by the first addition can also function as reducing agents, but this has little apparent effect on the relative reactivity of the carbonyl compounds. Table 8.3 presents some of the rate data obtained from these studies.

Reductions by $NaBH_4$ are characterized by low enthalpies of activation (8–13 kcal/mol) and large negative entropies of activation (-28 to -40 eu). Aldehydes are substantially more reactive than ketones, as can be seen by comparison of the rate data for benzaldehyde and acetophenone. This relative reactivity is characteristic of nearly all carbonyl addition reactions. The reduced reactivity of ketones is attributed primarily to steric effects. Not only does the additional substituent increase the steric restrictions to approach of the nucleophile, but it also causes larger steric interaction in the tetrahedral product as the hybridization changes from trigonal to tetrahedral.

Among the cyclic ketones, the reactivity of cyclobutanone is enhanced because the strain of the four-membered ring is decreased in going from sp^2 to sp^3 hybridization. The higher reactivity of cyclohexanone as compared to cyclopentanone is quite general for carbonyl addition reactions. The major factor responsible for the difference in this case is the change in torsional strain as addition occurs. As the hybridization goes from sp^2 to sp^3, the torsional strain is increased in cyclopentanone. The opposite is true for cyclohexanone. The equatorial hydrogens are nearly eclipsed with the carbonyl oxygen in cyclohexanone, but the chair structure of the addition product allows all bonds to attain staggered arrangements.

The borohydride reduction rate data are paralleled by the rate data for many other carbonyl addition reactions. In fact, for a series of ketones, most of which are cyclic, a linear free-energy correlation of the form

$$\log k = A \ \log k_0 + B$$

exists for nucleophiles such as NH_2OH, CN^-, $HOCH_2CH_2S^-$, and SO_3^-.[28] These nucleophiles span a wide range of reactivity and represent nitrogen, carbon, and sulfur atoms acting as the nucleophile. The free-energy relationship implies that in this series of ketones the same structural features govern reactivity toward each of the nucleophiles. To a good approximation, the parameter A is equal to 1, which reduces the correlation to

$$\log(k/k_0) = B$$

This equation implies that the relative reactivity is independent of the specific nucleophile and that relative reactivity is insensitive to changes in position of the transition state. Table 8.4 lists the B values for some representative ketones. The parameter B indicates relative reactivity on a log scale. Cyclohexanone is seen to be a particularly reactive ketone, being almost as reactive as cyclobutanone and more than 10 times as reactive as acetone.

The same structural factors come into play in determining the position of equilibria in reversible additions to carbonyl compounds. The best studied of such equilibrium processes is probably addition of cyanide to give cyanohydrins.

$$R_2C{=}O + HCN \rightleftharpoons N{\equiv}C{-}\underset{\underset{R}{|}}{\overset{\overset{R}{|}}{C}}{-}OH$$

For cyclopentanone, cyclohexanone, and cycloheptanone, the K values for addition are 48, 1000, and $8\,M^{-1}$, respectively.[29] For aromatic aldehydes, the equilibria are affected by the electronic nature of the aryl substituent. Electron donors disfavor addition by stabilizing the aldehyde whereas electron-accepting substituents have the opposite effect.

K is correlated by Hammett equation with σ^+, $\rho = 1.01$
k_a is correlated by Hammett equation with σ^+, $\rho = 1.18$

(Ref. 30)

Table 8.4. Relative Reactivity of Some Ketones toward Addition of Nucleophiles

Ketone	$B = \log$ relative reactivity[a]
Cyclobutanone	0.09
Cyclohexanone	0.00
4-t-Butylcyclohexanone	-0.008
Adamantanone	-0.46
Cycloheptanone	-0.95
Cyclopentanone	-1.18
Acetone	-1.19
Bicyclo[2.2.1]heptan-2-one	-1.48
3,3,5,5-Tetramethylcyclohexanone	-1.92

a. A. Finiels and P. Geneste, *J. Org. Chem.* **44**:1577 (1979); reactivity relative to cyclohexanone as a standard.

28. A. Finiels and P. Geneste, *J. Org. Chem.* **44**:1577 (1979).
29. V. Prelog and M. Kobelt, *Helv. Chim. Acta* **32**:1187 (1949).
30. W. M. Ching and R. G. Kallen, *J. Am. Chem. Soc.* **100**:6119 (1978).

There are large differences in reactivity among the various carboxylic acid derivatives, such as amides, esters, and acyl chlorides. One important factor is the resonance stabilization provided by the heteroatom. This decreases in the order $N > O > Cl$. Electron donation reduces the electrophilicity of the carbonyl group, and the corresponding stabilization is lost in the tetrahedral intermediate.

473

SECTION 8.4.
REACTIVITY OF
CARBONYL
COMPOUNDS TOWARD
ADDITION

The high reactivity of the acyl chlorides also reflects the polar electron-withdrawing effect of the chlorine, which more than outweighs the small π-donor effect.

Another factor which strongly affects the reactivity of these carboxylic acid derivatives is the leaving-group ability of the substituents. The order is $Cl > OAr > OR > NR_2 > O^-$ so that not only does the ease of forming the tetrahedral intermediate decrease in the order $Cl > OAr > OR > NR_2 > O^-$, but the tendency for subsequent elimination to occur is also in the same order. Because the two factors work together, there are large differences in reactivity toward the nucleophiles.

	Approximate relative reactivity toward hydrolysis
RCOCl	10^{11}
RCO$_2$R'	1
RCONR$_2'$	10^{-2}
RCO$_2^-$	$\ll 1$

Table 8.5 lists the enthalpies for a series of isodesmic reactions involving conversion of carbonyl derivatives to the methyl ketones. The ΔH of the reactions is given both from thermodynamic data (ΔH_{obs}) and as calculated at the MP3/6$-$31$++$G* level. These values show that NH$_2$, OH, and F all provide significant stabilization to carbonyl groups. A major component of this stabilization is π donation of a lone pair from the substituent. This effect is diminished for $X = Cl$ and is absent for $X = CN$. The cyano substituent is destabilizing because of a Coulombic repulsion of the adjacent positively charged carbons. The trifluoromethyl group has a similar effect.[31] Figure 8.5 shows bond orders and charge densities for a number of carbonyl derivatives.

Fig. 8.5. Charge density and bond orders for carbonyl derivatives.

31. K. B. Wiberg, C. M. Hadad, P. R. Rablen, and J. Cioslowski, *J. Am. Chem. Soc.* **114**:8644 (1992).

**Table 8.5. Substituent Effects on Carbonyl
Group Stabilization**

X	ΔH_{obs} (kcal/mol)	ΔH_{calc} (kcal/mol)[a]
H	−9.9	−9.3
NH$_2$	+19.6	+18.3
OH	+23.4	+22.3
F	+17.9	+16.4
Cl	+6.6	+6.7
CN	—	−11.0
CF$_3$	—	−12.4

a. MP3/G-31++G* calculations; K. B. Wiberg, C. M. Hadad,
P. R. Rablen, and J. Cioslowski, *J. Am. Chem. Soc.* **114**:8644
(1992).

**Table 8.6. Relative Stabilization of Carbonyl
Compounds by Substituent Groups**[a]

$$\underset{\|}{\overset{O}{XCY}} + 2CH_4 \longrightarrow H_2C{=}O + CH_3X + CH_3{-}Y$$

$$\underset{\underset{H}{|}}{\overset{O^-}{\underset{|}{XCY}}} + 2CH_4 \longrightarrow CH_3O^- + CH_3X + CH_3{-}Y$$

$$\underset{XCY}{\overset{O}{\|}} + \underset{\underset{H}{|}}{\overset{O^-}{\underset{|}{XCY}}} \rightleftharpoons H_2C = O + CH_3O^-$$

X	Y	Hydride affinity (ΔH_{RHA}) (kcal/mol)
H	H	0
H	CH$_3$	1.5
H	OCH$_3$	3.8
H	NH$_2$	9.7
H	F	−12.0
CH$_3$	CH$_3$	1.9
CH$_3$	NH$_2$	9.3
CH$_3$	F	−10.7

Table 8.6 shows the hydride affinities calculated for a related series of compounds at the G2 level of theory.[32] The computation was made on the basis of two isodesmic reactions that allow a comparison of the hydride affinity of the compound with that of formaldehyde. These results place the C=O stabilizing effect in the order F≪H < CH$_3$ < OCH$_3$ < NH$_2$.

8.5. Ester Hydrolysis

Esters can be hydrolyzed in either basic or acidic solution. In acidic solution, the reaction is reversible. The position of the equilibrium depends on the relative concentra-

32. R. E. Rosenberg, *J. Am. Chem. Soc.* **117**:10358 (1993).

tions of water and the alcohol. In aqueous solution, hydrolysis occurs. In alcoholic solution, the equilibrium is shifted in favor of the ester:

$$\underset{\text{RCOR}'}{\overset{\text{O}}{\|}} + H_2O \underset{H^+}{\rightleftharpoons} \underset{\text{RCOH}}{\overset{\text{O}}{\|}} + R'OH$$

In alkaline aqueous solution, ester hydrolysis is essentially irreversible:

$$\underset{\text{RCOR}'}{\overset{\text{O}}{\|}} + {}^-\text{OH} \longrightarrow RCO_2^- + R'OH$$

The carboxylic acid is converted to its anion under these conditions, and the position of the equilibrium lies far to the right. The mechanistic designations $A_{AC}2$ and $B_{AC}2$ are given to the acid- and base-catalyzed hydrolysis mechanisms, respectively. The A denotes acid catalysis, whereas B indicates base catalysis. The subscript AC indicates that acyl–oxygen bond cleavage occurs. The digit 2 has its usual significance, indicating the bimolecular nature of the rate-determining step.

$A_{AC}2$ mechanism

$$\underset{\text{RCOR}'}{\overset{\text{O}}{\|}} + H^+ \rightleftharpoons \underset{\text{RCOR}'}{\overset{{}^+\text{OH}}{\|}}$$

$$\underset{\text{RCOR}'}{\overset{{}^+\text{OH}}{\|}} + H_2O \rightleftharpoons \underset{\substack{| \\ H_2O^+}}{\overset{\text{OH}}{\underset{|}{\text{RCOR}'}}}$$

$$\underset{\substack{| \\ H_2O^+}}{\overset{\text{OH}}{\underset{|}{\text{RCOR}'}}} \rightleftharpoons \underset{\substack{| \text{H} \\ \text{HO}}}{\overset{\text{OH}}{\underset{|+}{\text{RCOR}'}}} \rightleftharpoons \underset{\text{RCOH}}{\overset{\text{O}}{\|}} + R'OH + H^+$$

$B_{AC}2$ mechanism

$$\underset{\text{RCOR}'}{\overset{\text{O}}{\|}} + {}^-\text{OH} \rightleftharpoons \underset{\substack{| \\ \text{OH}}}{\overset{\text{O}^-}{\underset{|}{\text{RCOR}'}}}$$

$$\underset{\substack{| \\ \text{OH}}}{\overset{\text{O}^-}{\underset{|}{\text{RCOR}'}}} \longrightarrow \underset{\text{RCOH}}{\overset{\text{O}}{\|}} + R'O^- \longrightarrow \underset{\text{RCO}^-}{\overset{\text{O}}{\|}} + R'OH$$

Esters without special structural features hydrolyze by these mechanisms. Among the evidence supporting these mechanisms are kinetic studies that show the expected dependence on hydrogen-ion or hydroxide-ion concentration and isotopic labeling studies that prove that it is the acyl–oxygen, not the alkyl–oxygen bond, that is cleaved during hydrolysis.[33] Acid-catalyzed hydrolysis of esters is accompanied by some exchange of oxygen from water into the carbonyl group. This exchange occurs by way of the tetrahedral intermediate because expulsion of water is competitive with expulsion of the alcohol.

$$\underset{\text{RCOR}'}{\overset{\text{O}}{\|}} + H_2\overset{*}{O} \rightleftharpoons \underset{\substack{| \\ {}^*\text{OH}}}{\overset{\text{OH}}{\underset{|}{\text{RCOR}'}}} \rightleftharpoons \underset{\substack{\| \\ {}^*\text{O}}}{\overset{\text{RCOR}'}{}} + H_2O$$

33. M. I. Bender, *Chem. Rev.* **60**:53 (1960).

Substituent effects come into play at several points in the ester hydrolysis mechanism. In the base-catalyzed reaction, electron-withdrawing substituents in either the acyl or alkoxy group facilitate hydrolysis. Because the tetrahedral intermediate formed in the rate-determining step is negatively charged, this intermediate, and the corresponding transition state, is stabilized by electron withdrawal. If the carbonyl group is conjugated with an electron-releasing group, reactivity is decreased by ground-state stabilization. The partitioning of the tetrahedral intermediate between reversion to starting material by loss of hydroxide ion and formation of product by expulsion of the alkoxide is strongly affected by substituents in the alkoxy group. Electron-withdrawing groups in the alkoxy group shift the partitioning in favor of loss of the alkoxide and favor hydrolysis. For this reason, exchange of carbonyl oxygen with solvent does not occur in basic hydrolyses when the alkoxy group is a good leaving group. This has been demonstrated, for example, for esters of phenols. Because phenols are stronger acids than alcohols, their conjugate bases are better leaving groups than alkoxide ions. Aryl esters are hydrolyzed faster than alkyl esters and without observable exchange of carbonyl oxygen with solvent:

$$
\underset{\substack{\parallel \\ RCOAr}}{O} + {}^-OH \underset{slow}{\overset{}{\rightleftharpoons}} \underset{\substack{| \\ OH}}{\overset{O^-}{\underset{|}{RCOAr}}} \xrightarrow{fast} \underset{\substack{\parallel \\ RCOH}}{O} + ArO^- \xrightarrow{{}^-OH} RCO_2^- + ArO^- + H_2O
$$

Even alkyl benzoate esters give only a small amount of exchange under basic hydrolysis conditions. This means that reversal of the hydroxide addition must be slow relative to the forward breakdown of the tetrahedral intermediate.[34]

These substituent effects can be summarized in a general way for the B_{AC} mechanism by noting the effect of substituents on each step of the mechanism:

$$
\underset{\substack{\parallel \\ RCOR'}}{O} + {}^-OH \rightleftharpoons \underset{\substack{| \\ OH}}{\overset{O^-}{\underset{|}{R-C-OR'}}} \qquad \text{favored by electron-attracting substituents in both R and } R'
$$

$$
\underset{\substack{| \\ OH}}{\overset{O^-}{\underset{|}{R-C-OR'}}} \longrightarrow \underset{\substack{\parallel}}{\overset{O}{R-C-O^-}} + HOR' \qquad \text{strongly favored by electron-attracting substituents in } R'
$$

It is possible to shift ester hydrolysis away from the normal $A_{AC}2$ or $B_{AC}2$ mechanisms by structural changes in the substrate. When the ester is derived from a tertiary alcohol, acid-catalyzed hydrolysis often occurs by a mechanism involving alkyl–oxygen fission. The change in mechanism is due to the stability of the tertiary carbocation that can be formed by alkyl–oxygen cleavage.[35] When this mechanism occurs, alkenes as well as alcohols may be produced, since the carbocation can react by substitution or elimination. This mechanism is referred to as $A_{AL}1$, reflecting the fact that the alkyl–oxygen bond is broken.

$$
\underset{\substack{\parallel \\ RCOCR_3'}}{O} + H^+ \rightleftharpoons \underset{\substack{\parallel \\ RCOCR_3'}}{\overset{+OH}{}} \xrightarrow{slow} RCO_2H + R_3C^+
$$

$$
R_3'C^+ + H_2O \longrightarrow R_3'COH \quad and \quad R_3'C^+ \longrightarrow alkene + H^+
$$

34. R. A. McClelland, *J. Am. Chem. Soc.* **106**:7579 (1984).
35. A. G. Davies and J. Kenyon, *Q. Rev. Chem. Soc.* **9**:203 (1955).

The change of mechanism with tertiary alkyl esters is valuable in synthetic methodology because it permits certain esters to be hydrolyzed very selectively. The usual situation involves the use of *t*-butyl esters, which can be cleaved to carboxylic acids by action of acids such as *p*-toluenesulfonic acid or trifluoroacetic acid under anhydrous conditions where other esters are stable.

In the preceding paragraphs, the ester hydrolysis mechanisms discussed pertained to aqueous solutions of strong acids and strong bases. These are conditions in which specific acid catalysis or specific base catalysis is expected to be dominant. In media in which other acids or bases are present, the possible occurrence of general acid-catalyzed and general base-catalyzed hydrolysis must be considered. General base catalysis has been observed in the case of esters in which the acyl group carries electron-attracting substituents.[36] The transition state for esters undergoing hydrolysis by a general base-catalyzed mechanism involves partial proton transfer from the attacking water molecule to the general base during formation of the tetrahedral intermediate:

$$
B^- \quad H-O \quad \overset{R}{\underset{OR'}{C}}=O \quad \xrightarrow{slow} \quad B-H + HO-\overset{R}{\underset{OR'}{C}}-O^- \quad \xrightarrow{fast} \quad RCOH + {}^-OR' \longrightarrow RCO_2^- + R'OH
$$

Ester hydrolysis can also be promoted by *nucleophilic catalysis*. If a component of the reaction system is a more effective nucleophile toward the carbonyl group than hydroxide ion or water under a given set of conditions, an acyl-transfer reaction can take place to form an intermediate:

$$
HNu: + RCOR' \longrightarrow RCNu + R'OH
$$

$$
RCNu + H_2O \longrightarrow RCO_2H + HNu:
$$

If this intermediate, in turn, is more rapidly attacked by water or hydroxide ion than the original ester, the overall reaction will be faster in the presence of the nucleophile than in its absence. These are the requisite conditions for nucleophilic catalysis. Esters of relatively acidic alcohols (in particular, phenols) are hydrolyzed by the nucleophilic catalysis mechanism in the presence of imidazole[37]:

$$
HN\underset{}{\diagup}N: + RCOAr \longrightarrow :N\underset{}{\diagup}N-CR + ArOH
$$

$$
:N\underset{}{\diagup}N-CR + H_2O \longrightarrow :N\underset{}{\diagup}NH + RCOH
$$

36. W. P. Jencks and J. Carriuolo, *J. Am. Chem. Soc.* **83**:1743 (1961); D. Stefanidis and W. P. Jencks, *J. Am. Chem. Soc.* **115**:6045 (1993).

37. T. C. Bruice and G. L. Schmir, *J. Am. Chem. Soc.* **79**:1663 (1967); M. L. Bender and B. W. Turnquest, *J. Am. Chem. Soc.* **79**:1652, 1656 (1957).

Carboxylate anions can also serve as nucleophilic catalysts.[38] In this case, an anhydride is the reactive intermediate:

$$RCOR' + R''CO_2^- \longrightarrow RCOCR'' + R'O^-$$

$$RCOCR'' + H_2O \longrightarrow RCO_2H + R''CO_2H$$

The nucleophilic catalysis mechanism only operates when the alkoxy group being hydrolyzed is not much more basic than the nucleophilic catalyst. This relationship can be understood by considering the tetrahedral intermediate generated by attack of the potential catalyst on the ester:

$$R''CO_2^- + RCOR' \longrightarrow \underset{OR'}{RC-OCR''}$$

The relative leaving-group abilities of $R'O^-$ and $R''CO_2^-$ are strongly correlated with the basicity of the two anions. If $R''CO_2^-$ is a much better leaving group than $R'O^-$, it will be eliminated preferentially from the tetrahedral intermediate and no hydrolysis will occur.

The preceding discussion has touched on the most fundamental aspects of ester hydrolysis mechanisms. Much effort has been devoted to establishing some of the finer details, particularly concerning proton transfers during the formation and breakdown of the tetrahedral intermediates. These studies have been undertaken in part because of the fundamental importance of hydrolytic reactions in biological systems. These biological hydrolytic reactions are catalyzed by enzymes. The detailed mechanistic studies of ester hydrolysis laid the groundwork for understanding the catalytic mechanisms of the hydrolytic enzymes. Discussions of the biological mechanisms and their relationship to the fundamental mechanistic studies are available in several books which discuss enzyme catalysis in terms of molecular mechanisms.[39]

Esters react with alcohols in either acidic or basic solution to exchange alkoxy groups (ester interchange) by mechanisms which parallel hydrolysis. The alcohol or alkoxide acts as the nucleophile:

$$RCOR' + R''OH \underset{R''O^-}{\rightleftarrows} RCOR'' + R'OH$$

$$RCOR' + R''OH \underset{H^+}{\rightleftarrows} RCOR'' + R'OH$$

As in the case of hydrolysis, there has been a good deal of study of substituent effects, solvent effects, isotopic exchange, kinetics, and the catalysis of these processes.[40] The alcoholysis reaction is reversible in both acidic and basic solution, in contrast to hydrolysis (see p. 475). The key intermediate is the tetrahedral addition product. Its fate is determined

38. V. Gold, D. G. Oakenfull, and T. Riley, *J. Chem. Soc., Perkin Trans. B* **1968**:515.

39. T. C. Bruice and S. J. Benkovic, *Bioorganic Mechanisms*, Vol. 1, W. A. Benjamin, New York, 1966, pp. 1–258; W. P. Jencks, *Catalysis in Chemistry and Enzymology*, McGraw-Hill, New York, 1969; M. L. Bender, *Mechanisms of Homogeneous Catalysis from Protons to Proteins*, Wiley-Interscience, New York, 1971; C. Walsh, *Enzymatic Reaction Mechanisms*, W. H. Freeman, San Francisco, 1979; A. Fersht, *Enzyme Structure and Mechanism*, 2nd ed., W. H. Freeman, New York, 1985.

40. C. G. Mitton, R. L. Schowen, M. Gresser, and J. Shapely, *J. Am. Chem. Soc.* **91**:2036 (1969); C. G. Mitton, M. Gresser, and R. L. Schowen, *J. Am. Chem. Soc.* **91**:2045 (1969).

mainly by the relative basicities of the two alkoxy groups. A tetrahedral intermediate generated by addition of methoxide ion to a *p*-nitrophenyl ester, for example, breaks down exclusively by elimination of the much less basic *p*-nitrophenoxide ion:

In general, the equilibrium in a base-catalyzed alcohol exchange reaction goes in the direction of incorporation of the less acidic alcohol in the ester. This is a reflection of both the kinetic factor, the more acidic alcohol being a better leaving group, and the greater stabilization provided to the carbonyl group by the more electron-donating alkoxy substituent.

8.6. Aminolysis of Esters

Esters react with ammonia and amines to give amides. The mechanism involves nucleophilic attack of the amine at the carbonyl group, followed by expulsion of alkoxide ion from the tetrahedral intermediate. The identity of the rate-determining step depends primarily on the leaving-group ability of the alkoxy group.[41] With relatively good leaving groups such as phenols or trifluoroethanol, the slow step is expulsion of the oxygen leaving group from a zwitterionic tetrahedral intermediate **A**. With poorer leaving goups, breakdown of the tetrahedral intermediate occurs only after formation of the anionic species **B**. For such systems, the deprotonation of **A** is rate-determining.

Aminolysis of esters often reveals general base catalysis and, in particular, a contribution to the reaction rate from terms that are second-order in the amine. The general base is believed to function by deprotonating the zwitterionic tetrahedral intermediate.[42] Deprotonation of the nitrogen facilitates breakdown of the tetrahedral intermediate, since the increased electron density at nitrogen favors expulsion of an anion:

41. F. M. Menger and J. H. Smith, *J. Am. Chem. Soc.* **94**:3824 (1972); A. C. Satterthwait and W. P. Jencks, *J. Am. Chem. Soc.* **96**:7018 (1974).
42. W. P. Jencks and M. Gilchrist, *J. Am. Chem. Soc.* **88**:104 (1966); J. F. Kirsch and A. Kline, *J. Am. Chem. Soc.* **91**:1841 (1969).

Detailed mechanistic studies have been carried out on aminolysis of substituted aryl acetates and aryl carbonates.[43] Aryl esters are considerably more reactive than alkyl esters because the phenoxide ions are better leaving groups than alkoxide ions. The tetrahedral intermediate formed in aminolysis can exist in several forms which differ in extent and site of protonation:

$$RNH_2 + R'COR'' \longrightarrow R\overset{+}{N}H_2-\underset{R'}{\overset{O^-}{\underset{|}{\overset{|}{C}}}}-OR'' \underset{A}{\rightleftharpoons} RNH-\underset{R'}{\overset{OH}{\underset{|}{\overset{|}{C}}}}-OR''$$

$$RNH-\underset{R'}{\overset{O^-}{\underset{|}{\overset{|}{C}}}}-OR'' \underset{B}{\overset{H^+}{\rightleftharpoons}} R\overset{+}{N}H_2-\underset{R'}{\overset{OH}{\underset{|}{\overset{|}{C}}}}-OR''$$

In **A** and **D**, the best leaving group is the neutral amine, whereas in **B** and **C** the group $R''O^-$ would be expected to be a better leaving group than RNH^-. Furthermore, in **B** and **C** the lone pair on nitrogen can assist in elimination. In **A**, the negatively charged oxygen also has increased capacity to "push" on the leaving group, with re-formation of the carbonyl group. Precisely how the intermediate proceeds to product depends upon pH and the identity of the groups RNH_2 and $R''O^-$. When $R''O^-$ is a relatively poor leaving group, as would be the case for alkyl esters, reaction usually occurs through **B** or **C**.

$$\underset{B}{\overset{H}{\underset{|}{R}}N-\underset{R'}{\overset{O^-}{\underset{|}{\overset{|}{C}}}}-OR'' \longrightarrow \overset{H}{\underset{|}{R}}N-\overset{O}{\overset{||}{C}}-R' + R''O^-}$$

$$\underset{C}{\overset{H}{\underset{|}{R}}\ddot{N}-\underset{R'}{\overset{OH}{\underset{|}{\overset{|}{C}}}}-OR'' \longrightarrow \overset{H}{\underset{|}{R}}N=\overset{OH}{\overset{+}{C}}-R' + {}^-OR'' \longrightarrow \overset{H}{\underset{|}{R}}N-\overset{O}{\overset{||}{C}}-R' + R''OH}$$

When the leaving group is better, breakdown can occur directly from **A**. This is the case when $R''O^-$ is a phenolate anion. The mechanism also depends upon the pH and the presence of general acids and bases because the position of the equilibria among the tetrahedral intermediates and their rates of breakdown are determined by these factors.

Insight into the factors that govern breakdown of tetrahedral intermediates has also been gained by studying the hydrolysis of amide acetals. If the amine is expelled, an ester is formed, whereas elimination of an alcohol gives an amide:

$$R-\underset{OR'}{\overset{OR'}{\underset{|}{\overset{|}{C}}}}-NR_2'' \overset{H_2O}{\longrightarrow} \overset{O}{\overset{||}{RCNR_2''}} + 2\ R'OH \qquad \text{alcohol elimination}$$

$$R-\underset{OR'}{\overset{OR'}{\underset{|}{\overset{|}{C}}}}-NR_2'' \overset{H_2O}{\longrightarrow} \overset{O}{\overset{||}{RCOR'}} + R'OH + R_2''NH \qquad \text{amine elimination}$$

43. W. P. Jencks and M. Gilchrist, *J. Am. Chem. Soc.* **90:**2622 (1968); A. Satterthwait and W. P. Jencks, *J. Am. Chem. Soc.* **96:**7018 (1974); A. Satterthwait and W. P. Jencks, *J. Am. Chem. Soc.* **96:**7031 (1974); M. J. Gresser and W. P. Jencks, *J. Am. Chem. Soc.* **99:**6970 (1977).

The pH of the solution is of overwhelming importance in determining the course of these hydrolyses.[44] In basic solution, oxygen elimination is dominant. This is because the unprotonated nitrogen substituent is a very poor leaving group and is also more effective at facilitating the alkoxide elimination by electron donation:

$$R_2''\overset{..}{N}-\overset{\overset{\displaystyle R}{|}}{\underset{\underset{\displaystyle OR'}{|}}{C}}-OR' \longrightarrow R_2''\overset{+}{N}=\overset{\overset{\displaystyle R}{|}}{\underset{\underset{\displaystyle OR'}{}}{C}} + {}^-OR' \xrightarrow{H_2O} \overset{\overset{\displaystyle O}{||}}{RCNR_2''} + 2R'OH$$

In acidic solution, the nitrogen is protonated and becomes a better leaving group and also loses its ability to assist in the elimination of the alkoxide. Under these circumstances, nitrogen elimination is favored:

$$R-\overset{\overset{\displaystyle \cdot OR'}{|}}{\underset{\underset{\displaystyle OR'}{|}}{C}}-\overset{+}{N}HR_2'' \longrightarrow R-\overset{\overset{\displaystyle {}^+OR'}{||}}{\underset{\underset{\displaystyle OR'}{|}}{C}} + R_2''NH \xrightarrow{H_2O} RCOR' + R'OH + R_2''NH$$

In analyzing the behavior of these types of tetrahedral intermediates, it should be kept in mind that proton-transfer reactions are usually fast relative to other steps. This circumstance permits the possibility that a minor species in equilibrium with the major species may be the major intermediate. Detailed studies of kinetics, solvent isotope effects, and the nature of catalysis are the best tools for investigating the various possibilities.

It is useful to recognize that the dissociation of tetrahedral intermediates in carbonyl chemistry is related to the generation of carbocations by ionization processes. Thus, the question of which substituent on a tetrahedral intermediate is the best leaving group is similar to the issues raised in comparing the reactivity of S_N1 reactants on the basis of leaving-group ability. Poorer leaving groups can function as leaving groups in the case of tetrahedral intermediates because of the assistance provide by the remaining oxygen or nitrogen substituents. For example, the iminium ions and protonated carbonyl compounds that are generated by breakdown of tetrahedral intermediates can be recognized as being examples of resonance-stabilized carbocations:

$$\overset{R}{\underset{R}{}}\overset{+}{N}=\overset{R}{\underset{R}{C}} \longleftrightarrow \overset{R}{\underset{R}{}}\overset{..}{N}=\overset{R}{\underset{R}{C^+}} \qquad H\overset{+}{\overset{..}{O}}=\overset{R}{\underset{R}{C}} \longleftrightarrow H\overset{..}{O}-\overset{R}{\underset{R}{C^+}}$$

Keeping these relationships in mind should be helpful in understanding the reactivity of tetrahedral intermediates.

8.7. Amide Hydrolysis

The hydrolysis of amides to carboxylic acids and amines requires considerably more vigorous conditions than ester hydrolysis.[45] The reason is that the electron-releasing

44. R. A. McClelland, *J. Am. Chem. Soc.* **100**:1844 (1978).
45. C. O'Connor, *Q. Rev. Chem. Soc.* **24**:553 (1970).

nitrogen substituent imparts a very significant ground-state stabilization, which is lost in the transition state leading to the tetrahedral intermediate:

$$
\underset{\substack{\| \\ RC-NR_2}}{O} \longleftrightarrow \underset{\substack{| \\ RC=\overset{+}{N}R_2}}{O^-}
$$

In basic solution, a mechanism similar to the $B_{AC}2$ mechanism for ester hydrolysis is believed to operate[46]:

$$
\underset{\substack{\| \\ RCNHR + {}^-OH}}{O} \rightleftarrows \underset{\substack{| \\ RCNHR \\ | \\ OH}}{O^-} \underset{{}^-OH}{\overset{H_2O}{\rightleftarrows}} \underset{\substack{| \\ RC\overset{+}{N}H_2R \\ | \\ OH}}{O^-} \longrightarrow \underset{\substack{\| \\ RCOH + H_2NR}}{O}
$$

$$
RCO_2H + {}^-OH \longrightarrow RCO_2^- + H_2O
$$

The principal difference lies in the poorer ability of amide ions to act as leaving groups, compared to alkoxides. As a result, protonation at nitrogen is required for breakdown of the tetrahedral intermediate. Also, exchange between the carbonyl oxygen and water is extensive because reversal of the tetrahedral intermediate to reactants is faster than its decomposition to products.

In some amide hydrolyses, the breakdown of the tetrahedral intermediate in the forward direction may require formation of a dianion[47]:

$$
\underset{\substack{\| \\ RCNHR'}}{O} + {}^-OH \rightleftarrows \underset{\substack{| \\ RCNHR' \\ | \\ OH}}{O^-}
$$

$$
\underset{\substack{| \\ RCNHR' \\ | \\ OH}}{O^-} + {}^-OH \longrightarrow \underset{\substack{| \\ RCNHR \\ | \\ O^-}}{O^-} \longrightarrow RCO_2^- + \bar{N}HR
$$

$$
R\bar{N}H + H_2O \longrightarrow RNH_2 + {}^-OH
$$

This variation from the ester hydrolysis mechanism also reflects the poorer leaving ability of amide ions as compared to alkoxide ions. The evidence for the involvement of the dianion comes from kinetic studies and from solvent isotope effects, which suggest that a rate-limiting proton transfer is involved.[48] The reaction is also higher than first-order in hydroxide ion under these circumstances, which is consistent with the dianion mechanism.

The mechanism for acid-catalyzed hydrolysis of amides involves attack by water on the protonated amide. An important feature of the chemistry of amides is that the most basic site in an amide is the carbonyl oxygen. Very little of the N-protonated form is present.[49] The major factor that contributes to the stability of the O-protonated form is the

46. M. L. Bender and R. J. Thomas, *J. Am. Chem. Soc.* **83**:4183 (1961); R. S. Brown, A. J. Bennet, and H. Slebocka-Tilk, *Acc. Chem. Res.* **25**:481 (1992).

47. R. M. Pollack and M. L. Bender, *J. Am. Chem. Soc.* **92**:7190 (1970).

48. R. L. Schowen, H. Jayaraman, L. Kershner, and G. W. Zuorick, *J. Am. Chem. Soc.* **88**:4008 (1966).

49. R. J. Gillespie and T. Birchall, *Can. J. Chem.* **41**:148, 2642 (1963); A. R. Fersht, *J. Am. Chem. Soc.* **93**:3504 (1971); R. B. Martin, *J. Chem. Soc., Chem. Commun.* **1972**:793; A. J. Kresge, P. H. Fitzgerald, and Y. Chiang, *J. Am. Chem. Soc.* **96**:4698 (1974).

π-electron delocalization over the O—C—N system. No such delocalization is possible in the N-protonated form.

Amides are weak bases with pK_a values in the range of 0 to -2.[50] When amide resonance is prevented, as in 1-azabicyclo[2.2.2]octanone, N protonation is preferred.[51]

The usual hydrolysis mechanism in strongly acidic solution involves addition of water to the O-protonated amide, followed by breakdown of the tetrahedral intermediate:

By using *N*-acyltrialkylammonium ions as models of the N-protonated amide, it has been possible to show that it would be kinetically impossible for acid-catalyzed hydrolysis to proceed via the N-protonated form.[52] There is almost no exchange of oxygen with water during acid-catalyzed hydrolysis of amides.[53] Because a tetrahedral intermediate is involved, the lack of exchange requires that the intermediate must dissociate exclusively by elimination of the nitrogen substituent. This requirement is reasonable, since the amino group is the most basic site and is the preferred site of protonation in the tetrahedral intermediate. The protonated amine is a much better leaving group than hydroxide ion.

Acylimidazoles and related amides in which the nitrogen atom is part of an aromatic ring hydrolyze much more rapidly than other amides. A major factor is the decreased resonance stabilization of the carbonyl group, which is opposed by the delocalization of the nitrogen lone pair as part of the aromatic sextet.

unfavorable

The acid-catalyzed hydrolysis of imidazolides can also be accelerated by protonation of N-3, which increases the leaving-group ability of the ring.[54] Accumulation of additional

50. R. A. Cox, L. M. Druet, A. E. Klausner, T. A. Modro, P. Wan, and K. Yates, *Can. J. Chem.* **59**:1568 (1981); A. Bagno, G. Lovato, and G. Scorrano, *J. Chem. Soc., Perkin Trans. 2* **1993**:1091.
51. A. Greenberg and C. A. Venanzi, *J. Am. Chem. Soc.* **115**:6951 (1993).
52. A. Williams, *J. Am. Chem. Soc.* **98**:5645 (1976).
53. R. A. McClelland, *J. Am. Chem. Soc.* **97**:5281 (1975); for cases in which some exchange does occur, see H. Slebocka-Tilk, R. S. Brown, and J. Olekszyk, *J. Am. Chem. Soc.* **109**:4620 (1987); A. J. Bennet, H. Slebocka-Tilk, R. S. Brown, J. P. Guthrie, and A. J. Jodhan, *J. Am. Chem. Soc.* **112**:8497 (1990).
54. T. H. Fife, *Acc. Chem. Res.* **26**:325 (1993).

nitrogens in the ring (triazoles, tetrazoles) further increases the leaving-group ability of the ring.[55]

8.8. Acylation of Nucleophilic Oxygen and Nitrogen Groups

The conversion of alcohols to esters by O-acylation and of amines to amides by N-acylation are fundamental organic reactions. These reactions are the reverse of the hydrolytic procedures discussed in the preceding sections. Section 3.4 in Part B discusses these reactions from the point of view of synthetic applications and methods.

Although the previous two sections of this chapter emphasized hydrolytic processes, two mechanisms that led to O- or N-acylation were considered. In the discussion of acid-catalyzed ester hydrolysis, it was pointed out that this reaction is reversible (p. 475). Thus, it is possible to acylate alcohols by reaction with a carboxylic acid. To drive the reaction forward, the alcohol is usually used in large excess, and it may also be necessary to remove water as it is formed. This can be done by azeotropic distillation in some cases.

$$RCOH + R'OH \xrightleftharpoons{H^-} RCOR' + H_2O$$

The second reaction that should be recalled is the aminolysis of esters (p. 479). This reaction leads to the formation of amides by N-acylation:

$$RCOR' + R''NH_2 \xrightleftharpoons{H^-} RCNHR'' + R'OH$$

The equilibrium constant for this reaction is ordinarily favorable, but the reactions are rather slow.

The most common O- and N-acylation procedures use acylating agents that are more reactive than carboxylic acids or their esters. Carboxylic acid chlorides and anhydrides react rapidly with most unhindered hydroxy and amino groups to give esters and amides, respectively:

$$R'OH + RCCl \longrightarrow RCOR' + HCl$$
$$R'OH + (RCO)_2O \longrightarrow RCOR' + RCO_2H$$
$$R'NH_2 + RCCl \longrightarrow RCNHR' + HCl$$
$$R'NH_2 + (RCO)_2O \longrightarrow RCNHR' + RCO_2H$$

The general mechanisms are well established.[56] The nucleophilic species undergoes addition at the carbonyl group, followed by elimination of the halide or carboxylate

55. J. F. Patterson, W. P. Huskey, and J. L. Hoggs, *J. Org. Chem.* **45**:4675 (1980); B. S. Jursic and Z. Zdravkovski, *THEOCHEM* **109**:177 (1994).
56. D. P. N. Satchell, *Q. Rev. Chem. Soc.* **17**:160 (1963).

group. Acyl halides and anhydrides are reactive acylating reagents because of a combination of the polar effect of the halogen or oxygen substituent, which enhances the reactivity of the carbonyl group, and the ease with which the tetrahedral intermediate can expel such relatively good leaving groups:

485

SECTION 8.8.
ACYLATION OF
NUCLEOPHILIC
OXYGEN AND
NITROGEN GROUPS

$$
\text{R'OH} + \underset{\substack{\| \\ \text{RCX}}}{\text{O}} \longrightarrow \underset{\substack{| \\ \text{RCX} \\ | \\ \text{R'O}^+\text{H}}}{\text{O}^-} \longrightarrow \underset{\substack{\| \\ \text{RCOR'}}}{\text{O}} + \text{HX}
$$

$$
\text{R}_2\text{NH} + \underset{\substack{\| \\ \text{RCX}}}{\text{O}} \longrightarrow \underset{\substack{| \\ \text{RCX} \\ | \\ \text{R}_2'\text{NH} \\ +}}{\text{O}^-} \longrightarrow \underset{\substack{\| \\ \text{RCNR}_2'}}{\text{O}} + \text{HX}
$$

X = halogen or carboxylate

Acylation of alcohols is often performed in the presence of an organic base such as pyridine. The base serves two purposes. It neutralizes the protons generated in the reaction and prevents the development of high acid concentrations. Pyridine also becomes directly involved in the reaction as a *nucleophilic catalyst* (see Section 8.5).

Pyridine is more nucleophilic than an alcohol toward the carbonyl center of an acyl chloride. The product that results, an acylpyridinium ion, is, in turn, more reactive toward an alcohol than the original acyl chloride. The conditions required for nucleophilic catalysis therefore exist, and acylation of the alcohol by acyl chloride is faster in the presence of pyridine than in its absence. Among the evidence that supports this mechanism is spectroscopic observation of the acetylpyridinium ion.[57] An even more effective catalyst is 4-dimethylaminopyridine (DMAP), which functions in the same way but is more reactive because of the electron-donating dimethylamino substituent.[58]

With more strongly basic tertiary amines such as triethylamine, another mechanism can come into play. It has been found that when methanol deuterated on oxygen reacts with acyl chlorides in the presence of triethylamine, some deuterium is found α to the carbonyl group in the ester:

$$
\underset{\substack{\| \\ \text{CH}_3\text{CH}_2\text{CH}_2\text{CCl}}}{\text{O}} + \text{CH}_3\text{OD} \xrightarrow[\text{octane}]{\text{Et}_3\text{N}} \underset{\substack{\| \\ \text{CH}_3\text{CH}_2\text{CH}_2\text{COCH}_3 \\ \\ 67\%}}{\text{O}} + \underset{\substack{\| \\ \text{CH}_3\text{CH}_2\text{CHCOCH}_3 \\ | \\ \text{D} \\ 33\%}}{\text{O}}
$$

57. A. R. Fersht and W. P. Jencks, *J. Am. Chem. Soc.* **92**:5432, 5442 (1970).
58. E. F. V. Scriven, *Chem. Soc. Rev.* **12**:129 (1983).

This finding suggests that some of the ester is formed via a ketene intermediate[59]:

$$RCH_2CCl + Et_3N \longrightarrow RCH{=}C{=}O + Et_3\overset{+}{N}H + Cl^-$$

$$RCH{=}C{=}O + CH_3OD \longrightarrow RCHCOCH_3$$
$$\hspace{5cm} D$$

Ketenes undergo rapid addition by nucleophilic attack at the *sp*-carbon atom. The reaction of tertiary amines and acyl halides, in the absence of nucleophiles, is a general preparation for ketenes.[60]

Kinetic studies of the reaction of alcohols with acyl chlorides in polar solvents in the absence of basic catalysts generally reveal terms both first-order and second-order in alcohol.[61] Transition states in which the second alcohol molecule acts as a proton acceptor have been proposed:

$$R{-}\overset{O}{\underset{|}{C}}{-}Cl \longrightarrow R'\overset{+}{O}H_2 + RCCl \longrightarrow RCOR' + Cl^-$$

There are alternatives to the addition–elimination mechanism for nucleophilic substitution of acyl chlorides. Certain acyl chlorides are known to react with alcohols by a dissociative mechanism in which acylium ions are intermediates. This mechanism is observed with aroyl halides having electron-releasing substituents.[62] Other acyl halides show reactivity indicative of mixed or borderline mechanisms.[63] The existence of the S_N1-like dissociative mechanism reflects the relative stability of acylium ions.

$$ArCCl \longrightarrow ArC{\equiv}O^+ \overset{ROH}{\longrightarrow} ArCOR + H^+$$

In addition to acyl chlorides and acid anhydrides, there are a number of other types of compounds that are reactive acylating agents. Many have been developed to facilitate the synthesis of polypeptides, in which mild conditions and high selectivity are required. An important group of reagents for converting carboxylic acids to active acylating agents is

59. W. E. Truce and P. S. Bailey, *J. Org. Chem.* **34**:1341 (1969).
60. R. N. Lacey, in *The Chemistry of Alkenes*, S. Patai, ed., Interscience Publishers, New York, 1964, pp. 1168–1170; W. E. Hanford and J. C. Sauer, *Org. React.* **3**:108 (1947).
61. D. N. Kevill and F. D. Foss, *J. Am. Chem. Soc.* **91**:5054 (1969); S. D. Ross, *J. Am. Chem. Soc.* **92**:5998 (1970).
62. T. W. Bentley, H. C. Harris, and I. S. Koo, *J. Chem. Soc., Perkin Trans. 2* **1988**:783; B. D. Song and W. P. Jencks, *J. Am. Chem. Soc.* **111**:8470 (1989).
63. T. W. Bentley, I. S. Koo and S. J. Norman, *J. Org. Chem.* **56**:1604 (1991); T. W. Bentley and C. S. Shim, *J. Chem. Soc., Perkin Trans. 2* **1993**:1659.

the carbodiimides, such as dicyclohexylcarbodiimide. The mechanism for carbodiimide-promoted amide bond formation is shown below:

487

SECTION 8.8.
ACYLATION OF
NUCLEOPHILIC
OXYGEN AND
NITROGEN GROUPS

$$
\underset{\text{RCOH}}{\overset{\text{O}}{\|}} + \text{R'N=C=NR'} \longrightarrow \underset{\text{RCO-C=NR'}}{\overset{\text{O} \quad \text{HNR'}}{\|}}
$$

$$
\underset{\underset{\text{R''NH}_2}{\downarrow}}{\underset{\text{RCOC=NR'}}{\overset{\text{O} \quad \text{NHR'}}{\|}}} \longrightarrow \underset{\underset{\text{R''NH}_2^+}{}}{\underset{\text{RC-O-C=NR'}}{\overset{\text{O} \quad \text{NHR'}}{}}} \longrightarrow \underset{\text{RCNHR''}}{\overset{\text{O}}{\|}} + \underset{\text{R'NHCNHR'}}{\overset{\text{O}}{\|}}
$$

The first step is addition of the carboxylic acid to the C=N bond of the carbodiimide, which generates an O-acyl urea derivative. This is a reactive acylating agent because there is a strong driving force for elimination of the urea unit, with formation of the very stable urea carbonyl group.[64] The amine reacts with the active acylating agent. In the absence of an amine, the acid would be converted to the anhydride, with a second molecule of the carboxylic acid serving as the nucleophile.

Carboxylic acids react with trifluoroacetic anhydride to give mixed anhydrides that are especially useful for the acylation of hindered alcohols and phenols:

The active acylating agent may be the protonated mixed anhydride,[65] or, alternatively, the anhydride may dissociate to the acylium and trifluoroacetate ions[66]:

$$
\underset{\text{RCOCCF}_3}{\overset{\text{O} \quad \text{O}}{\| \quad \|}} \overset{\text{H}^+}{\rightleftharpoons} \underset{\text{RCOCCF}_3}{\overset{\text{HO}^+ \quad \text{O}}{\| \quad \|}} \quad \text{or} \quad \underset{\text{RCOCCF}_3}{\overset{\text{O} \quad \text{O}}{\| \quad \|}} \overset{\text{H}^+}{\rightleftharpoons} \text{RC} \equiv \overset{+}{\text{O}} + \text{CF}_3\text{CO}_2\text{H}
$$

Either mechanism explains why trifluoroacetylation of the nucleophile does not occur. Protonation of the anhydride would occur selectively at the more electron-rich carbonyl oxygen, rather than at the carbonyl flanked by the very electron-withdrawing trifluoromethyl group. Similarly, cleavage of the unsymmetrical anhydride would occur to give the more stable acylium ion. The trifluoroacetylium ion would be less stable.

Enol esters are another useful family of acylating agents. The acetate of the enol form of acetone, isopropenyl acetate, is the most commonly used member of this group of

64. D. F. DeTar and R. Silverstein, *J. Am. Chem. Soc.* **88**:1013, 1020 (1966).
65. R. C. Parish and L. M. Stock, *J. Org. Chem.* **30**:927 (1965).
66. J. M. Tedder, *Chem. Rev.* **55**:787 (1955).

compounds. Enol esters act as acylating agents in the presence of a trace amount of an acid catalyst and are reactive toward weak nucleophiles such as hindered hydroxyl groups:

(Ref. 67)

94%

The active acylating agent is presumably the C-protonated enol ester:

$$(CH_3)_2C=\overset{+}{O}-\overset{O}{\overset{\|}{C}}CH_3 \rightleftharpoons CH_3\overset{O}{\overset{\|}{C}}CH_3 + \overset{+}{O}\equiv CCH_3$$

This species would be highly reactive owing to the presence of a positively charged oxygen. An alternative possibility is that the protonated enol ester decomposes to acetone and an acylium ion, which then acts as the acylating agent. Section 3.4 of Part B gives additonal examples of synthetically useful acylating reagents.

8.9. Intramolecular Catalysis

The reactions of carbonyl compounds have provided an important testing ground for developing an understanding of intramolecular catalysis. Studies in intramolecular catalysis have been designed to determine how much more efficiently a given functional group can act as a catalyst when it is part of the reacting molecule and located in such a position that encounter between the catalytic group and the reaction center is facilitated. These studies are important to understanding biological mechanisms because enzymes act as exceedingly efficient catalysts by bringing together, at the "active site," various basic, acidic, or nucleophilic groups in a geometry that is particularly favorable for reaction. This section will illustrate some of the facts that have emerged from these studies and the mechanistic conclusions that have been drawn.

It was pointed out in the mechanistic discussion concerning acetal and ketal hydrolysis that general acid catalysis occurs only for acetals and ketals having special structural features (see p. 453). Usually, specific acid catalysis operates. The question of whether general acid catalysis could be observed in intramolecular reactions has been of interest because intramolecular general acid catalysis is postulated to play a part in the mechanism of action of the enzyme lysozyme, which hydrolyzes the acetal linkage present in certain polysaccharides. One group of molecules that has been examined as a model system is acetals derived from o-hydroxybenzoic acid (salicylic acid):

67. W. S. Johnson, J. Ackerman, J. F. Eastham, and H. A. DeWalt, Jr., *J. Am. Chem. Soc.* **78**:6302 (1956).

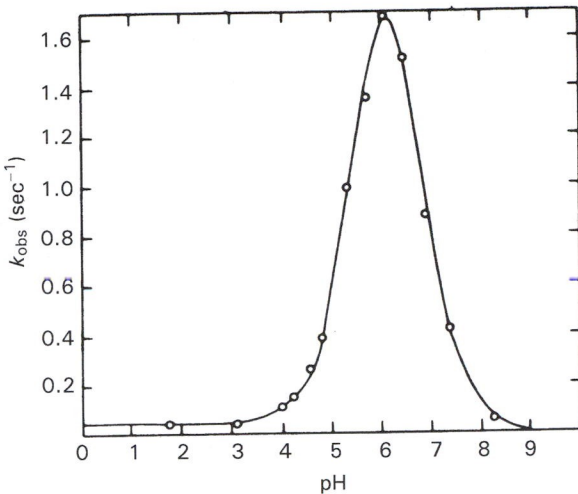

Fig. 8.6. pH–Rate profile for release of salicylic acid from benz-aldehyde disalicyl acetal. [Reproduced from E. Anderson and T. H. Fife, *J. Am. Chem. Soc.* **95:**6437 (1973) by permission of the American Chemical Society.]

The pH–rate profile (see Fig. 8.6) indicates that of the species that are available, the monoanion of the acetal is the most reactive. The reaction is fastest in the intermediate pH range, where the concentration of this species is at a maximum. The concentration of the neutral molecule decreases with increasing pH; the converse is true of the concentration of the dianion.

The transition state for the rapid hydrolysis of the monoanion has been depicted as involving an intramolecular general acid catalysis by the carboxylic acid group, with participation by the anionic carboxylate group, which becomes bound at the developing electrophilic center:

A mixed acetal of benzaldehyde, methanol, and salicylic acid has also been studied.[68] It, too, shows a marked rate enhancement attributable to intramolecular general acid catalysis:

+ CH$_3$OH

The case of intramolecular participation in ester hydrolysis has been extensively studied using acetylsalicylic acid (aspirin) and its derivatives. The kinetic data show that the anion is hydrolyzed more rapidly than the neutral species, indicating that the carboxylate group becomes involved in the reaction in some way. Three mechanisms can be considered:

I. Nucleophilic catalysis

+ CH$_3$CO$_2$H

II. General base catalysis

+ CH$_3$CO$_2$H

III. General acid catalysis of hydroxide ion attack

+ CH$_3$CO$_2$H

Mechanism III cannot be distinguished from the first two on the basis of kinetics alone, because the reactive species shown is in rapid equilibrium with the anion and therefore equivalent to it in terms of reaction kinetics.

68. T. H. Fife and E. Anderson, *J. Am. Chem. Soc.* **93**:6610 (1971).

Mechanism I was ruled out by an isotopic labeling experiment. The mixed anhydride of salicylic acid and acetic acid is an intermediate if nucleophilic catalysis occurs by mechanism I. This molecule is known to hydrolyze in water with about 25% incorporation of solvent water into the salicylic acid.

Hydrolysis of aspirin in $H_2^{18}O$ leads to no incorporation of ^{18}O into the product salicylic acid, ruling out the anhydride as an intermediate and thereby excluding mechanism I.[69] The general acid catalysis of mechanism III can be ruled out on the basis of failure of other nucleophiles to show evidence for general acid catalysis by the neighboring carboxylic acid group. Because there is no reason to believe hydroxide should be special in this way, mechanism III is eliminated. Thus, mechanism II, general base catalysis of hydroxide-ion attack, is believed to be the correct description of the hydrolysis of aspirin.

The extent to which intramolecular nucleophilic catalysis of the type depicted in mechanism I is important is a function of the leaving ability of the alkoxy group. This has been demonstrated by the study of the hydrolysis of a series of monoesters of phthalic acid:

Nucleophilic participation is important only for esters of alcohols that have $pK_a \leq 13$. Specifically, phenyl and trifluoroethyl esters show nucleophilic catalysis, but methyl and 2-chloroethyl esters do not.[70] This result reflects the fate of the tetrahedral intermediate that results from nucleophilic participation. For relatively acidic alcohols, the alkoxide group can be eliminated, leading to hydrolysis via nucleophilic catalysis:

For less acidic alcohols, nucleophilic participation is ineffective because of the low tendency for such alcohols to function as leaving groups. The tetrahedral intermediate formed by intramolecular addition simply returns to starting material because the carboxylate is a much better leaving group than the alkoxide. A similar observation is made for salicylate esters. In contrast to aspirin itself, acetylsalicylates with electron-withdrawing groups (o- and p-nitro analogs) hydrolyze via the nucleophilic catalysis

69. A. R. Fersht and A. J. Kirby, J. Am. Chem. Soc. **89**: 4857 (1967).
70. J. W. Thanassi and T. C. Bruice, J. Am. Chem. Soc. **88**: 747 (1966).

mechanism in which the phenolates act as leaving groups from the cyclic intermediate[71]:

H₂O   + CH₃CO₂H

Intramolecular catalysis of ester hydrolysis by nitrogen nucleophiles is also important. The role of imidazole rings in intramolecular catalysis has received particularly close scrutiny. There are two major reasons for this. One is that the imidazole ring of the histidine residue in proteins is believed to be frequently involved in enzyme-catalyzed hydrolyses. Secondly, the imidazole ring has several possible catalytic functions, which include acting as a general acid in the protonated form, acting as a general base in the neutral form, and acting as a nucleophile in the neutral form. A study of a number of derivatives of structure **1** was undertaken to distinguish between the importance of these various possible mechanisms as a function of pH.[72]

1

general acid
catalysis

general base
catalysis

nucleophilic
catalysis

The relative importance of the potential catalytic mechanisms depends on pH, which also determines the concentration of the other participating species such as water, hydronium ion, and hydroxide ion. At low pH, the general acid catalysis mechanism dominates, and comparison with analogous systems in which the intramolecular proton transfer is not available suggests that the intramolecular catalysis results in a 25- to 100-fold rate enhancement. At neutral pH, the intramolecular general base catalysis mechanism begins to operate. It is estimated that the catalytic effect for this mechanism is a factor of about 10^4. Although the nucleophilic catalysis mechanism was not observed in the parent compound, it occurred in certain substituted derivatives.

The change in mechanism with pH for compound **1** gives rise to the pH–rate profile shown in Fig. 8.7. The rates at the extremities pH < 2 and pH > 9 are proportional to [H⁺] and [⁻OH], respectively, and represent the specific proton-catalyzed and hydroxide-catalyzed mechanisms. In the absence of the intramolecular catalytic mechanisms, the

71. A. R. Fersht and A. J. Kirby, *J. Am. Chem. Soc.* **89**:5960 (1967); *J. Am. Chem. Soc.* **90**:5818 (1968).
72. G. A. Rogers and T. C. Bruice, *J. Am. Chem. Soc.* **96**:2463 (1974).

Fig. 8.7. pH–Rate profile for compound **1**. (Reproduced from Ref. 72 by permission of the American Chemical Society.)

rates of the H$^+$- and $^-$OH-catalyzed reactions would decrease in proportion to the concentration of the catalytic species and interesect at a minimum value representing the "uncatalyzed water hydrolysis." An estimate of the effectiveness of the intramolecular mechanisms can be made by extrapolating the lines which are proportional to [H$^+$] and [$^-$OH]. The pH range in which intramolecular general acid catalysis operates is pH 2–4. At pH 6–8, the intramolecular general base catalysis mechanism is dominant. The extent to which the actual rate lies above these extrapolated lines in the pH range 2–8 represents the contribution from the intramolecular catalysis.

Intramolecular participation of the *o*-hydroxy group in aminolysis of phenyl salicylate has been established by showing that such compounds are more reactive than analogs lacking the hydroxyl substituent. This reaction exhibits overall third-order kinetics, second-order in the reacting amine. Similar kinetics are observed in the aminolysis of simple esters. Both intermolecular general base catalysis (by the second amine molecule) and intramolecular general acid catalysis (by the hydroxyl group) apparently occur.[73]

This mechanism can reduce the overall activation energy of the reaction in at least two ways. The partial transfer of a proton to the carbonyl oxygen increases the electrophilicity of the carbonyl. Likewise, partial deprotonation of the amino group increases its nucleophilicity.

Certain molecules that can permit concerted proton transfers are efficient catalysts for reactions at carbonyl centers. An example is the catalytic effect that 2-pyridone has on the aminolysis of esters. Although neither a strong base (pK_{aH^+} = 0.75) nor a strong acid (pK_a = 11.6), 2-pyridone is an effective catalyst of the reaction of *n*-butylamine with 4-nitrophenyl acetate.[74] The overall rate is more than 500 times greater when 2-pyridone acts

73. F. M. Menger and J. H. Smith, *J. Am. Chem. Soc.* **91**:5346 (1969).

as the catalyst than when a second molecule of butylamine (acting as a general base) is present in the transition state. 2-Pyridone has been called a *tautomeric catalyst* to emphasize its role in proton transfer. Such molecules are also called *bifunctional catalysts*, because two atoms in the molecule are involved in the proton-transfer process.

2-Pyridone also catalyzes epimerization of the anomeric position of the tetramethyl ether of glucose. The mechanism involves two double proton transfers. The first leads to a ring-opened intermediate, and the second results in ring closure to the isomerized product:

Another type of bifunctional catalysis has been noted with α,ω-diamines in which one of the amino groups is primary and the other tertiary. These substituted diamines are from several times to as much as 100 times more reactive toward imine formation than similar monofunctional amines.[75] This is attributed to a catalytic intramolecular proton transfer.

The rate enhancement is greatest for $n = 2$ (1000) but still significant for $n = 3$ (a factor of 10). As the chain is lengthened to $n = 4$ and $n = 5$, the rate enhancement, if any, is minor. This relationship reflects the fact that when n is 4 or 5, the transition state for the intramolecular proton transfer would have to involve rings of 9 and 10 atoms, respectively, and would not be geometrically advantageous. The particularly rapid reaction when $n = 2$ corresponds to the possibility for a proton transfer via a seven-membered cyclic transition state. Assuming that the proton is transferred in a collinear fashion through a hydrogen

74. P. R. Rony, *J. Am. Chem. Soc.* **91**:6090 (1969).

75. J. Hine, R. C. Dempsey, R. A. Evangelista, E. T. Jarvi, and J. M. Wilson, *J. Org. Chem.* **42**:1593 (1977); J. Hine and Y. Chou, *J. Org. Chem.* **46**:649 (1981).

bond, this represents a favorable transition-state geometry:

These examples serve to illustrate the concept of intramolecular catalysis and the fact that favorable juxtaposition of acidic, nucleophilic, or basic sites can markedly accelerate some of the common reactions of carbonyl compounds. Nature, through evolution, has used optimal placement of functional groups to achieve the catalytic activity of enzymes. The functional groups employed to accomplish this are those present on the amino acid residues found in proteins. The acidic sites available include phenolic or carboxylic acid groups from tyrosine, glutamic acid, and aspartic acid. Basic sites include the imidazole ring in histidine and the ω-amino group of lysine. This latter group and the amidine group in arginine are normally protonated at physiological pH and can serve as cationic centers or general acids, as well. Thiol (cysteine) and hydroxyl (threonine and serine) groups and the deprotonated carboxyl groups of glutamic acid and aspartic acid are potential nucleophilic sites.

A good example of an enzyme active site is the "catalytic triad" found in various hydrolytic enzymes. Such an active site contains a hydroxyl group (from serine), a carboxylate group (from glutamic acid or aspartic acid), and an imidazole ring (from histidine). The three groups are aligned in such a way that the carboxylate group assists a proton transfer from the serine hydroxyl to the imidazole. This enhances the nucleophilicity of the serine toward the carbonyl group of the substrate undergoing hydrolysis. The acyl group is transferred to the serine via a tetrahedral intermediate. Breakdown of the tetrahedral intermediate is accompanied by transfer of a proton back to the leaving group. Subsequently, a water molecule is activated by the same mechanism to cleave the acyl enzyme intermediate.[76]

General References

M. L. Bender, *Mechanisms of Homogeneous Catalysis from Protons to Proteins*. Wiley-Interscience, New York, 1971.

T. C. Bruice and S. J. Benkovic, *Bioorganic Mechanisms*, W. A. Benjamin, New York, 1966.

H. Dugas and C. Penney, *Bioorganic Chemistry: A Chemical Approach to Enzyme Action*, 3rd ed., Springer-Verlag, New York, 1996.

76. D. M. Blow, *Acc. Chem. Res.* **9**:145 (1976); R. M. Garavito, M. G. Rossman, P. Argos, and W. Eventoff, *Biochemistry* **16**:5065 (1977); M. L. Bender, R. J. Bergeron, and M. Komiyama, *The Bioorganic Chemistry of Enzymatic Catalysis*, John Wiley & Sons, New York, 1984, pp. 121–123; C.-H. Hu, T. Brinck, and K. Hult, *Int. J. Quantum Chem.* **69**:89 (1998).

W. P. Jencks, *Catalysis in Chemistry and Enzymology*, McGraw-Hill, New York, 1969.

A. J. Kirby and A. R. Fersht, in *Progress in Bioorganic Chemistry*, Vol. 1, E. T. Kaiser and R. J. Kezdy, eds., Wiley-Interscience, New York, 1971, pp 1–82.

S. Patai, ed., *The Chemistry of Acid Derivatives*, Supplement B, Vol. 2, John Wiley & Sons, New York, 1992.

S. Patai, ed., *The Chemistry of the Carbonyl Group*, John Wiley & Sons, New York, 1969.

S. Patai, ed., *The Chemistry of Carboxylic Acids and Esters*, John Wiley & Sons, New York, 1969.

J. E. Zabricky, ed., *The Chemistry of Amides*, John Wiley & Sons, New York, 1970.

Problems

(References for these problems will be found on page 799.)

1. The hydrates of aldehydes and ketones are considerably more acidic than normal alcohols ($pK \approx 16$–19). How would you account for this fact? Some reported values are shown below. Explain the order of relative acidity.

Hydrate	pK
$H_2C(OH)_2$	13.4
$Cl_3CCH(OH)_2$	10.0
$PhC(OH)_2$ $|$ CF_3	10.0
O_2N—⟨⟩—$C(OH)_2$ $|$ CF_3	9.2

2. Suggest explanations for each of the following observations.

   a. The equilibrium constant for cyanohydrin formation from 3,3-dimethyl-2-butanone is 40 times larger than that for acetophenone.

   b. The rate of release of *p*-nitrophenoxide from compound **A** is independent of pH in aqueous solution of pH > 10.

A

   c. Ester **B** undergoes alkaline hydrolysis 8300 times faster than ester **C** in aqueous dioxane.

B   **C**

d. Under comparable conditions, the general base-catalyzed elimination of bisulfite ion from **D** is about 10 times greater than from **E**.

e. The rate of isotopic exchange of the carbonyl group in tropone (**F**) and 2,3-diphenylcyclopropenone (**G**) is much less than for acetophenone.

3. Arrange the carbonyl compounds in each group in order of decreasing rate of hydrolysis of their respective diethyl acetals or ketals. Explain your reasoning.

a. acetaldehyde, chloroacetaldehyde, crotonaldehyde
b. acetaldehyde, formaldehyde, acetone
c. cyclopentanone, cyclohexanone, camphor
d. acetone, methyl *t*-butyl ketone, methyl neopentyl ketone
e. benzaldehyde, *p*-methoxybenzaldehyde, butyraldehyde.

4. The acid-catalyzed hydrolysis of 2-alkoxy-2-phenyl-1,3-dioxolanes has been studied. The initial step is rate-determining under certain conditions and is described by the rate law given below, which reveals general acid catalysis.

$$k_{obs} = k_{H^+}[H^+] + k_{H_2O}[H_2O] + k_{HA}[HA]$$

By determining k_{HA} for several different buffer catalysts and each of several alkoxy leaving groups, it was determined that there was a relationship between the Brønsted coefficient α and the structure of the alkoxy leaving group. The data are given and show that α decreases as the alkoxy group becomes less basic.

RO–	α
Cl_2CHCH_2O-	0.69
$ClCH_2CH_2O-$	0.80
$CH_3OCH_2CH_2O-$	0.85
CH_3O-	0.90

What mechanistic information is provided by the fact that the Brønsted α decreases as the acidity of the alcohol increases? Discuss these results in terms of a three-dimensional potential energy diagram with the extent of O—H bond formation and the extent of C—O bond breaking taken as the reaction progress coordinates.

5. Each of the following molecules has been considered to be capable of some form of intramolecular catalysis of ester hydrolysis. For each reactant, indicate one or more mechanisms by which intramolecular catalysis might occur. Depict a transition-state arrangement that shows this catalysis.

(a)

(b)

(c)

(d)

(e)

(f)

6. Consider the alkaline pH region of the pH–rate profile in Fig. 8.4 (p. 459), which indicates a rate independent of pH. The rate-controlling reaction in this region is

$$HO^- + PhCH{=}\overset{+}{\underset{H}{N}}C(CH_3)_3 \longrightarrow PhCHNHC(CH_3)_3$$

with OH above PhCH.

Show that the rate of this reaction is pH-independent, despite the involvement of two species, the concentrations of which are pH-dependent.

7. Derive the general expression for the observed rate constant for hydrolysis of **A** as a function of pH. Assume, as is the case experimentally, that intramolecular general acid catalysis completely outweighs intermolecular catalysis by hydronium ion in the pH range of interest. Does the form of your expression agree with the pH–rate profile given for this reaction in Fig. 8.6 (p. 489)?

A

8. Enantiomerically pure dipeptide is obtained when the *p*-nitrophenyl ester of *N*-benzoyl-L-leucine is coupled with glycine ethyl ester in ethyl acetate:

$(CH_3)_2CHCH_2CHC-O-$〈aryl〉$-NO_2 + H_2NCH_2CO_2CH_2CH_3$
$\quad\quad\quad\quad C_6H_5CNH$

$\downarrow$

$(CH_3)_2CHCH_2CHCNHCH_2CO_2CH_2CH_3 + HO-$〈aryl〉$-NO_2$
$\quad\quad\quad\quad C_6H_5CNH$

If, however, the *p*-nitrophenyl ester of *N*-benzoyl-L-leucine is treated with 1-methyl-piperidine in chloroform for 30 min and then coupled with glycine ethyl ester, the dipeptide isolated is almost completely racemic. Furthermore, treatment of the *p*-nitrophenyl ester of *N*-benzoyl-L-leucine with 1-methylpiperidine alone leads to the formation of a crystalline material, $C_{13}H_{15}NO_2$, having strong IR bands at 1832 and 1664 cm^{-1}. Explain these observations, and suggest a reasonable structure for the crystalline product.

9. Offer as complete as possible explanations, based on structural and mechanistic concepts, of the following observations.

   a. The bicyclic lactam **A** hydrolyzes 10^7 times faster than the related monocyclic compound **B**.

A   B

   b. Leaving groups, X, solvolyze from structure **C** at a rate which is 10^{-13} less than that for monocyclic model **D**.

C   D

10. Analyze the factors which would determine stereoselectivity in the addition of organometallic compounds to the following carbonyl compounds. Predict the major product.

(a)

(b)

(c)

$$PhCHCCH(CH_3)_2 \xrightarrow{PhMgBr}$$

with CH$_3$ and O substituents

11. Indicate which compound in each of the following pairs will have the more negative standard free-energy change for hydrolysis at pH 7:

(a) $CH_3CO_2CH_3$ *or* CH_3CSCH_3 (with O)

(b) $CH_3CO_2CH_3$ *or* $CH_3CO_2PO_3H_2$

(c) $CH_3CO_2CH_3$ *or* $H_2NCH_2CO_2CH_3$

(d) $CH_3CCH_2CSCH_3$ *or* CH_3CSCH_3 (with O's)

(e) $CH_3CO_2CH_3$ *or* CH_3C (imidazolide)

(f) $CH_3CO_2CH_3$ *or* $CH_3CN(CH_3)_2$ (with O)

12. Sodium acetate reacts with *p*-nitrophenyl benzoates to give mixed anhydrides if the reaction is conducted in a polar aprotic solvent in the presence of a crown ether. The reaction is strongly accelerated by quarternary nitrogen groups substituted at the *ortho* position. Explain the basis for the enhanced reactivity of these compounds.

Introduction of $^+NR_3$ for H accelerates
the reaction by a factor of $>10^3$

13. The kinetics of the hydrolysis of some imines derived from benzophenone and primary amines revealed the normal dependence of mechanism on pH with rate-determining nucleophilic attack at high pH and rate-determining decomposition of the tetrahedral intermediate at low pH. The simple primary amines show a linear correlation between the rate of nucleophilic addition and the basicity of the amine. Several diamines which were included in the study, in particular **A**, **B**, and **C**, all showed a positive (more reactive) deviation from the correlation line for the simple amines. Why might these amines be more reactive than predicted on the basis of their

$$H_2NCH_2CH_2N(C_2H_5)_2 \qquad H_2NCH_2CH_2N\!\!\diagdown\!\!O \qquad H_2NCH_2CH_2\!\!-\!\!\text{(pyridine)}$$

| A | B | C |

14. The following data give the dissociation constants for several acids that catalyze hydration of acetaldehyde. Also given are the rate constants for the hydration reaction catalyzed by each acid. Treat the data according to the Brønsted equation, and comment on the mechanistic significance of the result.

Acid	K_a	k_{hydr}
Formic	1.77×10^{-4}	1.74
Phenylacetic	4.9×10^{-5}	0.91
Acetic	1.75×10^{-5}	0.47
Pivalic	9.4×10^{-6}	0.33

15. 1,1-Diphenylthioalkanes react with mercuric fluoride to give 1-fluoro-1-phenylthio-alkanes. Provide a detailed description of a likely mechanism for this reaction. Consider such questions as: (1) Is an S_N1 or an S_N2 process most likely to be involved? (2) Would NaF cause the same reaction to occur? (3) Why is only one of the phenylthio groups replaced?

$$\underset{PhS}{\overset{PhS}{\diagdown}}\!\!C\!\!\underset{H}{\overset{R}{\diagup}} \xrightarrow[\text{CH}_3\text{CN}]{\text{HgF}_2} \underset{F}{\overset{PhS}{\diagdown}}\!\!C\!\!\underset{H}{\overset{R}{\diagup}}$$

16. The acid-catalyzed hydrolysis of thioacetanilide can follow two courses.

$$\underset{\text{O}}{\overset{\text{S}}{\parallel}}$$

$$\text{(phenyl)}\!\!-\!\!NH\overset{S}{\overset{\parallel}{C}}CH_3 \xrightarrow[\text{H}_2\text{O}]{\text{H}^+} \underset{B}{\overset{A}{\diagup}}$$

Path A: $\overset{S}{\overset{\parallel}{CH_3COH}} \rightleftharpoons \overset{SH}{\overset{|}{CH_3C}}\!\!=\!\!O + H_2N\!\!-\!\!\text{(phenyl)}$

Path B: $\overset{}{CH_3CNH}\!\!-\!\!\text{(phenyl)} + H_2S$, with C=O below.

The product analysis permits determination of the amount of product formed by each path, as a function of the acidity of the solution. The results are as shown:

H_2SO_4 (% by weight)	3.2	6.1	12	18	36	48
% following path A	50	55	65	75	96	100

Provide a mechanism in sufficient detail to account for the change in product ratio with acid strength.

17. A comparison of the kinetics of hydrolysis and isotopic exchange of amides **A** and **B** was carried out. Some of the data are given below. An interesting observation is that there is more C=O exchange for **A** than for **B**. From this observation, and other data given, develop a stepwise mechanism for the hydrolysis of each amide and a qualitative comparison of the substituent effects on the various steps.

ΔG^{*}_{hydrol} (100°C):	29.6 kcal/mol	24.4 kcal/mol
ΔG^{*}_{ex} (100°C):	26.5 kcal/mol	27.1 kcal/mol
k_{hydrol} (1.0 M OH⁻ in D₂O):	3.06×10^{-5} s⁻¹	3.85×10^{-4} s⁻¹
k_{ex} (1.0 M OH⁻ in D₂O):	1.09×10^{-3} s⁻¹	1.53×10^{-5} s⁻¹

18. Adjacent formyl groups facilitate the hydrolysis of certain aromatic esters, as illustrated by the examples below. Indicate a mechanism for this rate enhancement.

Rate enhancement:	6400	400,000

19. The rates of hydrolysis of the ester group in compounds **A** and **B** have been compared. The effect of an added metal ion (Ni²⁺) on the rate of hydrolysis has been studied, and the observed rate constants for attack by ⁻OH are tabulated. Suggest the most favorable transition-state structure for the addition step of the hydrolysis reaction for each substrate under each set of conditions. Discuss the relationship between the structures of these transition states and the relative rates of attack by hydroxide ion.

k_{HO^-} : 3.0×10^2 M^{-1} s⁻¹
k_{HO^-} in presence of excess
  Ni²⁺: 2.8×10^6 M^{-1} s⁻¹

k_{HO^-} : 7.1 M^{-1} s⁻¹
k_{HO^-} at pH values where salicyclic acid group is
  not ionized: 2.7×10^5 M^{-1} s⁻¹
k_{HO^-} when salicyclic acid group is not ionized and
  excess Ni²⁺ is present: 2.7×10^7 M^{-1} s⁻¹

20. Data pertaining to substituent effects on the acid-catalyzed hydrolysis of mixed methyl aryl acetals of benzaldehyde are given below. The reactions exhibited general acid catalysis, and the Brønsted α values are tabulated. Discuss the information provided by these data about the transition state for the first hydrolysis step, making reference to a diagram showing the location of the transition state as a function of O—H bond formation and C—O bond breaking.

$$\underset{\underset{OCH_3}{\overset{\overset{OAr'}{|}}{Ar-CH}}}{} + H_2O \xrightarrow{H^+} ArCH=O + Ar'OH + CH_3OH$$

Series I, substituent in Ar			Series II, substituent in Ar'		
X	$k_{cat}{}^a$	α	X	$k_{cat}{}^a$	α
m-NO$_2$	2.7×10^{-4}	1.05	m-NO$_2$	8.85×10^{-2}	0.49
m-F	2.2×10^{-3}	0.92	m-Br	4.7×10^{-2}	0.65
m-CH$_3$O	9.6×10^{-3}	0.78	m-F	2.45×10^{-2}	0.67
H	1.3×10^{-2}	0.77	m-CH$_3$O	2.55×10^{-2}	0.71
p-CH$_3$	1.1×10^{-1}	0.72	H	1.3×10^{-2}	0.77
p-CH$_3$O	2.8×10^{-1}	0.68	p-CH$_3$	1.3×10^{-2}	0.88
			p-CH$_3$O	1.65×10^{-2}	0.96

a. Rate constant in s^{-1} for catalysis by acetic acid.

21. The hydrolysis of the lactone **A** shows a significant catalysis by acetate ion in acetate buffer, with the rate expression being

$$k_{obs} = 1.6 \times 10^{-6} + 6.4 \times 10^{-4}[H^+] + 2.08 \times 10^{-5}[^-OAc] + 49[^-OH]$$

This results in a pH–rate profile as shown in Fig. 8.P21, with the acetate catalysis being significant in the pH range 3–6. Discuss how this catalysis by acetate ion might occur. What are the most likely mechanisms for hydrolysis at pH < 2 and pH > 7, where the rates are linear in [H$^+$] and [$^-$OH], respectively?

22. Some data on substituent effects for the reaction of trifluoroacetanilides with

Fig. 8.P21. pH–Rate profile for hydrolysis of **A** in buffered aqueous solution at 70°C.

methoxide ion in methanol and methanol-O-*d* are given below. Calculate the isotope effect for each system. Plot the rate data against appropriate Hammett substituent constants. What facets of the data are in agreement with a normal addition–elimination sequence passing through a tetrahedral intermediate? What facets of the data indicate additional complications? Can you propose a mechanism that is consistent with all the data given?

X	k_{CH_3OH}[a]	k_{CH_3OD}[a]
m-NO$_2$	5.75	8.13
m-Br	0.524	0.464
p-Cl	0.265	0.274
p-Br	0.349	0.346
m-Cl	0.513	0.430
m-OCH$_3$	0.110	0.101
H	0.104	0.0899
m-CH$_3$	0.0833	0.0595
p-CH$_3$	0.0729	0.0451
p-OCH$_3$	0.0564	0.0321

a. Second-order rate constants in M^{-1} s^{-1}.

23. The halogenation of simple ketones such as acetone can proceed through the enol or enolate. By applying the steady-state condition to the enolate, derive a kinetic expression for reaction of acetone with any halogenating agent X−Y in a buffered solution where both C-protonation and O-protonation of the enolate can compete with halogenation. Show that this rate expression predicts that halogenation will be zero-order in halogenating agent under some conditions but first-order in halogenating agent under other conditions.

24. The order of reactivity toward hydrolysis of the cyclic acetals shown below is **A ≪ B ≪ C**. Offer an explanation for this order of reactivity.

25. Examine the structure of the reactants given and the pH–rate profiles (Figs. 8.P25a–d) of the reactions in question. Offer explanations for the response of the observed

reaction rate to the pH for each case.

(a)

(b)

(c)

(d) $(CH_3)_2CHCH=NCH_3$

hydrolysis

Fig. 8.P25a. (Reproduced from problem reference 25a by permission of the American Chemical Society.)

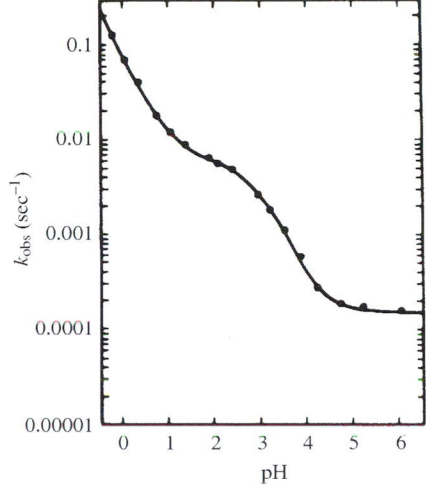

Fig. 8.P25b. (Reproduced from problem reference 25b by permission of the American Chemical Society.)

Fig. 8.P25c. (Reproduced from problem reference 25c by permission of the American Chemical Society.)

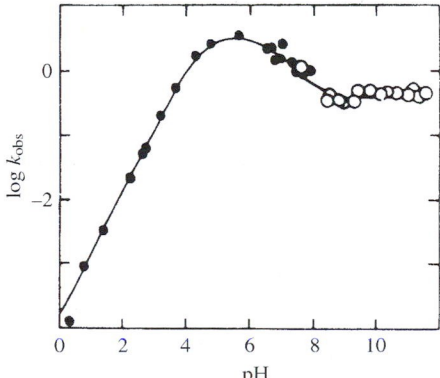

Fig. 8.P25d. (Reproduced from problem reference 25d by permission of the American Chemical Society.)

26. The rates of both formation and hydrolysis of dimethyl acetals of p-substituted benzaldehydes are substituent-dependent. Do you expect k_{form} to increase or decrease with increasing electron-attracting capacity of the *para* substituent? Do you expect the k_{hydrol} to increase or decrease with the electron-attracting power of the substituent? How do you expect K, the equilibrium constant for acetal formation, to vary with the nature of the substituent?

27. Consider the kinetic isotope effect that would be observed in the reaction of semicarbazide with benzaldehyde:

$$PhC\overset{*}{H}=O \ + \ H_2NNHCNH_2 \ \longrightarrow \ PhC\overset{*}{H}=NNHCNH_2$$

Would you expect to find k_H/k_D to be normal or inverse? Would you expect k_H/k_D to be constant, or would it vary with pH?

28. Figure 8.P28 gives the pH–rate profile for conversion of the acid **A** to the anhydride **B** in aqueous solution. The reaction shows no sensitivity to buffer concentration. Notice that the reaction rate increases with the size of the alkyl substituent, and, in fact, the derivative with $R^1 = R^2 = CH_3$ is still more reactive. Propose a mechanism which is consistent with the pH–rate profile and the structure of the initially formed product (which is subsequently hydrolyzed to the diacid). How do you account for the effect of the alkyl substituents on the rate?

29. Assume that the usual mechanism for hydrolysis of an imine, Im, is operative, i.e., that the hydrolysis occurs through a tetrahedral intermediate, TI:

$$R_2C=NR' \ + \ H_2O \ \rightleftharpoons \ \underset{\underset{\text{TI}}{}}{R_2\overset{\overset{OH}{|}}{C}NHR'} \ \rightleftharpoons \ R_2\overset{\overset{O^-}{|}}{C}N^+H_2R' \ \rightleftharpoons \ R_2C=O \ + \ H_2NR'$$

Assume that the steady-state approximation can be applied to the intermediate TI. Derive the kinetic expression for hydrolysis of the imine. How many variables must be determined to construct the pH–rate profile? What simplifying assumptions are justified at very high and very low pH values? What are the kinetic expressions that result from these assumptions?

Fig. 8.P28. pH–Ratio profiles for the hydrolysis of alkyl-*N*-methylmaleamic acids at 39°C and ionic strength 1.0. In increasing order of reactivity, R = H, Me, Et, *i*-Pr, *t*-Bu. Reproduced from problem reference 28 by permission of the Royal Chemical Society.

30. Give a specific structure, including all stereochemical features, for the product expected for each of the following reactions.

(a) $CH_3CH_2CH_2CH{=}O \xrightarrow[80°C]{1M\ NaOH} C_8H_{17}O$

(b)

$\xrightarrow[100°C,\ 1\ h]{5\%\ Na_2CO_3} C_{10}H_{14}O$

(c) $CH_3CH_2\overset{O}{\overset{\|}{C}}Ph \xrightarrow[\substack{2)\ PhCH{=}O,\ -78°C \\ \text{quench after 10 seconds}}]{1)\ LiN(i\text{-}Pr)_2,\ -78°C} C_{16}H_{16}O_2$

(d) $CH_3\overset{O}{\overset{\|}{C}}CH_2CH_3 + PhCH{=}O$ $\underset{HCl}{\overset{NaOH}{\diagdown}}$ $\begin{matrix} C_{11}H_{14}O_2 \\ C_{11}H_{12}O \end{matrix}$

31. Figure 8.P31 gives the pH–rate profile for hydrolysis of thioesters **A–D** and shows a dependence on the nature of the substituents in the alkylthio group. Propose a mechanism which would account for the pH–rate profile of each compound.

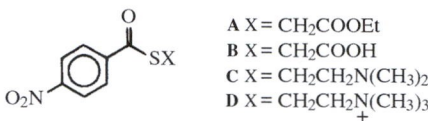

A X = CH$_2$COOEt
B X = CH$_2$COOH
C X = CH$_2$CH$_2$N(CH$_3$)$_2$
D X = CH$_2$CH$_2$N(CH$_3$)$_3$
        +

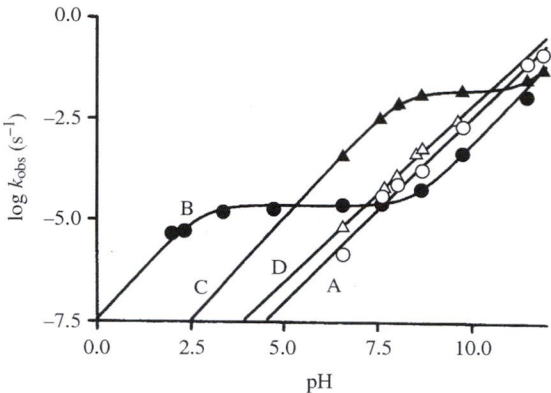

Fig. 8.P31. Plot of the pseudo-first-order rate constants for hydrolysis of thioesters **A** ($\bigcirc$), **B** ($\bullet$), **C** ($\blacktriangle$), **D** ($\triangle$) as a function of pH at 50°C and ionic strength 0.1 (KCl). Lines are from fits of the data to $k_{obs} = (k_{OH}(K_w/H^+)) + (k_{gb}K_a/K_a + [H^+]))$ where k_{OH} is the hydroxide term and k_{gb} is the intramolecular assistance term for **B** and **C** and from linear regression for **A** and **D**. Reproduced from problem reference 31 by permission of the American Chemical Society.

Aromaticity

9.1. The Concept of Aromaticity

The meaning of the word *aromaticity* has evolved as understanding of the special properties of benzene and other aromatic molecules has deepened.[1] Originally, aromaticity was associated with a special chemical reactivity. The aromatic hydrocarbons were considered to be those unsaturated systems that underwent substitution reactions in preference to addition. Later, the idea of special stability became more important. Benzene can be shown to be much lower in enthalpy than predicted by summation of the normal bond energies for the $C=C$, $C-C$, and $C-H$ bonds in the Kekulé representation of benzene. Aromaticity is now generally associated with this property of special stability of certain completely conjugated cyclic molecules. A major contribution to the stability of aromatic systems results from the delocalization of electrons in these molecules.

Aromaticity is usually described in MO terminology. Cyclic structures that have a particularly stable arrangement of occupied π molecular orbitals are called aromatic. A simple expression of the relationship between an MO description of structure and aromaticity is known as the *Hückel* rule. It is derived from Hückel molecular orbital (HMO) theory and states that *planar monocyclic completely conjugated hydrocarbons will be aromatic when the ring contains $4n + 2$ π electrons*. HMO calculations assign the π-orbital energies of the cyclic unsaturated systems of ring size 3–9 as shown in Fig. 9.1. (See Chapter 1, Section 1.4, p. 31, to review HMO theory.)

In Fig. 9.1, orbitals below the dashed reference line are bonding orbitals; when they are filled, the molecule is stabilized. The orbitals that fall on the reference line are nonbonding; placing electrons in these orbitals has no effect on the total bonding energy of the molecule. The orbitals above the reference line are antibonding; the presence of electrons in these orbitals destabilizes the molecule. The dramatic difference in properties of cyclobutadiene (extremely unstable) and benzene (very stable) is explicable in terms of

1. M. Glukhovtsev, *J. Chem. Educ.* **74**:132 (1997); D. Lloyd, *J. Chem. Inf. Comput. Sci.* **36**:442 (1996); Z. Zhou, *Int. Rev. Phys. Chem.* **11**:243 (1992); J. P. Snyder, *Nonbenzenoid Aromatics*, Vol. 1, Academic Press, New York, 1969, Chapter 1.

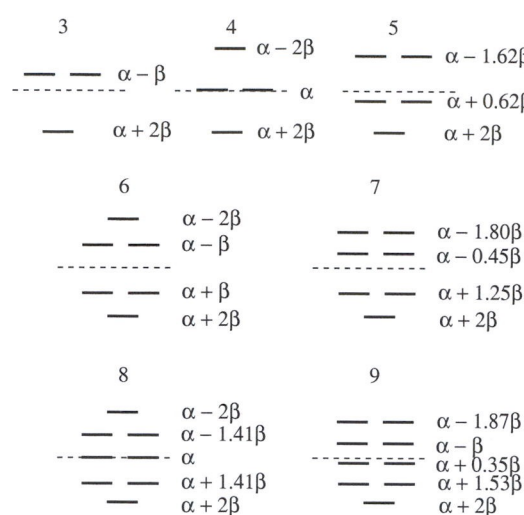

Fig. 9.1. HMO energies for conjugated ring systems of three to nine carbon atoms.

these energy level diagrams:

$$\begin{array}{cc} \text{cyclobutadiene} & \text{benzene} \\ E = 4\alpha + 4\beta & E = 6\alpha + 8\beta \end{array}$$

Cyclobutadiene has two bonding electrons, but the other two electrons are unpaired because of the degeneracy of the two nonbonding orbitals. The two electrons in the nonbonding levels do not contribute to the stabilization of the molecule. Furthermore, because these electrons occupy a high-energy orbital, they are particularly available for chemical reactions. As we shall see in a moment, experimental evidence indicates that cyclobutadiene is rectangular rather than square. This modifies somewhat the orbital picture from the simple Hückel pattern, which assumes a square geometry. The two nonbonding levels are no longer degenerate, so cyclobutadiene is not predicted to have unpaired electrons. Nevertheless, higher-level MO calculations agree with the Hückel concept in predicting cyclobutadiene to be an extremely unstable molecule with a high-energy occupied orbital.

Simple Hückel calculations on benzene, in contrast, place all the π electrons in bonding MOs. The π-electron energy of benzene is calculated by summing the energies of the six π electrons, which is $6\alpha + 8\beta$, lower by 2β than the value of $6\alpha + 6\beta$ for three isolated double bonds. Thus, the HMO method predicts a special stabilization for benzene.

The pattern of two half-filled degenerate levels persists for larger rings containing $4n$ π electrons. In contrast, all $4n + 2$ systems are predicted to have all electrons paired in bonding MOs with net stabilization relative to isolated double bonds. This pattern provides the theoretical basis of the Hückel rule.

As indicated in Chapter 1, the simple HMO theory is based on rather drastic assumptions. More elaborate MO treatments indicate that the most stable geometry for cyclobutadiene is rectangular.[2] Although several derivatives of cyclobutadiene are known and will be discussed shortly, cyclobutadiene itself has been observed only as a "matrix-isolated" species. Several compounds when photolyzed at very low temperature ($\sim 10\,K$) in solid argon give rise to cyclobutadiene. Analysis of the IR spectra of the product and deuterated analogs generated from labeled precursors has confirmed the theoretical conclusion that cyclobutadiene is a rectangular molecule.[3]

Attempts to describe just how stable a given aromatic molecule is in terms of simple HMO calculations have centered on the *delocalization energy*. The total π-electron energy of a molecule is expressed in terms of the energy parameters α and β, which arise in HMO calculations. This energy value can be compared to that for a hypothetical localized version of the same molecule. The HMO energy for the π electrons of benzene is $6\alpha + 8\beta$. The same quantity for the hypothetical localized model cyclohexatriene is $6\alpha + 6\beta$, the sum of three isolated C=C bonds. The difference of 2β is called the *delocalization energy* or *resonance energy*. Although this quantity is often useful for comparing related systems, it is not a measurable physical quantity; rather, it is obtained by comparing a real molecule and a hypothetical one. Most estimates of the stabilization of benzene are in the range of 20–40 kcal/mol and depend on the choice of properties assigned to the hypothetical cyclohexatriene reference point.

There have been two general approaches to determining the amount of stabilization that results from aromatic delocalization. One is to use experimental thermodynamic measurements. Bond energies, as was mentioned in Chapter 1, are nearly additive when there are no special interactions between the various bond types. Thus, it is possible to calculate such quantities as the heat of combustion or heat of hydrogenation of "cyclohexatriene" by assuming that it is a compound with no interaction between the conjugated double bonds. For example, a very simple calculation of the heat of hydrogenation for cyclohexatriene would be to multiply the heat of hydrogenation of cyclohexene by 3, i.e., $3 \times 28.6 = 85.8$ kcal/mol. The actual heat of hydrogenation of benzene is 49.8 kcal/mol, suggesting a total stabilization or delocalization energy of 36.0 kcal/mol. There are other, more elaborate, ways of approximating the thermodynamic properties of the hypothetical cyclohexatriene. The difference between the calculated and corresponding measured thermodynamic property of benzene is taken to be the aromatic stabilization. For benzene, the values obtained are usually around 30 kcal/mol, but the aromatic stabilization cannot be determined in an absolute sense because these values are established by the properties assigned to the cyclohexatriene model.

The second general approach to estimating aromatic stabilization is to use MO methods. This has already been illustrated by the discussion of benzene according to simple HMO theory, which assigns the stabilization energy a value of 2β units. More advanced MO methods can assign the stabilization energy in a more quantitative way. The most successful method is to perform calculations on the aromatic compound and on a linear, conjugated polyene containing the same number of double bonds.[4] This method

2. J. A. Jaffi and M. D. Newton, *J. Am. Chem. Soc.* **100**:5012 (1978); W. T. Borden, E. R. Davidson, and P. Hart, *J. Am. Chem. Soc.* **100**:388 (1978); H. Kollmar and V. Staemmler, *J. Am. Chem. Soc.* **99**:3583 (1977); M. J. S. Dewar and A. Kormornicki, *J. Am. Chem. Soc.* **99**:6174 (1977).
3. S. Masamune, F. A. Souto-Bachiller, T. Machiguchi, and J. E. Bertie, *J. Am. Chem. Soc.* **100**:4889 (1978); B. A. Hess, Jr., P. Carsky, and L. J. Schaad, *J. Am. Chem. Soc.* **105**:695 (1983).

assigns a resonance stabilization of zero to the polyene, even though it is known by thermodynamic criteria that conjugated polyenes do have some stabilization relative to isomeric compounds with isolated double bonds. Using this definition, semiempirical MO calculations assign a value of about 20 kcal/mol to the resonance energy of benzene, relative to 1,3,5-hexatriene. The use of polyenes as reference compounds has proven to give better agreement with experimental trends in stability than comparison with the sums of isolated double bonds.

The isodesmic reaction approach (see Section 4.1) has also been applied to calculation of the resonance stabilization of benzene.

$$\text{benzene} + 3\,CH_2{=}CH_2 \longrightarrow 3\,CH_2{=}CH{-}CH{=}CH_2$$

This approach can be taken using either experimental thermochemical data or energies obtained by MO calculations.[5] If the resonance energy of butadiene is assigned as zero, the above reaction gives the resonance energy of benzene as 21.2 kcal/mol. If butadiene is considered to have a delocalization energy, the computation must be modified to reflect that fact. Using 7.2 kcal/mol as the butadiene delocalization energy gives a value of 42.8 kcal/mol as the benzene resonance energy.

Both thermochemical and MO approaches agree that benzene is an especially stable molecule and are reasonably consistent with one another in the stabilization energy which is assigned. It is very significant that MO calculations also show a destabilization of certain conjugated cyclic polyenes, cyclobutadiene in particular. The instability of cyclobutadiene has precluded any thermochemical evaluation of the extent of destabilization. Compounds that are destabilized relative to conjugated noncyclic polyene models are called *antiaromatic*.[6]

Another characteristic of aromatic compounds is a relatively large HOMO–LUMO gap, which can be expressed in terms of hardness, η (see p. 21 for the definition of hardness)[7]:

$$\eta = (\varepsilon_{LUMO} - \varepsilon_{HOMO})/2$$

The numerical value of hardness obtained by MNDO-level calculations correlates with the stability of aromatic compounds.[8] The correlation can be extended to a wider range of compounds, including heterocyclic compounds, when hardness is determined experimentally on the basis of molar refractivity.[9] The relatively large HOMO–LUMO gap also indicates the absence of relatively high-energy, reactive electrons, in agreement with the reduced reactivity of aromatic compounds toward electrophilic reagents.

There are also physical measurements that can give evidence of aromaticity. The

4. M. J. S. Dewar and C. de Llano, *J. Am. Chem. Soc.* **91:**789 (1969).
5. P. George, M. Trachtman, C. W. Bock, and A. M. Brett, *J. Chem. Soc., Perkin Trans. 2* **1976:**1222; P. George, M. Trachtman, C. W. Bock, and A. M. Brett, *Tetrahedron* **32:**1357 (1976); M. N. Glukhovtsev, R. D. Bach, and S. Laiter, *THEOCHEM* **417:**123 (1997); A. Skanke, R. Hosmane, and J. F. Liebman, *Acta Chem. Scand.* **52:**967 (1998).
6. R. Breslow, *Acc. Chem. Res.* **6:**393 (1973).
7. Z. Zhou and R. G. Parr, *J. Am. Chem. Soc.* **111:**7371 (1989).
8. Z. Zhou and H. V. Navangul, *J. Phys. Org. Chem.* **3:**784 (1990); Z. Zhou, *Int. Rev. Phys. Chem.* **11:**243 (1992).
9. C. W. Bird, *Tetrahedron* **53:**3319 (1997).

determination of the bond lengths in benzene by electron diffraction is a classic example of use of the bond-length criterion of aromaticity. Spectroscopic methods or X-ray diffraction can also provide bond-length data. Aromatic hydrocarbons show carbon–carbon bond lengths in the range 1.38–1.40 Å, and the bond lengths are quite uniform around the ring. In contrast, localized polyenes show alternation between typical sp^3–sp^3 single-bond and sp^2–sp^2 double-bond lengths along the conjugated chain. The uniformity of bond lengths has been developed as a criterion of aromaticity.[10]

NMR spectroscopy also provides an experimental tool capable of assessing aromaticity. Aromatic compounds exhibit a *diamagnetic ring current*. Qualitatively, this ring current can be viewed as the migration of the delocalized electrons in the π system under the influence of the magnetic field in an NMR spectrometer. The ring current effect is responsible for a large magnetic anisotropy in aromatic compounds. The induced ring current gives rise to a local magnetic field that is opposed to the direction of the applied magnetic field. Nuclei in a cone above or below the plane of an aromatic ring are shielded by the induced field and appear at relatively high field in the NMR spectrum, whereas nuclei in the plane of the ring—i.e., the atoms bound directly to the ring—occur at downfield positions. Antiaromatic compounds show opposite effects. The occurrence of these chemical shift phenomena is evidence for aromaticity.[11] The chemical shift phenomena can be treated on a quantitative basis by quantum-mechanical calculation of the chemical shift at the center of the ring. The value of the chemical shift at a point in the center of the ring can be calculated. These values are referred to as the nucleus independent chemical shift (NICS). These values show excellent correlation with other manifestations of aromaticity.[12] Benzenoid hydrocarbons such as benzene, naphthalene, and anthracene show values of about −9 to −10 ppm. Heteroaromatic five-membered rings show slightly more negative values (pyrrole, −15.1; thiophene, −13.6; furan, −12.3). The values for aromatic ions such as cyclopentadienide (−14.3) and cycloheptatrienylium (−7.6) are also negative. Those for antiaromatic species, including cyclobutadiene (+27.6) and borole (+17.5), are positive. Saturated compounds such as cyclohexane have values near zero.

Another property associated with aromaticity is magnetic susceptibility. Magnetic susceptibility is determined by measuring the force exerted on the sample by a magnetic field.[13] Magnetic susceptibility can also be determined using an NMR spectrometer.[14] It is noted that aromatic compounds have enhanced magnetic susceptibility, relative to values predicted on the basis of the localized structural components.[15] Magnetic susceptibility can also be calculated by computational methods. Calculation of magnetic susceptibility by the B3LYP method correctly reproduces some of the trends in stability among the

10. C. W. Bird, *Tetrahedron* **48**:335 (1992); C. W. Bird, *Tetrahedron* **52**:9945 (1996); C. W. Bird, *Tetrahedron* **54**:4641 (1998).

11. R. C. Haddon, *J. Am. Chem. Soc.* **101**:1722 (1979); J. Aihara, *J. Am. Chem. Soc.* **103**:5704 (1981); R. C. Haddon and K. Raghavachari, *J. Am. Chem. Soc.* **107**:289 (1985); S. Kuwajima and Z. G. Soos, *J. Am. Chem. Soc.* **109**:107 (1987).

12. P. v. R. Schleyer, C. Maerker, A. Dransfeld, H. Jiao, and N. J. R. van Eikema Hommes, *J. Am. Chem. Soc.* **118**:6317 (1996).

13. E. A. Boudreaux and R. R. Gupta, *Physical Methods in Heterocyclic Chemistry*, R. R. Gupta, ed., Wiley-Interscience, New York, 1984, pp. 281–311.

14. K. Frei and H. J. Bernstein, *J. Chem. Phys.* **37**:1891 (1962).

15. P. v. R. Schleyer and H. Jiao, *Pure Appl. Chem.* **68**:209 (1996); P. Friedman and K. F. Ferris, *Int. J. Quantum Chem. Symp.* **24**:843 (1990).

benzo[*b*] and benzo[*c*] derivatives of five-membered heterocycles.[16]

It has been argued that there are two fundamental aspects of aromaticity, one reflecting structure and energy, and the other, magnetic properties and electron mobility.[17] Parameters of aromaticity such as bond length and stabilization appear to be largely independent of the magnetic criteria, such as diamagnetic ring current. However, there is a correlation between the two kinds of measurements. The more stabilized compounds exhibit the greatest magnetic susceptibility.[18] Aromaticity is thus best conceived as a single property resulting from cyclic delocalization that results in both stabilization and the magnetic phenomena associated with electron mobility.

9.2. The Annulenes

The term *annulene* was coined to refer to the completely conjugated monocyclic polyenes.[19] The synthesis of annulenes has been extended well beyond the first two members of the series [4]annulene (cyclobutadiene) and [6]annulene (benzene). The generality of the Hückel rule can be tested by considering the properties of members of the annulene series.

The smallest member, cyclobutadiene, was the objective of attempted synthesis for many years. The first success was achieved when cyclobutadiene released from a stable iron complex was trapped with various reagents.[20]

Dehalogenation of *trans*-3,4-dibromocyclobutene was shown to generate a species with the same reactivity.[21]

16. B. S. Jursic, *J. Heterocycl. Chem.* **33**:1079 (1996).
17. A. R. Katritzky, P. Barczyniski, G. Musumarra, D. Pisano, and M. Szafran, *J. Am. Chem. Soc.* **111**:7 (1989); A. R. Katritzky, M. Karelson, S. Sild, T. M. Krygowski, and K. Jug, *J. Org. Chem.* **63**:5228 (1998); V. I. Minkin, M. N. Glukhovtsev, and B. Ya. Simkin, *Aromaticity and Antiaromaticity*, John Wiley & Sons, New York, 1994.
18. P. v. R. Schleyer, P. K. Freeman, H. Jiao, and B. Goldfuss, *Angew. Chem. Int. Ed. Engl.* **34**:337 (1995); C. W. Bird, *Tetrahedron* **52**:9945 (1996).
19. F. Sondheimer, *Pure Appl. Chem.* **28**:331 (1971); *Acc. Chem. Res.* **5**:81 (1972).
20. L. Watts, J. D. Fitzpatrick, and R. Pettit, *J. Am. Chem. Soc.* **87**:3253 (1965).
21. E. K. G. Schmidt, L. Brener, and R. Pettit, *J. Am. Chem. Soc.* **92**:3240 (1970).

Various trapping agents react with cyclobutadiene to give Diels–Alder adducts[22]:

In the absence of trapping agents, a characteristic dimer is produced.

This dimerization is an extremely fast reaction and limits the lifetime of cyclobutadiene, except at very low temperatures.

Cyclobutadiene can also be prepared by photolysis of several different precursors at very low temperature in solid inert gases.[23] These methods provide cyclobutadiene in a form that is appropriate for spectroscopic study. Under these conditions, cyclobutadiene begins to dimerize at around 35 K. Whereas simple HMO theory assumes a square geometry for cyclobutadiene, most MO methods predict a rectangular structure as the minimum-energy geometry. With very high level calculations, good agreement is obtained between the calculated minimum-energy structure, which is rectangular, and observed spectroscopic properties.[24]

The rectangular structure is calculated to be strongly destabilized (antiaromatic) with respect to a polyene model.[25] With 6-31G* calculations, for example, cyclobutadiene is found to have a negative resonance energy of -54.7 kcal/mol, relative to 1,3-butadiene. In addition, 30.7 kcal of strain is found, giving a total destabilization of 85.4 kcal/mol.[26] G2 and MP4/G-31(d,p) calculations arrive at an antiaromatic destabilization energy of about 42 kcal/mol.[25,27]

22. L. Watts, J. D. Fitzpatrick, and R. Pettit, *J. Am. Chem. Soc.* **88**:623 (1966); J. C. Barborak, L. Watts, and R. Pettit, *J. Am. Chem. Soc.* **88**:1328 (1966); D. W. Whitman and B. K. Carpenter, *J. Am. Chem. Soc.* **102**:4272 (1980).

23. G. Maier and M. Scheider, *Angew. Chem. Int. Ed. Engl.* **10**:809 (1971); O. L. Chapman, C. L. McIntosh, and S. Pacansky, *J. Am. Chem. Soc.* **95**:614 (1973); O. L. Chapman, D. De La Cruz, R. Roth, and J. Pacansky, *J. Am. Chem. Soc.* **95**:1337 (1973); C. Y. Lin and A. Krantz, *J. Chem. Soc., Chem. Commun.* **1972**:111; G. Maier, H. G. Hartan, and T. Sayrac, *Angew. Chem. Int. Ed. Engl.* **15**:226 (1976); H. W. Lage, H. P. Reisenauer, and G. Maier, *Tetrahedron Lett.* **23**:3893 (1982).

24. B. A. Hess, Jr., P. Carsky, and L. J. Schaad, *J. Am. Chem. Soc.* **105**:695 (1983); H. Kollmar and V. Staemmler, *J. Am. Chem. Soc.* **100**:4304 (1978); C. van Wullen and W. Kutzelnigg, *Chem. Phys. Lett.* **205**:563 (1993).

25. M. N. Gluckhovtsev, S. Laiter, and A. Pross, *J. Phys. Chem.* **99**:6828 (1995).

26. B. A. Hess, Jr., and L. J. Schaad, *J. Am. Chem. Soc.* **105**:7500 (1983).

27. M. N. Glukhovtsev, R. D. Bach, and S. Laiter, *THEOCHEM* **417**:123 (1997).

A number of alkyl-substituted cyclobutadienes have been prepared by related methods.[28] Increasing alkyl substitution enhances the stability of the compounds. The tetra-*t*-butyl derivative is stable up to at least 150°C but is very reactive toward oxygen.[29] This reactivity reflects the high energy of the HOMO. The chemical behavior of the cyclobutadienes as a group is in excellent accord with that expected from the theoretical picture of the structure of these compounds.

[6]Annulene is benzene. Its properties are so familiar to students of organic chemistry that not much need be said here. It is the parent compound of a vast series of derivatives. The benzene ring shows exceptional stability, both in a thermodynamic sense and in terms of its diminished reactivity in comparison with conjugated polyenes. As was discussed earlier, a stabilization on the order of 30 kcal/mol is found by thermodynamic comparisons. Benzene is much less reactive toward electrophiles than are conjugated polyenes. This is in line with the HOMO of benzene being of lower energy than the HOMO of a conjugated polyene.

The next higher annulene, cyclooctatetraene, is nonaromatic.[30] The bond lengths around the ring alternate as expected for a polyene. The C=C bonds are 1.33 Å while the C−C bonds are 1.462 Å in length.[31] Thermodynamic data provide no evidence of any special stability.[32] Cyclooctatetraene is readily isolable, and its chemical reactivity is normal for a polyene. Structure determination shows that the molecule is tub-shaped,[19] and therefore is not a planar system to which the Hückel rule applies. There have been both experimental and theoretical studies aimed at trying to estimate the stability of the planar form of cyclooctatetraene.[33] The results of 6-31G* calculations indicate that the completely delocalized D_{8h} structure is about 4.1 kcal higher in energy than the conjugated planar D_{4h} structure, suggesting that delocalization leads to destabilization.[34] Similar results are obtained using CAS-SCF calculations.[35]

	D_{2d}	D_{4h}	D_{8h}
Relative energy	0	10.6 kcal/mol	14.7 kcal/mol

These two energies are, respectively, comparable to the experimental activation energies for conformation inversion of the tub conformer and bond shifting, suggesting that the two planar structures represent the transition states for those processes.

ring inversion   bond shift

28. G. Maier, *Angew. Chem. Int. Ed. Engl.* **13:**425 (1974); S. Masamune, *Tetrahedron* **36:**343 (1980).
29. G. Maier, S. Pfriem, U. Schafer, and R. Matusch, *Angew. Chem. Int. Ed. Engl.* **17:**520 (1978).
30. G. Schroder, *Cyclootatetraene*, Verlag Chemie, Weinheim, 1965; G. I. Fray and R. G. Saxton, *The Chemistry of Cyclooctatetraene and Its Derivatives*, Cambridge University Press, Cambridge, 1978.
31. M. Traetteberg, *Acta Chem. Scand.* **20:**1724 (1966).
32. R. B. Turner, B. J. Mallon, M. Tichy, W. v. E. Doering, W. Roth, and G. Schröder, *J. Am. Chem. Soc.* **95:**8605 (1973).
33. L. A. Paquette, *Acc. Chem. Res.* **26:**57 (1993).
34. D. A. Hrovat and W. T. Borden, *J. Am. Chem. Soc.* **114:**5879 (1992); P. Politzer, J. S. Murray, and J. M. Seminario, *Int. J. Quantum Chem.* **50:**273 (1994).
35. J. L. Andres, O. Castano, A. Morreale, R. Palmeiro, and R. Gomperts, *J. Chem. Phys.* **108:**203 (1998).

Larger annulenes permit the incorporation of *trans* double bonds into the rings. Beginning with [10]annulene, stereoisomeric structures are possible. According to the Hückel rule, [10]annulene should possess aromatic stabilization if it were planar. However, all the isomeric cyclodeca-1,3,5,7,9-pentaenes suffer serious steric strain that prevents the planar geometry from being adopted. The *Z,E,Z,Z,E*-isomer, which has minimal bond-angle strain, suffers a severe nonbonded repulsion between the two internal hydrogens.

The *Z,Z,Z,Z,Z*-isomer is required by geometry to have bond angles of 144° to maintain planarity and would therefore be enormously destabilized by distortion of the normal trigonal bond angle. The most stable structure is a twisted form of the *E,Z,Z,Z,Z*-isomer. MO (MP2/DZd) calculations suggest an aromatic stabilization of almost 18 kcal for a conformation of the *E,Z,Z,Z,Z*-isomer in which the inner hydrogens are twisted out of the plane by about 20°,[36] but other calculations point to a polyene structure.[37]

E,Z,Z,Z,Z   Z,Z,Z,Z,Z   Z,E,Z,Z,E

Experimental studies have indicated that all of the isomers prepared to date are quite reactive, but whether the most stable isomer has been observed is uncertain. Two of the isomeric [10]annulenes, as well as other products, are formed by photolysis of *cis*-9,10-dihydronaphthalene[38]:

Neither compound exhibits properties that would suggest aromaticity. The NMR spectra are consistent with polyene structures. Both compounds are thermally unstable and revert back to dihydronaphthalenes:

It appears that [10]annulene is sufficiently distorted from planarity that little stabilization is achieved.

36. H. M. Sulzbach, P. v. R. Schleyer, H. Jiao, Y. Xie, and H. F. Schaefer III, *J. Am. Chem. Soc.* **117**:1369 (1995).
37. H. M. Sulzbach, H. F. Schaefer III, W. Klopper, and H. P. Luthi, *J. Am. Chem. Soc.* **118**:3519 (1996).
38. S. Masamune, K. Hoko, G. Gigam, and D. L. Rabenstein, *J. Am. Chem. Soc.* **93**:4966 (19971); S. Masamune and N. Darby, *Acc. Chem. Res.* **5**:272 (1972).

A number of structures have been prepared that do not have the steric problems associated with the cyclodeca-1,3,5,7,9-pentaenes. In compound **1**, the steric problem is avoided with only a slight loss of planarity in the π system[39]:

The NMR spectrum of this compound shows a diamagnetic ring current of the type expected in an aromatic system.[40] X-ray crystal structures of **1**[41] and its carboxylic acid derivative **2**[42] are shown in Fig. 9.2. Both reveal a pattern of bond lengths very similar to that in naphthalene (see p. 534).[43]

Most MO methods find a bond alternation pattern in the minimum-energy structure, but calculations that include electron correlation lead to a delocalized minimum-energy structure.[44] Thus, although the π system in **1** is not completely planar, it appears to be sufficiently close to provide a delocalized 10-electron π system. A resonance energy of 17.2 kcal has been obtained on the basis of an experimental heat of hydrogenation.[45]

The deviation from planarity that is present in a structure such as **1** raises the question of how severely a conjugated system can be distorted from the ideal coplanar alignment of p orbitals and still retain aromaticity. This problem has been analyzed by determining the degree of rehybridization necessary to maximize p orbital overlap in **1**. It is found that rehybridization to incorporate fractional amounts of s character can improve orbital alignment substantially. Orbitals with about 6% s character are suggested to be involved

Fig. 9.2. X-ray crystal structures of 1,6-methanodeca-1,3,5,7,9-pentaene (A) and 1,6-methanodeca-1,3,5,7,9-pentaene-2-carboxylic acid (B). (Structures are reproduced from Refs. 41 and 42 by permission of the International Union of Crystallography and Verlag Helvetica Chimica Acta AG.)

39. E. Vogel and H. D. Roth, *Angew. Chem. Int. Ed. Engl.* **3**:228 (1964).
40. E. Vogel, *Pure Appl. Chem.* **20**:237 (1969).
41. R. Bianchi, T. Pilati and M. Simonetta, *Acta Crystallogr., Sect. B* **36**:3146 (1980).
42. M. Dobler and J. D. Dunitz, *Helv. Chim. Acta* **48**:1429 (1965).
43. O. Bastainsen and P. N. Skancke, *Adv. Chem. Phys.* **3**:323 (1961).
44. R. C. Haddon and K. Raghavchari, *J. Am. Chem. Soc.* **107**:289 (1985); L. Farnell and L. Radom, *Am. Chem. Soc.* **104**:2650 (1982).
45. W. R. Roth, M. Böhm, H. W. Lennartz, and E. Vogel, *Angew. Chem. Int. Ed. Engl.* **22**:1007 (1983).

for **1**.[46] Thus, a relatively small amount of rehybridization greatly improves orbital overlap in the twisted system.

[12]Annulene is a very unstable compound that undergoes cyclization to bicyclic isomers and can be kept only at very low temperature.[47] The NMR spectrum has been studied at low temperature.[48] Besides indicating the double-bond geometry shown in the structure below, the spectrum reveals a *paramagnetic* ring current, the opposite to what is observed for aromatic systems. This feature is quite characteristic of the [4*n*]annulenes and has been useful in characterizing the aromaticity or lack of it in annulenes.[49]

[14]Annulene was first prepared in 1960.[50] Its NMR spectrum has been investigated and shows that two stereoisomers are in equilibrium[51]:

Z,E,E,Z,E,Z,E   Z,E,Z,E,Z,Z,E

The spectrum also reveals a significant diamagnetic (aromatic) ring current. The signals of the internal hydrogens (C-3, C-6, C-10, and C-13) are very far upfield ($\delta = -0.61$ ppm).[50] The interconversion of the two forms involves a configurational change from E to Z at at least one double bond. The activation energy for this process is only about 10 kcal/mol. The crystal structure for [14]annulene shows the *Z,E,E,Z,E,Z,E*-form to be present in the solid.[52] The bond lengths around the ring range from 1.35 to 1.41 Å but do not show the alternating pattern of short and long bonds expected for a localized polyene. There is some distortion from planarity, especially at carbon atoms 3, 6, 10, and 13, which is caused by nonbonded repulsions between the internal hydrogens. MP2/6-31G* and B3LYP calculations, however, find the delocalized structure as the only minimum.[53] This discrepancy between experiment and theory awaits resolution.

A 14-electron π system can be generated in circumstances in which the steric problem associated with the internal hydrogens of [14]annulene are avoided. This can be achieved in 10b,10c-dihydropyrene systems in which the annulene ring is built around a saturated

46. R. C. Haddon, *Acc. Chem. Res.* **21**:243 (1988).
47. J. F. M. Oth, H. Rottele, and G. Schroder, *Tetrahedron Lett.* **1970**:61.
48. J. F. M. Oth, J.-M. Gilles, and G. Schroder, *Tetrahedron Lett.* **1970**:67.
49. R. C. Haddon, *Tetrahedron* **28**:3613, 3635 (1972).
50. F. Sondheimer and Y. Gaoni, *J. Am. Chem. Soc.* **82**:5765 (1960).
51. J. F. M. Oth, *Pure Appl. Chem.* **25**:573 (1971).
52. C. C. Chiang and I. C. Paul, *J. Am. Chem. Soc.* **94**:4741 (1972).
53. C. H. Choi, M. Kertesz, and A. Karpfen, *J. Am. Chem. Soc.* **119**:11994 (1997).

core:

Several derivatives of this ring system have been synthesized.[54,55] These compounds exhibit properties indicating that the conjugated system is aromatic. They exhibit NMR shifts characteristic of a diamagnetic ring current. Typical aromatic substitution reactions can be carried out.[56] An X-ray crystal structure ($R = C_2H_5$) shows that the bond lengths are in the aromatic range (1.39–1.40 Å), and there is no strong alternation around the ring.[57] The peripheral atoms are not precisely planar, but the maximum deviation from the average plane is only 0.23 Å. The dimethyl derivative is essentially planar, with bond lengths between 1.38 and 1.40 Å.[55]

Another family of 14-π-electron systems is derived from structure **3**[58]:

The *syn* isomer can achieve a conjugated system with angles of up to 35° between adjacent *p* orbitals. The *anti* isomer is much more twisted.[59] An X-ray crystal structure of the *syn* isomer shows C–C bond lengths between 1.368 and 1.418 Å for the conjugated system (Fig. 9.3).[60] The spectroscopic properties of the *syn* isomer are consistent with considering it to be a delocalized annulene.[61] B3LYP calculations indicate that both the *syn* and *anti* structures are stabilized by delocalization, the *syn* (17.6 kcal/mol) more so than the *anti* (8.1 kcal).[62]

Fig. 9.3. (A) Carbon framework from X-ray crystal structure of *syn*-tricyclo[8.4.1.1$^{3.8}$]hexadeca-1,3,5,7,9,11,13-heptaene. (B) Side view showing deviation from planarity of annulene ring. (Reproduced from Ref. 60 by permission from the International Union of Crystallography.)

54. R. H. Mitchell and V. Boekelheide, *J. Am. Chem. Soc.* **96**:1547 (1974); V. Boekelheide and T. A. Hylton, *J. Am. Chem. Soc.* **92**:3669 (1970); H. Blaschke, C. E. Ramey, I. Calder, and V. Boekelheide, *J. Am. Chem. Soc.* **92**:3675 (1970); V. Boekelheide and J. B. Phillips, *J. Am. Chem. Soc.* **89**:1695 (1967).
55. R. H. M. Mitchell, V. S. Iyer, N. Khalifa, R. Mahadevan, S. Venugopalan, S. A. Weerawarna, and P. Zhou, *J. Am. Chem. Soc.* **117**:1514 (1995).
56. J. B. Phillis, R. J. Molyneux, E. Sturm, and V. Boekelheide, *J. Am. Chem. Soc.* **89**:1704 (1967).
57. A. W. Hanson, *Acta Crystallogr.* **23**:476 (1967).
58. E. Vogel, *Pure Appl. Chem.* **28**:355 (1971).
59. E. Vogel, J. Sombroek, and W. Wagemann, *Angew. Chem. Int. Ed. Engl.* **14**:564 (1975); E. Vogel, U. Haberland, and H. Günther, *Angew. Chem. Int. Ed. Engl.* **9**:513 (1970).
60. R. Destro, T. Pilati, and M. Simonetta, *Acta Crystallogr. Sect. B.* **33**:940 (1977).
61. J. Dewey, H. M. Deger, W. Fröhlich, B. Dick, K. A. Klingensmith, G. Hohlneicher, E. Vogel, and J. Michl, *J. Am. Chem. Soc.* **102**:6412 (1980).
62. M. Nendel, K. N. Houk, L. M. Tolbert, E. Vogel, H. Jiao, and P. v. R. Schleyer, *Angew. Chem. Int. Ed. Engl.* **36**:748 (1997).

An isomeric system is related to the benzenoid hydrocarbon phenanthrene. Both the *syn* and *anti* stereoisomers have been synthesized.[63]

The NMR spectrum of the *syn* isomer shows evidence of a diamagnetic ring current, based on both the relatively low-field position of the vinylic hydrogens and the upfield shift of the methylene hydrogens. The *anti* isomer shows much less pronounced shifts. The X-ray crystal structure of the *syn* isomer shows a moderate level of bond alternation, ranging from 1.36 to 1.45 Å (Fig. 9.4A). In the *anti* isomer, bond alternation is more pronounced, with the double bond in the center ring being essentially a localized double bond (Fig. 9.4B).

Fig. 9.4. (A) X-ray crystal structure of *syn*-tricyclo[8.4.1.1^{4.9}]hexadeca-2,4,6,8,10,12,14-heptaene. (B) X-ray crystal structure of *anti* stereoisomer of tricyclo[8.4.1.1^{4.9}]hexadeca-2,4,6,8,10,12,14-heptaene-5-carboxylic acid. (Reproduced from Ref. 63 by permission of Wiley-VCH.)

The Hückel rule predicts nonaromaticity for [16]annulene. The compound has been synthesized and thoroughly characterized.[64] The bond lengths show significant alternation (C=C, 1.34 Å; C—C, 1.46 Å), and the molecule is less planar than [14]annulene.[65] These structural data are consistent with regarding [16]annulene as being nonaromatic.

[18]Annulene offers a particularly significant test of the Hückel rule. The internal cavity in [18]annulene is large enough to minimize steric interactions between the internal hydrogens in a geometry that is free of angle strain. Most MO calculations find the delocalized structure to be more stable than the polyene.[66]

63. E. Vogel, W. Püttmann, W. Duchatsch, T. Schieb, H. Schmickler, and J. Lex, *Angew. Chem. Int. Ed. Engl.* **25**:720 (1986); E. Vogel, T. Schieb, W. H. Schulz, K. Schmidt, H. Schmickler, and J. Lex, *Angew. Chem. Int. Ed. Engl.* **25**:723 (1986).
64. I. Calder, Y. Gaoni, and F. Sondheimer, *J. Am. Chem. Soc.* **90**:4946 (1968); G. Schröder and J. F. M. Oth, *Tetrahedron Lett.* **1966**:4083.
65. S. M. Johnson and I. C. Paul, *J. Am. Chem. Soc.* **90**:6555 (1968).
66. J. M. Shulman and R. L. Disch, *J. Mol. Struct.* **234**:213 (1991); K. Yoshizawa, T. Kato, and T. Yamabe, *J. Phys. Chem.* **100**:5697 (1996).

The properties of [18]annulene are consistent with its being aromatic. The X-ray crystal structure shows the molecule to be close to planarity, with the maximum deviation from the plane being 0.085 Å.[67] The bond lengths are in the range 1.385–1.405 Å, and the pattern is short, short, long, rather than alternating. The NMR spectrum is indicative of an aromatic ring current.[68] The chemical reactivity of the molecule would also justify its classification as aromatic.[69] Both MP2/6-31G* and DFT calculations find a delocalized structure with D_{6h} symmetry as the minimum-energy structure. The bond lengths are 1.39–1.42 Å, and a stabilization of 18 kcal/mol is indicated.[70]

There are also examples of [18]annulene systems constructed around a saturated central core, such as in compound **4**.[71] In this compound, the internal protons are at very high field (−6 to −8 ppm), whereas the external protons are far downfield (∼9.5 ppm).

4

The chemical shift data can be used as the basis for comparing the diamagnetic ring current with the maximum ring current expected for a completely delocalized π system. By this criterion, the flexible [18]annulene maintains only about half (0.56) of the maximum ring current, whereas the rigid ring in **4** gives a value of 0.88, indicating more effective conjugation in this system.

The synthesis of annulenes has been carried forward to larger rings as well. [20]Annulene,[72] [22]annulene,[73] and [24]annulene[74] have all been reported. The NMR spectrum of [22]annulene is consistent with regarding the molecule as aromatic, whereas those of the [20] and [24] analogs are not. In each case, there is some uncertainty as to the preferred conformation in solution, and the NMR spectra are temperature-dependent. Although the properties of these molecules have not been studied as completely as those of the smaller systems, they are consistent with the predictions of the Hückel rule.

Both clever synthesis and energetic processes leading to stable compounds have provided other examples of structures for which aromaticity might be important. Kekulene was synthesized in 1978.[75] How aromatic is this substance? Both by energy and magnetic criteria, it appears that it is primarily benzenoid in character. Its energy is close to that expected from isodesmic reactions summing smaller aromatic components. Magnetic

67. J. Bregman, F. L. Hirshfeld, D. Rabinovich, and G. M. J. Schmidt, *Acta Crystallogr.* **19**:227 (1965); F. L. Hirshfeld and D. Rabinovich, *Acta Crystallogr.* **19**:235 (1965); S. Gorter, E. Rutten-Keulemans, M. Krever, C. Romers, and D. W. J. Cruickshank, *Acta Crystallogr., Sect. B.* **51**:1036 (1995).
68. Y. Gaoni, A. Melera, F. Sondheimer, and R. Wolovsky, *Proc. Chem. Soc.* **1965**:397.
69. I. C. Calder, P. J. Garratt, H. C. Longuet-Higgins, F. Sondheimer, and R. Wolovsky, *J. Chem. Soc. C* **1067**:1041.
70. K. K. Baldridge and J. S. Siegel, *Angew. Chem. Int. Ed. Engl.* **36**:745 (1997); C. H. Choi, M. Kertesz, and A. Karpfen, *J. Am. Chem. Soc.* **119**:11994 (1974).
71. T. Otsubo, R. Gray, and V. Boekelheide, *J. Am. Chem. Soc.* **100**:2449 (1978).
72. B. W. Metcalf and F. Sondheimer, *J. Am. Chem. Soc.* **93**:6675 (1971).
73. R. M. McQuilkin, B. W. Metcalf, and F. Sondheimer, *J. Chem. Soc., Chem. Commun.* **1971**:338.
74. I. C. Calder and F. Sondheimer, *J. Chem. Soc., Chem. Commun.* **1966**:904.
75. F. Diederich and H. A. Staab, *Angew. Chem. Int. Ed. Engl.* **17**:372 (1978); H. A. Staab and F. Diederich, *Chem. Ber.* **116**:3487 (1983).

criteria, too, indicate that it is similar to the smaller polycyclic benzenoid hydrocarbons, such as phenanthrene and anthracene.[76] (See Problem 18 at the end of this chapter to consider this issue more thoroughly.)

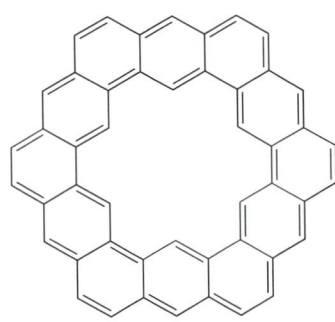

Fullerene, C_{60}, is a spherical cluster of carbon atoms which is produced by processes such as laser vaporization of graphite.[77] The structure consists of hexagons and pentagons, corresponding to the pattern of a soccer ball. There is bond-length variation, with the bonds shared by the hexagonal rings being shorter (1.40 ± 0.01 Å) than those of the pentagons (1.46 ± 0.01 Å). Unlike benzene, with its two Kekulé structures, there is only one valence bond structure for C_{60}. It has double bonds at all hexagon–hexagon edges and single bonds at the pentagonal edges. An isodesmic energy computation suggests that the π system is substantially less stable than for benzene on an atom-by-atom comparison.[78] Calculated chemical shift parameters suggest that the five-membered rings are antiaromatic whereas the hexagonal rings are aromatic.[79] Thus, it appears that fullerene is a delocalized molecule, but with both stabilizing and destabilizing components which are partially compensating in terms of stabilization energy.

It has been pointed out that a different array of atomic orbitals might be conceived of in large conjugated rings. The array, called a *Möbius twist*, results in there being one point in the ring at which the atomic orbitals would show a phase discontinuity.[80]

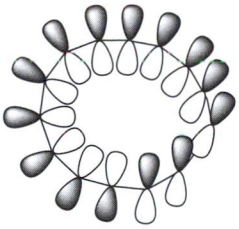

If the ring were sufficiently large that the twist between individual orbitals were small, such a system would not necessarily be less stable than the normal array of atomic orbitals. This

76. H. Jiao and P. v. R. Schleyer, *Angew. Chem. Int. Ed. Engl.* **35**:2383 (1996).
77. H. W. Kroto, J. R. Heath, S. C. O'Brien, R. F. Curl, and R. E. Smalley, *Nature* **318**:162 (1985).
78. P. W. Fowler, D. J. Collins, and S. J. Austin, *J. Chem. Soc., Perkin Trans. 2* **1993**:275.
79. P. v. R. Schleyer, C. Maerker, A. Dransfeld, H. Jiao, and N. J. R. van Eikema Hommes, *J. Am. Chem. Soc.* **118**:6317 (1996).
80. E. Heilbronner, *Tetrahedron Lett.* **1964**:1923.

same analysis points out that in such an array the Hückel rule is reversed and aromaticity is predicted for the $4n$ π-electron systems.

Hückel orbital aray	Möbius orbital array
$4n + 2$ aromatic	$4n$ aromatic
$4n$ antiaromatic	$4n + 2$ antiaromatic

So far, no ground-state molecule in which the twisted conjugation would exist has been made, so the prediction remains to be tested. It correctness is strongly suggested, however, by the fact that transition states with twisted orbital arrays appear to be perfectly acceptable in many organic reactions.[81] We will return to this topic in Chapter 11.

9.3. Aromaticity in Charged Rings

There are also striking stability relationships due to aromaticity in charged ring systems. The HMO energy levels that apply to fully conjugated, planar three- to nine-membered rings were given in Fig. 9.1 (p. 510). These energy levels are applicable to charged species as well as to the neutral annulenes. A number of cations and anions that have completely conjugated planar structures are shown in Scheme 9.1. Among these species, the Hückel rule predicts aromatic stability for cyclopropenium ion (**A**), cyclobutadiene dication (**C**), cyclobutadiene dianion (**D**), cyclopentadienide anion (**F**), cycloheptatrienyl cation (tropylium ion, **G**), the dications and dianions derived from cyclooctatetraene (**I** and **J**) and the cyclononatetraenide anion (**K**). The other species shown, having $4n$ π electrons, would be expected to be very unstable. Let us examine what is known about the chemistry of some of these species.

There is a good deal of information about the cyclopropenium ion that supports the idea that it is exceptionally stable. The cyclopropenium ion and a number of derivatives have been generated by ionization procedures:

Ref. 82

Ref. 83

The 1,2,3-tri-t-butylcyclopropenium cation is so stable that the perchlorate salt can be recrystallized from water.[84] An X-ray study of triphenylcyclopropenium perchlorate has verified the existence of the carbocation as a discrete species.[85] Quantitative estimation of the stability of the unsubstituted ion can be made in terms of its pK_{R^+} value of -7.4, which is intermediate between those of such highly stabilized ions as triphenylmethyl

81. H. E. Zimmerman, *J. Am. Chem. Soc.* **88**:1566 (1966); H. E. Zimmerman, *Acc. Chem. Res.* **4**:272 (1971).
82. S. W. Tobey and R. West, *J. Am. Chem. Soc.* **86**:1459 (1964); R. West, A. Sado, and S. W. Tobey, *J. Am. Chem. Soc.* **88**:2488 (1966).
83. R. Breslow, J. T. Groves, and G. Ryan, *J. Am. Chem. Soc.* **89**:5048 (1967).
84. J. Ciabattoni and E. C. Nathan III, *J. Am. Chem. Soc.* **91**:4766 (1969).
85. M. Sundaralingam and L. H. Jensen, *J. Am. Chem. Soc.* **88**:198 (1966).

Scheme 9.1. Completely Conjugated Cyclic Cations and Anions

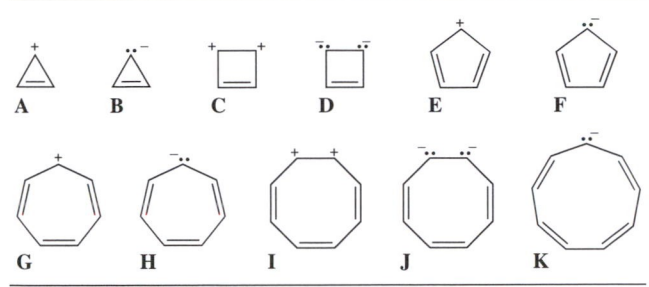

cation and that of the bis(4-methoxyphenyl)methyl cation[86] (see Section 5.4 for the definition of pK_{R+}). A 6-31G* MO calculation on the isodesmic reaction

$$\text{[cyclopropenium]}^{\oplus} + CH_3CH{=}CH_2 \longrightarrow \text{[cyclopropene]} + {}^+CH_2CH{=}CH_2$$

yields a ΔH of +38.2 kcal/mol, while experimental data on the heats of formation of the various species indicate $\Delta H = +31$ kcal/mol.[87] Both values imply that the cyclopropenium ion is much more stable than the allyl cation. G2 calculations indicate total aromatic stabilization of 60 kcal/mol based on the reaction:[88]

$$\text{[cyclopropenyl}^+] + \text{[cyclopropene]} \longrightarrow \text{[cyclopropene]} + \text{[cyclopropenyl}^+]$$

The heterolytic gas-phase bond dissociation energy to form cyclopropenium ion from cyclopropene is 225 kcal/mol. This compares with 256 kcal/mol for formation of the allyl cation from propene or 268 kcal/mol for formation of the 1-propyl cation from propane.[89]

In contrast, the less strained four-π-electron cyclopentadienyl cation is very unstable. It is calculated to have a negative stabilization energy of 56.7 kcal/mol.[90] The cyclopentadienyl cation is also found to be antiaromatic on the basis of magnetic susceptibility and chemical shift criteria.[91] Its pK_{R+} has been estimated as −40, from an electrochemical cycle.[92] The heterolytic bond dissociation energy to form the cation from cyclopentadiene is 258 kcal/mol, which is substantially more than for formation of an allylic cation from cyclopentene but only slightly more than the 252 kcal/mol for formation of an unstabilized secondary carbocation.[89] A rate retardation of 10^{-14}, relative to cyclopentyl analogs, has been estimated from solvolytic rate data.[93] Solvolysis of cyclopentadienyl halides assisted by silver ion is extremely slow, even though the cyclopentadienyl ring is doubly allylic.[94] When the bromide and antimony pentafluoride react at −78°C, an EPR spectrum is

86. R. Breslow and J. T. Groves, *J. Am. Chem. Soc.* **92**:984 (1970).
87. L. Radom, P. C. Hariharan, J. A. Pople, and P. v. R. Schleyer, *J. Am. Chem. Soc.* **98**:10 (1976).
88. M. N. Glukhovtsev, S. Laiter, and A. Pross, *J. Phys. Chem.* **100**:17801 (1996).
89. F. P. Lossing and J. L. Holmes, *J. Am. Chem. Soc.* **106**:6917 (1984).
90. P. v. R. Schleyer, P. K. Freeman, H. Jiao, and B. Goldfuss, *Angew. Chem. Int. Ed. Engl.* **34**:337 (1995); B. Reindl and P. v. R. Schleyer, *J. Comput. Chem.* **19**:1402 (1998).
91. H. Jiao, P. v. R. Schleyer, Y. Mo, M. A. McAllister, and T. T. Tidwell, *J. Am. Chem. Soc.* **119**:7075 (1997).
92. R. Breslow and S. Mazur, *J. Am. Chem. Soc.* **95**:584 (1973).
93. A. D. Allen, M. Sumonja, and T. T. Tidwell, *J. Am. Chem. Soc.* **119**:2371 (1997).
94. R. Breslow and J. M. Hoffman, Jr., *J. Am. Chem. Soc.* **94**:2110 (1972).

observed which indicates that the cyclopentadienyl cation is a triplet.[95] Similar studies indicate that the penta-*i*-propyl-[96] and pentachlorocyclopentadienyl cations are also triplets, but the ground state of the pentaphenyl derivative is a singlet.

The relative stability of the anions derived from cyclopropene and cyclopentadiene by deprotonation is just the reverse of the situation for the cations. Cyclopentadiene is one of the most acidic hydrocarbons known, with a pK_a of 16.0.[97] The pK's of triphenylcyclopropene and trimethylcyclopropene have been estimated as 50 and 62, respectively, from electrochemical cycles.[98] The unsubstituted compound would be expected to fall somewhere in between and thus must be about 40 powers of 10 less acidic than cyclopentadiene. MP2/6-31(d,p) and B3LYP calculations indicate a small destabilization, relative to the cyclopropyl anion.[99] Thus, the six-π-electron cyclopentadienide ion is enormously stabilized relative to the four-π-electron cyclopropenide ion, in agreement with the Hückel rule.

The Hückel rule predicts aromaticity for the six-π-electron cation derived from cycloheptatriene by hydride abstraction and antiaromaticity for the planar eight-π-electron anion that would be formed by deprotonation. The cation is indeed very stable, with a pK_{R+} of +4.7.[100] Salts containing the cation can be isolated as a product of a variety of preparative procedures.[101] On the other hand, the pK_a of cycloheptatriene has been estimated at 36.[102] This value is similar to those of normal 1,4-dienes and does not indicate strong destabilization. Thus, the seven-membered eight-π-electron anion is probably nonplanar. This would be similar to the situation in the nonplanar eight-π-electron hydrocarbon, cyclooctatetraene.

The cyclononatetraenide anion is generated by treatment of the halide **5** with lithium metal[103]:

An isomeric form of the anion that is initially formed is converted to the all-*cis* system rapidly at room temperature.[104] Data on the equilibrium acidity of the parent hydrocarbon are not available, so the stability of the anion cannot be judged quantitatively. The NMR spectrum of the anion, however, is indicative of aromatic character.[105]

95. M. Saunders, R. Berger, A. Jaffe, J. M. McBride, J. O'Neill, R. Breslow, J. M. Hoffman, Jr., C. Perchonock, E. Wasserman, R. S. Hutton, and V. J. Kuck, *J. Am. Chem. Soc.* **95**:3017 (1973).

96. H. Sitzmann, H. Bock, R. Boese, T. Dezember, Z. Havlas, W. Kaim, M. Moscherosch, and L. Zanathy, *J. Am. Chem. Soc.* **115**:12003 (1993).

97. A. Streitwieser, Jr., and L. L. Nebenzahl, *J. Am. Chem. Soc.* **98**:2188 (1976).

98. R. Breslow and W. Chu, *J. Am. Chem. Soc.* **95**:411 (1973).

99. G. N. Merrill and S. R. Kass, *J. Am. Chem. Soc.* **119**:12322 (1997).

100. W. v. E. Doering and L. H. Knox, *J. Am. Chem. Soc.* **76**:3202 (1954).

101. T. Nozoe, *Prog. Org. Chem.* **5**:132 (1961); K. M. Harmon, in *Carbonium Ions*, Vol. IV, G. A. Olah and P. v. R. Schleyer, eds., Wiley-Interscience, 1973, Chapter 2.

102. R. Breslow and W. Chu, *J. Am. Chem. Soc.* **95**:411 (1973).

103. T. J. Katz and P. J. Garratt, *J. Am. Chem. Soc.* **86**:5194 (1964); E. A. LaLancette and R. E. Benson, *J. Am. Chem. Soc.* **87**:1941 (1965).

104. G. Boche, D. Martens, and W. Danzer, *Angew. Chem. Int. Ed. Engl.* **8**:984 (1969).

105. S. Fliszar, G. Cardinal, and M. Bernaldin, *J. Am. Chem. Soc.* **104**:5287 (1982); S. Kuwajima and Z. G. Soos, *J. Am. Chem. Soc.* **108**:1707 (1986).

Several doubly charged ions are included in Scheme 9.1; some have been observed experimentally. Ionization of 3,4-dichloro-1,2,3,4-tetramethylcyclobutene in SbF_5–SO_2 at $-75°C$ results in an NMR spectrum attributed to the tetramethyl derivative of the cyclobutadienyl dication[106]:

It is difficult to choose a reference compound against which to judge the stability of the dication. That it can be formed at all, however, suggests special stabilization associated with the two-π-electron system. The dianion formed by adding two electrons to the π system of cyclobutadiene also meets the $4n + 2$ criterion. In this case, however, four of the six electrons would occupy nonbonding orbitals so high reactivity could be expected. There is some evidence that this species may have a finite existence.[107] Reaction of 3,4-dichlorocyclobutene with sodium naphthalenide, followed a few minutes later by addition of methanol-O-d, gives a low yield of 3,4-di-*deuterio*-cyclobutene. The inference is that the dianion $[C_4H_4{}^{2-}]$ is present. As yet, however, no direct experimental observation of this species has been accomplished.

Cyclooctatetraene is reduced by alkali metals to a dianion:

The NMR spectrum indicates a planar aromatic structure.[108] It has been demonstrated that the dianion is more stable than the radical anion formed by one-electron reduction, since the radical anion disproportionates to cyclooctatetraene and the dianion:

The crystal structure of the potassium salt of 1,3,5,7-tetramethylcyclooctatetraene dianion has been determined by X-ray diffraction.[109] The eight-membered ring is planar, with "aromatic" C–C bond lengths of about 1.41 Å without significant alternation. The spectroscopic and structural studies lead to the conclusion that the cyclooctatetraene dianion is a stabilized delocalized structure.

A dication derived from 1,3,5,7-tetramethylcyclooctatetraene is formed at $-78°C$ in SO_2Cl by reaction with SbF_5. Both the proton and carbon NMR spectra indicate that the ion is a symmetrical, diamagnetic species, and the chemical shifts are consistent with an

106. G. A. Olah, J. M. Bollinger, and A. M. White, *J. Am. Chem. Soc.* **91**:3667 (1969); G. A. Olah and G. D. Mateescu, *J. Am. Chem. Soc.* **92**:1430 (1970).
107. J. S. McKennis, L. Brener, J. R. Schweiger, and R. Pettit, *J. Chem. Soc., Chem. Commun.* **1972**:365.
108. T. J. Katz, *J. Am. Chem. Soc.* **82**:3784 (1960).
109. S. Z. Goldberg, K. N. Raymond, C. A. Harmon, and D. H. Templeton, *J. Am. Chem. Soc.* **96**:1348 (1974).

aromatic ring current. At about −20°C, this dication undergoes a chemical transformation to a more stable dication[110]:

Reduction of the nonaromatic polyene [12]annulene, either electrochemically or with lithium metal, generates a 14-π-electron dianion[111]:

The NMR spectrum of the resulting dianion shows a diamagnetic ring current indicative of aromatic character, even though steric interactions among the internal hydrogens must prevent complete coplanarity. In contrast to the neutral [12]annulene, which is thermally unstable above −50°C, the dianion remains stable at 30°C. The dianion of [16]annulene has also been prepared and shows properties consistent with its being regarded as aromatic.[112]

The pattern of experimental results on charged species with cyclic conjugated systems is summarized in Table 9.1. It is consistent with the applicability of Hückel's rule to charged, as well as neutral, conjugated planar cyclic structures.

Table 9.1. Hückel's Rule Relationships for Charged Species

Compound	π electrons
Aromatic species	
Cyclopropenium cation	2
Cyclopentadienide anion	6
Cycloheptatrienyl cation	6
Cyclooctatetraene dianion	10
Cyclononatetraenide anion	10
[12]Annulene dianion	14
Antiaromatic species	
Cyclopropenide anion	4
Cyclopentadienyl cation	4
Nonaromatic species	
Cycloheptatrienyl anion	8

110. G. A. Olah, J. S. Staral, G. Liang, L. A. Paquette, W. P. Melega, and M. I. Carmody, *J. Am. Chem. Soc.* **99:**3349 (1977).
111. J. F. M. Oth and G. Schröder, *J. Chem. Soc., B* **1971:**904.
112. J. F. M. Oth, G. Anthoine, and J.-M. Gilles, *Tetrahedron Lett.* **1968:**6265.

Homoaromaticity is a term used to describe systems in which a stabilized cyclic conjugated system is formed by bypassing one saturated atom.[113] The resulting stabilization would, in general, be expected to be reduced because of poorer overlap of the orbitals. The properties of several such cationic species, however, suggest that substantial stabilization does exist. The cyclooctatrienyl cation is an example:

A significant feature of the NMR spectrum of this cation is that the protons *a* and *b* exhibit sharply different chemical shifts. Proton *a* is 5.8 ppm upfield of *b*, indicating the existence of an aromatic ring current.[114] The fact that the two protons exhibit separate signals also establishes that there is a substantial barrier for the conformational process that interchanges H_a and H_b. MO calculations that include the effects of electron correlation indicate that the homoconjugated structure is a good description of the cation and find that there is a strong aromatic ring current.[115]

The cyclobutenyl cation is the homoaromatic analog of the very stable cyclopropenium cation. This ion can be prepared from 3-acetoxycyclobutene with the use of "superacid" conditions[116]:

The temperature-dependent NMR spectrum of the ion can be analyzed to show that there is a barrier (8.4 kcal/mol) for the ring flip that interchanges the two hydrogens of the methylene group. The ^{13}C-NMR chemical shift is also compatible with the homoaromatic structure. MO calculations are successful in reproducing the structural and spectroscopic characteristics of the cation and are consistent with a homoaromatic structure.[117]

The existence of stabilizing homoconjugation in anions has been more difficult to establish. Much of the discussion has revolved about anion **8**. This species was proposed to have aromatic character on the basis of the large upfield shift of the CH_2 group that would lie in the shielding region generated by a diamagnetic ring current.[118] The ^{13}C-NMR

113. S. Winstein, *Q. Rev. Chem. Soc.* **23**:141 (1969); L. A. Paquette, *Angew. Chem. Int. Ed. Engl.* **17**:106 (1978); R. V. Williams, *Adv. Phys. Org. Chem.* **29**:273 (1994).
114. P. Warner, D. L. Harris, C. H. Bradley, and S. Winstein, *Tetrahedron Lett.* **1970**:4013; C. E. Keller and R. Pettit, *J. Am. Chem. Soc.* **88**:604, 606 (1966); R. F. Childs, *Acc. Chem. Res.* **17**:347 (1984).
115. R. C. Haddon, *J. Am. Chem. Soc.* **110**:1108 (1988).
116. G. A. Olah, J. S. Staral, R. J. Spear, and G. Liang, *J. Am. Chem. Soc.* **97**:5489 (1975).
117. R. C. Haddon and K. Raghavachari, *J. Am. Chem. Soc.* **105**:1188 (1983); M. Schindler, *J. Am. Chem. Soc.* **109**:1020 (1987); S. Sieber, P. v. R. Schleyer, A. H. Otto, J. Gauss, F. Reichel, and D. Cremer, *J. Phys. Org. Chem.* **6**:445 (1993).
118. S. Winstein, M. Ogliaruso, M. Sakai, and J. M. Nicholson, *J. Am. Chem. Soc.* **89**:3656 (1967).

Fig. 9.5. Structure of TMEDA complex of lithium bicyclo[3.2.1]octa-2,6-dienide. (Reproduced from Ref. 121 by permission of Wiley-VCH.)

spectrum can also be interpreted in terms of homoaromaticity.[119] Both gas-phase and solution measurements suggest that the parent hydrocarbon is more acidic than would be anticipated if there were no special stabilization of the anion.[120] An X-ray crystal structure of the lithium salt has been done.[121] The structure is a monomeric TMEDA complex (Fig. 9.5). The lithium is not symmetrically disposed toward the anion but is closer to one carbon of the allyl system. There is no indication of flattening of the homoconjugated atoms, and the C(6)—C(7) bond distance is in the normal double-bond range (1.354 Å). In contrast to the results from MO calculations on the homoaromatic cations **6** and **7**, MO calculations fail to reveal substantial stabilization of the anion **8**.[122] The final reconciliation of the divergent indications of the degree of delocalization and stabilization of this anion will have to await further work.

9.5. Fused-Ring Systems

Many completely conjugated hydrocarbons can be built up from the annulenes and related structural fragments. Scheme 9.2 gives the structures, names, and stabilization energies of a variety of such hydrocarbons. Derivatives of these hydrocarbons having heteroatoms in place of one or more carbon atoms constitute another important class of organic compounds.

119. M. Cristl, H. Leininger, and D. Brückner, *J. Am. Chem. Soc.* **105:**4843 (1983).
120. R. E. Lee and R. R. Squires, *J. Am. Chem. Soc.* **108:**5078 (1986); W. N. Washburn, *J. Org. Chem.* **48:**4287 (1983).
121. N. Hertkron, F. H. Kohler, G. Müller, and G. Reber, *Angew. Chem. Int. Ed. Engl.* **25:**468 (1986).
122. J. B. Grutzner and W. L. Jorgenson, *J. Am. Chem. Soc.* **103:**1372 (1981); E. Kaufman, H. Mayr, J. Chandrasekhar, and P. v. R. Schleyer, *J. Am. Chem. Soc.* **103:**1375 (1981); R. Lindh, B. O. Roos, G. Jonsäll, and P. Ahlberg, *J. Am. Chem. Soc.* **108:**6554 (1986).

	Benzene	Naphthalene	Anthracene	Naphthacene
HMO	2.00β	3.698β	5.31β	6.93β
HMO'	0.39β	0.55β	0.66β	0.76β
RE	0.38β	0.59β	0.71β	0.83β
SCF-MO	$0.869\,eV$	$1.323\,eV$	$1.600\,eV$	$1.822\,eV$

	Phenanthrene	Triphenylene	Pyrene	Perylene
HMO	5.44β	7.27β	6.50β	8.24β
HMO'	0.77β	1.01β	0.82β	0.96β
RE	0.85β	1.13β	0.95β	1.15β
SCF-MO	$1.933\,eV$	$2.654\,eV$	$2.10\,eV$	$2.619\,eV$

	Butalene	Pentalene	Azulene	Heptalene
HMO	1.66β	2.45β	3.25β	3.61β
HMO'	-0.48β	-0.14β	0.23β	-0.048β
RE	-0.34β	-0.09β	0.27β	-0.01β
SCF-MO	$-0.28\,eV$	$-0.006\,eV$	$0.169\,eV$	$-0.004\,eV$

	Methylene-cyclopropene	Fulvene	Calicene	Fulvalene
HMO	0.96β	1.46β	—	2.80β
HMO'	0.02β	-0.012β	0.34β	-0.33β
RE	—	-0.01β	0.39β	-0.29β
SCF-MO	—	—	—	—

	Benzocyclobutadiene	Biphenylene	Acenaphthylene	
HMO	2.38β	4.50β	4.61β	—
HMO'	-0.22β	0.32β	0.47β	$-0.22\,\beta$
RE	-0.16β	0.42β	0.57β	—
SCF-MO	—	$1.346\,eV$	$1.335\,eV$	—

a. Stabilization energies given are from the following sources:
   HMO: C.A. Coulson, A. Streitwieser, Jr., M. D. Poole, and J. I. Brauman, *Dictionary of π-Electron Calculations*, W. H. Freeman, San Franscisco, 1965.
   HMO': B. A. Hess, Jr., and L. J. Schaad, Jr., *J. Am. Chem. Soc.* **93**:305, 2413 (1971); *J. Org. Chem.* **36**:3418 (1971); *J. Org. Chem.* **37**: 4179 (1972).
   RE: A. Moyano and J. C. Paniagua, *J. Org. Chem.* **51**:2250 (1986).
   SCF-MO: M. J. S. Dewar and C. de Llano, *J. Am. Chem. Soc.* **91**:789 (1969). 1 eV = 23 kcal/mol.

It is of interest to be able predict the stability of such fused-ring compounds. Because Hückel's rule applies only to monocyclic systems, it cannot be applied to the fused-ring compounds, and there have been many efforts to develop relationships which would predict their stability. The underlying concepts are the same as for monocyclic systems; stabilization should result from a particularly stable arrangement of MOs whereas instability would be associated with unpaired electrons or electrons in high-energy orbitals.

The same approximations discussed in Section 1.4 permit calculation of the HMOs for conjugated systems of the type shown in Scheme 9.2, and many of the results have been tabulated.[123] However, attempts to correlate stability with the Hückel delocalization energy relative to isolated double bonds give poor correlation with the observed chemical properties of the compounds. By choosing a polyene as the reference state, much better agreement between calculated stabilization energy and experimental chemical properties is achieved (see Section 1.1.2). A series of energy terms corresponding to the structural units in the reference polyene have been established empirically by Hess and Schaad.[124] The difference between the energy of the conjugated hydrocarbon by HMO calculation and the sum of the energies of the appropriate structural units gives a stabilization energy. For azulene, for example, the HMO calculation gives an energy of $10\alpha + 13.36\beta$. The energy for the polyene reference is obtained by summing contributions for the component bond types: $3(HC=CH) + 2(HC=C) + 3(HC-CH) + 2(HC-C) + 1(C-C) = 10\alpha + 13.13\beta$. The difference, 0.23β, is the stabilization or resonance energy assigned to azulene by this method. For comparison of nonisomeric molecules, the Hess–Schaad treatment uses resonance energy per electron, which is obtained simply by dividing the calculated stabilization energy by the number of π electrons. Although the resulting stabilization energies are based on a rudimentary HMO calculation, they are in good qualitative agreement with observed chemical stability. The stabilizations have been calculated for most of the molecules in Scheme 9.2 and are listed as HMO′.

The energy parameters used for the reference polyene by Hess and Schaad were developed on a strictly empirical basis. Subsequently, Moyano and Paniagua developed an alternative set of reference bond energies on a theoretical basis.[125] These values are shown along with the Hess–Schaad values in Table 9.2. In Scheme 9.2, the stabilizations calculated for the various hydrocarbons using this point of reference are those listed as RE (for resonance energy). The Hess–Schaad HMO′ and the RE values are in generally

Table 9.2. Energy Values for Reference Bond Types

Bond type	Hess–Schaad value (β)	Bond type	Moyano–Paniagua value (β)
$H_2C=CH$	2.000	$H_2C=CH-CH=$	2.2234
$HC=CH$	2.070	$H_2C=CH-C=$	2.2336
$H_2C=C$	2.000	$=CH-CH=CH-CH=$	2.5394
$HC=C$	2.108	$=CH-CH=CH-C=$	2.5244
$C=C$	2.172	$=C-CH=CH-C=$	2.4998
$HC-CH$	0.466	$H_2C=C-$	2.4320
$HC-C$	0.436	$-CH=CH-$	2.7524
$C-C$	0.436	$-C=C-$	2.9970

123. F. Heilbronner and P. A. Straub, *Hückel Molecular Orbitals*, Springer-Verlag, Berlin, 1966; C. A. Coulson and A. Streitwieser, *Dictionary of π-Electron Calculations*, W. H. Freeman, San Francisco, 1965.
124. B. A. Hess, Jr., and L. J. Schaad, *J. Am. Chem. Soc.* **93**:305, 2413 (1971); *J. Org. Chem.* **36**:3418 (1971); *J. Org. Chem.* **37**:4179 (1972).
125. A. Moyano and J. C. Paniagua, *J. Org. Chem.* **51**:2250 (1986).

good agreement with observed stability. Both calculations give negative stabilization for benzocyclobutadiene, for example.[126]

The values listed in Scheme 9.2 as SCF-MO are from an early semiempirical SCF calculation. This was the first instance in which a polyene was chosen as the reference state.[127]

All these approaches agree that benzene and the structures that can be built up by fusing together benzenoid rings are strongly stabilized relative to the reference polyenes. The structures with more rings tend to have lower resonance energies per π electron compared to benzene. This feature is in agreement with experimental trends in reactivity.[128] Because the structures with fewer rings are more stable, there is a tendency for species in which several rings are fused together to react by addition to an internal ring to give the smaller and more stable structures.

SCF stabilization energy = 41.9 kcal   SCF stabilization energy = 50.4 kcal

This trend is revealed, for example, by the rates of Diels–Alder addition reactions of anthracene, naphthacene, and pentacene, in which three, four, and five rings, respectively are linearly fused. The rate data are shown in Table 9.3. The same trend can be seen in the activation energy and the resonance energy gained when cycloreversion of the adducts **9–12** yields the aromatic compound, as shown in Scheme 9.3.

Benzene rings can also be fused in angular fashion, as in phenanthrene, chrysene, and picene. These compounds, while reactive toward additions in the center ring, retain most of the resonance energy per electron (REPE) stabilization of benzene and naphthalene.[129]

phenanthrene   chrysene   picene

Table 9.3. Rate of Diel-Alder Additions of Linear Polycyclic Aromatic Hydrocarbons[a]

Dienophile	k (M^{-1} s^{-1}) (80°C, in toluene)		
	Anthracene	Naphthacene	Pentacene
Benzoquinone		44	181
Maleic anhydride	5	294	4710
N-Phenylmaleimide	10	673	19,280

a. V. D. Samuilov, V. G. Uryadov, L. F. Uryadova and A. J. Konolova, *Zh. Org. Khim.* (Engl. Transl.), **21**:1137 (1985).

126. There are a number of other systems for comparing the stability of conjugated cyclic compounds with reference polyenes. For example, see L. J. Schaad and B. A. Hess, Jr., *Pure Appl. Chem.* **54**:1097 (1982); J. Aihara, *Pure Appl. Chem.* **54**:1115 (1982); K. Jug, *J. Org. Chem.* **48**:1344 (1983); W. Gründler, *Monatsh Chem.* **114**:155 (1983).

127. M. J. S. Dewar and C. de Llano, *J. Am. Chem. Soc.* **91**:789 (1969).

128. D. Biermann and W. Schmidt, *J. Am. Chem. Soc.* **102**:3163, 3173 (1980).

129. K. B. Wiberg, *J. Org. Chem.* **62**:5720 (1997).

Scheme 9.3. Correlation between E_a for Retro-Diels–Alder Reaction and Resonance Stabilization of Aromatic Products

	9	10	11	12
	2 benzene	benzene + naphthalene	benzene + anthracene	benzene + naphthacene
E_a	16 kcal	20 kcal	29 kcal	31 kcal
Gain in resonance energy	40 kcal	30 kcal	17 kcal	11 kcal

There is evidence that aromatic segments can exist as part of larger conjugated units, resulting in an aromatic segment in conjugation with a "localized" double bond. For example, in acenaphthylene, the double bond in the five-membered ring is both structurally and chemically similar to a normal localized double bond. The resonance energy given in Scheme 9.2, 0.57β, is slightly less than that for naphthalene (0.59β). The additional double bond of acenaphthylene has only a small effect on the stability of the conjugated system. The molecular structure determined at 80 K by neutron diffraction shows bond lengths for the aromatic portion that are very similar to those of naphthalene.[130] The double bond is somewhat longer than a normal double bond, but this may reflect the strain imposed by the naphthalene framework on the double bond.

naphthalene	acenaphthylene

The predictions of relative stability obtained by the various approaches diverge more widely when nonbenzenoid systems are considered. The simple Hückel method using total π delocalization energies relative to an isolated double-bond reference energy ($\alpha + \beta$) fails. This approach predicts stabilization of the same order of magnitude for such unstable systems as pentalene and fulvalene as it does for much more stable aromatics. The HMO′, RE, and SCF-MO methods, which use polyene reference energies, do much better. All show drastically reduced stabilization for such systems and, in fact, indicate destabilization of systems such as butalene and pentalene (Scheme 9.2).

It is of interest to consider at this point some of the specific molecules in Scheme 9.2 and compare their chemical properties with the calculated stabilization energies. Benzocyclobutadiene has been generated in a number of ways, including dehalogenation of

130. R. A. Wood, T. R. Welberry, and A. D. Rae, *J. Chem. Soc., Perkin Trans. 2* **1985**:451.

dibromobenzocyclobutene.[131] The compound is highly reactive, dimerizing or polymerizing readily[132]:

Benzocyclobutadiene is very reactive as a dienophile in the Diels–Alder reaction:

Ref. 133

Generation of benzocyclobutadiene by fluoride-induced elimination has permitted the NMR spectrum to be observed under flow conditions.[134]

All the peaks are somewhat upfield of the aromatic region, suggesting polyene character. This structure would also be consistent with the observed reactivity since the polyene has a quinodimethane structure (see Section 11.3). The implication of a nonaromatic structure is that the combination of ring strain and the antiaromaticity associated with the four-membered ring results in a localized system.[135]

Azulene is one of the few nonbenzenoid hydrocarbons that appears to have appreciable aromatic stabilization. There is some divergence on this point between the SCF-MO and HMO′ results in Scheme 9.2. The latter estimates a resonance energy about half that for the isomeric naphthalene, whereas the SCF-MO method assigns a resonance energy that is only about one-seventh that of naphthalene. Naphthalene is more stable than azulene by about 38.5 kcal/mol. Molecular mechanics calculations attribute about 12.5 kcal/mol of this difference to strain and about 26 kcal/mol to greater resonance stabilization of naphthalene.[136] Based on heats of hydrogenation, the stabilization energy of azulene is about 16 kcal/mol.[137] The parent hydrocarbon and many of its derivatives are well-characterized compounds with considerable stability. The structure of azulene has been determined by both X-ray crystallography and electron diffraction measurements.[138] The peripheral bond lengths are in the aromatic range and show no regular alternation. The

131. M. P. Cava and D. R. Napier, *J. Am. Chem. Soc.* **78**:500 (1956); *J. Am. Chem. Soc.* **79**:1701 (1957).
132. M. P. Cava and M. J. Mitchell, *Cyclobutadiene and Related Compounds*, Academic Press, New York, 1967, pp. 192–216; M. K. Shepherd, *Cyclobutaarenes: Chemistry of Benzocyclobutene, Biphenylene and Related Compounds*, Elsevier, New York, 1991; W. S. Trahanovsky and K. B. Arvidson, *J. Org. Chem.* **61**:9528 (1996); P. Gandhi, *J. Sci. Ind. Res.* **41**:495 (1982); M. P. Cava and D. R. Napier, *J. Am. Chem. Soc.* **80**:2255 (1958).
133. M. P. Cava and M. J. Mitchell, *J. Am. Chem. Soc.* **81**:5409 (1959)
134. W. S. Trahanovsky and D. R. Fischer, *J. Am. Chem. Soc.* **112**:4971 (1990).
135. P. B. Kardakov, J. Gerratt, D. L. Cooper, M. Raimondi, and M. Sironi, *Int. J. Quantum Chem.* **60**:545 (1996).
136. N. L. Allinger and Y. H. Yu, *Pure Appl. Chem.* **55:**, 191 (1983).
137. W. R. Roth, M. Boehm, H. W. Lennartz, and E. Vogel, *Angew. Chem. Int. Ed. Engl.* **22**:1007 (1983).
138. A. W. Hanson, *Acta Crystallogr.* **19**:19 (1965); O. Bastiansen, and J. L. Derissen, *Acta Chem. Scand.* **20**:1319 (1966).

bond shared by the two rings is significantly longer, indicating that it has predominantly single-bond character. Theoretical calculations indicate that the molecule has C_{2v} symmetry, indicating delocalization of the π electrons.[139]

| azulene | X-ray bond lengths | electron diffraction bond lengths |

An interesting structural question involves the contribution of a dipolar structure which pictures the molecule as the fusion of a cyclopentadienide anion and a cycloheptatrienyl cation:

Azulene does have an appreciable dipole moment (0.8 D).[140] The essentially single-bond nature of the shared bond indicates, however, that the conjugation is principally around the periphery of the molecule. Several MO calculations have been applied to azulene. At the MNDO and STO-3G levels, structures with considerable bond alternation are found as the minimum-energy structures. Calculations which include electron correlation effects give a delocalized π system as the minimum-energy structure.[141]

In contrast to the significant resonance stabilization of azulene, pentalene and heptalene are indicated to be destabilized relative to a reference polyene:

| pentalene | heptalene |

Preparation of pentalene is followed by immediate dimerization.[142] Low-temperature photolysis produces a new species believed to be pentalene, but the compound reverts to dimer at $-100°C$. The matrix-isolated monomer has been characterized spectroscopically.[143] The results are in accord with the predicted lack of stabilization.[144]

139. S. Grimme, *Chem. Phys. Lett.* **201**:67 (1993).
140. H. J. Tobler, A. Bauder, and H. H. Günthard, *J. Mol. Spectrosc.* **18**:239 (1965); G. W. Wheland and D. E. Mann, *J. Chem. Phys.* **17**:264 (1949).
141. C. Glidewell and D. Lloyd, *Tetrahedron* **40**:4455 (1984); R. C. Haddon and K. Raghavachari, *J. Am. Chem. Soc.* **104**:3516 (1982).
142. K. Hafner, R. Dönges, E. Goedecke, and R. Kaiser, *Angew. Chem. Int. Ed. Engl.* **12**:337 (1973); S. You and M. Neuenschwander, *Chimia* **50**:24 (1996).
143. T. Bally, S. Chai, M. Neuenschwander, and Z. Zhu, *J. Am. Chem. Soc.* **119**:1869 (1997).
144. T. K. Zywietz, H. Jiao, P. v. R. Schleyer, and A. de Meijere, *J. Org. Chem.* **63**:3417 (1998).

Heptalene readily polymerizes and is sensitive to oxygen. The NMR spectrum does not indicate the presence of an aromatic ring current. The conjugate acid of heptalene, however, is very stable (even at pH 7 in aqueous solution), reflecting the stability of the cation, which is a substituted tropylium ion.[145]

Another structure with a 10-π-electron conjugated system is bicyclo[6.2.0]deca-1,3,5,7,9-pentaene. The crystal structure of the 9,10-diphenyl derivative (Fig. 9.6) shows the conjugated system to be nearly planar.[146]

There is significant bond alternation, however. The bond at the ring fusion is quite long (1.539 Å). A molecular mechanics calculation on this molecule that included an SCF-MO treatment of the planar conjugated system found the molecule to be slightly destabilized (4 kcal/mol) relative to a polyene reference.[147]

The possibility of extra stabilization in conjugated systems that have conjugated components exocyclic to the ring has also been examined. The substituents complete conjugated rings but are not part of the cyclic system. Some representative structures are shown in Scheme 9.4.

Fig. 9.6. Crystal structure of 9,10-diphenylbicyclo[6.2.0]deca-1,3,5,7,9-pentaene. (Reproduced from Ref. 146 by permission from Elsevier Science.)

145. H. J. Dauben, Jr., and D. J. Bertelli, *J. Am. Chem. Soc.* **83**:4657, 4659 (1961).
146. C. Kabuto and M. Oda, *Tetrahedron Lett.* **1980**:103.
147. N. L. Allinger and Y. H. Yuh, *Pure Appl. Chem.* **55**:191 (1983).

**Scheme 9.4. Completely Conjugated Hydrocarbons
Incorporating Exocyclic Double Bonds**

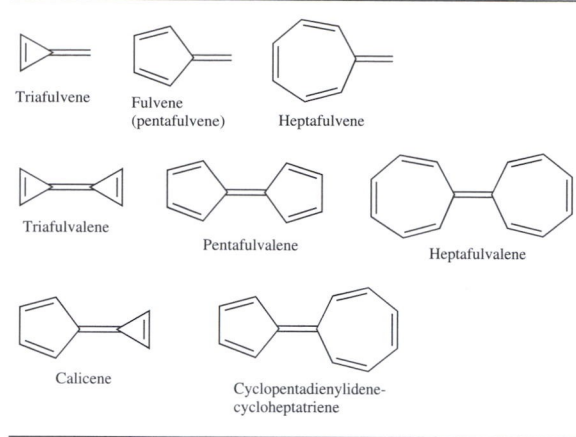

Triafulvene

Fulvene
(pentafulvene)

Heptafulvene

Triafulvalene

Pentafulvalene

Heptafulvalene

Calicene

Cyclopentadienylidene-
cycloheptatriene

Cyclopropenes and cyclopentadienes with exocyclic double bonds provide the possibility of dipolar resonance structures that suggest aromatic character in the cyclic structure:

fulvene

For methylenecyclopropene, a microwave structure determination has established bond lengths which show the strong alternation anticipated for a localized structure.[148] The molecule does have a significant dipole moment (1.90 D), implying a contribution from the dipolar resonance structure. The net stabilization calculated at the 6-31G* level is small and comparable to the stabilization of 1,3-butadiene. The molecular geometry of dimethylfulvene has been examined by electron diffraction methods. Strong bond-length alternation indicative of a localized structure is found[149]:

The fulvalene systems are not predicted to be aromatic by any of the theoretical estimates of stability. Even simple resonance considerations would suggest polyene behavior, since only dipolar resonance structures can be drawn in addition to the single nonpolar structure.

triafulvalene

pentafulvalene

heptafulvalene

Triafulvalene (cyclopropenylidenecyclopropene) has not been isolated. A substantial

148. T. D. Norden, S. W. Staley, W. H. Taylor, and M. D. Harmony, *J. Am. Chem. Soc.* **108**:7912 (1986).
149. J. F. Chiang and S. H. Bauer, *J. Am. Chem. Soc.* **92**:261 (1970).

number of pentafulvalene derivatives have been prepared.[150] The chemical properties of these molecules are those of reactive polyenes. The NMR spectrum of pentafulvalene is characteristic of a localized system.[151] Heptafulvalene (cycloheptatrienylidenecycloheptatriene) is a well-characterized compound with the properties expected for a polyene.[152]

Because the five-membered ring is a substituted cyclopentadienide anion in some dipolar resonance structures, it might be expected that exocyclic groups that could strongly stabilize a positive charge would lead to a larger contribution from dipolar structures and enhanced stability. The structures **13** and **14** are cases in which a dipolar contribution would be feasible.

13   **14**

The stability of such dipolar systems depends on the balance between the increase in energy required to separate unlike charges and the aromaticity associated with Hückel $4n + 2$ systems. The parent compound, triapentafulvalene, is unknown, but B3LYP and MP2/6-31G* calculations suggest some delocalization and a substantial dipole moment.[153] Phenyl-substituted analogs are known and the large measured dipole moments suggest considerable charge separation:

$\mu = 6.3$ D   Ref. 154

Some alkyl derivatives have been prepared. Their chemical behavior is that of highly reactive polyenes. One interesting property that is revealed by the NMR spectra is a reduced barrier to rotation about the double bond between the two rings.[155] This property suggests that rotation about this bond takes place easily through a transition state in which the two charged aromatic rings are twisted out of conjugation:

STO-3G and 3-21G MO calculations indicate a rotational barrier that is substantially reduced relative to the corresponding barrier in ethylene. The transition state for the rotation is calculated to have a charge separation of the type suggested by the dipolar

150. E. D. Bergmann, *Chem. Rev.* **68**:41 (1968).
151. E. Escher, P. Bönzil, A. Otter, and M. Neuenschwander, *Magn. Reson. Chem.* **24**:350 (1986).
152. T. Nozoe and I. Murata, *Int. Rev. Sci., Org. Chem. Ser. Two* **3**:197 (1976).
153. A. P. Scott, I. Agranat, P. U. Biedermann, N. V. Riggs, and L. Radom, *J. Org. Chem.* **62**:2026 (1997).
154. E. D. Bergmann and I. Agranat, *J. Chem. Soc., Chem. Commun.* **1965**:512.
155. A. S. Kende, P. T. Izzo, and W. Fulmor, *Tetrahedron Lett.* **1966**:3697; H. Prinzbach, *Pure Appl. Chem.* **28**:281 (1971).

Fig. 9.7. Hückel molecular orbitals for phenalenyl.

resonance structure.[156]

The hydrocarbon phenalene is the precursor of both a highly stabilized anion and a highly stabilized cation. The Hückel MO diagram is shown in Fig. 9.7. The single orbital at the nonbonding level is the LUMO in the cation and the HOMO in the anion. The stabilization energy calculated for both would be the same and is 0.41β by the HMO′ comparison.[157]

The pK for conversion of phenalene to its anion is 19.[158] The cation is estimated to have a pK_{R+} of about 0–2.[159] Several methods for generating the phenalenyl cation have been developed.[160] Because the center carbon is part of the conjugated system, the Hückel rule, which applies only to *monocyclic* conjugated systems, cannot be applied to just the peripheral conjugation. The nature of the phenalenyl system is considered further in Problem 12 at the end of this chapter.

In general conclusion, the HMO′ and SCF methods both appear able to make reasonably accurate predictions about the stabilization in conjugated molecules. The stabilization is general for benzenoid compounds but quite restricted in nonbenzenoid systems. Because the HMO′ method of estimating stability is based on the ideas of HMO theory, its general success vindicates the ability of this very simplified MO approach to provide insight into the structural nature of the annulenes and other conjugated polyenes. More sophisticated MO methods, of course, are now accessible and should be applied for more detailed analysis of the structures of these molecules.

9.6. Heterocyclic Rings

Certain structural units containing heteroatoms can be substituted into conjugated systems in such a way that the system remains conjugated and isoelectronic with the original hydrocarbon. The most common examples are −CH=N− and −N=N− double

156. B. A. Hess, Jr., L. J. Schaad, C. S. Ewig, and P. Carsky, *J. Comput. Chem.* **4:**53 (1982).
157. J. Aihara, *Bull. Chem. Soc. Jpn.* **51:**3540 (1978); P. Ilic and N. Trinjastic, *J. Org. Chem.* **45:**1738 (1980).
158. A. Streitwieser, Jr., J. M. Word, F. Guibe, and J. S. Wright, *J. Org. Chem.* **46:**2588 (1981); R. A. Cox and R. Stewart, *J. Am. Chem. Soc.* **98:**488 (1976).
159. D. Menche, H. Strauss, and E. Heilbronner, *Helv. Chim. Acta.* **41:**57 (1958).
160. I. Murata, in *Topics in Nonbenzenoid Aromatic Chemistry*, T. Nozoe, R. Breslow, K. Hafner, S. Ito, and I. Murata, eds., Hirokawa, Tokyo, 1976, pp. 159–190.

Scheme 9.5. Stabilization Energies and Index of Aromaticity for Heteroaromatic Structures Isoelectronic with Benzene or Naphthalene [a]

A. Structures isoelectronic with benzene

	Pyridine	Pyrimidine	Pyrazine	Pyridazine	s-Triazine
RE	43.3	40.6	40.9	32.7	44.9
HMO'	0.35	0.30	0.29		
SCF-MO	20.9	20.2	14.6		
AM1	25.6	25.0	24.6	22.6	
IA	86	84	89	79	100

	Furan	Pyyrole	Thiophene	Thiazole	Imidazole
RE	27.2	40.4	43.0	42.0	48.3
HMO'			0.19		15.4
SCF-MO	1.6	8.5			
AM1	12.1	22.5	16.5		
IA	53	90	81.5	79	79

B. Structures isoelectronic with naphthalene

	Quinoline	Isoquinoline	Indole	Benzimidazole
RE	81.0	81.0	73.8	78.9
HMO'	0.51	0.52		30.9
SCF-MO	32.9			
IA	134	133	146	148

	Benzofuran	Isobenzofuran	Isoindole	Benzothiophene
RE	55.4			
HMO'				0.44
SCF-MO	20.3			
IA	94			

RE: Thermochemical stabilization (in kcal/mol) based on difference between ΔH_f and summation of standard bond energies (benzene RE = 45.8 kcal/mol).[a]

HMO': Hückel MO stabilization in β relative to a localized model (benzene = 0.39β).[b]

SCF-MO: Difference in SCF-MO total energy (in kcal mol) for heterocycle and sum of localized polyene energies (benzene = 20 kcal/mol).[c]

AM1: Aromatic stabilization in kcal/mol based on semiempirical AM1 calculations.[d]

IA: Index of aromaticity based on bond-length variation (benzene = 100).[a]

a. C. W. Bird, *Tetrahedron* **48**:335 (1992); *Tetrahedron* **52**:9945 (1996).

b. B. A. Hess, Jr., L. J. Schaad, and C. W. Holyoke, *J. Org. Chem.* **31**:295 (1975); B. A. Hess, Jr., and L. J. Schaad, *J. Am. Chem. Soc.* **95**:3907 (1973).

c. M. J. S. Dewar, A.J. Harget, and N. Trinajstic, *J. Am. Chem. Soc.* **91**:6321 (1969).

d. M. J. S. Dewar and A. J. Holder, *Heterocycles* **28**:1135 (1989).

bonds and divalent sp^2 $-O-$, $-S-$, and $-NR-$ units. Each of these structural fragments can replace a $-CH=CH-$ unit in a conjugated system and contribute two π electrons.[161] These compounds are called *heteroaromatic* to recognize both the heterocyclic structure and the relationship to benzene and other aromatic structures. Scheme 9.5 gives some of the common structures that are isoelectronic with benzene and naphthalene.

MO calculations on compounds in which a $-CH=N-$ unit replaces $-CH=CH-$ indicate that the resonance stabilization is very similar to that of the original compound. For the $-O-$, $-S-$, and $-NR-$ fragments, the resonance stabilization is somewhat reduced but nevertheless high enough to consider the resulting compounds to be aromatic in character.[162] Various approaches have been used to estimate the aromaticity of these compounds. The Hess–Schaad HMO' values are available,[163] as are SCF comparisons with polyene models.[164] Generally speaking, the various approaches suggest that the aromatic stabilization of pyridine is similar to that of benzene. This is in agreement with thermochemical estimates of the pyridine stabilization energy.[165] Typically, the five-membered compounds are found to be somewhat less stabilized than benzene, with resonance energies in the range of one-half to three-quarters of that for benzene. Theoretical calculations at the MP2/6-31G* level have provided aromatic stabilization energies (ASE) for the five-membered heteroaromatic compounds. The ASE values correlate with magnetic susceptibility.[166] Magnetic and polarizability criteria put the order of aromaticity as thiophene>pyrrole>furan.[167]

Additional heteroaromatic structures can be built up by fusing benzene rings to the aromatic heterocyclic rings or by fusing together heterocyclic rings. Examples of this type are included in Scheme 9.5. When benzene rings are fused to the heterocyclic five-membered rings, the structures from fusion at the 2,3-positions are much more stable than those from fusion at the 3,4-positions. The π-electron system in the 3,4-fused compounds is more similar to a peripheral 10-π-electron system than to the 10-electron system of naphthalene. As a result, these compounds have a strong tendency to undergo reactions that restore benzene conjugation in the carbocyclic ring. The isobenzofuran structure **15** is known to be an exceptionally reactive diene, for example. Isoindole, **16**, readily tautomerizes to the benzenoid imine **17**.

161. B. Ya. Simkin, V. I. Minkin, and M. N. Glukhovtsev, *Adv. Heterocycl. Chem.* **56**:303 (1993).
162. L. Nyulaszi, P. Varnai, and T. Veszpremi, *THEOCHEM* **358**:55 (1995); P. Friedman and K. F. Ferris, *Int. J. Quantum Chem. Symp.* **24**:843 (1990); G. P. Bean, *J. Org. Chem.* **63**:2497 (1998).
163. B. A. Hess, Jr., L. J. Schaad, and C. W. Holyoke, *Tetrahedron* **31**:295 (1975); B. A. Hess and L. J. Schaad, *J. Am. Chem. Soc.* **95**:3907 (1973).
164. M. J. S. Dewar, A. J. Harget, N. Trinajstic, and S. D. Worley, *Tetrahedron* **28**:4505 (1970).
165. K. B. Wiberg, D. Nakaji, and K. M. Morgan, *J. Am. Chem. Soc.* **115**:3527 (1993).
166. P. v. R. Schleyer, P. K. Freeman, H. Jiao, and B. Goldfuss, *Angew. Chem. Int. Ed. Engl.* **34**:337 (1995).
167. M. Stolze and D. H. Sutter, *Z. Naturforsch. A* **42**:49 (1987); A. Hinchliffe and H. J. Soscun M., *THEOCHEM* **331**:109 (1995).

E. Clar, *Polycyclic Hydrocarbons*, Academic Press, New York, 1964.

P. J. Garratt, *Aromaticity*, John Wiley & Sons, New York, 1986.

I. Gutman and S. J. Cyvin, *Introduction to the Theory of Benzenoid Hydrocarbons*, Springer-Verlag, Berlin, 1989.

D. Lloyd, *Nonbenzenoid Conjugated Carbocyclic Compounds*, Elsevier, Amsterdam, 1984.

V. I. Minkin, M. N. Glukhovtsev, and B. Y. Simkin, *Aromaticity and Anti-aromaticity*, John Wiley & Sons, New York, 1994.

M. Sainsbury, *Aromatic Chemistry*, Oxford University Press, Oxford, U.K., 1992.

Problems

(References for these problems will be found on page 800.)

1. The reaction of *o*-diphenylcyclobutadiene (generated *in situ* by oxidation of its iron tricarbonyl complex) with *p*-benzoquinone yields **A** as the exclusive product. With tetracyanoethylene, however, **B** and **C** are formed in a 1 : 7 ratio. Discuss these results, and explain how they relate to the question of the square versus rectangular shape of cyclobutadiene.

2. A single resonance structure is shown below for each of several molecules. Consider other resonance structures. Comment on those that would be expected to make a major stabilizing contribution to the molecule in question.

3. (a) A synthesis of tropone (cycloheptatrienone) entails treating 1-methoxycyclohep-tatriene with bromine. A salt is produced that yields tropone on treatment with aqueous sodium bicarbonate. What is the salt? Write a mechanism for its formation.

   (b) The optically active dichlorophenylcyclobutenone **A** undergoes racemization in acetic acid at 100°C. Suggest an experiment to determine if the enol (a hydroxycyclobutadiene) is an intermediate.

A

4. Predict whether the following systems would be expected to show strong (aromatic or homoaromatic) stabilization, weak stabilization by conjugation (non-aromatic); or destabilization (antiaromatic) relative to localized model structures. Explain the basis for your prediction.

5. Bicyclo[6.2.0]deca-2,4,6,8,10-pentaene has been synthesized, and a number of molecular orbital and molecular mechanics calculations have been performed to determine whether it is aromatic or antiaromatic. Consider the structure and discuss the following points.

(a) What aspects of the structure suggest antiaromaticity might be observed?

(b) What aspects of the structure suggest aromaticity might be observed?

(c) What are some of the experimental and theoretical criteria which could be applied to assess aromaticity or antiaromaticity? Cite at least three such probes, and indicate the nature of the observation and how it would be interpreted.

6. Using the empirically chosen energy equivalents for contributing bond types given on p. 532 and a standard compilation of simple HMO calculations, calculate resonance energies for the following molecules according to the modified procedure of Hess and Schaad. Do you find any discrepancies between predicted and observed stability?

(a) (b) (c) (d)

7. The relative basicity of carbonyl oxygen atoms can be measured by studying strength of hydrogen bonding between the carbonyl compound and a hydrogen donor such as phenol. In carbon tetrachloride, values of K_{eq} for $1:1$ complex formation for the compounds shown have been measured. Rationalize the observed order of basicity.

K_{eq}:   6.2   31.2   83.2   117

8. One criterion of aromaticity is the "ring current," which is indicated by a chemical shift difference between protons. in the plane of the conjugated system and those above or below the plane. The chemical shifts of two isomeric hydrocarbons are given below. In qualitative terms, which appears to be more aromatic? (Because the chemical shift depends on the geometric relationship to the ring current, a quantitative calculation would be necessary to confirm the correctness of this qualitative impression.) Does Hückel MO theory predict a difference in the aromaticity of these two compounds?

−6.9   −8.0   −6.5   external protons at 9.3–9.5 ppm

2.7   −2.7   −3.0   external protons at 7.2–7.9 ppm

chemical shift in ppm from tetramethylsilane

9. Offer an explanation for the following observations.

(a) Hydrocarbon **A** (p$K \approx 14$) is much more acidic than **B** (p$K \approx 22$).

(b) The hydrocarbon **C** has an unusually small separation of its oxidation and reduction potentials, as established by electrochemical measurements. It is both easily reduced and easily oxidized. Both mono- and dications and mono- and dianions can be formed readily.

C

(c) The barrier for rotation about the marked bond in **D** is only about 14 kcal/mol.

D

(d) The hydrocarbon **E** is easily reduced to a dianion. The proton NMR spectrum of the dianion shows an average *downfield* shift, relative to the hydrocarbon. The center carbon shows a very large upfield shift in the ^{13}C-NMR spectrum.

E

10. The Hückel molecular orbitals for acenaphthylene are shown below. The atomic coefficients for the orbital which is the LUMO in the neutral compound and the

$$\alpha - 2.36\beta$$
$$\alpha - 1.92\beta$$
$$\alpha - 1.43\beta$$
$$\alpha - 1.31\beta$$
$$\alpha - 1.00\beta$$
$$\alpha - 0.28\beta$$
$$\alpha + 0.63\beta$$
$$\alpha + 0.83\beta$$
$$\alpha + 1.00\beta$$
$$\alpha + 1.69\beta$$
$$\alpha + 2.47\beta$$

Comment on the aromaticity, antiaromaticity, or nonaromaticity of acenaphthylene and its dianion on the basis of the following physical measurements.

(a) The bond lengths of acenaphthylene are shown below. Compare them with the bond lengths for naphthalene given on p. 534. What conclusions do you draw about the aromaticity of acenaphthylene?

(b) Both X-ray and NMR data indicate that the C(1)–C(2) bond lengthens significantly in the dianion as shown below. There is also a different pattern of bond-length alternation. What conclusions do you draw about the aromaticity of the acenaphthylene dianion?

acenaphthylene   acenaphthylene dianion

(c) The ^{1}H- and ^{13}C-NMR shifts for acenaphthylene and its dianion (Na$^+$ counterion) are given below. What conclusions about charge density and aromaticity can be drawn from these data?

		2,3	3,8	4,7	5,6	2a, 8a	5a	8b
^{1}H	neutral	7.04	7.65	7.50	7.78			
	dianion	4.49	4.46	5.04	3.34			
^{13}C	neutral	129.9	124.7	128.3	127.8	140.7	129.1	129.3
	dianion	86.1	97.0	126.8	82.6	123.4	149.3	137.7

11. There have been extensive physical and chemical studies of cyclopropenone, cyclopentadienone, and cycloheptatrienone (tropone). The results of these studies can be briefly summarized as follows:

(a) Cyclopropenone appears to be stabilized by 20 ± 5 kcal/mol relative to a localized model structure.
(b) Cyclopentadienone is a kinetically unstable molecule.
(c) Tropone is estimated to be stabilized by less than 10 kcal/mol relative to localized models. It is a nonplanar molecule.

Derive the π-MO patterns for these three molecules by treating them as derivatives of the three-, five-, and seven-membered cyclic conjugated systems. Explain the relationship between the derived MO pattern and the observed properties and stabilities of the molecules.

12. The orbital coefficients for the MO of energy α for the phenalenyl system described in Fig. 9.7 are as shown below. Predict the general appearance of the NMR spectra of the anion and cation derived from phenalene.

13. The ^{13}C-NMR spectrum of octalene is temperature-dependent. At $-150°C$, there are signals for 14 different carbons. At $-100°C$, these collapse to seven different signals. Above $80°C$, all but one of the remaining signals becomes broad. Although not attained experimentally, because of decomposition, it would be expected that only four different signals would be observed at still higher temperature. (1) Show that these data rule out structures **A** and **B** for the room temperature structure of octalene and favor structure **C**; (2) indicate the nature of the dynamic process that converts the 14-line spectrum to the 7-line spectrum; (3) indicate the nature of the process which would be expected to convert the 7-line spectrum to a 4-line spectrum.

A

B

C

14. When the alcohol **1** is dissolved in fluorosulfonic acid at $-136°C$ and then allowed to warm to $-110°C$, it gives rise to a cation having a ^{13}C-NMR spectrum consisting of five lines in the intensity ratio $2:1:2:2:2$. Suggest possible structures for this cation, and discuss any stabilizing features which might favor a particular structure.

1

15. (a) The heats of combustion, ΔH_c, the heats of hydrogenation for addition of one mole of H_2, ΔH_{H_2}, and the estimated stabilization energies (S.E.) for benzene and cyclooctatetraene are given below. The heat of combustion and the heat of

hydrogenation of [16]annulene are also given. Estimate the stabilization energy of [16]annulene. Does this value agree with the prediction of simple Hückel MO theory?

	Benzene	Cyclooctatetraene	[16]Annulene
ΔH_c (kcal/mol)	781	1086	2182
ΔH_{H_2} (kcal/mol)	− 5.16	25.6	28
S.E.[a] (kcal/mol)	36	4	?

[a] Estimated stabilization resulting from conjugation in kcal/mol.

(b) The enthalpies of the reaction of the cyclooctatetraene and [16]annulene dianions $(C_nH_n)^{2-}$ with water have been measured.

$$2\,Na^+ + (C_nH_n)^{2-} = 2\,H_2O_{(1)} \rightarrow C_nH_{n+2} + 2\,NaOH$$
$$\Delta H = -33.3 \,\text{kcal/mol for cyclooctatetraene}$$
$$\Delta H = -10.9 \,\text{kcal/mol for [16]annulene}$$

Using these data and the enthalpy value for the reaction

$$2\,Na_{(s)} + 2\,H_2O \rightarrow 2\,NaOH_{(aq)} + H_2, \qquad \Delta H = -88.2 \,\text{kcal/mol}$$

calculate ΔH for the raection

$$2\,Na_{(s)} + C_nH_n \rightarrow 2\,Na^+ + (C_nH_n)^{2-}$$

How do you interpret the difference in the heat of reaction for the two hydrocarbons in the reaction to form the respective dianions?

16. Use the calculated energies for the molecules shown below to calculate isodesmic reaction energies for the equation:

n = 1, 2, 3

$CH_3CH=CH_2$ $CH_3CH=C=O$ ⬭=C=O   ⬭=C=O   ⬭=C=O   ⬭=CH_2   ⬭=CH_2   ⬭=CH_2

−117.0715   −190.7592   −227.3303   −304.3387   −381.2164   −153.6698   −230.6444   −307.5368

What trends do your calculations reveal? How do you account for these results?

17. The diagram below shows the distribution of the molecular orbitals for phenalenyl (Fig. 9.7). On the basis of these orbitals, predict the charge distributions for the cation,

anion, and radical.

COEFFICIENTS

		1	2	3	4	5	6	7
E		2.44949	1.73205	1.73205	1.00000	1.00000	1.00000	0.00000
	1	0.22361	0.35355	0.20412	−0.00473	−0.35355	−0.27382	−0.40825
	2	0.18257	0.40825	0.00000	−0.00947	0.00000	−0.54764	−0.00000
	10	0.36515	0.20412	0.35355	0.00473	−0.35355	0.27382	0.00000
	11	0.44721	0.00000	0.00000	−0.40188	0.00000	0.37215	0.00000
	9	0.22361	−0.00000	0.40825	0.41134	0.00000	0.17549	0.40825
	8	0.18257	−0.20412	0.35355	0.40661	0.35355	−0.09833	0.00000
	7	0.22361	−0.35355	0.20412	−0.00473	0.35355	−0.27382	−0.40825
	13	0.36515	−0.40825	0.00000	−0.41134	0.00000	−0.17549	−0.00000

18. Consider the two structures shown for kekulene, one suggesting inner and outer annulenes, and the other a series of phenanthrene-like units. Indicate properties that you would expect to be associated with each structure. Proton NMR and bond-length data are given. How do they compare with your expectations?

10

Aromatic Substitution

10.1. Electrophilic Aromatic Substitution Reactions

Electrophilic aromatic substitution reactions are important for synthetic purposes and also are one of the most thoroughly studied classes of organic reactions from a mechanistic point of view. The synthetic aspects of these reactions are discussed in Chapter 11 of Part B. The discussion here will emphasize the mechanisms of several of the most completely studied reactions. These mechanistic ideas are the foundation for the structure–reactivity relationships in aromatic electrophilic substitution that will be discussed in Section 10.2

A wide variety of electrophilic species can effect aromatic substitution. Usually, it is a substitution of some other group for hydrogen that is of interest, but this is not always the case. Scheme 10.1 lists some of the specific electrophilic species that are capable of carrying out substitution for hydrogen. Some indication of the relative reactivity of the electrophiles is given as well. Most of these electrophiles will not be treated in detail until Part B. Nevertheless, it is important to recognize the very broad scope of electrophilic aromatic substitution.

The reactivity of a particular electrophile determines which aromatic compounds can be successfully substituted. Those electrophiles grouped in the first category in Scheme 10.1 are sufficiently reactive to attack almost all aromatic compounds, even those having strongly electron-withdrawing substituents. Those in the second group react readily with benzene and derivatives having electron-releasing substituents but are not generally reactive toward aromatic rings with electron-withdrawing substituents. Those classified in the third group are reactive only toward aromatic compounds that are much more reactive than benzene. These groupings can provide a general guide to the feasibility of a given electrophilic aromatic substitution.

Despite the wide range of electrophilic species and aromatic ring systems that can undergo substitution, a single broad mechanistic picture encompasses most electrophilic aromatic substitution reactions. The identity of the rate-determining step and the shape of the potential energy surface are specific to the reagents which are involved, but the series of steps and nature of the intermediates are very similar across a wide range of reactivity. This permits discussion of electrophilic aromatic substitution in terms of a general mechanism. This mechanism is outlined in Scheme 10.2.

Scheme 10.1. Electrophilic Species Active in Aromatic Substitution

Electrophile	Typical mode of generation	Reference

A. Electrophiles capable of substituting both activated and deactivated aromatic rings

$O{=}\overset{+}{N}{=}O$	$2\,H_2SO_4 + HNO_3 \rightleftharpoons NO_2^+ + 2\,HSO_4^- + H_3O^+$	a
Br_2 or $Br_2{-}MX_n$	$Br_2 + MX_n \rightleftharpoons Br_2{-}MX_n$	b
$Br\overset{+}{O}H_2$	$BrOH + H_3O^+ \rightleftharpoons Br\overset{+}{O}H_2 + H_2O$	b
Cl_2 or $Cl_2{-}MX_n$	$Cl_2 + MX_n \rightleftharpoons Cl_2{-}MX_n$	b
$Cl\overset{+}{O}H_2$	$ClOH + H_3O^+ \rightleftharpoons Cl\overset{+}{O}H_2 + H_2O$	b
SO_3	$H_2S_2O_7 \rightleftharpoons H_2SO_4 + SO_3$	c
RSO_2^+	$RSO_2Cl + AlCl_3 \rightleftharpoons RS_2O^+ + AlCl_4^-$	d

B. Electrophiles capable of substituting activated but not deactivated aromatic rings

R_3C^+	$R_3CX + MX_n \rightleftharpoons R_3C^+ + [MX_{n+1}]^-$	e
	$R_3COH + H^+ \rightleftharpoons R_3C^+ + H_2O$	f
	$R_2C{=}CR_2' + H^+ \rightleftharpoons R_2\overset{+}{C}CHR_2'$	g
$RCH_2X{-}MX_n$	$RCH_2X + MX_n \rightleftharpoons RCH_2X{-}MX_n$	e
$RC{\equiv}O^+$	$\overset{O}{\overset{\|}{RCX}} + MX_n \rightleftharpoons RC{\equiv}O^+ + [MX_{n+1}]^-$	h
$\overset{O}{\overset{\|}{RCX}}{-}MX_n$	$\overset{O}{\overset{\|}{RCX}} + MX_n \rightleftharpoons \overset{O}{\overset{\|}{RCX}}{-}MX_n$	h
$R\overset{+}{C}{=}\overset{}{O}H$	$\overset{O}{\overset{\|}{RCX}} + MX_n + H^+ \longrightarrow R\overset{+}{C}{=}\overset{}{O}H + [MX_{n+1}]^-$	i
H^+	$HX \rightleftharpoons H^+ + X^-$	j
$R_2C{=}\overset{+}{O}H$	$R_2C{=}O + H^+ \rightleftharpoons R_2C{=}\overset{+}{O}H$	k
$R_2C{=}\overset{+}{O}{-}\overset{-}{M}X_n$	$R_2C{=}O + MX_n \rightleftharpoons R_2C{=}\overset{+}{O}{-}\overset{-}{M}X_n$	k
$H\overset{+}{C}{=}\overset{+}{N}H_2$	$H{-}C{\equiv}N + 2\,H^+ \rightleftharpoons H\overset{+}{C}{=}\overset{+}{N}H_2$	i

C. Electrophiles capable of substituting only strongly activated aromatic rings

$H\overset{+}{C}{=}\overset{+}{N}H$	$HC{\equiv}N + HX \rightleftharpoons HC{\equiv}\overset{+}{N}H + X^-$	l
$N{\equiv}O^+$	$HNO_2 + H^+ \rightarrow N{\equiv}O^+ + H_2O$	m
$Ar\overset{+}{N}{\equiv}N$	$ArNH_2 + HNO_2 + H^+ \rightarrow Ar\overset{+}{N}{\equiv}N + 2\,H_2O$	n

a. G. A. Olah and S. J. Kuhn, in *Friedel–Crafts and Related Reactions.*, Vol. III, G. A. Olah, ed., Interscience, New York, 1964, Chapter XLIII.

b. H. P. Braendlin and E. T. McBee, In *Friedel–Crafts and Related Reactions.*, Vol. III, G. A. Olah, ed., Interscience, New York, 1964, Chapter XLVI.

c. K. L. Nelson in *Friedel–Crafts and Related Reactions.*, Vol. III, G. A. Olah, ed., Interscience, New York, 1964, Chapter XLVII.

d. F. R. Jensen and G. Goldman, in *Friedel–Crafts and Related Reactions.*, Vol. III, G. A. Olah, ed., Interscience, New York, 1964, Chapter XL.

e. F. A. Drahowzal, in *Friedel–Crafts and Related Reactions.*, Vol. II, G. A. Olah, ed., Interscience, New York, 1964, Chapter XVII.

f. A. Schrelsheim, in *Friedel–Crafts and Related Reactions.*, Vol. II, G. A. Olah, ed., Interscience, New York, 1964, Chapter XVIII.

g. S. H. Patinkin and B. S. Friedman, in *Friedel–Crafts and Related Reactions.*, Vol. II, G. A. Olah, ed., Interscience, New York, 1964, Chapter XIV.

h. P. H. Gore, in *Friedel–Crafts and Related Reactions.*, Vol. III, G. A. Olah, ed., Interscience, New York, 1964, Chapter XXXI.

i. Y. Sato, M. Yato, T. Ohwada, S. Saito, and K. Shudo, *J. Am. Chem. Soc.* **117**:3037 (1995).

j. R. O. C. Norman and R. Taylor, *Electrophilic Substitution in Benzenoid Compounds*, Elsevier, New York, 1965, Chapter 8.

k. J. E. Hofmann and A. Schriesheim, in *Friedel–Crafts and Related Reactions.*, Vol. II, G. A. Olah, ed., Interscience, New York, 1964, Chapter XIX.

l. W. Ruske, in *Friedel–Crafts and Related Reactions.*, Vol. III, G. A. Olah, ed., Interscience, New York, 1964, Chapter XXXII

m. B. C. Challis, R. J. Higgins, and A. J. Lawson, *J. Chem. Soc., Perkin Trans. 2* **1972**:1831.

n. H. Zollinger, *Azo and Diazo Chemistry*, translated by H. E. Nursten, Interscience, New York, 1961, Chapter 10.

553

SECTION 10.1.
ELECTROPHILIC
AROMATIC
SUBSTITUTION
REACTIONS

**Scheme 10.2. Generalized
Mechanism for Electrophilic
Aromatic Substitution**

In this mechanism, a complexation of the electrophile with the π-electron system of the aromatic ring is the first step. This species, called the π-complex, may or may not be involved directly in the substitution mechanism. π-Complex formation is, in general, rapidly reversible, and in many cases the equilibrium constant is small. The π-complex is a donor–acceptor type complex, with the π electrons of the aromatic ring donating electron density to the electrophile. No position selectivity is associated with the π-complex.

In order for a substitution to occur, a "σ-complex" must be formed. The term σ-complex is used to describe an *intermediate* in which the carbon at the site of substitution is bonded to both the electrophile and the hydrogen that is displaced. As the term implies, a σ bond is formed at the site of substitution. The intermediate is a cyclohexadienyl cation. Its fundamental structural characteristics can be described in simple MO terms. The σ-complex is a four-π-electron delocalized system that is electronically equivalent to a pentadienyl cation (Fig. 10.1). There is no longer cyclic conjugation. The LUMO has nodes at C-2 and C-4 of the pentadienyl structure, and these positions correspond to the positions *meta* to the site of substitution on the aromatic ring. As a result, the positive charge of the cation is located at the positions *ortho* and *para* to the site of substitution.

Formation of the σ-complex can be reversible. The partitioning of the σ-complex forward to product or back to reactants depends on the ease with which the electrophile can be eliminated relative to a proton. For most electrophiles, it is easier to eliminate the proton, in which case the formation of the σ-complex is essentially irreversible. Formation of the σ-complex is least likely to be reversible for the electrophiles in group A in Scheme 10.1, whereas those in group C are most likely to undergo reversible σ-complex formation. Formation of the σ-complex is usually, but not always, the rate-determining step in

Fig. 10.1. π-Molecular orbitals and energy levels for the pentadienyl cation.

electrophilic aromatic substitution. There may also be a π-complex involving the aromatic ring and the departing electrophile. This would be logical on the basis of the principle of overall reversibility of the process. There is little direct evidence on this point however.[1]

Let us now consider some of the evidence for this general mechanism. Such evidence has, of course, been gathered by study of specific reaction mechanisms. Only some of the most clear-cut examples are cited here. Additional evidence will be mentioned when individual mechanisms are discussed in Section 10.4. A good example of a reaction that has been the subject of studies focused on the identity and mode of generation of the electrophile is aromatic nitration. Primarily on the basis of kinetic studies, it has been possible to show that the active electrophile in nitration is often the nitronium ion, NO_2^+. In some cases, the generation of the electrophile is the rate-determining step. Several lines of evidence have been used to establish the role of the nitronium ion. Reaction of nitric acid with concentrated sulfuric acid leads to formation of the nitronium ion. It can be detected spectroscopically, and the freezing-point depression of the solution is consistent with the following equation:

$$2H_2SO_4 + HNO_3 \rightarrow NO_2^+ + H_3O^+ + 2HSO_4^-$$

Solid salts in which the nitronium ion is the cation can be prepared with unreactive anions such as BF_4^- and PF_6^-. These salts can act as nitrating reagents.

Two types of rate expressions have been found to describe the kinetics of most aromatic nitration reactions. With relatively unreactive substrates, second-order kinetics, first-order in the nitrating reagent and first-order in the aromatic, are observed. This second-order relationship corresponds to rate-limiting attack of the electrophile on the aromatic reactant. With more reactive aromatics, this step can be faster than formation of the active electrophile. When formation of the active electrophile is the rate-determining step, the concentration of the aromatic reactant no longer appears in the observed rate expression. Under these conditions, different aromatic substrates undergo nitration at the same rate, corresponding to the rate of formation of the active electrophile.

An important general point to be drawn from the specific case of nitration is that the active electrophile is usually some species that is more reactive than the added reagents.

1. For additional discussion of the role of σ and π complexes in aromatic substitution, see G. A. Olah, *Acc. Chem. Res.* **4**:240 (1971); J. H. Ridd, *Acc. Chem. Res.* **4**:248 (1971).

555

SECTION 10.1.
ELECTROPHILIC
AROMATIC
SUBSTITUTION
REACTIONS

The active electrophile is formed by a subsequent reaction, often involving a Lewis acid. As discussed above with regard to nitration, the formation of the active electrophile may or may not be the rate-determining step. Scheme 10.1 indicates the structure of some of the electrophilic species that are involved in typical electrophilic aromatic substitution processes and the reactions involved in their formation.

There are several lines of evidence pointing to formation of σ-complexes as intermediates in electrophilic aromatic substitution. One involves measurement of isotope effects on the rate of substitution. If removal of the proton at the site of substitution is concerted with introduction of the electrophile, a primary isotope effect would be observed in reactions in which electrophilic attack on the ring is rate-determining. This is not the case for nitration nor for several other types of aromatic substitution reactions. Nitration of aromatic substrates partially labeled by tritium shows no selectivity between protium- and tritium-substituted sites.[2] Similarly, the rate of nitration of nitrobenzene is identical to that of penta-*deuterio*-nitrobenzene.[3] The lack of an isotope effect indicates that the proton is lost in a fast step subsequent to the rate-determining step. This means that proton loss must occur from some intermediate that is formed before the cleavage of the C$-$H bond begins. The σ-complex intermediate fits this requirement. There are some electrophilic aromatic substitution reactions that show values of k_H/k_D between 1 and 2, and there are a few others for which the values are in the range indicating a primary isotope effect.[4] The existence of these isotope effects is compatible with the general mechanism if the proton removal is rate-limiting (or partially rate-limiting). Many of the modest kinetic isotope effects ($k_H/k_D \approx 1.2$–2.0) have been interpreted in terms of comparable rates for formation and reaction of the σ-complex intermediate.

The case for the generality of the σ-complex mechanism is further strengthened by numerous studies showing that benzenium ions (an alternative name for the σ-complex) can exist as stable entities under suitable conditions. Substituted benzenium ions can be observed by NMR techniques under stable-ion conditions. They are formed by protonation of the aromatic substrate[5]:

Ref. 6

Ref. 7

2. L. Melander, *Acta Chem. Scand.* **3**:95 (1949); *Ark. Kemi.* **2**:211 (1950).
3. T. G. Bonner, F. Bower, and G. Williams, *J. Chem. Soc.* **1953**:2650.
4. H. Zollinger, *Adv. Phys. Org. Chem.* **2**:163 (1964).
5. G. A. Olah, R. H. Schlosberg, R. D. Porter, Y. K. Mo, D. P. Kelly, and G. Mateescu, *J. Am. Chem. Soc.* **94**:2034 (1972).
6. G. A. Olah and T. E. Kiovsky, *J. Am. Chem. Soc.* **89**:5692 (1967).
7. G. A. Olah, *J. Am. Chem. Soc.* **87**:1103 (1965).

Salts formed by alkylation of benzene derivatives have also been characterized:

Ref. 8

Under normal conditions of electrophilic substitution, these benzenium ions are short-lived intermediates. The fact that the structures are stable in nonnucleophilic media clearly demonstrates the feasibility of such intermediates.

The existence of σ-complex intermediates can be inferred from experiments in which they are trapped by nucleophiles under special circumstances. For example, treatment of the acid **1** with bromine gives the cyclohexadienyl lactone **2**. This product results from capture of the σ-complex by intramolecular nucleophilic attack by the carboxylate group:

Ref. 9

A number of examples of nucleophilic capture of σ-complexes have also been uncovered in the study of nitration of alkylated benzenes in acetic acid. For example, nitration of **3** at 0°C leads to formation of **4** with acetate serving as the nucleophile[10]:

This type of addition process is particularly likely to be observed when the electrophile attacks a position that is already substituted, since facile rearomatization by deprotonation is then blocked. Reaction at a substituted position is called *ipso* attack. Addition products have also been isolated, however, when initial electrophilic attack has occurred at an unsubstituted position. The extent of addition in competition with substitution tends to increase on going to naphthalene and the larger polycyclic aromatic ring systems.[11]

The general mechanistic framework outlined in this section must be elaborated by other details to fully describe the mechanisms of the individual electrophilic substitutions. The question of the identity of the active electrophile in each reaction is important. We have discussed the case of nitration, in which, under many circumstances, the electrophile is the nitronium ion. Similar questions arise in most of the other substitution reactions.

8. G. A. Olah and S. J. Kuhn, *J. Am. Chem. Soc.* **80**:6541 (1958).
9. E. J. Corey, S. Barcza, and G. Klotmann, *J. Am. Chem. Soc.* **91**:4782 (1969).
10. R. C. Hahn and D. L. Strack, *J. Am. Chem. Soc.* **96**:4335 (1974).
11. P. B. de la Mare, *Acc. Chem. Res.* **7**:361 (1974).

Other matters that are important include the ability of the electrophile to select among the alternative positions on a substituted aromatic ring. The relative reactivity of different substituted benzenes toward various electrophiles has also been important in developing a firm understanding of electrophilic aromatic substitution. The next section considers some of the structure–reactivity relationships that have proven to be informative.

10.2. Structure–Reactivity Relationships

The effect that substituents on the aromatic reactant have on electrophilic aromatic substitution reactions is an area of structure–reactivity relationships that has been studied since about 1870. The classification of substituents as activating and *ortho–para*-directing or deactivating and *meta*-directing became clear from the early studies. An understanding of the origin of these substituent effects became possible when ideas about electronic interactions and resonance theory were developed. Activating, *ortho–para*-directing substituents are those that can serve as electron donors and stabilize the transition state leading to σ-complex formation. Both saturated and unsaturated hydrocarbon groups and substituents with an unshared electron pair on the atom adjacent to the ring fall in this group. The stabilizing effects of these types of substituents can be expressed in terms of resonance structures. Direct resonance stabilization is only possible when the substituent is *ortho* or *para* to the incoming electrophile. As a result, the transition states for *ortho* and *para* substitution are favored over that for *meta* substitution.

stabilization by alkyl groups
by hyperconjugation

stabilization by heteroatoms
by delocalization

stabilization by vinyl groups
by delocalization

Because the substituent groups have a direct resonance interaction with the charge that develops in the σ-complex, quantitative substituent effects exhibit a high resonance component. Hammett equations usually correlate best with the σ^+ substituent constants (see Section 4.3).[12]

Electron-attracting groups retard electrophilic substitution. Substituents falling in this group include substituents in which a carbonyl group is directly attached to the ring and substituents containing electronegative elements that do not have a lone pair on an atom

12. H. C. Brown and Y. Okamoto, *J. Am. Chem. Soc.* **80**:4979 (1958).

adjacent to the ring. The classification of specific substituents given in Table 4.7 on p. 214 indicates which are electron-attracting. Because of the direct conjugation with the *ortho* and *para* positions, electrophilic attack occurs primarily at the *meta* position, since it is *less* deactivated than the *ortho* and *para* positions.

destabilization by electron-attracting groups   destabilization by electronegative heteroatoms with no unshared pairs

A few substituents, most notably chlorine and bromine, decrease the rate of reaction but nevertheless direct incoming electrophiles to the *ortho* and *para* positions. This is the result of the opposing influence of polar and resonance effects for these substituents. The halogens are more electronegative than carbon, and the carbon–halogen bond dipole opposes the development of positive charge in the ring. Overall reactivity toward electrophiles is therefore reduced. The unshared electron pairs on the halogen, however, can preferentially stabilize the *ortho* and *para* transition states by resonance. As a result, the substituents are deactivating but *ortho–para*-directing.

stabilization resonance participation   no resonance interaction
by chlorine   with chlorine

The *ortho–para*- versus *meta*-directing and activating versus deactivating effect of substituents can also be described in terms of MO theory. The discussion can focus either on the σ-complex or on the aromatic reactant. According to the Hammond postulate, it would be most appropriate to focus on the intermediate in reactions that have relatively high energies of activation and a late transition state. In such cases, the transition state should closely resemble the σ-complex. For highly reactive electrophiles, where the activation energy is low, it may be more appropriate to regard the transition state as closely resembling the reactant aromatic. Let us examine the MO description of substituent effects from both these perspectives.

If the transition state resembles the intermediate σ-complex, the structure involved is a substituted cyclohexadienyl cation. The electrophile has localized one pair of electrons to form the new σ bond. The Hückel orbitals are those shown for the pentadienyl system in Fig. 10.1. A substituent can stabilize the cation by electron donation. The LUMO is ψ_3. This orbital has its highest coefficients at carbons 1, 3, and 5 of the pentadienyl system. These are the positions which are *ortho* and *para* to the position occupied by the electrophile. Electron-donor substituents at the 2- and 4-positions will stabilize the system much less because of the nodes at these carbons in the LUMO.

If we consider a π-acceptor substituent, we see that such a substituent will strongly destabilize the system when it occupies the 1-, 3-, or 5-position on the pentadienyl cation. The destabilizing effect would be less at the 2- or 4-position. The conclusions drawn from this MO interpretation are the same as those based on resonance arguments. Electron-donating substituents will be most *stabilizing* in the transition state leading to *ortho, para* substitution. Electron-withdrawing substituents will be least *destabilizing* in the transitions state leading to *meta* substitution.

The effect of the bond dipole associated with electron-withdrawing groups can also be expressed in terms of its interaction with the cationic σ-complex. The atoms with the highest coefficients in the LUMO ψ_3 are the most positive. The unfavorable interaction of the bond dipole will therefore be greatest at these positions. This effect operates with substituents such as carbonyl, cyano, and nitro groups. With ether and amino substituents, the unfavorable dipole interaction is overwhelmed by the stabilizing effect of the lone-pair electrons stabilizing ψ_3.

The effect of substituents has been probed by MO calculations at the STO-3G level.[13] An isodesmic reaction corresponding to transfer of a proton from a substituted σ-complex to an unsubstituted one will indicate the stabilizing or destabilizing effect of the substituent. The results are given in Table 10.1.

The calculated energy differences give a good correlation with σ^+. The ρ parameter ($\rho = -17$) is larger than that observed experimentally for proton exchange ($\rho \approx -8$). A physical interpretation of this is that the theoretical results pertain to the gas phase, where

Table 10.1. Energy Changes for Isodesmic Proton-Transfer Reactions of Substituted Benzenes[a]

Substituent	ΔE (kcal/mol)	
	meta	*para*
NO_2	−17.9	−22.1
CN	−14.0	−13.8
CF_3	−7.5	−8.4
F	−7.5	3.7
CH_3	2.0	8.5
OCH_3		15.7
OH	−5.3	16.0
NH_2	0.6	27.2

a. From STO-3G calculations reported by J. M. McKelvey, S. Alexandratos, A. Streitwiesser, Jr., J.-L. M. Abboud, and W. J. Hehre, *J. Am. Chem. Soc.* **98**:244 (1976).

13. J. M. McKelvey, S. Alexandratos, A. Streitwiesser, Jr., J.-L. M. Abboud, and W. H. Hehre, *J. Am. Chem. Soc.* **98**:244 (1976).

the effect of substituents is at a maximum because of the absence of any leveling effect due to solvation.

Both HMO calculations and more elaborate MO methods can be applied to the issue of the position of electrophilic substitution in aromatic molecules. The most direct approach is to calculate the localization energy. This is the energy difference between the aromatic molecule and the σ-complex intermediate. In simple Hückel calculations, the localization energy is just the difference between the energy calculated for the initial π system and that remaining after two electrons and the carbon atom at the site of substitution have been removed from the conjugated system:

$$\text{localization energy} = 2\alpha + 2.54\beta$$

$$6\alpha + 8\beta \qquad 4\alpha + 5.46\beta$$

Comparison of localization energies has frequently been applied to prediction of the relative positional reactivity in polycyclic aromatic hydrocarbons. Simple HMO calculations have only marginal success. CNDO/2 and SCF calculations give results which show good correlation with experimental data on the rate of proton exchange.[14]

Now let us turn to the case of a highly reactive electrophile, where we expect an early transition state. In this case, the charge density and coefficients of the HOMO characteristic of the aromatic reactant would be expected to be major features governing the orientation of electrophilic attack. The transition state should resemble the reactants, and, according to frontier orbital theory, the electrophile should attack the position that has the largest coefficient in the HOMO. Methoxybenzene (anisole) can be taken as an example of a reactive molecule. MO calculations place the lone-pair oxygen orbital lower in energy than the aromatic π orbitals, leading to the MO diagram in Fig. 10.2. The degeneracy of the two highest-lying occupied π orbitals is lifted because the methoxy group interacts preferentially with one of them. The other has a node at the site of methoxy substitution. Figure 10.3 gives the coefficients for the two highest occupied π orbitals, as calculated by the CNDO/2 method. We see that the HOMO has its highest coefficients at the *ipso*, *ortho*, and *para* positions. As indicated in Fig. 10.2, the energy of this orbital is raised by its interaction with the electron-donor substituent. Figure 10.4 shows the distribution of π electrons from all the orbitals, based on STO-3G calculations, for various substituted benzenes. Those having the electron-donating substituents show increased electron density at the *ortho* and *para* positions. Both the HOMO coefficients and the total charge distribution predict preferential attack by the electrophile at the positions *ortho* and *para* to donor substituents.

benzene   methoxy   anisole

Fig. 10.2. MO diagram for anisole by application of perturbation for a methoxy substituent.

14. A. Streitwieser, Jr., P. C. Mowery, R. G. Jesaitis, and A. Lewis, *J. Am. Chem. Soc.* **92**:6529 (1970).

Fig. 10.3. Orbital coefficients for HOMO and next highest π orbital for some substituted benzenes. (From CNDO/2 calculations. *Ortho* and *meta* coefficients have been averaged in the case of the unsymmetrical methoxy and formyl substituents. Orbital energies are given in atomic units.)

Some examples of electron-withdrawing substituents are also shown in Figs. 10.3 and 10.4. As expected, the acceptor substituents lower the energies of the π orbitals. The HOMO distribution remains highest at the *para* position, however. The total charge distribution shows greater depletion at the *ortho* and *para* positions than at the *meta* position. The lower energy of the HOMO is consistent with decreased reactivity for rings with an electron-accepting substituent. The distribution of the HOMO would, however, erroneously predict *para* substitution if frontier orbital theory were used. Aromatic rings with acceptor substituents are relatively unreactive and therefore unlikely to have early transition states. For such compounds, considerations of the stability of the σ-complex intermediate, which predict *meta* substitution, are more appropriate.

Substituents which are not directly bound to the aromatic ring can also influence the course of electrophilic aromatic substitution. Several alkyl groups bearing electron-

Fig. 10.4. Total π-electron density for some substituted benzenes. [From STO-3G calculations as reported by W. J. Hehre, L. Radom, and J. A. Pople, *J. Am. Chem. Soc.* **94**:1496 (1972).]

Table 10.2. Percent *meta* Nitration for Some Alkyl Groups with Electron-Withdrawing Substituents[a]

$CH_2CO_2C_2H_5$	$CHCl_2$	CH_2CCl_3	CH_2NO_2	CCl_3	$CH_2\overset{+}{N}(CH_3)_3$
11%	34%	37%	55%	64%	85%

a. From C. K. Ingold, *Structure and Mechanism in Organic Chemistry*, 2nd ed., Cornell University Press, Ithaca, New York, 1969, pp. 275, 281; F. DeSarlo, G. Grynkiewicz, A. Rici, and J. H. Ridd, *J. Chem. Soc., B,* **1971**:719.

attracting substituents are *meta*-directing and deactivating. Some examples are given in Table 10.2. In these molecules, stabilization of the *ortho* and *para* σ-complex by electron release from the alkyl group is opposed by the polar effect of the electronegative substituent. Both the reduced electron density at the alkyl substituent and the bond dipoles in the substituent would reduce electron donation by the methylene group.

The relationships between substituents and the typical electrophilic substitution reactions, such as those listed in Scheme 10.1, can be summarized as follows:

1. The hydroxy and amino groups are highly activating *ortho–para*-directing groups. Such compounds are attacked by all the electrophilic reagents tabulated in Scheme 10.1 (p. 552). With some electrophilic reagents, all available *ortho* and *para* positions are rapidly substituted.
2. The alkyl, amido, and alkoxy groups are activating and *ortho–para*-directing, but not as strongly so as hydroxyl or amino groups. Synthetically useful conditions for selective substitution are available for essentially all the electrophiles in Scheme 10.1 except for very weak electrophiles such as NO^+ or PhN_2^+.
3. The halogens, as mentioned earlier, are unusual substituents, being deactivating but *ortho–para*-directing. In general, halogenated aromatics will react successfully with electrophiles listed in categories A and B in Scheme 10.1
4. The carbonyl group in aldehydes, ketones, acids, esters, and amides is deactivating and *meta*-directing. There are distinct limitations on the types of substitution reactions that are satisfactory for these deactivating substituents. In general, only those electrophiles in category A in Scheme 10.1 react readily.
5. The cyano, nitro, and quaternary ammonium groups are strongly deactivating and *meta*-directing. Electrophilic substitutions of compounds with these substituents require especially vigorous conditions and fail completely with all but the most reactive electrophiles.

Because nitration has been studied for a wide variety of aromatic compounds, it is a useful reaction with which to illustrate the directing effect of substituent groups. Table 10.3 presents some of the data. A variety of reaction conditions are represented, so direct comparison is not always valid, but the trends are nevertheless clear. It is important to remember that other electrophiles, while following the same qualitative trends, show large quantitative differences in position selectivity.

The effect of substituents on electrophilic substitution can be placed on a quantitative basis by use of *partial rate factors*. The reactivity of each position in a substituted aromatic compound can be compared with that of benzene by measuring the overall rate, relative to benzene, and dissecting the total rate by dividing it among the *ortho*, *meta*, and *para*

**Table 10.3. Isomer Proportions in the Nitration of
Some Substituted Benzenes**[a]

Substituent	Product (%)		
	o	m	p
$\overset{+}{N}H_3$	3–5	35–50	50–60
$\overset{+}{N}(CH_3)_3$	0	89	11
$CH_2\overset{+}{N}(CH_3)_3$	0	85	15
$\overset{+}{S}(CH_3)_2$	4	90	6
NO_2	5–8	91–93	0–2
CO_2H	15–20	75–85	~1
$C{\equiv}N$	15–17	81–83	~2
$CO_2C_2H_5$	24–28	66–73	1–6
$COCH_3$	26	72	0–2
F	9–13	0–1	86–91
Cl	30–35	~1	64–70
Br	36–43	1	56–62
I	38–45	1–2	54–60
CCl_3	7	64	29
CF_3	6	91	3
$CH_2C{\equiv}N$	24	20	56
CH_2NO_2	22	55	23
CH_2OCH_3	51	7	42
CH_3	56–63	2–4	34–41
CH_2CH_3	46–50	2–4	46–51
OCH_3	30–40	0–2	60–70

a. Data are from Tables 9.1, 9.2, 9.3, 9.4, 9.5, and 9.6 in J. G. Hoggett, R.
B. Moodie, J. R. Penton, and K. Schofield, *Nitration and Aromatic
Reactivity*, Cambridge University Press, Cambridge, 1971.

products. Correction for the statistical factor arising from the relative number of available
positions permits the partial rate factors to provide meaningful comparisons between the
reactivity of each position on a substituted ring and the reactivity of benzene:

$$\text{partial rate factor} = f = \frac{(6)(k_{subs})(\text{fraction } z \text{ product})}{(y)(k_{benz})}$$

where y is the number of equivalent positions. A partial rate factor calculation for nitration
of toluene is given in Example 10.1.

Example 10.1. The nitration of toluene is 23 times faster than nitration of benzene in
nitric acid–acetic anhydride. The product ratio is 63% *ortho*, 34% *para*, and 3% *meta*.
Calculate the partial rate factors.

$$f_o = \frac{(6)}{(2)} \times \frac{(23)}{(1)} \times (0.63) = 43.5$$

$$f_m = \frac{(6)}{(2)} \times \frac{(23)}{(1)} \times (0.03) = 2.1$$

$$f_p = \frac{(6)}{(1)} \times \frac{(23)}{(1)} \times (0.34) = 46.9$$

Partial rate factors give insight into two related aspects of reactivity. They reveal the selectivity of a given electrophile for different *reactants*. Some reactions exhibit high selectivity; that is, there are large differences in rate of reaction depending on the identity of the ring substituent. In general, low substrate selectivity is correlated with high electrophile reactivity and vice versa. Clearly, when substrate selectivity is high, the partial rate factors for the substituted aromatic compound will be very different from unity. The partial rate factors also reveal *positional* selectivity within the substituted aromatic. This selectivity also varies for different electrophiles and provides some insight into the mechanism. In general, there is a correlation between position and substrate selectivity. High substrate selectivity is accompanied by high position selectivity. Electrophiles that show high substrate selectivity generally exhibit low *ortho* : *para* ratios and negligible amounts of *meta* substitution. Very reactive electrophiles tend to show low position and substrate selectivity. Table 10.4 gives some data on the selectivity of some representative aromatic substitution reactions. The most informative entry in terms of substrate selectivity is f_p, since the partial rate factors for *ortho* substitution contain variable steric components. Using f_p as the criterion, halogenation and Friedel–Crafts acylation exhibit high selectivity, protonation and nitration are intermediate, and Friedel–Crafts alkylation shows low selectivity.

Reactivity and selectivity are largely determined by the position of the transition state on the reaction coordinate. With highly reactive electrophiles, the transition state will come early on the reaction coordinate, as in Fig. 10.5A. The transition state then resembles the reactants more closely than the σ-complex. The positive charge on the ring is small, and, as a result, the interaction with the substituent group is relatively weak. With a less reactive electrophile, the transition state comes later, as in Fig. 10.5B. The bond to the electrophile is more completely formed, and a substantial positive charge is present on the ring. This situation results in stronger substituent effects. These arguments follow the general lines of Hammond's postulate (Section 4.4). MO calculations at the STO-3G level reproduce these

Table 10.4. Selectivity in Some Electrophilic Aromatic Substitution Reactions[a]

Reaction	Partial rate factors for toluene		
	f_o	f_m	f_p
Nitration			
$\quad$ HNO$_3$ (CH$_3$NO$_2$)	38.9	1.3	45.7
Halogenation			
$\quad$ Cl$_2$ (CH$_3$CO$_2$H)	617	5	820
$\quad$ Br$_2$ (CH$_3$CO$_2$H, H$_2$O)	600	5.5	2420
Protonation			
$\quad$ H$_2$O–H$_2$SO$_4$	83	1.9	83
$\quad$ H$_2$O–CF$_3$CO$_2$H, H$_2$SO$_4$	330	7.2	313
Acylation			
$\quad$ PhCOCl(AlCl$_3$, PhNO$_2$)	32.6	5.0	831
$\quad$ CH$_3$COCl (AlCl$_3$, ClCH$_2$CH$_2$Cl)	4.5	4.8	749
Alkylation			
$\quad$ CH$_3$Br (GaBr$_3$)	9.5	1.7	11.8
$\quad$ (CH$_3$)$_2$CHBr (GaBr$_3$)	1.5	1.4	5.0
$\quad$ PhCH$_2$Cl (AlCl$_3$)	4.2	0.4	10.0

a. From L. M. Stock and H. C. Brown, *Adv. Phys. Org. Chem.* **1**:35 (1963).

Fig. 10.5. Transition states for highly reactive (A) and less reactive (B) electrophiles.

qualitative expectations by revealing greater stabilization of the *ortho* and *para* positions in toluene with a closer approach of an electrophile.[15]

Hammett correlations also permit some insight into the reactivity and selectivity of electrophiles in aromatic substitution reactions. In general, the values of the standard Hammett σ substituent constant give poor correlations with reactions involving electrophilic aromatic substitution. The σ^+ values, which reflect an increased importance of direct resonance interaction (see Section 4.3), give better correlations and, indeed, were developed as a result of the poor correlations observed with σ in electrophilic aromatic substitution. It has been suggested that one could judge the position of a transition state on the reaction coordinate by examining the slope (ρ) of the correlation line between rate of substitution and σ^+.[16] The rationale is along the following lines. A numerically large value

15. C. Santiago, K. N. Houk, and C. L. Perrin, *J. Am. Chem. Soc.* **101**:1337 (1979).
16. P. Rys, P. Skrabal, and H. Zollinger, *Angew. Chem. Int. Ed. Engl.* **11**:874 (1972).

Table 10.5. Values of ρ for Some Electrophilic Aromatic Substitution Reactions[a]

Reaction	ρ
Bromination (CH_3CO_2H)	-13.1
Chlorination (CH_3NO_2)	-13.0
Chlorination (CH_3CO_2H, H_2O)	-8.8
Proton exchange (H_2SO_4, CF_3CO_2H, H_2O)	-8.6
Acetylation (CH_3COCl, $AlCl_3$, $C_2H_4Cl_2$)	-8.6
Nitration (H_2SO_4, HNO_3)	-6.4
Chlorination ($HOCl$, H^+)	-6.1
Alkylation (C_2H_5Br, $GaBr_3$)	-2.4

a. From P. Rys, P. Skrabal, and H. Zollinger, *Angew. Chem. Int. Ed. Engl.* **11**:874 (1972).

for the slope suggests a strong substituent effect, i.e., a late transition state that resembles the σ-complex. A small value indicates a weak substituent effect and implies an early transition state. Table 10.5 gives some ρ values for typical electrophilic substitution reactions. The data indicate that halogenation reactions show the characteristics of a highly selective electrophile, nitration and Friedel–Crafts acylation represent reactions of modest selectivity, and Friedel–Crafts alkylation is an example of a reaction of low selectivity. This is in general agreement with the selectivity as measured by f_p indicated in Table 10.4.

Isotope effects are also useful in providing insight into other aspects of the mechanisms of individual electrophilic aromatic substitution reactions. In particular, because primary isotope effects are expected only when the breakdown of the σ-complex to product is rate-determining, the observation of a substantial k_H/k_D points to a rate-determining deprotonation. Some typical isotope effects are summarized in Table 10.6. Whereas isotope effects are rarely observed for nitration and halogenation, Friedel–Crafts acylation, sulfonation, nitrosation, and diazo coupling provide examples in which the rate of proton loss can affect the rate of substitution. Only in the case of the reactions involving weak electrophiles, namely, nitrosation and diazo coupling, are isotope effects in the expected range for a fully rate-controlling deprotonation.

Figure 10.6 summarizes the general ideas that have been presented in this section. At least four types of energy profiles can exist for individual electrophilic aromatic substitution reactions. Case A is the case of rate-determining generation of the electrophile. It is most readily identified by kinetics. A rate law independent of the concentration of the aromatic is diagnostic of this case. Case B represents rate-determining σ-complex formation with an electrophile of low selectivity. The rate law in such a case should have terms in both the electrophile and the aromatic. Furthermore, low selectivity, as indicated by low ρ values and low partial rate factors, is expected when this energy profile is applicable. Case C is rate-determining σ-complex formation with a more selective electrophile having a later transition state. Finally, there is case D, in which the proton removal and rearomatization are rate-limiting. This case can be recognized by the observation of a primary kinetic isotope effect at the site of substitution.

Table 10.6. Kinetic Isotope Effects in Some Electrophilic Aromatic Substitution Reactions

Reaction and substrates	Electrophilic reagents	k_H/k_D or k_H/k_T	Reference
Nitration			
Benzene-t	$HNO_3-H_2SO_4$	<1.2	a
Toluene-t	$HNO_3-H_2SO_4$	<1.2	a
Nitrobenzene-d_5	$HNO_3-H_2SO_4$	1	a
Halogenation			
Benzene-d_6	$HOBr, HClO_4$	1	a
Anisole-d	Br_2	1.05	a
Acylation			
Benzene-d_6	$CH_3\equiv O^+SbF_6^-, CH_3NO_2$	2.25	b
Benzene-d_6	$PhC\equiv O^+SbF_6^-, CH_3NO_2$	1.58	b
Sulfonation			
Benzene-d_6	$ClSO_3H, CH_3NO_2$	1.7	c
Benzene-d_6	$ClSO_3H, CH_2Cl_2$	1.6	c
Nitrobenzene-d_5	$H_2SO_4-SO_3$	1.6–1.7	a
Nitrosation			
Benzene-d_6	HNO_2, D_2SO_4	8.5	d
Diazo coupling			
1-Naphthol-4-sulfonic acid-2-d	$PhN_2{}^+$	1.0	a
2-Naphthol-8-sulfonic acid-1-d	$PhN_2{}^+$	6.2	a

a. From a more extensive compilation by H. Zollinger, *Adv. Phys. Org. Chem.* **2**:163 (1964).
b. G. A. Olah, J. Lukas, and E. Lukas, *J. Am. Chem. Soc.* **91**:5319 (1969).
c. M. P. van Albada and H. Cerfontain, *Rev. Trav. Chim.* **91**:499 (1972).
d. B. C. Challis, R. J. Higgins, and A. J. Lawson, *J. Chem. Soc., Perkin Trans. 2* **1972**:1831.

(A) rate-controlling formation of the electrophile

(B) rate-controlling σ-complex formation (non-selective electrophile)

(C) rate-controlling σ-complex formation (selective electrophile)

(D) rate-controlling deprotonation

Fig. 10.6. Various potential energy profiles for electrophilic aromatic substitution.

10.3. Reactivity of Polycyclic and Heteroaromatic Compounds

The polycyclic aromatic hydrocarbons such as naphthalene, anthracene, and phenanthrene undergo electrophilic aromatic substitution and are generally more reactive than benzene. One reason is that the activation energy for formation of the σ-complex is lower than for benzene because more of the initial resonance stabilization is retained in intermediates that have a fused benzene ring.

CNDO calculations provide estimates of the localization energies. For benzene, naphthalene, and anthracene, these are, respectively, 36.3, 15.4, and 8.3 kcal/mol.[17]

The relative stability of the intermediates determines the position of substitution under kinetically controlled conditions. For naphthalene, the preferred site for electrophilic attack is the 1-position. Two factors can result in substitution at the 2-position. If the electrophile is very bulky, the hydrogen on the adjacent ring may cause a steric preference for attack at C-2. Under conditions of reversible substitution, where relative thermodynamic stability is the controlling factor, 2-substitution is frequently preferred. An example of this behavior is in sulfonation, where low-temperature reaction gives the 1-isomer but at elevated temperatures the 2-isomer is formed.[17]

Phenanthrene and anthracene both react preferentially in the center ring. This behavior is expected from simple resonance considerations. The σ-complexes that result from substitution in the center ring have two intact benzene rings. The total resonance stabilization of these intermediates is larger than that of the naphthalene system that results if substitution occurs at one of the terminal rings.[18]

(two isomers possible)   favored

(four isomers possible)   favored

17. H. Cerfontain, *Mechanistic Aspects in Aromatic Sulfonation and Desulfonation*, Interscience, New York, 1968, pp. 68–69.
18. D. Z. Wang and A. Streitwieser, *Theor. Chim. Acta* **102**:78 (1999).

Both phenanthrene and anthracene have a tendency to undergo addition reactions under the conditions involved in certain electrophilic substitutions.[19] For example, in the nitration of anthracene in the presence of hydrochloric acid, an intermediate addition product can be isolated.[20] This is a result of the relatively close balance in resonance stabilization to be regained by elimination (giving an anthracene ring) or addition (resulting in two benzenoid rings).

569

SECTION 10.3.
REACTIVITY OF
POLYCYCLIC AND
HETEROAROMATIC
COMPOUNDS

The heteroaromatic compounds can be divided into two broad groups, called π-excessive and π-deficient, depending on whether the heteroatom acts as an electron donor or an electron acceptor. Furan, pyrrole, thiophene, and other heterocyclics incorporating an oxygen, nitrogen, or sulfur atom that contributes two π electrons are in the π-excessive group. This classification is suggested by resonance structures and confirmed by various MO methods.[21]

The reactivity order is pyrrole > furan > thiophene, which indicates that electron-donating capacity decreases in the order $N > O > S$.[22] The order $N > O$ is as expected on the basis of electronegativity, and $O > S$ probably reflects the better overlap of the oxygen $2p$ orbital, as compared to the sulfur $3p$ orbital, with the carbon $2p$ orbitals of the ring.

Structures that incorporate the $-N=CH-$ unit, such as pyridine, are π-deficient and are deactivated to electrophilic attack. Again, a resonance interpretation is evident. The nitrogen, being more electronegative than carbon, is a net acceptor of π electron density.

19. P. B. D. de la Mare and J. H. Ridd, *Aromatic Substitution*, Academic Press, New York, 1959, p. 174.
20. C. E. Braun, C. D. Cook, C. Merritt, Jr., and J. F. Rousseau, *Org. Synth.* **IV:**711 (1965).
21. N. D. Epiotis, W. P. Cherry, F. Bernardi, and W. J. Hehre, *J. Am. Chem. Soc.* **98:**4361 (1976); W. Adam and A. Grimison, *Theor. Chim. Acta* **7:**342 (1967); D. W. Genson and R. E. Christoffersen, *J. Am. Chem. Soc.* **94:**6904 (1972); N. Bodor, M. J. S. Dewar, and A. J. Harget, *J. Am. Chem. Soc.* **92:**2929 (1970).
22. S. Clementi, F. Genel, and G. Marino, *J. Chem. Soc., Chem. Commun.* **1967:**498.

There is another important factor in the low reactivity of pyridine derivatives toward electrophilic substitution. The −N=CH− unit is basic because the electron pair on nitrogen is not part of the aromatic π system. The nitrogen is protonated or complexed with a Lewis acid under many of the conditions typical of electrophilic substitution reactions. The formal positive charge present at nitrogen in such species further reduces the reactivity toward electrophiles.

The position selectivity for electrophilic substitution in the simple five-membered heteroaromatic rings is usually 2 > 3. This reflects the more favorable conjugation in intermediate **A** than in intermediate **B**. In structure **A** the remaining C=C bond can delocalize the positive charge more effectively than in **B**. Substituents on the ring can easily override this directive influence, however.

For pyridine, the reactivity toward electrophilic substitution is 3 > 4, 2. The ring nitrogen acts as a strongly destabilizing "internal" electron-withdrawing substituent in the 2- and 4-intermediates. The nitrogen also deactivates the 3-position, but less so than the 2- and 4-positions.

Reactivity and orientation in electrophilic aromatic substitution can also be related to the concept of hardness (see Section 1.2.3). Ionization potential is a major factor in determining hardness and is also intimately related to the process of σ-complex formation when an electrophile interacts with the π HOMO to form a new σ bond. In MO terms, hardness is related to the gap between the LUMO and HOMO, $\eta = (\varepsilon_{LUMO} - \varepsilon_{HOMO})/2$.[23] Thus, the harder a reactant ring system is, the more difficult it is for an electrophile to complete σ-bond formation.

This idea can be quantitatively expressed by defining *activation hardness* as the difference between the LUMO–HOMO gap for the reactant and that for the σ-complex intermediate.[24]

$$\Delta\eta^* = \beta[(\chi^R_{LUMO} - \chi^R_{HOMO}) - (\chi^\sigma_{LUMO} - \chi^\sigma_{HOMO})]/2$$

where χ^R and χ^σ are the orbital energies of the reactant and σ-complex.

23. R. G. Pearson, *Proc. Natl. Acad. Sci. U.S.A.* **83**:8440 (1986).
24. Z. Zhou and R. G. Parr, *J. Am. Chem. Soc.* **112**:5720 (1990).

Scheme 10.3. **Activation Hardness for Aromatic and Heteroaromatic Compounds**[a]

571

SECTION 10.4.
SPECIFIC
SUBSTITUTION
MECHANISMS

Hydrocarbons

Heteroaromatics

————→ increasing reactivity ————→

————→ increasing reactivity ————→

Activated benzenes

Deactivated benzenes

a. Z. Zhou and R. G. Parr, *J. Am. Chem. Soc.*, **112**: 5720 (1990).

Using simple HMO theory, $\Delta\eta^*$ has been calculated for several benzenoid hydrocarbons, substituted benzenes, and heterocycles. The resulting values are in excellent qualitative agreement with reactivity trends. Scheme 10.3 gives some of the data. The less positive the number, the more reactive is the position. Although there are some discrepancies between structural groups, within groups the $\Delta\eta^*$ correlates well with position selectivity. The most glaring discrepancy is the smaller activation hardness for deactivated benzene than for activated benzenes. In particular, benzaldehyde and benzoic acid have $\Delta\eta^*$ values less than that of benzene, which is counter to their relative reactivity. However, the preference for *meta* substitution of the deactivated benzenes is correctly predicted.

10.4. Specific Substitution Mechanisms

At this point, attention can be given to specific electrophilic substitution reactions. The kinds of data that have been especially useful for determining mechanistic details include linear free-energy relationships, kinetic studies, isotope effects, and selectivity patterns. In general, the basic questions that need to be asked about each mechanism are: (1) What is the active electrophile? (2) Which step in the general mechanism for electrophilic aromatic substitution is rate-determining? (3) What are the orientation and selectivity patterns?

10.4.1. Nitration

A substantial body of data, including reaction kinetics, isotope effects, and structure–reactivity relationships, has permitted a thorough understanding of the steps in aromatic nitration.[25] As anticipated from the general mechanism for electrophilic substitution, there are three distinct steps:

25. J. G. Hoggett, R. B. Moodie, J. R. Penton, and K. Schofield, *Nitration and Aromatic Reactivity*, Cambridge University Press, Cambridge, U.K., 1971; L. M. Stock, *Prog. Phys. Org. Chem.* **12**:21 (1976); G. A. Olah, R. Malhotra, and S. C. Narang, *Nitration*, VCH Publishers, New York, 1989.

1. Generation of the electrophile

$$2\,H_2SO_4 + HNO_3 \rightleftharpoons NO_2^+ + 2\,HSO_4^- + H_3O^+$$

or

$$2\,HNO_3 \rightleftharpoons NO_2^+ + NO_3^- + H_2O$$

2. Attack on the aromatic ring

3. Deprotonation

Although conditions under which each of the first two steps is rate-determining have been recognized, the third step is usually very fast.

The existence of the nitronium ion in sulfuric–nitric acid mixtures was demonstrated both by cryoscopic measurements and by spectroscopy. An increase in the strong acid concentration increases the rate of reaction by shifting the equilibrium of step 1 to the right. Addition of a nitrate salt has the opposite effect by suppressing the preequilibrium dissociation of nitric acid. It is possible to prepare crystalline salts of nitronium ions, such as nitronium tetrafluoroborate. Solutions of these salts in organic solvents rapidly nitrate aromatic compounds.[26]

There are three general types of kinetic behavior that have been observed for aromatic nitration. Aromatics of modest reactivity exhibit second-order kinetics in mixtures of nitric acid with the stronger acids sulfuric or perchloric acid.[27] Under these conditions, the formation of the nitronium ion is a rapid preequilibrium, and step 2 of the nitration mechanism is rate-controlling. If nitration is conducted in inert organic solvents, such as nitromethane or carbon tetrachloride, in the absence of a strong acid, the rate of formation of nitronium ion is slowed and becomes rate-limiting.[28] Finally, some reactive aromatics, including alkylbenzenes, can react so rapidly under conditions where nitronium ion concentration is high that the rate of nitration becomes governed by encounter rates. Under these circumstances, mixing and diffusion control the rate of reaction, and there are no differences in reactivity among the substrates.

With very few exceptions, the final step in the nitration mechanism, the deprotonation of the σ-complex, is fast and therefore has no effect on the observed kinetics. The fast deprotonation can be confirmed by the absence of an isotope effect when deuterium or

26. S. J. Kuhn and G. A. Olah, *J. Am. Chem. Soc.* **83**:4564 (1961); G. A. Olah and S. J. Kuhn, *J. Am. Chem. Soc.* **84**:3684 (1962); C. M. Adams, C. M. Sharts, and S. A. Shackelford, *Tetrahedron Lett.* **34**:6669 (1993); C. L. Dwyer and C. W. Holzapfel, *Tetrahedron* **54**:7843 (1998).
27. J. G. Hoggett, R. B. Moodie, J. R. Penton, and K. Schofield, *Nitration and Aromatic Reactivity*, Cambridge University Press, Cambridge, U.K., 1971, Chapter 2.
28. E. D. Hughes, C. K. Ingold, and R. I. Reed, *J. Chem. Soc.* **1950**:2400; R. G. Coombes, *J. Chem. Soc. B* **1969**:1256.

tritium is introduced at the substitution site. Several compounds such as benzene, toluene, bromobenzene, and fluorobenzene have been subjected to this test and found not to exhibit isotope effects during nitration.[29] The only case in which a primary isotope effect indicating rate-controlling deprotonation has been seen is with 1,3,5-tri-t-butylbenzene, where steric hindrance evidently makes deprotonation the slow step.[30]

The question of what other species can be the active electrophile in nitration arises in the case of nitration using solutions of nitric acid in acetic anhydride. The solutions are very potent nitrating mixtures and effect nitrations at higher rates than solutions of nitric acid in inert organic solvents. Acetyl nitrate is formed in such solutions, and may be the actual nitrating agent.

$$HNO_3 + (CH_3CO)_2O \longrightarrow CH_3\overset{\overset{\displaystyle O}{\|}}{C}ONO_2 + CH_3CO_2H$$

A very convenient synthetic procedure for nitration involves the mixing of a nitrate salt with trifluoroacetic anhydride.[31] This presumably generates trifluoroacetyl nitrate.

$$NO_3^- + (CF_3CO)_2O \longrightarrow CF_3CO_2NO_2 + CF_3CO_2H$$

Benzene, toluene, and aromatics of similar reactivity can be nitrated using $Yb(O_3SCF_3)_3$ and 69% nitric acid in an inert solvent.[32]

The active nitrating agent under these conditions is uncertain but must be some complex of nitrate with the oxyphilic lanthanide.

$$Yb\overset{\underset{\displaystyle H^+}{|}}{-}O-\overset{\overset{\displaystyle }{\underset{\displaystyle \|}{N}}}{\underset{\displaystyle O}{}}-O^- \longrightarrow YbOH + O=\overset{+}{N}=O$$

The identification of a specific nitrating species can be approached by comparing selectivity with that of nitration under conditions known to involve the nitronium ion. Examination of part B of Table 10.7 shows that the position selectivity exhibited by acetyl nitrate toward toluene and ethylbenzene is not dramatically different from that observed with nitronium ion. The data for i-propylbenzene suggest a lower $ortho:para$ ratio for acetyl nitrate nitrations. This could indicate a larger steric factor for nitration by acetyl nitrate.

29. G. A. Olah, S. J. Kuhn, and S. H. Flood, *J. Am. Chem. Soc.* **83**:4571, 4581 (1961); H. Suhr and H. Zollinger, *Helv. Chim. Acta* **44**:1011 (1961); L. Melander, *Acta Chem. Scand.* **3**:95 (1949); L. Melander, *Ark. Kemi.* **2**:211 (1950).
30. P. C. Myhre, M. Beug, and L. L. James, *J. Am. Chem. Soc.* **90**:2105 (1968).
31. J. V. Crivello, *J. Org. Chem.* **46**:3056 (1981).
32. F. J. Waller, A. G. M. Barrett, D. C. Braddock, and D. Ramprasad, *J. Chem. Soc., Chem. Commun.* **1997**:613.

Table 10.7. Relative Reactivity and Position Selectivity for Nitration of Some Aromatic Compounds

A. Relative reactivity of some hydrocarbons

Substrate	$H_2SO_4-HNO_3-H_2O$[a]	$HNO_3-CH_3NO_2$[b]	$HNO_3-(CH_3CO)_2O$[c]
Benzene	1	1	1
Toluene	17	25	27
p-Xylene	38	139	92
m-Xylene	38	146	–
o-Xylene	38	139	–
Mesitylene	36	400	1750

B. Partial rate factors for some monoalkylbenzenes

Substrate	$H_2SO_4-HNO_3$ in sulfolane[d]			$HNO_3-CH_3NO_2$[e,f]			$HNO_3-(CH_3CO)_2O$[g]		
	f_o	f_m	f_p	f_o	f_m	f_p	f_o	f_m	f_p
Toluene	52.1	2.8	58.1	49	2.5	56	49.7	1.3	60.0
Ethylbenzene	36.2	2.6	66.4	32.7	1.6	67.1	31.4	2.3	69.5
i-Propylbenzene	17.9	1.9	43.3	–	–	–	14.8	2.4	71.6
t-Butylbenzene	–	–	–	5.5	3.7	71.4	4.5	3.0	75.5

C. Relative reactivity and isomer distribution for nitrobenzene and the nitrotoluenes[h]

Substrate	Relative reactivity	Product %		
		o	m	p
Nitrobenzene	1	7	92	1
o-Nitrotoluene	545	29	1	70
m-Nitrotoluene	138	38	1	60
p-Nitrotoluene	217	100	0	–

a. R. G. Coombes, R. B. Moodie, and K. Schofield, *J. Chem. Soc. B* **1968**:800.
b. J. G. Hoggett, R. B. Moodie, and K. Schofield, *J. Chem. Soc.* **1969**:1.
c. A. R. Cooksey, K. J. Morgan, and D. P. Morrey, *Tetrahedron* **26**:5101 (1970).
d. G. A. Olah, S. J. Kuhn, S. H. Flood and J. C. Evans, *J. Am. Chem. Soc.* **84**:3687 (1962).
e. L. M. Stock, *J. Org. Chem.* **26**:4120 (1961).
f. G. A. Olah and H. C. Lin, *J. Am. Chem. Soc.* **96**:549 (1974); *o, m, p* designations refer to the methyl substituent.
g. J. R. Knowles, R. O. C. Norman, and G. K. Radda, *J. Chem. Soc.* **1960**:4885.
h. G. A. Olah and H. C. Lin, *J. Am. Chem. Soc.* **96**:549 (1974); *o, m, p* designations for the nitrotoluenes refer to the methyl groups.

Relative reactivity data for nitration must be treated with special caution because of the possibility of diffusion control. As an example of this, Table 10.7 shows that no difference in reactivity between mesitylene and xylene is found in $H_2SO_4-HNO_3$ nitration, whereas in $HNO_3-CH_3NO_2$, the rates differ by a factor of ~3. Diffusion controls prevail in the former case. In general, nitration is a relatively unselective reaction, with toluene *fp* being about 50–70, as shown in Table 10.7. When the aromatic reactant carries an electron-attracting group, the selectivity increases because the transition state occurs later. For example, whereas toluene is ~20 times more reactive than benzene, *p*-nitrotoluene is ~200 times more reactive than nitrobenzene. Because of the later transition state, the effect of the methyl substituent is magnified.

One aspect of aromatic nitration that has received attention is the role of charge-transfer and electron-transfer intermediates on the path to the σ-complex intermediate. For some NO_2—X nitrating reagents, a general mechanism might involve formation of such intermediates prior to the formation of the σ-complex.[33]

The existence of charge-transfer complexes can be demonstrated for several reaction combinations which eventually lead to nitration, but the crucial question is whether a complete electron transfer to form a radical : radical-cation pair occurs as a distinct step in the mechanism. This has been a matter of continuing discussion, both *pro*[34] and *con*.[35] One interesting fact that has emerged about nitration is that the product composition from toluene is virtually invariant at $4 \pm 2\%$ *meta*, $33 \pm 3\%$ *para*, and $65 \pm 5\%$ *ortho*, i.e., close to a statistical *ortho : para* ratio, over a wide range of nitrating species.[36] This argues for a common product-forming step, and one interpretation is that this step is a collapse of a NO_2· : radical-cation pair, as in the electron-transfer mechanism. If the σ-complex were formed in a single step from different NO_2—X reagents, some variation of the *ortho : para* ratio for different X would be expected.

10.4.2. Halogenation

Substitution of hydrogen by halogen is a synthetically important electrophilic aromatic substitution reaction. The reactivity of the halogens increases in the order $I_2 < Br_2 < Cl_2$. The molecular halogens are only reactive enough to halogenate activated aromatics. Halogenation reactions are normally run in the presence of Lewis acids, in which case a complex of the halogen with the Lewis acid is probably the active electrophile.

Bromine and iodine form complexes with the corresponding halide ions. These anionic trihalide ions are less reactive than the free halogen but are capable of substituting highly reactive rings. They complicate kinetic studies, since the concentration of halide ion increases during the course of halogenation, and successively more of the halogen will be present as the trihalide ion. The hypohalous acids, ClOH, BrOH, and IOH, are weak halogenating agents but are much more reactive in acidic solution. Halogenation is also effected by the hypohalites of carboxylic acids such as acetyl hypochlorite and trifluoroacetyl hypobromite[37]:

$$Cl_2 + Hg(OAc)_2 \rightleftharpoons HgClOAc + CH_3CO_2Cl$$
$$Br_2 + Hg(O_2CCF_3)_2 \rightleftharpoons HgBr(O_2CCF_3) + CF_3CO_2Br$$

33. C. L. Perrin, *J. Am. Chem. Soc.* **99**:5516 (1977).
34. J. K. Kochi, *Acc. Chem. Res.* **25**:39 (1992); T. M. Bockman and J. K. Kochi, *J. Phys. Org. Chem.* **7**:325 (1994); A. Peluso and G. Del Re, *J. Phys. Chem.* **100**:5303 (1996).
35. L. Eberson and F. Radner, *Acc. Chem. Res.* **20**:53 (1987); L. Eberson, M. P. Hartshorn, and F. Radner, *Acta Chem. Scand.* **48**:937 (1994); M. Lehnig, *J. Chem. Soc., Perkin Trans.* **1996**:1943.
36. E. K. Kim, K. Y. Lee, and J. K. Kochi, *J. Am. Chem. Soc.* **114**:1756 (1992).
37. P. B. D. de la Mare, I. C. Hilton, and S. Varma, *J. Chem. Soc.* **1960**:4044; J. R. Barnett, L. J. Andrews, and R. M. Keefer, *J. Am. Chem. Soc.* **94**:6129 (1972).

The latter is an extremely reactive species. Trifluoroacetate is a good leaving group and facilitates cleavage of the O–Br bond. The acyl hypohalites are also the active halogenating species in solutions of the hypohalous acids in carboxylic acids, where they exist in equilibrium.

$$HOCl + RCO_2H \rightleftharpoons R\overset{\overset{\text{O}}{\|}}{C}OCl + H_2O$$

$$HOBr + RCO_2H \rightleftharpoons R\overset{\overset{\text{O}}{\|}}{C}OBr + H_2O$$

Molecular chlorine is believed to be the active electrophile in uncatalyzed chlorination of aromatic compounds. Simple second-order kinetics are observed in acetic acid.[38] The reaction is much slower in nonpolar solvents such as dichloromethane and carbon tetrachloride. Chlorination in nonpolar solvents is catalyzed by added acid. The catalysis by acids is probably the result of assistance by proton transfer during the cleavage of the Cl–Cl bond.[39]

Chlorination in acetic acid is characterized by a large ρ value (~ -9 to -10), and the partial rate factor f_p for toluene is 820. Both values indicate a late transition state, which would resemble the σ-complex intermediate.

For preparative purposes, a Lewis acid such as $AlCl_3$ or $FeCl_3$ is often used to catalyze chlorination. Chlorination of benzene by $AlCl_3$ is overall third-order.[40]

$$\text{rate} = k[ArH][Cl_2][AlCl_3]$$

This rate law could correspond to formation of a Cl_2–$AlCl_3$ complex that acts as the active halogenating agent but is also consistent with a rapid equilibrium involving formation of Cl^+:

$$Cl_2 + AlCl_3 \rightleftharpoons Cl_2\text{–}AlCl_3 \rightleftharpoons Cl^+ + AlCl_4^-$$

There is, however, no direct evidence for the formation of Cl^+, and it is much more likely that the complex is the active electrophile. The substrate selectivity under catalyzed conditions ($k_{tol} = 160 k_{benz}$) is lower than in uncatalyzed chlorinations, as would be expected for a more reactive electrophile. The effect of the Lewis acid is to weaken the Cl–Cl bond, which lowers the activation energy for σ-complex formation.

Hypochlorous acid is a weak chlorinating agent. In acidic solution, it is converted to a much more active chlorinating agent. Although early mechanistic studies suggested that Cl^+ might be formed under these conditions, it has since been shown that this is not the

38. L. M. Stock and F. W. Baker, *J. Am. Chem. Soc.* **84**:1661 (1962).
39. L. J. Andrews and R. M. Keefer, *J. Am. Chem. Soc.* **81**:1063 (1959); R. M. Keefer and L. J. Andrews, *J. Am. Chem. Soc.* **82**:4547 (1960).
40. S. Y. Caille and R. J. P. Corriu, *Tetrahedron* **25**:2005 (1969).

case. Detailed kinetic analysis of the chlorination of methoxybenzene has revealed a rather complex rate law[41]:

$$\text{rate} = k_1[\text{HOCl}]^2 + k_2[\text{H}_3\text{O}^+][\text{HOCl}]^2 + k_3[\text{ArH}][\text{H}_3\text{O}^+][\text{HOCl}]$$

Some of the terms are independent of the concentration of the aromatic reactant. This rate law can best be explained in terms of the formation of Cl_2O, the anhydride of hypochlorous acid.

$$2\,\text{HOCl} \xrightarrow{\text{H}^+} \text{Cl}_2\text{O} + \text{H}_2\text{O}$$

Both Cl_2O and $[\text{H}_2\text{OCl}]^+$ apparently are active electrophiles under these conditions. The terms involving Cl_2O are zero-order in the aromatic reactant because the rate of formation of Cl_2O is slower than the rate of the subsequent reaction with the aromatic. Thermodynamic considerations argue strongly against rate-determining cleavage of $[\text{H}_2\text{OCl}]^+$ to H_2O and Cl^+. The estimated equilibrium constant for this dissociation is so small that the concentration of Cl^+ would be around $10^{-40}\,M$, which is far too low to account for the observed reaction rate.[42]

Molecular bromine is believed to be the reactive brominating agent in uncatalyzed brominations. The brominations of benzene and toluene are first-order in both bromine and the aromatic substrate in trifluoroacetic acid solution,[43] but the rate expressions become more complicated when these reactions take place in the presence of water.[44] The bromination of benzene in aqueous acetic acid exhibits a first-order dependence on bromine concentration when bromide ion is present. The observed rate is dependent on bromide ion concentration, decreasing with increasing bromide ion concentration. The detailed kinetics are consistent with a rate-determining formation of the σ-complex when bromide ion concentration is low, but with a shift to reversible formation of the σ-complex with rate-determining deprotonation at high bromide ion concentration.[45]

Bromination is characterized by high substrate selectivity.[46] The data in Table 10.4 (p. 564) show that for toluene f_p is around 2400, as compared to about 45 for nitration. The very large stabilizing effect of electron-donor substituents is also evident in the large negative ρ value (-12).[47] The fact that substituents can strongly influence both the rate and orientation implies that the transition state comes late in the reaction and resembles the σ-complex.

41. C. G. Swain and D. R. Crist, *J. Am. Chem. Soc.* **94**:3195 (1972).
42. E. Berliner, *J. Chem. Educ.* **43**:124 (1966).
43. H. C. Brown and R. A. Wirkkala, *J. Am. Chem. Soc.* **88**:1447 (1966).
44. W. M. Schubert and D. F. Gurka, *J. Am. Chem. Soc.* **91**:1443 (1969).
45. E. Berliner and J. C. Powers, *J. Am. Chem. Soc.* **83**:905 (1961); W. M. Schubert and J. L. Dial, *J. Am. Chem. Soc.* **97**:3877 (1975).
46. L. M. Stock and H. C. Brown, *Adv. Phys. Org. Chem.* **1**:35 (1963).
47. H. C. Brown and L. M. Stock, *J. Am. Chem. Soc.* **79**:1421 (1957).

Bromination has been shown not to exhibit a primary kinetic isotope effect in the case of benzene,[48] bromobenzene,[49] toluene,[50] or methoxybenzene.[51] There are several examples of substrates which do show significant isotope effects, including substituted anisoles,[46] N,N-dimethylanilines,[52] and 1,3,5-trialkylbenzenes.[53] The observation of isotope effects in highly substituted systems seems to be the result of steric factors that can operate in two ways. There may be resistance to the bromine taking up a position coplanar with adjacent substituents in the aromatization step. This would favor return of the σ-complex to reactants. In addition, the steric bulk of several substituents may hinder solvent or other base from assisting in the proton removal. Either factor would allow deprotonation to become rate-controlling.

Bromination is catalyzed by Lewis acids, and a study of the kinetics of bromination of benzene and toluene in the presence of aluminum chloride has been reported.[54] Toluene is found to be about 35 times more reactive than benzene under these conditions. The catalyzed reaction thus shows a good deal less substrate selectivity than the uncatalyzed reaction, as would be expected on the basis of the greater reactivity of the aluminum chloride–bromine complex.

Bromination can also be carried out using solutions of acetyl hypobromite or trifluoroacetyl hypobromite.[55] Acetyl hypobromite is considered to be the active halogenating species in solutions of hypobromous acid in acetic acid:

$$CH_3CO_2H + HOBr \rightleftharpoons CH_3\overset{\overset{O}{\|}}{C}OBr + H_2O$$

This reagent can also be formed by reaction of bromine with mercuric acetate:

$$Hg(OAc)_2 + Br_2 \rightleftharpoons HgBr(O_2CCH_3) + CH_3CO_2Br$$

Both of the above equilibria lie to the left, but acetyl hypobromite is sufficiently reactive that it is the principal halogenating species in both solutions. The reactivity of the acyl hypohalites as halogenating agents increases with the ability of the carboxylate to function as a leaving group. This, of course, correlates with the acidity of the carboxylic acid. The estimated order of reactivity of Br_2, CH_3CO_2Br, and CF_3CO_2Br is $1 : 10^6 : 10^{10}$.[56] It is this exceptionally high reactivity of the hypobromites that permits them to be the reactive halogenating species in solutions where they are present in relatively low equilibrium concentration.

Molecular iodine is not a very powerful halogenating agent. Only very reactive aromatics such as anilines or phenolate anions are reactive toward iodine. Iodine monochloride can be used as an iodinating agent. The greater electronegativity of the

48. P. B. D. de la Mare, T. M. Dunn, and J. T. Harvey, J. Chem. Soc. 1957:923.
49. L. Melander, Acta Chem. Scand. 3:95 (1949); Ark. Kemi. 2:211 (1950).
50. R. Josephson, R. M. Keefer, and L. J. Andrews, J. Am. Chem. Soc. 83:3562 (1961).
51. J.-J. Aaron and J.-E. Dubois, Bull. Soc. Chim. Fr. 1971:603.
52. J.-E. Dubois and R. Uzan, Bull. Soc. Chim. Fr. 1968:3534; A. Nilsson, Acta Chem. Scand. 21:2423 (1967); A. Nilsson and K. Olsson, Acta Chem. Scand. 23:7 (1969).
53. P. C. Myhre, Acta Chem. Scand. 14:219 (1960).
54. S. Y. Caille and R. J. P. Corriu, Tetrahedron 25:2005 (1969).
55. P. B. D. de la Mare and J. L. Maxwell, J. Chem. Soc. 1962:4829; Y. Hatanaka, R. M. Keefer, and L. J. Andrews, J. Am. Chem. Soc. 87:4280 (1965).
56. P. B. D. de la Mare, I. C. Hilton, and S. Vanna, J. Chem. Soc. 1960:4044; J. R. Bennett, L. J. Andrews, and R. M. Keefer, J. Am. Chem. Soc. 94:6129 (1972).

chlorine ensures that the iodine will be the electrophilic entity in the substitution reaction. Iodination by iodine monochloride can be catalyzed by Lewis acids, such as $ZnCl_2$.[57] Iodination can also be carried out with acetyl hypoiodite and trifluoroacetyl hypoiodite. The methods of formation of these reagents are similar to those for the hypobromites.[58]

Direct fluorination of aromatics is not a preparatively important reaction because it can occur with explosive violence. Mechanistic studies have been done at very low temperatures and with low fluorine concentrations. For toluene, the f_p and f_m values are 8.2 and 1.55, respectively, indicating that fluorine is a very unselective electrophile. The ρ value in a Hammett correlation with σ^+ is -2.45. Thus, fluorination exhibits the characteristics that would be expected for a very reactive electrophile.[59] A number of reagents in which fluorine is bound to a very electronegative group also serve as fluorinating agents. These include CF_3OF, CF_3CO_2F, CH_3CO_2F and $HOSO_2OF$.[60] The synthetic applications of these reagents will be discussed in Section 11.1.2. of Part B.

10.4.3. Protonation and Hydrogen Exchange

Hydrogen exchange resulting from reversible protonation of an aromatic ring can be studied by the use of isotopic labels. Either deuterium or tritium can be used, and the experiment can be designed to follow either the incorporation or the release of the isotope. The study of the mechanism of electrophilic hydrogen exchange is somewhat simplified by the fact that a solvated proton must be the active electrophile. The principle of microscopic reversibility implies that the transition state must occur on a symmetrical potential energy surface, since the attacking electrophile is chemically identical to the displaced proton. The transition states involve partial transfer of a proton to (or from) a solvent molecule(s) from (or to) the aromatic ring. The intermediate σ-complex is a cyclohexadienyl cation. As mentioned earlier, these cations are stable in strongly acidic nonnucleophilic media and have been observed spectroscopically.

Partial rate factors for exchange for a number of substituted aromatic compounds have been measured. They reveal activation of *ortho* and *para* positions by electron-releasing groups. Some typical data are given in Table 10.8. The k_{tol}/k_{benz} ratio of around 300 indicates considerable substrate selectivity. The f_p value for toluene varies somewhat, depending on the reaction medium, but generally is about 10^2.[61] The ρ value for hydrogen exchange in $H_2SO_4-CF_3CO_2H-H_2O$ is -8.6.[62] A similarly large ρ value of -7.5 has been observed in aqueous sulfuric acid.[63] As seen for other electrophilic aromatic substitution reactions, the best correlation is with σ^+.

Among the many experimental results pertaining to hydrogen exchange, a most important one is that general acid catalysis has been demonstrated.[64] This finding is in accord with a rate-limiting step involving proton transfer. Because proton removal is partially rate-determining, hydrogen exchange exhibits an isotope effect. A series of experiments using both deuterium and tritium labels arrived at $k_H/k_D = 9.0$ for the proton-

57. R. M. Keefer and L. J. Andrews, *J. Am. Chem. Soc.* **78**:5623 (1956).
58. E. M. Chen, R. M. Keefer, and L. J. Andrews, *J. Am. Chem. Soc.* **89**:428 (1967).
59. F. Cacace, P. Giacomello, and A. P. Wolff, *J. Am. Chem. Soc.* **102**:3511 (1980).
60. A. Haas and M. Lieb, *Chimia* **39**:134 (1985).
61. L. M. Stock and H. C. Brown, *Adv. Phys. Org. Chem.* **1**:35 (1963).
62. P. Rys, P. Skrabal, and H. Zollinger, *Angew. Chem. Int. Ed. Engl.* **11**:874 (1972).
63. S. Clementi and A. R. Katritzky, *J. Chem. Soc., Perkin Trans. II* **1973**:1077.
64. A. J. Kresge and Y. Chiang, *J. Am. Chem. Soc.* **83**:2877 (1961); A. J. Kresge, S. Slae, and D. W. Taylor, *J. Am. Chem. Soc.* **92**:6309 (1970).

Table 10.8. Partial Rate Factors for Hydrogen Exchange in Some Substituted Aromatic Compounds

X	f_o	f_m	f_p	Reference
CH_3	330	7.2	313	a
F	0.136	–	1.79	b
Cl	0.035	–	0.161	b
OPh	6900	~0.1	31,000	c
Ph	133	<1	143	d

a. C. Eaborn and R. Taylor, *J. Chem. Soc.* **1961**:247.
b. C. Eaborn and R. Taylor, *J. Chem. Soc.* **1961**:2388.
c. R. Baker and C. Eaborn, *J. Chem. Soc.* **1961**:5077.
d. C. Eaborn and R. Taylor, *J. Chem. Soc.* **1961**:1012.

loss step for 1,3,5-trimethoxybenzene.[65] A substantial isotope effect has also been observed for the exchange process with azulene.[66]

10.4.4. Friedel–Crafts Alkylation and Related Reactions

The Friedel–Crafts reaction is a very important method for introducing alkyl substituents on an aromatic ring. It involves generation of a carbocation or related electrophilic species. The most common method of generating these electrophiles involves reaction between an alkyl halide and a Lewis acid. The usual Friedel–Crafts catalyst for preparative work is $AlCl_3$, but other Lewis acids such as SbF_5, $TiCl_4$, $SnCl_4$, and BF_3 can also promote reaction. Alternative routes to alkylating species include protonation of alcohols and alkenes.

There are relatively few kinetic data on the Friedel–Crafts reaction. Alkylation of benzene or toluene with methyl bromide or ethyl bromide with gallium bromide as catalyst is first-order in each reactant and in catalyst.[67] With aluminum bromide as catalyst, the rate of reaction changes with time, apparently because of heterogeneity of the reaction mixture.[68] The initial rate data fit the kinetic expression:

$$rate = k[EtBr][benzene][AlBr_3]^2$$

The reaction rates of toluene and benzene with i-propyl chloride in nitromethane fit a third-order rate law[69]:

$$rate = k[AlCl_3][i\text{-}PrCl][ArH]$$

A rate law of the same form pertains to t-butyl chloride.[70]

65. A. J. Kresge and Y. Chiang, *J. Am. Chem. Soc.* **89**:4411 (1967).
66. L. C. Gruen and F. A. Long, *J. Am. Chem. Soc.* **89**:1287 (1967).
67. S. U. Choi and H. C. Brown, *J. Am. Chem. Soc.* **85**:2596 (1963).
68. B. J. Carter, W. D. Covey, and F. P. DeHaan, *J. Am. Chem. Soc.* **97**:4783 (1975); cf. S. U. Choi and H. C. Brown, *J. Am. Chem. Soc.* **81**:3315 (1959); F. P. DeHaan and H. C. Brown, *J. Am. Chem. Soc.* **81**:4484 (1969); H. Jungk, C. R. Smoot, and H. C. Brown, *J. Am. Chem. Soc.* **78**:2185 (1956).
69. F. P. DeHaan, G. L. Delker, W. D. Covey, J. Ahn, R. L. Cowan, C. H. Fong, G. Y. Kim, A. Kumar, M. P. Roberts, D. M. Schubert, E. M. Stoler, Y. J. Suh, and M. Tang, *J. Org. Chem.* **51**:1587 (1986).
70. F. P. DeHaan, W. H. Chan, J. Chang, D. M. Ferrara, and L. A. Wamschel, *J. Org. Chem.* **51**:1591 (1986).

Rates that are *independent* of aromatic substrate concentration have been found for reaction of benzyl chloride catalyzed by $TiCl_4$ or SbF_5 in nitromethane.[71] This can be interpreted as resulting from rate-determining formation of the electrophile, presumably a benzyl cation. The reaction of benzyl chloride and toluene shows a second-order dependence on titanium tetrachloride concentration under conditions where there is a large excess of hydrocarbon.[72]

$$\text{rate} = k[PhCH_2Cl][TiCl_4]^2$$

All these kinetic results can be accommodated by a general mechanism that incorporates the following fundamental components: (1) complexation of the alkylating agent and the Lewis acid; (2) electrophilic attack on the aromatic substrate to form the σ-complex; and (3) deprotonation. In many systems, there may be an ionization of the complex to yield a discrete carbocation. This step accounts for the fact that rearrangement of the alkyl group is frequently observed during Friedel–Crafts alkylation.

Absolute rate data for Friedel–Crafts reactions are difficult to obtain. The reaction is complicated by sensitivity to moisture and heterogeneity. For this reason, most of the structure–reactivity trends have been developed using competitive methods, rather than by direct measurements. Relative rates are established by allowing the electrophile to compete for an excess of the two reagents. The product ratio establishes the relative reactivity. These studies reveal low substrate and position selectivity.

A study of alkylations with a group of substituted benzyl halides and a range of Friedel–Crafts catalysts has provided insight into the trends in selectivity and orientation that accompany changes in both the alkyl group and the catalysts.[73] There is a marked increase in substrate selectivity on going from *p*-nitrobenzyl chloride to *p*-methoxybenzyl chloride. For example, with titanium tetrachloride as the catalyst, k_{tol}/k_{benz} increases from 2.5 to 97. This increase in substrate selectivity is accompanied by an increasing preference for *para* substitution. With *p*-nitrobenzyl chloride, the *ortho*:*para* ratio is 2:1 (the

71. F. P. DeHaan, G. L. Delker, W. D. Covey, J. Ahn, M. S. Anisman, E. C. Brehm, J. Chang, R. M. Chicz, R. L. Cowan, D. M. Ferrara, C. H. Fong, J. D. Harper, C. D. Irani, J. Y. Kim, R. W. Meinhold, K. D. Miller, M. P. Roberts, E. M. Stoler, Y. J. Suh, M. Tang, and E. L. Williams, *J. Am. Chem. Soc.* **106**:7038 (1984); F. P. DeHaan, W. H. Chan, J. Chang, B. Cheng, D. A. Chiriboga, M. M. Irving, C. R. Kaufman, G. Y. Kim, A. Kumar, J. Na, T. T. Nguyen, D. T. Nguyen, B. R. Patel, N. P. Sarin, and J. H. Tidwell, *J. Am. Chem. Soc.* **112**:356 (1990).

72. F. P. DeHaan, W. D. Covey, R. L. Ezelle, J. E. Margetan, S. A. Pace, M. J. Sollenberger, and D. S. Wolf, *J. Org. Chem.* **49**:3954 (1984).

73. G. A. Olah, S. Kobayashi, and M. Tashiro, *J. Am. Chem. Soc.* **94**:7448 (1972).

statistically expected ratio), whereas with the *p*-methoxy compound, the *para* product dominates by 2.3 : 1. Selectivity by substituted benzyl cations has also been investigated using cations generated from benzyl diazonium ion intermediates.[74] This system removes potential complications of direct involvement of the Lewis acid in the substitution.

Toluene/benzene selectivity decreases in the order $X = CH_3 > H \sim Cl > NO_2$, in agreement with the expectation that the least stable (and most reactive) carbocation would be least selective. The reactions also show low position selectivity.

There is a clear trend within the family of substituted benzyl chlorides of increasing selectivity with the increasing electron-donor capacity of the benzyl substituent. All of the reactions, however, remain in a region that constitutes rather low selectivity. Thus, the position of the transition state for substitution by a benzylic cation must be quite early on the reaction coordinate. The substituents on the ring undergoing substitution have a relatively weak orienting effect on the attacking electrophile. With benzylic cations stabilized by donor substituents, the transition state comes later and the selectivity is somewhat higher. Toluene/benzene reactivity ratios under a number of Friedel–Crafts conditions are recorded in Table 10.9. As would be expected on the basis of the low substrate selectivity, position selectivity is also modest. As shown by the isomer ratios in

Table 10.9. Substrate and Position Selectivity in Friedel-Crafts Alkylation Reactions

	Electrophilic reagents	k_{tol}/k_{benz}	Toluene $o:p$ ratio	Reference
1	$CH_3Br-AlBr_3$	2.5–4.1	1.9	a
2	$C_2H_5Br-GaBr_3$	6.5	–	b
3	$(CH_3)_2CHBr-AlCl_3$	1.9	1.2	c
4	$(CH_3)_2CHCl-AlCl_3$	2.0	1.5	d
5	$(CH_3)_3CCl-AlCl_3$	25	0	e
6	$(CH_3)_3CBr-SnCl_4$	16.6	0	f
7	$(CH_3)_3CBr-AlCl_3$	1.9	0	f
8	$PhCH_2Cl-AlCl_3$	3.2	0.82	g
9	$PhCH_2Cl-AlCl_3$	2–3	0.9	h
10	$PhCH_2Cl-TiCl_4$	6.3	0.74	i
11	p-$MeOPhCH_2Cl-TiCl_4$	97	0.40	i
12	p-$NO_2PhCH_2Cl-TiCl_4$	2.5	1.7	j

a. H. C. Brown and H. Jungk, *J. Am. Chem. Soc.* **77**:5584 (1955).
b. S. U. Choi and H. C. Brown, *J. Am. Chem. Soc.* **85**:2596 (1963).
c. G. A. Olah, S. H. Flood, S. J. Kuhn, M. E. Moffatt, and N. A. Overchuck, *J. Am. Chem. Soc.* **86**:1046 (1964).
d. F. P. DeHaan, G. L. Delker, W. D. Covey, J. Ahn, R. L. Cowan, C. H. Fong, G. Y. Kim, A. Kumar, M. P. Roberts, D. M. Schubert, E. M. Stoler, Y. J. Suh, and M. Tang, *J. Org. Chem.* **51**:1587 (1986).
e. F. P. DeHaan, W. H. Chan, J. Chang, D. M. Ferrara, and L. A. Wainschel, *J. Org. Chem.* **51**:1591 (1986).
f. G. A. Olah, S. H. Flood, and M. E. Moffatt, *J. Am. Chem. Soc.* **86**: 1060 (1964).
g. G. A. Olah, S. J. Kuhn, and S. H. Flood, *J. Am. Chem. Soc.* **84**:1688 (1962).
h. F. P. DeHaan, W. D. Covey, R. L. Ezelle, J. E. Margetan, S. A. Pace, M. J. Sollenberger, and D. S. Wolf, *J. Org. Chem.* **49**:3954 (1984).
i. G. A. Olah, S. Kobayashi, and M. Tashiro, *J. Am. Chem. Soc.* **94**:7448 (1972).

74. E. H. White, R. W. Darbeau, N. R. Darbean, Y. Chen, S. Chen, and D. Chen, *J. Org. Chem.* **61**:7986 (1996); E. H. White, *Tetrahedron Lett.* **38**:7649 (1997).

Table 10.9, the amount of *ortho* product is often comparable to the amount of *para* product.

Steric effects play a major role in determining the *ortho* : *para* ratio in Friedel–Crafts alkylations. The amount of *ortho* substitution of toluene decreases as the size of the entering alkyl group increases along the series methyl, ethyl, *i*-propyl.[75] No *ortho* product is found when the entering group is *t*-butyl.[76]

A good deal of experimental care is often required to ensure that the product mixture at the end of a Friedel–Crafts reaction is determined by *kinetic control*. The strong Lewis acid catalysts can catalyze the isomerization of alkylbenzenes, and if isomerization takes place, the product composition is not informative about the position selectivity of electrophilic attack. Isomerization increases the amount of the *meta* isomer in the case of dialkylbenzenes, because this isomer is thermodynamically the most stable.[77]

Alcohols and alkenes can also serve as sources of electrophiles in Friedel–Crafts reactions in the presence of strong acids:

$$R_3COH + H^+ \rightarrow R_3C\overset{+}{O}H_2 \rightarrow R_3C^+ + H_2O$$

$$R_2C = CHR' + H^+ \rightarrow R_2\overset{+}{C}CH_2R'$$

The generation of carbocations from these sources is well documented (see Section 5.4). The reaction of aromatics with alkenes in the presence of Lewis acid catalysts is the basis for the industrial production of many alkylated aromatic compounds. Styrene, for example, is prepared by dehydrogenation of ethylbenzene made from benzene and ethylene.

Benzyl and allyl alcohols which can generate stabilized carbocations give Friedel–Crafts alkylation products with mild Lewis acid catalysts such as scandium triflate.[78]

Scandium triflate and lanthanide triflates also catalyze alkylation by secondary methanesulfonates.[79]

10.4.5. Friedel–Crafts Acylation and Related Reactions

Friedel–Crafts acylation usually involves the reaction of an acyl halide, a Lewis acid catalyst, and the aromatic substrate. Several species may function as the active electrophile, depending on the reactivity of the aromatic compound. For activated aromatics, the electrophile can be a discrete positively charged acylium ion or the complex formed

75. R. H. Allen and L. D. Yats, *J. Am. Chem. Soc.* **83**:2799 (1961).
76. G. A. Olah, S. H. Flood, and M. E. Moffatt, *J. Am. Chem. Soc.* **86**:1060 (1964).
77. D. A. McCaulay and A. P. Lien, *J. Am. Chem. Soc.* **74**:6246 (1952).
78. T. Tsuchimoto, K. Tobita, T. Hiyama, and S. Fukuzawa, *Synlett.* **1996**:557; T. Tsuchimoto, K. Tobita, T. Hiyama, and S. Fukuzawa, *J. Org. Chem.* **62**:6997 (1997).
79. H. Kotsuki, T. Oshisi, and M. Inoue, *Synlett.* **1998**:255.

between the acyl halide and the Lewis acid catalyst. For benzene and less reactive aromatics, it is believed that the active electrophile is a protonated acylium ion or an acylium ion complexed by a Lewis acid.[80]

$$RCX + MX_n \rightleftharpoons RC{\equiv}O^+ + [MX_{n+1}]^-$$

$$RC{\equiv}O^+ + \text{[benzene ring with X]} \longrightarrow \text{[arenium intermediate]} \longrightarrow \text{[acylated product]} + H^+$$

or

$$RCX + MX_n \rightleftharpoons RC{-}\overset{+}{X}{-}\bar{M}X_n$$

$$\text{[benzene ring with X]} + RC{-}\overset{+}{X}{-}\bar{M}X_n \longrightarrow \text{[arenium intermediate]} + [MX_{n+1}]^- \longrightarrow \text{[product]} + H^+$$

or

$$\text{[benzene ring]} + RC{=}\overset{+}{O}H \longrightarrow \text{[arenium intermediate]} \longrightarrow \text{[product]} + 2H^+$$

The formation of acyl halide–Lewis acid complexes have been observed by several methods. For example, both 1 : 1 and 1 : 2 complexes of acetyl chloride, with $AlCl_3$ can be observed by NMR spectroscopy.[81] The existence of acylium ions has been demonstrated by X-ray diffraction studies on crystalline salts. For example, crystal structure determinations have been reported for p-methylphenylacylium[82] and acetylium[83] ions as SbF_6^- salts. There is also a good deal of evidence from NMR measurements which demonstrates that acylium ions can exist in nonnucleophilic solvents.[84] The positive charge on acylium ions is delocalized onto the oxygen atom.[85] This delocalization is demonstrated in particular by the short O−C bond lengths in acylium ions, which imply a major contribution from the structure having a triple bond:

$$R\overset{+}{C}{=}O \longleftrightarrow RC{\equiv}O^+$$

Aryl acylium ions have substantial charge delocalization into the aromatic ring.

80. M. Vol'pin, I. Akhrem, and A. Orlinkov, *New J. Chem.* **13**:771 (1989); Y. Sato, M. Yato, T. Ohwada, S. Saito, and K. Shudo, *J. Am. Chem. Soc.* **117**:3037 (1995).
81. B. Glavincevski and S. Brownstein, *J. Org. Chem.* **47**:1005 (1982).
82. B. Chevrier, J.-M. LeCarpentier, and R. Weiss, *J. Am. Chem. Soc.* **94**:5718 (1972).
83. F. P. Boer, *J. Am. Chem. Soc.* **90**:6706 (1968).
84. N. C. Deno, C. U. Pittman, Jr., and M. J. Wisotsky, *J. Am. Chem. Soc.* **86**:4370 (1964); G. A. Olah and M. B. Comisarow, *J. Am. Chem. Soc.* **88**:4442 (1966).
85. T. Xu, D. H. Barich, P. D. Torres, J. B. Nicholas, and J. F. Haw, *J. Am. Chem. Soc.* **119**:396 (1997).

As is the case with Friedel–Crafts alkylations, direct kinetic measurements are difficult, and not many data are available. Rate equations of the form

$$\text{rate} = k_1[\text{RCOCl–AlCl}_3][\text{ArH}] + k_2[\text{RCOCl–AlCl}_3]^2[\text{ArH}]$$

have been reported for reaction of benzene and toluene with both acetyl and benzoyl chloride.[86] The available kinetic data usually do not permit unambiguous conclusions about the identity of the active electrophile. Direct kinetic evidence for acylium ions acting as electrophiles has been obtained using aroyl triflates which can ionize without assistance from a Lewis acid.[87] Either formation of the acylium ion or formation of the σ-complex can be rate-determining, depending on the reactivity of the substrate.

This provides unequivocal evidence that the acylium ion can act as the active electrophile.

Most mechanistic discussions have depended on competitive rate data and on structure–reactivity relationships. Selectivity in Friedel–Crafts acylation, with regard to both substrate and position, is moderate. Some representative data are collected in Table 10.10. It can be seen that the toluene : benzene reactivity ratio is generally between 100 and 200. A progression from low substrate selectivity (entries 5 and 6) to higher substrate selectivity (entries 8 and 9) has been demonstrated for a series of aroyl halides.[88] Electron-attracting groups on the aroyl chloride lead to low selectivity, presumably because of the increased reactivity of such electrophiles. Electron-releasing groups diminish reactivity and increase selectivity. For the more selective electrophiles, the selectivity for *para*

Table 10.10. Substrate and Position Selectivity in Friedel–Crafts Acylation Reactions

Electrophilic reagents	$\dfrac{k_{\text{tol}}}{k_{\text{benz}}}$	Toluene $o:p$ ratio	References
1 Acetyl chloride—AlCl$_3$	134	0.012	a
2 Propionyl chloride—AlCl$_3$	106	0.033	b
3 CH$_3$C≡O$^+$ SbF$_6$$^-$	125	0.014	c
4 Formyl fluoride—BF$_3$	35	0.82	d
5 2,4-Dinitrobenzoyl chloride—AlCl$_3$	29	0.78	d
6 Pentafluorobenzoyl chloride—AlCl$_3$	16	0.61	d
7 Benzoyl chloride—AlCl$_3$	153	0.09	d
8 *p*-Methylbenzoyl chloride—AlCl$_3$	164	0.08	d
9 *p*-Methoxybenzoyl chloride—AlCl$_3$	233	0.2	d

a. G. A. Olah, M. E. Moffatt, S. J. Kuhn, and B. A. Hardie, *J. Am. Chem. Soc.* **86**:2198 (1964).
b. G. A. Olah, J. Lukas, and E. Lukas, *J. Am. Chem. Soc.* **91**:5319 (1969).
c. G. A. Olah, S. J. Kuhn, S. H. Flood, and B. A. Hardie, *J. Am. Chem. Soc.* **86**:2203 (1964).
d. G. A. Olah and S. Kobayashi, *J. Am. Chem. Soc.* **93**:6964 (1971).

86. R. Corriu, M. Dore, and R. Thomassin, *Tetrahedron* **27**:5601, 5819 (1971).
87. F. Effenberger, J. K. Ebehard, and A. H. Maier, *J. Am. Chem. Soc.* **118**:12572 (1996).
88. G. A. Olah and S. Kobayashi, *J. Am. Chem. Soc.* **93**:6964 (1971).

substitution is unusually high. Friedel–Crafts acylation is, in general, a more selective reaction than Friedel–Crafts alkylation. The implication is that acylium ions are less reactive electrophiles than the cationic intermediates involved in the alkylation process.

Steric factors clearly enter into determining the *ortho* : *para* ratio. The hindered 2,4,6-trimethylbenzoyl group is introduced with a 50 : 1 preference for the *para* position.[88] Similarly, in the benzoylation of alkylbenzenes by benzoyl chloride–aluminum chloride, the amount of *ortho* product decreases (10.3%, 6.0%, 3.1%, 0.6%, respectively) as the branching of the alkyl group is increased along the series methyl, ethyl, *i*-propyl, *t*-butyl.[89]

One other feature of the data in Table 10.10 is worthy of further comment. Notice that alkyl substituted acylium ions exhibit a smaller *ortho* : *para* ratio than the various aroyl systems. If steric factors were dominating the position selectivity, one would expect the opposite result. A possible explanation for this feature of the data could be that the aryl compounds are reacting via free acylium ions, whereas the alkyl systems may involve more bulky acyl chloride–catalyst complexes.

Friedel–Crafts acylation sometimes shows a modest kinetic isotope effect.[90] This observation suggests that the proton removal is not much faster than the formation of the σ-complex and that the formation of the σ-complex may be reversible under some conditions.

Although the Lewis acids used as co-reagents in Friedel–Crafts acylations are often referred to as "catalysts," they are, in fact, consumed in the reaction, with the generation of strong acids. There has been considerable interest in finding materials which could function as true catalysts.[91] Considerable success has been achieved using lanthanide triflates.[92]

$$X-\bigcirc + ArCOCl \xrightarrow[5\% \ CF_3SO_3H]{5\% \ Hf(O_3SCF_3)_4} X-\bigcirc-\overset{\overset{O}{\|}}{C}Ar$$

These reactions are presumed to occur through aroyl triflate intermediates which dissociate to aryl acylium ions. Lithium perchlorate and scandium triflate also promote acylation.[93]

$$CH_3O-\bigcirc + (CH_3CO)_2O \xrightarrow[LiClO_4]{Sc(O_3SCF_3)_3} CH_3O-\bigcirc-\overset{\overset{O}{\|}}{C}CH_3$$
90%

A number of variations of the Friedel–Crafts reaction conditions are possible. Acid anhydrides can serve as the acylating agent in place of acid chlorides. Also, the carboxylic acid can be used directly, particularly in combination with strong acids. For example, mixtures of carboxylic acids with polyphosphoric acid, in which a mixed anhydride is presumably formed *in situ*, are reactive acylating agents.[94] Similarly, carboxylic acids dissolved in trifluoromethanesulfonic acid can carry out Friedel–Crafts acylation. The reactive electrophile under these conditions is believed to be the protonated mixed anhydride.[95] In these procedures, the leaving group from the acylating agent is different,

89. G. A. Olah, J. Lukas, and E. Lukas, *J. Am. Chem. Soc.* **91**:5319 (1969).

90. G. A. Olah, S. J. Kuhn, S. H. Flood, and B. A. Hardie, *J. Am. Chem. Soc.* **86**:2203 (1964); D. B. Denney and P. P. Klemchuk, *J. Am. Chem. Soc.* **80**:3285, 6014 (1958).

91. K. Smith, *J. Chem. Technol. Biotechnol.* **68**:432 (1997).

92. I. Hachiya, K. Morikawa, and S. Kobayashi, *Tetrahedron Lett.* **36**:409 (1995); S. Kobayashi and S. Iwamoto, *Tetrahedron Lett.* **39**:4697 (1998).

93. A. Kawada, S. Mitamura, and S. Kobayashi, *J. Chem. Soc., Chem. Commun.* **1996**:183.

94. T. Katuri and K. M. Damodaran, *Can. J. Chem.* **47**:1529 (1969).

but other aspects of the reaction are similar to those under the usual conditions. Synthetic applications of Friedel–Crafts acylation are discussed in Chapter 11 of Part B.

10.4.6. Coupling with Diazonium Compounds

Among the reagents that are classified as weak electrophiles, the best studied are the aromatic diazonium ions, which reagents react only with aromatic substrates having strong electron-donor substituents. The products are azo compounds. The aryl diazonium ions are usually generated by diazotization of aromatic amines. The mechanism of diazonium ion formation is discussed more completely in Section 11.2.1 of Part B.

Aryl diazonium ions are stable in solution only near room temperature or below, and this also limits the range of compounds that can be successfully substituted by diazonium ions.

Kinetic investigations have revealed second-order kinetic behavior for substitution by diazonium ions in a number of instances. In the case of phenols, it is the conjugate base that undergoes substitution.[96] This finding is entirely reasonable, since the deprotonated oxy group is a better electron donor than the neutral hydroxy substituent. The reactivity of the diazonium ion depends on the substituent groups which are present. Reactivity is increased by electron-attracting groups and decreased by electron donors.[97]

A unique feature of the mechanism for diazonium coupling is that proton loss can be clearly demonstrated to be the rate-determining step in some cases. This feature is revealed in two ways. First, diazonium couplings of several naphthalenesulfonate ions exhibit primary isotope effects in the range 4–6 when deuterium is present at the site of substitution, clearly indicating that cleavage of the C−H bond is rate-determining. Second, these reactions can also be shown to be general-base-catalyzed. This, too, implies that proton removal is rate-determining.[98]

Because of the limited range of aromatic compounds that react with diazonium ions, selectivity data comparable to those discussed for other electrophilic substitutions are not available. Because diazotization involves a weak electrophile, it would be expected to reveal high substrate and position selectivity.

95. R. M. G. Roberts and A. R. Sardi, *Tetrahedron* **39**:137 (1983).
96. R. Wistar and P. D. Bartlett, *J. Am. Chem. Soc.* **63**:413 (1941).
97. A. F. Hegarty, in *The Chemistry of the Diazonium and Diazo Groups*, S. Patai, ed., John Wiley & Sons, 1978, Chapter 12; H. Mayr, M. Hartnagel, and K. Grimm, *Liebigs Ann.* **1997**:55.
98. H. Zollinger, *Azo and Diazo Chemistry*, translated by H. E. Nursten, Interscience, New York, 1961, Chapter 10; H. Zollinger, *Adv. Phys. Org. Chem.* **2**:163 (1964); H. Zollinger, *Helv. Chim. Acta* **38**:1597 (1955).

10.4.7. Substitution of Groups Other Than Hydrogen

The general mechanism for electrophilic substitution suggests that groups other than hydrogen could be displaced, provided the electrophile attacked at the substituted carbon. Substitution at a site already having a substituent is called *ipso* substitution and has been observed in a number of circumstances. The ease of removal of a substituent depends on its ability to accommodate a positive charge. This factor determines whether the newly attached electrophile or the substituent is eliminated from the σ-complex on rearomatization:

One example of substituent replacement involves cleavage of a highly branched alkyl group. The alkyl group is expelled as a carbocation, and for this reason, substitution is most common for branched alkyl groups. The nitration of 1,4-bis(*i*-propyl)benzene provides an example:

Ref. 99

Cleavage of *t*-butyl groups has also been observed in halogenation reactions. Minor amounts of dealkylated products are formed during chlorination and bromination of *t*-butylbenzene.[100] The amount of dealkylation increases greatly in the case of 1,3,5-tri-*t*-butylbenzene, and the principal product of bromination is 3,5-dibromo-*t*-butylbenzene.[101]

The replacement of bromine and iodine during aromatic nitration has also been observed. *p*-Bromoanisole and *p*-iodoanisole, for example, both give 30–40% of *p*-nitroanisole, a product resulting from displacement of halogen on nitration.

Ref. 102

Because of the greater resistance to elimination of chlorine as a positively charged species, *p*-chloroanisole does not undergo dechlorination under similar conditions.

99. G. A. Olah and S. J. Kuhn, *J. Am. Chem. Soc.* **86**:1067 (1964).
100. P. R. D. de la Mare and J. T. Harvey, *J. Chem. Soc.* **1957**:131; P. B. D. de la Mare, J. T. Harvey, M. Hassan, and S. Varma, *J. Chem. Soc.* **1958**:2756.
101. P. D. Bartlett, M. Roha, and R. M. Stiles, *J. Am. Chem. Soc.* **76**:2349 (1954).
102. C. L. Perrin and G. A. Skinner, *J. Am. Chem. Soc.* **93**:3389 (1971).

589

SECTION 10.5.
NUCLEOPHILIC
AROMATIC
SUBSTITUTION BY THE
ADDITION–
ELIMINATION
MECHANISM

The most useful group of aromatic substitutions involving replacement of a substituent group in preference to a hydrogen are electrophilic substitutions of arylsilanes.

$$Ar-SiR_3 + E^+ + X^- \rightarrow Ar-E + R_3SiX$$

The silyl group directs electrophiles to the substituted position. That is, it is an *ipso*-directing group. Because of the polarity of the carbon–silicon bond, the substituted position is relatively electron-rich. The ability of silicon substituents to stabilize carbocation character at β-carbon atoms (see Section 6.10, p. 393) also promotes *ipso* substitution. The silicon substituent is easily removed from the σ-complex by reaction with a nucleophile. The desilylation step probably occurs through a pentavalent silicon species:

The reaction exhibits other characteristics typical of an electrophilic aromatic substitution.[103] Examples of electrophiles that can effect substitution for silicon include protons and the halogens, as well as acyl, nitro, and sulfonyl groups.[104] The fact that these reactions occur very rapidly has made them attractive for situations where substitution must be done under very mild conditions.[105]

Trialkyltin substituents are also powerful *ipso*-directing groups. The overall electronic effects are similar to those in silanes, but the tin substituent is a better electron donor. The electron density at carbon is increased, as is the stabilization of β-carbocation character. Acidic cleavage of arylstannanes is formulated as an electrophilic aromatic substitution proceeding through an *ipso*-oriented σ-complex.[106]

10.5. Nucleophilic Aromatic Substitution by the Addition–Elimination Mechanism

Neither the S_N1 nor S_N2 mechanism is accessible for nucleophilic substitution on aromatic rings. A back-side S_N2-type reaction is precluded by the geometry of the benzene ring. The back lobe of the sp^2 orbital is directed toward the center of the ring. Any inversion mechanism is precluded by the geometry of the ring. An S_N1 mechanism is very costly in terms of energy because a cation directly on a benzene ring is very unstable. From the data in Table 5.2, (p. 279), it is clear that a phenyl cation is less stable than even a

103. F. B. Deans and C. Eaborn, *J. Chem. Soc.* **1959**:2299.
104. F. B. Deans, C. Eaborn, and D. F. Webster, *J. Chem. Soc.* **1959**:3031; C. Eaborn, Z. Lasocki, and D. E. Webster, *J. Chem. Soc.* **1959**:3034; C. Eaborn, *J. Organomet. Chem.* **100**:43 (1975); J. D. Austin, C. Eaborn, and J. D. Smith, *J. Chem. Soc.* **1963**:4744; F. B. Deans and C. Eaborn, *J. Chem. Soc.* **1957**:498; R. W. Bott, C. Eaborn, and T. Hashimoto, *J. Chem. Soc.* **1963**:3906.
105. S. R. Wilson and L. A. Jacob, *J. Org. Chem.* **51**:4833 (1986).
106. C. Eaborn, I. D. Jenkins, and D. R. M. Walton, *J. Chem. Soc., Perkin Trans.* **1974**:596.

primary carbocation. This is again a consequence of the geometry and hybridization of the aromatic carbon atoms. An aryl carbocation is localized in an sp^2 orbital. This orbital is orthogonal to the π system so there is no stabilization available from the π electrons.

back-side approach of
nucleophile with
inversion is possible

phenyl cation is
highly unstable

There are several mechanisms by which net nucleophilic aromatic substitution can occur. In this section we will discuss the addition–elimination mechanism and the elimination–addition mechanism. Substitutions via organometallic intermediates and via aryl diazonium ions will be considered in Chapter 11 of Part B.

The addition–elimination mechanism[107] uses one of the vacant π^* orbitals for bonding interaction with the nucleophile. This permits addition of the nucleophile to the aromatic ring without displacement of any of the existing substituents. If attack occurs at a position occupied by a potential leaving group, net substitution can occur by a second step in which the leaving group is expelled.

The π-electron system of the addition intermediate is isoelectronic with that of a pentadienyl anion.

$$\psi_5 \quad\rule{1cm}{0.4pt}\quad \alpha - 1.7\beta$$
$$\psi_4 \quad\rule{1cm}{0.4pt}\quad \alpha - \beta$$
$$\psi_3 \quad\text{—}\!\!\uparrow\!\downarrow\!\!\text{—}\quad \alpha$$
$$\psi_2 \quad\text{—}\!\!\uparrow\!\downarrow\!\!\text{—}\quad \alpha + \beta$$
$$\psi_1 \quad\text{—}\!\!\uparrow\!\downarrow\!\!\text{—}\quad \alpha + 1.7\beta$$

The HOMO of the pentadienyl anion is ψ_3, which has its electron density primarily at the carbons *ortho* and *para* to the position of substitution. The intermediate is therefore strongly stabilized by an electron-accepting group *ortho* or *para* to the site of substitution. Such substituents activate the ring to nucleophilic substitution. The most powerful effect is exerted by a nitro group, but cyano and carbonyl groups are also favorable. Generally speaking, nucleophilic aromatic substitution is an energetically demanding reaction, even when electron-attracting substituents are present. The addition disrupts the aromatic π system. Without electron-attracting groups present, nucleophilic aromatic substitution occurs only under extreme reaction conditions.

The role of the leaving group in determining the reaction rate is somewhat different from its role in S_N2 and S_N1 substitution at alkyl groups. In those cases, the bond strength is usually the dominant factor so that the order of reactivity of the halogens is

107. For reviews, see, C. F. Bernasconi, in *MTP International Review of Science, Organic Series One*, Vol. 3, H. Zollinger, ed., Butterworths, London 1973; J. A. Zoltewicz, *Top. Curr. Chem.* **59**:33 (1975); J. Miller, *Aromatic Nucleophilic Substitution*, Elsevier, Amsterdam, 1968.

591

SECTION 10.5.
NUCLEOPHILIC
AROMATIC
SUBSTITUTION BY THE
ADDITION–
ELIMINATION
MECHANISM

$I > Br > Cl > F$. In nucleophilic aromatic substitution, the formation of the addition intermediate is usually the rate-determining step so the ease of C–X bond breaking does not affect the rate. When this is the case, the order of reactivity is often $F > Cl > Br > I$.[108] This order is the result of the polar effect of the halogen. The stronger bond dipoles associated with the more electronegative halogens favor the addition step and thus increase the overall rate of reaction.

Groups other than halogen can serve as leaving groups. Alkoxy groups are very poor leaving groups in S_N2 reactions but can act as leaving groups in aromatic substitution. The reason is the same as for the inverted order of reactivity for the halogens. The rate-determining step is the addition, and the alkoxide can be eliminated in the energetically favorable rearomatization. Nitro[109] and sulfonyl[110] groups can also be displaced.

Ref. 111

Ref. 112

The addition intermediates can frequently be detected spectroscopically and sometimes can be isolated.[113] They are called *Meisenheimer complexes*. Especially in the case of adducts stabilized by nitro groups, the intermediates are often strongly colored.

The range of nucleophiles which have been observed to participate in nucleophilic aromatic substitution is similar to that for S_N2 reactions and includes alkoxides,[114] phenoxides,[115] sulfides,[116] fluoride ion,[117] and amines.[118] Substitutions by carbanions are somewhat less common. This may be because there are frequently complications resulting from electron-transfer processes with nitroaromatics. Solvent effects on nucleophilic aromatic substitutions are similar to those discussed for S_N2 reactions. Dipolar

108. G. P. Briner, J. Miller, M. Liveris, and P. G. Lutz, *J. Chem. Soc.* **1954**:1265; G. Bartoli and P. E. Todesco, *Acc. Chem. Res.* **10**:125 (1977).
109. J. R. Beck, *Tetrahedron* **34**:2057 (1978).
110. A. Chisari, E. Maccarone, G. Parisi, and G. Perrini, *J. Chem. Soc., Perkin Trans. II* **1982**:957.
111. J. F. Bunnett, E. W. Garbisch, Jr., and K. M. Pruitt, *J. Am. Chem. Soc.* **79**:385 (1957).
112. J. R. Beck, R. L. Sobczak, R. G. Suhr, and J. A. Vahner, *J. Org. Chem.* **39**:1839 (1974).
113. E. Buncel, A. R. Norris, and K. E. Russel, *Q. Rev. Chem. Soc.* **22**:123 (1968); M. J. Strauss, *Chem. Rev.* **70**:667 (1970); C. F. Bernasconi, *Acc. Chem. Res.* **11**:147 (1978).
114. J. P. Idoux, M. L. Madenwald, B. S. Garcia, D. L. Chu, and J. T. Gupton, *J. Org. Chem.* **509**:1876 (1985).
115. R. O. Brewster and T. Groening, *Org. Synth.* **II**:445 (1943).
116. M. T. Bogert and A. Shull, *Org. Synth.* **I**:220 (1941); N. Kharasch and R. B. Langford, *Org. Synth.* **V**:474 (1973); W. P. Reeves, T. C. Bothwell, J. A. Rudis, and J. V. McClusky, *Synth. Commun.* **12**:1071 (1982).
117. W. M. S. Berridge, C. Crouzel, and D. Comar, *J. Labelled Compd. Radiopharm.* **22**:687 (1985).
118. H. Bader, A. R. Hansen, and F. J. McCarty, *J. Org. Chem.* **31**:2319 (1966); F. Pietra and F. Del Cima, *J. Org. Chem.* **33**:1411 (1968); J. F. Pilichowski and J. C. Gramain, *Synth. Commun.* **14**:1247 (1984).

aprotic solvents,[119] crown ethers,[120] and phase-transfer catalysts[121] can all enhance the rate of substitution by providing the nucleophile in a reactive state with weak solvation.

One of the historically most significant examples of aromatic nucleophilic substitution is the reaction of amines with 2,4-dinitrofluorobenzene. This reaction was used by Sanger[122] to develop a method for identification of the N-terminal amino acid in proteins and the process opened the way for structural characterization of proteins and other biopolymers.

The pyridine family of heteroaromatic nitrogen compounds is reactive toward nucleophilic substitution at the C-2 and C-4 positions. The nitrogen atom serves to activate the ring toward nucleophilic attack by stabilizing the addition intermediate. This kind of substitution reaction is especially important in the chemistry of pyrimidines.

Ref. 123

Ref. 124

119. F. Del Cima, G. Biggi, and F. Pietra. *J. Chem. Soc., Perkin Trans. II* **1973**:55; M. Makosza, M. Jagusztyn-Grochowska, M. Ludwikow, and M. Jawdosiuk, *Tetrahedron* **30**:3723 (1974); M. Prato, U. Quintily, S. Salvagno, and G. Scorrano, *Gazz. Chim. Ital.* **114**:413 (1984).

120. J. S. Bradshaw, E. Y. Chen, R. H. Holes, and J. A. South, *J. Org. Chem.* **37**:2051 (1972); R. A. Abramovitch and A. Newman, *J. Org. Chem.* **39**:2690 (1974).

121. M. Makosza, M. Jagusztyn-Grochowska, M. Ludwikow, and M. Jawdosiuk, *Tetrahedron* **30**:3723 (1974).

122. F. Sanger, *Biochem. J.* **45**:563 (1949).

123. J. A. Montgomery and K. Hewson, *J. Med. Chem.* **9**:354 (1966).

124. D. J. Brown, B. T. England, and J. M. Lyall, *J. Chem. Soc. C* **1966**:226.

593

SECTION 10.6.
NUCLEOPHILIC
AROMATIC
SUBSTITUTION BY THE
ELIMINATION–
ADDITION
MECHANISM

A variation of the aromatic nucleophilic substitution process in which the leaving group is part of the entering nucleophile has been developed and is called *vicarious nucleophilic aromatic substitution*.

The combinations $Z = CN$, RSO_2, CO_2R, or SR and $X = F$, Cl, Br, I, ArO, ArS, or $(CH_3)_2NCS_2$ are among those that have been demonstrated.[125]

10.6. Nucleophilic Aromatic Substitution by the Elimination–Addition Mechanism

The elimination–addition mechanism involves a highly unstable intermediate, which is referred to as *dehydrobenzene* or *benzyne*.[126]

A characteristic feature of this mechanism is the substitution pattern in the product. The entering nucleophile need not always enter at the carbon to which the leaving group was bound.

Benzyne can be observed spectroscopically in an inert matrix at very low temperatures.[127] For these studies, the molecule is generated photolytically.

125. M. Makosza and J. Winiarski, *J. Org. Chem.* **45**:1574 (1980); M. Makosza, J. Golinski, and J. Baron, *J. Org. Chem.* **49**:1488 (1984); M. Makosza and J. Winiarski, *J. Org. Chem.* **49**:1494 (1984); M. Makosza, and J. Winiarski, *J. Org. Chem.* **49**:5272 (1984).
126. R. W. Hoffmann, *Dehydrobenzene and Cycloalkynes*, Academic Press, New York, 1967.
127. O. L. Chapman, K. Mattes, C. L. McIntosh, J. Pacansky, G. V. Calder, and G. Orr, *J. Am. Chem. Soc.* **95**:6134 (1973); J. W. Laing and R. S. Berry, *J. Am. Chem. Soc.* **98**:660 (1976); J. G. Radziszewski, B. A. Hess, Jr., and R. Zahradnik, *J. Am. Chem. Soc.* **114**:52 (1992).

There have been several representations of the bonding in benzyne. The one most generally used pictures benzyne as being similar to benzene but with an additional weak bond in the plane of the ring, formed by overlap of the two sp^2 orbitals.[128] Comparison of the NMR characteristics[129] with MO calculations indicate that the conjugation is maintained and that benzyne is a strained but aromatic molecule.[130]

An early case in which the existence of benzyne as a reaction intermediate was established was in the reaction of chlorobenzene with potassium amide. Carbon-14 label in the starting material was found to be distributed in the aniline as expected for a benzyne intermediate.[131]

The elimination–addition mechanism is facilitated by electronic effects that favor removal of a hydrogen from the ring by strong base. Relative reactivity also depends on the halide. The order Br > I > Cl > F has been established in the reaction of aryl halides with KNH_2 in liquid ammonia.[132] This order has been interpreted as representing a balance between two effects. The polar order favoring proton removal would be F > Cl > Br > I, but this is largely overwhelmed by the order of leaving-group ability I > Br > Cl > F, which reflects bond strengths. With organometallic compounds as bases in aprotic solvents, the acidity of the *ortho* hydrogen is the dominant factor, and the reactivity order is F > Cl > Br > I because of the bond polarity effect.[133]

Addition of nucleophiles such as ammonia or alcohols or their conjugate bases to benzynes takes place very rapidly. These nucleophilic additions are believed to involve capture of the nucleophile, followed by protonation to give the substituted benzene.[134]

128. H. E. Simmons, *J. Am. Chem. Soc.* **83**:1657 (1961).
129. R. Warmuth, *Angew. Chem. Int. Ed. Engl.* **36**:1347 (1997).
130. H. Jiao, P. v. R. Schleyer, B. R. Beno, K. N. Houk, and R. Warmuth, *Angew. Chem. Int. Ed. Engl.* **36**:2761 (1997).
131. J. D. Roberts, D. A. Semenow, H. E. Simmons, Jr., and L. A. Carlsmith *J. Am. Chem. Soc.* **78**:601 (1956).
132. F. W. Bergstrom, R. E. Wright, C. Chandler, and W. A. Gilkey, *J. Org. Chem.* **1**:170 (1936).
133. R. Huisgen and J. Sauer, *Angew. Chem.* **72**:91 (1960).
134. J. F. Bunnett, D. A. R. Happer, M. Patsch, C. Pyun, and H. Takayama, *J. Am. Chem. Soc.* **88**:5250 (1966); J. F. Bunnett and J. K. Kim, *J. Am. Chem. Soc.* **95**:2254 (1973).

The regiochemistry of the nucleophilic addition is influenced by ring substituents. Electron-attracting groups tend to favor addition of the nucleophile at the more distant end of the "triple bond," because this permits maximum stabilization of the developing negative charge. Electron-donating groups have the opposite effect. These directive effects probably arise through interaction of the substituent with the electron pair which is localized on the *ortho* carbon by the addition step.

595

SECTION 10.6.
NUCLEOPHILIC
AROMATIC
SUBSTITUTION BY THE
ELIMINATION–
ADDITION
MECHANISM

EWG, electron-withdrawing group
ERG, electron-releasing group

Selectivity is usually not high, however, and formation of both possible products from monosubstituted benzynes is common.[135]

There are several methods for generation of benzyne in addition to base-catalyzed elimination of hydrogen halide from a halobenzene, and some of these are more generally applicable for preparative work. Probably the most convenient method is diazotization of *o*-aminobenzoic acid.[136] Concerted loss of nitrogen and carbon dioxide follows diazotization and generates benzyne. Benzyne can be formed in this manner in the presence of a variety of compounds with which it reacts rapidly.

Oxidation of 1-aminobenzotriazole also yields benzyne under mild conditions. An oxidized intermediate decomposes with loss of two molecules of nitrogen.[137]

Another heterocyclic molecule that can serve as a benzyne precursor is benzothiadiazole-1,1-dioxide, which decomposes with elimination of sulfur dioxide and nitrogen.[138]

135. E. R. Biehl, E. Nieh, and K. C. Hsu, *J. Org. Chem.* **34**:3595 (1969).
136. M. Stiles, R. G. Miller, and U. Burckhardt, *J. Am. Chem. Soc.* **85**:1792 (1963); F. M. Logullo, A. H. Seitz, and L. Friedman, *Org. Synth.* **V.**54. (1973); P. C. Buxton, M. Fensome, F. Heaney, and K. G. Mason, *Tetrahedron* **51**:2959 (1995).
137. C. D. Campbell and C. W. Rees, *J. Chem. Soc. C* **1969**:742, 745.
138. G. Wittig and R. W. Hoffmann, *Org. Synth.* **47**:4 (1967); G. Wittig and R. W. Hoffmann, *Chem. Ber.* **95**:2718, 2729 (1962).

Benzyne can also be generated from *o*-dihaloaromatics. Reaction of lithium amalgam or magnesium results in formation of a transient organometallic compound that decomposes with elimination of lithium halide. 1-Bromo-2-fluorobenzene is the usual starting material in this procedure.[139]

Benzyne is capable of dimerizing, so that in the absence of either a nucleophile or a reactive unsaturated compound, biphenylene is formed.[140] The lifetime of benzyne is estimated to be on the order of a few seconds in solution near room temperature.[141]

When benzyne is generated in the presence of potential dienes, additions at the highly strained "triple bond" occur. Among the types of compounds that give Diels–Alder addition products are furans, cyclopentadienones, and anthracene.

Ref. 142

Ref. 143

Ref. 144

139. G. Wittig and L. Polmer, *Chem. Ber.* **89**:1334 (1956); G. Wittig, *Org. Synth.* **IV**:964 (1963).
140. F. M. Logullo, A. H. Seitz, and L. Friedman, *Org. Synth.* **48**:12 (1968).
141. F. Gavina, S. V. Luis, and A. M. Costero, *Tetrahedron* **42**:155 (1986).
142. G. Wittig and L. Pohmer, *Angew. Chem.* **67**:348 (1955).
143. L. F. Fieser and M. J. Haddadin, *Org. Synth.* **46**:107 (1966).
144. L. Friedman and F. M. Logullo, *J. Org. Chem.* **34**:3089 (1969).

L. F. Albright, R. V. C. Carr, and R. J. Schmitt, *Nitration: Recent Laboratory and Industrial Developments*, American Chemical Society, Washington, D.C., 1996.

R. W. Hoffmann, *Dehydrobenzene and Cycloalkynes*, Academic Press, New York, 1967.

J. G. Hoggett, R. B. Moodie, J. R. Penton, and K. S. Schofield, *Nitration and Aromatic Reactivity*, Cambridge University Press, Cambridge, U.K., 1971.

C. K. Ingold, *Structure and Mechanism in Organic Chemistry*, Cornell University Press, Ithaca, New York, 1969, Chapter VI.

R. O. C. Norman and R. Taylor, *Electrophilic Substitution in Benzenoid Compounds*, Elsevier, Amsterdam, 1965.

G. A. Olah, *Friedel–Crafts Chemistry*, John Wiley & Sons, New York, 1973.

S. Patai, ed., *The Chemistry of Diazonium and Diazo Groups*, John Wiley & Sons, New York, 1978.

R. M. Roberts and A. A. Khalaf, *Friedel–Crafts Alkylation Chemistry*, Marcel Dekker, New York, 1984.

R. Taylor, *Electrophilic Aromatic Substitution*, John Wiley & Sons, Chichester, U.K., 1990.

F. Terrier, *Nucleophilic Aromatic Substitution*, VCH Publishers, New York, 1991.

Problems

(References for these problems will be found on page 800.)

1. Predict qualitatively the isomer ratio for the nitration of each of the following compounds.

(a) CH_2F

(c) CH_2OCH_3

(e)   F

(b) CF_3

(d) $^+N(CH_3)_3$

(f) $O_2SCH_2CH_3$

2. While *N,N*-dimethylaniline is an extremely reactive aromatic substrate and is readily attacked by such weak electrophiles as aryl diazonium ions and nitrosonium ion, this reactivity is greatly diminished by introduction of an alkyl substituent in the *ortho* position. Explain.

3. Toluene is 17 times more reactive than benzene and isopropylbenzene is 14 times more reactive than benzene when nitration is carried out in the organic solvent sulfolane. The $o:m:p$ ratio for toluene is $62:3:35$, and for isopropylbenzene it is $43:5:52$. Calculate the partial rate factors for each position in toluene and isopropylbenzene. Discuss the significance of the partial rate factors. Compare the reactivity at the various positions of each molecule, and explain any differences you consider to be significant.

4. Some bromination rate constants are summarized below. Compare the correlation of the rate data with σ and σ^+ substituent constants. What is the value of ρ? What is the mechanistic significance of these results?

X	k $(M^{-1} s^{-1})$
H	2.7×10^{-6}
CH_3	1.5×10^{-2}
OCH_3	9.8×10^3
OH	4.0×10^4
$N(CH_3)_2$	2.2×10^8

5. Compare the results given below for the alkylation of p-xylene under a variety of conditions. Explain the reasons for the variation in product composition with temperature and with the use of n- versus i-propyl chloride.

		A	B	C
n-propyl chloride	0°C	27%	73%	0%
n-propyl chloride	50°C	31%	53%	16%
i-propyl chloride	0°C	100%	0%	0%
i-propyl chloride	50°C	62%	0%	38%

6. The table below gives first-order rate constants for reaction of substituted benzenes with m-nitrobenzenesulfonyl peroxide. From these data, calculate the overall relative reactivity and partial rate factors. Does this reaction fit the pattern of an electrophilic aromatic substitution? If so, does the active electrophile exhibit low, moderate, or high substrate and position selectivity?

		Product composition		
X	k (s^{-1})	o	m	p
H	8.6×10^{-5}	–	–	–
Br	4.8×10^{-5}	21	3	76
CH_3	1.7×10^{-3}	32	3	65
CH_3O	4.3×10^{-2}	14	0	86
CH_3O_2C	9.1×10^{-6}	24	67	9

7. Give the products to be expected from each of the following reactions.

(a)

(b)

(c) $(CH_3)_3Si$

(d)

8. In 100% sulfuric acid, the cyclization shown below occurs.

When one of the *ortho* hydrogens is replaced by deuterium, the rate drops from $1.53 \times 10^{-4}\,s^{-1}$ to $1.38 \times 10^{-4}\,s^{-1}$. What is the kinetic isotope effect? The product from such a reaction contains 60% of the original deuterium. Give a mechanism for this reaction that is consistent with both the kinetic isotope effect and the deuterium retention data.

9. Reaction of 3,5,5-trimethyl-2-cyclohexen-1-one with $NaNH_2$ (3 equiv) in THF generates its enolate. When bromobenzene is then added to this solution and stirred for 4 h, the product **A** is isolated in 30% yield. Formulate a mechanism for this transformation.

10. Various phenols can be selectively hydroxymethylated at the *ortho* position by heating with paraformaldehyde and phenylboronic acid.

An intermediate **A**, having the formula $C_{14}H_{13}O_2B$ for the case above, can be isolated after the first step. Postulate a structure for the intermediate and comment on its role in the reaction.

11. When compound **B** is dissolved in FSO_3H at $-78°C$, the NMR spectrum shows that a carbocation is formed. If the solution is then allowed to warm to $-100°C$, a different ion forms. The first ion gives compound **C** when quenched with base, while the second gives **D**. What are the structures of the two carbocations, and why do they give different products on quenching?

12. Alkyl groups which are *para* to strong π-donor substituents such as hydroxy or methoxy can be removed from aromatic rings under acidic conditions if the alkyl group is capable of forming a stable carbocation:

For the equation above, when $R = CH_3$, the solvent isotope effect is $k_H/k_D = 0.1$. When $R = Ph$, $k_H/k_D = 4.3$. How do you account for the difference in the isotope effect for the two systems, and, particularly, what is the probable cause of the inverse isotope effect in the case of $R = CH_3$?

13. Acylation of 1,4-dimethoxynaphthalene with acetic anhydride (1.2 equiv) and aluminum chloride (2.2 equiv) in ethylene dichloride (60°C, 3 h) gives two products, 6-acetyl-1,4-dimethoxynaphthalene (30%) and 1-hydroxy-2-acetyl-4-methoxynaphthalene (50%). Suggest a rationalization for the formation of these two products and, in particular, for the differing site of substitution in the two products.

14. The solvolysis of 4-arylbutyl arenesulfonates in nonnucleophilic media leads to the formation of tetralins:

Two σ-intermediates are conceivable. **A** would lead directly to product on deprotonation, while **B** could give product by rearrangement to **A**, followed by deprotonation.

Devise an experiment that would permit one to determine how much product is formed via **A** and how much via **B**. How would you expect the relative importance of the alternative routes to be related to the identity of the substituent group X?

15. The complex kinetic expression for chlorination of anisole by hypochlorous acid (p. 577) becomes simpler for both less reactive and more reactive substrates. For benzene, the expression is

$$\text{rate} = k[\text{benzene}][\text{HOCl}][\text{H}^+]$$

For *p*-dimethoxybenzene, it is

$$\text{rate} = k[\text{HOCl}][\text{H}^+]$$

What is the reason for this dependence of the form of the rate expression on the reactivity of the aromatic compound?

16. When acetyl nitrate is the nitration reagent, the reactivities of chlorobenzene and bromobenzene relative to that of benzene are 0.033 and 0.030, respectively. The product ratios are: chlorobenzene, $o:m:p$, 30%, 1%, 69%; bromobenzene, $o:m:p$ 37%, 1%, 62%. Calculate the partial rate factors.

17. The chlorination of a series of compounds having electron-withdrawing substituents has been studied. The relative rates of chlorination and the isomer distributions are known. The data give a satisfactory correlation with the Hammett equation using σ^+, but no rate measurement for benzene under precisely comparable conditions is possible. How could you estimate f_o, f_m, and f_p for chlorination from the available data?

$$\rho = -6.6$$

$$o:m:p \text{ ratio for benzonitrile} = 34:55:11$$

18. *Ipso* substitution, in which the electrophile attacks a position already carrying a substituent, is relatively rare in electrophilic aromatic substitution and was not explicitly covered in Section 10.2 in the discussion of substituent effects on reactivity and selectivity. Using qualitative MO concepts, discuss the effect of the following types of substituents on the energy of the transition state for *ipso* substitution.

(a) A π-donor substituent which is more electronegative than carbon, e.g., F or CH_3O.

(b) A π-acceptor substituent which is more electronegative than carbon, e.g., NO_2 or CN.

(c) A group without a strong π-conjugating capacity which is more electronegative than carbon, e.g., $^+N(CH_3)_3$.

(d) A group without a strong π-conjugating capacity which is less electronegative than carbon, e.g., $(CH_3)_3Si$.

According to this analysis, which types of groups will most favor *ipso* substitution? Can you cite any experimental evidence to support this conclusion?

19. The nitration of 2,4,6-tri-*t*-butyltoluene gives rise to three products. The distribution is changed when the 3- and 5-positions are deuterated:

*H = H	40.3%
*H = D	42.4%

*H = H	51.0%	*H = H	8.7%
*H = D	54.6%	*H = D	2.7%

Indicate mechanisms that would account for the formation of each product. Show how the isotopic substitution could cause a change in product composition. Does your mechanism predict that the isotopic substitution would give rise to a primary or secondary deuterium kinetic isotope effect? Calculate the magnitude of the kinetic isotope effect from the data given.

20. (a) Under several reaction conditions designed to determine the products of cyclization under Friedel-Crafts conditions, six-membered cyclic products were found to be favored over seven-membered ring products. Write a detailed mechanism for each of the reactions shown below, and comment on the significance of the apparently general preference for formation of a six-membered

ring over a seven-membered ring.

(b) Examine the data below for cyclization of a variety of phenylalkanols in 85% H_3PO_4 at elevated temperatures. What general conclusions do you draw about the preferences for ring closure (as a function of ring size) under these conditions?

(1) $PhCH_2CH_2CH_2OH$ $\xrightarrow[>200°C]{H_3PO_4}$ mainly isomeric phenylpropenes (89%)

(2) $\underset{\underset{CH_2}{|}}{\overset{\overset{CH_2}{|}}{PhCCH_2CH_2OH}}$ $\xrightarrow[>200°C]{H_3PO_4}$ mainly 2-methyl-3-phenyl-2-butene (82%) plus some 1,1-dimethylindane (18%)

(3) $PhCH_2CH_2CH_2CH_2OH$ $\xrightarrow[>200°C]{H_3PO_4}$ mainly tetralin (80%)

(4) $\underset{\underset{OH}{|}}{PhCH_2CH_2CHCH_3}$ $\xrightarrow[>200°C]{H_3PO_4}$ phenylbutenes (100%)

(5) $\underset{\underset{OH}{|}}{m\text{-}CH_3PhCH_2CH_2C(CH_3)_2}$ $\xrightarrow[>200°C]{H_3PO_4}$ mainly 1,1,5- and 1,1,7-trimethylindane

21. Explain each of the following reaction processes by presenting a detailed stepwise mechanism to show how the observed products are formed.

(a) The reaction of 2,6-di-*t*-butylphenoxide with *o*-nitroaryl halides gives 2,6-di-*t*-butyl-4-(*o*-nitrophenyl)phenols in 60%–90% yield. 1,4-Dinitrobenzene reacts under similar conditions to give 2,6-di-*t*-butyl-4-(*p*-nitrophenyl)phenol.

(b) 2-(3-Chlorophenyl)-4,4-dimethyloxazoline reacts with alkyllithium reagents to give 2-(2-alkylphenyl)-4,4-dimethyloxazolines.

(c) Nitrobenzene reacts with cyanomethyl phenyl sulfide in the presence of sodium hydroxide in dimethyl sulfoxide to give a mixture of 2- and 4-nitrophenylaceto-nitrile.

(d)

(e) Reaction of benzene with 3,3,3-trifluoropropene in the presence of aluminum chloride and a trade of moisture gives 3,3,3-trifluoropropylbenzene.

22. Reaction of several 3-bromobenzoic acids with excess LDA, followed by addition of a benzyl cyanide, gave the product mixtures shown. Suggest a mechanism for the formation of each of these products.

X	Ar	A	B	C
4-MeO	Ph	56	9	11
4-MeO	4-MePh	70	8	12
4-MeO	2-MePh	44	5	10
4-Me	Ph	53	<2	7
4-Me	4-MePh	43	<2	8

Concerted Pericyclic Reactions

Introduction

There are many reactions in organic chemistry that give no evidence of involving intermediates when subjected to the usual probes for studying reaction mechanisms. Highly polar transition states do not seem to be involved either, since the rates of the reactions are often insensitive to solvent polarity. Efforts to detect free-radical intermediates in these reactions by spectroscopic or chemical means have not been successful, and the reaction rates are neither increased by initiators nor decreased by inhibitors of free-radical reactions. This absence of evidence of intermediates leads to the conclusion that the reactions are single-step processes in which bond making and bond breaking both contribute to the structure at the transition state, although not necessarily to the same degree. Such processes are called *concerted reactions*. There are numerous examples of both unimolecular and bimolecular concerted reactions.

An important group of concerted reactions are the *concerted pericyclic reactions*.[1] A pericyclic reaction is characterized as a change in bonding relationship that takes place as a continuous concerted reorganization of electrons. The word "concerted" specifies that there is a single transition state and that there are no intermediates in the process. To maintain continuous electron flow, pericyclic reactions must occur through *cyclic transition states*. Furthermore, the cyclic transition state must correspond to an arrangement of the participating orbitals that can maintain a bonding interaction between the reacting atoms throughout the course of the reaction. We shall see shortly that these requirements make pericyclic reaction highly predictable in terms of such features as relative reactivity, stereospecificity, and regioselectivity.

The key to understanding the mechanism of the concerted pericyclic reactions was the recognition by Woodward and Hoffmann[2] that the pathways of such reactions were determined by the symmetry properties of the orbitals that were directly involved. Their recognition that the symmetry of each participating orbital must be conserved during the

1. R. B. Woodward and R. Hoffmann, *The Conservation of Orbital Symmetry*, Academic Press, New York, 1970.
2. R. B. Woodward and R. Hoffmann, *J. Am. Chem. Soc.* **87**:395 (1965).

reaction process dramatically transformed the understanding of this family of reactions and stimulated much experimental work to test and extend their theories.[3] Woodward and Hoffmann's approach led to other related interpretations of orbital properties which are also successful in predicting and explaining the course of concerted pericyclic reactions.[4]

There has been a great deal of effort at modeling the *transition states* of concerted pericyclic reactions.[5] Each of the major theoretical approaches—semiempirical MO, *ab initio* MO, and density functional theory—have been applied to the problem, and some comparisons made.[6] These approaches generally parallel the orbital symmetry rules in their prediction of stereochemistry and provide insight into relative activation energies and substituent effects. We will refer to specific examples for several reactions.

11.1. Electrocyclic Reactions

There are several general classes of pericyclic reactions for which orbital symmetry factors determine both the stereochemistry and relative reactivity. The first class that we will consider are *electrocyclic reactions*. An electrocyclic reaction is defined as the formation of a single bond between the ends of a linear conjugated system of π electrons and the reverse process. An example is the thermal ring opening of cyclobutenes to butadienes:

Ref. 7

It is not surprising that thermolysis of cyclobutenes leads to ring opening, since the strain in the four-membered ring is relieved. The activation energy for simple alkyl-substituted cyclobutenes is in the range of 30–35 kcal/mol.[8] *What is particularly significant about these reactions is that they are stereospecific. cis*-3,4-Dimethylcyclobutene is converted to *E,Z*-2,4-hexadiene, whereas *trans*-3,4-dimethylcyclobutene yields the *E,E*-isomer. The stereospecificity of such processes is very high. In the ring opening of *cis*-3,4-dimethylcyclobutene, for example, only 0.005% of the minor product *E,E*-2,4-hexadiene is formed, even though it is more stable than the *E,Z*-isomer.[9]

3. For reviews of several concerted reactions within the general theory of pericyclic reactions, see A. P. Marchand and R. E. Lehr, eds., *Pericyclic Reactions*, Vols. I and II, Academic Press, New York, 1977.
4. H. C. Longuet-Higgins and E. W. Abrahamson, *J. Am. Chem. Soc.* **87**:2045 (1965); M. J. S. Dewar, *Angew. Chem. Int. Ed. Engl.* **10**:761 (1971); M. J. S. Dewar, *The Molecular Orbital Theory of Organic Chemistry*, McGraw-Hill, New York, 1969; H. E. Zimmerman, *Acc. Chem. Res.* **4**:272 (1971); K. N. Houk, Y. Li, and J. D. Evanseck, *Angew. Chem. Int. Ed. Engl.* **31**:682 (1992).
5. O. Wiest, D. C. Montiel, and K. N. Houk, *J. Phys. Chem.* **101**:8378 (1997).
6. D. Sperling, H. U. Reissig, and J. Fabian, *Liebigs Ann. Chem.* **1997**:2443; B. S. Jursic, *THEOCHEM* **358**:139 (1995); H. Y. Yoo and K. N. Houk, *J. Am. Chem. Soc.* **119**:2877 (1997); V. Aviente, Y. H. Yoo, and K. N. Houk, *J. Org. Chem.* **62**:6121 (1997); K. N. Houk, B. R. Bono, M. Nendal, K. Black, H. Y. Yoo, S. Wilsey, and J. K. Lee, *THEOCHEM* **398**:169 (1997); J. E. Carpenter and C. P. Sosa, *THEOCHEM* **311**:325 (1994); B. Jursic, *THEOCHEM* **423**:189 (1998); V. Brachadell, *Int. J. Quantum Chem.* **61**:381 (1997).
7. R. F . K. Winter, *Tetrahedron Lett.* **1965**:1207.
8. W. Kirmse, N. G. Rondon, and K. N. Houk, *J. Am. Chem. Soc.* **106**:7989 (1984).
9. J. I. Brauman and W. C. Archie, Jr., *J. Am. Chem. Soc.* **94**:4262 (1972).

The reason for the observed stereospecificity is that the groups bonded to the breaking bond all rotate in the same sense during the ring-opening process. Such motion, in which either all the substitutents rotate clockwise or all rotate counterclockwise, is called the *conrotatory* mode. When such motion is precluded by some structural feature, ring opening requires a higher temperature. In the bicycloheptene shown below, the five-membered ring prevents completion of a conrotatory ring opening because it would lead to a *cis,trans*-cycloheptadiene. The reaction takes place only at very high temperature, 400°C, and probably involves the diradical shown as an intermediate.

The principle of microscopic reversibility requires that the reverse process, ring closure of a butadiene to a cyclobutene, must also be a conrotatory process. Usually, this is thermodynamically unfavorable, but a case in which the ring closure is energetically favorable is conversion of *E,Z*-2,4-cyclooctadiene (**1**) to bicyclo[4.2.0]oct-7-ene (**2**). The ring closure is favorable in this case because of the strain associated with the *trans* double bond. The ring closure occurs by a conrotatory process.

Electrocyclic reactions of 1,3,5-trienes lead to 1,3-cyclohexadienes. These ring closures also exhibit a high degree of stereospecificity. The ring closure is normally the favored reaction in this case, because the cyclic compound, which has six σ bonds and two π bonds, is thermodynamically more stable than the triene, which has five σ and three π bonds. The stereospecificity is illustrated with octatrienes **3** and **4**. *E,Z,E*-2,4,6-Octatriene (**3**) cyclizes only to *cis*-5,6-dimethyl-1,3-cyclohexadiene, whereas the *E,Z,Z*-2,4,6-octatriene (**4**) leads exclusively to the *trans* cyclohexadiene isomer.[12] A point of particular importance regarding the stereochemistry of this reaction is that the groups at the termini of the triene system rotate in the opposite sense during the cyclization process. This mode

10. R. Criegee and H. Furr, *Chem. Ber.* **97**:2949 (1964).
11. K. M. Schumate, P. N. Neuman, and G. J. Fonken, *J. Am. Chem. Soc.* **87**:3996 (1965); R. S. H. Liu, *J. Am. Chem. Soc.* **89**:112 (1967).
12. E. N. Marvell, G. Caple, and B. Schatz, *Tetrahedron Lett.* **1965**:385; E. Vogel, W. Grimme, and E. Dinne, *Tetrahedron Lett.* **1965**:391; J. E. Baldwin and V. P. Reddy, *J. Org. Chem.* **53**:1129 (1988).

of electrocyclic reaction is called *disrotatory*.

A complete mechanistic description of these reactions must explain not only their high degree of stereospecificity, but also why four-π-electron systems undergo conrotatory reactions whereas six-π-electron systems undergo disrotatory reactions. Woodward and Hoffmann proposed that the stereochemistry of the reactions is controlled by the symmetry properties of the HOMO of the reacting system.[13] The idea that the HOMO should control the course of the reaction is an example of frontier orbital theory, which holds that it is the electrons of highest energy, i.e., those in the HOMO, that are of prime importance.[14] The symmetry characteristics of the occupied orbitals of 1,3-butadiene are shown in Fig. 11.1.

Why do the symmetry properties of ψ_2 determine the stereochemistry of the electrocyclic reaction? For convenience, let us examine the microscopic reverse of the ring opening. The stereochemical features of the reaction are the same in either the forward or the reverse direction. For bonding to occur between C-1 and C-4 of the conjugated system, the positive lobe on C-1 must overlap with the positive lobe on C-4 (or negative with negative, since the signs are arbitrary). This overlap of lobes of the same sign can be accomplished only by a conrotatory motion. Disrotatory motion causes overlap of orbitals of opposite sign. This is an antibonding overlap and would preclude bond formation. Other conjugated dienes have identical orbital symmetries so that the conrotatory mode is preferred for all thermal electrocyclic processes of 1,3-dienes.

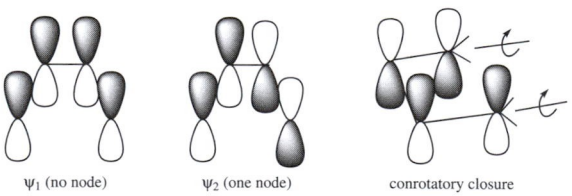

ψ_1 (no node) ψ_2 (one node)   conrotatory closure

Fig. 11.1. Symmetry properties for the π system of a conjugated diene.

13. See Ref. 2.
14. K. Fukui and H. Fujimoto, in *Mechanisms of Molecular Migrations*, Vol. 2, B. S. Thyagarajan, ed., Interscience, New York, 1968, p. 117; K. Fukui, *Acc. Chem. Res.* **4**:57 (1971); K. Fukui, *Angew. Chem. Int. Ed. Engl.* **21**:801 (1982).

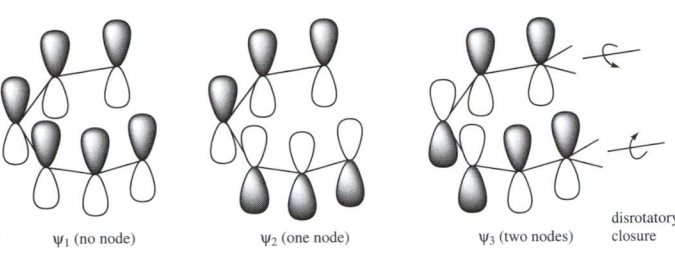

ψ_1 (no node) ψ_2 (one node) ψ_3 (two nodes)   disrotatory closure

Fig. 11.2. Symmetry properties of hexatriene molecular orbitals.

The π orbitals for the 1,3,5-triene system are shown in Fig. 11.2. The analysis according to frontier orbital theory proceeds in the same way but leads to the conclusion that a bonding interaction betwen C-1 and C-6 of the triene will require a disrotatory motion. This is because the HOMO, ψ_3, has positive lobes on the same face of the π system, and these must come together to permit bond formation. The symmetry properties of other six-π-electron conjugated triene systems are the same, so we would expect disrotatory ring closure (or opening) to be general for conjugated trienes.

When we recall the symmetry patterns for linear polyenes that were discussed in Chapter 1 (see p. 33), we can further generalize the predictions based on the symmetry of the polyene HOMO. Systems with $4n$ π electrons will undergo electrocyclic reactions by conrotatory motion, whereas systems with $4n + 2$ π electrons will react by the disrotatory mode.

An additional dimension was introduced into the analysis of concerted reactions with the use of orbital correlation diagrams.[15] This approach focuses attention on the orbital symmetries of both reactants and products and considers the symmetry properties of all the orbitals. In any concerted process, the orbitals of the starting material must be smoothly transformed into orbitals of product having the same symmetry. If this process of orbital conversion leads to the ground-state electronic configuration of the product, the process should have a relatively low activation energy and is called an *allowed* process. If, on the other hand, the orbitals of the reactant are transformed into a set of orbitals that does not correspond to the ground state of the product, a high-energy transition state occurs and the reaction is called *forbidden*, since it would lead to an excited state of the product.

The cyclobutene–butadiene interconversion can serve as an example of the reasoning employed in construction of an orbital correlation diagram. For this reaction, the four π orbitals of butadiene are converted smoothly into the two π and two σ orbitals of the ground state of cyclobutene. The analysis is done as shown in Fig. 11.3. The π orbitals of butadiene are ψ_1, ψ_2, ψ_3, and ψ_4. For cyclobutene, the four orbitals are σ, π, σ^*, and π^*. Each of the orbitals is classified with respect to the symmetry elements that are maintained in the course of the transformation. The relevant symmetry features depend on the structure of the reacting system. The most common elements of symmetry to be considered are planes of symmetry and rotation axes. An orbital is classified as symmetric (S) if it is unchanged by reflection in a plane of symmetry or by rotation about an axis of symmetry. If the orbital changes sign (phase) at each lobe as a result of the symmetry operation, it is called antisymmetric (A). Proper MOs must be either symmetric or antisymmetric. If an orbital is not sufficiently symmetric to be either S or A, it must be adapted by combination with other orbitals to meet this requirement.

15. R. Hoffmann and R. B. Woodward, *J. Am. Chem. Soc.* **87**:2046 (1965).

Figure 11.3 illustrates the classification of the MOs of butadiene and cyclobutene. There are two elements of symmetry that are common to both *s-cis*-butadiene and cyclobutene. These are a plane of symmetry and a twofold axis of rotation. The plane of symmetry is maintained during a disrotatory transformation of butadiene to cyclobutene. In the conrotatory transformation, the axis of rotation is maintained throughout the process. Therefore, to analyze the disrotatory process, the orbitals must be classified with respect to the plane of symmetry, and to analyze the conrotatory process, they must be classified with respect to the axis of rotation.

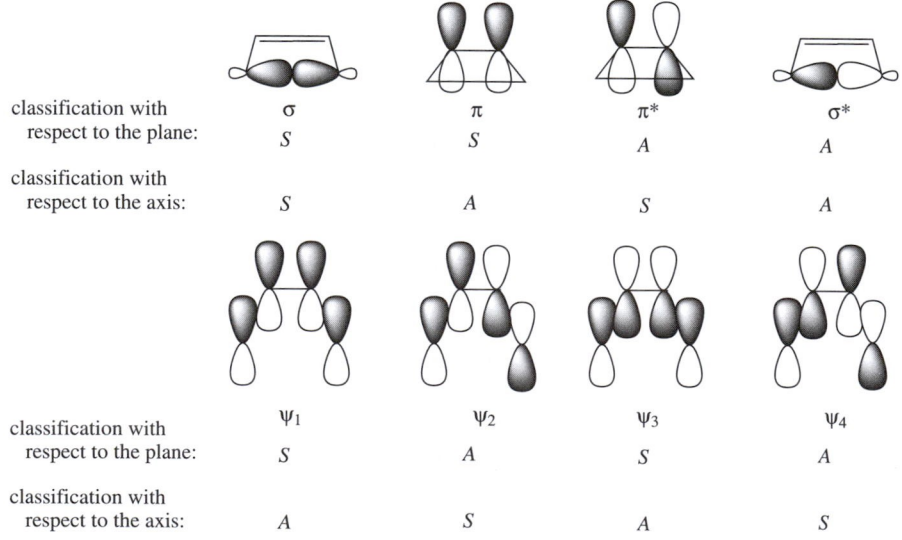

Both the disrotatory process and the conrotatory process can be analyzed by comparing the symmetry classifications of the reactant and product orbitals given in Fig. 11.3. The orbtitals are arranged according to energy in Fig. 11.4, and the states of like symmetry for the disrotatory process are connected. It is evident that not all of the ground-state orbitals of cyclobutene correlate with ground-state orbitals of butadiene. The bonding π orbital of cyclobutene is transformed into an antibonding orbital (ψ_3) of butadiene. Considering the reverse process, ψ_2 of butadiene is transformed into the antibonding π^* orbital of cyclobutene. Because of the failure of the orbitals of the ground-state molecules to correlate, the transformation would have to attain a very high-energy transition state, and the disrotatory reaction is said to be *symmetry forbidden*.

Fig. 11.3. Symmetry properties of cyclobutene (top) and butadiene (bottom) orbitals.

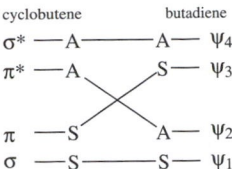

Fig. 11.4. Correlation diagram for cyclobutene and butadiene orbitals (symmetry-forbidden disrotatory reaction).

Analysis of the conrotatory process is carried out in exactly the same fashion. In this case, the element of symmetry that is maintained throughout the reaction process is the twofold rotation axis. The resulting correlation diagram is shown in Fig. 11.5. The conrotatory reaction is *symmetry allowed*, since the bonding orbitals of butadiene correlate with the bonding orbitals of cyclobutene and vice versa. Detailed MO analysis of the transition state for conrotatory ring opening using 3-21G and SCF(CI) level MO methods fully supports the conclusion that the reaction proceeds by a concerted process.[16]

Correlation diagrams can be constructed in an analogous fashion for the disrotatory and conrotatory modes for interconversion of hexatriene and cyclohexadiene. They lead to the prediction that the disrotatory mode is an allowed process whereas the conrotatory reaction is forbidden. This is in agreement with the experimental results on this reaction. Other electrocyclizations can be analyzed by the same method. Substituted derivatives of polyenes obey the orbital symmetry rules, even in cases in which the substitution pattern does not correspond in symmetry to the orbital system. It is the symmetry of the participating orbitals, not of the molecule as a whole, that is crucial to the analysis.

There is another useful viewpoint of concerted reactions that is based on the idea that transition states can be classified as aromatic or antiaromatic, just as is the case for ground-state molecules.[17] A stabilized aromatic transition state will lead to a low activation energy, i.e., an allowed reaction. An antiaromatic transition state will result in a high energy barrier and correspond to a forbidden process. The analysis of concerted reactions by this process consists of examining the array of orbitals that would be present in the transition state and classifying the system as aromatic or antiaromatic.

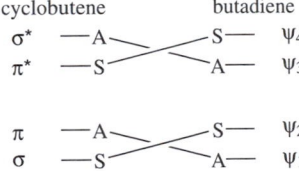

Fig. 11.5. Correlation diagram for cyclobutene and butadiene orbitals (symmetry-allowed conrotatory reaction).

16. N. G. Rondan and K. N. Houk, *J. Am. Chem. Soc.* **107**:2099 (1985); J. Breulet and H. F. Schaefer III, *J. Am. Chem. Soc.* **106**:1221 (1984).
17. H. E. Zimmerman, *Acc. Chem. Res.* **4**:272 (1971); M. J. S. Dewar, *Angew. Chem. Int. Ed. Engl.* **10**:761 (1971).

For the butadiene–cyclobutene interconversion, the transition states for conrotatory and disrotatory interconversion are shown below. The array of orbitals represents the *basis set orbitals*, i.e., the total set of 2*p* orbitals involved in the reaction process, not the individual MOs. Each of the orbitals is π in character, and the phase difference is represented by shading. The tilt at C-1 and C-4 as the butadiene system rotates toward the transition state is different for the disrotatory and conrotatory modes. The dashed line represents the σ bond that is being broken (or formed).

basis set orbitals
for conrotatory closure

basis set orbitals
for disrotatory closure

We will consistently assign phases to the basis set orbitals in such a way as to minimize the phase changes in the transition state. It has, however, been shown that this is not necessary and that the correct classification of the transition state will be achieved for any combination of basis set orbitals. When the array of orbitals corresponding to the transition state being analyzed has been drawn, two features must be determined to decide if it is aromatic or antiaromatic. These features are the topology of the orbital array and the number of electrons involved. The topology can be of the *Hückel type* or the *Möbius type*. A Hückel system has zero (or any even number) of phase changes around the orbital array. A Möbius system has one (or any odd number) of phase changes. For the cyclobutene–butadiene transition state, the conrotatory closure results in a Möbius system, whereas a disrotatory closure gives a Hückel system. The second important feature of the transition state is the number of participating electrons. In the present case, it is four. The same rules of aromaticity apply as for ground-state molecules. A Hückel system is aromatic when it has $4n + 2$ electrons. A Möbius system is aromatic when it has $4n$ electrons. In the case of the cyclobutene–butadiene interconversion, it is then the conrotatory transition state that is the favored aromatic transition state.

There are two stereochemically distinct possibilities for the conrotatory process. A substituent group might move toward or away from the breaking bond:

Steric factors would ordinarily be expected to induce a preference for the larger group to move outward and thus generate the *E*-isomer. It was observed, however, that in the case of 1,2,3,4-tetrafluoro-*trans*-3,4-bis(trifluoromethyl)cyclobutene, ring opening occurred with an inward rotation of the trifluoromethyl groups.[18]

18. W. R. Dolbier, Jr., H. Koroniak, D. J. Burton, and P. Heinze, *Tetrahedron Lett.* **27**:4387 (1986).

MO calculation for the case of Y = CH=O found that the formyl group preferred to rotate inward, and this was confirmed experimentally.[19] A general theoretical analysis indicates that the preference is for donor substituents to rotate outward, whereas acceptor substituents prefer to rotate inward.[20] A qualitative understanding of this stereoselectivity can be based on analysis of the interaction of substituents with the orbitals of the C(3)−C(4) σ bond which is breaking. Outward rotation of donor groups stabilizes the LUMO (**A**) whereas inward rotation leads to a repulsive interaction between two filled orbitals (**B**). For acceptor substituents, a π* substituent orbital can provide a stabilizing interaction with the C(3)−C(4) HOMO (**C**). More quantitative analyses have been done by MO computation of transition-state energies. Table 11.1 shows some of the results. The trends observed experimentally are successfully reproduced, with donor substituents stabilizing outward rotation by as much as 9 kcal/mol for alkoxy groups.[21]

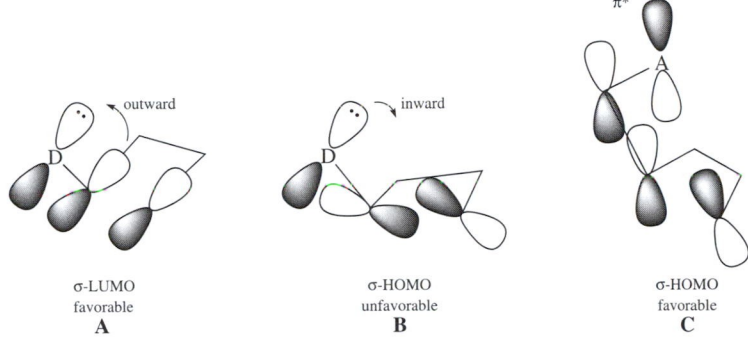

σ-LUMO
favorable
A

σ-HOMO
unfavorable
B

σ-HOMO
favorable
C

Table 11.1 Directive Effect of Substituents on Inward/Outward Conrotation in Cyclobutene Ring Opening[a]

Donor substituent	$\Delta_{\text{in-out}}$	Acceptor substituent	$\Delta_{\text{in-out}}$
NH_2	17.5	NO_2	7.3
OH	17.2	CH=O	− 4.6
F	16.9	N=O	− 2.6
Cl	13.6	CF_3	2.6
CH_3	6.8	CN	4.3

a. $\Delta_{\text{in-out}}$ = Energy difference, in kcal/mol, between transition states for inward and outward rotation; S. Niwayama, E. A. Kallel, D. C. Spellmeyer, C. Sheu, and K. N. Houk, *J. Org. Chem.* **61**:2813 (1996).

19. K. Rudolf, D. C. Spellmeyer, and K. N. Houk, *J. Org. Chem.* **52**:3708 (1987).

20. N. G. Rondan and K. N. Houk, *J. Am. Chem. Soc.* **107**:2099 (1985); W. Kirmse, N. G. Rondan, and K. N. Houk, *J. Am. Chem. Soc.* **106**:7989 (1984); D. C. Spellmeyer and K. N. Houk, *J. Am. Chem. Soc.* **110**:3412 (1988).

21. E. A. Kallel, Y. Wang, D. C. Spellmeyer, and K. N. Houk, *J. Am. Chem. Soc.* **112**:6759 (1990); W. R. Dolbier, Jr., H. Koroniak, K. N. Houk, and C. Sheu, *Acc. Chem. Res.* **29**:471 (1996).

Analysis of the hexatriene–cyclohexadiene system leads to the conclusion that the disrotatory process will be favored. The basis set orbitals for the conrotatory and disrotatory transition states are shown below.

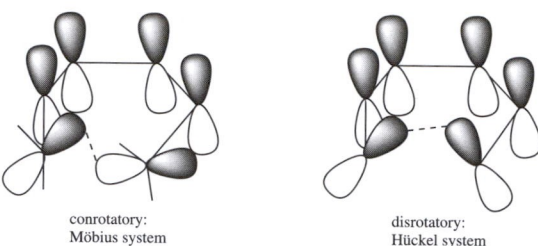

conrotatory:
Möbius system

disrotatory:
Hückel system

Here, with six electrons involved, it is the disrotatory mode (Hückel system) in which the transition state is stabilized. There are numerous examples of interconversion of 1,3,5-trienes and 1,3-cyclohexadiene systems by the electrocyclic mechanism.[22] The chart which follows gives a general summary of the relationship between transition-state topology, the number of electrons, and the stability of the transition state.

Electrons	Hückel (disrotatory)	Möbius (conrotatory)
2	aromatic	antiaromatic
4	antiaromatic	aromatic
6	aromatic	antiaromatic
8	antiaromatic	aromatic

We have now considered three viewpoints from which thermal electrocyclic processes can be analyzed: symmetry characteristics of the frontier orbitals, orbital correlation diagrams, and transition-state aromaticity. All arrive at the same conclusions about stereochemistry of electrocyclic reactions. *Reactions involving 4n + 2 electrons will be disrotatory and involve a Hückel-type transition state, whereas those involving 4n electrons will be conrotatory and the orbital array will be of the Möbius type.* These general principles serve to explain and correlate many specific experimental observations made both before and after the orbital symmetry rules were formulated. We will discuss a few representative examples in the following paragraphs.

The bicyclo[2.2.0]hexa-2,5-diene ring system is a valence isomer of the benzene ring and is often referred to as *Dewar benzene*. After many attempts to prepare Dewar benzene derivatives failed, a pessimistic opinion existed that all such efforts would be fruitless because Dewar benzene would be so unstable as to immediately revert to benzene. Then, in 1962, van Tamelen and Pappas isolated a stable Dewar benzene derivative by photolysis of 1,2,4-tri(*t*-butyl)benzene.[23]

5

22. V. A. Bakulev, *Russ. Chem. Rev.* **64**:99 (1995).
23. E. E. van Tamelen. S. P. Pappas, and K. L. Kirk, *J. Am. Chem. Soc.* **93**:6092 (1971); this paper contains references to the initial work and describes subsequent studies.

The compound was reasonably stable, reverting to the aromatic starting material only on heating. Part of the stability of this particular Dewar benzene derivative can be attributed to steric factors. The *t*-butyl groups are farther apart in the Dewar benzene structure than in the aromatic structure. The unsubstituted Dewar benzene was successfully prepared in 1963.

This compound is less stable than **5** and reverts to benzene with a half-life of about 2 days at 25°C, with $\Delta H^{\ddagger} = 23$ kcal/mol.[24] The observed kinetic stability of Dewar benzene is surprisingly high when one considers that its conversion to benzene is exothermic by 71 kcal/mol. The stability of Dewar benzene is intimately related to the orbital symmetry requirements for concerted electrocyclic transformations. The concerted thermal pathway should be conrotatory, since the reaction is the ring opening of a cyclobutene and therefore leads not to benzene, but to a highly strained *Z,Z,E*-cyclohexatriene. A disrotatory process, which would lead directly to benzene, is forbidden.[25]

Theoretical treatment of the reaction as a conrotatory process proceeding through the very unstable *Z,Z,E*-isomer of benzene satisfactorily accounts for the observed activation barrier.[26]

An especially interesting case of the hexatriene–cyclohexadiene type interconversion is the rapid equilibrium between cycloheptatrienes and bicyclo[4.1.0]hepta-2,4-dienes[27]:

The energy requirement for this electrocyclic transformation is so low that the process is very rapid at room temperature. Low-temperature NMR measurements give a value of about 7 kcal/mol for the activation energy in the case where $R=CO_2CH_3$.[28] This transformation is an example of *valence tautomerism*, a rapid process involving only reorganization of bonding electrons. The reason the reaction is much more rapid than electrocyclization of acyclic trienes is that the ring holds the reacting termini together, reducing the negative entropy of activation. In contrast to the ring opening of Dewar

24. M. J. Goldstein and R. S. Leight, *J. Am. Chem. Soc.* **99**:8112 (1977).
25. E. E. van Tamelen, *Acc. Chem. Res.* **5**:186 (1972)
26. R. P. Johnson and K. J. Daoust, *J. Am. Chem. Soc.* **118**:7381 (1996).
27. G. Maier, *Angew. Chem. Int. Ed. Engl.* **6**:402 (1967).
28. M. Görlitz and H Günther, *Tetrahedron*, **25**:4467 (1969).

benzene, disrotatory opening of the bicyclo[4.1.0]hepta-2,4-diene is allowed by orbital symmetry rules and is easily accommodated by the ring geometry. For unsubstituted bicyclo[4.1.0]hepta-2,4-diene, the equilibrium constant for ring closure is small, about 3×10^{-3} at $100°C$. Alkyl groups do not have much of an effect on the position of the equilibrium, but electron-withdrawing groups such as cyano and trifluoromethyl shift the equilibrium more in favor of the bicyclic ring.[29]

The prediction on the basis of orbital symmetry analysis that cyclization of eight-π-electron systems will be conrotatory has been confirmed by study of isomeric 2,4,6,8-decatetraenes. Electrocyclic reaction occurs near room temperature and establishes an equilibrium that favors the cyclooctatriene product. At slightly more elevated temperatures, the hexatriene system undergoes a subsequent disrotatory cyclization, establishing equilibrium with the corresponding bicyclo[4.2.0]octa-2,4-diene[30]:

a = d = CH₃;   b = c = H
b = c = CH₃;   a = d = H
b = d = CH₃;   a = c = H

Theoretical calculations (MP2/6-31G*) on the transition states indicated they are helical and conform to the expected conrotatory mode.[31] This is a Möbius-type transition state. The chemical shift and magnetic properties attributed to the transition state by MO calculation are in agreement with the idea that the orbitals are in a Möbius arrangement.[32]

The Woodward–Hoffmann orbital symmetry rules are not limited in application to the neutral polyene systems that have been discussed up to this point. They also apply to charged systems, just as the Hückel aromaticity rules can be applied to charged cyclic systems. The conversion of a cyclopropyl cation to an allyl cation is the simplest possible case of an electrocyclic process, since it involves only two π electrons.[33] Because of the strain imposed by the small internuclear angle in the cyclopropyl ring, cyclopropyl cations do not form easily, and cyclopropyl halides and sulfonates are quite unreactive under ordinary solvolytic conditions. For example, solvolysis of cyclopropyl tosylate in acetic acid requires a temperature of $180°C$. The product is allyl acetate rather than cyclopropyl acetate.[34] This transformation might occur by formation of the cyclopropyl cation,

29. P. Warner and S.-L. Lu, *J. Am. Chem. Soc.* **95**:5099 (1973); P. M. Warner and S.-L. Lu, *J. Am. Chem. Soc.* **102**:331 (1980); K. Takeuchi, H. Fujimoto, and K. Okamoto, *Tetrahedron Lett.* **22**:4981 (1981); T.-H. Tang, C. S. Q. Lew, Y.-P. Cui, B. Capon, and I. G. Csizmadia, *THEOCHEM* **305**:49 (1994); Y. Güzel, E. Saripinar and L. Yildirim, *Monatsh. Chem.* **127**:513 (1996).

30. R. Huisgen, A. Dahmen, and H. Huber, *Tetrahedron Lett.* **1969**:1461; R. Huisgen, A. Dahmen, and H. Huber, *J. Am. Chem. Soc.* **89**:7130 (1967); A. Dahmen and R. Huisgen, *Tetrahedron Lett.* **1969**:1465.

31. B. E. Thomas IV, J. D. Evanseck, and K. N. Houk, *J. Am. Chem. Soc.* **115**:4165 (1993); B. E. Thomas, J. D. Evanseck, and K. N. Houk, *Isr. J. Chem.* **33**:287 (1993).

32. H. Jiao and P. v. R. Schleyer, *J. Chem. Soc., Perkin Trans. 2* **1994**:407.

33. P. v. R. Schleyer, W. F. Sliwinski, G. W. Van Dine, U. Schöllkopf, J. Paust, and K. Fellenberger, *J. Am. Chem. Soc.* **94**:125 (1972); W. F. Sliwinski, T. M. Su, and P. v. R. Schleyer, *J. Am. Chem. Soc.* **94**:133 (1972).

34. J. D. Roberts and V. C. Chambers, *J. Am. Chem. Soc.* **73**:5034 (1951).

followed by ring opening to the allyl cation:

Formation of allylic products is characteristic of solvolytic reactions of other cyclopropyl halides and sulfonates. Similarly, diazotization of cyclopropylamine in aqueous solution gives allyl alcohol.[35] The ring opening of a cyclopropyl cation is an electrocyclic process of the $4n + 2$ type, where n equals zero. It should therefore be a disrotatory process. There is another facet to the stereochemistry in substituted cyclopropyl systems. Note that for a *cis*-2,3-dimethylcyclopropyl cation, for example, two different disrotatory modes are possible, leading to conformationally distinct allyl cations:

inward rotation   outward rotation

The disrotatory mode, in which the methyl groups move away from each other, would be more favorable for steric reasons. If the ring opening occurs through a discrete cyclopropyl cation, the W-shaped allylic cation should be formed in preference to the sterically less favorable U-shaped cation. This point was investigated by comparing the rates of solvolysis of the cyclopropyl tosylates **6–8**:

	6	**7**	**8**
relative rate:	1	4	41,000

Some very significant conclusions can be drawn from the data. If formation of the cyclopropyl cation were the rate-determining step, **7** should be more reactive than **8**, because the steric interaction between the tosylate leaving group and the methyl substituents in **7** will be relieved as ionization occurs. Since **7** is 10,000 times less reactive than **8**, some other factor must be determining the relative rates of reaction, and it is doubtful that rate-limiting ionization to a cyclopropyl cation is occurring. The results can be explained, as proposed by DePuy,[36] if the ionization and ring opening are part of a single, concerted process. In such a process, the ionization would be assisted by the electrons in the cleaving C(2)–C(3) bond. These electrons would provide maximum assistance if positioned toward the back side of the leaving group. This, in turn, requires that the substituents *anti* to the leaving group rotate outward as the ionization proceeds. This concerted process would explain why **7** reacts more slowly than **8**. In **7**, such a

35. P. Lipp, J. Buchkremer, and H. Seeles, *Justus Liebigs Ann. Chem.* **499**:1 (1932); E. J. Corey and R. F. Atkinson, *J. Org. Chem.* **29**:3703 (1964).
36. C. H. DePuy, *Acc. Chem. Res.* **1**:33 (1968).

rotation would move the methyl groups together, resulting in increased steric interaction and the formation of the U-shaped allylic anion. In **8**, the methyl groups would move away from one another and form the W-shaped allylic ion.

This interpretation is supported by results on the acetolysis of the bicyclic tosylates **9** and **10**. With **9**, after three months in acetic acid at 150°C, 90% of the starting material was recovered. This means that both ionization to a cyclopropyl cation and a concerted ring opening must be extremely slow. The preferred disrotatory ring-opening process would lead to an impossibly strained structure, the *trans*-cyclohexenyl cation. In contrast, the stereoisomer **10** reacts at least 2×10^6 more rapidly because it can proceed to a stable *cis*-cyclohexenyl cation[37]:

When the size of the fused ring permits ring opening to a *trans*-allylic cation, as in the case of compound **11**, solvolysis proceeds at a reasonable rate[38]:

An example of preferred conrotatory cyclization of four-π-electron pentadienyl cation systems can be found in the acid-catalyzed cyclization of the dienone **12**, which proceeds through the 3-hydroxypentadienyl cation **13**. The stereochemistry is that expected for a conrotatory process.[39]

37. See Ref. 33.
38. G. H. Whitham and M. Wright, *J. Chem. Soc., C* **1971**:883.
39. R. B. Woodward, in *Aromaticity*, Chemical Society Special Publication No. 21, Chemical Society, London, 1969, p. 217; K. L. Habermas, S. E. Denmark, and T. K. Jones, *Org. React.* **45**:1 (1994).

While most synthetic examples of this cyclization have involved protonation of divinyl ketones to give 3-hydroxy-1,4-pentadienyl cations, theoretical studies suggest that the cyclization would occur even more readily with alternative substituents at C-3.[40]

There are also examples of electrocyclic processes involving anionic species. Since the pentadienyl anion is a six-π-electron system, thermal cyclization to a cyclopentenyl anion should be disrotatory. Examples of this electrocyclic reaction are rare. NMR studies of pentadienyl anions indicate that they are stable and do not tend to cyclize.[41] Cyclooctadienyllithium provides an example where cyclization of a pentadienyl anion fragment does occur, with the first-order rate constant being 8.7×10^{-3} min^{-1}. The stereochemistry of the ring closure is consistent with the expected disrotatory nature of the reaction.

In contrast to pentadienyl anions, heptatrienyl anions have been found to cyclize readily to cycloheptadienyl anions.[42] The transformation of heptatrienyl anion to cyclo-heptadienyl anion proceeds with a half-life of 13 min at $-13°$C. The Woodward–Hoffmann rules predict that this would be a conrotatory closure.[43]

11.2. Sigmatropic Rearrangements

Sigmatropic rearrangements are another important class of concerted pericyclic reactions governed by orbital symmetry.[44] Sigmatropic rearrangements involve a concerted reorganization of electrons during which a group attached by a σ bond migrates to the terminus of an adjacent π-electron system. There is a simultaneous shift of the π electrons. Sigmatropic rearrangements are further described by stating the relationship between the reacting centers in the migrating fragment and the π system. The order $[i, j]$ specifies the number of atoms in the migrating fragment and the number of atoms in the π system that are directly involved in the bonding changes. This classification system is illustrated by the examples in Scheme 11.1. As with other concerted reactions, the topology of the

40. D. A. Smith and C. W. Ulmen II, *J. Org. Chem.* **62**:5110 (1997).
41. R. B. Bates, D. W. Gosselink, and J. A . Kaczynski, *Tetrahedron Lett.* **1967**:199, 205; R. B. Bates and D. A. McCombs, *Tetrahedron Lett.* **1969**:977.
42. E. A. Zuech, D. L. Crain, and R. F. Kleinschmidt, *J. Org. Chem.* **33**:771 (1968); R. B. Bates, W. H. Deines, D. A. McCombs, and D. E. Potter, *J. Am. Chem. Soc.* **91**:4608 (1969).
43. S. W. Staley, in *Pericyclic Reactions*, Vol. 1, A. P. Marchand and R. E. Lehr, eds., Academic Press, New York, 1977, Chapter. 4.
44. R. B. Woodward and R. Hoffmann, *J. Am. Chem. Soc.* **87**:2511 (1965).

Scheme 11.1. Examples of Sigmatropic Rearrangements

interacting orbitals dictates the facility of the various sigmatropic rearrangements and the stereochemistry of the process. First, it must be recognized that there are two topologically distinct processes by which a sigmatropic migration can occur. If the migrating group remains associated with the same face of the conjugated π system throughout the process, the migration is termed *suprafacial*. The alternative mode involves a process in which the migrating group moves to the opposite face of the π system during the course of the migration and is called *antarafacial*.

The orbital symmetry requirements of sigmatropic reactions are analyzed by considering the interactions between the orbitals of the π system and those of the migrating fragment. The simplest case, 1,3-sigmatropic shift of a hydrogen, is illustrated in the first entry in Scheme 11.1. A frontier orbital analysis of this process treats the system as a hydrogen atom interacting with an allyl radical. The frontier orbitals are the hydrogen $1s$ and allyl ψ_2 orbitals. These interactions are depicted below for both the suprafacial and antarafacial modes.

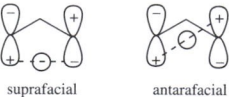

A bonding interaction can be maintained only in the antarafacial mode. The 1,3-suprafacial shift of hydrogen is therefore forbidden by orbital symmetry considerations. The allowed antarafacial process is symmetry allowed, but it involves such a contorted geometry that this shift, too, would be expected to be energetically difficult. As a result, orbital symmetry considerations indicate that concerted 1,3-shifts of hydrogen are unlikely processes.

A similar analysis of the 1,5-sigmatropic shift of hydrogen leads to the opposite conclusion. The relevant frontier orbitals in this case are the hydrogen $1s$ orbital and ψ_3 of the pentadienyl radical. The suprafacial mode is allowed whereas the antarafacial mode is forbidden. The suprafacial shift corresponds to a favorable six-membered ring.

thermally allowed 1,5-suprafacial hydrogen shift in 1,3-pentadiene

An alternative analysis of sigmatropic reactions involves drawing the basis set atomic orbitals and classifying the resulting system as Hückel or Möbius in character. When this classification has been done, the electrons involved in the process are counted to determine if the transition state is aromatic or antiaromatic. This analysis is illustrated in Fig. 11.6. The conclusions reached are the same as for the frontier orbital approach. The suprafacial 1,3-shift of hydrogen is forbidden, but the suprafacial 1,5-shift is allowed. Proceeding to a 1,7-shift of hydrogen, it is found that the antarafacial shift is allowed. These conclusions based on transition-state aromaticity are supported by 6-31G* MO calculations, which conclude that 1,5-shifts should be suprafacial whereas 1,7-shifts should be antarafacial.[45] Theoretical calculations also find that the 1,3-shift of hydrogen should be antarafacial, but, in agreement with expectations based on molecular geometry, the transition state is so energetic that it is close to a stepwise bond dissociation process.[46]

Sigmatropic migration involving alkyl group shifts can also occur:

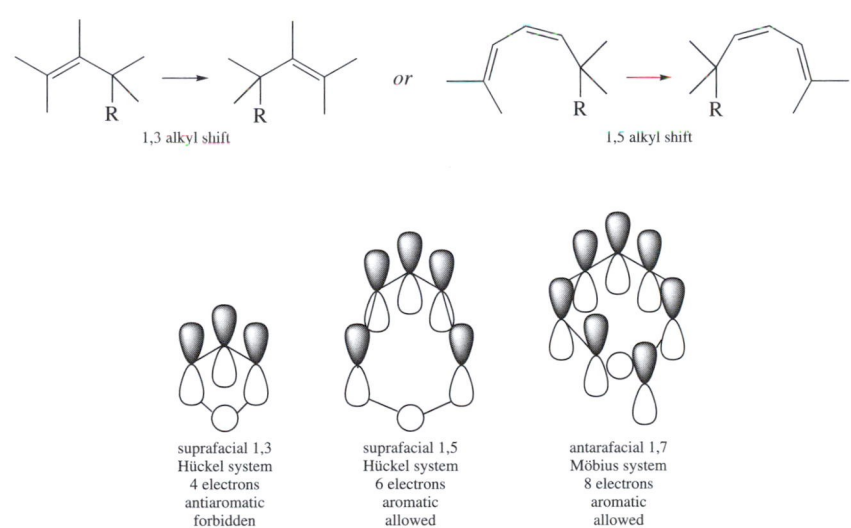

1,3 alkyl shift   or   1,5 alkyl shift

suprafacial 1,3	suprafacial 1,5	antarafacial 1,7
Hückel system	Hückel system	Möbius system
4 electrons	6 electrons	8 electrons
antiaromatic	aromatic	aromatic
forbidden	allowed	allowed

Fig. 11.6. Classification of sigmatropic hydrogen shifts with respect to basis set orbitals.

45. B. A Hess, Jr., L. J. Schaad, and J. Pancir, *J. Am. Chem. Soc.* **107**:149 (1985).
46. F. Bernardi, M. A. Robb, H. B. Schlegel, and G. Tonachini, *J. Am. Chem. Soc.* **106**:1198 (1984).

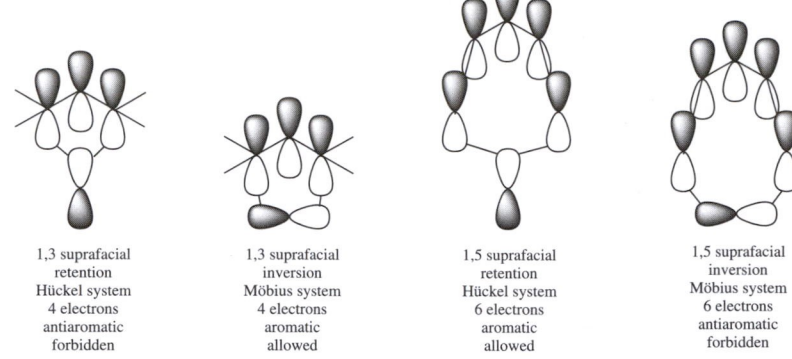

1,3 suprafacial
retention
Hückel system
4 electrons
antiaromatic
forbidden

1,3 suprafacial
inversion
Möbius system
4 electrons
aromatic
allowed

1,5 suprafacial
retention
Hückel system
6 electrons
aromatic
allowed

1,5 suprafacial
inversion
Möbius system
6 electrons
antiaromatic
forbidden

Fig. 11.7. Classification of sigmatropic shifts of alkyl groups with respect to basis set orbitals.

When an alkyl group migrates, there is an additional stereochemical feature to consider. The shift can occur with retention or inversion at the migrating center. The analysis of sigmatropic shifts of alkyl groups is illustrated in Fig. 11.7. The allowed processes include the suprafacial 1,3-shift with inversion and the suprafacial 1,5-shift with retention.

Sigmatropic rearrangements of order [3,3] are very common:

The transition state for such processes is represented as two interacting allyl fragments. When the process is suprafacial in both groups, an aromatic transition state results, and the process is thermally allowed. Usually, a chairlike transition state is involved, but a boatlike conformation is also possible.

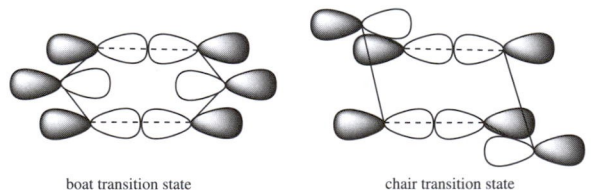

boat transition state   chair transition state

Generalization of the Woodward–Hoffmann rules for sigmatropic processes leads to the selection rules in Scheme 11.2.[47] With these generalized rules as a unifying theoretical framework, we can consider specific examples of sigmatropic rearrangements. In accord with the theoretical concepts, there are many examples of sigmatropic 1,5-hydrogen migrations in molecules that incorporate a pentadienyl fragment. The activation energies for such reactions are usually in the vicinity of 35 kcal/mol so the reactions require

47. See Ref. 44.

Selection rules for sigmatropic shifts of order $[i, j]$			
A. Order $[1, j]$ $1 + j$	supra/retention	supra/inversion	antara/retention	antara/inversion
$4n$	forbidden	allowed	allowed	forbidden
$4n+2$	allowed	forbidden	forbidden	allowed
B. Order $[i, j]$ $i + j$	supra/supra	supra/antara	antara/antara	
$4n$	forbidden	allowed	forbidden	
$4n + 2$	allowed	forbidden	allowed	

moderately elevated temperatures.[48] Two examples are given:

Ref. 49

Ref. 50

The conversion of **14** to **15** provides a test of the prediction that 1,3-alkyl shifts should occur with inversion of configuration[51]:

In the absence of the orbital symmetry considerations, one might assume that the C(3)−C(7) bond would form as the C(1)−C(7) bond breaks by simply "sliding over." This would be a violation of orbital symmetry rules because, as indicated in Scheme 11.2, a 1,3-suprafacial shift of an alkyl group must proceed with inversion at the migrating center. This point was investigated using **14** labeled with deuterium at C-7. In the starting material, the deuterium was *trans* to the acetoxy group, whereas in the product it was found to be exclusively *cis*. This result establishes that inversion of configuration occurred

48. W. R. Roth and J. König, *Justus Liebigs Ann. Chem.* **699**:24 (1966).
49. A. P. ter Borg, H. Kloosterziel, and N. Van Meurs, *Proc. Chem. Soc.* **1962**:359.
50. J. Wolinsky, B. Chollar, and M. D. Baird, *J. Am. Chem. Soc.* **84**:2775 (1962).
51. J. A. Berson, *Acc. Chem. Res.* **1**:152 (1968); J. A. Berson and G. L. Nelson, *J. Am. Chem. Soc.* **89**:5503 (1967).

at C-7 during the migration, in accord with the stereochemistry required by the Wood-ward–Hoffmann rules.

This particular reaction has quite a high activation energy, which suggests that a nonconcerted mechanism might be involved. An alternative explanation for the observed stereochemistry has been advanced.[52]

Suprafacial 1,3-shifts with inversion of configuration at the migrating carbon have also been observed in the thermal conversion of bicyclo[2.1.1]hexenes to bicyclo[3.1.0]-hexenes.[53]

$$X = H, \quad Y = CH_3$$
$$X = CH_3, \quad Y = H$$
$$X = H \quad Y = OAc$$

The thermal rearrangements of methyl-substituted cycloheptatrienes have been proposed to proceed by sigmatropic migration of the norcaradiene valence tautomer.[54] The first step is an electrocyclization analogous to those discussed in Section 11.1.

These are suprafacial sigmatropic shifts of order [1,5] and should occur with retention of configuration at the migrating carbon. This stereochemical course has been established for the 1,5-alkyl shift that converts **16** to **17**.[55] The product which is isolated, **18**, results from a subsequent 1,5-hydrogen shift, but this does not alter the stereochemistry at the migrating

52. B. K. Carpenter, *J. Am. Chem. Soc.* **117**:6336 (1995).

53. W. R. Roth and A. Friedrich, *Tetrahedron Lett.* **1969**:2607; S. Masamune, S. Takada, N. Nakatasuka, R. Vukov, and E. N. Cain, *J. Am. Chem. Soc.* **91**:4322 (1969).

54. J. A. Berson and M. R. Willcott, III, *Recent Chem. Prog.* **27**:139 (1966); *J. Am. Chem. Soc.* **88**:2494 (1966).

55. M. A. M. Boersma, J. W. de Haan, H. Kloosterziel, and L. J. M. van de Ven, *J. Chem. Soc., Chem. Commun.* **1970**:1168.

carbon. The configuration of the migrating carbon is retained, as expected.

16   **17**   **18**

Like the thermal 1,3-hydrogen shift, a 1,7-hydrogen shift is allowed when antarafacial but forbidden when suprafacial. Because a π system involving seven carbon atoms is more flexible than one involving only three carbon atoms, the geometrical restrictions on the antarafacial transition state are not as severe as in the 1,3-case. For the conversion of Z,Z-1,3,5-octatriene to Z,Z,E-2,4,6-octatriene, the energy of activation is 20.2 kcal/mol. The Z,Z,Z-isomer is also formed, but with a slightly higher activation energy. The primary kinetic isotope effect for the transferred hydrogen is around 7, consistent with C$-$H bond breaking being involved in the rate-determining step.[56] A similar activation energy has been measured for the unsubstituted Z,Z-1,3,5-heptatriene.[57]

More complex structures, such as **19**, exhibit similar activation energies. Compound **19** has also been used to demonstrate that the stereochemistry is *antarafacial*, as predicted.[58]

19   **20**   **21**

An especially important case is the thermal equilibrium between precalciferol (pre-vitamin D$_2$, **22**) and calciferol (vitamin D$_2$, **23**).[59,60]

22   **23**

56. J. E. Baldwin and V. P. Reddy, *J. Am. Chem. Soc.* **109**:8051 (1987).
57. M. Gurski, I. D. Gridnev, Y. V. Il'ichev, A. V. Ignatenko, and Y. N. Bubnov, *Angew. Chem. Int. Ed. Engl.* **31**:781 (1992).
58. C. A. Hoeger, A. D. Johnston, and W. H. Okamura, *J. Am. Chem. Soc.* **109**:4690 (1987); W. H. Okamura, C. A. Hoeger, K. J. Miller, and W. Reischl, *J. Am. Chem. Soc.* **110**:973 (1988).
59. J. L. M. A. Schlatmann, J. Pot, and E. Havinga, *Recl. Trav. Chim.* **83**:1173 (1964).
60. For a historical review of this reaction, see L. Fieser and M. Fieser, *Steroids*, Reinhold, New York, 1959, Chapter 4.

The most important sigmatropic rearrangements from the synthetic point of view are the [3,3] processes involving carbon–carbon bonds. The thermal rearrangement of 1,5-dienes by [3,3] sigmatropy is called the *Cope rearrangement*. The reaction establishes equilibrium between the two 1,5-dienes and proceeds in the thermodynamically favored direction. The conversion of **24** to **25** provides an example:

Ref. 61

24   **25**

The equilibrium in this case is controlled by the conjugation present in the product.

The rearrangement of the simplest possible case, 1,5-hexadiene, has been studied using deuterium labeling. The activation enthalpy is 33.5 kcal/mol, and the entropy of activation is -13.8 eu.[62] The substantially negative entropy reflects the formation of the cyclic transition state.

$E_a = 33.5$ kcal/mol

Conjugated substituents at C-2, C-3, C-4, or C-5 accelerate the rearrangement.[63] Donor substituents at C-2 and C-3 have an accelerating effect.[64] The effect of substituents can be rationalized in terms of the stabilization of the transition state by depicting their effect on two interacting allyl systems.

The transition state involves six partially delocalized electrons being transformed from one 1,5-diene system to another. The transition state could range in character from a 1,4-diradical to two nearly independent allyl radicals, depending on whether bond making or bond breaking is more advanced.[65] The general framework for understanding the substituent effects is that the reactions are concerted with a relatively late transition state with well-developed C(1)−C(6) bonds.

61. A. C. Cope and E. M. Hardy, *J. Am. Chem. Soc.* **62:**441 (1940).

62. W. v. E. Doering, V. G. Toscano, and G. H. Beasley, *Tetrahedron* **27:**5299 (1971); K. A. Black, S. Wilsey, and K. N. Houk, *J. Am. Chem. Soc.* **120:**5622 (1998).

63. M. J. S. Dewar and L. E. Wade, *J. Am. Chem. Soc.* **95:**290 (1973); *J. Am. Chem. Soc.* **99:**4417 (1977); R. Wehrli, H. Schmid, D. E. Bellus, and H. J. Hansen, *Helv. Chim. Acta*, **60:**1325 (1977).

64. M. Dollinger, W. Henning, and W. Kirmse, *Chem. Ber.* **115:**2309 (1982).

65. J. J. Gajewski and N. D. Conrad, *J. Am. Chem. Soc.* **100:**6268 (1978); J. J. Gajewski and K. E. Gilbert, *J. Org. Chem.* **49:**11 (1984).

Table 11.2. Effect of Phenyl Substituents on Activation Enthalpy and Entropy of Cope Rearrangements

R^1	R^2	R^3	R^4	R^5	R^6	$\Delta H^{\ddagger}$ (kcal/mol)	$\Delta S^{\ddagger}$ (eu)	Reference
H	H	H	H	H	H	33.5	−13.8	a
H	Ph	H	H	H	H	29.3	−11.3	b
H	H	Ph	H	H	H	31.6	−16.6	b
Ph	H	H	Ph	H	H	30.1	−14.4	c
H	Ph	H	H	Ph	H	21.3	−20.7	d
H	H	Ph	Ph	H	II	24.0	−12.4	e

a. W. v. E. Doering, V. G. Toscano, and G. H. Beasley, *Tetrahedron*, **27**:5299 (1971).
b. M. J. S. Dewar, and L. E. Wade, Jr., *J. Am. Chem. Soc.* **99**:4417 (1977).
c. E. G. Foster, A. C. Cope, and F. Daniels, *J. Am. Chem. Soc.* **69**:1893 (1947).
d. W. R. Roth, H.-W. Lennartz, W. v. E. Doering, L. Birladeanu, C. A. Guyton, and T. Kitagawa, *J. Am. Chem. Soc.* **112**:1722 (1990).
e. R. P. Lutz and H. A. J. Berg, *J. Org. Chem.* **45**:3915 (1980).

The most advanced MO and DFT calculations support the idea of an aromatic transition state.[66] The net effect on reaction rate of any substituent is determined by whether it stabilizes the transition state or the ground state more effectively.[67] The aromatic concept of the transition state predicts that it would be stabilized by substituents at all positions, and this is true for phenyl substituents, as shown in Table 11.2.

The Cope rearrangement usually proceeds through a chairlike transition state. The stereochemical features of the reaction can usually be predicted and analyzed on the basis of a chair transition state that minimizes steric interactions between the substituents. Thus, compound **26** reacts primarily through transition state **27a** to give **28** as the major product. Minor product **29** is formed through the less sterically favorable transition state **27b**.

When enantiomerically pure **26** is used, the product is $> 95\%$ optically pure and has the chirality shown above.[68] This result establishes that chirality is maintained throughout the

66. D. A. Hrovat, W. T. Borden, R. L. Vance, N. G. Rondan, K. N. Houk, and K. Morokuma, *J. Am. Chem. Soc.* **112**:2018 (1990); D. A. Hrovat, K. Morokuma, and W. T. Borden, *J. Am. Chem. Soc.* **116**:1072 (1994); O. Wiest, K. A. Black, and K. N. Houk, *J. Am. Chem. Soc.* **116**:10336 (1994); M. D. Davidson, I. H. Hillier, and M. A. Vincent, *Chem. Phys. Lett.* **246**:536 (1995); S. Yamabe, S. Okumoto, and T. Hayashi, *J. Org. Chem.* **61**:6218 (1996); W. T. Borden and E. R. Davidson, *Acc. Chem. Res.* **29**:67 (1995); P. M. Kozlowski, M. Dupuis, and E. R. Davidson, *J. Am. Chem. Soc.* **117**:774 (1995); K. N. Houk, B. R. Beno, M. Nendel, K. Block, H. Y. Yoo, S. Wiley, and J. K. Lee, *THEOCHEM* **398**:169 (1997). E. R. Davidson, *Chem. Phys. Lett.* **284**:301 (1998).
67. For analysis of substituent effects in molecular orbital terminology, see B. K. Carpenter, *Tetrahedron* **34**:1877 (1978); F. Delbecq and N. T. Anh, *Nouv. J. Chim.* **7**:505 (1983).
68. R. K. Hill and N. W. Gilman, *J. Chem. Soc., Chem. Commun.* **1967**:619.

course of the reaction. This stereospecificity is a general feature of [3,3] sigmatropic shifts and has made them valuable reactions in enantiospecific syntheses.[69]

There is a second possible transition state for the Cope rearrangement in which the transition state adopts a boatlike geometry:

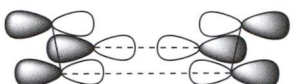

It is generally assumed that the boat transition state is higher in energy than the chair transition state. There have been several studies aimed at determining the energy difference between the two transition states. One study involved 1,1,1,8,8,8-*deuterio*-4,5-dimethyl-2,6-octadienes. Different stereoisomeric products would be predicted for the chair and boat transition states:

chair   threo   boat   erythro

Although the process is further complicated by *cis–trans* isomerizations not considered in the above structures, it was possible by analysis of the product ratio to determine that the boat transition state is about 6 kcal/mol less stable than the chair.[70] This study also demonstrated that dissociation to two independent allyl radicals is only slightly higher in energy than the boat transition state. Related experiments on deuterated 1,5-hexadiene itself indicated a difference of 5.8 kcal in $\Delta G^{\ddagger}$ for the chair and boat transition states.[71]

Another approach to determining the energy difference between the chair and boat transition states is based on measurement of the activation parameters for the isomeric alkenes **30** and **31**.[72] These two compounds are diastereomeric. Whereas **30** can attain a chairlike transition state, **30ch**, **31** can achieve bonding between the 1,6-carbons only in a boatlike transition state, **31bt**.

30   30ch   31   31bt

Comparison of the rates of rearrangement of **30** and **31** showed **30** to react faster by a factor of 18,000. This corresponds to about 14 kcal/mol in the measured $\Delta H^{\ddagger}$, but is partially compensated for by a more favorable $\Delta S^{\ddagger}$ for **31**. In the corresponding

69. R. K. Hill, in *Asymmetric Synthesis*, Vol. 3, J. D. Morrison, ed., Academic Press, New York, 1984, Chapter 8; D. Enders, M. Knopp, and R. Schiffers, *Tetrahedron Asymmetry* **7**:1847 (1996).

70. J. J. Gajewski, C. W. Benner, and C. M. Hawkins, *J. Org. Chem.* **52**:5198 (1987).

71. M. J. Goldstein and M. S. Benzon, *J. Am. Chem. Soc.* **94**:7147 (1972).

72. K. J. Shea and R. B. Phillips, *J. Am. Chem. Soc.* **102**:3156 (1980).

methylenecyclohexane analogs, the $\Delta H^{\ddagger}$ favors the chairlike transition state by 16 kcal/mol.

Some particularly striking examples of Cope rearrangement can be found in the rearrangement of *cis*-divinylcyclopropanes. However, before we go into these, let us examine vinylcyclopropane itself, which is known to rearrange thermally to cyclopentene[73]:

4 electrons, 0 nodes
antiaromatic

The most geometrically accessible transition state corresponds to a forbidden 1,3-suprafacial alkyl shift with retention of configuration. The rearrangement requires a temperature of at least 200–300°C.[74] The measured activation energy is about 50 kcal/mol, which is high enough to suggest a stepwise reaction beginning with rupture of a cyclopropane bond.[75] Support for a nonconcerted sequence comes from the observation that *cis–trans* isomerization of **32** occurs faster than the rearrangement. This isomerization presumably occurs by a reversible cleavage of the C(1)–C(2) cyclopropane bond.

Ref. 76

When this prior stereoisomerization is accounted for, the rearrangement is found to have resulted from a mixture of all possible suprafacial, antarafacial, inversion, and retention combinations in roughly equal amounts, indicating that no stereochemical pathway is strongly preferred.[77] Substituted systems, however, show higher stereoselectivity.[78] Theoretical modeling of the reaction finds no intermediate, but the transition state is diradical in character.[79]

A dramatic difference in reactivity is evident when *cis*-divinylcyclopropane is compared with vinylcyclopropane.[80] *cis*-Divinylcyclopropane can only be isolated at low temperature because it very rapidly undergoes Cope rearrangement to 1,4-cyclohepta-

73. C. G. Overberger and A. E. Borchert, *J. Am. Chem. Soc.* **82**:1007 (1960).

74. T. Hudlicky, T. M. Kutchan, and S. M. Naqui, *Org. React.* **33**:247 (1984); T. Hudlicky and J. D. Price, *Chem. Rev.* **89**:1467 (1989); J. E. Baldwin, in *Chemistry of the Cyclopropyl Group*, Vol. 2, Z. Rappoport, ed., John Wiley & Sons, Chichester, U.K., 1995, pp. 469–494.

75. D. K. Lewis, D. J. Charney, B. L. Kalra, A.-M. Plate, M. H. Woodard, S. J. Cianciosi, and J. E. Baldwin, *J. Phys. Chem.* **101**:4097 (1997).

76. M. R. Willcott and V. H. Cargle, *J. Am. Chem. Soc.* **89**:723 (1967).

77. J. J. Baldwin, K. A. Villarica, D. I. Freedberg, and F. A. L. Anet, *J. Am. Chem. Soc.* **116**:10845 (1994).

78. J. J. Gajewski, L. P. Olson, and M. R. Willcott III, *J. Am. Chem. Soc.* **118**:299 (1996); J. E. Baldwin and S. J. Bonacorsi, Jr., *J. Am. Chem. Soc.* **118**:8258 (1996).

79. E. R. Davidson and J. J. Gajewski, *J. Am. Chem. Soc.* **119**:10543 (1997); K. N. Houk, M. Nendal, O. Wiest, and J. W. Storer, *J. Am. Chem. Soc.* **119**:10545 (1997); J. E. Baldwin, *J. Comput. Chem.* **19**:222 (1998).

80. T. Hudlicky, R. Fan, J. W. Reed, and K. G. Gadamasetti, *Org. React.* **41**:1 (1992).

triene.[81] At 0°C, $\Delta H^{\ddagger}$ is 18.8 kcal/mol and $\Delta S^{\ddagger}$ is -9.4 eu:

Because of unfavorable molecular geometry, the corresponding rearrangement of *trans*-divinylcyclopropane to cycloheptatriene cannot be concerted, and it requires temperatures on the order of 190°C to occur at a significant rate. The very low energy requirement for the Cope rearrangement of *cis*-divinylcyclopropane reflects several favorable circumstances. The *cis*-orientation provides a favorable geometry for interaction of the diene termini so the loss in entropy in going to the transition state is smaller than for an acyclic diene. The bond being broken is strained, and this reduces the enthalpy of activation. The importance of the latter factor can be appreciated by comparison with *cis*-divinylcyclobutane and *cis*-divinylcyclopentane. The former compound has $\Delta H^{\ddagger} = 23$ kcal/mol for rearrangement to cyclooctadiene.[82] *cis*-Divinylcyclopentane does not rearrange to cyclononadiene, even at 250°C.[83] In the latter case, the rearrangement is presumably thermodynamically disfavored.

Divinylcyclopropane rearrangements can proceed with even greater ease if the entropy of activation is made less negative by incorporating both vinyl groups into a ring. An example of this is found in the degenerate homotropilidene rearrangement. A *degenerate rearrangement* is a reaction process in which no overall change in structure occurs. The product of rearrangement is structurally identical to the starting material. Depending on the rate at which the reaction occurs, the existence of a degenerate rearrangement can be detected by use of isotopic labels or by interpretation of the temperature dependence of NMR spectra. In the case of homotropilidene, **36**, the occurrence of a dynamic equilibrium is evident from the NMR spectrum. At low temperature, the rate of interconversion is slow and the spectrum is consistent with the presence of four vinyl protons, two allylic protons, and four cyclopropyl protons. As the temperature is raised and the rate of the rearrangement increases, it is observed that two of the vinyl protons remain essentially unchanged with respect to their chemical shift, whereas the signals of the other two vinyl protons coalesce with those of two of the cyclopropyl protons. Coalescence is also observed between the signals of the allylic protons and the two remaining cyclopropyl protons.[84] This means that sets of protons whose signals coalesce are undergoing sufficiently rapid interchange with one another to result in an averaged signal.

homotropilidene

36

81. J. M. Brown, B. T. Bolding, and J. F. Stofko, Jr., *J. Chem. Soc., Chem. Commun.* **1973**:319; M. Schneider, *Angew. Chem. Int. Ed. Engl.* **14**:707 (1975); M. P. Schneider and A. Rau, *J. Am. Chem. Soc.* **101**:4426 (1979).
82. E. Vogel, *Justus Liebigs Ann. Chem.* **615**:1 (1958); G. S. Hammond and C. D. DeBoer, *J. Am. Chem. Soc.* **86**:899 (1964).
83. E. Vogel, W. Grimme, and E. Dinne, *Angew. Chem.* **75**:1103 (1963).
84. G. Schröder, J. F. M. Oth, and R. Merenyi, *Angew. Chem. Int. Ed. Engl.* **4**:752 (1965); H. Günther, J. B. Pawliczek, J. Ulmen, and W. Grimme, *Angew. Chem. Int. Ed. Engl.* **11**:517 (1972); W. v. E. Doering and W. R. Roth, *Tetrahedron* **19**:715 (1963).

Many other examples of this type of rearrangement are known, one of the most interesting being the case of bullvalene, which is converted into itself with a first-order rate constant of $3.4 \times 10^3\,s^{-1}$ at 25°C.[85] At 100°C, the ^{1}H NMR spectrum of bullvalene exhibits a single peak at 4.22 ppm. This result indicates the "fluxional" nature of the molecule. Because of the threefold axis of symmetry present in bullvalene, the degenerate rearrangement results in all of the carbons having an identical averaged environment. The energy of activation for the rearrangement has been determined to be 13.9 kcal/mol.[86]

structures indicating changing environment of carbons in bullvalene

Substituted bullvalenes have also been studied.[87]

Other degenerate rearrangements that are even faster than that of bullvalene have been discovered. Barbaralane (**37**), rearranges to itself with a rate constant of $1.7 \times 10^7\,s^{-1}$ at 25°C.[88] The energy of activation of this rearrangement is only 7.7 kcal/mol. The lowered energy requirement can be attributed to an increase in ground-state energy due to strain. Barbaralane is less symmetrical than bullvalene. There are four different kinds of carbons and protons in the averaged structure. Only the methylene group labeled d is not affected by the degenerate rearrangement.

37

A further reduction in the barrier and increase in rate is seen with semibullvalene (**38**), in which strain is increased still more. The $\Delta G^{\ddagger}$ for this rearrangement is 5.5 kcal/mol at -143°C.[89]

38

The [3,3] sigmatropic reaction pattern is quite general for other systems that incorporate one or more heteroatoms in place of carbon in the 1,5-hexadiene unit. The

85. G. Schörder and J. F. M. Oth, *Angew. Chem. Int. Ed. Engl.* **6**:414 (1967).
86. R. Poupko, H. Zimmermann, and Z. Luz, *J. Am. Chem. Soc.* **106**:5391 (1984).
87. R. Poupko, H. Zimmermann, K. Müller, and Z. Luz, *J. Am. Chem. Soc.* **118**:7995 (1996).
88. W. v. E. Doering, B. M. Ferrier, E. T. Fossel, J. H. Hartenstein, M. Jones, Jr., G. Klumpp, R. M. Rubin, and M. Saunders, *Tetrahedron* **23**:3943 (1967); H. Günther, J. Runsink, H. Schmickler, and P. Schmitt, *J. Org. Chem.* **50**:289 (1985).
89. A. K. Cheng, F. A. L. Anet, J. Mioduski, and J. Meinwald, *J. Am. Chem. Soc.* **96**:2887 (1974); D. Moskau, R. Aydin, W. Leber, H. Günther, H. Quast, H.-D. Martin, K. Hassenruck, L. S. Miller, and K. Grohmann, *Chem. Ber.* **122**:925 (1989).

most synthetically useful and widely studied of these reactions is the Claisen rearrangement, in which an oxygen atom is present in the reacting system.[90] The simplest example of a Claisen rearrangement is the thermal conversion of allyl vinyl ether to 4-pentenal:

This reaction occurs with an energy of activation of 30.6 kcal/mol and an entropy of activation of -7.7 eu at $180°C$.[91] Both theoretical modeling of the transition-state structure and analysis of kinetic isotope effects are in accord with a concerted mechanism in which C—O bond cleavage is more advanced than C—C bond formation.[92] Claisen rearrangements show a considerable sensitivity to solvent polarity, the reaction rates increasing with solvent polarity.[93] Water is an especially favorable solvent.[94] The solvent effect is believed to be due to differential hydration of the reactants and the transition state. Hydrogen bonding contributes to stabilization of the transition state.[95]

Extensive studies on Claisen rearrangements of allyl ethers of phenols have provided further evidence bearing on [3,3] sigmatropic rearrangements.[96] For example, an important clue as to the mechanism of the Claisen rearrangement was obtained by use of ^{14}C-labeled allyl phenyl ether. It was found that the rearrangement was specific with respect to which carbon atom of the allyl group became bonded to the ring and led to the proposal of the following mechanism[97]:

The intramolecular nature of the rearrangement was firmly established by a crossover experiment in which **39** and **40** were heated simultaneously and found to yield the same products as when they were heated separately. There was no evidence for the formation of the crossover products **43** and **44**.[98] This indicates that the rearrangement must be

90. G. B. Bennett, *Synthesis* **1977**:589; S. J. Rhoads and N. R. Raulins, *Org. React.* **22**:1 (1975).
91. F. W. Schuler and G. W. Murphy, *J. Am. Chem. Soc.* **72**:3155 (1950).
92. J. J. Gajewski and N. D. Conrad, *J. Am. Chem. Soc.* **101**:6693 (1979); R. L. Vance, N. G. Rondan, K. N. Houk, F. Jensen, W. T. Borden, A. Komornicki, and E. Wimmer, *J. Am. Chem. Soc.* **110**:2314 (1988); L. Kupczyk-Subotkowska, W. H. Saunders, Jr., H. J. Shine, and W. Subotkowski, *J. Am. Chem. Soc.* **115**:5957 (1993).
93. B. Ganem, *Angew. Chem. Int. Ed. Engl.* **35**:937 (1996).
94. P. A. Grieco, E. B. Brandes, S. McCann, and J. D. Clark, *J. Org. Chem.* **54**:5849 (1989); A. Lubineau, J. Auge, N. Bellanger, and S. Caillebourdin, *J. Chem. Soc., Perkin Trans. 2*, **1992**:1631.
95. D. L. Severance and W. L. Jorgensen, *J. Am. Chem. Soc.* **114**:10966 (1992); M. M. Davidson and I. H. Hillier, *J. Phys. Chem.* **99**:6748 (1995); J. J. Gajewski, *Acc. Chem. Res.* **30**:219 (1997).
96. D. S. Tarbell, *Org. React.* **2**:1 (1944).
97. J. P. Ryan and P. R. O'Connor, *J. Am. Chem. Soc.* **74**:5866 (1952).
98. C. D. Hurd and L. Schmerling, *J. Am. Chem. Soc.* **59**:107 (1937).

The stereochemical features of the Claisen rearrangement are very similar to those described for the Cope rearrangement, and reliable stereochemical predictions can be made on the basis of the preference for a chairlike transition state. The major product has the E-configuration at the newly formed double bond because of the preference for placing the larger substituent in the pseudoequatorial position in the transition state.[99]

Studies of chiral substrates have also demonstrated that chirality is maintained in the reaction.[100] Further examples of the synthetic application of the Claisen rearrangement are discussed in Section 6.5 of Part B.

Like the Cope rearrangement, the Claisen rearrangement is sensitive to substituents on the reacting system. Cyano groups promote the rearrangement by a factor of 10^2 at positions 2 and 4 and have smaller effects at the other positions, as shown below.[101] Data are also available for methoxy groups at positions 2, 4, 5, and 6.[102]

A donor substituent, e.g., trimethylsilyloxy, at C-2 has a strongly accelerating effect.[103]

99. R. Marbet and G. Saucy, *Helv. Chim. Acta* **50**:2095 (1967); A. W. Burgstahler, *J. Am. Chem. Soc.* **82**:4681 (1960).
100. H. L. Goering and W. I. Kimoto, *J. Am. Chem. Soc.* **87**:1748 (1965).
101. C. J. Burrows and B. K. Carpenter, *J. Am. Chem. Soc.* **103**:6983 (1981).
102. R. M. Coates, B. D. Rogers, S. J. Hobbs, D. R. Peck, and D. P. Curran, *J. Am. Chem. Soc.* **109**:1160 (1987).

This effect is the basis of the synthetic importance of ester enolate Claisen rearrangements in which enolates or silyl enol ethers of allylic esters are rearranged into 4-pentenoate esters.[104]

As in the case of the Cope rearrangement, the interpretation of these substituent effects is best approached by considering the effect on transition-state stability. The effect on the transition state for the Claisen rearrangement by hydroxy substituents has been probed using both MP2/6-31G* and B3LYP calculations.[105] The effects of cyano and amino groups have also been calculated.[106] The calculated stabilization or destabilization energies are shown below and are in qualitative agreement with experimental results.

A model has been developed which treats the transition state as intermediate in character between a 2,5-diradical with strong 3–4 and 6–1 bonding and a pair of weakly interacting allyl-like radicals. By estimating the effect of the substituent on each structure and introducing a contribution for the overall free-energy change of the process, the following equation gives a good correlation with experimental activation energies.[107]

$$\Delta G^{\ddagger} = \frac{\Delta G_x^{\ddagger} \Delta G_y^{\ddagger}}{\Delta G_x^{\ddagger} + \Delta G_y^{\ddagger} + \Delta G_r}$$

where $\Delta G_x{}^{\ddagger}$ and $\Delta G_y{}^{\ddagger}$ represent substituent effects on bond formation and bond breaking, respectively, and ΔG_r is the overall free-energy change of the reaction. The correlation can be improved by including a weighting factor whose physical significance is related to the extent of bond making versus bond breaking in the transition state. This relationship can be conceptualized with reference to a More O'Ferrall diagram showing 1,6-bond formation and 3,4-bond breaking as the two coordinates (Fig. 11.8). A concerted path would proceed more or less diagonally to the product, with the transition state located along this path. The transition state should reflect the stabilizing effect of both the bond breaking (primarily at C-1, C-3, C-4, and C-6) and bond formation (primarily at C-2 and C-6).

There are also concerted rearrangements that exhibit the [2,3] sigmatropic reactivity pattern. The most well developed of these reactions are rearrangements of nitrogen[108] and sulfur[109] ylides and of allyl sulfoxides[110] and selenoxides.[111] One requirement for a [2,3] sigmatropic process is that the atom at the allylic position be able to act as a leaving group when the adjacent atom begins bonding to the allyl system, so X is normally an

103. J. J. Gajewski and J. Emrani, *J. Am. Chem. Soc.* **106**:5733 (1984); S. E. Denmark and M. A. Harmata, *J. Am. Chem. Soc.* **104**:4972 (1982).
104. S. Pereira and M. Srebnik, *Aldrichimica Acta* **26**:17 (1993).
105. H. Y. Yoo and K. N. Houk, *J. Am. Chem. Soc.* **119**:2877 (1997).
106. V. Aviyente, K. Y. Yoo, and K. N. Houk, *J. Org. Chem.* **62**:6121 (1997).
107. J. J. Gajewski and K. E. Gilbert, *J. Org. Chem.* **49**:11 (1984).
108. E. Vedejs, J. P. Hagen, B. L. Roach, and K. L. Spear, *J. Org. Chem.* **43**:1185 (1978).
109. B. M. Trost and L. S. Melvin, Jr., *Sulfur Ylides*, Academic Press, New York, 1975.
110. D. A. Evans and G. C. Andrews, *Acc. Chem. Res.* **7**:147 (1974).
111. K. B. Sharpless and R. F. Lauer, *J. Am. Chem. Soc.* **95**:2697 (1973); D. L. J. Clive, *Tetrahedron* **34**:1049 (1978).

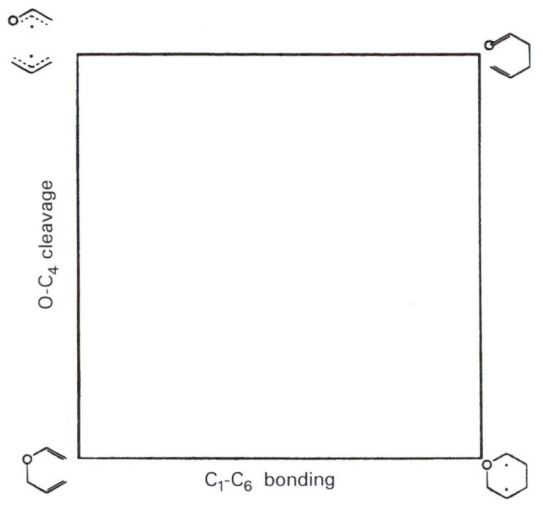

Fig. 11.8. More O'Ferrall representation of Claisen rearrangement. Adapted from J. J. Gajewski, *Acc Chem. Res.* **30**, 219 (1997).

electronegative element:

$$\text{[structure]}$$

This reaction is most facile in systems where the atoms X and Y bear formal charges, as in the case of ylides and sulfoxides:

$$CH_2=CHCH_2-\overset{+}{\underset{O^-}{S}}-R \longrightarrow RSO-CH_2CH=CH_2$$

$$CH_2=CHCH_2-\overset{+}{\underset{^-CHR}{S}}-R' \longrightarrow R'\underset{R}{SCHCH_2CH=CH_2}$$

$$CH_2=CHCH_2-\overset{+}{\underset{^-CHR}{N}}(R')_2 \longrightarrow (R')_2\underset{R}{NCHCH_2CH=CH_2}$$

Other examples of [2,3] sigmatropic rearrangements involve amine oxides and diazenes:

$$PhCH=CH-CHCH_3 \xrightarrow[\text{24 days}]{-20^\circ C} PhCH-CH=CHCH_3 \qquad \text{Ref. 112}$$
$$\overset{|}{\underset{O^-}{N}}(CH_3)_2 \qquad\qquad \overset{|}{O}-N(CH_3)_2$$

$$CH_3N-CH_2CH=CH_2 \xrightarrow[\text{Et}_3N]{(CH_3)_3COCl} CH_3-\overset{+}{\underset{\underset{-}{\parallel}}{N}}-CH_2-CH=CHCH_3 \longrightarrow CH_3-N=N-\overset{\overset{CH_3}{|}}{CHCH}=CH_2$$
$$\overset{|}{NH_2} \qquad\qquad :N: \qquad\qquad\qquad \text{Ref. 113}$$

112. Y. Yamamoto, J. Oda, and Y. Inouye, *J. Org. Chem.* **41**:303 (1976).
113. J. E. Baldwin, J. E. Brown, and G. Höfle, *J. Am. Chem. Soc.* **93**:788 (1971).

Some of the most useful synthetic applications of these reactions are for ring expansion:

Ref. 114

Further examples of synthesis using [2,3] sigmatropic reactions are given in Section 6.6 of Part B.

11.3. Cycloaddition Reactions

Cycloaddition involves the combination of two molecules in such a way that a new ring is formed. The principles of conservation of orbital symmetry also apply to concerted cycloaddition reactions and to the reverse, concerted fragmentation of one molecule into two or more smaller components (cycloreversion). The most important cycloaddition reaction from the point of view of synthesis is the *Diels–Alder reaction*. This reaction has been the object of extensive theoretical and mechanistic study, as well as synthetic application.[115] The Diels–Alder reaction is the addition of an alkene to a diene to form a cyclohexene. It is called a $[4\pi + 2\pi]$-cycloaddition reaction because four π electrons from the diene and the two π electrons from the alkene (which is called the *dienophile*) are directly involved in the bonding change. For most systems, the reactivity pattern, regioselectivity, and stereoselectivity are consistent with describing the reaction as a concerted process. In particular, the reaction is a stereospecific *syn* (suprafacial) addition with respect to both the alkene and the diene. This stereospecificity has been demonstrated with many substituted dienes and alkenes and also holds for the simplest possible example of the reaction, that of ethylene with butadiene[116]:

The issue of the concertedness of the Diels–Alder reaction has been debated extensively. It has been argued that there might be an intermediate which is diradical in

114. V. Ceré, C. Paolucci, S. Pollicino, E. Sandri, and A. Fava, *J. Org. Chem.* **43**:4826 (1978).
115. M. C. Kloetzel, *Org. React.* **4**:1 (1948); H. L. Holmes, *Org. React.* **4**:60 (1958); S. Seltzer, *Adv. Alicyclic Chem.* **2**:1 (1968); J. G. Martin and R. K. Hill, *Chem. Rev.* **61**:537 (1961); A. Wasserman, *Diels–Alder Reactions*, Elsevier, London, 1965; J. Hamer, ed., *1,4-Cycloaddition Reactions—The Diels–Alder Reaction in Heterocyclic Syntheses*, Academic Press, New York, 1967; J. Sauer and R. Sustmann, *Angew. Chem. Int. Ed. Engl.* **19**:779 (1980); R. Gleiter and M. C. Böhm, *Pure Appl. Chem.* **55**:237 (1983); R. Gleiter and M. C. Böhm, in *Stereochemistry and Reactivity of Systems Containing π Electrons*, W. H. Watson, ed., Verlag Chemie, Deerfield Beach, Florida, 1983.
116. K. N. Houk, Y.-T. Lin, and F. K. Brown, *J. Am. Chem. Soc.* **108**:554 (1986).
117. M. J. S. Dewar, S. Olivella, and J. P. Stewart, *J. Am. Chem. Soc.* **108**:5771 (1986).

character.[117] It is recognized that in the reaction between unsymmetrical alkenes and dienes, bond formation might be more advanced at one pair of termini than at the other. This can be described as being a *nonsynchronous* process. Loss of stereospecificity is to be expected, however, only if there is a intermediate in which one bond is formed and the other is not so that there can be rotation or inversion at the unbound termini:

product of suprafacial,
suprafacial addition

product of nonstereoselective
addition

Diels–Alder reactions are almost always stereospecific, which implies that if an intermediate exists, it cannot have a lifetime sufficient to permit rotation or inversion. The prevailing view is that the majority of Diels–Alder reactions are concerted processes, and most current theoretical analyses agree with this view.[118]

Another stereochemical feature of the Diels–Alder reaction is addressed by the *Alder rule*. The empirical observation is that if two isomeric adducts are possible, the one that has an unsaturated substituent(s) on the alkene oriented toward the newly formed cyclohexene double bond is the preferred product. The two alternative transition states are referred to as the *endo* and *exo* transition states:

endo addition   *exo* addition

For example, the addition of dienophiles to cyclopentadiene usually favors the *endo* stereoisomer, even though this is the sterically more congested product.

The preference for the *endo* mode of addition is not restricted to cyclic dienes such as

118. J. J. Gajewski, K. B. Peterson, and J. R. Kagel, *J. Am. Chem. Soc.* **109**:5545 (1987); K. N. Houk, Y.-T. Lin, and F. K. Brown, *J. Am. Chem. Soc.* **108**:554 (1986); E. Goldstein, B. Beno, and K. N. Houk, *J. Am. Chem. Soc.* **118**:6036 (1996); V. Branchadell, *Int. J. Quantum Chem.* **61**:381 (1997).

cyclopentadiene. By using deuterium labels, it has been found that in the addition of maleic anhydride to 1,3-butadiene, 85% of the product arises from the *endo* transition state[119]:

In general, stereochemical predictions based on the Alder rule can be made by aligning the diene and dienophile in such a way that the unsaturated substituent on the dienophile overlaps the diene π system. The stereoselectivity predicted by the Alder rule is independent of the requirement for suprafacial–suprafacial cycloaddition, since both the *endo* and *exo* transition states meet this requirement.

There are probably several factors which contribute to determining the *endo* : *exo* ratio in any specific case. These include steric effects, dipole–dipole interactions, and London dispersion forces. MO interpretations emphasize *secondary orbital interactions* between the π orbitals on the dienophile substituent(s) and the developing π bond between C-2 and C-3 of the diene. There are quite a few exceptions to the Alder rule, and in most cases the preference for the *endo* isomer is relatively modest. For example, whereas cyclopentadiene reacts with methyl acrylate in decalin solution to give mainly the *endo* adduct (75%), the ratio is solvent-sensitive and ranges up to 90% *endo* in methanol. When a methyl substituent is added to the dienophile (methyl methacrylate), the *exo* product predominates.[120]

R = H	75–90%	25–10%
R = CH₃	22–40%	78–60%

How do orbital symmetry requirements relate to $[4\pi + 2\pi]$ and other cycloaddition reactions? Let us construct a correlation diagram for the addition of butadiene and ethylene to give cyclohexene. For concerted addition to occur, the diene must adopt an *s-cis* conformation. Because the electrons that are involved are the π electrons in both the diene and dienophile, it is expected that the reaction must occur via a face-to-face rather than edge-to-edge orientation. When this orientation of the reacting complex and transition state is adopted, it can be seen that a plane of symmetry perpendicular to the planes of the

119. L. M. Stephenson, D. E. Smith, and S. P. Current, *J. Org. Chem.* **47**:4170 (1982).
120. J. A. Berson, Z. Hamlet, and W. A. Mueller, *J. Am. Chem. Soc.* **84**:297 (1962).

reacting molecules is maintained during the course of the cycloaddition:

An orbital correlation diagram can be constructed by examining the symmetry of the reactant and product orbitals with respect to this plane. The orbitals are classified by symmetry with respect to this plane in Fig. 11.9. For the reactants ethylene and butadiene, the classifications are the same as for the consideration of electrocyclic reactions on p. 610. An additional feature must be taken into account in the case of cyclohexene. The cyclohexene orbitals σ_1, σ_2, σ_1^*, and σ_2^* are called *symmetry-adapted orbitals*. We might be inclined to think of the σ and σ^* orbitals as localized between specific pairs of carbon

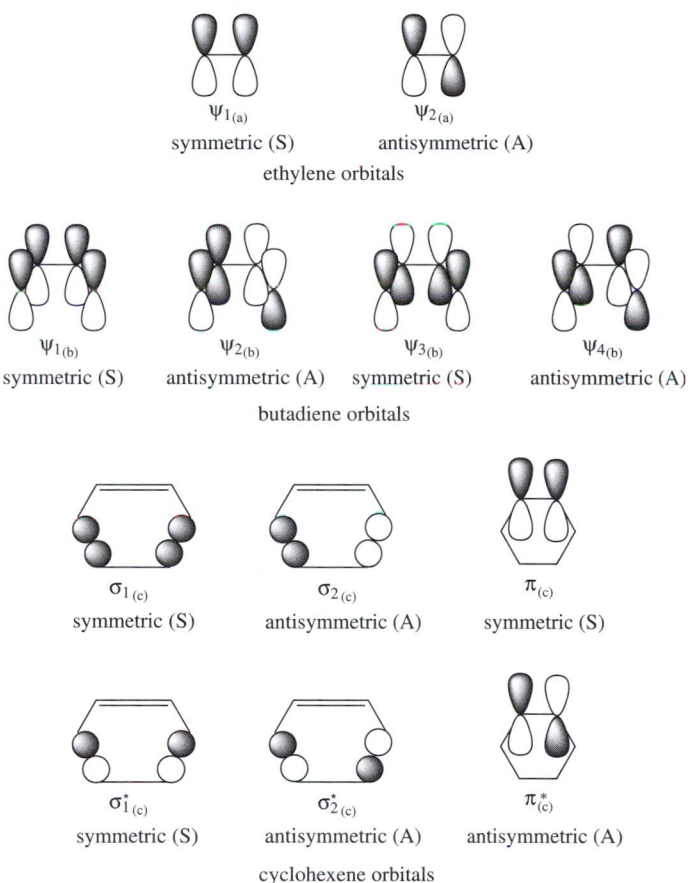

Fig. 11.9. Symmetry properties of ethylene, butadiene, and cyclohexene orbitals with respect to cycloaddition.

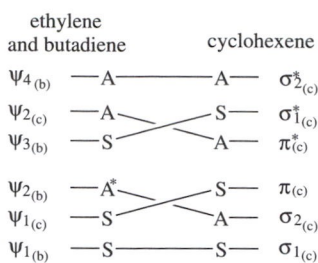

Fig. 11.10. Orbital correlation diagram for ethylene, butadiene, and cyclohexene orbitals.

atoms. This is not the case for the MO treatment because localized orbitals would fail the test of being either symmetric or antisymmetric with respect to the plane of symmetry. In the construction of orbital correlation diagrams, *all* orbitals involved must be either symmetric or antisymmetric with respect to the element of symmetry being considered.

When the orbitals have been classified with respect to symmetry, they can be arranged according to energy and the correlation lines can be drawn as in Fig. 11.10. From the orbital correlation diagram, it can be concluded that the thermal concerted cycloadditon reaction between butadiene and ethylene is allowed. All bonding levels of the reactants correlate with product ground-state orbitals. Extension of orbital correlation analysis to cycloaddition reactions involving other numbers of π electrons leads to the conclusion that the suprafacial–suprafacial addition is allowed for systems with $4n + 2$ π electrons but forbidden for systems with $4n$ π electrons.

The same conclusions are drawn by analysis of the frontier orbitals involved in cycloadditions. For the most common case of the Diels–Alder reaction, which involves dienophiles with electron-attracting substituents, the frontier orbitals are ψ_2 of the diene (which is the HOMO) and π^* of the dienophile (which is the LUMO). Reaction occurs by interaction of the HOMO and LUMO, which can be seen from the illustration below to be allowed.

The selection rules for cycloaddition reactions can also be derived from consideration of the aromaticity of the transition state. The transition states for $[2\pi + 2\pi]$ and $[4\pi + 2\pi]$ cycloadditions are depicted in Fig. 11.11. For the $[4\pi + 2\pi]$ suprafacial–suprafacial cycloaddition, the transition state is aromatic. For $[2\pi + 2\pi]$ cycloaddition, the suprafacial–suprafacial mode is antiaromatic, but the suprafacial–antarafacial mode is aromatic. In order to specify the topology of cycloaddition reactions, subscripts are added to the numerical classification. Thus, a Diels–Alder reaction is a $[4\pi_s + 2\pi_s]$ cycloaddition. The

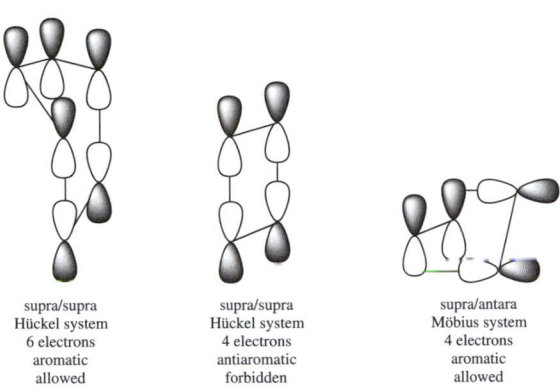

supra/supra	supra/supra	supra/antara
Hückel system	Hückel system	Möbius system
6 electrons	4 electrons	4 electrons
aromatic	antiaromatic	aromatic
allowed	forbidden	allowed

Fig. 11.11. Classification of cycloaddition reactions with respect to basis set orbitals.

allowed $[2\pi + 2\pi]$ cycloaddition process would be more completely described by the desiganation $[2\pi_s + 2\pi_a]$, where the subscripts a and s indicate the antarafacial or suprafacial nature of the addition at each component. The generalized Woodward–Hoffmann rules for cycloadditon are summarized below. These can be arrived at by orbital correlation analysis, frontier orbital theory, or classification of the transition states.

Selection rules for $m + n$ cycloadditions			
$m + n$	supra/supra	supra/antara	antara/antara
$4n$	forbidden	allowed	forbidden
$4n + 2$	allowed	forbidden	allowed

It has long been known that the Diels–Alder reaction is particularly efficient and rapid when the dienophile contains one or more electron-attracting groups. These substituent effects are illustrated by the data in Table 11.3. In the case of the diene, reactivity is increased by electron-releasing substituents. Some illustrative data are given in Table 11.4.

The regioselectivity of the Diels–Alder reaction is also sensitive to the nature of

**Table 11.3 Relative Reactivity toward Cyclopentadiene
in the Diels–Alder Reaction**

Dienophile	Relative rate[a]
Tetracyanoethylene	4.3×10^7
1,1-Dicyanoethylene	4.5×10^5
Maleic anhydride	5.6×10^4
p-Benzoquinone	9.0×10^3
Maleonitrile	91
Fumaronitrile	81
Dimethyl fumarate	74
Dimethyl maleate	0.6
Methyl acrylate	1.2
Acrylonitrile	1.0

a. From second-order rate constants in dioxane at 20°C as reported by J. Sauer, H. Wuest, and A. Mielert, *Chem. Ber.* **97**:3183 (1964).

642

CHAPTER 11
CONCERTED
PERICYCLIC
REACTIONS

Table 11.4 Relative Reactivity of Substituted Butadienes in the Diels–Alder Reaction[a]

	Dienophile	
Diene	Tetracyanoethylene	Maleic anhydride
$CH_2=CHCH=CH_2$	1	1
$CH_3CH=CHCH=CH_2$	103	3.3

$CH_2=CCH=CH_2$
    $\vert$
    CH_3
	45	2.3

$CH_3CH=CHCH=CHCH_3$	1.66×10^3	
$PhCH=CHCH=CH_2$	385	1.65

$CH_2=CCH=CH_2$
    $\vert$
    Ph
	191	8.8

$CH_3OCH=CHCH=CH_2$	5.09×10^4	12.4

$CH_2=CCH=CH_2$
    $\vert$
    OCH_3
	1.75×10^3	

$CH_3OCH=CHCH=CHOCH_3$	4.98×10^4	
Cyclopentadiene	2.1×10^6	1.35×10^3

a. C. Rücker, D. Lang, J. Sauer, H. Friege, and R. Sustmann, *Chem. Ber.* **113**:1663 (1980).

substituents on the diene and dienophile. The combination of an electron donor in the diene and an electron acceptor in the dienophile gives rise to cases A and B. The preferred orientations are shown in Scheme 11.3. There are also examples where the substituents are reversed so that the electron-donor substituents are on the dienophile and the electron-accepting substituent is on the diene. These are called *inverse-electron-demand Diels–Alder reactions* and give rise to combinations C and D in Scheme 11.3. Cases in which both the diene and the dienophile have the same type of substituent lead to poor reactivity and are rare.

Scheme 11.3. Regioselectivity of the Diels–Alder Reaction

a. ERG, electron-releasing group.
b. EWG, electron-withdrawing group.

Both the reactivity data in Tables 11.3 and 11.4 and the regiochemical relationships in Scheme 11.3 can be understood on the basis of frontier orbital theory. In reactions of types A and B illustrated in Scheme 11.3, the frontier orbitals will be the diene HOMO and the dienophile LUMO. This is illustrated in Fig. 11.12. This will be the strongest interaction because the donor substituent on the diene will raise the diene orbitals in energy whereas the acceptor substituent will lower the dienophile orbitals. The strongest interaction will be between ψ_2 and π^*. In reactions of types C and D, the pairing of diene LUMO and dienophile HOMO will be expected to be the strongest interaction because of the substituent effects, as illustrated in Fig. 11.12.

The regiochemical relationships summarized in Scheme 11.3 can be understood by considering the atomic coefficients of the frontier orbitals. Figure 11.13 gives the approximate energies and orbital coefficients for the various classes of dienes and dienophiles. The regiochemistry can be predicted by the generalization that the strongest interaction will be between the centers on the frontier orbitals having the largest orbital coefficients. For dienophiles with electron-withdrawing substituents, π^* has its largest coefficient on the β-carbon atom. For dienes with electron-donor substituents at C-1, the HOMO has its largest coefficient at C-4. This is the case designated A in Scheme 11.3. The prediction of regiochemistry which follows is that C-2 of the dienophile will interact most strongly with C-4 of the diene. This is the preferred regiochemistry for the type A Diels–Alder addition. A similar analysis of each of the other combinations in Scheme 11.3 using the orbitals in Fig. 11.13 leads to the prediction of the favored regiochemistry. Notice that in the type A and C reactions, preferential formation of the more sterically congested 1,2-disubstituted cyclohexene is predicted. These frontier orbital relationships provide an excellent interpretation of the regiochemistry of Diels–Alder reactions.[121]

Frontier orbital concepts can also serve to explain the very strong catalysis of certain Diels–Alder reactions by Lewis acids.[122] A variety of Lewis acids are effective catalysts, including $SnCl_4$, $ZnCl_2$, $AlCl_3$, and derivatives of $AlCl_3$ such as $(C_2H_5)_2AlCl$. The types of dienophiles which are subject to catalysis are typically those with carbonyl activating groups. Lewis acids are known to form complexes at the carbonyl oxygen, and this has the

Fig. 11.12. Frontier orbital interactions in Diels–Alder reactions.

121. For general discussion on the development and application of frontier orbital concepts in cycloaddition reactions, see K. N. Houk, *Acc. Chem. Res.* **8**:361 (1975); K. N. Houk, *J. Am. Chem. Soc.* **95**:4092 (1973); K. N. Houk, *Top. Curr. Chem.* **79**:1 (1979); R. Sustmann and R. Schubert, *Angew. Chem. Int. Ed. Engl.* **11**:840 (1972); J. Sauer and R. Sustmann, *Angew. Chem. Int. Ed. Engl.* **19**:779 (1980).
122. P. Laszlo and J. Lucche, *Actual. Chim.* **1984**:42.

Orbital energies are given in electron volts. The size of the circles give relative indication of orbital coefficients at each carbon. Z = conjugated electron-withdrawing substituent, e.g., $C=O$, $C\equiv N$, NO_2. C = conjugating group with modest electron-releasing capacity. X = electron-donating substituent, e.g., OCH_3, NH_2.

Fig. 11.13. Coefficients and relative energies of dienophile and diene molecular orbitals. [From K. N. Houk, *J. Am. Chem. Soc.* **95:**4092 (1973).]

effect of increasing the electron-withdrawing capacity of the substituent group:

This complexation accentuates both the energy and orbital distortion effects of the substituent and enhances both the reactivity and selectivity of the dienophile, relative to the uncomplexed compound.[123] The effects are well modeled by 3-21G-level computations on the transition-state structures.[124]

More complete interpretations of Diels–Alder regioselectivity have been developed. MO results can be analyzed from an electrostatic perspective by calculating potentials at the various atoms in the diene and dienophile. These results give a more quantitatively accurate estimate of the substituent effects.[125] Diels–Alder regioselectivity can also be accounted for in terms of HSAB theory (see Section 1.2.3). The expectation would be that the most polarizable (softest) atoms would lead to bond formation and that regioselectivity would reflect the best match between the diene and dienophile termini.[126] These ideas have been applied using 3-21G computations. The results are in agreement with the "ortho rule" for normal-electron-demand Diels–Alder reactions.[127]

Cycloadditions are not restricted to the reactions of combinations of neutral dienes and dienophiles. There are examples of corresponding reactions involving ionic intermediates. The addition of 2-methylallyl cation to cyclopentadiene is an example[128]:

A similar transformation results when trimethylsilyloxy-substituted allylic halides react with silver perchlorate in nitromethane. The resulting allylic cation gives cycloaddition reactions with dienes such as cyclopentadiene. The isolated products result from desilylation of the initial adducts[129]:

123. K. N. Houk and R. W. Strozier, *J. Am. Chem. Soc.* **95**:4094 (1973).
124. D. M. Birney and K. N. Houk, *J. Am. Chem. Soc.* **112**:4127 (1990); M. I. Menendez, J. Gonzalez, J. A. Sordo, and T. L. Sordo, *THEOCHEM* **120**:241 (1994).
125. S. D. Kahn, C. F. Pau, L. E. Overman, and W. J. Hehre, *J. Am. Chem. Soc.* **108**:7381 (1986).
126. J. L. Gazquez and F. Mendez, *J. Phys. Chem.* **98**:4591 (1994).
127. S. Damoun, G. Van de Woude, F. Mendez, and P. Geerlings, *J. Phys. Chem., A* **101**:886 (1997).
128. H. M. R. Hoffmann, D. R. Joy, and A. K. Suter, *J. Chem. Soc., B* **1968**:57; H. M. R. Hoffmann and D. R. Joy, *J. Chem. Soc., B* **1968**:1182.
129. N. Shimizu, M. Tanaka, and Y. Tsuno, *J. Am. Chem. Soc.* **104**:1330 (1982).

Scheme 11.4. Some 1,3-Dipoles

Nitrile oxide	$R—C\equiv\overset{+}{N}—\overset{..}{\overset{..}{O}}:^-$ ⟷ $R—\overset{..}{\overset{..}{C}}=\overset{+}{N}=O:$
Azides	$R—N=\overset{+}{N}=\overset{..}{N}:^-$ ⟷ $R—\overset{..}{N}—\overset{+}{N}\equiv N:$
Diazomethane	$R_2C=\overset{+}{N}=\overset{..}{N}:^-$ ⟷ $^-:CH_2—\overset{+}{N}\equiv N:$
Nitrones	$R_2C=\overset{+}{N}(R)—\overset{..}{\overset{..}{O}}:^-$ ⟷ $R_2\overset{..}{\overset{..}{C}}—\overset{+}{N}=\overset{..}{\overset{..}{O}}$
	$\overset{\textstyle |}{R}$
Nitrile imines	$R—\overset{..}{\overset{..}{C}}=\overset{+}{N}=N—R$ ⟷ $R—C\equiv\overset{+}{N}=\overset{..}{N}—R$

There is a large class of reactions known as *1,3-dipolar cycloaddition reactions* that are analogous to the Diels–Alder reaction in that they are concerted $[4\pi + 2\pi]$ cycloadditions.[130] These reactions can be represented as in the following diagram. The entity a—b—c is called the *1,3-dipolar molecule* and d—e is the *dipolarophile*.

The 1,3-dipolar molecules are isoelectronic with the allyl anion and have four electrons in a π system encompassing the 1,3-dipole. Some typical 1,3-dipolar species are shown in Scheme 11.4. It should be noted that all have one or more resonance structures showing the characteristic 1,3-dipole. The dipolarophiles are typically alkenes or alkynes, but all that is essential is a π bond. The reactivity of dipolarophiles depends both on the substituents present on the π bond and on the nature of the 1,3-dipole involved in the reaction. Because of the wide range of structures that can serve either as a 1,3-dipole or as a dipolarophile, the 1,3-dipolar cycloaddition is a very useful reaction for the construction of five-membered heterocyclic rings.

The stereochemistry of the 1,3-dipolar cycloaddition reaction is analogous to that of the Diels–Alder reaction and is a stereospecific *syn* addition. Diazomethane, for example, adds stereospecifically to the diesters **43** and **44** to yield the pyrazolines **45** and **46**, respectively.

Ref. 131

130. R. Huisgen, *Angew. Chem. Int. Ed. Engl.* **2**:565 (1963); R. Huisgen, R. Grashey, and J. Sauer, in *The Chemistry of the Alkenes,* S. Patai, ed., Interscience Publishers, London, 1965, pp. 806–878; A. Padwa, *1,3-Dipolar Addition Chemistry*, John Wiley & Sons, New York, 1984.
131. K. v. Auwers and E. Cauer, *Justus Liebigs Ann. Chem.* **470**:284 (1929); K v. Auwers and F. König, *Justus Liebigs Ann. Chem.* **496**:252 (1932); T. L. van Auken and K. L. Rinehart, *J. Am. Chem. Soc.* **84**:3736 (1962).

When both the 1,3-dipole and the dipolarophile are unsymmetrical, there are two possible orientations for addition. Both steric and electronic factors play a role in determining the regioselectivity of the addition. The most generally satisfactory interpretation of the regiochemistry of dipolar cycloadditions is based on frontier orbital concepts. As with the Diels–Alder reaction, the most favorable orientation is that which involves complementary interaction between the frontier orbitals of the 1,3-dipole and the dipolarophile. Although most dipolar cycloadditions are of the type in which the LUMO of the dipolarophile interacts with the HOMO of the 1,3-dipole, there are a significant number of systems in which the relationship is reversed. There are also some in which the two possible HOMO–LUMO interactions are of comparable magnitude.

The analysis of the regioselectivity of a 1,3-dipolar cycloaddition therefore requires information about the energy and atomic coefficients of the frontier orbitals of the 1,3-dipole and the dipolarophile. A range of 1,3-dipoles has been examined using CNDO/2 calculations. Some of the results are shown in Fig. 11.14. By using these orbital coefficients and by calculation or estimation of the relative energies of the interacting orbitals, it is possible to predict the regiochemistry of 1,3-dipolar cycloaddition reactions.[132] The most important dipolarophiles are the same types of compounds which are dienophiles. The orbital coefficients given in Fig. 11.13 can be used in the analysis of 1,3-dipolar cycloaddition reactions. Figure 11.15 gives estimates of the energies of the HOMO and LUMO orbitals of some representative 1,3-dipoles. In conjunction with the orbital

Fig. 11.14. Orbital coefficients for HOMO and LUMO π MOs of some common 1,3-dipoles. [From K. N. Houk, J. Sims, R. E. Duke, Jr., R. W. Strozier, and J. K. George, *J. Am. Chem. Soc.* **95**: 7287 (1973).]

132. K. N. Houk, J. Sims, R. E. Duke, Jr., R. W. Strozier, and J. K. George, *J. Am. Chem. Soc.* **95**:7287 (1973); R. Sustmann and H. Trill, *Angew. Chem. Int. Ed. Engl.* **11**:838 (1972); K. N. Houk, *Top. Curr. Chem.* **79**:1 (1979).

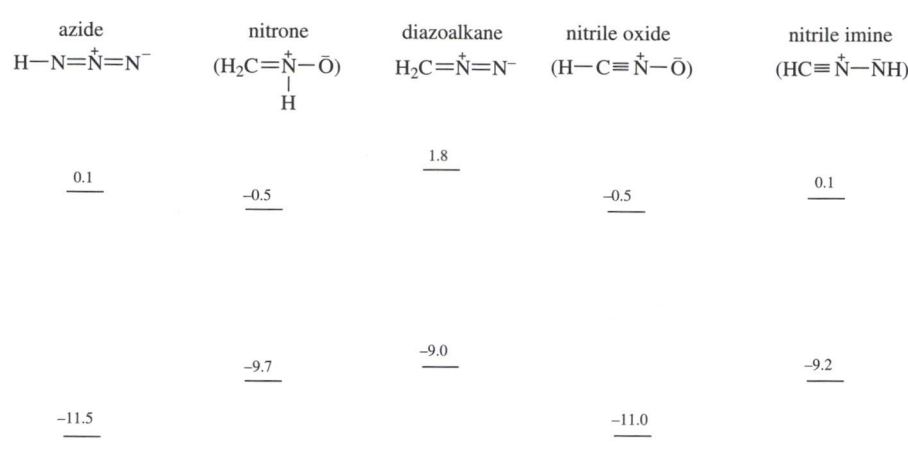

azide	nitrone	diazoalkane	nitrile oxide	nitrile imine
$H-N=\overset{+}{N}=N^-$	$(H_2C=\overset{+}{N}-\bar{O})$ $\vert$ H	$H_2C=\overset{+}{N}=N^-$	$(H-C\equiv\overset{+}{N}-\bar{O})$	$(HC\equiv\overset{+}{N}-\bar{N}H)$

Fig. 11.15. Estimated energies of frontier π orbitals for some common 1,3-dipoles. [From K. N. Houk, J. Sims, R. E. Duke, Jr., R. W. Strozier, and J. K. George, *J. Am. Chem. Soc.* **95**: 7287 (1973).]

energies given in Fig. 11.13, this information allows conclusions as to which HOMO–LUMO combination will interact most strongly for a given pair of reactants.

The regioselectivity of 1,3-dipolar cycloadditions can also be analyzed by MO calculations on transition-state models. For example, there are two possible regioisomers from the reaction of diazomethane and methyl vinyl ether, but only the 3-methoxy isomer is formed.

$$CH_3OCH=CH_2 + CH_2=\overset{+}{N}=\bar{N} \longrightarrow$$

not observed

Calculations at several levels of theory (AM1, 6-31G, and MP2/6-31G*) find lower activation energies for the transition state leading to the observed product.[133] The transition-state calculations presumably reflect the same structural features as the frontier orbital approach. The greatest transition-state stabilization should arise from the most favorable orbital interactions. As discussed earlier for Diels–Alder reactions, the HSAB theory can also be applied to interpretation of the regiochemistry of 1,3-dipolar cycloaddition reactions.[134]

$[2\pi + 2\pi]$ Additions are allowed only when the reaction is antarafacial for one of the components. There are relatively few systems which can meet the geometrical limits imposed by this restriction. Most examples of $[2\pi_s + 2\pi_a]$ additions are reactions involving ketenes.[135] Ketenes are ideal components in reactions of this type, because

133. A. Rastelli, M. Bagatti, R. Gandolfi, and M. Burdisso, *J. Chem. Soc., Faraday Trans.* **90**:1077 (1994); Y. L. Pascal, J. Chanet-Ray, R. Vessiere, and A. Zeroual, *Tetrahedron* **48**:7197 (1992).

134. F. Mendez, J. Tamariz, and P. Geerlings, *J. Phys. Chem., A* **102**:6292 (1998); A. K. Chandra and M. T. Nguyen, *J. Comput. Chem.* **19**:195 (1998).

135. W. T. Brady and R. Roe, *J. Am. Chem. Soc.* **93**:1662 (1971); W. T. Brady, in *The Chemistry of Ketenes, Allenes and Related Compounds*, S. Patai, ed., John Wiley & Sons, Chichester, U.K., 1980, Chapter. 8.

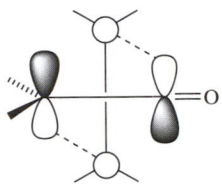

Fig. 11.16. Concerted cycloaddition of a ketene and an olefin. The orbitals represented are the HOMO of the olefin and the LUMO of the ethylenic portion of the ketene.

the linear geometry associated with the *sp*-hybridized carbon atom leads to a minimum of steric repulsion in the antarafacial transition state. MO modeling of the reaction suggests a concerted but nonsynchronous process, with bond formation at the *sp* carbon of the ketene leading bond formation at the terminal carbon.[136] The idealized geometry for the antarafacial addition of a ketene to an alkene is shown in Fig. 11.16.

The observed stereochemistry of $[2\pi + 2\pi]$ cycloadditions is in accord with the stereochemical predictions that can be made on the basis of this model. For example, *E*- and *Z*-2-butene give stereoisomeric products with ethoxyketene.[137] For monosubstituted alkenes, the substituent is vicinal and *cis* to the ethoxy group in the cyclobutanone product. This is exactly the stereochemistry predicted by the model in Fig. 11.16, since it maximizes the separation of the alkyl and ethoxy substituents in the transition state.

There is an alternative description of the $[2\pi + 2\pi]$ cycloaddition reactions of ketenes. This formulation is $[2\pi_s + (2\pi_s + 2\pi_s)]$.[138] The basis set orbital array is shown below. This system has Hückel-type topology, and with six π electrons involved, it would be an allowed process.

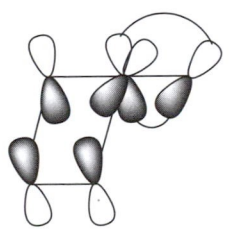

This analysis is equally compatible with available data, and some predictions of stereoselectivity and reactivity based on this model are in better accord with experimental results

136. L. A. Burke, *J. Org. Chem.* **50**:3149 (1985).
137. T. DoMinh and O. P. Strausz, *J. Am. Chem. Soc.* **92**:1766 (1970).
138. D. J. Pasto, *J. Am. Chem. Soc.* **101**:37 (1979).

than predictions derived from the $[2\pi_s + 2\pi_a]$ transition state. The transition state found for this mode of addition is very nonsynchronous, with a strong initial interaction of the ketene C-2 sp carbon with both carbons of the alkene, and has considerable polar character.[139]

Charges and atomic distances in MP2/6-31G transition
state structure for ketene + ethene cycloaddition

The prediction of the Woodward–Hoffmann rules that thermal concerted cycloadditions are allowed for combinations in which $4n + 2$ π electrons are involved has stimulated the search for combinations with 10 and larger numbers of participating electrons. An example of a [6 + 4] cycloaddition is the reaction of tropone with 2,5-dimethyl-3,4-diphenylcyclopentadienone[140]:

Heptafulvene derivatives undergo [8 + 2] cycloadditions.

Ref. 141

Ref. 142

139. X. Wang and K. N. Houk, *J. Am. Chem. Soc.* **112**:1754 (1990).
140. K. N. Houk and R. B. Woodward, *J. Am. Chem. Soc.* **92**:4145 (1970).
141. W. v. E. Doering and D. W. Wiley, *Tetrahedron* **11**:183 (1960).
142. J. Bindl, T. Burgemeister, and J. Daub, *Justus Liebigs Ann. Chem.* **1985**:1346.

Flexible six- and eight-π-electron systems would impose an entropic barrier to concerted 10-π-electron concerted reactions. Most of the examples, as in the cases above, involve cyclic systems in which the two termini of the conjugated system are held close together.

General References

M. J. S. Dewar and R. C. Dougherty, *The PMO Theory of Organic Chemistry*, Plenum Press, New York, 1975.

I. Fleming, *Frontier Orbitals and Organic Chemical Reactions*, Wiley-Interscience, New York, 1976.

T. L. Gilchrist and R. C. Storr, *Organic Reactions and Orbital Symmetry*, 2nd ed.. Cambridge University Press, Cambridge, U.K., 1979.

A. P. Marchand and R. E. Lehr, *Pericyclic Reactions*, Vols. I and II, Academic Press, New York, 1977.

E. N. Marvell, *Thermal Electrocyclic Reactions*, Academic Press, New York, 1980.

S. J. Rhoads and N. R. Raulins, "The Claisen and Cope Rearrangements," *Org. React.* **22**:1 (1975).

L. Salem, *Electrons in Chemical Reactions*, John Wiley & Sons, New York, 1982.

R. B. Woodward and R. Hoffmann, *The Conservation of Orbital Symmetry*, Academic Press, New York, 1970.

Problems

(References for these problems will be found on page 801.)

1. Show, by constructing a correlation diagram, whether each of the following disrotatory cyclizations is symmetry allowed:

   (a) pentadienyl cation to cyclopentenyl cation
   (b) pentadienyl anion to cyclopentenyl anion

2. Which of the following reactions are allowed according to the orbital symmetry conservation rules? Explain.

(a)

(b)

(c)

(d)

(e)

(f)

3. All-*cis*-cyclononatetraene undergoes a spontaneous electrocyclic ring closure at 25°C to afford a single product. Suggest a structure for this product. Also, describe an alternative symmetry-allowed electrocyclic reaction that would lead to an isomeric bicyclononatriene. Explain why the product of this alternative reaction pathway is not formed.

4. Offer a mechanistic explanation of the following observations.

(a) The 3,5-dinitrobenzoate esters of the epimers shown below both yield 3-cyclopenten-1-ol on hydrolysis in dioxane–water. The relative rates, however, differ by a factor of 10 million! Which is more reactive and why?

(b) Optically active **A** racemizes on heating at 50°C with a half-life of 24 h.

(c) The anions of various 2-vinylcyclopropanols undergo very facile vinylcyclopropane rearrangement to give 3-cyclopenten-1-ols. For example:

Offer an explanation for the facility of the reaction, as compared to the vinylcyclopropane rearrangement of hydrocarbons, which requires a temperature above 200°C. Consider concerted reaction pathways which would account for the observed stereospecificity of the reaction.

(d) On being heated to 320–340°C, compound **B** produces 1,4-dimethoxynaphthalene and 1-acetoxybutadiene.

(e) It has been found that compounds **C** and **D** are opened at $-25°C$ to allylic anions in the presence of strong bases such as lithium di-*t*-butylamide. In contrast, **E** opens only slowly at 25°C.

(f) When the 1,6-methanodeca-1,3,5,7,9-pentaene structure is modified by fusion of two benzene rings as in **Fa**, it exists as the valence tautomer **Fb**.

5. Suggest a mechanism by which each transformation could occur. More than one step is involved in each case.

(a)

(b)

(c)

(d)

PhCH$_2$N=CHPh $\xrightarrow[\substack{3) \text{H}_2\text{O}}]{\substack{1) \text{LiN}(i\text{-C}_3\text{H}_7)_2/\text{THF} \\ 2) E\text{-PhCH}=\text{CHPh}}}$

(e)

(f)

(g)

+ CH$_3$O$_2$CC≡CCO$_2$CH$_3$ $\xrightarrow{\Delta}$

(h)

(i)

6. Predict which is the more likely mode of ring closure for oxonin:

   (a) cyclization of the tetraene unit to a bicyclo[6.1.0] system, or
   (b) cyclization of a triene unit to a bicyclo [4.3.0] system.

oxonin

7. Give the structure, including stereochemistry, of the products expected for the
   following reactions.

(a)

+ Ph$_2$C=C=O ⟶

(b) (CH$_3$)$_2$C=C=O + C$_2$H$_5$OC≡CH $\xrightarrow{0°C}$

(c)

$\xrightarrow[25°]{KH}$

(d) PhCH$_2$CHSCH=CH$_2$ $\xrightarrow[\substack{DME, reflux \\ 12\ h}]{H_2O}$ (C$_{11}$H$_{14}$O)

(e)

$$(CH_3)_3CCHC=C(CH_3)_2 + \text{} \xrightarrow[\substack{CaCO_3,\ CH_3NO_2}]{AgClO_4}$$

with OSi(CH$_3$)$_3$ and Cl substituents

(f)

$\xrightarrow{K^+ {}^-OC(CH_3)_3}$

(g)

+ H$_3$CO$_2$CC≡CCO$_2$CH$_3$

(h)

$\xrightarrow{\Delta}$

(i)

$\xrightarrow{\Delta}$

(j)

+ $\overset{(CH_3)_3C}{\underset{N≡C}{>}}$C=C=O ⟶

(k)

$\xrightarrow{(C_2H_5)_3N}$

(l)

$\xrightarrow{heat}$

8. In the series of oxepins **A** ($n = 3$, 4, and 5), only the compound with $n = 5$ undergoes rearrangement (at 60°C) to the isomeric oxepin **B**. The other two compounds ($n = 3$ or 4) are stable, even at much higher temperature. When **B** ($n = 3$) was synthesized by an alternate route, it showed no tendency to revert to **A** ($n = 3$). Explain these observations.

9. Bromocyclooctatetraene rearranges to E-β-bromostyrene. The rate of the rearrangement is solvent-dependent, the first-order rate constant increasing from $\sim 10^{-7}\ s^{-1}$ in cyclohexane to $\sim 10^{-3}\ s^{-1}$ in acetonitrile at 80°C. In the presence of lithium iodide, the product is E-β-iodostyrene, although E-β-bromostyrene is unaffected by lithium iodide under the conditions of the reaction. Suggest a mechanism for this rearrangement.

10. The BF$_3$-catalyzed rearrangement of phenyl pentadienyl ether has been shown to proceed strictly intramolecularly and with the isotopic pattern shown. Analyze the

possibilities for concerted rearrangement in frontier orbital terms, and specify those mechanisms which are consistent with the observed result. Discuss the role of the ctalyst in the reaction.

11. At room temperature, the diene **1** and maleic anhydride react to give **A** and **B**. When the reaction mixture is heated to 160°C, **C** and **D** are formed, and as heating is continued, the **C** : **D** ratio decreases. Explain the reason for the different products at the two reaction temperatures and the reason for the change of product ratio with time at the higher temeprature.

12. Classify the following reactions as electrocyclizations, sigmatropic rearrangements, cycloadditions, etc., and give the correct symbolism for the electrons involved in each concerted process. Some of the reactions proceed by two sequential processes.

(a)

(b)

R = Si(Ph)₂C(CH₃)₃

(c)

(d)

(e)

(f)

13. The "ene" reaction is a concerted reaction in which addition of an alkene to an electrophilic olefin occurs with migration of a hydrogen and the alkene double bond. For example:

Depict the orbital array through which this process could occur in a concerted manner.

14. Predict the regiochemistry and stereochemistry of the following cycloaddition reactions and indicate the basis for your prediction.

(a)

(b)

(c)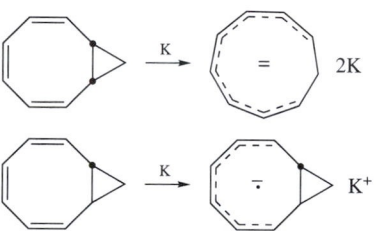

(d)

CH_2＝$CHCH$＝CH_2 +

(e)

15. On treatment with potassium metal, *cis*-bicyclo[6.1.0]nona-2,4,6-triene gives a monocyclic dianion. The *trans* isomer under similar conditions gives only a bicyclic
monoanion (radical anion). Explain how the stereochemistry of the ring junction
can control the course of these reductions.

16. The following compounds are capable of degenerate rearrangement at the temperature
given. Identify chemical processes which are consistent with the temperature and
which would lead to degenerate rearrangement. Indicate by an appropriate labeling
scheme the carbons and hydrogens which become equivalent as a result of the
rearrangement process you have suggested.

17. On heating at 225°C, 5-allylcyclohexa-1,3-diene, **A**, undergoes intramolecular cycloaddition to give the tricyclic nonene **B**. The mechanism of formation of **B** was probed using the deuterium-labeled sample of **A** which is shown. Indicate the position of deuterium labels in product **B** if the reaction proceeds by (a) a [2 + 2] cycloaddition or (b) a [4 + 2] cycloaddition.

18. The thermal behavior of **A** and **B** above 150°C has been studied. Both in the gas phase and in solution, each compound yields a 3 : 5 mixture of E,Z-1,5-cyclooctadiene (**C**) and Z,Z-1,5-cyclooctadiene (**D**). When hexachlorocyclopentadiene is present, compound **E** is found in place of **C**, but the amount of **D** formed is about the same as in its absence. Formulate a description of the thermolysis mechanism that is consistent with these facts and the general theory of thermal electrolytic reactions.

19. Suggest mechanisms for the following reactions. Classify the orbital symmetry-controlled process as clearly as you can with respect to type.

(a) $CH_2{=}CHCH_2CH_2CH_2CHO$

(b) $(CH_2{=}CH)_2CHOH + CH_2{=}CHOEt \xrightarrow{Hg^{2+}} CH_2{=}CHCH{=}CHCH_2CH_2CHO$

(c)

(d)

(e)

(f)

(g)

(h)

(i)

$(t\text{-Bu})_2C{=}C{=}CH_2 \ + \ CH_3CO_3H \longrightarrow$

(j)

(k)

(l)

(m)

(n)

(o)

(p)

(q)

20. Compound **1** undergoes rearrangement to **2** in SO_2 at $-66°C$. The deuterium label becomes uniformly scrambled among all the carbon atoms in **2**.

1   **2** X = Cl

It has been suggested that the cyclononatetraenyl cation might be an intermediate, and several structures have been calculated to determine their relative energy. Structure **3** is the lowest-energy structure. It has an *E*-configuration at one double bond. Structure **4** is also an energy minimum, but it is 21.6 kcal higher in energy than **3**. The all *Z*-isomer is not an energy minimum. By the NMR shift criterion, **5** is strongly antiaromatic whereas **3** is aromatic. The calculated nucleus-independent chemical shift (NICS) are given. (See section 9.1 to review NICS as a criterion of aromaticity.) Deduce a mechanism which accounts for the deuterium scrambling in the product.

3, C_2 (NICS = −13.4)

6, C_s (NICS = −0.9)   **7**, C_1 (NICS = −6.0)

4, C_s (NICS = 8.6)   **5**, C_{2v} (NICS = 42.0)   **8**, C_s (NICS = −11.8)

12

Free-Radical Reactions

12.1. Generation and Characterization of Free Radicals

12.1.1. Background

A free-radical reaction is a chemical process which involves molecules having unpaired electrons. The radical species could be a starting compound or a product, but the most common cases are reactions that involve radicals as intermediates. Most of the reactions discussed to this point have been heterolytic processes involving polar intermediates and/or transition states in which all electrons remained paired throughout the course of the reaction. In radical reactions, *homolytic* bond cleavages occur. The generalized reactions shown below illustrate the formation of alkyl, vinyl, and aryl free radicals by hypothetical homolytic processes.

$$Y\cdot + R_3C\!-\!X \longrightarrow R_3C\cdot + X\!-\!Y$$

$$R_2C\!=\!C\overset{R}{\underset{X}{\diagup}} \longrightarrow R_2C\!=\!C\overset{R}{\diagup}\cdot + X\cdot$$

$$\text{Ph}\!-\!X\!-\!Y\!-\!Z \longrightarrow \text{Ph}\!-\!X\!-\!Y\cdot + Z\cdot \longrightarrow \text{Ph}\cdot + X\!=\!Y$$

The idea that trivalent carbon atoms having only seven valence electrons could be involved in organic reactions took firm hold in the 1930s. Two studies have special historical significance in the development of the concept of free-radical reactions. First, the work of Gomberg around 1900 provided evidence that when triphenylmethyl chloride was treated with silver metal, the resulting solution contained $\text{Ph}_3\text{C}\cdot$ in equilibrium with a less reactive molecule. It was originally thought that the more stable molecule was hexaphenylethane, but eventually this was shown not to be so. The dimeric product is actually a cyclohexadiene derivative.[1]

The dissociation constant is small, only about $2 \times 10^{-4} M$ at room temperature. The presence of the small amount of the radical at equilibrium was deduced from observation of reactions that could not be reasonably attributed to a normal hydrocarbon.

The second set of experiments was carried out in 1929 by Paneth. The decomposition of tetramethyllead was carried out in such a way that the decomposition products were carried by a flow of inert gas over a film of lead metal. The lead was observed to disappear with re-formation of tetramethyllead. The conclusion was reached that methyl radicals must exist long enough in the gas phase to be transported from the point of decomposition to the lead film, where they were reconverted to tetramethyllead.

$$Pb(CH_3)_{4(g)} \xrightarrow{450°C} Pb_{(s)} + 4CH_3\cdot_{(g)}$$

$$4CH_3\cdot_{(g)} + Pb_{(s)} \xrightarrow{100°C} Pb(CH_3)_{4(g)}$$

Since these early experiments, a great deal of additional information about the existence and properties of free-radical intermediates has been developed. In this chapter, we will discuss the structure of free radicals and some of the special properties associated with free radicals. We will also discuss some of the key chemical reactions in which free-radical intermediates are involved.

12.1.2. Stable and Persistent Free Radicals

Most organic free radicals have very short lifetimes, but certain structural features enhance stability. Radicals without special stabilization rapidly dimerize or disproportionate. The usual disproportionation process for alkyl radicals involves transfer of a hydrogen from the carbon β to the radical site, leading to formation of an alkane and an alkene:

1. H. Lankamp, W. Th. Nauta, and C. MacLean, *Tetrahedron Lett.* **1968**:249; J. M. McBride, *Tetrahedron* **30**:2009 (1974); K. J. Skinner, H. S. Hochster, and J. M. McBride, *J. Am. Chem. Soc.* **96**:4301 (1974).

Radicals also rapidly abstract hydrogen atoms from many types of solvents, and most radicals are highly reactive toward oxygen.

Hydrogen atom abstraction
$$—\overset{|}{\underset{|}{C}}\cdot\ +\ H—Y\ \longrightarrow\ —\overset{|}{\underset{|}{C}}—H\ +\ Y\cdot$$

Addition to oxygen
$$—\overset{|}{\underset{|}{C}}\cdot\ +\ O_2\ \longrightarrow\ —\overset{|}{\underset{|}{C}}—O—O\cdot$$

A few free radicals are indefinitely stable. Entries 1, 4, and 6 in Scheme 12.1 are examples. These molecules are just as stable under ordinary conditions of temperature and atmosphere as typical closed-shell molecules. Entry 2 is somewhat less stable to oxygen, although it can exist indefinitely in the absence of oxygen. The structures shown in entries 1, 2, and 4 all permit extensive delocalization of the unpaired electron into aromatic rings. These highly delocalized radicals show no tendency toward dimerization or disproportionation. Radicals that have long lifetimes and are resistant to dimerization or other routes for bimolecular self-annihilation are called *stable free radicals*. The term *inert free radical* has been suggested for species such as entry 4, which is unreactive under ordinary conditions and is thermally stable even at 300°C.[2]

Entry 3 has only alkyl substituents and yet has a significant lifetime in the absence of oxygen. The tris(*t*-butyl)methyl radical has an even longer lifetime, with a half-life of about 20 min at 25°C.[3] The steric hindrance provided by the *t*-butyl substituents greatly retards the rates of dimerization and disproportionation of these radicals. They remain highly reactive toward oxygen, however. The term *persistent radicals* is used to describe these species, because their extended lifetimes have more to do with kinetic factors than with inherent stability.[4] Entry 5 is a sterically hindered perfluorinated radical and is even more long-lived than similar alkyl radicals.

There are only a few functional groups that contain an unpaired electron and yet are stable in a wide variety of structural environments. The best example is the nitroxide group, and numerous specific nitroxide radicals have been prepared and characterized. The unpaired electron is delocalized between nitrogen and oxygen in a structure with an N—O bond order of 1.5.

$$\underset{R}{\overset{R}{\diagdown}}\overset{+}{N}—\ddot{\underset{..}{O}}{:}^{-}\ \longleftrightarrow\ \underset{R}{\overset{R}{\diagdown}}\dot{N}—\ddot{\underset{..}{O}}\cdot\ \ or\ \ \underset{R}{\overset{R}{\diagdown}}\cdot N\text{∺}\ddot{\underset{..}{O}}\cdot$$

Many nitroxides are very stable under normal conditions, and heterolytic reactions can be carried out on other functional groups in the molecule without destroying the nitroxide group.[5] Nitroxides are very useful in biochemical studies by virtue of being easily detected paramagnetic probes.[6]

2. M. Ballester, *Acc. Chem. Res.* **18**:380 (1985).
3. G. D. Mendenhall, D. Griller, D. Lindsay, T. T. Tidwell, and K. U. Ingold, *J. Am. Chem. Soc.* **96**:2441 (1974).
4. For a review of various types of persisent radicals, see D. Griller and K. U. Ingold, *Acc. Chem. Res.* **9**:13 (1976).
5. For reviews of the preparation, reactions, and uses of nitroxide radicals, see J. F. W. Keana, *Chem. Rev.* **78**:37 (1978); L. J. Berliner, ed., *Spin-Labelling*, Vol. 2, Academic Press, New York, 1979; S. Banerjee and G. K. Trivedi, *J. Sci. Ind. Res.* **54**:623 (1995); L. B. Volodarsky, V. A. Reznikov, and V. I. Ovcharenko, *Synthetic Chemistry of Stable Nitroxides*, CRC Press, Boca Raton, Florida, 1994.
6. G. L. Millhauser, W. R. Fiori, and S. M. Miick, *Methods Enzymol.* **246**:589 (1995).

Scheme 12.1. Stability of Some Free Radicals

Structure	Conditions for stability
1[a]	Indefinitely stable as a solid, even in the presence of air.
2[b]	Crystalline substance is not rapidly attacked by oxygen, although solutions are air-sensitive; the compound is stable to high temperature in the absence of oxygen.
3[c] $(CH_3)_3C$ $\overset{\cdot}{\underset{H}{C}}-C(CH_3)_3$	Stable in dilute solution ($< 10^{-5}\ M$) below $-30°C$ in the absence of oxygen, $t_{1/2}$ of 50 s at 25°C.
4[d]	Stable in solution for days, even in the presence of air. Indefinitely stable in solid state. Thermally stable up to 300°C.
5[e] $(CF_3)_2CF$ $\overset{\cdot}{C}-CF_2CF_3$ $(CF_3)_2CF$	Stable to oxygen; thermally stable to 70°C.
6[e] galvinoxyl	Stable to oxygen; stable to extended storage as a solid. Slowly decomposes in solution.
7[f] $(CH_3)_3C$ $\overset{}{\underset{(CH_3)_3C}{N-O^\cdot}}$	Stable to oxygen even above 100°C.

a. C. F. Koelsch, *J. Am. Chem. Soc.* **79:**4439 (1957).
b. K. Ziegler and B. Schnell, *Justus Liebigs Ann. Chem.* **445:**266 (1925).
c. G. D. Mendenhall, D. Griller, D. Lindsay, T. T. Tidwell, and K. U. Ingold, *J. Am. Chem. Soc.* **96:**2441 (1974).
d. M. Ballester, J. Riera, J. Castañer, C. Badia, and J .M. Monsó, *J. Am. Chem. Soc.* **93:**2215 (1971).
e. K. V. Scherer, Jr., T. Ono, K. Yamanouchi, R. Fernandez, and P. Henderson, *J. Am. Chem. Soc.* **107:**718 (1985).
f. G. M. Coppinger, *J. Am. Chem. Soc.* **79:**501 (1957); P. D. Bartlett and T. Funahashi, *J. Am. Chem. Soc.* **84:**2596 (1962).
g. A. K. Hoffmann and A. T. Henderson, *J. Am. Chem. Soc.* **83:**4671 (1961).

Although the existence of the stable and persistent free radicals is of significance in establishing that free radicals can have extended lifetimes, most free-radical reactions involve highly reactive intermediates that have fleeting lifetimes and are present at very low concentrations. The techniques for study of radicals under these conditions are the subject of the next section.

12.1.3. Direct Detection of Radical Intermediates

The distinguishing characteristic of free radicals is the presence of an unpaired electron. Species with an unpaired electron are said to be *paramagnetic*. The most useful method for detecting and characterizing unstable radical intermediates is *electron paramagnetic resonance* (EPR) spectroscopy. *Electron spin resonance* (ESR) spectroscopy is synonymous. EPR spectroscopy detects the transition of an electron between the energy levels associated with the two possible orientations of electron spin in a magnetic field. An EPR spectrometer records the absorption of energy when an electron is excited from the lower to the higher state. The energy separation is very small on an absolute scale and corresponds to the energy of microwaves. EPR spectroscopy is a highly specific tool for detecting radical species because only molecules with unpaired electrons give rise to EPR spectra. As with other spectroscopic methods, detailed analysis of the absorption spectrum can give structural information. One feature that is determined is the g value, which specifies the separation of the two spin states as a function of the magnetic field strength of the spectrometer:

$$hv = E = g\mu_B H$$

where μ_B is a constant, the Bohr magneton ($=9.273$ erg/G), and H is the magnetic field in gauss. The measured value of g is a characteristic of the particular type of radical, just as the line positions in IR and NMR spectra are characteristic of the absorbing species.

A second type of structural information can be deduced from the *hyperfine splitting* in EPR spectra. The origin of this splitting is closely related to the factors that cause spin–spin splitting in proton NMR spectra. Certain nuclei have a magnetic moment. Those which are of particular interest in organic chemistry include ^{1}H, ^{13}C, ^{14}N, ^{19}F, and ^{31}P. Interaction of the electron with one or more of these nuclei splits the signal arising from the unpaired electron. The number of lines is given by the following equation:

$$\text{number of lines} = 2nI + 1$$

where I is the nuclear spin quantum number and n is the number of equivalent interacting nuclei. For ^{1}H, ^{13}C, ^{19}F, and ^{31}P, $I = \frac{1}{2}$. Thus, a single hydrogen splits a signal into a doublet. Interaction with two equivalent hydrogens, as in a methylene group, gives rise to splitting that produces three lines. This splitting is illustrated in Fig. 12.1. Nitrogen (^{14}N) with $I = 1$ splits each energy level into three lines. Neither ^{12}C nor ^{16}O has a nuclear magnetic moment, and, just as they cause no splitting in NMR spectra, they have no effect on the multiplicity in EPR spectra.

A great deal of structural information can be obtained by analysis of the hyperfine splitting pattern of a free radical. If we limit our discussion for the moment to radicals without heteroatoms, the number of lines indicates the number of interacting hydrogens, and the magnitude of the splitting, given by the hyperfine coupling constant a, is a measure of the unpaired electron density in the hydrogen $1s$ orbital. For planar systems in which the

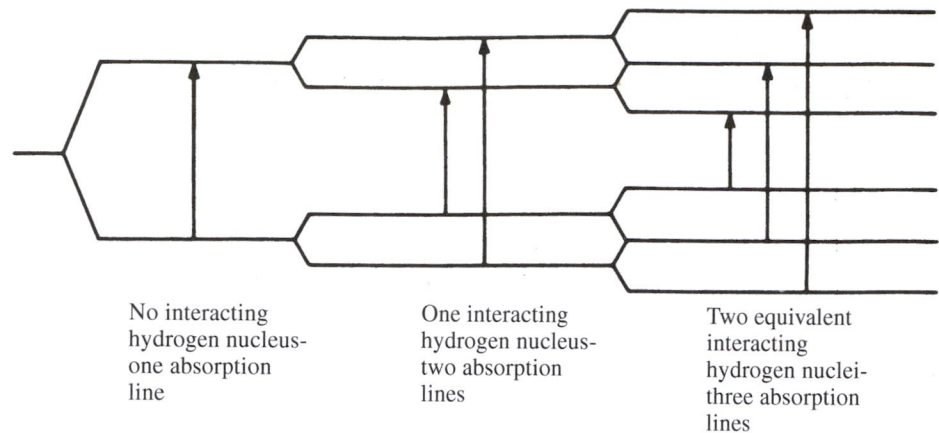

| No interacting hydrogen nucleus-one absorption line | One interacting hydrogen nucleus-two absorption lines | Two equivalent interacting hydrogen nuclei-three absorption lines |

Fig. 12.1. Hyperfine splitting in EPR spectra.

unpaired electron resides in a π-orbital system, the relationship between electron spin density and the splitting constant is given by the McConnell equation:[7]

$$a = \rho Q$$

where a is the hyperfine coupling constant for a proton, Q is a proportionality constant (about 23 G), and ρ is the spin density on the carbon to which the hydrogen is attached. For example, taking $Q = 23.0\,G$, the hyperfine splitting in the benzene radical anion may be readily calculated by taking $\rho = \frac{1}{6}$, since the one unpaired electron is distributed equally among the six carbon atoms. The calculated value of $a = 3.83$ is in good agreement with the observed value. The spectrum of benzene radical anion (Fig. 12.2a) consists of seven lines separated by 3.75 G.

The EPR spectrum of the ethyl radical presented in Fig. 12.2b is readily interpreted, and the results are relevant to the distribution of unpaired electron density in the molecule. The 12-line spectrum is a triplet of quartets resulting from unequal coupling of the electron spin to the α and β protons. The two coupling constants are $a_\alpha = 22.38\,G$ and $a_\beta = 26.87\,G$ and imply extensive delocalization of spin density through the σ bonds. Note that EPR spectra, unlike NMR and IR spectra, are displayed as the derivative of absorption rather than as absorption.

EPR spectra have been widely used in the study of reactions to detect free-radical intermediates. An interesting example involves the cyclopropylmethyl radical. Much chemical experience has indicated that this radical is unstable, giving rise to 3-butenyl radical rapidly after being generated.

$$\triangleright\!\!-CH_2\cdot \longrightarrow \begin{array}{c} CH_2 \\ | \\ \dot{C}H_2 \end{array}\!\!CH\!=\!CH_2$$

7. H. M. McConnell, *J. Chem. Phys.* **24**:764 (1956).

(a)

5 Gauss

(b)

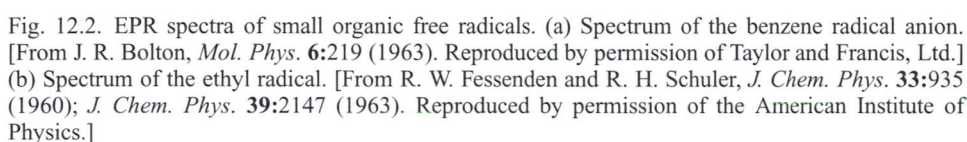

Fig. 12.2. EPR spectra of small organic free radicals. (a) Spectrum of the benzene radical anion. [From J. R. Bolton, *Mol. Phys.* **6**:219 (1963). Reproduced by permission of Taylor and Francis, Ltd.] (b) Spectrum of the ethyl radical. [From R. W. Fessenden and R. H. Schuler, *J. Chem. Phys.* **33**:935 (1960); *J. Chem. Phys.* **39**:2147 (1963). Reproduced by permission of the American Institute of Physics.]

The radical is generated by photolytic decomposition of di-*t*-butyl peroxide in methylcy-clopropane, a process that leads to selective abstraction of a methyl hydrogen from methylcyclopropane:

$$(CH_3)_3COOC(CH_3)_3 \xrightarrow{h\nu} 2\ (CH_3)_3C—O\cdot$$

$$(CH_3)_3C—O\cdot + \triangleright—CH_3 \longrightarrow \triangleright—CH_2\cdot + (CH_3)_3COH$$

Below $-140°C$, the EPR spectrum observed was that of the cyclopropylmethyl radical. If the photolysis was done above $-140°C$, however, the spectrum of a second species was seen, and above $-100°C$, this was the only spectrum observed. This second spectrum could be shown to be that of the 3-butenyl radical.[8] This study also established that the 3-butenyl radical did not revert to the cyclopropylmethyl radical on being cooled back to $-140°C$. The conclusion is that the ring opening of the cyclopropyl radical is a very facile process and that the lifetime of the cyclopropyl radical above $-100°C$ is very short. Even though the equilibrium favors the 3-butenyl radical, the reversible ring closure can be detected by isotopic labeling experiments, which reveal the occurrence of deuterium migration:

$$CH_2=CHCH_2CD_2\cdot \rightleftharpoons \underset{H_2C—CD_2}{\overset{\overset{\displaystyle CH_2\cdot}{H}}{\triangle}} \rightleftharpoons CH_2=CHCD_2CH_2\cdot$$

8. J. K. Kochi, P. J. Krusic, and D. R. Eaton, *J. Am. Chem. Soc.* **91**:1877 (1969).

The rates of both the ring opening ($k = 1 \times 10^8$ s^{-1} at 25°C) and the ring closure ($k = 3 \times 10^3$ s^{-1}) have been measured and confirm that only a very small amount of the cyclopropylmethyl radical is present at equilibrium, in agreement with the EPR results.[9]

It is important to emphasize that direct studies such as those carried out on the cyclopropylmethyl radical can be done with low *steady-state* concentrations of the radical. In the case of the study of the cyclopropylmethyl radical, removal of the source of irradiation leads to rapid disappearance of the EPR spectrum, because the radicals react rapidly and are not replaced by continuing radical formation. Under many conditions, the steady-state concentration of a radical intermediate may be too low to permit direct detection. Failure to observe an EPR signal, therefore, cannot be taken as conclusive evidence against a radical intermediate.

A technique called *spin trapping* can also be used to study radicals. A diamagnetic molecule that has the property of reacting rapidly with radicals to give a stable paramagnetic species is introduced into the reaction system being studied. As radical intermediates are generated, they are trapped by the reactive molecule to give more stable, detectable radicals. The most useful spin traps are nitroso compounds. They rapidly react with radicals to give stable nitroxides.[10] Analysis of the EPR spectrum of the nitroxide can provide information about the structure of the original radical.

$$R\cdot + R'N{=}O \longrightarrow \begin{matrix} R \\ \diagdown \\ N{-}\dot{O} \\ \diagup \\ R' \end{matrix}$$

Another technique that is highly specific for radical processes is known as CIDNP, an abbreviation for *chemically induced dynamic nuclear polarization*.[11] The instrumentation required for such studies is a normal NMR spectrometer. CIDNP is observed as a strong perturbation of the intensity of NMR signals in products formed in certain types of free-radical reactions. The variation in intensity results when the normal population of nuclear spin states dictated by the Boltzmann distribution is disturbed by the presence of an unpaired electron. The magnetic moment associated with an electron causes a redistribution of the nuclear spin states. Individual nuclei can become overpopulated in either the lower or the upper spin state. If the lower state is overpopulated, an enhanced absorption signal is oberved. If the upper state is overpopulated, an emission signal is observed. The CIDNP method is not as general as EPR spectroscopy because not all free-radical reactions can be expected to exhibit the phenomenon.[12]

Figure 12.3 shows the observation of CIDNP during the decomposition of benzoyl peroxide in cyclohexanone:

$$\underset{\substack{|| || \\ PhCOOCPh}}{\overset{O O}{}} \longrightarrow 2\ Ph\cdot + 2\ CO_2$$

$$Ph\cdot + S{-}H \longrightarrow C_6H_6 + S\cdot$$

9. A. Effio, D. Griller, K. U. Ingold, A. L. J. Beckwith, and A. K. Serelis, *J. Am. Chem. Soc.* **102**:1734 (1980); L. Mathew and J. Warkentin, *J. Am. Chem. Soc.* **108**:7981 (1986); M. Newcomb and A. G. Glenn, *J. Am. Chem. Soc.* **111**:275 (1989); A. L. J. Beckwith and V. W. Bowry, *J. Org. Chem.* **54**:2681 (1989); D. C. Nonhebel, *Chem. Soc. Rev.* **22**:347 (1993).

10. E. G. Janzen, *Acc. Chem. Res.* **4**:31 (1971); E. G. Janzen, in *Free Radicals in Biology*, Vol. 4, W. A. Pryor, ed., Academic Press, New York, 1980, pp. 115–154.

11. H. R. Ward, *Acc. Chem. Res.* **5**:18 (1972); R. G. Lawler, *Acc. Chem. Res.* **5**:25 (1972).

12. For a discussion of the theory of CIDNP and the conditions under which spin polarization occurs, see G. L. Closs, *Adv. Magn. Res.* **7**:157 (1974); R. Kaptein, *Adv. Free Radical Chem.* **5**:318 (1975); G. L. Closs, R. J. Miller, and O. D. Redwine, *Acc. Chem. Res.* **18**:196 (1985).

**BPO 0.05 *M* in cyclohexanone
100 MHz 110°C**

t = 12 min

t = 8 min

t = 4 min

t = 0

Fig. 12.3. NMR spectra recorded during thermal decomposition of dibenzoyl peroxide. Singlet at high field is due to benzene; other signals are due to dibenzoyl peroxide. [From H. Fischer and J. Bargon, *Acc. Chem. Res.* **2**:110 (1969). Reproduced by permission of the American Chemical Society.]

The emission signal corresponding to benzene confirms that it is formed by a free-radical process. As in steady-state EPR experiments, the enhanced emission and absorption are observed only as long as the reaction is proceeding. When the reaction is complete or is stopped in some way, the signals rapidly return to their normal intensity, because equilibrium population of the two spin states is rapidly reached.

One aspect of both EPR and CIDNP studies that should be kept in mind is that either is capable of detecting very small amounts of radical intermediates. This sensitivity makes both techniques quite useful, but it can also present a pitfall. The most prominent features of either EPR or CIDNP spectra may actually be due to radicals that account for only minor amounts of the total reaction process. An example of this was found in a study of the decomposition of trichloroacetyl peroxide in alkenes.

$$Cl_3CCOOCCCl_3 \longrightarrow 2\ Cl_3C\cdot + 2\ CO_2$$

$$Cl_3C\cdot + CH_2{=}C(CH_3)_2 \longrightarrow Cl_3CCH_2\dot{C}(CH_3)_2$$

$$Cl_3C\cdot + Cl_3CCH_2\dot{C}(CH_3)_2 \longrightarrow Cl_3CH + Cl_3CCH_2C{=}CH_2$$
$$\underset{CH_3}{|}$$

In addition to the emission signals of $CHCl_3$ and $Cl_3CCH_2C(CH_3){=}CH_2$, which are the major products, a strong emission signal for Cl_3CCHCl_2 was identified. However, this compound is a very minor product of the reaction, and when the signals have returned to their normal intensity, Cl_3CCHCl_2 is present in such a small amount that it cannot be detected.[13]

12.1.4. Sources of Free Radicals

There are several reactions that are frequently used to generate free radicals, both for the study of radical structure and reactivity and also in synthetic processes. Some of the most general methods are outlined here. These reactions will be encountered again when specific examples are discussed. For the most part, we will defer discussion of the reactions of the radicals until then.

Peroxides are a common source of radical intermediates. An advantage of the generation of radicals from peroxides is that reactions generally occur at relatively low temperature. The oxygen–oxygen bond in peroxides is weak (~ 30 kcal/mol), and activation energies for radical formation are low. Dialkyl peroxides are decomposed thermally to give two alkoxy radicals[14]:

$$C_2H_5OOC_2H_5 \xrightarrow{80^\circ C} 2C_2H_5O\cdot$$

Diacyl peroxides are sources of alkyl radicals because the carboxyl radicals that are intitially formed lose CO_2 very rapidly.[15] In the case of aroyl peroxides, products may be derived from the carboxyl radical or the radical formed by decarboxylation.[16] The decomposition of peroxides can also be accomplished by photochemical excitation.

$$\underset{\text{O O}}{CH_3COOCCH_3} \xrightarrow{80-100^\circ C} 2\ CH_3\overset{O}{\overset{\|}{C}}O\cdot \longrightarrow 2\ CH_3\cdot + 2\ CO_2$$

$$\underset{\text{O O}}{PhCOOCPh} \xrightarrow{80-100^\circ C} 2\ Ph\overset{O}{\overset{\|}{C}}O\cdot \longrightarrow 2\ Ph\cdot + 2\ CO_2$$

Peresters are also sources of radicals. The acyloxy portion normally loses carbon dioxide, so peresters yield an alkyl (or aryl) and an alkoxy radical[17]:

$$\underset{\text{O}}{R\overset{O}{\overset{\|}{C}}OOC(CH_3)_3} \longrightarrow R\cdot + CO_2 + \cdot OC(CH_3)_3$$

The thermal decompositions described above are unimolecular reactions that should exhibit first-order kinetics. Under many conditions, peroxides decompose at rates faster than expected for unimolecular thermal decomposition and with more complicated kinetics. This behavior is known as *induced decomposition* and occurs when part of the peroxide decomposition is the result of bimolecular reactions with radicals present in solution, as illustrated below specifically for diethyl peroxide.

$$X\cdot + CH_3CH_2OOCH_2CH_3 \rightarrow CH_3\dot{C}HOOCH_2CH_3 + H-X$$
$$CH_3\dot{C}HOOCH_2CH_3 \rightarrow CH_3CH{=}O + \cdot OCH_2CH_3$$

13. H. Y. Loken, R. G. Lawler, and H. R. Ward, *J. Org. Chem.* **38**:106 (1973).
14. W. A. Pryor, D. M. Huston, T. R. Fiske, T. L. Pickering, and E. Ciuffarin, *J. Am. Chem. Soc.* **86**:4237 (1964).
15. J. C. Martin, J. W. Taylor, and E. H. Drew, *J. Am. Chem. Soc.* **89**:129 (1967); F. D. Greene, H. P. Stein, C.-C. Chu, and F. M. Vane, *J. Am. Chem. Soc.* **86**:2080 (1964).
16. D. F. DeTar, R. A. J. Long, J. Rendleman, J. Bradley, and P. Duncan, *J. Am. Chem. Soc.* **89**:4051 (1967).
17. P. D. Bartlett and R. R. Hiatt, *J. Am. Chem. Soc.* **80**:1398 (1958).

The amount of induced decomposition that occurs depends on the concentration and reactivity of the radical intermediates and the susceptibility of the substrate to radical attack. The radical X· may be formed from the peroxide, but it can also be derived from subsequent reactions with the solvent. For this reason, both the structure of the peroxide and the nature of the reaction medium are important in determining the extent of induced decomposition, relative to unimolecular homolysis.

Alkyl hydroperoxides give alkoxy radicals and the hydroxyl radical. *t*-Butyl hydroperoxide is often used as a radical source. Detailed studies on the mechanism of the decomposition indicate that it is a more complicated process than simple unimolecular decomposition.[18] The alkyl hydroperoxides are also sometimes used in conjunction with a transition-metal salt. Under these conditions, an alkoxy radical is produced, but the hydroxyl portion appears as hydroxide ion as the result of one-electron reduction by the metal ion.[19]

$$(CH_3)_3COOH + M^{2+} \rightarrow (CH_3)_3CO\cdot + {}^-OH + M^{3+}$$

Another quite general source of free radicals is the decomposition of azo compounds. The products are molecular nitrogen and the radicals derived from the substituent groups:

$$R-N{=}N-R' \xrightarrow[h\nu]{\Delta \text{ or}} R\cdot + N{\equiv}N + \cdot R'$$

Both symmetrical and unsymmetrical azo compounds can be made, so that a single radical or two different ones may be generated. The energy for the decomposition can be either thermal or photochemical.[20] In the thermal decomposition, it has been established that the temperature at which decomposition occurs depends on the nature of the substituent groups. Azomethane does not decompose to methyl radicals and nitrogen until temperatures above 400°C are reached. Azo compounds that generate relatively stable radicals decompose at much lower temperatures. Azo compounds derived from allyl groups decompose somewhat above 100°C; for example:

$$CH_3CH_2CH_2N{=}NCH_2CH{=}CH_2 \xrightarrow{130°C} CH_3CH_2CH_2\cdot + N_2 + CH_2{=}CHCH_2\cdot \qquad \text{Ref.21}$$

Unsymmetrical azo compounds must be used to generate phenyl radicals because azobenzene is very stable thermally. Phenylazotriphenylmethane decomposes readily because of the stability of the triphenylmethyl radical:

$$PhN{=}NC(Ph)_3 \xrightarrow{60°C} Ph\cdot + Ph_3C\cdot + N_2 \qquad \text{Ref.22}$$

Azo compounds having functional groups that stabilize the radical products are especially reactive. The stabilizing effect of the cyano substituent is responsible for the easy decomposition of azobis(isobutyronitrile) (AIBN), which is frequently used as a radical initiator.

$$\underset{\underset{CN}{|}}{(CH_3)_2C}-N{=}N-\underset{\underset{CN}{|}}{C(CH_3)_2} \longrightarrow 2\,(CH_3)_2\overset{\cdot}{C}CN + N_2$$

18. R. Hiatt, T. Mill, and F. R. Mayo, *J. Org. Chem.* **33**:1416 (1968), and accompanying papers.
19. W. H. Richardson, *J. Am. Chem. Soc.* **87**:247 (1965).
20. P. S. Engel, *Chem. Rev.* **80**:99 (1980).
21. K. Takagi and R. J. Crawford, *J. Am. Chem. Soc.* **93**:5910 (1971).
22. R. F. Bridger and G. A. Russell, *J. Am. Chem. Soc.* **85**:3754 (1963).

Many azo compounds also generate radicals when photolyzed. This can occur by a thermal decomposition of the *cis*-azo compounds that are formed in the photochemical step.[23] The *cis* isomers are thermally much more labile than the *trans* isomers.

N-Nitrosoanilides are an alternative source of aryl radicals. There is a close mechanistic relationship to the decomposition of azo compounds. The *N*-nitrosoanilides rearrange to intermediates that have a nitrogen–nitrogen double bond. The intermediate then decomposes to generate aryl radicals.[24]

A technique that is a convenient source of radicals for study by EPR involves photolysis of a mixture of di-*t*-butyl peroxide, triethylsilane, and the alkyl bromide corresponding to the radical to be studied.[25] Photolysis of the peroxide gives *t*-butoxy radicals, which selectively abstract hydrogen from the silane. This reactive silicon radical in turn abstracts bromine, generating the alkyl radical at a steady-state concentration suitable for EPR study.

$$(CH_3)_3COOC(CH_3)_3 \xrightarrow{h\nu} 2(CH_3)_3CO\cdot$$
$$(CH_3)_3CO\cdot + (C_2H_5)_3SiH \longrightarrow (CH_3)_3COH + (C_2H_5)_3Si\cdot$$
$$(C_2H_5)_3Si\cdot + RBr \longrightarrow (C_2H_5)_3SiBr + R\cdot$$

The acyl derivatives of *N*-hydroxypyridine-2-thione are a synthetically versatile source of free radicals.[26] These compounds are readily prepared from reactive acylating agents, such as acyl chlorides, and a salt of *N*-hydroxypyridine-2-thione:

Radicals react at the sulfur, and decomposition generating an acyloxy radical ensues. The acyloxy radical undergoes decarboxylation. Usually, the radical then gives product and another radical which can continue a chain reaction. The process can be illustrated by the reactions with tri-*n*-butylstannane and bromotrichloromethane.

23. M. Schmittel and C. Rüchardt, *J. Am. Chem. Soc.* **109**:2750 (1987).
24. C. Rüchardt and B. Freudenberg, *Tetrahedron Lett.* **1964**:3623; J. I. G. Cadogan, *Acc. Chem. Res.* **4**:186 (1971).
25. A. Hudson and R. A. Jackson, *J. Chem. Soc., Chem. Commun.* **1969**:1323; D. J. Edge and J. K. Kochi, *J. Am. Chem. Soc.* **94**:7695 (1972).
26. D. H. R. Barton, D. Crich, and W. B. Motherwell, *Tetrahedron* **41**:3901 (1985).

(*a*) Reductive decarboxylation by reaction with tri-n-butylstannane

675

SECTION 12.1.
GENERATION AND
CHARACTERIZATION
OF FREE RADICALS

(*b*) Conversion of aromatic carboxylic acid to aryl bromide by reaction with bromotrichloromethane

12.1.5. Structural and Stereochemical Properties of Radical Intermediates

EPR studies and other physical methods have provided the basis for some insight into the detailed geometry of radical species.[29] Deductions about structure can also be drawn from the study of the stereochemistry of reactions involving radical intermediates. Several structural possibilities must be considered. If discussion is limited to alkyl radicals, the possibilities include a rigid pyramidal structure, rapidly inverting pyramidal structures, or a planar structure.

rigid pyramidal   flexible pyramidal   planar

Precise description of the pyramidal structures would also require that the bond angles be specified. The EPR spectrum of the methyl radical leads to the conclusion that its structure could be either planar or a very shallow pyramid.[30] The IR spectrum of the methyl radical has been recorded at very low tempertures in frozen argon.[31] This IR study puts a maximum of $\sim 5°$ on the deviation from planarity. A microwave study has also indicated that the methyl radical is planar.[32] Various MO calculations indicate a planar structure.[33]

27. D. H. R. Barton, D. Crich, and W. B. Motherwell, *J. Chem. Soc., Chem. Commun.* **1983**:939.
28. D. H. R. Barton, B. Lacher, and S. Z. Zard, *Tetrahedron Lett.* **26**:5939 (1985).
29. For a review, see J. K. Kochi, *Adv. Free Radicals* **5**:189 (1975).
30. M. Karplus and G. K. Fraenkel, *J. Chem. Phys.* **35**:1312 (1961).
31. L. Andrews and G. C. Pimentel, *J. Chem. Phys.* **47**:3637 (1967).
32. E. Hirota, *J. Phys. Chem.* **87**:3375 (1983).
33. F. M. Bickelhaupt, T. Ziegler, and P. v. R. Schleyer, *Organometallics* **15**:1477 (1996).

The *t*-butyl radical has been studied extensively. Whereas experimental results have been interpreted in terms of both planar and slightly pyramidal structures, theoretical calculations favor a pyramidal structure.[34] It appears that simple alkyl radicals are generally pyramidal, although the barrier to inversion is very small. *Ab initio* MO calculations suggest that two factors are of principal importance in favoring a pyramidal structure. One is a torsional effect in which the radical center tends to adopt a staggered conformation of the radical substituents. There is also a hyperconjugative interaction between the half-filled orbital and the hydrogen that is aligned with it. This hyperconjugation is stronger in the conformation in which the pyramidalization is in the same direction as to minimize eclipsing.[35] The theoretical results also indicate that the barrier to inversion is no more than 1–2 kcal/mol, so rapid inversion will occur. Note that this hyperconjugative interaction can account for the substantial hyperfine coupling with the β hydrogen that was discussed in Section 12.1.3. The β C−H bond is also greatly weakened by the hyperconjugation. MP4/6-311G(d,p) calculations assign a bond energy of only about 36 kcal/mol.[36]

Radical geometry is significantly affected by substituent groups that can act as π donors. Addition of a fluorine or oxygen substituent, in particular, favors a pyramidal structure. Analysis of the EPR spectra of the mono-, di-, and trifluoromethyl radicals indicates a progressive distortion from planarity.[37] Both EPR and IR studies of the trifluoromethyl radical show it to be pyramidal.[38] The basis of this structural effffect has been probed by MO calculations, and it is considered to result from interactions of both the σ and the π type. There is a repulsive interaction between the singly occupied p orbital and the filled orbitals occupied by "lone-pair" electrons on the fluorine or oxygen substituents. This repulsive interaction is minimized by adoption of a pyramidal geometry. The tendency for pyramidal geometry is reinforced by an interaction between the p orbital on carbon and the σ^* antibonding orbital associated with the C−F or C−O bonds. The energy of the p orbital can be lowered by interaction with the σ^* orbital. This interaction increases electron density on the more electronegative fluorine or oxygen atom. This stabilizing p–σ^* interaction is increased by pyramidal geometry.

stabilizing interaction with σ^*

There have been many studies aimed at deducing the geometry of radical sites by examining the stereochemistry of radical reactions. The most direct kind of study involves the generation of a radical at a carbon which is a stereogenic center. A planar or rapidly inverting radical would lead to racemization, whereas a rigid pyramidal structure should

34. L. Bonazolla, N. Leroy, and J. Roncin, *J. Am. Chem. Soc.* **99**:8348 (1977); D. Griller, K. H. U. Ingold, P. J. Krusic, and H. Fischer, *J. Am. Chem. Soc.* **100**:6750 (1978); J. Pacansky and J. S. Chang, *J. Phys. Chem.* **74**:5539 (1981); B. Schrader, J. Pacansky, and U. Pfeiffer, *J. Phys. Chem.* **88**:4069 (1984).

35. M. N. Paddon-Row and K. N. Houk, *J. Am. Chem. Soc.* **103**:5046 (1981); M. N. Paddon-Row and K. N. Houk, *J. Phys. Chem.* **89**:3771 (1985); J. Pacansky, W. Koch, and M. D. Miller, *J. Am. Chem. Soc.* **113**:317 (1991).

36. J. A. Seetula, *J. Chem. Soc., Faraday Trans.* **94**:1933 (1998).

37. P. J. Krusic and R. C. Bingham, *J. Am. Chem. Soc.* **98**:230 (1976); F. Bernardi, W. Cherry, S. Shaik, and N. D. Epiotis, *J. Am. Chem. Soc.* **100**:1352 (1978).

38. R. W. Fessenden and R. H. Schuler, *J. Chem. Phys.* **43**:2704 (1965); G. A. Carlson and G. C. Pimentel, *J. Chem. Phys.* **44**:4053 (1966).

Scheme 12.2. Stereochemistry of Radical Reactions at Chiral Carbon Atoms

a. H. C. Brown, M. S. Kharasch, and T. H. Chao, *J. Am. Chem. Soc.* **62**:3435 (1940).
b. W. v. E. Doering, M. Farber, M. Sprecher, and K. B. Wiberg, *J. Am. Chem. Soc.* **74**:3000 (1952).
c. F. D. Greene, *J. Am. Chem. Soc.* **81**:2688 (1959); D. B. Denney and W. F. Beach, *J. Org. Chem.* **24**:108 (1959).
d. H. J. Dauben, Jr., and L. L. McCoy, *J. Am. Chem. Soc.* **81**:5404 (1959).

lead to product of retained configuration. Some examples of reactions that have been subjected to this kind of study are shown in Scheme 12.2. In each case, racemic product is formed, indicating that alkyl radicals do not retain the tetrahedral geometry of their precursors.

Cyclic molecules permit deductions about stereochemistry without the necessity of using resolved chiral compounds. The stereochemistry of a number of reactions of 4-substituted cyclohexyl radicals has been investigated.[39] In general, reactions starting from pure *cis* or *trans* stereoisomers give mixtures of *cis* and *trans* products. This result indicates that the radical intermediates do not retain the stereochemistry of the precursor. Radical reactions involving cyclohexyl radicals are not usually very stereoselective, but some show a preference for formation of the *cis* product. This has been explained in terms of a torsional effect. If the cyclohexyl radical is planar or a shallow pyramid, equatorial attack leading to *trans* product causes the hydrogen at the radical site to become eclipsed with the two neighboring equatorial hydrogens. Axial attack does not suffer from this strain, because the hydrogen at the radical site moves away from the equatorial hydrogens toward the staggered conformation that is present in the chair conformation of the ring. The pyramidalization of the radical would be expected to be in the direction favoring axial attack.[40]

39. F. R. Jensen, L. H. Gale, and J. E. Rodgers, *J. Am. Chem. Soc.* **90**:5793 (1968).
40. W. Damm, B. Giese, J. Hartung, T. Hasskerl, K. N. Houk, O. Hüter and H. Zipse, *J. Am. Chem. Soc.* **114**:4067 (1992).

Another approach to obtaining information about the geometric requirements of free radicals has been to examine bridgehead systems. It will be recalled that small bicyclic rings strongly resist formation of carbocations at bridgehead centers because the skeletal geometry prevents attainment of the preferred planar geometry by the carbocation. In an early study, the decarbonylation of bridgehead aldehydes was found to proceed without special difficulty.[41]

Subsequent rate studies have shown that there is significant rate retardation for reactions in which the norbornyl radical is generated in a rate-determining step.[42] Typically, such reactions proceed 500–1000 times slower than the corresponding reaction generating the *t*-butyl radical. This is a much smaller rate retardation than that of 10^{14} found in S_N1 solvolysis. Rate retardations are even smaller for less strained bicyclic systems. The EPR spectra of the bridgehead radicals **A** and **B** are consistent with pyramidal geometry at the bridgehead carbon atoms.[43]

The EPR spectra of a number of bridgehead radicals have been measured and the hyperfine couplings measured (see Section 12.2.3). Both the H_α and $^{13}C_\beta$ couplings are sensitive to the pyramidal geometry of the radical.[44] The reactivity of bridgehead radicals increases with increased pyramidal character.[45]

Radical	H_α	$^{13}C_\beta$	ϕ^a
Adamantyl	6.58	132	113.6
Bicyclo[2.2.2]octyl	6.64	143	113.2
Bicyclo[2.2.1]heptyl	2.35	151	112.9
Bicyclo[2.1.1]hexyl	0	174	111.9
Bicyclo[1.1.1]pentyl	−1.2	223	110.3

a. ϕ = the C—C—C bond angle at the bridgehead radical.

41. W. v. E. Doering, M. Farber, M. Sprecher, and K. B. Wiberg, *J. Am. Chem. Soc.* **74**:3000 (1952).
42. A. Oberlinner and C. Rüchardt, *Tetrahedron Lett.* **1969**:4685; L. B. Humphrey, B. Hodgson, and R. E. Pincock, *Can. J. Chem.* **46**:3099 (1968); D. E. Applequist and L. Kaplan, *J. Am. Chem. Soc.* **87**:2194 (1965).
43. P. J. Krusic, T. A. Rettig, and P. v. R. Schleyer, *J. Am. Chem. Soc.* **94**:995 (1972).
44. C. J. Rhodes, J. C. Walton, and E. W. Della, *J. Chem. Soc., Perkin Trans. 2* **1993**:2125; G. T. Binmore, J. C. Walton, W. Adcock, C. I. Clark, and A. R. Krstic, *Magn. Reson. Chem.* **33**(Suppl.):S53 (1995).
45. F. Recupero, A. Bravo, H. R. Bjorsvik, F. Fontana, F. Minisci, and M. Piredda, *J. Chem. Soc., Perkin Trans. 2* **1997**:2399; K. P. Dockery and W. G. Bentrude, *J. Am. Chem. Soc.* **119**:1388 (1997).

The broad conclusion of all these studies is that alkyl radicals are shallow pyramids and that the barrier to inversion of the pyramidal structures is low. Radicals also are able to tolerate some geometric distortion associated with strained ring systems.

The allyl radical would be expected to be planar in order to maximize π delocalization. Molecular structure parameters have been obtained from EPR, IR, and electron diffraction measurements and confirm that the radical is planar.[46]

The vinyl radical, $CH_2=CH\cdot$, is found by both experiment and theory to be bent, with a C−C−H bond angle of about $137°$.[47] Substituents affect the preferred geometry of vinyl radicals. Conjugation with π-acceptor substituents favors a planar geometry, whereas σ-donor substituents favor a bent geometry.[48] For σ donors the barriers (kcal/mol) for isomerization are in the order CH_3 (3.1) < OH (13.3) < F (19.5) according to B3LYP calculations. While these barriers have not been measured experimentally, reaction stereoselectivity is in agreement with the results. For the π-acceptor substituents, the preferred geometry is one in which the substituent is aligned with the singly occupied p orbital, not the π bond.

$C=Z = -CH=CH_2, -CH=O, -C\equiv N$

The stereochemistry of reactions involving substituted alkenyl free radicals indicates that radicals formed at trigonal centers rapidly undergo interconversion with the geometric isomer.[49] Reactions proceeding through alkenyl radical intermediates usually give rise to the same mixture from both the E- and the Z-precursor:

In this particular case, there is evidence from EPR spectra that the radical is not linear in its ground state but is an easily inverted bent species.[51] The barrier to inversion is very low

46. R. W. Fessenden and R. H. Schuler, *J. Chem. Phys.* **39**:2147 (1963); A. K. Maltsev, V. A. Korolev, and O. M. Nefedov, *Izv. Akad. Nauk SSSR, Ser. Khim.* **1984**:555; E. Vajda, J. Tremmel, B. Rozandai, I. Hargittai, A. K. Maltsev, N. D. Kagramanov, and O. M. Nefedov, *J. Am. Chem. Soc.* **108**:4352 (1986).

47. J. H. Wang, H.-C. Chang, and Y.-T. Chen, *Chem. Phys.* **206**:43 (1996).

48. C. Galli, A. Guarnieri, H. Koch, P. Mencarelli, and Z. Rappoport, *J. Org. Chem.* **62**:4072 (1997).

49. For reviews of the structure and reactivity of vinyl radicals, see W. G. Bentrude, *Annu. Rev. Phys. Chem.* **18**:283 (1967); L. A. Singer, in *Selective Organic Transformations*, Vol. II, B. S. Thyagarajan, ed., John Wiley & Sons, New York, 1972, p. 239; O. Simamura, *Top. Stereochem.* **4**:1 (1969).

50. L. A. Singer and N. P. Kong, *J. Am. Chem. Soc.* **88**:5213 (1966); J. A. Kampmeier and R. M. Fantazier, *J. Am. Chem. Soc.* **88**:1959 (1966).

51. R. W. Fessenden and R. H. Schuler, *J. Chem. Phys.*, **39**:2147 (1963).

(~2 kcal), so that the lifetime of the individual isomers is very short ($\sim 10^{-9}$ s). The transition state for inversion approximates sp hybridization.[52]

12.1.6. Charged Radical Species

Unpaired electrons can be present in charged species as well as in the neutral systems that have been considered up to this point. There have been many studies of such radical cations and radical anions, and we will consider some representative examples in this section.

Various aromatic and conjugated polyunsaturated hydrocarbons undergo one-electron reduction by alkali metals.[53] Benzene and naphthalene are examples. The EPR spectrum of the benzene radical anion was shown in Fig. 12.2a (p. 669). These reductions must be carried out in aprotic solvents, and ethers are usually used. The ease of formation of the radical anion increases as the number of fused rings increases. The electrochemical reduction potentials of some representative compounds are given in Table 12.1. The potentials correlate with the energy of the LUMO as calculated by simple HMO theory.[54] Notice that polycyclic aromatics are both easier to reduce and easier to oxidize than benzene. This is because the HOMO–LUMO gap decreases with increasing size of the molecule, with the HOMO being higher and the LUMO lower in energy than in benzene. A correlation which includes a more extensive series of compounds can be observed using somewhat more sophisticated MO methods.[55]

(many resonance structures)

Table 12.1. Oxidation and Reduction Potentials for Some Aromatic Hydrocarbons[a]

Hydrocarbon	Ar–H $\xrightarrow{+e^-}$ Ar–H$\bar{\cdot}$	Ar–H $\xrightarrow{-e^-}$ Ar–H$\overset{+}{\cdot}$
Benzene	-3.42^b	+2.06
Naphthalene	−2.95	+1.33
Anthracene	−2.36	+0.89
Phenanthrene	−2.87	+1.34
Tetracene	−1.92	+0.57

a. Except where noted otherwise, the data are from C. Madec and J. Courtot-Coupez, *J. Electroanal. Chem. Interfacial Electrochem.* **84**:177 (1977).
b. J. Mortensen and J. Heinze, *Angew. Chem. Int. Ed. Eng.*, **23**:84 (1984).

52. P. R. Jenkins, M. C. R. Symons, S. E. Booth, and C. J. Swain, *Tetrahedron Lett.* **33**:3543 (1992).
53. D. E. Paul, D. Lipkin, and S. I. Weissman, *J. Am. Chem. Soc.* **78**:116 (1956); T. R. Tuttle, Jr., and S. I. Weissman, *J. Am. Chem. Soc.* **80**:5342 (1958).
54. E. S. Pysh and N. C. Yang, *J. Am. Chem. Soc.* **85**:2124 (1963); D. Bauer and J. P. Beck, *Bull. Soc. Chim. Fr.* **1973**:1252; C. Madec and J. Courtot-Coupez, *J. Electroanal. Chem. Interfacial Electrochem.* **84**:177 (1977).
55. C. F. Wilcox, Jr., K. A. Weber, H. D. Abruna, and C. R. Cabrera, *J. Electroanal. Chem. Interfacial Electrochem.* **198**:99 (1986).

In the presence of a proton source, the radical anion is protonated and further reduction occurs (the Birch reduction; Part B, Section 5.5.1). In general, when no proton source is present, it is more difficult to add a second electron. Solutions of the radical anions of aromatic hydrocarbons can be maintained for relatively long periods in the absence of oxygen or protons.

Cyclooctatetraene provides a significant contrast to the preference of aromatic hydrocarbons for one-electron reduction. It is converted to a diamagnetic dianion by addition of two electrons.[56] It is easy to understand the ease with which the cyclooctatetraene radical accepts a second electron because of the aromaticity of the 10-π-electron aromatic system which results (Section 9.3).

$$\text{(cyclooctatetraene)} + 2\,K \longrightarrow \text{(cyclooctatetraene dianion)} + 2\,K^+$$

Radical cations can be derived from aromatic hydrocarbons or alkenes by one-electron oxidation. Antimony trichloride and pentachloride are among the chemical oxidants that have been used.[57] Photodissociation or γ-radiation can generate radical cations from aromatic hydrocarbons.[58] Most radical cations derived from hydrocarbons have limited stability, but EPR spectral parameters have permitted structural characterization.[59] The radical cations can be generated electrochemically, and some oxidation potentials are included in Table 12.1. The potentials correlate with the HOMO levels of the hydrocarbons. The higher the HOMO, the more easily oxidized is the hydrocarbon.

Two classes of charged radicals derived from ketones have been well studied. *Ketyls* are radical anions formed by one-electron reduction of carbonyl compounds. The formation of the benzophenone radical anion by reduction with sodium metal is an example. This radical anion is deep blue in color and is very reactive toward both oxygen and protons. Many detailed studies on the structure and spectral properties of this and related radical anions have been carried out.[60] A common chemical reaction of the ketyl radicals is coupling to form a diamagnetic dianion. This occurs reversibly for simple aromatic ketyls. The dimerization is promoted by protonation of one or both of the ketyls because the electrostatic repulsion is then removed. The coupling process leads to reductive dimerization of carbonyl compounds, a reaction that will be discussed in detail in Section 5.5.3 of Part B.

$$\text{Na} + \text{Ar}_2\text{C}=\text{O} \longrightarrow \text{Ar}_2\overset{\bullet}{\text{C}}-\text{O}^-\,\text{Na}^+ \rightleftharpoons \text{Na}^+\,^-\text{O}-\underset{\underset{\text{Ar}}{|}}{\overset{\overset{\text{Ar}}{|}}{\text{C}}}-\underset{\underset{\text{Ar}}{|}}{\overset{\overset{\text{Ar}}{|}}{\text{C}}}-\text{O}^-\,^+\text{Na}$$

$$\downarrow \text{H}^+$$

$$\text{Ar}_2\overset{\bullet}{\text{C}}-\text{OH} \qquad \text{Na}^+\,^-\text{O}-\underset{\underset{\text{Ar}}{|}}{\overset{\overset{\text{Ar}}{|}}{\text{C}}}-\underset{\underset{\text{Ar}}{|}}{\overset{\overset{\text{Ar}}{|}}{\text{C}}}-\text{OH}$$

56. T. J. Katz, *J. Am. Chem. Soc.* **82**:3784 (1960).
57. I. C. Lewis and L. S. Singer, *J. Chem. Phys.* **43**:2712 (1965); R. M. Dessau, *J. Am. Chem. Soc.* **92**:6356 (1970).
58. R. Gschwind and E. Haselbach, *Helv. Chim. Acta* **62**:941 (1979); T. Shida, E. Haselbach, and T. Bally, *Acc. Chem. Res.* **17**:180 (1984); M. C. R. Symons, *Chem. Soc. Rev.* **13**:393 (1984).
59. J. L. Courtneidge and A. G. Davies, *Acc. Chem. Res.* **20**:90 (1987).
60. For a summary, see N. Hirota, in *Radical Ions*, E. T. Kaiser and L. Kevan, eds., Interscience, New York, 1968, pp. 35–85.

One-electron reduction of α-dicarbonyl compounds gives radical anions known as *semidiones*.[61] Closely related are the products of one-electron reduction of aromatic quinones, the *semiquinones*. Both semidiones and semiquinones can be protonated to give neutral radicals which are relatively stable.

semidione radical anion

neutral semidione radical

semiquinone radical anion

Reductants such as zinc or sodium dithionite generate the semidione from diketones. Electrolytic reduction can also be used. Enolates can reduce diones to semidiones by electron transfer.

The radicals that are formed from the enolate in this process are rapidly destroyed so that only the stable semidione species remains detectable for EPR study. Semidiones can also be generated oxidatively from ketones by reaction with oxygen in the presence of base.[62] The diketone is presumably generated oxidatively and then reduced to the semidione via reduction by the enolate derived from the original ketone.

The EPR spectra of semidione radical anions can provide information on the spin density at the individual atoms. The semidione derived from butane-2,3-dione, for example, has a spin density of 0.22 at each oxygen and 0.23 at each carbonyl carbon. The small amount of remaining spin density is associated with the methyl groups. This extensive delocalization is consistent with the resonance picture of the semidione radical anion.

61. G. A. Russell, in *Radical Ions*, E. T. Kaiser and L. Kevan, eds., Interscience, New York, 1968, pp. 87–150.
62. G. A. Russell and E. T. Strom, *J. Am. Chem. Soc.* **86**:744 (1964).

12.2. Characteristics of Reaction Mechanisms Involving Radical Intermediates

683

SECTION 12.2.
CHARACTERISTICS OF
REACTION
MECHANISMS
INVOLVING RADICAL
INTERMEDIATES

12.2.1. Kinetic Characteristics of Chain Reactions

Certain kinetic aspects of free-radical reactions are unique in comparison with the kinetic characteristics of other reaction types that have been considered to this point. The underlying difference is that many free-radical reactions are chain reactions; that is, the reaction mechanism consists of a cycle of repetitive steps which form many product molecules for each initiation event. The hypothetical mechanism below illustrates a chain reaction.

$$
\begin{array}{lcll}
& \text{A}\!-\!\text{A} & \xrightarrow{k_i} & 2\,\text{A}\cdot & \text{initiation} \\
\left\{ \begin{array}{l} \text{A}\cdot + \text{B}\!-\!\text{C} \\ \text{C}\cdot + \text{A}\!-\!\text{A} \\ \text{A}\cdot + \text{B}\!-\!\text{C} \\ \text{C}\cdot + \text{A}\!-\!\text{A} \end{array} \right. & \begin{array}{l} \xrightarrow{k_{p1}} \\ \xrightarrow{k_{p2}} \\ \xrightarrow{k_{p1}} \\ \xrightarrow{k_{p2}} \end{array} & \left. \begin{array}{l} \text{A}\!-\!\text{B} + \text{C}\cdot \\ \text{A}\!-\!\text{C} + \text{A}\cdot \\ \text{A}\!-\!\text{B} + \text{C}\cdot \\ \text{A}\!-\!\text{C} + \text{A}\cdot \end{array} \right\} & \text{propagation} \\
\begin{array}{l} 2\,\text{A}\cdot \\ 2\,\text{C}\cdot \\ \text{A}\cdot + \text{C}\cdot \end{array} & \begin{array}{l} \xrightarrow{k_{t1}} \\ \xrightarrow{k_{t2}} \\ \xrightarrow{k_{t3}} \end{array} & \left. \begin{array}{l} \text{A}\!-\!\text{A} \\ \text{C}\!-\!\text{C} \\ \text{A}\!-\!\text{C} \end{array} \right\} & \text{termination} \\
\text{A}\!-\!\text{A} + \text{B}\!-\!\text{C} & \longrightarrow & \text{A}\!-\!\text{B} + \text{A}\!-\!\text{C} &
\end{array}
$$

(repeated many times; overall reaction)

The step in which the reactive intermediate, in this case A·, is generated is called the *initiation step*. In the next four equations in the example above, a sequence of two reactions is repeated; this is the *propagation phase*. Chain reactions are characterized by a *chain length*, which is the number of propagation steps that take place per initiation step. Finally, there are *termination steps*, which include any reactions that destroy one of the reactive intermediates necessary for the propagation of the chain. Clearly, the greater the frequency of termination steps, the lower the chain length will be.

The overall rate of a chain process is determined by the rates of initiation, propagation, and termination reactions. Analysis of the kinetics of chain reactions normally depends on application of the steady-state approximation (see Section 4.2) to the radical intermediates. Such intermediates are highly reactive, and their concentrations are low and nearly constant throughout the course of the reaction:

$$
\frac{d[\text{A}\cdot]}{dt} = \frac{d[\text{C}\cdot]}{dt} = 0
$$

The result of the steady-state condition is that the overall rate of initiation must equal the total rate of termination. The application of the steady-state approximation and the resulting equality of the initiation and termination rates permits formulation of a rate law for the reaction mechanism above. The overall stoichiometry of a free-radical chain reaction is independent of the initiating and termination steps because the reactants are consumed and products formed almost entirely in the propagation steps.

$$
\text{A}_2 + \text{B}\!-\!\text{C} \rightarrow \text{A}\!-\!\text{B} + \text{A}\!-\!\text{C}
$$

The overall reaction rate is given by:

$$\text{rate} = \frac{d[\text{A}-\text{B}]}{dt} = \frac{d[\text{A}-\text{C}]}{dt} = \frac{-d[\text{A}_2]}{dt} = \frac{-d[\text{B}-\text{C}]}{dt}$$

Setting the rate of initiation equal to the rate of termination and assuming that k_{t2} is the dominant termination process gives:

$$k_i[\text{A}_2] = 2k_{t2}[\text{C}\cdot]^2$$

$$[\text{C}\cdot] = \left(\frac{k_i}{2k_{t2}}\right)^{1/2} [\text{A}_2]^{1/2}$$

In general, the rate constants for termination reactions involving coupling of two radicals are very large. Because the concentration of the reactive intermediates is very low, the overall rate of temination is low enough that the propagation steps can compete, since these steps involve the reactants, which are present at much higher concentration. The rate of the overall reaction is that of either propagation step:

$$\text{rate} = k_{p2}[\text{C}\cdot][\text{A}_2] = k_{p1}[\text{A}\cdot]\,[\text{B}-\text{C}]$$

Following the steady-state approximation, both propagation steps must proceed at the same rate; otherwise, the concentration of A· or C· would build up. By substituting for the concentration of the intermediate C·, we obtain:

$$\text{rate} = k_{p2}\left(\frac{k_i}{2k_{t2}}\right)^{1/2} [\text{A}_2]^{3/2} = k_{\text{obs}}[\text{A}_2]^{3/2}$$

The observed rate law is then three-halves order in the reagent A_2. In most real systems, the situation is complicated because more than one termination reaction makes a contribution to the total termination rate. A more complete discussion of the effect of termination steps on the form of the rate law has been given by Huyser.[63]

The overall rates of chain reactions can be greatly modified by changing the rate at which initiation or termination steps occur. The idea of initiation was touched on in Section 12.1.4, where sources of free radicals were discussed. Many radical reactions of interest in organic chemistry depend on the presence of an *initiator*, which serves as a source of free radicals to start chain sequences. Peroxides are frequently used as initiators, since they give radicals by thermal decomposition at relatively low temperatures. Azo compounds are another very useful class of initiators, with azobis(isobutyronitrile), AIBN, being the most commonly used example. Initiation by irradiation of a photosensitive compound that generates radical products is also a common procedure. Conversely, chain reactions can be greatly retarded by *inhibitors*. A compound can act as an inhibitor if it is sufficiently reactive toward a radical involved in the chain process that it effectively traps the radical, thus terminating the chain. Certain stable free radicals, for example, galvinoxyl (see entry 6, Scheme 12.1), are used in this way. Because they contain an unpaired electron, they are usually very reactive toward radical intermediates. The sensitivity of the rates of free-radical chain reactions to both initiators and inhibitors can be used in

63. E. S. Huyser, *Free Radical Chain Reactions*, Wiley-Interscience, New York, 1970, pp. 39–54.

mechanistic studies to distinguish radical chain reactions from polar or concerted processes.

685

SECTION 12.2.
CHARACTERISTICS OF
REACTION
MECHANISMS
INVOLVING RADICAL
INTERMEDIATES

Free-radical chain inhibitors are of considerable economic importance. The term *antioxidant* is commonly applied to inhibitors that retard the free-radical chain oxidations, termed autoxidations, that can cause relatively rapid deterioration of many commercial materials derived from organic molecules, including foodstuffs, petroleum products, and plastics. The chain mechanism for autoxidation of hydrocarbons is:

$$\text{initiation} \quad \text{In}\cdot + \text{R}\text{—}\text{H} \longrightarrow \text{R}\cdot + \text{In}\text{—}\text{H}$$

$$\text{propagation} \quad \text{R}\cdot + \text{O}_2 \longrightarrow \text{ROO}\cdot$$

$$\text{ROO}\cdot + \text{R}\text{—}\text{H} \longrightarrow \text{ROOH} + \text{R}\cdot$$

The function of an antioxidant is to divert the peroxy radicals and thus prevent a chain process. Other antioxidants function by reacting with potential initiators and thus retard oxidative degradation by preventing the initiation of autoxidation chains. The hydroperoxides generated by autoxidation are themselves potential chain initiators, and autoxidations therefore have the potential of being autocatalytic. Certain antioxidants function by reducing such hydroperoxides and thereby preventing their accumulation.

The presence of oxygen can modify the course of a free-radical chain reaction if a radical intermediate is diverted by reaction with molecular oxygen. The oxygen molecule, with its two unpaired electrons, is extremely reactive toward most free-radical intermediates. The product which is formed is a reactive peroxyl radical, which can propagate a chain reaction leading to oxygen-containing products.

$$\text{R}\cdot + \text{O}_2 \rightarrow \text{R}\text{—}\text{O}\text{—}\text{O}\cdot$$

12.2.2. Structure–Reactivity Relationships

Structure–reactivity relationships can be probed by measurements of rates and equilibria, as was discussed in Chapter 4. Direct comparison of reaction rates is used relatively less often in the study of radical reactions than for heterolytic reactions. Instead, *competition methods* have frequently been used. The basis of competition methods lies in the rate expression for a reaction, and the results can be just as valid a comparison of relative reactivity as directly measured rates, *provided the two competing processes are of the same kinetic order*. Suppose that it is desired to compare the reactivity of two related compounds, $B\text{—}X$ and $B\text{—}Y$, in a hypothetical sequence:

$$A\text{—}A \rightarrow 2A\cdot$$
$$A\cdot + B\text{—}X \xrightarrow{k_X} A\text{—}B + X\cdot$$
$$X\cdot + A\text{—}A \rightarrow A\text{—}X + A\cdot$$

and

$$A\cdot + B\text{—}Y \xrightarrow{k_Y} A\text{—}B + Y\cdot$$
$$Y\cdot + A\text{—}A \rightarrow A\text{—}Y + A\cdot$$

The data required are the relative magnitudes of k_X and k_Y. When both B—X and B—Y are present in the reaction system, they will be consumed at rates that are a function of their reactivity and concentration:

$$\frac{-d[\text{B—X}]}{dt} = k_X[\text{A·}][\text{B—X}]$$

$$\frac{-d[\text{B—Y}]}{dt} = k_Y[\text{A·}][\text{B—Y}]$$

$$\frac{k_X}{k_Y} = \frac{d[\text{B—X}]/[\text{B—X}]}{d[\text{B—Y}]/[\text{B—Y}]}$$

Integration of this expression with the limits $[\text{B—X}] = [\text{B—X}]_0$ to $[\text{B—X}]_t$, where t is a point in time during the course of the reaction, gives

$$\frac{k_X}{k_Y} = \frac{\ln([\text{B—X}]_0/[\text{B—X}]_t)}{\ln([\text{B—Y}]_0/[\text{B—Y}]_t)}$$

This relationship permits the measurement of the ratio k_X/k_Y. The initial concentrations $[\text{B—X}]_0$ and $[\text{B—Y}]_0$ are known from the conditions of the experiment. The reaction can be stopped at some point when some of both B—X and B—Y remain unreacted, or an excess of B—X and B—Y can be used so that neither is completely consumed when A—A has completely reacted. Determination of $[\text{B—X}]_t$ and $[\text{B—Y}]_t$ then provides the information needed to calculate k_X/k_Y. It is clear that the reactions being compared must be of the same order. If they were not, division of the two rate expressions would leave uncanceled concentration terms.

Another experiment of the competition type involves the comparison of the reactivity of different atoms in the same molecule. For example, gas-phase chlorination of butane can lead to 1- or 2-chlorobutane. The relative reactivity (k_p/k_s) of the primary and secondary hydrogens is the sort of information that helps to characterize the details of the reaction process.

$$\text{Cl·} + \text{CH}_3\text{CH}_2\text{CH}_2\text{CH}_3 \xrightarrow{k_p} \text{HCl} + \text{·CH}_2\text{CH}_2\text{CH}_2\text{CH}_3$$

$$\text{Cl}_2 + \text{·CH}_3\text{CH}_2\text{CH}_2\text{CH}_3 \longrightarrow \text{Cl·} + \text{ClCH}_2\text{CH}_2\text{CH}_2\text{CH}_3$$

$$\text{Cl·} + \text{CH}_3\text{CH}_2\text{CH}_2\text{CH}_3 \xrightarrow{k_s} \text{HCl} + \text{CH}_3\overset{.}{\text{C}}\text{HCH}_2\text{CH}_3$$

$$\text{CH}_3\overset{.}{\text{C}}\text{HCH}_2\text{CH}_3 + \text{Cl}_2 \longrightarrow \text{Cl·} + \underset{\overset{|}{\text{Cl}}}{\text{CH}_3\text{CHCH}_2\text{CH}_3}$$

The value of k_p/k_s can be determined by measuring the ratio of the products 1-chlorobutane : 2-chlorobutane during the course of the reaction. A statistical correction must be made to take account of the fact that the primary hydrogens outnumber the secondary ones by 3 : 2. This calculation provides the relative reactivity of chlorine atoms toward the primary and secondary hydrogens in butane:

$$\frac{k_p}{k_s} = \frac{2[\text{1-chlorobutane}]_t}{3[\text{2-chlorobutane}]_t}$$

Recent development of techniques for measuring the rates of very fast reactions has permitted absolute rates to be measured for some fundamental types of free-radical reactions. Some examples of absolute rates and E_a values are given in Table 12.2.

687

SECTION 12.2.
CHARACTERISTICS OF
REACTION
MECHANISMS
INVOLVING RADICAL
INTERMEDIATES

Table 12.2. Absolute Rates of Some Free-Radical Reactions[a]

Reaction		Reference

A. Hydrogen-atom abstraction

1. Ph· + (tetrahydrofuran) ⟶ Ph—H + (tetrahydrofuranyl radical). $k = 4.8 \times 10^6 \ M^{-1} \ s^{-1}$ b

2. $(CH_3)_3CO·$ + $(CH_3)_2C(H)$—(phenyl) ⟶ $(CH_3)_3COH$ + $(CH_3)_2C·$—(phenyl)   c

   $k = 8.7 \times 10^5 \ M^{-1} \ s^{-1}$

3. $(CH_3)_3CO·$ + (tetrahydrofuran) ⟶ $(CH_3)_3COH$ + (tetrahydrofuranyl radical). $k = 8.3 \times 10^6 \ M^{-1} \ s^{-1}$ c

4. Cl· + (cyclopentane) ⟶ H—Cl + (cyclopentyl radical). $k = 4.7 \times 10^9 \ M^{-1} \ s^{-1}$ d
   (free)

5. Cl· + (cyclopentane) ⟶ H—Cl + (cyclopentyl radical). $k = 4.3 \times 10^7 \ M^{-1} \ s^{-1}$ d

   (benzene complex)

6. $CH_3·$ + Bu_3Sn—H ⟶ CH_4 + $Bu_3Sn·$ $k = 1.0 \times 10^7 \ M^{-1} \ s^{-1}$ e
   $E_a = 3.2$ kcal/mol

7. $(CH_3)_3C·$ + Bu_3Sn—H ⟶ $(CH_3)_3CH$ + $Bu_3Sn·$ $k = 1.8 \times 10^6 \ M^{-1} \ s^{-1}$ e
   $E_a = 2.95$ kcal/mol

8. Ph· + Bu_3Sn—H ⟶ Ph—H + $Bu_3Sn·$ $k = 7.8 \times 10^8 \ M^{-1} \ s^{-1}$ f

9. $CF_3CF_2CF_2·$ + Bu_3Sn—H ⟶ $CF_3CF_2CF_2H$ + $Bu_3Sn·$ $k = 2.0 \times 10^9 \ M^{-1} \ s^{-1}$ g

10. $PhCH_2·$ + PhSH ⟶ $PhCH_3$ + PhS· $k = 3.1 \times 10^5 \ M^{-1} \ s^{-1}$ h

11. (cyclopropyl)—$CH_2·$ + PhSeH ⟶ (cyclopropyl)—CH_3 + PhSe· $k = 2.1 \times 10^9 \ M^{-1} \ s^{-1}$ i

12. $(CH_3)_3C\overset{O}{\overset{\|}{C}}·$ + Bu_3SnH ⟶ $(CH_3)_3CCH{=}O$ + $Bu_3Sn·$ $k = 3.0 \times 10^5 \ M^{-1} \ s^{-1}$ j

13. $R\overset{O}{\overset{\|}{C}}·$ + Bu_3SnH ⟶ RCH$=$O + $Bu_3Sn·$ (80°C) $k = 1.3 \times 10^6 \ M^{-1} \ s^{-1}$ k

14. $R\overset{O}{\overset{\|}{C}}·$ + $[(CH_3)_3Si)_3]_3SiH$ ⟶ RCH$=$O + $[(CH_3)_3Si)_3]_3Si·$ (80°C) $k = 1.0 \times 10^5 \ M^{-1} \ s^{-1}$ k

B. Additions to alkenes and aromatics

15. $CH_3·$ + $CH_2{=}CHPh$ ⟶ $CH_3CH_2\dot{C}HPh$ $k = 2.6 \times 10^5 \ M^{-1} \ s^{-1}$ l
    $E_a = 4.9$ kcal/mol

16. $CH_3·$ + $CH_2{=}CHCN$ ⟶ $CH_3CH_2\dot{C}HCN$ $k = 6.1 \times 10^5 \ M^{-1} \ s^{-1}$ l

17. $CF_3·$ + $CH_2{=}CHPh$ ⟶ $CF_3CH_2\dot{C}HPh$ $k = 5.3 \times 10^6 \ M^{-1} \ s^{-1}$ m

18. Ph· + $CH_2{=}CHPh$ ⟶ $PhCH_2\dot{C}HPh$ $k = 1.1 \times 10^8 \ M^{-1} \ s^{-1}$ b

(continued)

Table 12.2. (*continued*)[a]

Reaction		Reference

19 Ph· + (cyclohexene) ⟶ (Ph-cyclohexyl radical) $k = 2.1 \times 10^7\ M^{-1}\ s^{-1}$ n

20 $PhCH_2· + CH_2{=}C(CH_3)_2 \longrightarrow PhCH_2CH_2\overset{\cdot}{C}(CH_3)_2$ $k = 18\ M^{-1}\ s^{-1}$ o

21 $(CH_3)_2\overset{\cdot}{C}CN + CH_2{=}CHPh \longrightarrow (CH_3)_2\overset{\cdot}{C}\overset{\cdot}{C}HPh$ (CN) $k = 7.0 \times 10^4\ M^{-1}\ s^{-1}$ $E_a = 6.4\ kcal/mol$ p

22 $CH_3·$ + (benzene) ⟶ (methylcyclohexadienyl radical, CH_3, H) $k = 46\ M^{-1}\ s^{-1}$ $E_a = 9.1\ kcal/mol$ q

23 Ph· + (benzene) ⟶ (Ph cyclohexadienyl radical, H) $k = 4.5 \times 10^5\ M^{-1}\ s^{-1}$ b

24 $(CH_3)_3CO· + CH_2{=}CH(CH_2)_5CH_3 \longrightarrow (CH_3)_3COCH_2\overset{\cdot}{C}H(CH_2)_5CH_3$ $k = 1.5 \times 10^6\ M^{-1}\ s^{-1}$ c

25 $PhS· + CH_2{=}CHPh \longrightarrow PhSCH_2\overset{\cdot}{C}HPh$ $k = 2.0 \times 10^7\ M^{-1}\ s^{-1}$ r

C. Cyclization and ring opening

26 (cyclopropylcarbinyl radical) ⟶ (allylcarbinyl radical) $k = 9.4 \times 10^7\ s^{-1}$ $E_a = 7.6\ kcal/mol$ s

27 (cyclopropyl)$-\overset{\cdot}{C}(CH_3)_2 \longrightarrow ·CH_2CH_2CH{=}C(CH_3)_2$ $k = 1.8 \times 10^7\ s^{-1}$ t

28 (hexenyl radical) ⟶ (cyclopentylmethyl radical) $k = 2.4 \times 10^5\ s^{-1}$ $E_a = 6.2\ kcal/mol$ u

29 X-(chain, Ph, Ph)· ⟶ (cyclopentane, $-\overset{\cdot}{C}Ph_2$, X)   (20°C)

X = H	$k = 4 \times 10^7\ s^{-1}$	v
X = CH₃	$k = 2 \times 10^7\ s^{-1}$	
X = OCH₃	$k = 5.4 \times 10^7\ s^{-1}$	
X = COC₂H₅	$k = 4 \times 10^7\ s^{-1}$	

30 (aryl radical with allyl chain) ⟶ (indane radical) $k = 4 \times 10^8\ s^{-1}$ $E_a = 3.6\ kcal/mol$ w

31 (benzyl-type radical) ⟶ (indane radical) $k = 1.5 \times 10^5\ s^{-1}$ $E_a = 7.3\ kcal/mol$ x

32 (aryl radical) ⟶ (indane radical) $k = 2 \times 10^{-1}\ s^{-1}$ $E_a = 16.3\ kcal/mol$ y

33 (alkynyl radical) ⟶ (cyclopentylidene radical) $k = 2.8 \times 10^4\ s^{-1}$ $E_a = 8.3\ kcal/mol$ z

Table 12.2. (*continued*)[a]

689

SECTION 12.2.
CHARACTERISTICS OF
REACTION
MECHANISMS
INVOLVING RADICAL
INTERMEDIATES

Reaction		Reference

34 $O{=}CHCH_2CH_2CH_2CH_2{\cdot} \rightleftharpoons$ [cyclopentyl-O·] $\overset{k = 1.4 \times 10^5\ s^{-1}}{\underset{k = 9.1 \times 10^7\ s^{-1}}{\rightleftarrows}}$ $E_a = 6.8$ kcal/mol   aa

35 $O{=}CHCH_2CH_2CH_2CH_2CH_2{\cdot} \longrightarrow$ [cyclohexyl-O·] $k = 2.5 \times 10^5\ s^{-1}$ $E_a = 5.4$ kcal/mol   bb

D. Other reactions

36 $(CH_3)_3C{\cdot} + O_2 \longrightarrow (CH_3)_3C{-}O{-}O{\cdot}$ $k = 4.9 \times 10^9\ M^{-1}\ s^{-1}$ w

37 $PhCH_2{\cdot} + O_2 \longrightarrow PhCH_2{-}O{-}O{\cdot}$ $k = 2.4 \times 10^9\ M^{-1}\ s^{-1}$ w

38 $(CH_3)_3C{-}\overset{\cdot}{\underset{\underset{O}{\|}}{C}} \longrightarrow (CH_3)_3C{\cdot} + C{\equiv}O$ $k = 3.0 \times 10^5\ s^{-1}$ j

39 $PhCH_2\overset{\cdot}{\underset{\underset{O}{\|}}{C}} \longrightarrow PhCH_2{\cdot} + C{\equiv}O$ $k = 5.2 \times 10^7\ s^{-1}$ $E_a = 7.2$ kcal/mol   cc

40 [cyclohexyl with $\overset{\overset{O}{\|}}{C}{\cdot}$ and CH_3] $\longrightarrow$ [cyclohexyl·CH_3] $+ C{\equiv}O$ $k = 5.2 \times 10^5\ s^{-1}$ $E_a = 10.0$ kcal/mol   dd

41 $Ph{-}\overset{\overset{CH_3}{|}}{\underset{\underset{CH_3}{|}}{C}}{-}\overset{\overset{O}{\diagup}}{\underset{}{C}}{\cdot} \longrightarrow Ph\overset{\cdot}{C}(CH_3)_2 + C{\equiv}O$ (0°C) $k = 11 \times 10^7\ s^{-1}$ ee

42 $Ph{\cdot} + CCl_4 \longrightarrow PhCl + {\cdot}CCl_3$ $k = 2.3 \times 10^6\ M^{-1}\ s^{-1}$ w

43 $CH_2{=}CHCH_2{\cdot} + ClOC(CH_3)_3 \longrightarrow CH_2{=}CHCH_2Cl + (CH_3)_3CO{\cdot}$ $k = 2.6 \times 10^9\ M^{-1}\ s^{-1}$ ff

44 $C_8H_{19}{\cdot} + PhSeCH_2CO_2C_2H_5 \longrightarrow C_8H_{19}SePh + {\cdot}CH_2CO_2C_2H_5$ (50°C) $k = 1.0 \times 10^5\ M^{-1}\ s^{-1}$ gg

45 $C_8H_{19}{\cdot} + PhSe\overset{\overset{}{\underset{\underset{CH_3}{|}}{C}}}{C}(CO_2C_2H_5)_2 \longrightarrow C_8H_{19}SePh + CH_3\overset{\cdot}{C}(CO_2C_2H_5)_2$ $k = 8 \times 10^5\ M^{-1}\ s^{-1}$ gg

a. Rates, unless otherwise noted, are for temperatures near 25°C. The original literature should be consulted for precise temperature and other conditions.
b. J. C. Scaiano and L. C. Stewart, *J. Am. Chem. Soc.* **105**:3609 (1983).
c. A. Baignee, J. A. Howard, J. C. Scaiano, and L. C. Stewart, *J. Am. Chem. Soc.* **105**:6120 (1983).
d. N. J. Bunce, K. U. Ingold, J. P. Landers, J. Lusztyk, and J. Scaiano, *J. Am. Chem. Soc.* **107**:564 (1985).
e. C. Chatgilialoglu, K. U. Ingold, and J. C. Scaiano, *J. Am. Chem. Soc.* **103**:7739 (1981).
f. S. J. Garden, D. V. Avila, A. L. J. Beckwith, V. W. Bowry, K. U. Ingold, and J. Lusztyk, *J. Org. Chem.* **61**:805 (1996).
g. D. V. Avila, K. U. Ingold, J. Lusztyk, W. R. Dolbier, Jr., H.-Q. Pan, and M. Muir, *J. Am. Chem. Soc.* **116**:99 (1994).
h. J. A. Franz, N. K. Suleman, and M. S. Alnajjar, *J. Org. Chem.* **51**:19 (1986).
i. M. Newcomb, T. R. Varick, C. Ha, M. B. Manek, and X. Yue, *J. Am. Chem. Soc.* **114**:8158 (1992).
j. C. E. Brown, A. G. Neville, D. M. Rayner, K. U. Ingold, and J. Lusztyk, *Aust. J. Chem.* **48**:363 (1995).
k. C. Chatgilialoglu and M. Lucarini, *Tetrahedron Lett.* **36**:1299 (1995).
l. T. Zytowski and H. Fischer, *J. Am. Chem. Soc.* **118**:437 (1996).
m. D. V. Avila, K. U. Ingold, J. Lusztyk, W. R. Dolbier, Jr., and H.-Q. Pan, *Tetrahedron* **52**:12351 (1996).
n. D. Weldon, S. Holland, and J. C. Scaiano, *J. Org. Chem.* **61**:8544 (1996).
o. K. Heberger. M. Walbiner, and H. Fischer, *Angew. Chem. Int. Ed. Engl.* **31**:635 (1992).
p. K. Heberger and H. Fischer, *Int. J. Chem. Kinet.* **25**:249 (1993).
q. T. Zytowski and H. Fischer, *J. Am. Chem. Soc.* **119**:12869 (1997).
r. O. Ito, S. Tamura, K. Murakami, and M. Matsuda, *J. Org. Chem.* **53**:4758 (1988).
s. M. Newcomb and A. G. Glem, *J. Am. Chem. Soc.* **111**:275 (1989); A. L. J. Beckwith and V. W. Bowry, *J. Org. Chem.* **54**:2681 (1989).
t. P. S. Engel, S.-L. He, J. T. Banks, K. U. Ingold, and J. Lusztyk, *J. Org. Chem.* **62**:1210 (1997).
u. A. L. J. Beckwith, C. J. Easton, T. Lawrence, and A. K. Serelis, *Aust. J. Chem.* **36**:545 (1983).

(*continued*)

v. C. Ha, J. H. Horner, M. Newcomb, T. R. Varick, B. R. Arnold, and J. Lusztyk, *J. Org. Chem.* **58**:1194 (1993); M. Newcomb, J. H. Horner, M. A. Filipkowski, C. Ha, and S.-U. Park, *J. Am. Chem. Soc.* **117**:3674 (1995).
w. L. J. Johnson, J. Lusztyk, D. D. M. Wayner, A. N. Abeywickreyma, A. L. J. Beckwith, J. C. Scaiano, and K. U. Ingold, *J. Am. Chem. Soc.* **107**:4594 (1985).
x. J. A. Franz, R. D. Barrows, and D. M. Camaioni, *J. Am. Chem. Soc.* **106**:3964 (1984).
y. J. A. Franz, M. S. Alnajjar, R. D. Barrows, D. L. Kaisaki, D. M. Camaioni, and N. K. Suleman, *J. Org. Chem.* **51**:1446 (1986).
z. A. L. J. Beckwith and C. H. Schiesser, *Tetrahedron Lett.* **26**:373 (1985).
aa. A. L. J. Beckwith and B. P. Hay, *J. Am. Chem. Soc.* **111**:230 (1989).
bb. A. L. J. Beckwith and B. P. Hay, *J. Am. Chem. Soc.* **111**, 2674 (1989).
cc. D. Griller and K. U. Ingold, *Acc. Chem. Res.* **13**:317 (1980).
dd. C. Chatgilialoglu, C. Ferreri, M. Lucarini, P. Pedrielli, and G. F. Pedulli, *Organometallics* **14**:2672 (1995).
ee. R. Hany and H. Fischer, *Chem. Phys.* **172**:131 (1993).
ff. J. M. Tanko and J. F. Blackert, *J. Chem. Soc., Perkin Trans. 2* **1996**:1175.
gg. D. P. Curran, A. A. Martin-Esker, S.-B. Ko, and M. Newcomb, *J. Org. Chem.* **58**:4691 (1993).

In hydrogen-atom abstraction reactions, the strength of the bond to the reacting hydrogen is a major determinant of the rate at which reaction occurs. It is for this reason that trisubstituted stannanes are among the most reactive reagents as hydrogen-atom donors. As indicated by entries 6–8 in Table 12.2, hydrogen abstraction from stannanes proceeds with rates greater than 10^6 M^{-1} s^{-1} and has very low activation energies. This high reactivity correlates with the low bond strength of the Sn—H bond (74 kcal, Table 12.6, p. 697). For comparison, entries 1–3 give the rates of hydrogen abstraction from two of the more reactive C—H hydrogen-atom donors, tetrahydrofuran and isopropylbenzene. For the directly comparable reaction with the phenyl radical (entries 1 and 8), tri-*n*-butylstannane is about 100 times more reactive than tetrahydrofuran as a hydrogen-atom donor. Phenylselenol is an even more reactive hydrogen-atom donor than tri-*n*-butylstannane (see entry 11). Entries 10 and 14 provide abstraction rates from two other kinds of hydrogen-atom donors, thiols and silanes.

Entries 4 and 5 point to another important aspect of free-radical reactivity. The data given illustrate that the observed reactivity of the chlorine atom is strongly influenced by the presence of benzene. Evidently, a complex is formed which attenuates the reactivity of the chlorine atom. This is probably a general feature of radical chemistry, but there are relatively few data available on solvent effects on either absolute or relative reactivity of radical intermediates.

Part B of Table 12.2 gives some addition reaction rates. Comparison of entries 19 and 20 shows that the phenyl radical is much more reactive toward addition than the benzyl radical. Comparison of entries 22 and 23 shows that methyl radicals are less reactive than phenyl radicals in additions to an aromatic ring. Note that additions to aromatic rings are much slower than additions to alkenes.

Part C of Table 12.2 shows some reactions involving ring opening or closure. Cyclization of unsaturated radicals has become an important application of free-radical chemistry in synthesis and will be discussed more thoroughly in Section 10.3 of Part B. Rates of cyclization reactions have also proved useful in mechanistic studies, where they can serve as reference points for comparison with other reaction rates. Entry 26 is the case of ring opening of the cyclopropylmethyl radical, which was discussed on p. 668. Notice that the activation energy is slightly higher than that for a normal single-bond rotation but less than that for cyclohexane inversion.

Part D of Table 12.2 gives rates of some other important kinds of radical reactions, including reaction with O_2 (entries 36 and 37), decarbonylation (entries 38–41), and group transfer for the phenylselenenyl group (entries 44 and 45).

691

SECTION 12.2.
CHARACTERISTICS OF
REACTION
MECHANISMS
INVOLVING RADICAL
INTERMEDIATES

Among the entries in part C of Table 12.2 are examples of ring closures of unsaturated radicals. Free-radical cyclizations show a preference for *exo* cyclization with formation of five-membered rings over *endo* cyclization with formation of six-membered rings.[64] (Review Section 3.9.) This trend is observed even though it results in formation of a less stable primary radical. The cause for this preference has been traced to stereoelectronic effects. In order for a bonding interaction to occur, the radical center must interact with the π^* orbital of the alkene. Bond formation takes place as the result of initial interaction with the LUMO, which is the π^* orbital. According to MO calculations, the preferred direction of attack is from an angle of about 70° with respect to the plane of the double bond.[65]

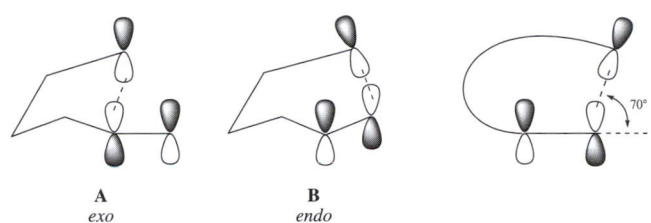

A
exo

B
endo

When this stereoelectronic requirement is combined with a calculation of the steric and angle strain imposed on the transition state, as determined by MM-type calculations, preferences for the *exo* versus *endo* modes of cyclization are predicted to be as summarized in Table 12.3. The observed results show the expected qualitative trend. The observed preferences for ring formation are $5 > 6$, $6 > 7$, and $8 > 7$, in agreement with the calculated preferences. The relationship only holds for terminal double bonds. An additional alkyl substituent at either end of the double bond reduces the relative reactivity as a result of a steric effect.

The relatively low rate and high activation energy noted for entry 32 also reflect a stereoelectronic effect. The preference for delocalization at the radical center requires

Table 12.3. Regioselectivity of Radical Cyclization as a Function of Ring Size[a]

Ring size	exo : endo Ratio	
	Calc.	Found
5:6	10:1	50:1
6:7	>100:1	10:1
7:8	1:5.8	<1:100

a. D. C. Spellmeyer and K. N. Houk, *J. Org. Chem.* **52**:959 (1987).

64. A. L. J. Beckwith, C. J. Eaton, and A. K. Serelis, *J. Chem. Soc., Chem. Commun.* **1980**:482; A. L. J. Beckwith, T. Lawrence, and A. K. Serelis, *J. Chem. Soc., Chem. Commun.* **1980**:484; A. L. J. Beckwith, *Tetrahedron* **37**:3073 (1981).

65. M. J. S. Dewar and S. Olivella, *J. Am. Chem. Soc.* **100**:5290 (1978); D. C. Spellmeyer and K. N. Houk, *J. Org. Chem.* **52**:959 (1987); A. L. J. Beckwith and C. H. Schiesser, *Tetrahedron* **41**:3925 (1985).

coplanarity of the substituents at the radical site:

poor alignment in
planar conformation

optimal alignment disrupts
benzylic conjugation

In view of the restrictions on the mode of approach of the radical to the double bond, significant strain develops at the transition state, and this requires rotation of the benzylic methylene group out of its preferred coplanar alignment.[66]

Some general remarks about structure–reactivity relationships in radical reactions can be made at this point. The reactivity of C$-$H groups toward radicals that abstract hydrogen is usually *pri* < *sec* < *tert*. Vinyl and phenyl substituents at a reaction site increase the reactivity toward radicals. This reactivity order reflects the bond dissociation energies of the various C$-$H bonds, which are in the order allyl $\sim$ benzyl < *tert* < *sec* < *pri*, and parallels the stability of the product radical.[67] The relative reactivity of primary, secondary, and tertiary positions in aliphatic hydrocarbons toward hydrogen abstraction by the methyl radical is 1 : 4.3 : 46.[68] The relative reactivity toward the *t*-butoxy radical is 1 : 10 : 44.[69] An allylic or benzylic hydrogen is more reactive toward a methyl radical by a factor of about 9, compared to a corresponding unactivated hydrogen.[68] Data for other types of radicals have been obtained and tabulated.[69] The trend of reactivity *tert* > *sec* > *pri* is consistently observed, but the range of reactivity is determined by the nature of the reacting radical. In the gas phase, the bromine atom, for example, is very selective, with relative reactivities of 1 : 250 : 6300 for abstraction of primary, secondary, and tertiary hydrogens.[70]

The stabilizing effects of vinyl groups (in allylic radicals) and phenyl groups (in benzyl radicals) are very significant and can be satisfactorily rationalized in resonance terminology:

$$CH_2{=}CH{-}CH_2\cdot \quad \longleftrightarrow \quad \cdot CH_2{-}CH{=}CH_2$$

For the vinyl substituent, we can analyze the stabilization in terms of simple HMO theory. The interaction of a p orbital with an adjacent vinyl group creates the allyl radical. In Hückel calculations, the resulting orbitals have energies of $\alpha + 1.4\beta$, α, and $\alpha - 1.4\beta$. Thus, the interaction of the p orbital with both the π and π^* orbitals leaves it at the same energy level, but the π and π^* levels are transformed to ψ_1 and ψ_3 of the allyl radical. There is a net stabilization of 0.4β. A measure of the stabilization energy is the barrier to rotation at a terminal methylene group. A value of 16.8 kcal/mol has been calculated.[71] The measured barrier is 15.7 kcal/mol.

66. J. A. Franz, N. K. Suleman, and M. S. Alnajjar, *J. Org. Chem.* **51**:19 (1986).
67. J. A. Kerr, *Chem. Rev.* **66**:465 (1966).
68. W. A. Pryor, D. L. Fuller, and J. P. Stanley, *J. Am. Chem. Soc.* **94**:1632 (1972).
69. C. Walling and B. B. Jacknow, *J. Am. Chem. Soc.* **82**:6108 (1960).
70. A. F. Trotman-Dickenson, *Adv. Free Radical Chem.* **1**:1 (1965).
71. D. A. Hrovat and W. T. Borden, *J. Phys. Chem.* **98**:10460 (1994).

693

SECTION 12.2.
CHARACTERISTICS OF
REACTION
MECHANISMS
INVOLVING RADICAL
INTERMEDIATES

The stabilizing role of other functional groups can also be described in resonance terms. Both electron-attracting groups such as carbonyl and cyano and electron-donating groups such as methoxy and dimethylamino have a stabilizing effect on a radical intermediate at an adjacent carbon. The resonance structures which depict these interactions indicate delocalization of the unpaired electron onto the adjacent substituents:

A description of the radical-stabilizing effect of both types of substituents can also be presented in MO terms. In this case, the question we ask is how will the unpaired electron in a p orbital interact with the orbitals of the adjacent substituent, such as a vinyl, carbonyl, or methoxy group? Figure 12.4 presents a qualitative description of the situation. The basic tenet of PMO theory (see Section 1.6) that the orbitals which are closest in energy will interact most strongly is employed in the analysis of the effect of electron-withdrawing and electron-donating substituents. In the case of an electron-accepting substituent, such as a carbonyl group, the strongest interaction is with the carbonyl LUMO, π^*. This results in a lowering of the energy of the orbital containing the unpaired electron; i.e., the radical is stabilized. For an electron-donating substituent, the strongest interaction is between the electron in the p orbital and a nonbonding electron pair on the electron donor. This interaction results in a lowering of the energy of the orbital occupied by the electron pair and a raising of the energy of the orbital occupied by the single electron. The net effect is still stabilizing because there are two electrons in the stabilized orbital and only one in the destabilized one. Some results of calculations of substituent effects on radical stability are presented in Table 12.4.

Allylic and benzylic radicals are also stabilized by both acceptor and donor substituents. As shown in Table 12.5, theoretical calculations at the MP2 level indicate that substituents at the 2-position are only slightly less effective than 1-substituents in the

Fig. 12.4. PMO representation of p-orbital interactions with C=C (a), C=O (b), and ÖR (c) substituents. Form of SOMO (singly occupied molecular orbital) is shown.

Table 12.4. Substituent Effects on Radical Stability from Measurements of Bond Dissociation Energies and Theoretical Calculations of Radical Stabilization Energies

Substituent	ΔBDE^a	RSE^b	RSE^c
H	0.0	0.0	0.0
CH_3	7	3.2	4.8
$CH_2=CH$	19	12.6	18.8
C_6H_5	17		
$HC=O$		9.5	
$CH_3C=O$	11		9.2
$C_6H_5C=O$	12		
$CO_2C_2H_5$	10		
CN	12	6.7	10.4
NO_2	7		
F	3	5.0	
CH_3O	12	8.9 (OH)	9.8
NH_2	22	11.1	13.3
$(CH_3)_2N$	21		

a. ΔBDE, Difference in bond dissociation energy relative to methane. Values from F. G. Bordwell, X.-M. Zhang, and M. S. Alnajjar, *J. Am. Chem. Soc.* **114**:7623 (1992); F. G. Bordwell and X.-M. Zhang, *Acc. Chem. Res.* **26**:570 (1993).
b. AUMP2 calculations; M. Lehd and F. Jensen, *J. Org. Chem.* **56**:884 (1991).
c. B3LYP/6-31 + G; B. S. Jursic, J. W. Timberlake, and P. S. Engel, *Tetrahedron Lett.* **37**:6473 (1996).

stabilization of allylic radicals. Table 12.5 also includes results of calculations that have been done on the stabilizing effect of *para*-substituents on benzylic radicals.

Radicals are particularly strongly stabilized when both an electron-attracting and an electron-donating substituent are present at the radical site. This has been called "*mero-stabilization*"[72] or "*capto-dative stabilization*."[73] This type of stabilization results from mutual reinforcement of the two substituent effects.[74] Scheme 12.3 gives some information on the stability of this type of radical.

stabilization
by π acceptor

stabilization
by π donor

combined capto-dative
stabilization by π donor
and π acceptor

A comparison of the rotational barriers in allylic radicals **A–D** provides evidence for the stabilizing effect of the capto-dative combination:

A	B	C	D
15.7	14.5	10.2	6.0

72. R. W. Baldock, P. Hudson, A. R. Katritzky, and F. Soti, *J. Chem. Soc., Perkin Trans. I* **1974**:1422.
73. H. G. Viehe, R. Merenyi, L. Stella, and Z. Janousek, *Angew. Chem. Int. Ed. Engl.* **18**:917 (1979).
74. R. Sustmann and H.-G. Korth, *Adv. Phys. Org. Chem.* **26**:131 (1990).

695

SECTION 12.2.
CHARACTERISTICS OF
REACTION
MECHANISMS
INVOLVING RADICAL
INTERMEDIATES

Table 12.5. Substituent Effects on Stability of Allylic and Benzylic Radicals from Theoretical Calculations of Radical Stabilization Energies

Substituent	Allylic[a]		
	1-Position	2-Position	p-Position
H	0	0	0
CH_3	5.6	4.3	0.3
CN	9.9	3.0	1.4
CH=O	11.7	11.6	
F	8.3	11.0	−0.1
OH	12.8	12.6	
CH_3O			0.7
NH_2	13.7	9.4	
$(CH_3)_2N$			1.8

a. AUMP2 calculation; M. Lehd and F. Jensen, *J. Org. Chem.* **56**:884 (1991).
b. BLYP-6-31G*; Y.-D. Wu, C.-L. Wong, K. W. K. Chan, G.-Z. Ji, and X.-K. Jang, *J. Org. Chem.* **61**:746 (1996).

The decreasing barrier at the formal single bond along the series **A** to **D** implies decreasing π-allyl character of this bond. The decrease in the importance of the π bonding in turn reflects a diminished degree of interaction of the radical center with the adjacent double bond. The fact that the decrease from **C** to **D** is greater than from **A** to B indicates a synergistic effect, as implied by the capto-dative formulation. The methoxy group is more stabilizing when it can interact with the cyano group than as an isolated substituent, as it is in **B**.[75]

The capto-dative effect has also been demonstrated by studying the bond dissociation process in a series of 1,5-dienes substituted at C-1, C-3, C-4, and C-6.

X	Y	X′	Y′	ΔH
CO_2R	CO_2R	CO_2R	CO_2R	38.1
CO_2R	CO_2R	CO_2R	OR	28.2
CN	OR	CN	OR	24.5
CN	NR_2	CN	NR_2	8.1

When the combination X, Y or X′, Y′ is of the capto-dative type, as is the case for an alkoxy and an ester group, the enthalpy of bond dissociation is 10–15 kcal lower than when all four groups are electron-attracting. When the capto-dative combination CN/NR_2 occupies both the X, Y and the X′, Y′ positions, the enthalpy for dissociation of the C(3)−C(4) bond is less than 10 kcal/mol.[76]

The radical stabilization provided by various functional groups results in reduced bond dissociation energies for bonds to the stabilized radical center. Some bond dissociation energy values are given in Table 12.6. As an example of the effect of substituents on bond dissociation energies, it can be seen that the primary C−H bonds in acetonitrile (86 kcal/mol) and acetone (92 kcal/mol) are significantly weaker than a primary C−H

75. H.-G. Korth, P. Lommes, and R. Sustmann, *J. Am. Chem. Soc.* **106**:663 (1984).
76. M. Van Hoecke, A. Borghese, J. Penelle, R. Merenyi, and H. G. Viehe, *Tetrahedron Lett.* **27**:4569 (1986).

Scheme 12.3. Radicals with Capto-dative Stabilization

1^a	Wurster's salts. Generated by one-electron oxidation of the diamine. Indefinitely stable.
2^b	Generated by one-electron reduction of the pyridinium salt. Stable, distillable, and only moderately reactive to oxygen.
3^c	Stable and distillable. A small amount of dimer is present in equilibrium with the radical.
4^d	Generated by spontaneous dissociation of the dimer. In equilibrium with dimer.
5^e	Generated by spontaneous dissociation of the dimer. Stable for several days at room temperature. Oxidized by oxygen.
6^f	Generated spontaneously from dimethylamino-malononitrile at room temperature. Observed to be persistent over many hours by EPR.
7^g	Total radical stabilization energy of 19.8 kcal/mol implies $\simeq 10$ kcal/mol of excess radical stabilization relative to the combined substituents. $CH-N(CH_3)_2$ rotational barrier is > 17 kcal/mol, implying strong resonance interaction.
8^h	Stabilization of about 6.3 kcal/mol based on thermochemical analysis of dimerization reaction.

a. A. R. Forrester, J. M. Hay, and R. H. Thompson, *Organic Chemistry of Stable Free Radicals*, Academic Press, New York, 1968, pp. 254–261.
b. J. Hermolin, M. Levin, and E. M. Kosower, *J. Am. Chem. Soc.* **103**:4808 (1981).
c. J. Hermolin, M. Levin, Y. Ikegami, M. Sawayangai, and E. M. Kosower, *J. Am. Chem. Soc.* **103**:4795 (1981).
d. T. H. Koch, J. A. Oleson, and J. DeNiro, *J. Am. Chem. Soc.* **97**:7285 (1975).
e. J. M. Burns, D. L. Wharry, and T. H. Koch, *J. Am. Chem. Soc.* **103**:849 (1981).
f. L. de Vries, *J. Am. Chem. Soc.* **100**:926 (1978).
g. F. M. Welle, H.-D. Beckhaus, and C. Rüchardt, *J. Org. Chem.* **62**:552 (1997).
h. F. M. Welle, S. P. Verevkin, H.-D. Beckhaus, and G. Rüchardt, *Liebigs Ann.* **1997**:155.

bond in ethane (100.1 kcal/mol). Differences in bond dissociation energies relative to methane can be taken as a measure of the stabilizing effect of the substituent on the radical.

The bond dissociation energy method assumes that all the bond strength changes are due to effects in the radical product. By analysis of heats of formation of compounds incorporating radical fragments and assignment of standard sets of bond energies, it is

697

SECTION 12.2.
CHARACTERISTICS OF
REACTION
MECHANISMS
INVOLVING RADICAL
INTERMEDIATES

Table 12.6. Bond Dissociation Energies (kcal/mol)^a

Bond	D.E.	Bond	D.E.
CH_3—H	104.3^b	H—Br	87.5
CH_3CH_2—H	100.1^b	H—I	71.3
$(CH_3)_2CH$—H	96.3^b	$HOCH_2$—H	92
$(CH_3)_3C$—H	93.6^b	$CH_3CH_2OCHCH_3$	92^c
CH_2=CH—H	103^b	$\underset{H}{\mid}$	
$\underset{CH_2\text{-}CH\text{-}H}{\overset{CH_2}{\diagup \diagdown}}$	101	$\underset{CH_3CCH_2\text{-}H}{\overset{O}{\overset{\parallel}{}}}$	92
$PhCH_2$—H	87.9^b	$N{\equiv}CCH_2$—H	86
CH_2=$CHCH_2$—H	86.8^b	CH_3S—H	91.8^b
		PhS—H	82^d
		$(CH_3)_3Si$—H	90^e
	73^c	$(CH_3)_3Ge$—H	82^e
		$(C_4H_9)_3Sn$—H	74^c
F_3C—H	106		
Cl_3C—H	96	$\underset{CH_3CO\text{-}OCCH_3}{\overset{O\quad\ \ O}{\overset{\parallel\quad\ \ \parallel}{}}}$	30
C_2H_5—F	106	$(CH_3)_3CO$—OH	44
C_2H_5—Cl	81	F—F	38
C_2H_5—Br	69	Cl—Cl	58
C_2H_5—I	53	Br—Br	46
H—F	136.3	I—I	36
H—Cl	103.1		

a. Except where noted otherwise, data are from J. A. Kerr, *Chem. Rev.* **66**:465 (1966); S. W. Benson, *J. Chem. Educ.* **42**:502 (1965).
b. D. D. M. Wagner and D. Griller, *Adv. Free Radical Chem.* **1**:159 (1990).
c. T. J. Burkey, M. Majewski, and D. Griller, *J. Am. Chem. Soc.* **108**:2218 (1986).
d. S. W. Benson, *Chem. Rev.* **78**:23 (1978).
e. R. A. Jackson, *J. Organomet. Chem.* **166**:17 (1979).

possible to arrive at energies corresponding to the stabilization of the radical fragment. This energy then reflects the stabilization or destabilization of the radical center in the particular structural environment. Table 12.7 lists values assigned by two such approaches. Although the two sets of values differ by a few kilocalories, the trends are consistent with the qualitative predictions of radical stability based on MO theory. These data reveal the familiar stabilization of allyl and benzyl radicals, which appears as positive stabilization energies. The trend of increasing stability of alkyl radicals in the order *tert* > *sec* > *pri* > methyl is also apparent. According to this analysis, donor substituents, such as hydroxy, are not as strongly stabilizing as acceptor substituents, such as the carbonyl group. The assignment of significantly negative stabilization (i.e., destabilization) energies implies that the substituent will increase the bond energy of adjacent bonds. The phenyl and vinyl radicals are examples of radicals that have negative stabilization energies, implying an increase in the strength of the bonds to substituents. These negative values are in agreement with the relative instability of phenyl and vinyl radicals and the difficulty in abstraction of a hydrogen atom from aromatic rings or double bonds.

Bond dissociation energies such as those in Table 12.6 are also useful for estimation of the energy balance in individual steps in a free-radical reaction sequence. This is an

Table 12.7. Thermochemical Stabilization Energies for Some Substituted Radicals

	SE^{0a}	SE^b
$\dot{C}H_3$	−1.67	−3.9
$CH_3\dot{C}H_2$	2.11	−1.5
$CN\dot{C}H_2$	6.50	
$NH_2\dot{C}H_2$	3.92	
$CH_3NH\dot{C}H_2$	12.18	
$(CH_3)_2N\dot{C}H_2$	14.72	
$HO\dot{C}H_2$	3.13	1.7
$CH_3O\dot{C}H_2$	3.64	
$F\dot{C}H_2$	−1.89	−0.7
$(CH_3)_2\dot{C}H$	2.57	
$(CN)_2\dot{C}H$	5.17	
$(HO)_2\dot{C}H$	−2.05	
$(CN)(NH_2)\dot{C}H$	12.94	
$(CN)(OH)\dot{C}H$	2.15	
$(CN)F\dot{C}H$	−2.79	
$F_2\dot{C}H$	−4.11	1.7
$(CH_3)_3\dot{C}$	4.35	
$CH_3\dot{C}(CN)_2$	3.92	
$CH_3\dot{C}(OH)_2$	0.15	
$CH_3\dot{C}(CN)(OH)$	2.26	
$\dot{C}F_3$	−4.17	
$\dot{C}Cl_3$	−13.79	
$CH_2{=}CH{-}\dot{C}H_2$	13.28	10.5
$C_6H_5{-}\dot{C}H_2$	12.08	13.9
$CH_2{=}\dot{C}H$	−6.16	−10.0
$HC{\equiv}C\cdot$	−15.57	
$(C_6H_5)\cdot$	−10.27	−12.3
$(C_5H_5)\cdot$	19.24	
$CH_3\overset{\cdot}{C}{=}O$	7.10	
$H\overset{\cdot}{C}{=}O$		5.8

a. Stabilization energy as defined by C. Leroy, D. Peeters, and C. Wilante, *THEOCHEM* **5**:217 (1982).
b. Reorganization energy as defined by R. T. Sanderson, *J. Org. Chem.* **47**:3835 (1982) but given here as a stabilization energy of opposite sign.

important factor in assessing the feasibility of chain reaction sequences, since only reactions with low activation energies are fast enough to sustain a chain process. If individual steps are identified as being endothermic by more than a few kilocalories, it is very unlikely that a chain mechanism can operate.

Example 12.1. Calculate the enthalpy for each step in the bromination of ethane by bromine atoms from molecular bromine. Determine the overall enthalpy of the reaction.

699

SECTION 12.2.
CHARACTERISTICS OF
REACTION
MECHANISMS
INVOLVING RADICAL
INTERMEDIATES

Initiation   Br—Br $\longrightarrow$ 2 Br·   Br—Br +46 kcal/mol

propagation {

Br· + CH$_3$—CH$_3$ $\longrightarrow$ H—Br + ·CH$_2$—CH$_3$ H—Br -87.5
                                                               H—C $+100.1$
                                                               _____
                                                                        $+12.6$

Br—Br + ·CH$_2$—CH$_3$ $\longrightarrow$ BrCH$_2$—CH$_3$ + Br·   Br—Br $+46$
                                                                 Br—C -69
                                                                 _____
                                                                          -23

total -10.4 kcal/mol

The enthalpy of the reaction is given by the sum of the enthalpies for the propagation steps and is -10.4 kcal/mol. Analysis of the enthalpy of the individual steps indicates that the first step is somewhat endothermic ($+12.6$ kcal). This endothermicity is the lower limit of the activation energy for the step. Radical-chain processes depend on a series of rapid steps that maintain the reactive intermediates at low concentrations. Because termination reactions are usually very fast, the presence of an endothermic step in a chain sequence means that the chains will be short. The value for ethane is borderline and suggests that radical bromination of ethane would exhibit only short chain lengths. Because the enthalpy of the corresponding steps for abstraction of secondary or tertiary hydrogens is less positive, the bromination selects for tertiary > secondary > primary in compounds with more than one type of hydrogen. Enthalpy calculations cannot give a direct evaluation of the activation energy of either exothermic or endothermic steps. These will depend on the energy of the transition state. The bond dissociation energies can therefore provide only permissive, not definitive, conclusions.

Radical stability is reflected in a variety of ways in addition to the bond dissociation energy of the corresponding C–H bond. It has already been indicated that radical structure and stability determines the temperature at which azo compounds undergo decomposition with elimination of nitrogen (Section 12.1.4). Similar trends have been established in other radical-forming reactions. Rates of thermal decomposition of t-butyl peresters, for example, vary over a wide range, depending on the structure of the carbonyl group.[77] This clearly indicates that the bonding changes involved in the rate-determining step are not completely localized in the O–O bond. Radical character must also be developing at the alkyl group by partial cleavage of the alkyl–carbonyl bond:

$$R-\overset{\overset{O}{\|}}{C}-O-O-C(CH_3)_3 \longrightarrow R- -\overset{\overset{O}{\|}}{C} \doubleequals O - -OC(CH_3)_3 \longrightarrow R\cdot + CO_2 + \cdot OC(CH_3)_3$$

R	Relative rate at 60°C
CH$_3$	1
Ph	17
PhCH$_2$	290
(CH$_3$)$_3$C	1,700
Ph$_2$CH	19,300
Ph(CH$_3$)$_2$C	41,500
PhCHCH=CH$_2$	125,000

77. P. D. Bartlett and R. R. Hiatt, *J. Am. Chem. Soc.* **80**:1398 (1958).

The same is true for decarbonylation of acyl radicals. The rates of decarbonylation have been measured over a very wide range of structural types.[78] There is a very strong dependence of the rate on the stability of the radical that results from decarbonylation. For example, rates for decarbonylations giving tertiary benzylic radicals are on the order of 10^8 s^{-1} whereas the benzoyl radical decarbonylates to phenyl radical with a rate on the order of 1 s^{-1}.

Free-radical reactions written in the simplest way imply no separation of charge. The case of toluene bromination can be used to illustrate this point:

$$Br\cdot + H—CH_2Ph \longrightarrow Br\text{-}\text{-}\text{-}H\text{-}\text{-}CH_2Ph \longrightarrow Br—H + \cdot CH_2Ph$$

$$PhCH_2\cdot + Br—Br \longrightarrow PhCH_2\text{-}\text{-}\text{-}Br\text{-}\text{-}Br \longrightarrow PhCH_2Br + Br\cdot$$

Nevertheless, many free-radical processes respond to introduction of polar substituents, just as do heterolytic processes that involve polar or ionic intermediates. The substituent effects on toluene bromination, for example, are correlated by the Hammett equation, which gives a ρ value of -1.4, indicating that the benzene ring acts as an electron donor in the transition state.[79] Other radicals, for example the t-butyl radical, show a positive ρ for hydrogen abstraction reactions involving toluene.[80]

Why do free-radical reactions involving neutral reactants and intermediates respond to substituent changes that modify electron distribution? One explanation has been based on the idea that there would be some polar character in the transition state because of the electronegativity differences of the reacting atoms[81]:

This idea receives general support from the fact that the most negative ρ values are found for more electronegative radicals such as Br$\cdot$, Cl$\cdot$ and Cl$_3$C$\cdot$. There is, however, no simple correlation with a single property, and this probably reflects the fact that the *selectivity* of the radicals is also different. Furthermore, in hydrogen abstraction reactions, for which many of the quantitative measurements have been done, the C—H bond dissociation energy is also subject to a substituent effect.[82] Thus, the extent of bond cleavage and formation at the transition state may be different for various radicals. Successful interpretation of radical reactions therefore requires consideration of factors such as the electronegativity and polarizability of the radicals and the bond energy of the reacting C—H bond. The relative importance of these effects may vary from system to system. As a result, substituent effect trends in radical reactions appear to be more complicated than

78. H. Fischer and H. Paul, *Acc. Chem. Res.* **20**:200 (1987).

79. J. Hradil and V. Chvalovsky, *Collect. Czech. Chem. Commun.* **33**:2029 (1968); S. S. Kim, S. Y. Choi, and C. H. Kong, *J. Am. Chem. Soc.* **107**:4234 (1984); G. A. Russell, C. DeBoer, and K. M. Desmond, *J. Am. Chem. Soc.* **85**:365 (1963); C. Walling, A. L. Rieger, and D. D. Tanner, *J. Am. Chem. Soc.* **85**:3129 (1963).

80. W. A. Pryor, F. Y. Tang, R. H. Tang, and D. F. Church, *J. Am. Chem. Soc.* **104**:2885 (1982); R. W. Henderson and R. O. Ward, Jr., *J. Am. Chem. Soc.* **96**:7556 (1974); W. A. Pryor, D. F. Church, F. Y. Tang, and R. H. Tang, *Frontiers of Free Radical Chemistry*, W. A. Pryor, ed., Academic Press, New York, 1980, pp. 355–380.

81. E. S. Huyser, *Free Radical Chain Reactions*, Wiley-Interscience, New York, 1970, Chapter 4; G. A. Russell, in *Free Radicals*, Vol. 1, J. Kochi, ed., John Wiley & Sons, New York, 1973, Chapter 7.

82. A. A. Zavitsas and J. A. Pinto, *J. Am. Chem. Soc.* **94**:7390 (1972); W. M. Nau, *J. Phys. Org. Chem.* **10**:445 (1997).

those for heterolytic reactions, for which substituent effects are usually dominated by the electron-releasing or electron-donating capacity of the substituent.[83]

Despite their overall electrical neutrality, carbon-centered radicals show pronounced electrophilic or nucleophilic character, depending on the substituents present. This electrophilic or nucleophilic character is then reflected in rates of reaction with nonradical species, for example, in additions to substituted alkenes. Unsubstituted alkyl radicals and α-alkoxyalkyl radicals are nucleophilic in character and react most rapidly with alkenes having electron-attracting substituents.[84] Radicals having electron-attracting groups, such as the radicals derived from malonate esters, react preferentially with double bonds having electron-releasing substituents.[85] Some representative data are given in Table 12.8.

Simple alkyl radicals such as methyl are considered to be nonnucleophilic. Methyl radicals are somewhat more reactive toward alkenes bearing electron-withdrawing substituents than towards those with electron-releasing substituents. However, much of this effect can be attributed to the stabilizing effect that these substituents have on the product radical. There is a strong correlation of reaction rate with the overall exothermicity of the reaction.[86] Hydroxymethyl and 2-hydroxy-2-propyl radicals show nucleophilic character.[87] The hydroxymethyl radical shows a slightly enhanced reactivity toward acrylonitrile and acrolein, but a sharply decreased reactivity toward ethyl vinyl ether. Table 12.9 gives some of the reactivity data.

α-Fluoro substituents enhance reactivity toward alkene addition. The effect of polyfluorination is more than cumulative. The relative rates for RCH_2 (1), $RCHF$ (3.5),

Table 12.8. Relative Rates of Radical Addition as a Function of Alkene Substitution[a]

| A. Addition to substituted ethenes, $CH_2=CH-X$ | | | |
X	$CH_3\cdot$	$C_2H_5\cdot$	$c-C_6H_{11}\cdot$
CN	2.2	5.1	24
$COCH_3$	2.3		13
CO_2CH_3	1.3	1.9	6.7
Ph	1.0	1.0	1.0
O_2CCH_3		0.05	0.016

| B. Addition to α-substituted styrenes, | | |
X	$c-C_6H_{11}$	$\cdot CH(CO_2C_2H_5)_2$
CN	122	
$CO_2C_2H_5$	11.7	0.28
Ph	1.0	1.0
CH_3	0.28	1.06
OCH_3		0.78
$N(CH_3)_2$		6.6

a. Data from B. Giese, H. Horler, and M. Leising, *Chem. Ber.* **119**:444 (1986); B. Giese, *Angew. Chem. Int. Ed. Engl.* **22**:753 (1983).

83. W. H. Davis, Jr., and W. A. Pryor, *J. Am. Chem. Soc.* **99**:6365 (1977); W. H. Davis, Jr., J. H. Gleason, and W. A. Pryor, *J. Org. Chem.* **42**:7 (1977); W. A. Pryor, G. Gojon, and D. F. Church, *J. Org. Chem.* **43**:793 (1978).
84. B. Giese, *Angew. Chem. Int. Ed. Engl.* **22**:753 (1983).
85. B. Giese, H. Horler, and M. Leising, *Chem. Ber.* **119**:444 (1986).
86. M. W. Wong, A. Pross, and L. Radom, *J. Am. Chem. Soc.* **115**:11050 (1993); R. Arnaud, N. Bugaud, V. Vetere, and V. Barone, *J. Am. Chem. Soc.* **120**:5733 (1998).
87. J. Q. Wu and H. Fischer, *Int. J. Chem. Kinet.* **27**:167 (1995); S. N. Batchelor and H. Fischer, *J. Phys. Chem.* **100**:9794 (1996).

Table 12.9. Absolute Rates of Addition Reactions of Methyl, Cyanomethyl, and Hydroxymethyl Radicals toward Substituted Alkenes, $CH_2=CHX$

X	k (M^{-1} s^{-1})		
	$\cdot CH_3{}^a$	$\cdot CH_2CN{}^b$	$\cdot CH_2OH{}^c$
H	3.5×10^3	3.3×10^3	4.1×10^2
Ph	2.6×10^5	3.8×10^5	2.3×10^4
CN	6.1×10^5	1.1×10^5	1.1×10^6
CH=O	7.4×10^5	2.5×10^4	2.1×10^6
CO_2CH_3	3.4×10^5	1.1×10^5	7.1×10^5
OC_2H_5	1.4×10^4	1.2×10^4	1.8×10^2
CH_3	4.3×10^3	1.2×10^4	2.7×10^2

a. T. Zytowski and H. Fischer, *J. Am. Chem. Soc.* **118**:437 (1996); T. Zytowski and H. Fischer, *J. Am. Chem. Soc.* **119**:12869 (1997).
b. J. Q. Wu and H. Fischer, *Int. J. Chem. Kinet.* **27**:167 (1995).
c. J. Q. Wu, I. Beranck, and H. Fischer, *Helv. Chim. Acta* **78**:194 (1995).

RCF_2 (20), and CF_3 (300) show this trend.[88] Further accumulation of fluorine enhances this effect still more, and the perfluoro-*t*-butyl radical is typically 8–10 times more reactive than $CF_3\cdot$ toward alkenes.

These reactivity trends are in line with what would be expected on the basis of a frontier orbital interpretation.[89] As shown in Fig. 12.5, electron-releasing substituents will raise the energy of the singly occupied molecular orbital (SOMO) and increase the strength of interaction with the relatively low-lying LUMO of alkenes having electron-withdrawing groups. When the radical site is substituted by an electron-attracting group, the SOMO is lower, and the strongest interaction will be with the HOMO of the alkene. This interaction is strenghtened by donor substituents on the alkene.

Unsubstituted | Donor-substituted radical and acceptor-substituted alkene | Acceptor-substituted radical and donor-substituted alkene

Fig. 12.5. Frontier orbital interactions between different combinations of substituted radicals and alkenes.

88. D. V. Avila, K. U. Ingold, J. Lusztyk, W. R. Dolbier, Jr., and H.-Q. Pan, *J. Org. Chem.* **61**:2027 (1996); D. Avila, K. U. Ingold, J. Lusztyk, W. R. Dolbier, Jr., and H.-Q. Pan, *Tetrahedron* **52**:12351 (1996).
89. U. Berg, E. Butkus, and A. Stoncius, *J. Chem. Soc., Perkin Trans. 2* **1995**:97; M. W. Wong, A. Pross, and L. Radom, *J. Am. Chem. Soc.* **116**:6284 (1994).

12.3. Free-Radical Substitution Reactions

12.3.1. Halogenation

Free-radical bromination is an important method of selective functionalization of hydrocarbons.[90] The process is a chain reaction involving the following steps:

$$\text{initiation} \qquad \text{Br—Br} \longrightarrow 2\,\text{Br·}$$
$$\text{propagation} \qquad \text{Br·} + R_3CH \longrightarrow R_3C\text{·} + \text{HBr}$$
$$R_3C\text{·} + Br_2 \longrightarrow R_3CBr + \text{Br·}$$

The reaction is often initiated by photolysis of bromine. The hydrogen-atom abstraction step is rate-limiting, and the product composition is governed by the selectivity of this step. The enthalpy requirements for abstraction of hydrogen from methane, ethane (primary), propane (secondary), and isobutane (tertiary) by bromine atoms are $+16.5$, $+10.5$, $+7.0$, and $+3.5$ kcal/mol, respectively.[91] These bond-energy differences are reflected in the activation energies, and there is a substantial kinetic preference for hydrogen abstraction in the order *tert* > *sec* > *pri*. Substituents that promote radical stability, such as phenyl, vinyl, or carbonyl groups, also lead to kinetic selectivity in radical brominations. Bromination at benzylic positions is a particularly efficient process, as illustrated by entries 2 and 4 in Scheme 12.4.

Important differences are seen when the reactions of the other halogens are compared to bromination. In the case of chlorination, although the same chain mechanism is operative as for bromination, there is a key difference in the *greatly diminished selectivity of the chlorination*. For example, the *pri*:*tert* selectivity in 2,3-dimethylbutane for chlorination is 1:3.6 in typical solvents.[92] Because of the greater reactivity of the chlorine atom, abstractions of primary, secondary, and tertiary hydrogens are all *exothermic*. As a result of this exothermicity, the stability of the product radical has less influence on the activation energy. In terms of Hammond's postulate (Section 4.4.2), the transition state would be expected to be more *reactant-like*. As an example of the low selectivity, ethylbenzene is chlorinated at both the methyl and the methylene positions, despite the much greater stability of the benzyl radical[93]:

4.25 : 1
Ratio is solvent dependent

Radical chlorination reactions show a substantial polar effect. Positions substituted by electron-withdrawing groups are relatively unreactive toward chlorination, even though the substituents may be potentially capable of stabilizing the free-radical intermediate[94]:

$$CH_3CH_2CH_2CN + Cl_2 \longrightarrow \underset{\underset{69\%}{|}}{CH_3CHCH_2CN} + ClCH_2CH_2CH_2CN$$
$$Cl \qquad\qquad 31\%$$

90. W. A. Thaler, *Methods Free Radical Chem.* **2**:121 (1969); A. Nechvatal, *Adv. Free Radicals* **4**:175 (1972).
91. E. S. Huyser, *Free Radical Chain Reactions*, Wiley-Interscience, New York, 1970, p. 91.
92. K. D. Raner, J. Lusztyk, and K. U. Ingold, *J. Org. Chem.* **53**:5220 (1988).
93. G. A. Russell, A. Ito, and D. G. Hendry, *J. Am. Chem. Soc.* **85**:2976 (1963).
94. A. Bruylants, M. Tits, C. Dieu, and R. Gauthier, *Bull. Soc. Chim. Belg.* **61**:266 (1952); A. Bruylants, M. Tits, and R. Danby, *Bull. Soc. Chim. Belg.* **58**:210 (1949); M. S. Kharasch and H. C. Brown, *J. Am. Chem. Soc.* **62**:925 (1940).

Scheme 12.4. Radical Halogenation

Molecular bromine

1^a $(CH_3)_2CHCH_2CH_2CH_3$ $\xrightarrow[h\nu]{Br_2}$ $(CH_3)_2CCH_2CH_2CH_3$ 90%
  |
  Br

2^b $\xrightarrow[h\nu]{Br_2}$ 48%–53%

N-Bromosuccinimide

3^c $CH_3(CH_2)_3CH{=}CHCH_3$ $\xrightarrow[(PhCO_2)_2]{NBS}$ $CH_3(CH_2)_2CHCH{=}CHCH_3$ 58%–64%
  |
  Br

4^d $\xrightarrow[(PhCO_2)_2]{NBS}$ 79%

Other halogenating agents

5^e $\xrightarrow[40°C]{(CH_3)_3COCl}$

  65% 25% 10%

6^f $CH_2{=}CHCH_2CH_3 + (CH_3)_3COCl$ $\xrightarrow[-78\,°C]{h\nu}$ $CH_3CH{=}CHCH_2Cl + CH_2{=}CHCHCH_3$
  74% Cl 26%

7^g $\xrightarrow[(PhCO_2)_2]{SO_2Cl_2}$ 50%

a. G. A. Russell and H. C. Brown, *J. Am. Chem. Soc.* **77**:4025 (1955).
b. E. F. M. Stephenson, *Org. Synth.* **IV**:984 (1963).
c. F. L. Greenwood, M. D. Kellert, and J. Sedlak, *Org. Synth.* **IV**:108 (1963).
d. E. Campaigne and B. F. Tullar, *Org. Synth.* **IV**:921 (1963).
e. C. Walling and B. B. Jacknow, *J. Am. Chem. Soc.* **82**:6108 (1960).
f. C. Walling and W. Thaler, *J. Am. Chem. Soc.* **83**:3877 (1961).
g. H. C. Brown and A. B. Ash, *J. Am. Chem. Soc.* **77**:4019 (1955).

Similarly, carboxylic acid and ester groups tend to direct chlorination to the β and γ positions, because attack at the α position is electronically disfavored. The polar effect is attributed to the fact that the chlorine atom is an electrophilic species, and the relatively electron-poor position adjacent to an electron-withdrawing group is avoided. The effect of an electron-withdrawing substituent is to decrease the electron density at the potential radical site. Because the chlorine atom is highly reactive, the reaction would be expected to have a very early transition state, and this electrostatic effect predominates over the stabilizing substituent effect on the intermediate. The substituent effect dominates the *kinetic selectivity* of the reaction, and the relative stability of the radical intermediate has relatively little influence.

Radical substitution reactions by iodine are not practical because the abstraction of hydrogen from hydrocarbons by iodine is endothermic, even for stable radicals. The enthalpy of the overall reaction is also slightly endothermic. Thus, because of both the kinetic problem excluding a chain reaction and an unfavorable equilibrium constant for substitution, iodination cannot proceed by a radical-chain mechanism.

Fluorination presents problems of the other extreme. Both steps in the substitution chain reaction are so exothermic that the reaction is violent if not performed under carefully controlled conditions. Furthermore, fluorine atoms are capable of cleaving carbon–carbon bonds:

$$F\cdot + CH_3CH_3 \longrightarrow CH_3F + CH_3\cdot$$

Saturated hydrocarbons such as neopentane, norbornane, and cyclooctane have been converted to the corresponding perfluoro derivatives in 10–20% yield by gas-phase reaction with fluorine gas diluted with helium at $-78°C$.[95] Simple ethers can be completely fluorinated under similar conditions.[96] Crown polyethers can be fluorinated by passing an F_2/He stream over a solid mixture of sodium fluoride and the crown ether.[97] Liquid-phase fluorination of hydrocarbons has also been observed, but the reaction is believed to be ionic, rather than radical, in character.[98] A variety of milder fluorination agents have been developed for synthetic purposes and will be discussed in Chapter 6 of Part B.

Halogenations of organic molecules can also be carried out with several other reagents besides the molecular halogens. N-Bromosuccinimide (NBS) has been used extensively, especially for allylic and benzylic bromination. Mechanistic investigations have established that molecular bromine is the active halogenating agent under the conditions used for NBS bromination.[99] Molecular bromine is maintained at a low concentration throughout the course of the reaction by formation from NBS and hydrogen bromide.

The fact that the bromine concentration remains at very low levels is important to the success of the allylic halogenation process. The allylic bromination of alkenes must

95. N. J. Maraschin, D. B. Catsikis, L. H. Davis, G. Jarvinen, and R. J. Lagow, *J. Am. Chem. Soc.* **97**:513 (1975).
96. D. F. Persico, H.-N. Huang, R. J. Lagow, Jr., and L. C. Clark, *J. Org. Chem.* **50**:5156 (1985).
97. W.-H. Lin, W. I. Bailey, Jr., and R. J. Lagow, *J. Chem. Soc., Chem. Commun.* **1985**:1350.
98. C. Gal and S. Rozen, *Tetrahedron Lett.* **25**:449 (1984).
99. R. E. Pearson and J. C. Martin, *J. Am. Chem. Soc.* **85**:354, 3142 (1963); G. A. Russell, C. DeBoer, and K. M. Desmond, *J. Am. Chem. Soc.* **85**:365 (1963); J. H. Incremona and J. C. Martin, *J. Am. Chem. Soc.* **92**:627 (1970); J. C. Day, M. J. Lindstrom, and P. S. Skell, *J. Am. Chem. Soc.* **96**:5616 (1974).

compete with polar addition of bromine via a bromonium ion intermediate. The addition reaction frequently follows a second-order dependence on $[Br_2]$ (see Section 6.3). Therefore, a low concentration of Br_2 favors substitution.[100]

NBS can also be used to brominate alkanes. For example, cyclopropane, cyclopentane, and cyclohexane give the corresponding bromides when irradiated in a solution of NBS in dichloromethane.[101] Under these conditions, the succinimidyl radical appears to be involved as the hydrogen-abstracting intermediate:

$$Br\cdot + R{-}H \longrightarrow R\cdot + HBr$$

Another reagent which effects chlorination by a radical mechanism is t-butyl hypochlorite. The hydrogen-abstracting species in the chain mechanism is the t-butoxy radical.

$$(CH_3)_3CO\cdot + H{-}R \rightarrow (CH_3)_3COH + R\cdot$$
$$R\cdot + (CH_3)_3COCl \rightarrow (CH_3)_3CO\cdot + RCl$$

The selectivity and product composition is different from that obtained for direct chlorination. The selectivity of the t-butoxy radical is intermediate between that of chlorine and bromine atoms. The selectivity is also solvent- and temperature-dependent. A typical ratio, in chlorobenzene as solvent, is $tert : sec : pri; = 60 : 10 : 1$.[102] Scheme 12.4 (p. 704) gives examples of specific halogenation reactions that proceed by radical-chain mechanisms.

12.3.2. Oxidation

Free-radical chain oxidation of organic molecules by molecular oxygen is often referred to as *autoxidation* (see Section 12.2.1). The general mechanism is outlined below.

$$
\begin{aligned}
\text{initiation} \quad & In\cdot + R{-}H \longrightarrow In{-}H + R\cdot \\
\text{propagation} \quad & R\cdot + O_2 \longrightarrow R{-}O{-}O\cdot \\
& R{-}O{-}O\cdot + R{-}H \longrightarrow R{-}O{-}O{-}H + R\cdot
\end{aligned}
$$

The rate of reaction of oxygen with most radicals is very rapid because of the triplet character of molecular oxygen. The ease of autoxidation is therefore largely governed by

100. D. W. McMillen and J. B. Grutzner, *J. Org. Chem.* **59**:4516 (1994).
101. J. G. Traynham and Y.-S. Lee, *J. Am. Chem. Soc.* **96**:3590 (1974).
102. C. Walling and P. J. Wagner, *J. Am. Chem. Soc.* **86**:3368 (1964).

**Table 12.10. Relative Reactivities of Some
Aromatic Hydrocarbons toward Oxygen[a]**

PhCH(CH$_3$)$_2$	1.0	PhCH$_2$CH$_3$	0.18
PhCH$_2$CH=CH$_2$	0.8	PhCH$_3$	0.015
(Ph)$_2$CH$_2$	0.35		

a. G. A. Russell, *J. Am. Chem. Soc.* **78**:1047 (1956).

the rate of hydrogen abstraction in the second step of the propagation sequence. The alkylperoxyl radicals that act as the chain carrier are fairly selective. Substrates that are relatively electron-rich or that provide particularly stable radicals are the most easily oxidized. Benzylic, allylic, and tertiary positions are especially susceptible to oxidation. This selectivity makes radical-chain oxidation a preparatively useful reaction in some cases.

The reactivity of a series of hydrocarbons toward oxygen measured under a standard set of conditions can give some indication of the susceptibility of various structural units to autoxidation.[103] Table 12.10 gives the results for a series of hydrocarbons. These data indicate the activating effect of alkyl, vinyl, and phenyl substituents.

The best preparative results from autoxidation are encountered when only one relatively reactive hydrogen is available for abstraction. The oxidation of isopropylbenzene (cumene) is carried out on an industrial scale, with the ultimate products being acetone and phenol:

The benzylic position in tetralin can be selectively oxidized to the hydroperoxide[104]:

Functional groups that stabilize radicals would be expected to increase susceptibility to autoxidation. This is illustrated by two cases that have been relatively well studied. Aldehydes, in which abstraction of the aldehyde hydrogen is facile, are easily autoxidized. The autoxidation initially forms a peroxycarboxylic acid, but usually the corresponding carboxylic acid is isolated because the peroxy acid oxidizes additional aldehyde in a

103. G. A. Russell, *J. Am. Chem. Soc.* **78**:1047 (1956).
104. H. B. Knight and D. Swern, *Org. Synth.* **IV**:895 (1963).

parallel heterolytic reaction:

$$\underset{\parallel}{\overset{O}{RCH}} + In\cdot \longrightarrow \underset{\parallel}{\overset{O}{RC\cdot}} + InH$$

$$\underset{\parallel}{\overset{O}{RC\cdot}} + |O_2 \longrightarrow \underset{\parallel}{\overset{O}{RCOO\cdot}}$$

$$\underset{\parallel}{\overset{O}{RCOO\cdot}} + \underset{\parallel}{\overset{O}{RCH}} \longrightarrow \underset{\parallel}{\overset{O}{RCOOH}} + \underset{\parallel}{\overset{O}{RC\cdot}}$$

$$\underset{\parallel}{\overset{O}{RCOOH}} + \underset{\parallel}{\overset{O}{RCH}} \longrightarrow 2\,RCO_2H$$

The final step is not a radical reaction, but an example of the Baeyer–Villiger reaction, which will be discussed in Section 12.5.2 of Part B.

Similarly, the α position in ethers is autoxidized quite readily to give α-hydroperoxy ethers.

This reaction is the cause of a widely recognized laboratory hazard. The peroxides formed from several commonly used ethers, such as diethyl ether and tetrahydrofuran, are explosive. Appreciable amounts of such peroxides can build up in ether samples that have been exposed to the atmosphere. Because the hydroperoxides are less volatile than the ethers, they are concentrated by evaporation or distillation, and the concentrated peroxide solutions may explode. For this reason, extended storage of ethers that have been exposed to oxygen is extremely hazardous.

12.4. Free-Radical Addition Reactions

12.4.1. Addition of Hydrogen Halides

The anti-Markownikoff addition of hydrogen bromide to alkenes was one of the earliest free-radical reactions to be put on a firm mechanistic basis. In the presence of a suitable initiator, such as a peroxide, a radical-chain mechanism becomes competitive with the ionic mechanism for addition of hydrogen bromide:

$$In\cdot + HBr \longrightarrow Br\cdot + HIn$$
$$Br\cdot + RCH{=}CH_2 \longrightarrow R\overset{\cdot}{C}HCH_2Br$$
$$R\overset{\cdot}{C}HCH_2Br + HBr \longrightarrow RCH_2CH_2Br + Br\cdot$$

Because the bromine adds to the less substituted carbon atom of the double bond, generating the more stable radical intermediate, the regioselectivity of radical-chain hydrobromination is opposite to that of ionic addition. The early work on the radical mechanism of addition of hydrogen bromide was undertaken to understand why Markownikoff's rule was violated under certain circumstances. The cause was found to be conditions that initiated the radical-chain process, such as peroxide impurities or light.

Some examples of radical-chain additions of hydrogen bromide to alkenes are included in Scheme 12.5.

The stereochemistry of radical addition of hydrogen bromide to alkenes has been studied with both acyclic and cyclic alkenes.[105] *Anti* addition is favored.[106, 107] This is contrary to what would be expected if the sp^2 carbon of the radical were rapidly rotating or inverting with respect to the remainder of the molecule:

The stereoselectivity of the radical addition can be explained in terms of a bridged structure similar to that involved in discussion of ionic bromination of alkenes[108]:

Further evidence for a bromine-bridged radical comes from radical substitution of optically active 2-bromobutane. Most of the 2,3-dibromobutane which is formed is racemic, indicating that the stereogenic center is involved in the reaction. A bridged intermediate that can react at either carbon can explain the racemization. When the 3-deuterated reagent is used, it can be shown that the hydrogen (or deuterium) that is abstracted is replaced by bromine with *retention of stereochemistry.*[109] These results are also consistent with a bridged bromine radical.

105. B. A. Bohm and P. I. Abell, *Chem. Rev.* **62**:599 (1962).
106. P. S. Skell and P. K. Freeman, *J. Org. Chem.* **29**:2524 (1964).
107. H. L. Goering and D. W. Larsen, *J. Am. Chem. Soc.* **81**:5937 (1959).
108. P. S. Skell and J. G. Traynham, *Acc. Chem. Res.* **17**:160 (1984).
109. P. S. Skell, R. R. Pavlis, D. C. Lewis, and K. J. Shea, *J. Am. Chem. Soc.* **95**:6735 (1973).

Scheme 12.5. Free-Radical Chain Additions to Alkenes

Hydrogen bromide

1^a $\quad$ $CH_3CH{=}C(CH_3)_2$ $\xrightarrow{HBr}$ $CH_3CHCH(CH_3)_2$ $\quad$ 55%
$\qquad\qquad\qquad\qquad\qquad\qquad\quad$ |
$\qquad\qquad\qquad\qquad\qquad\qquad\quad$ Br

2^b $\quad$ $CH_2{=}CHCO_2CH_3$ $\xrightarrow{HBr}$ $BrCH_2CH_2CO_2CH_3$ $\quad$ 80–84%

3^c $\quad$ $CH_2{=}CHCH_2\overset{\overset{\displaystyle CH_3}{|}}{\underset{\underset{\displaystyle CH_3}{|}}{C}}{-}\overset{\overset{\displaystyle CH_3}{|}}{\underset{\underset{\displaystyle O}{\|}}{C}}$ $\xrightarrow[hv]{HBr}$ $BrCH_2CH_2CH_2\overset{\overset{\displaystyle CH_3}{|}}{\underset{\underset{\displaystyle CH_3}{|}}{C}}{-}\overset{\overset{\displaystyle CH_3}{|}}{\underset{\underset{\displaystyle O}{\|}}{C}}$ $\quad$ 76%

Addition of halomethanes

4^d $\quad$ $CCl_4 + H_2C{=}CH(CH_2)_5CH_3$ $\xrightarrow{(PhCO_2)_2}$ $Cl_3CCH_2\underset{\underset{\displaystyle Cl}{|}}{CH}(CH_2)_5CH_3$ $\quad$ 75%

5^e $\quad$ $BrCCl_3 + CH_2{=}C(C_2H_5)_2$ $\xrightarrow{hv}$ $Cl_3CCH_2\underset{\underset{\displaystyle Br}{|}}{C}(C_2H_5)_2$ $\quad$ 91%

6^f $+ CCl_4$ $\xrightarrow{(PhCO_2)_2}$ $+$
$\qquad\qquad\qquad\qquad\qquad\qquad\qquad\qquad\qquad\qquad$ 73% $\qquad\qquad\qquad\qquad$ 4%

7^g $\quad$ $CCl_4 + CH_2{=}\underset{\underset{\displaystyle OH}{|}}{CH}C(CH_3)_2$ $\xrightarrow[80°C]{(PhCO_2)_2}$ $Cl_3CCH_2\underset{\underset{\displaystyle Cl}{|}}{CH}\overset{\overset{\displaystyle OH}{|}}{C}(CH_3)_2$ $\quad$ 70%

Addition of other carbon radicals

8^h $\quad$ $CH_3CH{=}O + CH_2{=}CH(CH_2)_5CH_3$ $\xrightarrow{hv}$ $CH_3\overset{\overset{\displaystyle O}{\|}}{C}(CH_2)_7CH_3$ $\quad$ 64%

9^i $\quad$ $CH_3(CH_2)_5CH{=}O + H_5C_2O_2CCH{=}CHCO_2C_2H_5$ $\xrightarrow{(PhCO_2)_2}$ $CH_3(CH_2)_5\overset{\overset{\displaystyle O}{\|}}{C}\underset{\underset{\displaystyle CH_2CO_2C_2H_5}{|}}{CH}CO_2C_2H_5$
$\qquad\qquad\qquad\qquad\qquad\qquad\qquad\qquad\qquad\qquad\qquad\qquad\qquad\qquad\qquad\qquad\qquad\qquad\qquad$ 76%

10^j $\quad$ $CH_2{=}CHCH_2CH_2CO_2CH_3 + H\overset{\overset{\displaystyle O}{\|}}{C}NH_2$ $\xrightarrow[acetone]{hv}$ $H_2N\overset{\overset{\displaystyle O}{\|}}{C}(CH_2)_4CO_2CH_3$ $\quad$ 58%

11^k $+ H_3CO_2CCH{=}CHCO_2CH_3$ $\xrightarrow[acetone]{hv}$

12^l $+ CH_3\overset{\overset{\displaystyle O}{\|}}{C}CH_3$ $\xrightarrow{hv}$

13^m $\quad$ $CH_2{=}CH(CH_2)_5CH_3 + CH_2(CO_2C_2H_5)_2$ $\xrightarrow{ROOR}$ $CH_3(CH_2)_7CH(CO_2C_2H_5)_2$ $\quad$ 67–85%

14^n $\quad$ $CH_2{=}CH(CH_2)_5CH_3 + N{\equiv}CCH_2CO_2C_2H_5$ $\xrightarrow{CuO}$ $C_2H_5O_2CCH(CH_2)_7CH_3$ $\quad$ 75%
$\qquad\qquad\qquad\qquad\qquad\qquad\qquad\qquad\qquad\qquad\qquad\qquad\qquad\qquad\qquad\qquad\quad$ |
$\qquad\qquad\qquad\qquad\qquad\qquad\qquad\qquad\qquad\qquad\qquad\qquad\qquad\qquad\qquad\qquad\quad$ $C{\equiv}N$

15° $\quad$ $CH_2{=}CHCO_2C(CH_3)_3 + CH_3CH{=}O$ $\xrightarrow{hv}$ $CH_3\overset{\overset{\displaystyle O}{\|}}{C}CH_2CH_2CO_2C(CH_3)_3$ $\quad$ 99%

Scheme 12.5. (*continued*)

711

SECTION 12.4.
FREE-RADICAL
ADDITION REACTIONS

Addition of thiols and thio acids

16^p $CH_3\overset{\overset{O}{\|}}{C}SH + C_6H_5CH\!=\!CHCH\!=\!O \longrightarrow CH_3\overset{\overset{O}{\|}}{C}S\overset{\overset{C_6H_5}{|}}{C}HCH_2CH\!=\!O$ 90%

17^q $CH_3CH_2CH_2CH_2SH + CH_2\!=\!CH(CH_2)_4CH_3 \xrightarrow{(PhCO_2)_2} CH_3(CH_2)_3S(CH_2)_6CH_3$ 68%

a. W. J. Bailey and S. S. Hirsch, *J. Org. Chem.* **28**:2894 (1963).
b. R. Mozingo and L. A. Patterson, *Org. Synth.* **III**:576 (1955).
c. H. O. House, C.-Y. Chu, W. V. Phillips, T. S. B. Sayer, and C.-C. Yau, *J. Org. Chem.* **42**:1709 (1977).
d. M. S. Kharasch, E. V. Jensen, and W. H. Urry, *J. Am. Chem. Soc.* **69**:1100 (1947).
e. M. S. Kharasch and M. Sage, *J. Org. Chem.* **14**:537 (1949).
f. C. L. Osborn, T. V. Van Auken, and D. J. Trecker, *J. Am. Chem. Soc.* **90**:5806 (1968).
g. P. D. Klemmensen, H. Kolind-Andersen, H. B. Madsen, and A. Svendsen, *J. Org. Chem.* **44**:416 (1979).
h. M. S. Kharasch, W. H. Urry, and B. M. Kuderna, *J. Org. Chem.* **14**:248 (1949).
i. T. M. Patrick, Jr., and F. B. Erikson, *Org. Synth.* **IV**:430 (1963).
j. D. Elad and J. Rokach, *J. Org. Chem.* **29**:1855 (1964).
k. I. Rosenthal and D. Elad, *J. Org. Chem.* **33**:805 (1968).
l. W. Reusch, *J. Org. Chem.* **27**:1882 (1962).
m. J. C. Allen, J. I. G. Cadogan, B. W. Harris, and D. H. Hey, *J. Chem. Soc.* **1962**:4468.
n. A. Hajek and J. Malek, *Synthesis* **1977**:454.
o. F. C. Macias, J. M. G. Molinillo, G. M. Massanet, and F. Rodriguez-Luis, *Tetrahedron* **48**:3345 (1992).
p. R. Brown, W. E. Jones, and A. R. Pinder, *J. Chem. Soc* **1951**:2123.
q. D. W. Grattan, J. M. Locke, and S. R. Wallis, *J. Chem. Soc. Perkin Trans. 1* **1973**:2264.

Other mechanisms must also operate, however, to account for the fact that 5–10% of the product is formed with retained configuration at the chiral center. Isotopic labeling studies have also demonstrated that the 3-bromo-2-butyl radical undergoes reversible loss of bromine atom to give 2-butene at a rate which is competitive with that of the bromination reaction:

$$CH_3\overset{\overset{Br}{|}}{\underset{\cdot}{C}}HCHCH_3 \rightleftharpoons CH_3CH\!=\!CHCH_3 + Br\cdot$$

This process can account for some of the racemization by a mechanism that does not involve the bridged intermediate.[110]

    trans-Diaxial addition is the preferred stereochemical mode for addition to cyclohexene and its derivatives[111]:

This stereochemistry can be explained in terms of a bromine-bridged intermediate.

    Product mixtures from radical-chain addition of hydrogen chloride to alkenes are much more complicated than is the case for addition of hydrogen bromide. The problem is that the rate of abstraction of hydrogen from hydrogen chloride is not fast relative to the rate of addition of the alkyl radical to the alkene. This results in the formation of low-

110. D. D. Tanner, E. V. Blackburn, Y. Kosugi, and T. C. S. Rao, *J. Am. Chem. Soc.* **99**:2714 (1977).
111. H. L. Goering and L. L. Sims, *J. Am. Chem. Soc.* **77**:3465 (1955); N. A. LeBel, R. F. Czaja, and A. DeBoer, *J. Org. Chem.* **34**:3112 (1969); P. D. Readio and P. S. Skell, *J. Org. Chem.* **31**:753 (1966); H. L. Goering, P. I. Abell, and B. F. Aycock, *J. Am. Chem. Soc.* **74**:3588 (1952).

molecular-weight polymers in competition with simple addition:

$$Cl\cdot + RCH{=}CH_2 \longrightarrow R\overset{\centerdot}{C}HCH_2Cl \xrightarrow[RCH=CH_2]{HCl} RCHCH_2(\overset{R}{\underset{|}{C}}HCH_2)_n\overset{\centerdot}{C}l \xrightarrow{HCl} RCH_2CH_2(\overset{R}{\underset{|}{C}}HCH_2)_nCl + Cl\cdot$$

with branch: $RCH_2CH_2Cl + Cl\cdot$

Radical-chain additions of hydrogen fluoride and hydrogen iodide to alkenes are not observed. In the case of hydrogen iodide, the addition of an iodine atom to an alkene is an endothermic process and is too slow to permit a chain reaction, even though the hydrogen abstraction step would be favorable. In the case of hydrogen fluoride, the abstraction of hydrogen from hydrogen fluoride is energetically prohibitive.

12.4.2. Addition of Halomethanes

One of the older preparative free-radical reactions is the addition of polyhalomethanes to alkenes. Examples of addition of carbon tetrabromide, carbon tetrachloride, and bromoform have been recorded.[112] The reactions are chain processes that depend on facile abstraction of halogen or hydrogen from the halomethane:

$$In\cdot + CBr_4 \longrightarrow InBr + \cdot CBr_3$$
$$\cdot CBr_3 + CH_2{=}CHR \longrightarrow Br_3CCH_2\overset{\centerdot}{C}HR$$
$$Br_3CCH_2\overset{\centerdot}{C}HR + CBr_4 \longrightarrow Br_3CCH_2\overset{Br}{\underset{|}{C}}HR + \cdot CBr_3$$

or

$$In\cdot + CHBr_3 \longrightarrow InH + \cdot CBr_3$$
$$\cdot CBr_3 + CH_2{=}CHR \longrightarrow Br_3CCH_2\overset{\centerdot}{C}HR$$
$$Br_3CCH_2\overset{\centerdot}{C}HR + HCBr_3 \longrightarrow Br_3CCH_2CH_2R + \cdot CBr_3$$

Bromotrichloromethane can also be used effectively in the addition reaction. Because of the preferential abstraction of bromine, a trichloromethyl unit is added to the less substituted carbon atom of the alkene:

$$BrCCl_3 + CH_2{=}CHR \xrightarrow[\Delta]{peroxide} Cl_3CCH_2\overset{Br}{\underset{|}{C}}HR$$

The efficiency of the halomethane addition process depends on the relative rate of halogen-atom abstraction versus that of addition to the alkene:

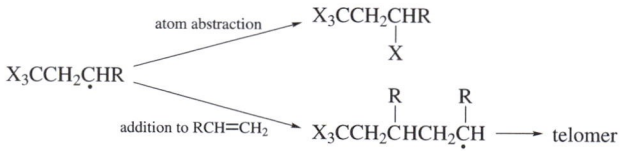

$$X_3CCH_2\overset{\centerdot}{C}HR \xrightarrow{atom\ abstraction} X_3CCH_2\overset{X}{\underset{|}{C}}HR$$
$$X_3CCH_2\overset{\centerdot}{C}HR \xrightarrow{addition\ to\ RCH=CH_2} X_3CCH_2\overset{R}{\underset{|}{C}}HCH_2\overset{R}{\underset{|}{C}}H \longrightarrow telomer$$

112. E. Sosnovsky, *Free Radical Reactions in Preparative Organic Chemistry*, Macmillan, New York, 1964, Chapter 2.

For a given alkene, the order of reactivity of the halomethanes is $CBr_4 > CBrCl_3 > CCl_4 > CH_2Cl_2 > CHCl_3$. The efficiency of 1 : 1 addition for a given alkene depends on the ease with which it undergoes radical-chain polymerization, since polymerization can compete with the halogen-atom abstraction step in the chain mechanism. Polymerization is usually most rapid for terminal alkenes bearing stabilizing substituents such as a phenyl or ester group. Several specific examples of additions of polyhaloalkanes are included in Scheme 12.5.

The addition of bromotrichloromethane to cyclohexene gives a nearly 1 : 1 mixture of the two possible stereoisomers[113]:

This result shows than the initially added trichloromethyl group has little influence on the stereochemistry of the subsequent bromine atom-abstraction. The intermediate 2-(trichloromethyl)cyclohexyl radical presumably relaxes to the equatorial conformation faster than bromine-atom abstraction occurs. In contrast with addition to $\Delta^{2,3}$-octahydronaphthalene, the addition is exclusively *trans*-diaxial:

The *trans*-fused decalin system is conformationally rigid, and the stereochemistry of the product indicates that the initial addition of the trichloromethyl radical is from the axial direction. This would be expected on stereoelectronic grounds, because the radical should initially interact with the π^* orbital. The axial trichloromethyl group then shields the adjacent radical position enough to direct the bromine abstraction in the *trans* sense.

Addition of bromotrichloromethane to norbornene is also *anti*:

This is again the result of steric shielding by the trichloromethyl group, which causes the bromine atom to be abstracted from the *endo* face of the intermediate radical.

12.4.3. Addition of Other Carbon Radicals

Other functional groups provide sufficient stabilization of radicals to permit successful chain additions to alkenes. Acyl radicals are formed by abstraction of the formyl hydrogen from aldehydes. As indicated in Table 12.7, the resulting acyl radicals are

113. J. G. Traynham, A. G. Lane, and N. S. Bhacca, *J. Org. Chem.* **34**:1302 (1969).

somewhat stabilized. The chain process results in formation of a ketone by addition of the aldehyde to an alkene:

$$In\cdot + \overset{O}{\overset{\|}{RCH}} \longrightarrow \overset{O}{\overset{\|}{RC}}\cdot + InH$$

$$\overset{O}{\overset{\|}{RC}}\cdot + CH_2{=}CHR' \longrightarrow \overset{O}{\overset{\|}{RCCH_2\overset{\cdot}{C}HR'}}$$

$$\overset{O}{\overset{\|}{RCCH_2\overset{\cdot}{C}HR'}} + \overset{O}{\overset{\|}{RCH}} \longrightarrow \overset{O}{\overset{\|}{RCCH_2CH_2R'}} + \overset{O}{\overset{\|}{RC}}\cdot$$

Some specific examples are included in Scheme 12.5.

The chain addition of formamide to alkenes is a closely related reaction. It results in the formation of primary amides.[114] The reaction is carried out with irradiation in acetone. The photoexcited acetone initiates the chain reaction by abstracting hydrogen from formamide:

$$CH_3COCH_3 \xrightarrow{h\nu} CH_3COCH_3{*}$$

$$CH_3COCH_3{*} + HCONH_2 \longrightarrow \cdot CONH_2 + (CH_3)_2\overset{\cdot}{C}OH$$

$$\cdot CONH_2 + RCH{=}CH_2 \longrightarrow R\overset{\cdot}{C}HCH_2CONH_2$$

$$R\overset{\cdot}{C}HCH_2CONH_2 + HCONH_2 \longrightarrow RCH_2CH_2CONH_2 + \cdot CONH_2$$

12.4.4. Addition of Thiols and Thiocarboxylic Acids

The addition of S—H compounds to alkenes by a radical-chain mechanism is a quite general and efficient reaction.[115] The mechanism is analogous to that for hydrogen bromide addition. The energetics of both the hydrogen abstraction and addition steps are favorable. Entries 16 and 17 in Scheme 12.5 are examples.

$$In\cdot + R'SH \longrightarrow In{-}H + R'S\cdot$$

$$R'S\cdot + CH_2{=}CHR \longrightarrow R'SCH_2\overset{\cdot}{C}HR$$

$$R'SCH_2\overset{\cdot}{C}HR + R'SH \longrightarrow R'SCH_2CH_2R + R'S\cdot$$

The preferred stereochemistry of addition to cyclic alkenes is *anti*.[116] The additions are not as highly stereoselective as hydrogen bromide addition, however.

12.5. Halogen, Sulfur, and Selenium Group Transfer Reactions

Besides hydrogen, other atoms and groups are susceptible to abstraction by free radicals. The most important from a synthetic point of view are bromine, iodine, sulfur,

114. D. Elad and J. Rokach, *J. Org. Chem.* **29**:1855 (1964).
115. K. Griesbaum, *Angew. Chem. Int. Ed. Engl.* **9**:273 (1970).
116. N. A. LeBel, R. F. Czaja, and A. DeBoer, *J. Org. Chem.* **34**:3112 (1969); P. D. Readio and P. S. Skell, *J. Org. Chem.* **31**:759 (1966); F. G. Bordwell, P. S. Landis, and G. S. Whitney, *J. Org. Chem.* **30**:3764 (1965); E. S. Huyser, H. Benson, and H. J. Sinnige, *J. Org. Chem.* **32**:622 (1967).

and selenium substituents. Group transfer reactions can occur intermolecularly or intramolecularly. Indeed, we have already encountered one example in the addition of polyhalogenated methanes to alkenes. The chain is propagated by a halogen-atom abstraction.

Cyclizations involving iodine-atom transfers have been developed. Among the most effective examples are reactions involving the cyclization of 6-iodohexene derivatives.[117] The 6-hexenyl radical generated by iodine-atom abstraction rapidly cyclizes to a cyclopentylmethyl radical. The chain is propagated by iodine-atom transfer.

Various functionalized derivatives can be cyclized, including α-iodoesters, α-iodoketones, and α-iodomalonates.[118]

Similarly, 6-iodoalkyne derivatives cyclize to iodomethylenecyclopentanes.[119]

117. D. P. Curran and D. Kim, *Tetrahedron Lett.* **27**:5821 (1986).
118. D. P. Curran and C. T. Chang, *Tetrahedron Lett.* **28**:2477 (1987).
119. D. P. Curran, M.-H. Chen, and D. Kim, *J. Am. Chem. Soc.* **108**:2489 (1986).

The chain is propagated by abstraction of iodine by the cyclized vinyl radical intermediate.

$$\dot{R}CHCH_2CH_2CH_2C\equiv CH \longrightarrow \quad\quad \xrightarrow{\quad RCHCH_2CH_2CH_2C\equiv CH \atop \quad\quad\quad I \quad\quad} \quad \dot{R}CHCH_2CH_2CH_2C\equiv CH$$

The hexabutyldistannane used in this reaction is not involved in the propagation sequence but may be involved in initiation or scavenging of potential chain-termination radicals.

Intermolecular additions of alkyl radicals to alkynes have also been observed.

$$(CH_3)_2CHI + HC\equiv CCO_2CH_3 \longrightarrow (CH_3)_2CHCH=C \begin{array}{c} I \\ CO_2CH_3 \end{array}$$

With 4-iodoalkynes, the intermediate radicals can be trapped by activated alkenes and lead to cyclized products.[120]

$$ICH_2CH_2C\equiv CH + CH_2=CHCO_2CH_3 \longrightarrow$$

$$ICH_2CH_2C\equiv CH + CH_2=CHCN \longrightarrow$$

$$HC\equiv CCH_2CH_2\cdot + CH_2=CHZ \longrightarrow HC\equiv CCH_2CH_2CH_2\dot{C}HZ \longrightarrow$$

$$\cdots=CH\cdot + ICH_2CH_2C\equiv CH \longrightarrow \cdots=CHI + \cdot CH_2CH_2C\equiv CH$$

For all of these reactions, the reagents and reaction conditions must be chosen to meet the fundamental requirement for successful chain reactions. Each step in the sequence must be exothermic to permit chain propagation.[121]

Mixed aryl selenides have also proven to be excellent reagents for group transfer reactions.[122] Photolysis of selenides in an inert solvent such as benzene can initiate chain reactions. Substituted radicals can be generated in this manner, from α-seleno-

120. D. P. Curran and M.-H. Chen, J. Am. Chem. Soc. 109:6558 (1987).
121. D. P. Curran, Synthesis 1988:417, 489; C. P. Jasperse, D. P. Curran, and T. L. Fevig, Chem. Rev. 91:1237 (1991).
122. L. Castle and M. J. Perkins, in The Chemistry of Organic Selenium and Tellurium Compounds, Part 2, S. Patai, ed., John Wiley & Sons, Chichester, U.K., 1987, Chapter 16.

esters,[123] α-selenonitriles,[124] α-selenomalonates,[125] α-methoxyesters,[126] and α-selenophosphonates.[127] The resulting radicals undergo addition to substituted alkenes to generate γ-seleno derivatives.

$$RCHX + CH_2{=}CHZ \longrightarrow RCHCH_2CHZ$$

(with SePh on the left reactant, and X and SePh substituents on the product)

Appropriately substituted selenides can undergo cyclization reactions via a group transfer process.[128]

If selenide additions are carried out in the presence of tri-n-butylstannane, the radical generated by addition can be reduced by hydrogen abstraction. The chain is then continued by selenide abstraction by the stannyl radical. This leads to nonselenated addition and cyclization products.

Intermolecular addition

$$RCH{-}SePh + R_3Sn{\cdot} \longrightarrow RCH_2 + R_3SnSePh$$
(with X on both RCH species)

$$RCH{\cdot} + CH_2{=}CHZ \longrightarrow RCHCH_2CHZ$$
(with X substituent)

$$RCHCH_2CHZ + R_3SnH \longrightarrow RCHCH_2CH_2Z + R_3Sn{\cdot}$$
(with X substituent)

These reactions can also be used to achieve cyclization by intramolecular addition.

Intramolecular addition

Ref. 129

123. J. H. Byers and B. C. Harper, *Tetrahedron Lett.* **33**:6953 (1992).
124. D. L. J. Clive, T. L. B. Boivin, and A. G. Angoh, *J. Org. Chem.* **52**:4943 (1987); D. P. Curran and G. Thoma, *J. Am. Chem. Soc.* **114**:4436 (1992).
125. J. H. Byers and G. C. Lane, *J. Org. Chem.* **58**:3355 (1993).
126. P. Renaud and S. Abazi, *Synthesis* **1996**:253.
127. P. Balczewski, W. M. Pietrzykowski, and M. Mikolajczyk, *Tetrahedron* **51**:7727 (1995); J. H. Byers, J. G. Thissell, and M. A. Thomas, *Tetrahedron Lett.* **36**:6403 (1995).
128. J. H. Byers, T. G. Gleason, and K. S. Knight, *J. Chem. Soc., Chem. Commun.* **1991**:354.
129. D. L. J. Clive, T. L. B. Boivin, and A. G. Angoh, *J. Org. Chem.* **52**:4943 (1982).

12.6. Intramolecular Free-Radical Reactions

Substitution, addition, and group transfer reactions can occur intramolecularly. Intramolecular substitution reactions that involve hydrogen abstraction have some important synthetic applications, since they permit functionalization of carbon atoms relatively remote from the initial reaction site.[130] The preference for a six-membered cyclic transition state in the hydrogen abstraction step imparts position selectivity to the process:

There are several reaction sequences which involve such intramolecular hydrogen abstraction steps. One example is the photolytically initiated decomposition of N-haloamines in acidic solution, which is known as the *Hofmann–Löffler reaction*.[131] The reaction leads initially to γ-haloamines, but these are usually converted to pyrrolidines by intramolecular nucleophilic substitution:

There are related procedures involving N-haloamides which lead to lactones via iminolactone intermediates[132]:

130. G. Majetich and K. Wheless, *Tetrahedron* **51**:7095 (1995).
131. M. E. Wolff, *Chem. Rev.* **63**:55 (1963).
132. D. H. R. Barton, A. L. J. Beckwith, and A. Goosen, *J. Chem. Soc.* **1965**:18; R. S. Neale, N. L. Marcus, and R. G. Schepers, *J. Am. Chem. Soc.* **88**:3051 (1966).

719

SECTION 12.7.
REARRANGEMENT
AND FRAGMENTATION
REACTIONS OF FREE
RADICALS

Intramolecular hydrogen abstraction reactions have also been observed in medium-sized rings. The reaction of cyclooctene with carbon tetrachloride is an interesting case. As shown in the equation below, whereas bromotrichloromethane adds to cyclooctene in a completely normal manner, carbon tetrachloride gives some 4-chloro-1-trichloromethyl-cyclooctane as well as the expected product[133]:

In the case of carbon tetrachloride, the radical intermediate undergoes two competing reactions; intramolecular hydrogen abstraction is competitive with abstraction of a chlorine atom from carbon tetrachloride:

No product derived from the transannular hydrogen abstraction is observed in the addition of bromotrichloromethane because bromine-atom abstraction is sufficiently rapid to prevent effective competition by the intramolecular hydrogen abstraction.

The selectivity observed in most intramolecular functionalizations depends on the preference for a six-membered transition state in the hydrogen-atom abstraction step. Appropriate molecules can be constructed in which steric or conformational effects dictate a preference for selective abstraction of a hydrogen that is more remote from the reactive radical.

Intramolecular addition reactions are quite common when radicals are generated in molecules with unsaturation in a sterically favorable position.[134] Cyclization reactions based on intramolecular addition of radical intermediates have become synthetically useful, and several specific cases will be considered in Section 10.3.4 of Part B.

12.7. Rearrangement and Fragmentation Reactions of Free Radicals

12.7.1. Rearrangement Reactions

Compared with rearrangements of carbocations, rearrangements of radical intermediates are quite rare. However, for specific structural types, migrations can be expected. The groups that are usually involved in migration in free-radical intermediates include aryl, vinyl, acyl, and other unsaturated substituents. Migration of saturated groups is unusual, and there is a simple structural reason for this. In carbocations, migration occurs through a

133. J. G. Traynham, T. M. Couvillon, and N. S. Bhacca, *J. Org. Chem.* **32**:529 (1967); J. G. Traynham and T. M. Couvillon, *J. Am. Chem. Soc.* **87**:5806 (1965); J. G. Traynham and T. M. Couvillon, *J. Am. Chem. Soc.* **89**:3205 (1967).
134. A. L. J. Beckwith, *Tetrahedron* **37**:3073 (1981).

bridged transition state (or intermediate) that involves a three-center, two-electron bond:

In a free radical, there is a third electron in the system. It cannot occupy the same orbital as the other two electrons and must instead be in an antibonding level. As a result, the transition state for migration is much less favorable than for the corresponding carbocation.

The more facile migration of aryl and other unsaturated groups involves bridged intermediates formed by an addition process. In the case of aryl migration, the intermediate is a cyclohexadienyl radical:

Aryl migrations are promoted by steric crowding in the initial radical site. This trend is illustrated by data from the thermal decomposition of a series of diacyl peroxides. The amount of product derived from rearrangement increases with the size and number of substituents[135]:

R	R^1	Rearrangement
CH$_3$	H	39%
Ph	H	63%
Ph	Ph	100%

It has been possible to measure absolute rates and activation energies for rearrangement of the substituents in a series of 2-substituted 2,2-dimethylethyl radicals. The rates at 25°C and the E_a for several substituents are indicated below.[136]

X	k (s^{-1})	E_a (kcal/mol)
Ph	7.6×10^2	11.8
CH$_2$=CH	10^7	5.7
(CH$_3$)$_3$CC≡C	9.3	12.8
(CH$_3$)$_3$CC=O	1.7×10^5	7.8
C≡N	9.0	16.4

135. W. Rickatson and T. S. Stevens, *J. Chem. Soc.* **1963**:3960.
136. D. A. Lindsay, J. Lusztyk, and K. U. Ingold, *J. Am. Chem. Soc.* **106**:7087 (1984).

The rapid rearrangement of vinyl and acyl substituents can be explained as proceeding through cyclopropyl intermediates:

721

SECTION 12.7.
REARRANGEMENT
AND FRAGMENTATION
REACTIONS OF FREE
RADICALS

The substantially slower rearrangement of alkynyl and cyano substituents can be attributed to the reduced stability of the intermediate derived by cyclization of the triply bonded substituents:

Scheme 12.6 gives some examples of reactions in which free-radical rearrangements have been observed.

Scheme 12.6. Free-Radical Rearrangements

a. S. Winstein and F. H. Seubold, Jr., *J. Am. Chem. Soc.* **69**:2916 (1947).
b. J. W. Wilt and H. P. Hogan, *J. Org. Chem.* **24**:441 (1959).
c. P. S. Skell, R. G. Allen, and N. D. Gilmour, *J. Am. Chem. Soc.* **83**:504 (1961).
d. L. H. Slaugh, *J. Am. Chem. Soc.* **87**:1522 (1965).
e. H. Meislich, J. Costanza, and J. Strelitz, *J. Org. Chem.* **33**:3221 (1968).

12.7.2. Fragmentation

Earlier sections have already provided several examples of radical fragmentation reactions, although this terminology was not explicitly used. The facile decarboxylation of acyloxy radicals is an example.

$$\underset{\substack{\| \\ O}}{RCOOC(CH_3)_3} \longrightarrow \underset{\substack{\| \\ O}}{RCO\cdot} \longrightarrow R\cdot + CO_2$$

For the acetoxy radical, the E_a for decarboxylation is about 6.5 kcal/mol and the rate is about 10^9 s^{-1} at 60°C and 10^6 s^{-1} at −80°C.[137] Thus, only very rapid reactions can compete with decarboxylation. As would be expected because of the lower stability of aryl radicals, the rates of decarboxylation of aroyloxy radicals are slower. The rate for p-methoxybenzoyloxy radical has been determined to be 3×10^5 s^{-1} near room temperature.[137] Hydrogen donation by very reactive hydrogen-atom donors such as triethylsilane can compete with decarboxylation at moderate temperatures.

Acyl radicals can fragment with loss of carbon monoxide. Decarbonylation is slower than decarboxylation, but the rate also depends on the stability of the radical that is formed.[138] For example, when reaction of isobutyraldehyde with carbon tetrachloride is initiated by t-butyl peroxide, both isopropyl chloride and isobutyroyl chloride are formed. Decarbonylation is competitive with the chlorine-atom abstraction.

$$\underset{\substack{\| \\ O}}{(CH_3)_2CHCH} + In\cdot \longrightarrow \underset{\substack{\| \\ O}}{(CH_3)_2CHC\cdot}$$

$$\underset{\substack{\| \\ O}}{(CH_3)_2CHC\cdot} \overset{CCl_4}{\longrightarrow} \underset{\substack{\| \\ O}}{(CH_3)_2CHCCl} + \cdot CCl_3$$

$$\overset{-CO}{\longrightarrow} (CH_3)_2CH\cdot + CO \overset{CCl_4}{\longrightarrow} (CH_3)_2CHCl + \cdot CCl_3$$

Another common fragmentation reaction is the cleavage of an alkoxy radical to an alkyl radical and a carbonyl compound[139]:

$$(CH_3)_3CO\cdot \rightarrow CH_3\cdot + (CH_3)_2C{=}O$$

This type of fragmentation is involved in the chain decomposition of alkyl hypochlorites[140]:

$$RCH_2OCl \longrightarrow RCH_2O\cdot + Cl\cdot$$
$$RCH_2O\cdot \longrightarrow R\cdot + CH_2{=}O$$
$$R\cdot + RCH_2OCl \longrightarrow RCl + RCH_2O\cdot$$

In this reaction, too, the stability of the radical being eliminated is the major factor in determining the rate of fragmentation.

137. J. Chateauneuf, J. Lusztyk, and K. U. Ingold, *J. Am. Chem. Soc.* **109**:897 (1987).
138. D. E. Applequist and L. Kaplan, *J. Am. Chem. Soc.* **87**:2194 (1965); W. H. Urry, D. J. Trecker, and H. D. Hartzler, *J. Org. Chem.* **29**:1663 (1964); H. Fischer and H. Paul, *Acc. Chem. Res.* **20**:200 (1987).
139. P. Gray and A. Williams, *Chem. Rev.* **59**:239 (1959).
140. F. D. Greene, M. L. Savitz, F. D. Osterholtz, H. H. Lau, W. N. Smith, and P. M. Zanet, *J. Org. Chem.* **28**:55 (1963); C. Walling and A. Padwa, *J. Am. Chem. Soc.* **85**:1593, 1597 (1963).

723

SECTION 12.7.
REARRANGEMENT
AND FRAGMENTATION
REACTIONS OF FREE
RADICALS

In cyclic systems, the fragmentation of alkoxy radicals can be a reversible process. The 10-decalyloxy radical can undergo fragmentation of either the C(1)–C(9) or the C(9)–C(10) bond:

By using various trapping reagents, it has been deduced that the transannular fragmentation is rapidly reversible. The cyclization of the fragmented radical **C** is less favorable, and it is trapped at rates which exceed that for recyclization under most circumstances.[141]

Radicals derived from ethers and acetals by hydrogen abstraction are subject to fragmentation, with formation of a ketone or ester, respectively.

$$R_2\dot{C}OR' \longrightarrow R_2C{=}O + R'\cdot$$

$$R\dot{C}(OR')_2 \longrightarrow R\overset{\overset{\displaystyle O}{\|}}{C}OR' + R'\cdot$$

These fragmentations are sufficiently slow that the initial radicals can undergo reactions such as addition to alkenes at rates which are competitive with the rate of fragmentation.

A special case of fragmentation is that of 1,4-diradicals where fragmentation can lead to two stable molecules. In the case of 1,4-diradicals without functional-group stabilization, reclosure to cyclobutanes is normally competitive with fragmentation to two molecules of alkene.

Theoretical calculations on the simplest such radical, 1,4-butadinyl, indicate that both processes are exothermic and can proceed with little, if any, barrier.[142]

A study of the lifetimes of the triplet biradicals **D**, **E**, and **F**, which were generated from the corresponding azo compounds, found that lifetime decreased in the order **D > E > F**. The lifetime of **D** is on the order of 2.6×10^{-7} s.[143]

141. A. L. J. Beckwith, P. Kazlauskas, and M. R. Syner-Lyons, *J. Org. Chem.* **48**:4718 (1983).
142. C. Doubleday, Jr., R. N. Camp, H. F. King, J. W. McIver, Jr., D. Mullaly, and M. Page, *J. Am. Chem. Soc.* **106**:447 (1984).
143. W. Adam, K. Hanneman, and R. M. Wilson, *J. Am. Chem. Soc.* **106**:7646 (1984); W. Adam, K. Hanneman, and R. M. Wilson, *Angew. Chem. Int. Ed. Engl.* **24**:1071 (1985); W. Adam, H. Platsch, J. Sendelbach, and J. Wirz, *J. Org. Chem.* **58**:1477 (1993).

The major factor identified as controlling the lifetimes of these radicals is the orientation of the singly occupied orbitals with respect to one another. One factor determining the lifetime is the rate of conversion (intersystem crossing) to the singlet biradical. The rate of conversion is dependent on the orientation of the orbital with unpaired electrons. In **D**, they are essentially parallel. This is a poor orientation for intersystem crossing. Diradical **E** is more flexible, and the triplet is converted more rapidly to the singlet diradical, which reacts rapidly to give the products. The geometry of the bicyclic ring system in radical **F** directs the half-filled orbitals toward one another, and the lifetime of **F** is less than 1×10^{-10} s.[144]

12.8. Electron-Transfer Reactions Involving Transition-Metal Ions

Most of the free-radical mechanisms discussed thus far have involved some combination of homolytic bond dissociation, atom abstraction, and addition steps. In this section, we will discuss reactions that include discrete electron-transfer steps. Addition to or removal of one electron from a diamagnetic organic molecule generates a radical. Organic reactions that involve electron-transfer steps are often mediated by transition-metal ions. Many transition-metal ions have two or more relatively stable oxidation states differing by one electron. Transition-metal ions therefore frequently participate in electron-transfer processes.

The decomposition of peroxyesters has been shown to be strongly catalyzed by Cu(I). The process is believed to involve oxidation of the copper to Cu(II):

An example of this reaction is the reaction of cyclohexene with *t*-butyl perbenzoate, which is mediated by Cu(I).[145] The initial step is the reductive cleavage of the perester. The *t*-butoxy radical then abstracts hydrogen from cyclohexene to give an allylic radical. The radical is oxidized by Cu(II) to the carbocation, which captures benzoate ion. The net effect is an allylic oxidation.

144. W. Adam, S. Grabowski, and R. M. Wilson, *Acc. Chem. Res.* **23**:165 (1990).
145. K. Pedersen, P. Jakobsen, and S.-O. Lawesson, *Org. Synth.* **V**:71 (1973).

725

SECTION 12.8.
ELECTRON-TRANSFER
REACTIONS
INVOLVING
TRANSITION-METAL
IONS

The reactions of copper salts with diacyl peroxides have been investigated quite thoroughly, and the mechanistic studies indicate that both radicals and carbocations are involved as intermediates. The radicals are oxidized to carbocations by Cu(II), and the final products can be recognized as having arisen from carbocations because characteristic patterns of substitution, elimination, and rearrangement can be discerned[146]:

$$[(CH_3)_3CCH_2CO_2]_2 + Cu(I) \longrightarrow (CH_3)_3CCH_2CO_2^- + (CH_3)_3CCH_2\cdot + CO_2 + Cu(II)$$

$$(CH_3)_3CCH_2 + Cu(II) \longrightarrow (CH_3)_3CCH_2^+ + Cu(I)$$

$$(CH_3)_3CCH_2^+ \longrightarrow (CH_3)_3\overset{+}{C}CH_2CH_3 \xrightarrow{-H^+} (CH_3)_2C{=}CHCH_3$$

$$\downarrow {}^-O_2CR$$

$$(CH_3)_2\underset{\underset{O_2CR}{|}}{C}CH_2CH_3$$

When the radicals have β hydrogens, alkenes are formed by a process in which carbocations are probably bypassed. Instead, the oxidation and the elimination of a proton probably occur in a single step through an alkylcopper species. The oxidation state of copper in such an intermediate is Cu(III).

$$R_2CHCH_2\cdot + Cu(II) \to R_2CHCH_2Cu(III) \to R_2C{=}CH_2 + Cu(I) + H^+$$

When halide ions or anions such as thiocyanate or azide are present, these anions are incorporated into the organic radical generated by decomposition of the peroxide. This anion transfer presumably occurs in the same step as the redox interaction with Cu(II), and such reactions have been called *ligand-transfer* reactions.[147]

$$(RCO_2)_2 + Cu(I)X \to RCO_2^- + R\cdot + CO_2 + Cu(II)X$$
$$R\cdot + Cu(II)X \to R{-}Cu(III)X$$
$$R{-}Cu(III)X \to R{-}X + Cu(I)$$

These reactions do not appear to involve free carbocations, because they proceed effectively in nucleophilic solvents that would successfully compete with halide or similar anions for free carbocations. Also, rearrangements are unusual under these conditions, although they have been observed in special cases.

A unified concept of these reactions is provided by the proposal that alkylcopper intermediates are involved in each of these reactions at the stage of the oxidation of the radical[148]:

146. J. K. Kochi, *J. Am. Chem. Soc.* **85**:1958 (1963); J. K. Kochi and A. Bemis, *J. Am. Chem. Soc.* **90**:4038 (1968).
147. C. L. Jenkins and J. K. Kochi, *J. Am. Chem. Soc.* **94**:856 (1972).
148. C. L. Jenkins and J. K. Kochi, *J. Am. Chem. Soc.* **94**:843 (1972).

The organocopper intermediate has three possible fates, and the preferred path is determined by the structure of the group R and the identity of the copper ligand X. If R is potentially a very stable carbocation, the intermediate breaks down to generate the carbocation, and the products are derived from it. When X is a halide or a pseudohalide such as $^-CN, ^-SCN$, or $^-N_3$, the preferred pathway is ligand transfer, leading to the alkyl halide or pseudohalide. If the R group is not capable of sustaining formation of a carbocation and no easily transferred anion is present, the organocopper intermediate is converted primarily to alkene by elimination of a proton.

One-electron oxidation of carboxylate ions generates acyloxy radicals, which undergo decarboxylation. Such electron-transfer reactions can be effected by strong one-electron oxidants, such as Mn(III), Ag(II), Ce(IV), and Pb(IV).[149] These metal ions are also capable of oxidizing the radical intermediate, so the products are those expected from carbocations. The oxidative decarboxylation by Pb(IV) in the presence of halide salts leads to alkyl halides.[150] For example, oxidation of pentanoic acid with lead tetraacetate in the presence of lithium chloride gives 1-chlorobutane in 71% yield:

$$CH_3(CH_2)_3CO_2H \xrightarrow[\text{LiCl}]{\text{Pb(OAc)}_4} CH_3(CH_2)_3Cl + CO_2$$

A chain mechanism is proposed for this reaction. The first step is oxidation of a carboxylate ion coordinated to Pb(IV), with formation of alkyl radical, carbon dioxide, and Pb(III). The alkyl radical then abstracts halogen from a Pb(IV) complex, generating a Pb(III) species that decomposes to Pb(II) and an alkyl radical. This alkyl radical can continue the chain process. The step involving abstraction of halide from a complex with a change in metal-ion oxidation state is a ligand-transfer type reaction.

$$RCO_2-\overset{\overset{\displaystyle X}{|}}{Pb(IV)}- \longrightarrow R\cdot + \overset{\overset{\displaystyle X}{|}}{Pb(III)}- + CO_2$$

$$R\cdot + X-\overset{|}{Pb(IV)}- \longrightarrow R-X + \overset{|}{Pb(III)}-$$

$$\overset{|}{Pb(III)}- + RCO_2^- \longrightarrow R\cdot + CO_2 + Pb(II)$$

In the absence of halide salts, the principal products may be alkanes, alkenes, or acetate esters.

A classic reaction involving electron transfer and decarboxylation of acyloxy radicals is the Kolbe electrolysis, in which an electron is abstracted from a carboxylate ion at the anode of an electrolysis system. This reaction gives products derived from coupling of the decarboxylated radicals.

$$RCO_2^- \longrightarrow RCO_2\cdot + e^-$$
$$RCO_2\cdot \longrightarrow R\cdot + CO_2$$
$$2\,R\cdot \longrightarrow RR$$

149. J. M. Anderson and J. K. Kochi, *J. Am. Chem. Soc.* **92**:2450 (1970); J. M. Anderson and J. K. Kochi, *J. Am. Chem. Soc.* **92**:1651 (1970); R. A. Sheldon and J. K. Kochi, *J. Am. Chem. Soc.* **90**:6688 (1968); W. A. Mosher and C. L. Kehr, *J. Am. Chem. Soc.* **75**:3172 (1953); J. K. Kochi, *J. Am. Chem. Soc.* **87**:1811 (1965).
150. J. K. Kochi, *J. Org. Chem.* **30**:3265 (1965); R. A. Sheldon and J. K. Kochi, *Org. React.* **19**:279 (1972).

Other transformations of the radicals are also possible. For example, the 5-hexenyl radical partially cyclizes in competition with coupling[151]:

Carbocations can also be generated during the electrolysis, and they give rise to alcohols and alkenes. The carbocations are presumably formed by an oxidation of the radical at the electrode before it reacts or diffuses into solution. For example, an investigation of the electrolysis of phenylacetic acid in methanol has led to the identification of benzyl methyl ether (30%), toluene (1%), benzaldehyde dimethylacetal (1%), methyl phenylacetate (6%), and benzyl alcohol (5%), in addition to the coupling product bibenzyl (26%).[152]

12.9. $S_{RN}1$ Substitution Processes

Electron-transfer processes are also crucially involved in a group of reactions that are designated by the mechanistic description $S_{RN}1$. This designation refers to a nucleophilic substitution via a radical intermediate that proceeds by unimolecular decomposition of a radical anion derived from the substrate. There are two families of such reactions which have been developed to a stage of solid mechanistic understanding and also synthetic utility. The common mechanistic pattern involves electron transfer to the substrate, generating a radical anion which then expels the leaving group. This becomes a chain process if the radical generated by expulsion of the leaving group can react with the nucleophile to give a radical anion capable of sustaining a chain reaction:

A mechanism of this type permits substitution of certain aromatic and aliphatic nitro compounds by a variety of nucleophiles. These reactions were discovered as the result of efforts to explain the mechanistic basis for high-yield carbon alkylation of the 2-nitropropane anion by p-nitrobenzyl chloride. p-Nitrobenzyl bromide and iodide and benzyl halides that do not contain a nitro substituent give mainly the unstable oxygen alkylation product with this ambient anion[153]:

151. R. F. Garwood, C. J. Scott, and B. C. L. Weedon, *J. Chem. Soc., Chem. Commun.* **1965**:14.
152. S. D. Ross and M. Finkelstein, *J. Org. Chem.* **34**:2923 (1969).
153. N. Kornblum, *Angew. Chem. Int. Ed. Engl.* **14**:734 (1975); N. Kornblum, in *The Chemistry of Amino, Nitroso and Nitro Compounds and Their Derivatives*, S. Patai, ed., Interscience, New York, 1982, Chapter 10.

A mixture of carbon and oxygen alkylation would be expected for an S_N2 substitution process. The high preference for carbon alkylation suggested that a different mechanism operates with p-nitrobenzyl chloride. This conclusion was further strengthened by the fact that the chloride is more reactive than would be predicted by application of the usual $I > Br > Cl$ reactivity trend for leaving groups in S_N2 reactions. The involvement of a free-radical process was indicated by EPR studies and by demonstrating that typical free-radical inhibitors decrease the rate of the carbon alkylation process. The mechanism proposed is a free-radical chain process initiated by electron transfer from the nitronate anion to the nitroaromatic compound.[154] This process is the principal reaction only for the chloride, because, with the better leaving groups bromide and iodide, a direct S_N2 process is more rapid.

The absolute rate of dissociation of the radical anion of p-nitrobenzyl chloride has been measured as 4×10^3 s^{-1}. The m-nitro isomer does not undergo a corresponding reaction.[155] This is because the *meta* nitro group provides no resonance stabilization of the benzylic radical.

The synthetic value of the $S_{RN}1$ arylation reaction has been developed from this mechanistic understanding. The reaction has been shown to be capable of providing highly substituted carbon skeletons that would be inaccessible by normal S_N2 processes. For example, tertiary p-nitrocumyl halides can act as alkylating agents in high yield. The nucleophile need not be a nitroalkane anion; the same mechanism operates for such anions as thiolate or phenolate or carbanions such as those derived from malonate esters.[156] Furthermore, the leaving group need not be a halide. Displacement of nitrite ion from α,p-dinitrocumene occurs with good efficiency[157]:

154. N. Kornblum, R. E. Michel, and R. C. Kerber, *J. Am. Chem. Soc.* **88**:5662 (1966); G. A. Russell and W. C. Danen, *J. Am. Chem. Soc.* **88**:5663 (1966).

155. R. K. Norris, S. D. Baker, and P. Neta, *J. Am. Chem. Soc.* **106**:140 (1984).

156. N. Kornblum, T. M. Davies, G. W. Earl, N. L. Holy, R. C. Kerber, M. T. Musser, and D. H. Snow, *J. Am. Chem. Soc.* **89**:725 (1967); M. Kornblum, L. Cheng, T. M. Davies, G. W. Earl, N. L. Holy, R. C. Kerber, M. M. Kestner, J. W. Manthey, M. T. Musser, H. W. Pinnick, D. H. Snow, F. W. Stuchal, and R. T. Swiger, *J. Org. Chem.* **52**:196 (1987).

157. N. Kornblum, T. M. Davies, G. W. Earl, G. S. Greene, N. L. Holy, R. C. Kerber, J. W. Manthey, M. T. Musser, and D. H. Snow, *J. Am. Chem. Soc.* **89**: 5714 (1967).

Azide, sulfonyl, and quaternary nitrogen groups can also be displaced by this mechanism.

A similar mechanism has been proposed for the alkylation of amines by p-nitrocumyl chloride.[158,159]

Clearly, the tertiary nature of the chloride would make an S_N2 mechanism highly unlikely. Furthermore, the nitro substituent is essential to the success of these reactions. Cumyl chloride itself undergoes elimination of HCl on reaction with amines.

A related process constitutes a method of carrying out alkylation reactions to give highly branched alkyl chains that could not easily be formed by an S_N2 mechanism. The alkylating agent must contain a nitro group and a second electron-attracting group. Such compounds react with nitronate anions to effect displacement of the nitro group[160]:

$$X = C{\equiv}N, CO_2C_2H_5, \overset{O}{\overset{\|}{C}}R, NO_2$$

Experiments in which radical scavengers are added indicate that a chain reaction is involved, because the reaction is greatly retarded in the presence of the scavengers. The mechanism shown below indicates that one of the steps in the chain process is an electron transfer and that none of the steps involves atom abstraction. The elimination of nitrite occurs as a unimolecular decomposition of the radical anion intermediate, and the $S_{RN}1$ mechanistic designation would apply.

This reaction can also be applied to tertiary nitroalkanes lacking any additional functional group. The reactions with nitro compounds lacking additional anion-stabilizing groups are carried out in DMSO solution[161]:

$$R_3CNO_2 + {}^-CH_2NO_2 \xrightarrow{\text{DMSO}} R_3CCH_2NO_2$$

These reactions also appear to be chain reactions that proceed through the $S_{RN}1$ mechanism. DMSO is a particularly favorable solvent for this reaction, probably because

158. N. Kornblum and F. W. Stuchal, *J. Am. Chem. Soc.* **92**:1804 (1970).
159. W. R. Bowman, *Chem. Soc. Rev.* **17**:283 (1988).
160. N. Kornblum and S. D. Boyd, *J. Am. Chem. Soc.* **92**:5784 (1970).
161. N. Kornblum and A. S. Erickson, *J. Org. Chem.* **46**:1037 (1981).

its conjugate base acts as an efficient chain initiator by transferring an electron to the nitroalkane.

Although the nitro group plays a crucial role in most of these $S_{RN}1$ reactions, reactions of this type have synthetic application beyond the area of nitro compounds. The nitromethyl groups can be converted to other functional groups, including aldehydes and carboxylic acids.[162] Nitro groups at tertiary positions can be reductively removed by reaction with the methanethiol anion.[163] This reaction also appears to be of the electron-transfer type, with the methanethiolate anion acting as the electron donor:

$$R_3C-NO_2 + CH_3S^- \longrightarrow R_3C-NO_2^{\cdot -} + CH_3S\cdot$$
$$R_3C-NO_2^{\cdot -} \longrightarrow R_3C\cdot + NO_2^-$$
$$R_3C\cdot + CH_3S^- \longrightarrow R_3C-H + H_2\overset{\cdot}{C}S^-$$
$$H_2\overset{\cdot}{C}S^- + R_3C-NO_2 \longrightarrow H_2C{=}S + R_3C-NO_2^{\cdot -}$$

The unique feature of the $S_{RN}1$ reactions of substituted alkyl nitro compounds is the facility with which carbon–carbon bonds between highly branched centers can be formed. This point is illustrated by several of the examples in Scheme 12.7.

A second general reaction that proceeds by an $S_{RN}1$ mechanistic pattern involves aryl halides. Aryl halides undergo substitution by certain nucleophiles by a chain mechanism of the $S_{RN}1$ class.[164] Many of the reactions are initiated photochemically, and most have been conducted in liquid ammonia solution.

The reactions can also be initiated by a strong chemical reductant or electrochemically.[165] There are several lines of evidence which support the operation of a chain mechanism, one of the most general observations being that the reactions are stopped or greatly retarded by radical traps. The reaction is not particularly sensitive to the aromatic ring substituents. Both electron-releasing groups such as methoxy and electron-attracting groups such as benzoyl can be present.[166] Groups which easily undergo one-electron reduction, especially the nitro group, cause the reaction to fail. The nucleophiles that have been used successfully include sulfide and phosphide anions, dialkyl phosphite anions, and certain enolates. Scheme 12.8 illustrates some typical reactions.

162. N. Kornblum, A. S. Erickson, W. J. Kelly, and B. Henggeler, *J. Org. Chem.* **47**:4534 (1982).

163. N. Kornblum, S. C. Carlson, and R. G. Smith, *J. Am. Chem. Soc.* **101**:647 (1979).

164. J. F. Bunnett, *Acc. Chem. Res.* **11**:413 (1978); R. A. Rossi and R. H. deRossi, *Aromatic Substitution by the $S_{RN}1$ Mechanism*, American Chemical Society Monograph No. 178, Washington, D.C., 1983; J.-M. Saveant, *Adv. Phys. Org. Chem.* **26**:1 (1990).

165. C. Amatore, J. Chaussard, J. Pinson, J.-M.Saveant, and A. Thiebault, *J. Am. Chem. Soc.* **101**:6012 (1979).

166. J. F. Bunnett, and J. E. Sundberg, *Chem. Pharm. Bull.* **23**:2620 (1975); R. A. Rossi, R. H. deRossi, and A. F. Lopez, *J. Org. Chem.* **41**:3371 (1976).

Scheme 12.7. Carbon Alkylation via Nitroalkane Radical Anions Generated by Electron Transfer

1[a]

$(CH_3)_2CCl$—[aryl, para-NO_2] + $(CH_3)_2\bar{C}NO_2$ Li^+ ⟶ $(CH_3)_2C-C(CH_3)_2$ (with NO_2)—[aryl, para-NO_2]   95%

2[b]

$(CH_3)_2CNO_2$—[aryl, para-NO_2] + $^-CH(CO_2C_2H_5)_2$ ⟶ $(CH_3)_2CCH(CO_2C_2H_5)_2$—[aryl, para-$NO_2$]   95%

3[c]

[cyclohexane ring with O_2N NO_2] + $(CH_3)_2\bar{C}NO_2$ ⟶ [cyclohexane ring with NO_2 and $-C(CH_3)_2$ bearing NO_2]   85%

4[d]

$(CH_3)_2CCO_2C_2H_5$ (with NO_2) + $(CH_3)_2\bar{C}NO_2$ ⟶ $(CH_3)_2C\underset{CO_2C_2H_5}{\overset{NO_2}{C}}(CH_3)_2$ 95%

5[e]

$N\equiv C$—[aryl]—$\underset{CH_3}{\overset{CH_3}{C}}-NO_2$ + $CH_3\bar{C}HNO_2$ $\xrightarrow{HMPA}$ $N\equiv C$—[aryl]—$\underset{CH_3}{\overset{CH_3}{C}}-\underset{CH_3}{CHNO_2}$ 94%

6[f]

$(CH_3)_3CCH_2\underset{CH_3}{\overset{CH_3}{C}}-NO_2$ + $^-CH_2NO_2$ $\xrightarrow{DMSO}$ $(CH_3)_3CCH_2\underset{CH_3}{\overset{CH_3}{C}}CH_2NO_2$ 60%

a. N. Kornblum, T. M. Davies, G. W. Earl, N. L. Holy, R. C. Kerber, M. T. Musser, and D. H. Snow, *J. Am. Chem. Soc.* **89**:725 (1967).
b. N. Kornblum, T. M. Davies, G. W. Earl, G. S. Greene, N. L. Holy, R. C. Kerber, J. W. Manthey, M. T. Musser and D. H. Snow, *J. Am. Chem. Soc.* **89**:5714 (1967).
c. N. Kornblum, S. D. Boyd, and F. W. Stuchal, *J. Am. Chem. Soc.* **92**:5783 (1970).
d. N. Kornblum, S. C. Carlson, J. Widmer, N. Fifolt, B. N. Newton, and R. G. Smith, *J. Org. Chem.* **43**:1394 (1978).
e. N. Kornblum and A. S. Erickson, *J. Org. Chem.* **46**:1037 (1981).

Kinetic studies have shown that the enolate and phosphorus nucleophiles all react at about the same rate.[167] This suggests that the only step directly involving the nucleophile (step 2 of the propagation sequence) occurs at essentially the diffusion-controlled rate so that there is little selectivity among the individual nucleophiles.[168] The synthetic potential of the reaction lies in the fact that other substituents which activate the halide to substitution are not required in this reaction, in contrast to aromatic nucleophilic substitutions that proceed by an addition–elimination mechanism (see Section 10.5).

167. X.-M. Zhang, D.-L. Yang, X.-Q. Jia, and Y.-C. Liu, *J. Org. Chem.* **58**:7350 (1993).
168. C. Galli and J. F. Bunnett, *J. Am. Chem. Soc.* **103**:7140 (1981); R. G. Scamehorn, J. M. Hardacre, J. M. Lukanich, and L. R. Sharpe, *J. Org. Chem.* **49**:4881 (1984).

Scheme 12.8. Aromatic Substitution by the $S_{RN}1$ Mechanism

a. J. E. Swartz and J. F. Bunnett, *J. Org. Chem.* **44**:340 (1979).
b. J. F. Bunnett and X. Creary, *J. Org. Chem.* **39**:3611 (1974).
c. J. F. Bunnett and X. Creary, *J. Org. Chem.* **39**:3612 (1974).
d. M. F. Semmelhack and T. Bargar, *J. Am. Chem. Soc.* **102**:7765 (1980).
e. R. A. Rossi and R. A. Alonso, *J. Org. Chem.* **45**:1239 (1980).

An alternative reaction mechanism has been suggested for nitroarylation of enolates. An impetus for considering other mechanisms is the fact that the by-products which might be expected from aryl radicals, such as reduction products from hydrogen abstraction from the solvent or biaryls from coupling, are not observed. One alternative is that, rather than being a chain process, the reaction may involve recombination whereby the radicals combine more rapidly than they separate.

Kinetics of the reaction of *p*-nitrochlorobenzene with the sodium enolate of ethyl cyanoacetate are consistent with this mechanism. Also, radical scavengers have no effect on the reaction, contrary to what would be expected for a chain mechanism in which aryl radicals would need to encounter the enolate in a propagation step. The reactant, *p*-nitrophenyl chloride, however, is one which might also react by the addition–elimination mechanism, and the postulated mechanism is essentially the stepwise electron-transfer version of this mechanism. The issue then becomes the question of whether the postulated radical pair is a distinct intermediate.

Instances of substitution of hindered alkyl halides by the $S_{RN}1$ mechanism have also been documented.[173] Some examples are shown below.

$$(CH_3)_3CCH_2Br + PhS^- \xrightarrow[NH_3]{h\nu} (CH_3)_3CCH_2SPh$$
60%

Ref. 174

Ref. 175

87%

Ref. 176

The mechanism is the same as for aryl halides:

initiation $\quad e^- + R—X \longrightarrow R—X^{\cdot -} + R\cdot + X^-$

propagation $\quad R\cdot + \mathbf{Nu:} \longrightarrow R—\mathbf{Nu}^{\cdot -}$

$\quad\quad\quad\quad R—\mathbf{Nu}^{\cdot -} + R—X \longrightarrow R—\mathbf{Nu} + R—X^{\cdot -}$

$\quad\quad\quad\quad R—X^{\cdot -} \longrightarrow R\cdot + X^-$

Acetophenone enolate and nitromethane anions have also been used successfully in alkyl substitution.

85%

Ref. 177

$$(CH_3)_3CCH_2I + {}^-CH_2NO_2 \longrightarrow (CH_3)_3CCH_2CH_2NO_2$$
69%

Ref. 178

General References

Reactions and Mechanisms

A. L. J. Beckwith and K. U. Ingold, in *Rearrangements in Ground and Excited States*, P. de Mayo, ed., Academic Press, New York, 1980, Chapter 4.

M. Birkhofer, H.-D. Beckhaus, and C. Ruchardt, *Substituent Effects in Radical Chemistry*, Reidel, Boston, 1986.

J. Fossey, D. Lefort, and J. Sorba, *Free Radicals in Organic Chemistry*, John Wiley & Sons, Chichester, U.K., 1995.

B. Giese, *Radicals in Organic Synthesis; Formation of Carbon–Carbon Bonds*, Permagon Press, Oxford, U.K., 1986.

E. S. Huyser, *Free Radical Chain Reactions*, Wiley-Interscience, New York, 1970.

J. E. Leffler, *An Introduction to Free Radicals*, John Wiley & Sons, New York, 1993.

173. S. M. Palacios, A. N. Santiago, and R. A. Rossi, *J. Org. Chem.* **49:**4609 (1984).

174. A. B. Pierini, A. B. Penenory, and R. A. Rossi, *J. Org. Chem.* **50:**2739 (1985).

175. R. A. Rossi, S. M. Palacios, and A. N. Santiago, *J. Org. Chem.* **47:**4654 (1982).

176. R. A. Rossi, A. N. Santiago, and S. M. Palacios, *J. Org. Chem.* **49:**3387 (1984).

177. M. A. Nazareno and R. A. Rossi, *J. Org. Chem.* **61:**1645 (1996).

178. A. B. Penenory and R. A. Rossi, *Gazz. Chim. Ital.* **125:**605 (1995).

W. B. Motherwell and D. Crich, *Free Radical Chain Reactions in Organic Synthesis*, Academic Press, London, 1992.

W. A. Pryor, *Free Radicals*, McGraw-Hill, New York, 1966.

Spectroscopic Methods

M. Bersohn and J. C. Baird, *An Introduction to Electron Paramagnetic Resonance*, W. A. Benjamin, New York, 1966.

N. Hirota and H. Ohya-Nishiguchi, in *Investigation of Rates and Mechanisms of Reactions*, C. Bernasconi, ed., *Techniques of Chemistry*, 4th ed., Vol. VI, Part 2, Wiley-Interscience, New York, 1986, Chapter XI.

A. G. Lawler and H. R. Ward, in *Determination of Organic Structures by Physical Methods*, Vol 5, F. C. Nachod and J. J. Zuckerman, eds., Academic Press, New York, 1973, Chapter 3.

Problems

(References for these problems will be found on page 803.)

1. Predict the products of the following reactions.

(a)

$\xrightarrow[\text{HBr}]{h\nu}$

(b) + $(CH_3)_2CHOH$ $\xrightarrow[130°C]{(CH_3)_3COOC(CH_3)_3}$

(c) $CH_3(CH_2)_6CH_2OCl$ $\xrightarrow[1\,M\text{ in }CCL_4]{h\nu}$

(d) $CH_3CH{=}O + H_2C{=}CHCH(OCH_3)_2$ $\xrightarrow[(PhCO_2)_2]{\Delta}$

(e) + $PhCO_2C(CH_3)_3$ $\xrightarrow[80°C]{CuBr}$

(f) + $HCCl_3$ $\xrightarrow[65°C]{(PhCO_2)_2}$

(g) $O_2N\overset{\overset{\displaystyle CH_3}{|}}{\underset{\underset{\displaystyle CH_3}{|}}{C}}CH_2CH_2CO_2^-$ $\xrightarrow{\text{electrolysis}}$

(h) $CH_2{=}CH\underset{\underset{\displaystyle OH}{|}}{C}(CH_3)_2 + CCl_4$ $\xrightarrow{(PhCO_2)_2}$

(i) + K^+ $^-CH_2\overset{\overset{\displaystyle O}{||}}{C}CH_3$ $\xrightarrow[NH_3]{h\nu}$

(j) $C_4H_9SH + CH_3(CH_2)_4CH{=}CH_2$ $\xrightarrow{(PhCO_2)_2}$

2. Using Table III in Ref. 68 (p. 692), calculate the expected product composition from the gas-phase photochemical chlorination and bromination of 3-methylpentane under conditions (excess hydrocarbon) in which only monohalogenation would occur.

3. A careful study of the photoinitiated addition of HBr to 1-hexene established the following facts. (1) The quantum yield is 400, (2) The products are 1-bromohexane, 2-bromohexane, and 3-bromohexane. The amounts of 2- and 3-bromohexane formed are always nearly identical and increase from about 8% each at 4°C to about 22% at 63°C, (3) During the course of the reaction, small amounts of 2-hexene can be detected. Write a mechanism that could accommodate all these facts.

4. The irradiation of 1,3-dioxolane in the presence of alkenes and a photochemically activated initiator at 30°C leads to 2-alkyldioxolanes:

The reaction is particularly effective with alkenes with electron-attracting substituents such as diethyl maleate. When the reaction is conducted thermally with a peroxide initiator at 160°C, the product mixture is much more complex:

(a) Provide a mechanism for the formation of the observed products.
(b) Why is diethyl maleate an especially good reactant?
(c) Why does the photochemical method lead to a different product distribution than the peroxide-catalyzed reaction?

5. Provide a detailed mechanistic explanation of the following results.

(a) Photochemical bromination of **1**, $\alpha_D + 4.21°$, affords **2**, which is optically active, $\alpha_D - 3.23°$, but **3** under the same conditions gives **4**, which is racemic.

1: X = Br
3: X = F

2: X = Br
4: X = F

(b) The stereoisomerization shown below proceeds efficiently, with no other chemical change occurring at a comparable rate, when the compound is warmed with N-bromosuccinimide and a radical chain initiator.

(c) There is a substantial difference in the reactivity of the two stereoisomeric compounds shown below toward abstraction of a hydrogen atom by the *t*-butoxy radical.

is 7–10 times more reactive than

(d) Free-radical chlorination of optically active 1-chloro-2-methylbutane yields six dichloro derivatives, of which four are optically active and two are not. Identify the optically active and optically inactive products and provide an explanation for the origin of each product.

(e) Irradiation of a mixture of the hydrocarbon **A** and di-*t*-butyl peroxide generates a free radical which can be identified as the 2-phenylethyl radical by its EPR spectrum. This is the only spectrum which can be observed, even when the photolysis is carried out at low temperatures (−173°C).

A

(f) Among the products from heating 1,6-heptadiene with 1-iodoperfluoropropane in the presence of azobis(isobutyronitrile) are two saturated 1:1 adducts. Both adducts gave the same product on dehydrohalogenation, and this product was shown by spectroscopic means to contain a $CH_2=C$ unit. Give the structures of the two adducts and propose a mechanism for their formation.

6. Write mechanisms which satisfactorily account for the following reactions.

(a)

(b) $Ph(CH_2)_5CH=CH_2$ $\xrightarrow{ROOH, 140°C}$

(c)

(d) $CH_3(CH_2)_5CH=CH_2 + Cl\overset{H}{N}CO_2C_2H_5 + Cr(II)$ $\xrightarrow{-78°C}$ $CH_3(CH_2)_5\underset{\underset{Cl}{|}}{C}HCH_2NHCO_2C_2H_5$

(e) reaction with $NH_3(l)$

(f) $C_6H_5CCH(CH_3)_2$ with OH and CH$_3$ substituents, electrolysis $R_4N^+BF_4^-$, $CH_3C\equiv N$ → $C_6H_5CCH_3$ (O) + $CH_3CNHCH(CH_3)_2$ (O)

(g) reaction at 100°C

(h) $CH_2=CHCH_2CH=C(CH_3)_2$ $\xrightarrow[O_2]{PhSH}$ $PhSCH_2$— ... —$C(CH_3)_2$, with O—O and OOH

7. The decarbonylation of the two labeled pentenals shown below has been studied. Write a mechanism that could explain the distribution of deuterium label found in the two products.

$CH_2=CHCH_2CD_2CH=O$ $\xrightarrow{ROOR}$ $CH_2=CHCD_2CH_3$ + $CH_2=CHCH_2CHD_2$

1 : 1 ratio in dilute solution, increasing
to 1 : 1.5 in concentrated solution

$\xrightarrow{ROOR}$ products

1 : 1 ratio in dilute solution, increasing
to 1 : 1.4 in concentrated solution

8. Decomposition of the *trans*-decalyl perester **A** gives a 9:1 ratio of *trans*:*cis* hydroperoxide product at all oxygen pressures studied. The product ratio from the *cis* isomer is dependent on the oxygen pressure. At 1 atm O_2, it is 9:1 *trans*:*cis*, as with the *trans* substrate, but this ratio decreases and eventually inverts with increasing O_2 pressure. It is 7:3 *cis*:*trans* at 545 atm oxygen pressure. What deduction about the stereochemistry of the decalyl free radical can be made from these data?

A $\xrightarrow{O_2}$ OOH + OOH

9 : 1 ratio at
all O_2 pressures

COOPh $\xrightarrow{O_2}$ (low pressure) / (high pressure) OOH + OOH

3 : 7 ratio at
545 atm

9. (a) Trichloromethanesulfonyl chloride, Cl_3CSO_2Cl, can chlorinate hydrocarbons as described in the stoichiometric equation below. The reaction is a chain process. Write at least two possible sequences for chain propagation. Suggest some likely termination steps.

$$\text{C}_6\text{H}_5\text{—CH}_3 + Cl_3CSO_2Cl \xrightarrow[\Delta]{(PhCO_2)_2} \text{C}_6\text{H}_5\text{—CH}_2Cl + HCCl_3 + SO_2$$

(b) Given the following additional information, choose between the chain propagation sequences you have postulated in part (a).
   (1) In the reaction

$$R-H + BrCCl_3 \xrightarrow{h\nu} RBr + HCCl_3$$

   the reactivity of cyclohexane is about one-fifth that of toluene.
   (2) In the chlorination by trichloromethanesulfonyl chloride, cyclohexane is about three times as reactive as toluene.

10. A highly selective photochemical chlorination of esters, amides, and alcohols can be effected in 70%–90% H_2SO_4 using N-chlorodialkylamines as chlorinating agents. Mechanistic studies indicate that a chain reaction is involved:

$$R_2\overset{+}{N}H\underset{Cl}{|} \xrightarrow{\text{initiation}} R_2\overset{+}{N}H + Cl\cdot$$

$$R_2\overset{+}{\dot{N}}H + CH_3CH_2(CH_2)_nX \longrightarrow R_2\overset{+}{N}H_2 + CH_3\overset{\cdot}{C}H(CH_2)_nX$$

$$CH_3\overset{\cdot}{C}H(CH_2)_nX + R_2\overset{+}{N}H\underset{Cl}{|} \longrightarrow CH_3\underset{Cl}{\overset{|}{C}}H(CH_2)_nX + R_2\overset{+}{\dot{N}}H$$

where $X = CO_2CH_3$, CH_2OH, or $CONH_2$. A very interesting feature of the reaction is that the chlorine atom is introduced on the next-to-terminal carbon atom for reactant molecules with $n = 4$ or 6. In contrast, chlorination of these same compounds with chlorine in nonpolar solvents shows little position selectivity. Rationalize the observed selectivity.

11. Analyze the hyperfine coupling in the spectrum of the butadiene radical anion given in Fig. 12.P11. What is the spin density at each carbon atom according to the McConnell equation?

(b) A representation of the EPR spectrum of allyl radical is presented below. Interpret the splitting pattern and determine the values of the hyperfine splitting constants.

Fig. 12.P11. Spectrum of the butadiene radical anion. [From D. H. Levy and R. J. Meyers, *J. Chem. Phys.* **41**:1062 (1964).]

12. The oxidation of norbornadiene by *t*-butyl perbenzoate and Cu(I) leads to 7-*t*-butoxynorbornadiene. Similarly, oxidation with dibenzoyl peroxide and CuBr leads to 7-benzyloxynorbornadiene. In both cases, when a 2-monodeuterated sample of norbornadiene is used, the deuterium is found distributed at all seven carbons in the product. Provide a mechanism which could account for this result. In what ways does this mechanism differ from the general mechanism discussed on pp. 724–725?

13. A very direct synthesis of certain lactones can be achieved by heating an alkene, a carboxylic acid, and the Mn(III) salt of the acid. Suggest a mechanism by which this reaction might proceed.

$$CH_3(CH_2)_5CH{=}CH_2 \ + \ Mn(O_2CCH_3)_3 \quad \xrightarrow{CH_3CO_2H} \quad$$

14. Indicate mechanisms that would account for each of the products observed in the thermal decomposition of compound **A**:

$$\underset{\substack{| \\ Ph \\ \mathbf{A}}}{CH_3\overset{O}{\overset{||}{C}}\overset{CH_3}{\overset{|}{C}}CH_2\overset{O}{\overset{||}{C}}OOC(CH_3)_3} \quad \xrightarrow{130°C}$$

$$\underset{\substack{| \\ Ph \\ 26\%}}{CH_3\overset{O}{\overset{||}{C}}\overset{CH_3}{\overset{|}{C}}CH_2OC(CH_3)_3} \ + \ \underset{15\%}{CH_3\overset{O}{\overset{||}{C}}CH_2\overset{Ph}{\overset{|}{C}}HCH_3} \ + \ \underset{9\%}{CH_3\overset{O}{\overset{||}{C}}CH{=}\overset{Ph}{\overset{|}{C}}CH_3} \ + \ \underset{\substack{| \\ Ph \\ 9\%}}{CH_3\overset{O}{\overset{||}{C}}CH_2C{=}CH_2}$$

15. The *spiro* peroxide **A**, which is readily prepared from cyclohexanone and hydrogen peroxide, decomposes thermally to give substantial amounts of cyclodecane (**B**) and

11-undecanolactone (**C**). Account for the efficient formation of these macrocyclic compounds.

16. Methylcyclopropane shows strikingly different reactivity toward chlorination and bromination under radical-chain conditions. With chlorine, cyclopropyl chloride (56%) is the major product, along with small amounts of 1,3-dichlorobutane (7%). Bromine gives a quantitative yield of 1,3-dibromobutane. Offer an explanation for the difference.

17. Electrolysis of 3,3-diphenylpropanoic acid in acetic acid–acetate solution gives the products shown below. Propose mechanisms for the formation of each of the major products.

18. Write a mechanism to account for the formation of the observed product of each of the following reactions.

19. The *N*-benzoyl methyl esters of the amino acids valine, alanine, and glycine have been shown to react with *N*-bromosuccinimide to give monobromination products contain-

ing bromine at the α carbon of the amino acid structure. The order of reactivity is glycine > alanine > valine $(23:8:1)$. Account for the observed trend in eactivity.

20. By measurements in an ion cyclotron resonance spectrometer, the proton affinity (PA) of free radicals can be measured.

$$RH^+ \rightarrow R\cdot + H^+, \quad \Delta H_{298} = PA$$

These data can be combined with ionization potential (IP) data according to the scheme below to determine bond dissociation energies (BDE).

$$
\begin{array}{ll}
RH^+ \rightarrow R^\cdot + H^+ & \Delta H_{298} = PA \\
RH \rightarrow RH^+ + e^- & \Delta H_{298} = IP \\
e^- + H^+ \rightarrow H^\cdot & \Delta H_{298} = -313.6\,\text{kcal/mol} \\
\hline
RH \rightarrow R^\cdot + H^\cdot & \Delta H_{298} = BDE
\end{array}
$$

Data for PA and IP are given for several hydrocarbons of interest.

	IP	PA
PhCH$_2$—H	203	198
⬡—H	190	200
⬠—H	198	199
▷—H	224	180
CH$_2$=CHCH$_2$—H	224	180
▷—H	232	187
CH$_2$=CH—H	242	183

According to these data, which structural features provide stabilization of radial centers? Determine the level of agreement between these data and the "radical stabilization energies" given in Table 12.7 if the standard C—H bond dissociation energy is taken to be 98.8 kcal/mol. (Compare the calculated and observed bond dissociation energies for the benzyl, allyl, and vinyl systems.)

21. Provide a stepwise mechanism for the following reactions.

(a)

(b)

(c)

Photochemistry

The photochemical reactions of organic compounds attracted great interest in the 1960s. As a result, many useful and fascinating reactions were uncovered, and photochemistry is now an important synthetic tool in organic chemistry. A firm basis for mechanistic description of many photochemical reactions has been developed. Some of the more general types of photochemical reactions will be discussed in this chapter. In Section 13.2, the relationship of photochemical reactions to the principles of orbital symmetry will be considered. In later sections, characteristic photochemical reactions of alkenes, dienes, carbonyl compounds, and aromatic rings will be introduced.

13.1. General Principles

A broad recognition of the fundamental elements of a photochemical reaction is not difficult. The first condition that must be met is that the compound absorb light emitted by the radiation source. For light to be absorbed, the molecule must have an energy level corresponding in energy to that of the radiation. Organic photochemical reactions involve excited electronic states. Depending on functionality, organic compounds can have electronic absorption bands from the far ultraviolet to the visible region of the spectrum. Table 13.1 lists the general regions of absorption for the classes of organic molecules that we will discuss in this chapter. A number of light sources can be used. The most common sources for preparative-scale work are mercury vapor lamps, which emit mainly at 254, 313, and 366 nm. The composition of the radiation reaching the sample can be controlled by use of filters. For example, if the system is constructed so that light must pass through borosilicate glass (Pyrex), only wavelengths longer than 300–310 nm will reach the sample, because the glass absorbs below this wavelength. Pure fused quartz, which transmits down to 200 nm, must be used if the 254-nm radiation is desired. Other materials have cutoff points between those of quartz and Pyrex. Filter solutions that absorb in specific wavelength ranges can also be used to control the energy of the light reaching the

**Table 13.1. General Wavelength Ranges for
Lowest-Energy Absorption Band of Some
Classes of Photochemical Substrates**

Substrates	Absorption maxima (nm)
Simple alkenes	190–200
Acyclic dienes	220–250
Cyclic dienes	250–270
Styrenes	270–300
Saturated ketones	270–280
α, β-Unsaturated ketones	310–330
Aromatic ketones and aldehydes	280–300
Aromatic compounds	250–280

sample.[1] The energy supplied by a particular wavelength of light can be calculated from the fundamental equation

$$E = h\nu$$

The energy in kilocalories per mole is

$$E = 2.62 \times 10^4 / \lambda$$

where λ is wavelength in nanometers. Thus, the energy supplied by light of $\lambda 254$ nm equals 113 kcal/mol, an energy sufficient to rupture most chemical bonds.

When a molecule absorbs a quantum of light, the electronic configuration changes to correspond to an excited state. Three general points about this process should be emphasized:

1. The excitation promotes an electron from a filled orbital to an empty one. In most cases, the promotion is from the HOMO to the LUMO, which is usually an antibonding orbital.
2. At the instant of excitation, only electrons are reorganized; the heavier nuclei retain their ground-state geometry. The statement of this condition is referred to as the *Franck–Condon principle*. A consequence is that the initially generated excited state will have a non-minimal-energy geometry.
3. The electrons do not undergo spin inversion at the instant of excitation. Inversion is forbidden by quantum-mechanical selection rules, which require that there be conservation of spin during the excitation process. Although a subsequent spin-state change may occur, it is a separate step from excitation.

Thus, in the very short time (10^{-15} s) required for excitation, the molecule does not undergo changes in nuclear position or in the spin state of the promoted electron. After the excitation, however, these changes can occur very rapidly. The new minimum-energy

1. Detailed information on the emission characteristics of various sources and the transmission properties of glasses and filter solutions can be found in A. J. Gordon and R. A. Ford, *The Chemist's Companion*, Wiley-Interscience, New York, 1972, pp. 348–368 and in S. L. Murov, I. Carmichael, and G. L. Hug, *Handbook of Photochemistry*, 2nd ed., Marcel Dekker, New York, 1993.

geometry associated with the excited state is rapidly achieved by vibrational processes, which transfer thermal energy to the solvent. Sometimes, chemical reactions of the excited molecule are fast relative to this vibrational relaxation, but this is rare in solution. When reaction proceeds faster than vibrational relaxation, the reaction is said to involve a "hot excited state," that is, one with excess vibrational energy. The excited state can also undergo *intersystem crossing*, the inversion of spin of an electron in a half-filled orbital to give a *triplet state*, in which both unpaired electrons have the same spin. The triplet state will also adopt a new minimum-energy molecular geometry.

The general situation can be represented for a hypothetical molecule using a potential energy diagram (Fig. 13.1). The designations S and T are used for singlet and triplet states, respectively. The excitation is a "vertical transition"; that is, it involves no distortion of the molecular geometry. Horizontal displacement on the diagram corresponds to motion of the atoms relative to one another. Because the potential energy curves of the excited states are displaced from that of the ground state, the species formed by excitation is excited both electronically and vibrationally. The energy wells corresponding to the triplet states also correspond to a different minimum-energy molecular geometry. Vibrational relaxation corresponds to dissipation of the vibrational energy as the molecule moves to the bottom of the potential energy well for a particular excited state. One of the central issues in the description of any photochemical reaction is the question of whether a singlet or triplet excited state is involved. This depends on the rate of intersystem crossing in comparison with the rate of chemical reaction of the singlet excited state. If intersystem crossing is fast relative to reaction, reaction will occur through the triplet excited state. If reaction is faster than intersystem crossing, the reaction will occur from the singlet state.

Photosensitization is an important alternative to direct excitation of molecules, and this method of excitation usually results in reactions that occur via triplet excited states. If

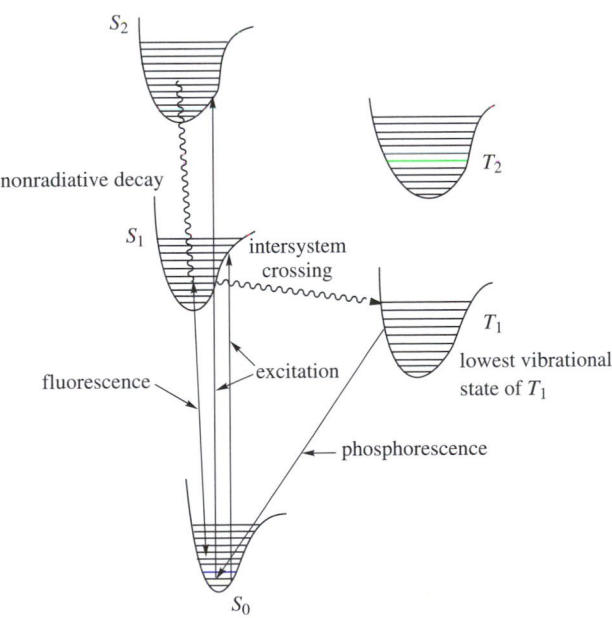

Fig. 13.1. Energy level diagram and summary of photochemical processes.

a reaction is to be carried out by photosensitization, a substance, the *sensitizer*, is present in the system. This substance is chosen to meet the following criteria:

1. It must be excited by the irradiation to be used.
2. It must be present in sufficient concentration and absorb more strongly than the other reactants under the conditions of the experiment so that it is the major light absorber.
3. The energy of the triplet state of the sensitizer (^{3}Sens*) must be greater than that of the reactant. If this condition is not met, the energy transfer becomes endothermic and cannot compete with other transformations of ^{3}Sens*.
4. It must be able to transfer energy to the desired reactant.

The common case, and the one that will be emphasized here, is triplet sensitization. In this case, the intersystem crossing of the sensitizer must be faster than energy transfer to the reactant or solvent from the singlet excited state.

The transfer of energy must proceed with net conservation of spin. In the usual case, the acceptor molecule is a ground-state singlet, and its reaction with the triplet state of the sensitizer will produce the triplet state of the acceptor. The mechanism for triplet photosensitization is outlined below:

$$\text{Sens} \longrightarrow {}^1\text{Sens*} \qquad \text{Sensitizer singlet formed}$$
$$^1\text{Sens*} \longrightarrow {}^3\text{Sens*} \qquad \text{Intersystem crossing of sensitizer}$$
$$^3\text{Sens*} + \text{React} \longrightarrow \text{Sens} + {}^3\text{React*} \qquad \text{Energy transfer to reactant molecule}$$

Once the excited state of the reactant has been formed, either by direct or sensitized energy transfer, the stage is set for a photochemical reaction. There are still, however, competitive processes that can occur and result in the return of unreacted starting material. The excited state can decay to the ground state by emission of light, a *radiative transition*. The rate of emission is very fast ($k = 10^5 - 10^9\ \text{s}^{-1}$) for radiative transitions between electronic states of the same multiplicity, and somewhat slower ($k = 10^3 - 10^5\ \text{s}^{-1}$) between states of different multiplicity. The two processes are known as *fluorescence* and *phosphorescence*, respectively. If energy is emitted as light, the reactant is no longer excited, of course, and a photochemical reaction will not occur.

Excited states can also be *quenched*. Quenching is the same physical process as sensitization, but the word "quenched" is used when a photoexcited state of the reactant is deactivated by transferring its energy to another molecule in solution. This substance is called a *quencher*.

Finally, *nonradiative decay* can occur. This name is given to the process by which the energy of the excited state is transferred to the surrounding molecules as vibrational (thermal) energy without light emission. The processes that can occur after photochemical excitation are summarized in Fig. 13.1.

Because of the existence of these competing processes, not every molecule that is excited undergoes a photochemical reaction. The fraction of molecules that react, relative to those that are excited, is called the *quantum yield*, ϕ. The quantum yield is a measure of the efficiency of the absorption of light in producing reaction product. A quantum yield of unity means that each molecule excited (which equals the number of quanta of light absorbed) goes to product. If the quantum yield is 0.01, then only 1 of every 100 molecules that are excited undergoes photochemical reaction. The quantum yield can vary widely, depending on the structure of the reactants and the reaction conditions. Quantum yields can be greater than 1 in chain reactions, in which a single photoexcitation initiates a repeating series of reactions leading to many molecules of product per initiation step.

747

SECTION 13.2.
ORBITAL SYMMETRY
CONSIDERATIONS
RELATED TO
PHOTOCHEMICAL
REACTIONS

Because photochemical processes are very fast, special techniques are required to obtain rate measurements. One method is flash photolysis. The excitation is effected by a short pulse of light in an apparatus designed to monitor very fast spectroscopic changes. The rate characteristics of the reactions following radiation can be determined from these spectroscopic changes.

Another useful technique for measuring the rates of certain reactions involves measuring the quantum yield as a function of quencher concentration. A plot of the inverse of the quantum yield versus quencher concentration is then made (*Stern–Volmer plot*). Because the quantum yield indicates the fraction of excited molecules that go on to product, it is a function of the rates of the processes that result in other fates for the excited molecule. These processes are described by the rate constants k_q (quenching) and k_n (other nonproductive decay to ground state).

$$\phi = \frac{k_r}{k_r + k_q[Q] + k_n}$$

A plot of $1/\phi$ versus quencher concentrations, $[Q]$, then gives a line with the slope k_q/k_r. It is usually possible to assume that quenching is diffusion-controlled, permitting assignment of a value to k_q. The rate of photoreaction, k_r, for the excited intermediate can then be calculated.

In this chapter, the discussion will center on the reactions of excited states, rather than on the other routes available for dissipation of excess energy. The chemical reactions of photoexcited molecules are of interest primarily for three reasons:

1. Excited states have a great deal of energy and can therefore undergo reactions that would be highly endothermic if initiated from the ground state. For example, we can calculate from the relationship $E = h\nu$ that excitation by 350-nm light corresponds to an energy transfer of 82 kcal/mol.
2. The population of an antibonding orbital in the excited state allows the occurrence of chemical transformations that are electronically not available to ground-state species.
3. Either the singlet or the triplet state may be involved in a photochemical reaction, whereas only singlet species are involved in most thermal processes. This permits the formation of intermediates that are unavailable under thermal conditions.

13.2. Orbital Symmetry Considerations Related to Photochemical Reactions

The complementary relationship between thermal and photochemical reactions can be illustrated by considering some of the same reaction types discussed in Chapter 11 and applying orbital symmetry considerations to the photochemical mode of reaction. The case of $[2\pi + 2\pi]$ cycloaddition of two alkenes can serve as an example. This reaction was classified as a forbidden thermal reaction (Section 11.3) The correlation diagram for cycloaddition of two ethylene molecules (Fig. 13.2) shows that the ground-state molecules would lead to an excited state of cyclobutane and that the cycloaddition would therefore involve a prohibitive thermal activation energy.

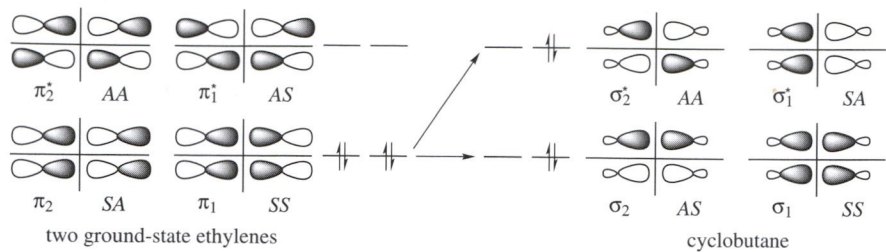

π_2^* AA π_1^* AS σ_2^* AA σ_1^* SA

π_2 SA π_1 SS σ_2 AS σ_1 SS

two ground-state ethylenes   cyclobutane

Fig. 13.2. Orbital correlation for two ground-state ethenes and cyclobutane. The symmetry designations apply, respectively, to the horizontal and vertical planes for two ethylene molecules approaching one another in parallel planes.

How does the situation change when a photochemical reaction involving one ground-state alkene and one excited-state alkene is considered? We can assume the same symmetrical approach as in the thermal reaction, so the same array of orbitals is involved. The occupation of the orbitals is different however; the π_1 (SS) orbital is doubly occupied, but π_2 (SA) and π_1^* (AS) are singly occupied. The reaction is therefore allowed. Although the correlation diagram illustrated in Fig. 13.3 might suggest that the product would initially be formed in an excited state, this is not necessarily the case. The concerted process can involve a transformation of the reactant excited state to the ground state of product. This transformation will be discussed shortly.

Consideration of the HOMO–LUMO interactions also indicates that the $[2\pi + 2\pi]$ additions would be allowed photochemically. The HOMO in this case is the excited alkene π^* orbital. The LUMO is the π^* of the ground-state alkene, and a bonding interaction is present between the carbons where new bonds must be formed:

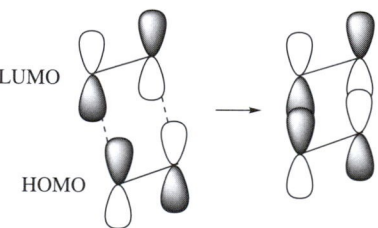

LUMO

HOMO

A striking illustration of the relationship between orbital symmetry considerations and the outcome of photochemical reactions can be found in the stereochemistry of electrocyclic reactions. In Chapter 11, the distinction between the conrotatory and the disrotatory mode of reaction as a function of the number of electrons in the system was

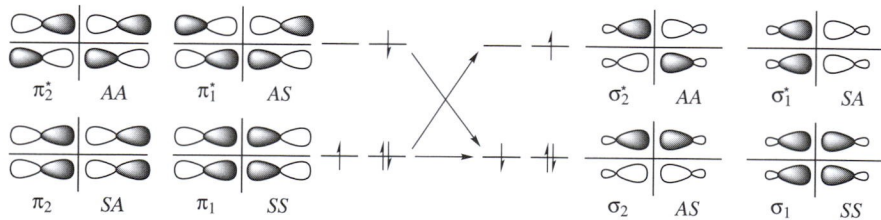

π_2^* AA π_1^* AS σ_2^* AA σ_1^* SA

π_2 SA π_1 SS σ_2 AS σ_1 SS

Fig. 13.3. Orbital correlation diagram for one ground-state ethene and one excited-state ethene. The symmetry designations apply, respectively, to the horizontal and vertical planes for two ethene molecules approaching one another in parallel planes.

described. Orbital symmetry considerations predict that photochemical electrocyclic reactions will show a reversal of stereochemistry[2]:

749

SECTION 13.2.
ORBITAL SYMMETRY
CONSIDERATIONS
RELATED TO
PHOTOCHEMICAL
REACTIONS

Number of π electrons	Thermal	Photochemical
2	disrotatory	conrotatory
4	conrotatory	disrotatory
6	disrotatory	conrotatory
8	conrotatory	disrotatory

One way of making this prediction is to construct an electronic energy state diagram for the reactant and product molecules and observe the correlation between the states.[3] Those reactions will be permitted in which the reacting state correlates with a state of the product that is not appreciably higher in energy[4]. The states involved in the photochemical butadiene-to-cyclobutene conversion are ψ_1, ψ_2, and ψ_3 for the first excited state of butadiene and σ, π, and π^* for cyclobutene. The appropriate elements of symmetry are the plane of symmetry for the disrotatory process and the axis of symmetry for the conrotatory process. The correlation diagram for this reaction is shown in Fig. 13.4.

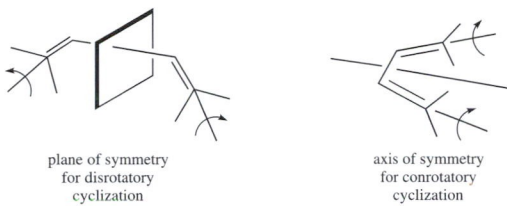

plane of symmetry
for disrotatory
cyclization

axis of symmetry
for conrotatory
cyclization

This analysis shows that disrotatory cyclization is allowed, whereas conrotation would lead to a highly excited σ^1, π^2, σ^{*1} configuration of cyclobutene. The same conclusion is reached if it is assumed that the frontier orbital will govern reaction stereochemistry[2]:

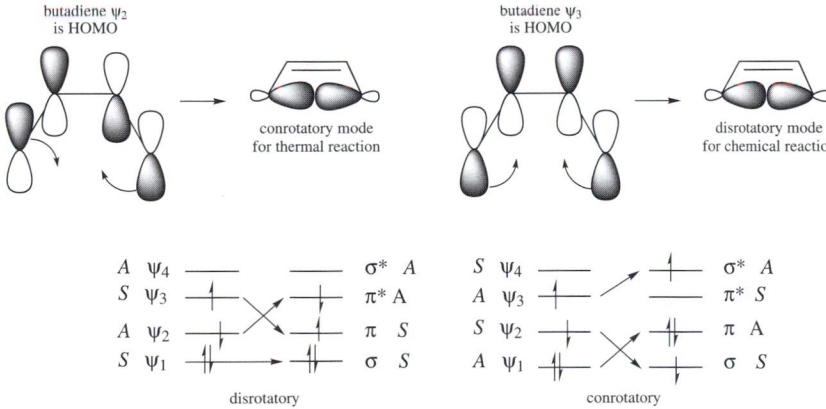

Fig. 13.4. Orbital correlation of energy states involved in the photochemical butadiene-to-cyclobutene conversion.

2. R. B. Woodward and R. Hoffmann, *J. Am. Chem. Soc.* **87**:395 (1965).
3. H. C. Longuet-Higgins and E. W. Abrahamson, *J. Am. Chem. Soc.* **87**:2045 (1965).
4. R. B. Woodward and R. Hoffmann, *The Conservation of Orbital Symmetry*, Academic Press, New York, 1970.

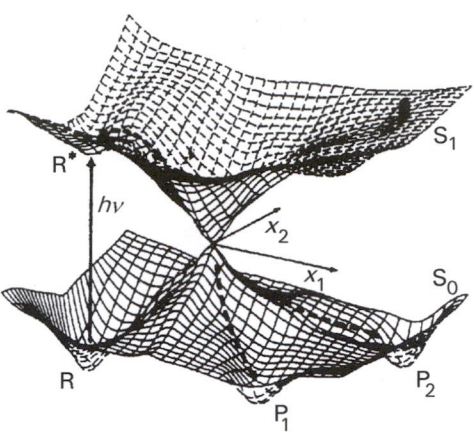

Fig. 13.5. Schematic representation of the potential energy surfaces of the ground state (S_0) and the excited state (S_1) of a nonadiabatic photoreaction of reactant R. Depending on the way the classical trajectories enter the conical intersection region, different ground-state valleys, which lead to products P_1 and P_2, can be reached. Reproduced from *Angew. Chem. Int. Ed. Engl.* **34**:549 (1995) by permission of Wiley-VCH.

In fact, it is a general result that the Woodward–Hoffmann rules predict that photochemical reactions will be complementary to thermal reactions. What is allowed photochemically is forbidden thermally, and vice versa. The physical basis for this complementary relationship is that the high barrier associated with forbidden thermal reactions provides a point for strong interaction of the ground-state and excited-state species. Because the two states are close in energy and of the same symmetry, they "mix" and allow the excited molecule to reach the ground state. This interaction is necessay for efficient photochemical reactions.[5] The point of intersection of an excited state with the ground state is called a *conical intersection*.[6] A diagram illustrating a conical intersection is shown in Fig. 13.5.

The interaction between excited states and the ground states of butadiene and cyclobutene is shown in Fig. 13.6. The energies are based on quantum-chemical calculations of the molecules in the geometries traversed during interconversion. The calculations find a minimum in the S_2 excited state.[7] They also show that with a small activation energy (calculated to be about 5 kcal/mol), the molecule in the S_1 state can reach points where the S_1 and S_2 energy surfaces cross. The excited molecules, whether generated from cyclobutene or butadiene, would be expected to follow the S_2 surface to the minimum located above the S_0 transition state. By loss of energy to the surrounding medium, the excited-state molecule can drop to the S_0 surface and then be transformed to butadiene or cyclobutene. This diagram also provides an explanation of how the excited state shown in Fig. 13.4, which has one singly occupied antisymmetric orbital, returns to

5. H. E. Zimmerman, *J. Am. Chem. Soc.* **88**:1566 (1966); W. Th. A. M. van der Lugt and L. J. Oosterhoff, *J. Chem. Soc. Chem. Commun.* **1968**:1235; W. Th. A. M. Van der Lugt and L. J. Oosterhoff, *J. Am. Chem. Soc.* **91**:6042 (1969); R. C. Dougherty, *J. Am. Chem. Soc.* **93**:7187 (1971); J. Michl, *Top. Curr. Chem.* **46**:1 (1974); J. Michl, *Photochem. Photobiol.* **25**:141 (1977).

6. M. Klessinger, *Angew. Chem. Int. Ed. Engl.* **34**:549 (1995).

7. D. Grimbert, G. Segal, and A. Devaquet, *J. Am. Chem. Soc.* **97**:6629 (1975).

751

SECTION 13.2.
ORBITAL SYMMETRY
CONSIDERATIONS
RELATED TO
PHOTOCHEMICAL
REACTIONS

Fig. 13.6. Energy diagram showing potential energy curves for interconversion of ground-state (S_0) and first and second singlet (S_1 and S_2) and first triplet (T_1) excited states. The angle ϕ is the C—C—C bond angle at C-2 and C-3. [From D. Grimbert, G. Segal, and A. Devaquet, *J. Am. Chem. Soc.* **97**:6629 (1975).]

ground-state cyclobutene, in which only symmetric orbitals are occupied. The S_2 state has the same symmetry as the ground state and can therefore decay directly to the ground state.

The experimental results on the photochemical cyclobutene-to-butadiene ring opening are not as straightforward as for the thermal reaction. For simple alkylcyclobutenes, the photolysis must be done in the vacuum ultraviolet ($\lambda < 200\,\text{nm}$) because of the high energy of the absorption maximum. Cyclobutene ring opening is accompanied by fragmentation to acetylene and ethylene and by rearrangement to methylenecyclopropene.[8] Both *cis*- and *trans*-3,4-dimethylcyclobutene give mixtures of all three isomers of 2,4-hexadiene on direct irradiation.[9]

8. W. Adam, T. Oppenländer, and G. Zans, *J. Am. Chem. Soc.* **107**:3921 (1985).
9. K. B. Clark and W. J. Leigh, *J. Am. Chem. Soc.* **109**:6086 (1987).

In contrast to the prediction that the disrotatory product should be formed, the conrotatory product is somewhat favored in each case. We will return to the mechanism of the butadiene–cyclobutene interconversion in Section 13.4.

Mention should be made of issues that can complicate the mechanistic interpretation of photochemical reactions:

1. *What is the geometry of the excited state?* In contrast to the well-defined information about ground-state geometries for most organic molecules, the excited-state geometries are often much less certain. Incorrect assumptions about the molecular geometry may lead to an erroneous basis for assignment of elements of molecular symmetry and orbital symmetry.

2. *Is the reaction concerted?* As was emphasized in Chapter 11, orbital symmetry considerations apply only to concerted reactions. The possible involvement of triplet excited states and, as a result, a nonconcerted process is much more common in photochemical reactions than in the thermal processes. A concerted mechanism must be established before the orbital symmetry rules can be applied.

Scheme 13.1. Some Examples of Photochemical Cycloaddition and Electrocyclic Reactions

a. O. L. Chapman and W. R. Adams, *J. Am. Chem. Soc.* **90**:2333 (1968).
b. A. G. Anastassiou, F. L. Setliff, and G. W. Griffin, *J. Org. Chem.* **31**: 2705 (1966).
c. A. Padwa, S. Goldstein, and M. Pulwer, *J. Org. Chem.* **47**:3893 (1982).
d. O. L. Chapman, D. L. Pasto, G. W. Borden, and A. A. Griswold, *J. Am. Chem. Soc.* **84**:1220 (1962).
e. G. J. Fonken, *Tetrahedron Lett.* **1962**:549.

Scheme 13.1 lists some example of photochemical cycloaddition and electrocyclic reactions of the type that are consistent with the predictions of orbital symmetry considerations. We will discuss other examples in Section 13.4.

13.3. Photochemistry of Carbonyl Compounds

The photochemistry of carbonyl compounds has been extensively studied, both in solution and in the gas phase. It is not surprising that there are major differences between the photochemical reactions in the two phases. In the gas phase, the energy transferred by excitation cannot be lost rapidly by collision, whereas in the liquid phase the excess energy is rapidly transferred to the solvent or to other components of the solution. Solution photochemistry will be emphasized here, since both mechanistic study and preparative applications of organic reactions usually involve solution processes.

The reactive excited state of saturated ketones usually is the $n-\pi^*$ state. On excitation, an electron from an oxygen nonbonding orbital is transferred to the π-antibonding orbital of the carbonyl group. The singlet excited state is initially formed, but intersystem crossing to the triplet can occur. For saturated ketones, the singlet lies about 80–85 kcal/mol above the ground state. The triplet state is about 75–80 kcal/mol above the ground state. The first excited singlet S_1 and triplet T_1 can be described structurally from spectroscopic data available for the simplest analog, formaldehyde. In both excited states, the molecule is pyramidal, the $C-O$ bond is lengthened, and the dipole moment is reduced from 2.34 D to 1.56 D.[10] The reduction of the dipole moment results from the transfer of electron density from an orbital localized on oxygen to one that also encompasses the carbon atom. An alternative excited state involves promotion of a bonding π electron to the antibonding π^* orbital. This is called the $\pi-\pi^*$ excited state and is most likely to be involved when the carbonyl group is conjugated with an extensive π-bonding system, as is the case for aromatic ketones.

It is not possible to draw unambiguous Lewis structures for excited states of the sort that are so useful in depicting ground-state chemistry. Instead, it is common to asterisk the normal carbonyl structure and provide information about the nature and multiplicity of the excited state:

$$[H_2C=O]^* \qquad {}^3[H_2C=O]^* \qquad {}^1[H_2C=O]^*$$

| excited state, multiplicity unspecified | excited-state triplet | excited-state singlet |

The excited carbonyl groups have radical character at both carbon and oxygen, and the oxygen is rather similar in its reactivity to alkoxyl radicals. The MO diagrams for the $n-\pi^*$ and $\pi-\pi^*$ states can be readily depicted:

π^*	π^*	π^*	π^*	π^*
n	n	n	n	n
π	π	π	π	π
ground state	$n \to \pi^*$ singlet	$n \to \pi^*$ triplet	$\pi \to \pi^*$ singlet	$\pi \to \pi^*$ triplet

10. J. C. D. Brand and D. G. Williamson, *Adv. Phys. Org. Chem.* **1**:365 (1963); D. E. Freeman and W. Klemperer, *J. Chem. Phys.* **45**:52 (1966).

For conjugated carbonyl compounds, such as α,β-enones, the orbital diagram would be similar, except for the recognition that the HOMO of the ground state is ψ_2 of the enone system, rather than the oxygen lone-pair orbital. The excited states can sometimes be usefully represented as dipolar or diradical intermediates:

Although Lewis structures of this type are not entirely adequate descriptions of the structure of the excited states, they do correspond to the MO picture by indicating distortion of charge and the presence of polar or radical-like centers. The excited states are much more reactive than the corresponding ground-state molecules. In addition to the increased energy content, this high reactivity is associated with the presence of half-filled orbitals. The two SOMO orbitals in the excited states have enhanced radical, cationic, or anionic character.

One of the most common reactions of photoexcited carbonyl groups is hydrogen-atom abstraction from solvent or some other hydrogen donor. A second common reaction is cleavage of the carbon–carbon bond adjacent to the carbonyl group:

The intermediates which are generated are free radicals. The hydrogen-atom abstraction can be either intramolecular or intermolecular. Many aromatic ketones react by hydrogen-atom abstraction, and the stable products are diols formed by coupling of the resulting α-hydroxybenzyl radicals:

These reactions usually occur via the triplet excited state T_1. The intersystem crossing of the initially formed singlet excited state is so fast ($k \approx 10^{10}$ s^{-1}) that reactions of the S_1 state are usually not observed. The reaction of benzophenone has been particularly closely studied. Some of the facts that have been established in support of the general mechanism outlined above are as follows:

1. For a given hydrogen donor S–H, replacement by S–D leads to a decreased rate of reduction, relative to nonproductive decay to the ground state.[11] This decreased rate is consistent with a primary isotope effect in the hydrogen abstraction step.
2. The photoreduction can be quenched by known triplet quenchers. The effective quenchers are those which have T_1 states less than 69 kcal/mol above S_0. Quenchers with higher triplet energies are ineffective because the benzophenone π–π^* triplet is then not sufficiently energetic to effect energy transfer.

---

11. W. M. Moore, G. S. Hammond, and R. P. Foss, *J. Am. Chem. Soc.* **83**:2789 (1961); G. S. Hammond, W. P. Baker, and W. M. Moore, *J. Am. Chem. Soc.* **83**:2795 (1961).

3. The intermediate diphenylhydroxymethyl radical has been detected after generation by flash photolysis.[12] Photolysis of benzophenone in benzene solution containing potential hydrogen donors results in the formation of two intermediates that are detectable, and their rates of decay have been measured. One intermediate is the $Ph_2\dot{C}OH$ radical. It disappears by combination with another radical in a second-order process. A much shorter-lived species disappears with first-order kinetics in the presence of excess amounts of various hydrogen donors. The pseudo-first-order rate constants vary with the structure of the donor; with 2,2-diphenylethanol, for example, $k = 2 \times 10^6 \, s^{-1}$. The rate is much less with poorer hydrogen-atom donors. The rapidly reacting intermediate is the triplet excited state of benzophenone.

4. In 2-propanol, the quantum yield for photolytic conversion of benzophenone to the coupled reduction product is 2.0.[13] The reason is that the radical remaining after abstraction of a hydrogen atom from 2-propanol transfers a hydrogen atom to ground-state benzophenone in a nonphotochemical reaction. Because of this process, two molecules of benzophenone are reduced for each one that is photoexcited:

$$Ph_2C=O \xrightarrow{h\nu} [Ph_2C=O]^*$$

$$[Ph_2C=O]^* + (CH_3)_2CHOH \longrightarrow Ph_2\dot{C}OH + (CH_3)_2\dot{C}OH$$

$$(CH_3)_2\dot{C}OH + Ph_2C=O \longrightarrow Ph_2\dot{C}OH + (CH_3)_2C=O$$

$$2\,Ph_2\dot{C}OH \longrightarrow \underset{\underset{HO}{|}\quad\underset{OH}{|}}{Ph_2C-CPh_2}$$

The efficiency of reduction of benzophenone derivatives is greatly diminished when an *ortho* alkyl substituent is present because a new photoreaction, intramolecular hydrogen-atom abstraction, then becomes the dominant process. The abstraction takes place from the benzylic position on the adjacent alkyl chain, giving an unstable enol that can revert to the original benzophenone without photoreduction. This process is known as *photoenolization*.[14] Photoenolization can be detected, even though no net transformation of the reactant occurs, by photolysis in deuterated hydroxylic solvents. The proton of the enolic hydroxyl is rapidly exchanged with solvent, so deuterium is introduced at the benzylic position. Deuterium is also introduced if the enol is protonated at the benzylic carbon by solvent:

12. J. A. Bell and H. Linschitz, *J. Am. Chem. Soc.* **85**:528 (1963).
13. N. J. Turro, *Molecular Photochemistry*, W. A. Benjamin, New York, 1965, pp. 143, 144.
14. P. G. Sammes, *Tetrahedron* **32**:405 (1976).

For some aromatic ketones the reactive dienols can also undergo electrocyclization to cyclobutenols.[15]

The dominant photochemical reaction of ketones in the gas phase is cleavage of one of the carbonyl substituents, which is followed by decarbonylation and subsequent reactions of the alkyl free radicals that result:

$$RCR' \xrightarrow{h\nu} R\dot{C}{=}O + \cdot R' \longrightarrow R\cdot + CO + \cdot R' \longrightarrow R{-}R'$$

This reaction is referred to as the *type-I* or *α-cleavage* reaction of carbonyl compounds. This type of reaction is not so common in solution, although some cyclic ketones do undergo decarbonylation:

Ref. 16

The facility with which α-cleavage occurs in solution depends on the stability of the radical fragments that can be ejected. Dibenzyl ketone, for example, is readily cleaved photolytically.[16] Similarly, *t*-butyl ketones undergo α-cleavage quite readily on photolysis in solution.[17]

$$(CH_3)_3CCCH_3 \xrightarrow{h\nu} (CH_3)_3CH + (CH_3)_2C{=}CH_2 + CH_3CH{=}O$$

Ref. 17

$$PhCH_2CCHPh_2 \xrightarrow{h\nu} PhCH_2CH_2Ph + PhCH_2CHPh_2 + Ph_2CHCHPh_2$$

Ref. 16

With cyclic ketones, the α-cleavage can also be followed by intramolecular hydrogen abstraction that leads eventually to an unsaturated ring-opened aldehyde[18]:

15. P. J. Wagner, D. Subrahmanyam and B.-S. Park, *J. Am. Chem. Soc.* **113**:709 (1991).
16. G. Quinkert, K. Opitz, W. W. Wiersdorff, and J. Weinlich, *Tetrahedron Lett.* **1963**:1863.
17. N. C. Yang and E. D. Feit, *J. Am. Chem. Soc.* **90**:504 (1968).
18. W. C. Agosta and W. L. Schreiber, *J. Am. Chem. Soc.* **93**:3947 (1971); P. J. Wagner and R. W. Spoerke, *J. Am. Chem. Soc.* **91**:4437 (1969).

An alternative reaction involves formation of a ketene.

The competition between these two reactions is determined by the effect of substituents on the conformation and reactivity of the diradical intermediate.[19]

In ketones having propyl or longer alkyl groups as a carbonyl substituent, intramolecular hydrogen abstraction can be followed by either cleavage of the bond between the α and β carbon atoms or by formation of a cyclobutanol:

Ketones such as 2,2,5,5-tetraphenylcyclopentanone and 2,2,6,6-tetraphenylcyclohexanone decarbonylate rapidly because of the stabilization afforded by the phenyl groups. The products result from recombination, disproportionation, or fragmentation of the diradical intermediate.[20]

19. P. J. Wagner, *Top. Curr. Chem.* **61**:1 (1976).
20. D. H. R. Barton, B. Charpiot, K. U. Ingold, L. J. Johnston, W. B. Motherwell, J. C. Scaiano, and S. Stanforth, *J. Am. Chem. Soc.* **107**:3607 (1985).

Cleavage between C_α and C_β is referred to as *type-II* photoelimination to distinguish it from α-cleavage. Type-II photoeliminations are observed for both aromatic and aliphatic ketones. Studies aimed at establishing the identity of the reactive excited state indicate that both S_1 and T_1 are involved for aliphatic ketones, but when one of the carbonyl substituents is aryl, intersystem crossing is very fast, and T_1 is the reactive state. Usually, cleavage is the dominant reaction, with cyclobutanol yields being rather low, but there are exceptions. The lifetime of the 1,4-diradical intermediate generated by intramolecular hydrogen abstraction is very short, probably not more than 10^{-7}–10^{-9} s.[21] The competition between these two processes can be broadly understood in terms of stability of the various intermediates.[22] Activation energies for the hydrogen abstraction process from a methylene group are about 4 kcal. The activation energy for carbon–carbon bond cleavage is 8–12 kcal. The nature of substituents on the aryl ring can affect the balance between the two competing reactions, but the overall reactivity pattern remains the same.[23]

Intramolecular hydrogen-atom abstraction is also an important process for acyclic α,β-unsaturated ketones.[24] The intermediate diradical then cyclizes to give the enol of a cyclobutyl ketone. Among the by-products of such photolyses are cyclobutanols resulting from alternative modes of cyclization of the diradical intermediate:

21. J. E. Rudzki, J. L. Goodman, and K. S. Peters., *J. Am. Chem. Soc.* **107**:7849 (1985); R. A. Caldwell, T. Majima and C. Pac, *J. Am. Chem. Soc.* **104**:629 (1982).
22. P. J. Wagner, *Acc. Chem. Res.* **4**:168 (1971).
23. M. V. Encina, E. A. Lissi, E. Lemp, A. Zanocco, and J. C. Scaiano, *J. Am. Chem. Soc.* **105**:1856 (1983).
24. R. A. Cormier, W. L. Schreiber, and W. C. Agosta, *J. Am. Chem. Soc.* **95**:4873 (1973); R. A. Cormier and W. C. Agosta, *J. Am. Chem. Soc.* **96**:618 (1974).

α,β-Unsaturated ketones with γ hydrogens can undergo hydrogen-atom transfer, resulting in formation of a dienol. Because the hydrogen-atom transfer occurs through a cyclic transition state, the originally formed product has Z-stereochemistry:

The dienol is unstable, and two separate processes have been identified for ketonization. These are a 1,5-sigmatropic shift of hydrogen leading back to the enone and a base-catalyzed proton transfer which leads to the β,γ-enone.[25] The deconjugated enone is formed because of the kinetic preference for reprotonation of the dienolate at the α carbon. Photochemical deconjugation is a synthetically useful way of effecting isomerization of α,β-unsaturated ketones and esters to the β,γ-isomers.

Cyclic α,β-unsaturated ketones present a rich array of photochemical reactions, some of which are of considerable synthetic value (see Section 6.4 of Part B). For cyclohexenones, two prominent reactions are the di-π-methane rearrangement (path A) and the lumiketone rearrangement (path B).

Both reactions proceed via triplet excited species and, to some extent, are controlled by whether the π–π^* (path A) or n–π^* states are involved. The di-π-methane rearrangement pathway is restricted to 4-aryl- or 4-vinylcyclohexenones. At the most basic level of

25. R. Ricard, P. Sauvage, C. S. K. Wan, A. C. Weedon, and D. F. Wong, *J. Org. Chem.* **51**:62 (1986).

mechanism, the reactions can be depicted as involving the following steps.

Path A

Path B

4,4-Dialkylcyclohexenones undergo the lumiketone rearrangement which involves the shift of the C(4)–C(5) bond to C-3 and formation of a new C(2)–C(4) bond[26]:

The reaction is stereospecific and can be described as a $[\pi 2_a + \sigma 2_a]$ cycloaddition. This mechanism requires that inversion of configuration occur at C-4 as the new σ bond is formed at the back lobe of the reacting C(4)–C(5) σ bond:

It has been demonstrated in several systems that the reaction is, in fact, stereospecific, with the expected inversion occurring at C-4. The ketone **1** provides a specific example. The

26. For a review of this reaction, see D. I. Schuster, in *Rearrangements in Ground and Excited States*, Vol. 3, P. de Mayo, ed., Academic Press, New York, 1980, Chapter 17.

stereoisomeric products **2** and **3** are both formed, but in each product inversion has occurred at C-4:

Ref. 27

This reaction is quite general and also proceeds in the case of some 4-alkyl-4-arylcyclo-hexenones:

Ref. 28

With 4,4-diarylcyclohexenones, the di-π-methane rearrangement occcurs. In compounds in which the two aryl groups are substituted differently, it is found that substituents which stabilize radical character favor migration. Thus, the *p*-cyanophenyl substituent migrates in preference to the phenyl substituent in **4**:

Ref. 29

This rearrangement can be considered to occur via a transition state in which C(2)−C(4) bridging is accompanied by a $4 \rightarrow 3$ aryl migration[30]:

27. D. I. Schuster and J. M. Rao, *J. Org. Chem.* **46**:1515 (1981); D. I. Schuster, R. H. Brown, and B. M. Resnick, *J. Am. Chem. Soc.* **100**:4504 (1978).

28. O. L. Chapman, J. B. Sieja, and W. J. Welstead, Jr., *J. Am. Chem. Soc.* **88**:161 (1966).

29. H. E. Zimmerman, R. D. Rieke, and J. R. Scheffer, *J. Am. Chem. Soc.* **89**:2033 (1967).

30. H. E. Zimmerman, *Tetrahedron* **30**:1617 (1974).

Note that the *endo* product is predicted by the concerted mechanism. It is the major product, even though it is sterically more congested than the *exo* isomer. This stereospecificity is characteristic of the reaction.

With other ring sizes, the photochemistry of cyclic enones may take different courses. For cyclopentenones, the principal products result from hydrogen abstraction processes. Irradiation of cyclopentenone in cyclohexane gives a mixture of 2- and 3-cyclohexylcyclopentanone.[31] These products can be formed by intermolecular hydrogen abstraction, followed by recombination of the resulting radicals:

If a substituent chain is present on the cyclopentenone ring, an intramolecular hydrogen abstraction can take place:

The bicyclic product is formed by coupling of the two radical sites, while the alkene results from an intramolecular hydrogen-atom transfer. These reactions can be sensitized by aromatic ketones and quenched by typical triplet quenchers and are therefore believed to proceed via triplet excited states.

In the case of cycloheptenone and larger rings, the main initial photoproducts are the *E*-cycloalkenones produced by photoisomerization. In the case of the seven- and eight-membered rings, the double bonds are sufficiently strained that rapid reactions follow. In nonnucleophilic solvents dimerization occurs, whereas in nucleophilic solvents addition occurs.[32]

31. S. Wolff, W. L. Schreiber, A. B. Smith, III, and W. C. Agosta, *J. Am. Chem. Soc.* **94**:7797 (1972).
32. H. Hart, B. Chen, and M. Jeffares, *J. Org. Chem.* **44**:2722 (1979).

Ketones in which the double bond is located in the β,γ position are likely candidates for α-cleavage because of the stability of the allyl radical that is formed. This is an important process on direct irradiation. Products then arise by recombination of the radicals or by recombination after decarbonylation.

$$RCCH_2CH{=}CHR' \xrightarrow{h\nu} R\dot{C}{=}O + \cdot CH_2CH{=}CHR' \longrightarrow \text{products}$$

In cyclic ketones, the diradical intermediates can recombine, leading to isomerized ketones:

Ref. 33

Ref. 34

Excitation of acyclic β,γ-unsaturated ketones by triplet photosensitization often gives cyclopropyl ketones.[35]

This reaction is known as the oxa-π-methane rearrangement.

Another class of carbonyl compounds that has received much attention is the cyclohexadienones.[36] The primary photolysis product of 4,4-diphenylcyclohexadienone, for example, is **5**.[37] Quenching and photosensitization experiments have indicated that the

33. H. Sato, N. Furutachi, and K. Nakanishi, *J. Am. Chem. Soc.* **94**:2150 (1972); L. A. Paquette, R. F. Eizember, and O. Cox, *J. Am. Chem. Soc.* **90**:5153 (1968).
34. P. S. Engel and M. A. Schexnayder, *J. Am. Chem. Soc.* **97**:145 (1975).
35. W. G. Dauben, M. S. Kellog, J. I. Seeman, and W. A. Spitzer, *J. Am. Chem. Soc.* **92**:1786 (1970).
36. H. E. Zimmerman, *Angew. Chem. Int. Ed. Engl.* **8**:1 (1969); K. Schaffner and M. Demuth, in *Rearrangements in Ground and Excited States*, Vol. 3, P. de Mayo, ed., Academic Press, New York, 1980, Chapter 18; D. I. Schuster, *Acc. Chem. Res.* **11**:65 (1978).
37. H. E. Zimmerman and D. I. Schuster, *J. Am. Chem. Soc.* **83**:4486 (1961).

reaction proceeds through a triplet excited state. A scheme that delineates the bonding changes is outlined below:

It is believed that a reactive ground-state species, the zwitterion **A**, is an intermediate and that it rearranges to the observed product.[38] To test this mechanism, generation of species **A** by nonphotochemical means was undertaken.[39] α-Haloketones, when treated with strong base, ionize to such dipolar intermediates. Thus, the bromoketone **6** is a potential precursor of intermediate **A**:

The zwitterion prepared by this route did indeed lead to **5**, as required if it is an intermediate in the photochemical reaction. Further study of this process established another aspect of the reaction mechanism. The product could be formed by a process involving inversion at C-4 or by one involving a pivot about the C(3)–C(4) bond:

The two mechanisms require the formation of stereochemically different products when the aryl groups at C-4 are different. When the experiment was carried out with **7**, only the product corresponding to inversion of configuration at C-4 was observed.[40]

The rearrangement step itself is a ground-state thermal process and may be classified as a [1,4] sigmatropic shift of carbon across the face of a 2-oxybutenyl cation. The

38. H. E. Zimmerman and J. S. Swenton, *J. Am. Chem. Soc.* **89**:906 (1967).
39. H. E. Zimmerman, D. Döpp, and P. S. Huyffer, *J. Am. Chem. Soc.* **88**:5352 (1966).
40. H. E. Zimmerman and D. S. Crumrine, *J. Am. Chem. Soc.* **90**:5612 (1968).

Woodward–Hoffmann rules require a sigmatropic shift of this type to proceed with inversion of configuration. The orbitals involved in a [1,4] sigmatropic shift are shown in Fig. 13.7.

As is clear from the preceding examples, there are a variety of overall reactions that can be initiated by photolysis of ketones. The course of photochemical reactions of ketones is very dependent on the structure of the reactant. Despite the variety of overall processes that can be observed, the number of individual steps involved is limited. For ketones, the most important are inter- and intramolecular hydrogen abstraction, cleavage α to the carbonyl group, and substituent migration to the β-carbon atom of α,β-unsaturated ketones. Reexamination of the mechanisms illustrated in this section will reveal that most of the reactions of carbonyl compounds that have been described involve combinations of these fundamental processes. The final products usually result from rebonding of reactive intermediates generated by these steps.

Some ketones undergo a cycloaddition reaction with alkenes to form oxetanes:

$$Ph_2C{=}O \ + \ CH_3CH{=}CHCH_3 \ \xrightarrow{h\nu} \ \text{(minor)} \ + \ \text{(major)}$$

(E or Z)

The reaction is ordinarily stereoselective, favoring the more stable adduct for either alkene isomer, and a long-lived triplet diradical intermediate is implicated.[41]

$$^3[Ph_2C{=}O]^* \ + \ CH_3CH{=}CHCH_3 \ \longrightarrow \ Ph_2\overset{\cdot}{C}OCH\overset{\cdot}{C}HCH_3 \ \longrightarrow \ \text{products}$$

The diradical is believed to be preceded on the reaction path by a complex of the alkene with excited-state ketone. This reaction, particularly its stereochemistry and regioselectivity,[42] will be considered in more detail in Chapter 6 of Part B.

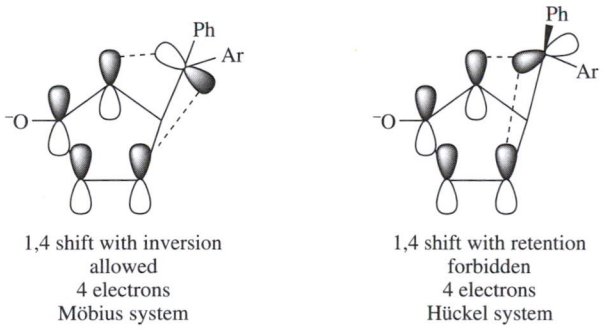

1,4 shift with inversion
allowed
4 electrons
Möbius system

1,4 shift with retention
forbidden
4 electrons
Hückel system

Fig. 13.7. Symmetry properties for [1,4] sigmatropic shifts with inversion and retention.

41. R. A. Caldwell, G. W. Sovocool, and R. P. Gajewski, *J. Am. Chem. Soc.* **95**:2549 (1973).
42. T. Bach, *Synthesis*, **1998**:683.

13.4. Photochemistry of Alkenes and Dienes

The photochemistry of alkenes and dienes has already been mentioned in connection with the principles of orbital symmetry control in electrocyclic and cycloaddition processes in Section 13.2. Cycloadditions are considered, from a synthetic viewpoint, in Chapter 6 of Part B. This section will emphasize unimolecular photoreactions of alkenes and dienes.

A characteristic photochemical reaction of alkenes is interconversion of *cis* and *trans* isomers. Usually, the *trans* isomer is thermodynamically more stable, and photolysis establishes a mixture that is richer in the *cis* isomer. Irradiation therefore provides a means of converting a *trans* alkene to the *cis* isomer. After a certain extent of irradiation the *cis-trans* ratio no longer changes. This is called the photostationary state. The composition of the photostationary state depends on the absorption spectra of the isomeric alkenes. A hypothetical case is illustrated in Fig. 13.8. Assume that the vertical line at 265 nm is the lower-wavelength limit of the light impinging on the system. This wavelength can be controlled by use of appropriate sources and filters. Because of its longer wavelength maximum and higher extinction coefficient, the *trans* isomer will be absorbing substantially more light than the *cis* isomer. The relative amounts of light absorbed by the *cis* and *trans* isomers at any wavelength will be proportional to their extinction coefficients, ε_c and ε_t, at that wavelength. Assuming that the quantum yield for conversion of *cis* $\rightarrow$ *trans* is approximately equal to that for *trans* $\rightarrow$ *cis*, the conversion of *trans* alkene to *cis* will occur faster than the converse process when the two isomers are in equal concentration. On continued photolysis, a photostationary state will be achieved at which the rate of *trans* $\rightarrow$ *cis* conversion is equal to that of *cis* $\rightarrow$ *trans* conversion. At that point, the concentration of the *cis* isomer will be greater than that of the *trans* isomer. The relationship can be expressed quantitatively for monochromatic light as

$$\frac{[t]_s}{[c]_s} = \left(\frac{\varepsilon_c}{\varepsilon_t}\right)\left(\frac{\phi_{c \rightarrow t}}{\phi_{t \rightarrow c}}\right)$$

The isomerization of alkenes is believed to take place via an excited state in which the two sp^2 carbons are twisted 90° with respect to one another. This state is referred to as the *p* (perpendicular) state. This geometry is believed to be the minimum-energy geometry for both the singlet and triplet excited states.

The perpendicular geometry for the excited state permits the possibility for returning to either the *cis* or the *trans* configuration of the ground state.

Especially detailed study of the mechanism of photochemical configurational isomerism has been done on *Z*- and *E*-stilbene.[43] Spectroscopic data have established

43. J. Saltiel, J. T. D'Agostino, E. D. Megarity, L. Metts, K. R. Neuberger, M. Wrighton, and O. C. Zafiriow, *Org. Photochem.* **3**:1 (1973); J. Saltiel and J. L. Charlton, in *Rearrangements in Ground and Excited States*, Vol. 3, P. de Mayo, ed., Academic Press, New York, 1980, Chapter 14; D. H. Waldeck, *Chem. Rev.* **91**:415 (1991); U. Mazzucato, G. G. Aloisi, G. Bartocci, F. Elisei, G. Galiazzo, and A. Spalletti, *Med. Biol. Environ.* **23**:69 (1995).

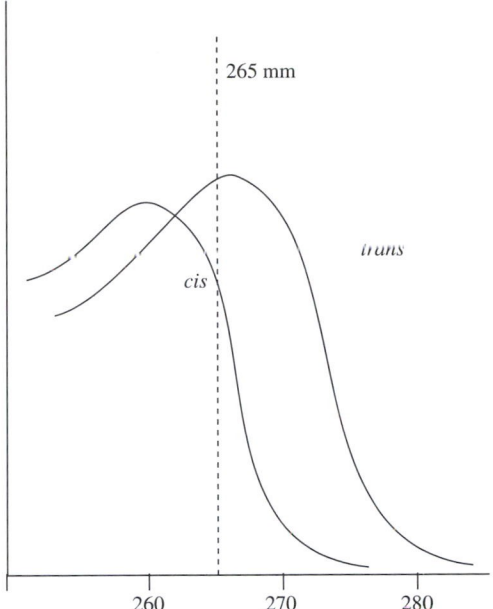

265 mm

cis

trans

260   270   280

Fig. 13.8. Absorption spectra of a *cis–trans* isomer pair.

the energies of the singlet and triplet states of both *Z*- and *E*-stilbene and of the twisted excited states that are formed from both isomers. The excited states derived from both *Z*- and *E*-stilbene can readily attain the *p* states. This information is summarized in Fig. 13.9. Direct irradiation leads to isomerization via singlet-state intermediates.[44] The isomerization presumably involves a twisted singlet state that can be achieved from either the *Z* or the *E* isomer. The temperature dependence of the isomerization further reveals that the process of formation of the twisted state involves a small activation energy. This energy is required for conversion of the initial excited state to the perpendicular geometry associated with the ^1S state. The energy barrier is greater from the *E* than from the *Z* isomer. Among the pieces of evidence indicating that a triplet intermediate is not involved in direct irradiation is the fact that azulene, which is known to intercept stilbene triplets, has only a minor effect on the efficiency of the direct photoisomerization.[45]

As shown in Fig. 13.10, the absorption spectra of *Z*- and *E*-stilbene differ substantially, with the *E* isomer absorbing more strongly at $\lambda > 280$ nm. Thus, the composition of the photostationary state can be controlled by altering the wavelength of the irradiation, as indicated by the equation on p. 766.[46]

The photoisomerization can also be carried out by photosensitization. Under these conditions, the composition of the photostationary state depends on the triplet energy of the sensitizer. With sensitizers having triplet energies above 60 kcal/mol, [*cis*]/[*trans*] is slightly more than 1, but a range of sensitizers having triplet energies of 52–58 kcal/mol

44. J. Saltiel, *J. Am. Chem. Soc.* **89**:1036 (1967); *J. Am. Chem. Soc.* **90**:6394 (1968).
45. J. Saltiel, E. D. Megarity, and K. G. Kneipp, *J. Am. Chem. Soc.* **88**:2336 (1966); J. Saltiel and E. D. Megarity, *J. Am. Chem. Soc.* **91**:1265 (1969).
46. J. Saltiel, A. Marinari, D. W.-L. Chang, J. C. Mitchener, and E. D. Megarity, *J. Am. Chem. Soc.* **101**:2982 (1979).

Fig. 13.9. Schematic reaction profile for the *EZ* isomerization of stilbene. The reaction coordinate θ is the torsion bond about the double bond. [From H. Meier, *Angew. Chem. Int. Ed. Engl.* **31**:1399 (1992)]. Reproduced by permission of Wiley-VCH.

afford much higher $Z:E$ ratios in the photostationary state.[47] The high $Z:E$ ratio in this region results from the fact that the energy required for excitation of *E*-stilbene is less than for excitation of *Z*-stilbene (see Fig. 13.9). Thus, sensitizers having triplet energies in the range of 52–58 kcal/mol selectively excite the *E* isomer. Since the rate of *trans* → *cis* conversion is increased, the composition of the photostationary state is then enriched in *cis* isomer.

Z-Stilbene also undergoes photocyclization to 4a,4b-dihydrophenanthrene via an electrocyclization.[48]

The cyclization product is thermally unstable relative to *Z*-stilbene and reverts to starting material unless trapped by an oxidizing agent.[49] The extent of cyclization is solvent-dependent, with nonpolar solvents favoring cyclization more than polar ones.[50] Whereas the quantum yield for *Z–E* isomerization is nearly constant at about 35%, the cyclization

47. G. S. Hammond, J. Saltiel, A. A. Lamola, N. J. Turro, J. S. Bradshaw, D. O. Cowan, R. C. Counsell, V. Vogt, and C. Dalton, *J. Am. Chem. Soc.* **86**:3197 (1964); S. Yamauchi and T. Azumi, *J. Am. Chem. Soc.* **95**:2709 (1973).
48. T. Wismonski-Knittel, G. Fischer, and E. Fischer, *J. Chem. Soc., Perkin Trans. 2* **1974**:1930.
49. L. Liu, B. Yang, T. J. Katz, and M. K. Poindexter, *J. Org. Chem.* **56**:3769 (1991).
50. J.-M. Rodier and A. B. Myers, *J. Am. Chem. Soc.* **115**:10791 (1993).

Fig. 13.10. UV spectra of *E*-and *Z*-stilbene in hexane at room temperature. [From H. Meier *Angew. Chem. Int. Ed. Engl.* **31**:1399 (1992). Reproduced by permission of Wiley-VCH.

quantum yield is in the range of 15–18% in hydrocarbons, as compared with 5–8% in acetonitrile or methanol. The detailed interpretation of the system involves the effect of solvent on the lifetime and dynamics of the excited states.[51] Figure 13.11 incorporates the reaction coordinate for cyclization into the *Z–E* energy diagram.

Direct photochemical excitation of unconjugated alkenes requires light with $\lambda \leq 230$ nm. There have been relatively few studies of direct photolysis of alkenes in solution because of the experimental difficulties imposed by this wavelength restriction. A study of *Z*- and *E*-2-butene diluted with neopentane demonstrated that *Z–E* isomerization was competitive with the photochemically allowed $[2\pi + 2\pi]$ cycloaddition that occurs in pure liquid alkene.[52] The cycloaddition reaction is completely stereospecific for each isomer, which requires that the *excited intermediates involved in cycloaddition must retain a geometry which is characteristic of the reactant isomer.* As the ratio of neopentane to butene is increased, the amount of cycloaddition decreases relative to that of *Z* : *E* isomerization. This effect presumably is the result of the very short lifetime of the intermediate responsible for cycloaddition. When the alkene is diluted by inert hydrocarbon, the rate of encounter with a second alkene molecule is reduced, and the unimolecular isomerization becomes the dominant reaction.

Aromatic compounds such as toluene, xylene, and phenol can photosensitize *cis–trans* interconversion of simple alkenes. This is a case in which the sensitization process must be somewhat endothermic because of the energy relationships between the excited states of the alkene and the sensitizers. The photostationary state obtained under these conditions favors the less strained of the alkene isomers. The explanation for this effect can be summarized with reference to Fig. 13.12. Isomerization takes place through a twisted triplet state. This state is achieved by a combination of energy transfer from the sensitizer and thermal activation. Because the *Z* isomer is somewhat higher in energy, its requirement for activation to the excited state is somewhat less than for the *E* isomer. If it is also assumed that the excited state forms the *Z*- and *E*-isomers with equal ease, the rate of

51. R. J. Sension, S. T. Repinec, A. Z. Szarka, and R. M. Hochstrasser, *J. Chem. Phys.* **98**:6291 (1993).
52. H. Yamazaki and R. J. Cventanovic, *J. Am. Chem. Soc.* **91**:520 (1969); H. Yamazaki, R. J. Cventanovic, and R. S. Irwin, *J. Am. Chem. Soc.* **98**:2198 (1976).

Fig. 13.11. A schematic drawing of the potential energy surfaces for the S_1 photochemical reactions of stilbene. Approximate branching ratios and quantum yields for the important processes are indicated. In this figure, the ground- and excited-state barrier heights are drawn to scale representing the best available values, as are the relative energies of the ground states of Z- and E-stilbene 4a,4b-dihydrophenanthrene (DHP). [Reproduced from R. J. Sension, S. T. Repinec, A. Z. Szarka, and R. M. Hochstrasser, *J. Chem. Phys.* **98**:6291 (1993) by permission of the American Institute of Physics.]

$Z \to E$ conversion exceeds that for $E \to Z$ conversion ($k_{Z \to E} > k_{Z \to E}$) and, at the photostationary state, and therefore [*trans*] > [*cis*].[53]

The reaction course taken by photoexcited cycloalkenes in hydroxylic solvents depends on ring size. 1-Methylcyclohexene, 1-methylcycloheptene, and 1-methylcyclooctene all add methanol, but neither 1-methylcyclopentene nor norbornene does so. The key intermediate in the addition reactions is believed to be the highly reactive E-isomer of the cycloalkene.

It appears that the E-cycloalkenes can be protonated exceptionally easily, because of the enormous relief of strain that accompanies protonation[54,55]. The E-isomers of cyclopen-

53. J. J. Snyder, F. P. Tise, R. D. Davis, and P. J. Kropp, *J. Org. Chem.* **46**:3609 (1981).
54. P. J. Kropp, E. J. Reardon, Jr., Z. L. F. Gaibel, K. F. Williard, and J. H. Hattaway, Jr., *J. Am. Chem. Soc.* **95**:7058 (1973).
55. J. A. Marshall, *Acc. Chem. Res.* **2**:33 (1969).

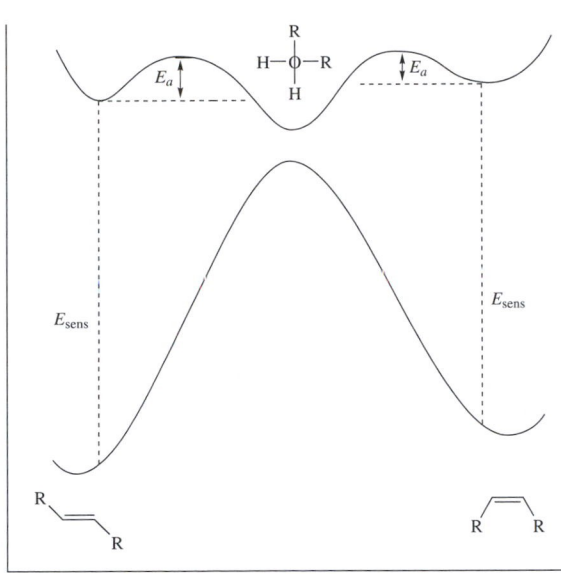

Fig. 13.12. Energy diagram illustrating differential in energy deficit for photosensitized isomerization of *cis* and *trans* isomers.

tene and norbornene are too strained to be formed. Cyclopentene and norbornene instead give products that result from hydrogen abstraction processes.[56] The reactivity of the excited state is that of a diradical species:

As was mentioned in Section 13.2, the $[2\pi + 2\pi]$ photocycloaddition of alkenes is an allowed reaction according to orbital symmetry considerations. Among the most useful reactions in this category, from a synthetic point of view, are intramolecular $[2\pi + 2\pi]$ cycloadditions of dienes and intermolecular $[2\pi + 2\pi]$ cycloadditions of alkenes with cyclic α,β-unsaturated carbonyl compounds. These reactions will be discussed in more detail in Section 6.4 of Part B.

The case of butadiene–cyclobutene interconversion, which one might expect to provide a simple case illustrating photochemical electrocyclization, is actually quite complex, especially when substituted systems are involved. Interpretation of the resonance Raman spectra of cyclobutene provides some insight into the dynamics of ring opening and is consistent with a disrotatory ring opening.[57] Theoretical analysis of the disrotatory and conrotatory processes predicts that the disrotatory process would be much more favorable, in accord with the Woodward–Hoffmann rules.[58] The experimental results for a number of cyclobutene photolyses reveal a more complicated picture. Cyclobutene gives ethylene, acetylene, and methylenecyclopropane, in addition to butadiene.[59] Furthermore,

56. P. J. Kropp, *J. Am. Chem. Soc.* **91:**5783 (1969).
57. M. K. Lawless, S. D. Wickham, and R. A. Mathies, *J. Am. Chem. Soc.* **116:**1593 (1994).
58. K. Morihashi and O. Kikuchi, *Theor. Chim. Acta* **67:**293 (1985).
59. W. Adam, T. Oppenländer, and G. Zang, *J. Am. Chem. Soc.* **107:**3921 (1985).

the ring opening of *cis-* and *trans*-3,4-dimethylcycylobutene is not stereospecific.[60] A number of other nonstereospecific photolytic ring openings of substituted cyclobutenes have also been reported.[61] These results point to a nonconcerted mechanism for the ring opening and several descriptions of the mechanism have been provided.[62] One possibility is that the ring opening proceeds stereospecifically to an *excited state* of the diene which then decays to the ground state with a stereoselectivity that is independent of the stereochemistry of the original cyclobutene.[63]

The reverse reaction, closure of butadiene to cyclobutene, has also been explored computationally, using CAS-SCF calculations. The distrotatory pathway is found to be favored, although the interpretation is somewhat more complex than the simplest Woodward–Hoffmann formulation. It is found that as disrotatory motion occurs, the singly excited state crosses the doubly excited state, which eventually leads to the ground state via a conical intersection.[64] A conrotatory pathway also exists, but it requires an activation energy.

Conjugated dienes can undergo a variety of photoreactions, depending on whether excitation is direct or photosensitized. The benzophenone-sensitized excitation of 1,3-pentadiene, for example, results in stereochemical isomerization and dimerization[65]:

Alkyl derivatives of 1,3-butadiene usually undergo photosensitized *Z–E* isomerism when photosensitizers that can supply at least 60 kcal/mol are used. Two conformers of the dienes, the *s-cis* and *s-trans*, exist in equilibrium, so there are two nonidentical ground states from which excitation can occur. Two triplet excited states that do not readily interconvert are derived from the *s-trans* and *s-cis* conformers. Theoretical calculations suggest that at their energy minimum the excited states of conjugated dienes can be described as an alkyl radical and an orthogonal allyl system called an allylmethylene diradical:

60. K. B. Clark and W. J. Leigh, *J. Am. Chem. Soc.* **109**:6086 (1987).

61. W. G. Dauben and J. E. Haubrich, *J. Org. Chem.* **53**:600 (1988); W. J. Leigh, K. Zheng, and K. B. Clark, *Can. J. Chem.* **68**:1988 (1990); W. J. Leigh and K. Zheng, *J. Am. Chem. Soc.* **113**:2163 (1991); G. Maier and A. Bothur, *Eur. J. Org. Chem.* **1998**:2063.

62. W. J. Leigh, *Can. J. Chem.* **71**:147 (1993); W. J. Leigh, K. Zheng, and K. B. Clark, *J. Org. Chem.* **56**:1574 (1991); W. J. Leigh, K. Zheng, N. Nguyen, N. H. Werstiuk, and J. Ma, *J. Am. Chem. Soc.* **113**:4993 (1991); F. Bernardi, M. Olivucci, and M. A. Robb, *Acc. Chem. Res.* **23**:405 (1990); F. Bernardi, M. Olivucci, I. N. Ragazos, and M. A. Robb, *J. Am. Chem. Soc.* **114**:2752 (1992).

63. W. J. Leigh, J. A. Postigo, and K. C. Zheng, *Can. J. Chem.* **74**:951 (1996).

64. P. Celani, F. Bernardi, M. Olivucci, and M. A. Robb, *J. Chem. Phys.* **102**:5733 (1995); F. Bernardi, M. Olivucci, and M. A. Robb, *J. Photochem. Photobiol.* **105**:365 (1997).

65. J. Saltiel, D. E. Townsend, and A. Sykes, *J. Am. Chem. Soc.* **95**:5968 (1973).

Such a structure implies that there would be a barrier to rotation about the C(2)–C(3) bond and would explain why the *s-trans* and *s-cis* conformers lead to different excited states.[66]

Another result that can be explained in terms of the two noninterconverting excited states is the dependence of the ratio of [2 + 2] and [2 + 4] addition products on sensitizer energy. The *s-cis* geometry is suitable for cyclohexene formation, but the *s-trans* is not. The excitation energy for the *s-cis* state is slightly lower than that for the *s-E*. With low-energy sensitizers, therefore, the *s-cis* excited state is formed preferentially, and the ratio of cyclohexene to cyclobutane product increases.[67]

Sensitizer	E_T	Butadiene dimers		α-Acetoxy acrylonitrile adducts	
		2 + 2	2 + 4	2 + 2	2 + 4
Acetophenone	73.6	97	3	98	2
Benzophenone	68.5	96	4	98	2
Triphenylene	66.6			97	3
Anthraquinone	62.4			95	5
Biacetyl	54.9	71	29	76	24
Benzil	53.7	66	34	74	26
Pyrene	48.7			69	31
Anthracene	42.5			87	13

The structure of the excited state of 1,3-dienes is also significant with respect to *Z–E* isomerization. If the excited state is an allylmethylene diradical, only one of the two double bonds would be isomerized in any single excitation event:

On the other hand, if the two possible allylmethylene diradicals interconvert rapidly, isomerization could take place at both double bonds. In this case, excitation could lead to isomerization at both double bonds. It is this latter situation that apparently exists in the

66. R. Hoffmann, *Tetrahedron* **22**:521 (1966); N. C. Baird and R. M. West, *J. Am. Chem. Soc.* **93**:4427 (1971).
67. R. S. Liu, N. J. Turro, Jr., and G. S. Hammond, *J. Am. Chem. Soc.* **87**:3406 (1965); W. L. Dilling, R. D. Kroening, and J. C. Little, *J. Am. Chem. Soc.* **92**:928 (1970).

triplet state. The triplet state has a high bond order between C-2 and C-3 and resists rotation about this bond, but the barrier to rotation at either of the terminal carbons is low.[53,68] Both double bonds can isomerize through this excited state. In contrast, it has been shown that in direct irradiation of 2,4-hexadiene, only one of the double bonds isomerizes on excitation.[69] The singlet state apparently retains a substantial barrier to rotation about the bonds in the allyl system:

An alternative description of the singlet excited state is a cyclopropylmethyl singlet diradical. Only one of the terminal carbons would be free to rotate in such a structure.

Orbital symmetry control of subsequent ring opening could account for isomerization at only one of the double bonds. Taking ψ_3 as the controlling frontier orbital, it can be seen that a concerted return to ψ_2 leads to rotation at only one terminus of the diene:

On direct irradiation of 1,3-pentadiene, Z–E isomerization is accompanied by cyclization to 1,3-dimethylcyclopropene and 3-methylcyclobutene:

The latter product is an example of a product formed by a concerted, photochemically allowed, electrocyclic reaction. A hydrogen-atom migration from a cyclopropyldimethyl

68. J. Saltiel, A. D. Rousseau, and A. Sykes, *J. Am. Chem. Soc.* **94**:5903 (1972).
69. J. Saltiel, L. Metts, and M. Wrighton, *J. Am. Chem. Soc.* **92**:3227 (1970).

diradical intermediate can account for the cyclopropene formation.[70] This product, then, is suggestive of a ring structure in the excited state:

The cyclohexadiene–hexatriene system seems to be less complicated than the cyclobutene–butadiene system. Cyclohexadiene undergoes photochemical electrocyclic ring opening:

The reaction occurs within 10^{-8} s of excitation. It is believed that the first step after excitation is attainment of a planar geometry, which occurs within 10^{-11} s. The planar geometry optimizes the orientation of the orbitals for the pericyclic ring opening.[71]

Ring closure of more highly substituted cyclohexadienes also follows the Woodward–Hoffmann rules and, indeed, provided the initial examples of the dichotomy between thermal and photochemical processes that led to development of the concepts underlying the Woodward–Hoffmann rules.[72]

5,5-Diphenylcyclohexadiene shows divergent photochemical behavior, depending on whether the reaction is induced by direct irradiation or by photosensitization. On direct irradiation, the electrocyclic conversion to 1,1-diphenylhexatriene is dominant, whereas a

70. S. Boué and R. Srinivasan, *J. Am. Chem. Soc.* **92**:3226 (1970).
71. P. J. Reid, M. K. Lawless, S. D. Wickham, and R. A. Mathies, *J. Phys. Chem.* **98**:5597 (1994); W. Fuss, T. Höfer, P. Hering, K. L. Kompa, S. Lochbrunner, T. Schikarski, and W. E. Schmid, *J. Phys. Chem.* **100**:921 (1996).
72. R. B. Woodward, in *Aromaticity*, Chemical Society Special Publication No. 21, Chemical Society, London, 1967, p. 217; E. Havinga and J. L. M. A. Schlatmann, *Tetrahedron* **16**:146 (1961); J. A. Berson, *Tetrahedron* **48**:3 (1992).

rearrangement involving one of the aromatic rings is the major reaction of the triplet excited state formed by photosensitization[73]:

The latter reaction is an example of the *di-π-methane rearrangement.*[74] This rearrangement is a very general reaction for 1,4-dienes and other systems that have two π systems separated by an sp^3-hybridized carbon atom:

The transformation can be formulated in terms of bonding between C-2 and C-4 involving a cyclopropyl diradical intermediate:

An alternative mechanism involves a direct migration of one of the vinyl groups.[75]

It has been found that the di-π-methane rearrangement can proceed through either a singlet or a triplet excited state.[76] The reaction can be formulated as a concerted process,

73. H. E. Zimmerman and G. A. Epling, *J. Am. Chem. Soc.* **94**:8749 (1972); J. S. Swenton, J. A. Hyatt, T. J. Walker, and A. L. Crumrine, *J. Am. Chem. Soc.* **93**:4808 (1971).
74. For reviews of the di-π-methane rearrangement, see S. S. Hixson, P. S. Mariano, and H. E. Zimmerman, *Chem. Rev.* **73**:531 (1973); H. E. Zimmerman, in *Rearrangements in Ground and Excited States*, Vol. 3, P. de Mayo, ed., Academic Press, New York, 1980, Chapter 16; H. E. Zimmerman, in *Organic Photochemistry and Photobiology*, W. M. Horspool and P.-S. Song, eds., CRC Press, Boca Raton, Florida, 1995, pp. 184–193; H. E. Zimmerman and D. Armesto, *Chem. Rev.* **96**:3065 (1996).
75. L. A. Paquette and E. Bay, *J. Org. Chem.* **47**:4597 (1982).
76. H. E. Zimmerman and P. S. Mariano, *J. Am. Chem. Soc.* **91**:1718 (1969); P. S. Mariano, R. B. Steitle, D. G. Watson, M. J. Peters, and E. Bay, *J. Am. Chem. Soc.* **98**:5899 (1976).

and this mechanism is followed in the case of some acyclic dienes and for cyclic systems in which a concerted process is sterically feasible.

orbital array for concerted
di-π-methane rearrangement

Notice that the orbital array is of the Möbius topology with a phase change depicted between the C-1 and C-2 positions. This corresponds to an allowed photochemical process since there are six electrons involved in bonding changes.

The di-π-methane rearrangement has been studied in a sufficient number of cases to develop some of the patterns regarding substituent effects. When the central sp^3 carbon is unsubstituted, the di-π-methane mechanism becomes less favorable. The case of 1,1,5,5-tetraphenyl-1,4-pentadiene is illustrative. Although one of the products has the expected structure for a product of the di-π-methane rearrangement, labeling with deuterium proves that an alternative mechanism operates:

The cyclopropane bridge is formed only after hydrogen-atom migration. The driving force for this migration may be the fact that a more stable allylic radical results:

The resistance of the unsubstituted system to the di-π-methane rearrangement probably occurs at the second step of the rearrangement.[77] If the central carbon is unsubstituted, this step results in the formation of a primary radical and would be energetically unfavorable.

77. H. E. Zimmerman and J. O. Pincock, *J. Am. Chem. Soc.* **95**:2957 (1973).

Photolysis of 1,5-diphenyl-1,4-pentadiene is another example of a reaction that takes this course[78]:

$$PhCH{=}CHCH_2CH{=}CHPh \xrightarrow{hv}$$

Intramolecular cycloaddition is also observed for 1,4-cyclooctadiene[79]:

The groups at the termini of the 1,4-pentadiene system also affect the efficiency and direction of the the di-π-methane reaction. The general trend is that cyclization occurs at the diene terminus that best stabilizes radical character. Thus, a terminus substituted with aryl groups will cyclize in preference to an unsubstituted or alkyl-substituted terminus:

Ref. 80

Ref. 81

This result can be rationalized in terms of a cyclopropyl diradical structure by noting that the bond cleavage of the intermediate will occur to give the more stable of the two possible 1,3-diradicals.[82] The cyclopropane ring in the final product will then incorporate this terminus:

The one-step vinyl migration mechanism makes the same regiochemical prediction. The less substituted vinyl group should migrate, because the substituent(s) will stabilize the residual radical.

The di-π-methane rearrangement is a stereospecific reaction. There are several elements of stereochemistry to be considered. It is known that the double bond that remains uncyclized retains the E or Z configuration present in the starting material. This result excludes any intermediate with a freely rotating terminal radical. The concerted

78. E. Block and H. W. Orf, *J. Am. Chem. Soc.* **94**:8438 (1972).
79. S. Moon and C. R. Ganz, *Tetrahedron Lett.* **1968**:6275.
80. H. E. Zimmerman and A. C. Pratt, *J. Am. Chem. Soc.* **92**:1409 (1970).
81. H. E. Zimmerman and A. A. Baum, *J. Am. Chem. Soc.* **93**:3646 (1971).
82. H. E. Zimmerman and A. C. Pratt, *J. Am. Chem. Soc.* **92**:6259, 6267 (1970).

transition state implies that C-3 would undergo inversion of configuration because the new C(3)−C(5) bond is formed using the back lobe of the C(2)−C(3) σ bond. This inversion of configuration has been confirmed[83]:

Thus, the transition state depicted on p. 777 for the concerted reaction correctly predicts the stereochemical course of the di-π-methane rearrangement.

13.5. Photochemistry of Aromatic Compounds

Irradiation of benzene and certain of its derivatives results in bond reorganization and formation of nonaromatic products.[84] Irradiation of liquid benzene with light of 254-nm wavelength results in the accumulation of fulvene (see p. 538 for structure) and a very small amount of tricyclo[3.1.0.0^{2,6}]hex-3-ene, also known as benzvalene.[85] The maximum conversion to this product in liquid benzene is about 0.05%. The key intermediate is believed to be a biradical formed by 1,3-bonding.

Because of the low photostationary concentration of benzvalene, photolysis is not an efficient way of accumulating this compound. The highly reactive molecule can be trapped, however, if it is generated in the presence of other molecules with which it reacts. Irradiation of benzene in acidic hydroxylic solvents gives products formally resulting from 1,3-bonding in the benzene ring and addition of a molecule of solvent:

These compounds are not direct photoproducts, however. The compounds of structure **9** arise by solvolysis of benzvalene, the initial photoproduct. Products of type **10** are secondary photoproducts derived from **9**.[86]

83. H. E. Zimmerman, J. D. Robbins, R. D. McKelvey, C. J. Samuel, and L. R. Sousa, *J. Am. Chem. Soc.* **96**:1974, 4630 (1974).
84. D. Bryce-Smith and A. Gilbert, *Tetrahedron* **32**:1309 (1976); A. Gilbert, in *Organic Photochemistry and Photobiology*, W. M. Horspool and P.-S. Song, eds., CRC Press, Boca Raton, Florida, 1995, pp. 229–236.
85. K. E. Wilzbach, J. S. Ritscher, and L. Kaplan, *J. Am. Chem. Soc.* **89**:1031 (1967).
86. L. Kaplan, D. J. Rausch, and K. E. Wilzbach, *J. Am. Chem. Soc.* **94**:8638 (1972).

The photoisomerization of aromatic rings has also been studied using 1,3,5-tri-*t*-butylbenzene. The composition of the photostationary state is shown below[87]:

R = *t*-butyl

7.3%   <0.7%   20.6%   7.1%   64.8%

These various photoproducts are all valence isomers of the normal benzenoid structure. These alternative bonding patterns are reached from the excited state, but it is difficult to specify a precise mechanism. The presence of the *t*-butyl groups introduces a steric factor that works in favor of the photochemical valence isomerism. Whereas the *t*-butyl groups are coplanar with the ring in the aromatic system, the geometry of the bicyclic products results in reduced steric interactions between adjacent *t*-butyl groups.

Irradiation of solutions of alkenes in benzene or substituted benzenes gives primarily 1 : 1 adducts in which the alkene bridges *meta* positions of the aromatic ring.[88]

These reactions are believed to proceed through a complex of the alkene with a singlet excited state of the aromatic compound (an exciplex). The alkene and aromatic ring are presumed to be oriented in such a manner that the alkene π system reacts with *p* orbitals on 1,3-carbons of the aromatic. The structure of the excited-state species has been probed in more detail using CAS-SCF *ab initio* calculations.[89]

87. K. E. Wilzbach and L. Kaplan, *J. Am. Chem. Soc.* **87**:4004 (1965).
88. K. E. Wilzbach and L. Kaplan, *J. Am. Chem. Soc.* **88**:2066 (1966); J. Cornelisse, V. Y. Merritt, and R. Srinivasan, *J. Am. Chem. Soc.* **95**:6197 (1973); A. Gilbert and P. Yianni, *Tetrahedron* **37**:3275 (1981); D. Bryce-Smith and A. Gilbert, *Tetrahedron* **33**:2459 (1977); T. Wagner-Jauregg, *Synthesis* **1980**:165, 769; J. Cornelisse, *Chem. Rev.* **93**:615 (1993).
89. S. Clifford, M. J. Bearpack, F. Bernardi, M. Olivucci, M. A. Robb, and B. R. Smith, *J. Am. Chem. Soc.* **118**:7353 (1996).

This addition to the aromatic ring is believed to be concerted, since the relative geometry of the substituents on the alkene is retained in the product. Lesser amounts of products involving addition to 1,2- or 1,4-positions of the aromatic ring are also formed in such photolyses.[90] This type of addition reaction has also been realized intramolecularly when the distance between the alkene and the phenyl substituent is sufficient to permit interaction.

General References

D. R. Arnold, N. C. Baird, J. R. Bolton, J. C. D. Brand, P. W. M. Jacobs, P. de Mayo, and W. R. Ware, *Photochemistry; An Introduction*, Academic Press, New York, 1974.

D. O. Cowan and R. L. Drisko, *Elements of Organic Photochemistry*, Plenum, New York, 1976.

J. M. Coxon and B. Halton, *Organic Photochemistry*, Cambridge University Press, London, 1974.

P. de Mayo, ed., *Rearrangements in Ground and Excited States*, Vol. 3, Academic Press, New York, 1980.

A. Gilbert and J. Baggott, *Essentials of Molecular Photochemistry*, CRC Press, Boca Raton, Florida, 1991.

W. Horspool and D. Armester, *Organic Photochemistry, A Comprehensive Treatment*, Ellis Horwood/Prentice Hall, 1992.

W. M. Horspool and P.-S. Song, eds., *Organic Photochemistry and Photobiology*, CRC Press, Boca Raton, Florida, 1995.

J. Kagan, *Organic Photochemistry*, Principles and Applications, Academic Press, San Diego, 1993.

M. Klessinger and J. Michl, *Excited States and Photochemistry of Organic Molecules*, VCH, New York, 1995.

J. Kopecky, *Organic Photochemistry: A Visual Approach*, VCH, Weinhim, Germany, 1992.

N. J. Turro, *Modern Molecular Photochemistry*, Benjamin-Cummings, Menlo Park, California, 1978.

M. A. West, in *Investigation of Rates and Mechanisms of Reactions*, C. F. Bernasconi, ed., *Techniques of Chemistry*, 4th ed., Vol. VI, Part 2, Wiley-Interscience, New York, 1986. Chapter VIII.

H. E. Zimmerman, Topic in Photochemistry, *Top. Curr. Chem.* **100**:45 (1982).

Problems

(References for these problems will be found on page 804).

1. The bridged radical **A** has been suggested as a possible intermediate in the photochemical decarbonylation of 3-phenylpropanal. Suggest an experiment to test this hypothesis.

A H_2C-CH_2

90. K. E. Wilzbach and L. Kaplan, *J. Am. Chem. Soc.* **93**:2073 (1971).

2. Predict the structure, including all aspects of stereochemistry, for the product expected to result from direct irradiation of each compound.

(a)

(e)

(b) H_2C=CH HC=CH_2

(f)

(c)

—CO_2CH_3

(g)

(d)

(h)

3. Suggest reasonable explanations for the following observations.

(a) Optically active 2,3-pentadiene is racemized rapidly under conditions of toluene-sensitized photolysis.

(b) Direct photolysis of diene **A** at 254 nm produces a photostationary state containing 40% **A** and 60% triene **B**. When the irradiation is carried out at 300 nm, no **B** is produced, and **C** is the observed product.

(c) Photochemical cyclization of the acetal **D** to **E** in benzene is efficiently catalyzed by benzoic acid.

4. Provide a mechanistic rationalization for each of the following reactions.

(a)

(b)

(c)

(d)

(e)

(f)

(g)

(h)

(i)

(j)

(k)

(l)

(m)

$(CH_3)_2C{=}CHCH{=}CHCH_2CO_2H$

(n)

(o)

$H_3CO_2CC{\equiv}CCO_2CH_3 \xrightarrow[\text{benzene}]{hv}$

(p)

$PhCH{=}O + H_3CC{\equiv}CCH_3 \xrightarrow{hv} PhCH{=}\overset{O}{\overset{||}{C}}\underset{CH_3}{C}CH_3$

5. Benzene-sensitized photolysis of methyl 3-cyclohexene-1-carboxylate in acetic acid leads to addition of acetic acid to the double bond. Only the *trans* adducts are formed. What factor(s) is (are) responsible for the reaction stereochemistry? Which of the two possible addition products, **A** or **B**, do you expect to be the major product?

6. Photolysis of bicyclo[2.2.2]octan-2-one (**A**) gives **B** in good yield. When **A** labeled as shown is used, the aldehyde group carries deuterium to the extent of 51.7%. Write a mechanism to account for the overall transformation. Calculate the isotope effect for the step in which hydrogen-atom transfer occurs. What mechanistic conclusion do you

7. The photolysis of benzobarrelene, **A**, has been studied in considerable detail. Direct photolysis gives **C**, but when acetone is used as a photosensitizer, the di-π-methane rearrangement product **B** is formed.

(A small amount of D is at C-3,8.)

A deuterium labeling study has been performed with the results shown. Discuss the details of the mechanism that are revealed by these results. Is there a feasible mechanism that would have led to **B** having an alternative label distribution?

8. Quantum yield data for several processes that occur on photolysis of *S*-4-methyl-1-phenyl-1-hexanone have been determined. The results are tabulated below for benzene as solvent.

Process	Quantum yield
Type-II elimination	0.23
Cyclobutanol formation	0.03
Racemization	0.78

What information about the mechanism operating under these conditions can be drawn from these data?

9. Show by a diagram why the energy of radiation emitted from an excited electronic state (by fluorescence or phosphorescence) is of lower energy than the exciting radiation. Would you expect the shift to lower energy to be more pronounced for fluorescence or phosphorescence? Explain.

10. *cis*-2-Propyl-4-*t*-butylcyclohexanone is photolyzed to 4-*t*-butylcyclohexanone. The *trans* isomer is converted to the *cis* isomer, which then undergoes cleavage. Offer a rationale for this pronounced stereochemical effect.

11. The quantum yield for formation of 3-methylcyclobutene from *E*-1,3-pentadiene is 10

times greater than for the cyclization of Z-1,3-pentadiene. Can you offer an explanation?

12. The irradiation of 2-methoxytropone (**A**) leads to methyl 4-oxo-2-cyclopentenyl-acetate (**D**). The reaction can be followed by analytical gas chromatography and two intermediates are observed. These have the structures **B** and **C**. Indicate a mechanism by which each of the three successive reactions might occur. The first two steps are photochemical, while the third is probably an acid-catalyzed reaction which occurs under the photolysis conditions.

13. When an aryl substituent is placed at C-5 of a 4-substituted cyclohexenone, a new product type containing a cyclobutanone ring is formed.

The reaction products are the same for both direct irradiation and acetophenone sensitization. When the reactant **B** is used in homochiral form, the product **D** is nearly racemic (6% e.e.). Relate the formation of the cyclobutanones to the more normal products of type **E** and **F**. Why does the 5-aryl substituent favor formation of the cyclobutanones? Give a complete mechanism for formation of **D** which is consistent with the stereochemical result.

14. In the rearrangement of 4,4-diphenylcyclohexenone to 5,6-diphenylbicyclo-[3.1.0]hexan-2-one, there is a strong preference for formation of the *endo* phenyl

stereoisomer. Offer an explanation.

15. Compound **A** is photochromic; that is, it becomes colored on exposure to light. The process is reversible, giving back the starting material in the dark. Suggest a structure for the colored photoisomer.

16. A study of the photolysis of **A**, **B**, and **C** has been reported. **A** gives both **D** and the cleavage product benzaldehyde. **B** gives only **E**. **C** gives benzaldehyde and the stereoisomer **B**. Discuss the ways in which the presence and configuration of the remote *t*-butyl group can control the product composition, and account for the formation of the observed products.

17. The azo compounds **A** and **B** were prepared and the thermal and photochemical behavior of these materials was investigated. The results are summarized in the equations below. Discuss how these results may relate to the photochemical di-π-methane rearrangement. (See Section 12.1.4 for some indications of the reactivity of

azo compounds.)

18. Irradiation of 1,1,3,3-tetraphenyl-5-methyl-1,4-hexadiene gives the two products shown below. When the reaction is carried out by photosensitization, **B** is not formed. Suggest mechanisms for the formation of **A** and **B**. What other products might have been expected? Can you rationalize their absence?

19. Suggest a reasonable pathway for the formation of each of the photoproducts formed on irradiation of the Diels–Alder adduct of 2,3-dimethylbutadiene and quinone:

20. The direct irradiation of **A** gives predominantly **B**, but the photosensitized reaction gives more **C**. Explain.

	C	B
Direct:	28%	72%
Sensitized:	68%	32%

21. 9,10-Ethenoanthracenes undergo the di-π-methane photorearrangement. Analyze the substituent effects that are observed in this reaction.

X	Y	% A	% B
H	CO$_2$CH$_3$	0	100
H	CON(CH$_3$)$_2$	0	100
CON(CH$_3$)$_2$	CO$_2$CH$_3$	3	97
CO$_2$CH$_3$	CSOCH$_3$	0	100
CH$_3$	CO$_2$CH$_3$	0	100
Ph	CO$_2$CH$_3$	100	0

References to Problems

Chapter 1

2a. H. Prinzbach, *Pure Appl. Chem.* **28:**281 (1971).

b. E. S. Gould, *Mechanism and Structure in Organic Chemistry,* Holt-Dryden, New York, 1959, pp. 69–70.

c. K. E. Laidig, P. Speers, and A. Streitwieser, *Can. J. Chem.* **74:**1215 (1996).

3a. R. L. Reeves and W. F. Smith, *J. Am. Chem. Soc.* **85:**724 (1963).

b. R. B. Martin, *J. Chem. Soc., Chem. Commun.* **1972:**793.

c, d. J. A. Joule and G. F. Smith, *Heterocyclic Chemistry,* Van Nostrand Reinhold, London, 1972, c: pp. 194–195; d: p. 64.

4a. H. L. Ammon and G. L. Wheeler, *J. Am. Chem. Soc.* **97:**2326 (1975).

b. R. Breslow, T. Eicher, A. Krebs, R. A. Peterson, and J. Posner, *J. Am. Chem. Soc.* **87:**1320 (1965).

c. R. B. Woodward, *J. Am. Chem. Soc.* **63:**1123 (1941).

5. R. T. Sanderson, *J. Am. Chem. Soc.* **97:**1367 (1975).

6b. B. B. Ross and N. S. True, *J. Am. Chem. Soc.* **106:**2451 (1984).

c. H. Slebocka-Tilk and R. S. Brown, *J. Org. Chem.* **52:**805 (1987); V. Somayaji and R. S. Brown, *J. Org. Chem.* **51:**2676 (1986).

7a. R. C. Bingham, *J. Am. Chem. Soc.* **98:**535 (1976).

b. W. L. Jorgensen and L. Salem, *The Organic Chemist's Book of Orbitals,* Academic Press, New York, 1973, pp. 179–184.

c. H. Stafast and H. Bock, *Tetrahedron* **32:**855 (1976).

d. I. Fleming, *Frontier Orbitals and Organic Chemical Reactions,* Wiley-Interscience, New York, 1976, pp. 51–57.

e. L. Libit and R. Hoffmann, *J. Am. Chem. Soc.* **96:**1370 (1974).

f. W. J. Hehre, L. Radom, and J. A. Pople, *J. Am. Chem. Soc.* **94:**1496 (1972).

9. W. L. Jorgensen and L. Salem, *The Organic Chemist's Book of Orbitals,* Academic Press, New York, 1973, pp 93–94.

11a. R. E. Lehr and A. P. Marchand, *Orbital Symmetry: A Problem-Solving Approach,* Academic Press, New York, 1972, p. 37.

b. C. A. Coulson and A. Streitwieser, Jr., *Dictionary of π-Electron Calculations,* W. H. Freeman, San Francisco, 1965, p. 184.

12. T. Ohwada and K. Shudo, *J. Am. Chem. Soc.* **11:**34 (1989).

13. C. A. Coulson and A. Streitwieser, Jr., *Dictionary of π-Electron Calculations,* W. H. Freeman, San Francisco, 1965, p. 1.

15. N. L. Allinger and N. A. Pamphilis, *J. Org. Chem.* **38:**316 (1973).

16. K. B. Wiberg and F. H. Walker, *J. Am. Chem. Soc.* **104:**5239 (1982).

17. R. B. Turner, B. J. Mallon, M. Tichy, W. v. E. Doering, W. R. Roth, and G. Schröder, *J. Am. Chem. Soc.* **95:**8605 (1973).

20a. C. H. DePuy, V. M. Bierbaum, and R. Damrauer, *J. Am. Chem. Soc.* **106:**4051 (1984); C. H. DePuy, S. Gronert, S. E. Barlow, U. M. Bierbaum, and R. Damrauer, *J. Am. Chem. Soc.* **111:**1968 (1989).

   b. P. C. Redfern, J. S. Murray, and P. Politzer, *Can. J. Chem.* **70:**636 (1992).

   c. C. F. Rodriguez, S. Sirois, and A. C. Hopkinson, *J. Org. Chem.* **57:**4869 (1992).

21a. R. Breslow and J. M. Hoffmann, Jr., *J. Am. Chem. Soc.* **94:**2110 (1972).

   b. R. Breslow and P. Dowd, *J. Am. Chem. Soc.* **85:**2729 (1963).

   c. R. Breslow and J. M. Hoffmann, *J. Am. Chem.* **94:**2110 (1972); M. Saunders, R. Berger, A. Jaffe, M. McBride, J. O'Neill, R. Breslow, J. M. Hoffmann, Jr., C. Perchonock, E. Wasserman, R. S. Hutton, and V. J. Kuck, *J. Am. Chem. Soc.* **95:**3017 (1973); R. Breslow and S. Mazur, *J. Am. Chem. Soc.* **95:**584 (1973).

22. W. Kirmse, K. Kund, E. Pitzer, A. Dorigo, and K. N. Houk, *J. Am. Chem. Soc.* **108:**6045 (1986).

23. K. N. Houk, N. G. Rondan, M. N. Paddon-Row, C. W. Jefford, P. T. Huy, P. D. Burrow, and K. N. Jordan, *J. Am. Chem. Soc.* **105:**5563 (1983).

24. A. Pross, L. Radom, and N. V. Riggs, *J. Am. Chem. Soc.* **102:**2252 (1980).

25. M. A. McAllister and T. T. Tidwell, *J. Chem. Soc., Perkin Trans. 2* **1994:**2239.

Chapter 2

2. M. Sprecher and D. B. Sprinson, *J. Org. Chem.* **28:**2490 (1963).

3. K. L. Marsi, *J. Org. Chem.* **39:**265 (1974).

4a. J. A. Pettus, Jr., and R. E. Moore, *J. Am. Chem. Soc.* **93:**3087 (1971).

   b. K. T. Black and H. Hope, *J. Am. Chem. Soc.* **93:**3053 (1971).

   c. J. Dillon and K. Nakanishi, *J. Am. Chem. Soc.* **96:**4055 (1974).

   d. M. Miyano and C. R. Dorn, *J. Am. Chem. Soc.* **95:**2664 (1973).

   e. M. Koreeda, G. Weiss, and K. Nakanishi, *J. Am. Chem. Soc.* **95:**239 (1973).

   f. P. A. Apgar and M. L. Ludwig, *J. Am. Chem. Soc.* **94:**964 (1972).

   g. M. R. Jones and D. J. Cram, *J. Am. Chem. Soc.* **96:**2183 (1974).

5a. J. W. deHaan and L. J. M. van den Ven, *Tetrahedron Lett.* **1971:**2703.

   b. T. Sone, S. Terashima, and S. Yamada, *Synthesis* **1974:**725.

   c. B. Witkop and C. M. Foltz, *J. Am. Chem. Soc.* **79:**197 (1957).

   e. E. W. Yankee, B. Spencer, N. E. Howe, and D. J. Cram, *J. Am. Chem. Soc.* **95:**4220 (1973).

   f. R. S. Lenox and J. A. Katzenellenbogen, *J. Am. Chem. Soc.* **95:**957 (1973).

   g. R. Bausch, B. Bogdanovic, H. Dreeskamp, and J. B. Koster, *Justus Liebigs Ann. Chem.* **1974:**1625.

6. J. A. Marshall, T. R. Konicek, and K. E. Flynn, *J. Am. Chem. Soc.* **102:**3287 (1980).

7. E. Vogel, W. Tückmantel, K. Schlög, M. Widham, E. Kraka, and D. Cremer, *Tetrahedron Lett.* **25:**4925 (1984).

8. R. A. Pascal, Jr., and R. B. Grossman, *J. Org. Chem.* **52:**4616 (1987); A. Rici, R. Danieli, and S. Rossini, *J. Chem. Soc., Perkin Trans. 1* **1976:**1691.

9. K. Koyama, S. Natori, and Y. Iitaka, *Chem. Pharm. Bull.* **35:**4049 (1987).

10. K. J. Thottathil, C. Przybyla, M. Malley, and J. Z. Gougoutas, *Tetrahedron Lett.* **27:**1533 (1986).

12. H. W. Gschwend, *J. Am. Chem. Soc.* **94:**8430 (1972).

13a, b. M. Kainosho, K. Akjisaka, W. H. Pirkle, and S. D. Beare, *J. Am. Chem. Soc.* **94:**5924 (1972).

   c. M. Rabin and K. Mislow, *Tetrahedron Lett.,* **1966:**3961.

   d. R. K. Hill, S. Yan, and S. M. Arfin, *J. Am. Chem. Soc.* **95:**7857 (1973).

   e. V. J. Morlino and R. B. Martin, *J. Am. Chem. Soc.* **89:**3107 (1967).

   f. J. A. Elvidge and R. G. Foster, *J. Chem. Soc.* **1964:**981; L. S. Rattet and J. H. Goldstein, *Org. Magn. Reson.* **1:**229 (1969).

14a. G. Helmchen and G. Staiger, *Angew. Chem. Int. Ed. Engl.* **16:**116 (1977).

   b. M. Farina and C. Morandi, *Tetrahedron* **30:**1819 (1974).

   c. D. T. Longone and M. T. Reetz, *J. Chem. Soc., Chem. Commun.* **1967:**46.

   d. I. T. Jacobson, *Acta Chem. Scand.* **21:**2235 (1967).

   e. R. J. Ternansky, D. W. Balogh, and L. A. Paquette, *J. Am. Chem. Soc.* **104:**4503 (1982).

   f. C. Ganter and K. Wicker, *Helv. Chim. Acta* **53:**1693 (1970).

g. M. S. Newman and D. Lednicer, *J. Am. Chem. Soc.* **78**:4765 (1956).

h. T. Otsubo, R. Gray, and V. Boekelheide, *J. Am. Chem. Soc.* **100**:2449 (1978).

i. J. H. Brewster and R. S. Jones, Jr., *J. Org. Chem.* **34**:354 (1969).

j. H. Gerlach, *Helv. Chim. Acta* **51**:1587 (1968).

k, l. H. Gerlach, *Helv. Chim. Acta* **68**:1815 (1985).

m. R. K. Hill, G. H. Morton, J. R. Peterson, J. A. Walsh, and L. A. Paquette, *J. Org. Chem.* **50**:5528 (1985).

15. T. C. Bruice and S. Benkovic, *Bioorganic Mechanisms*, Vol. 1, W. A. Benjamin, New York, 1966, p. 305.

16. M. Cohn, J. E. Pearson, E. L. O'Connell, and I. A. Rose, *J. Am. Chem. Soc.* **92**:4095 (1970).

17. R. K. Hill, S. Yan, and S. M. Arfin, *J. Am. Chem. Soc.* **95**:7857 (1973).

18. D. J. Cram and F. A. Abd Elhafez, *J. Am. Chem. Soc.* **74**:5828 (1952).

19. Y.-F. Cheung and C. Walsh, *J. Am. Chem. Soc.* **98**:3397 (1976).

20. D. M. Jerina, H. Selander, H. Yagi, M. C. Wells, J. F. Davey, V. Magadevan, and D. T. Gibson, *J. Am. Chem. Soc.* **98**:5988 (1976).

21. Y. Fujimoto, F. Irreverre, J. M. Karle, I. L. Karle, and B. Witkop, *J. Am. Chem. Soc.* **98**:5988 (1976).

22a M. Raban, S. K. Lauderback, and D. Kost, *J. Am. Chem. Soc.* **97**:5178 (1975).

b. P. Finocchiaro, D. Gust, and K. Mislow, *J. Am. Chem. Soc.* **96**:2165 (1974).

23. F.-C. Huang, L. F. H. Lee, R. S. D. Mittal, P. R. Ravikumar, J. A. Chan, C. J. Sih, E. Capsi, and C. R. Eck, *J. Am. Chem. Soc.* **97**:4144 (1975).

24. J. Dominguez, J. D. Dunitz, H. Gerlach, and V. Prelog, *Helv. Chim. Acta* **45**:129 (1962); H. Gerlach and V. Prelog, *Justus Liebigs Ann. Chem.* **669**:121 (1963); B. T. Kilbourn, J. D. Dunitz, L. A. R. Pioda, and W. Simon, *J. Mol Biol.* **30**: 559 (1967).

Chapter 3

1a. R. L. Lipnick and E. W. Garbisch, Jr., *J. Am. Chem. Soc.* **95**:6375 (1973).

b. L. Lunazzi, D. Macciantelli, F. Bemardi, and K. U. Ingold, *J. Am. Chem Soc.* **99**:4573 (1977).

c. B. Rickborn and M. T. Wuesthoff, *J. Am. Chem. Soc.* **92**:6894 (1970).

2. H. C. Brown and W. C. Dickason, *J. Am. Chem. Soc.* **91**:1226 (1969).

3. E. L. Eliel and S. H. Schroeter, *J. Am. Chem. Soc.* **87**:5031 (1965).

4. N. L. Allinger and M. T. Tribble, *Tetrahedron Lett.* **1971**:3259; W. F. Bailey, H. Connor, E. L. Eliel, and K. B. Wiberg, *J. Am. Chem. Soc.* **100**:2202 (1978).

5a. C. L Stevens, J. B. Filippi, and K. G. Taylor, *J. Org. Chem* **31**:1292 (1966).

b. M. Miyamoto, Y. Kawamatsu, M. Shinohara, Y. Nakadaira, and K. Nakanishi, *Tetrahedron* **22**:2785 (1966).

c. D. H. Williams and J. R. Kalman, *J. Am. Chem. Soc.* **99**:2768 (1977).

6a. J. E. Baldwin and J. A. Reiss, *J. Chem. Soc., Chem. Commun.* **1977**:77.

b. H. C. Brown, J. H. Kawakami, and S. Ikegami, *J. Am. Chem. Soc.* **92**:6914 (1970).

c. R. E. Lyle, E. W. Southwick, and J. J. Kaminski, *J. Am. Chem. Soc.* **94**:1413 (1972).

d. E. C. Ashby and S. A. Noding, *J. Am. Chem. Soc.* **98**:2010 (1976).

e. R. D. G. Cooper, P. V. DeMarco, and D. O. Spry, *J. Am. Chem. Soc.* **91**:1528 (1969).

f. J. E. Baldwin and M. J. Lusch, *Tetrahedron* **38**:2939 (1982).

g. S. K. Taylor, G. H. Hockerman, G. L Karnick, S. B. Lyle, and S. B. Schramm, *J. Org. Chem.* **48**:2449 (1983).

7a. H. Tanida, S. Yamamoto, and K. Takeda, *J. Org. Chem* **38**:2792 (1973).

b. J. E. Baldwin and L I. Kruse, *J. Chem. Soc., Chem. Commun.*, **1977**:233.

c. N. L. Allinger and J. C. Graham, *J. Org. Chem.* **36**:1688 (1971).

d. C. Galli, G. Illuminati, L Mandolini, and P. Tamborra, *J. Am. Chem. Soc.* **99**:2591 (1977).

e. C. M. Evans, R. Glenn, and A. J. Kirby, *J. Am. Chem Soc.* **104**:4706 (1982).

f. J. F. Bunnett, S. Sekiguchi, and L A. Smith, *J. Am. Chem. Soc.* **103**:4865 (1981).

g. P. Müller and J.-C. Perlberger, *J. Am. Chem. Soc.* **98**:8407 (1976).

8a. J. C. Little, Y.-L. C. Tong, and J. P. Heeschen, *J. Am. Chem. Soc.* **91**:7090 (1969).

b. D. J. Pasto and D. R. Rao, *J. Am. Chem. Soc.* **92**:5151 (1970).

c. P. E. Pfeffer and S. F. Osman, *J. Org. Chem.* **37**:2425 (1972).

d. P. L Durrette and D. Horton, *Carbohydr. Res.* **18**:57 (1971).

e. B. Fuchs and A. Ellencweig, *J. Org. Client* **44:**2274 (1979).

f. M. J. Anteunis, D. Tavernier, and F. Borremans, *Heterocyclics* **4:**293 (1976).

g. P. Deslongchamps, D. D. Rowan, N. Pothier, T. Sauvé, and J. K. Saunders, *Can. J. Chem.* **59:**1105 (1981).

h. P. v. R. Schleyer and A. J. Kos, *Tetrahedron* **39:**1141 (1983).

i. H. Senderowitz, P. Aped, and B. Fuchs, *Helv. Chim. Acta* **73:**2113 (1990).

9. K. B. Wiberg, *.J. Am. Chem. Soc.* **108:**5817 (1986).

10. L. Lunazzi, D. Macciantelli, F. Bernardi, and K. U. Ingold, *J. Am. Chem. Soc.* **99:**4573 (1977).

11a. E. N. Marvell and R. S. Knutson, *J. Org. Chem.* **35:**388 (1970).

b. R. J. Oulette, J. D. Rawn, and S. N. Jreissaty, *J. Am. Chem. Soc.* **93:**7117 (1971).

c. V. S. Mastryukov, E. L Olsina, L V. Vilkov, and R. L. Hilderbrandt, *J. Am. Chem. Soc.* **99:**6855 (1977).

d. D. D. Danielson and K. Hedberg, *J. Am. Chem. Soc.* **101:**3730 (1979).

12. S. E. Denmark, M. S. Dappen, N. L. Sear, and R. T. Jacobs, *J. Am. Chem. Soc.* **112:**3466 (1990)

13. M. Traetteberg, P. Bakken, H. Hopfa, and R. Hänel, *Chem. Ber.* **127:**1469 (1994).

14. J. B. Lambert, R. R.Clikeman, and E. S. Magyar, *J. Am. Chem. Soc.* **96:**2265 (1974).

15. J. G. Vintner and H. M. R. Hoffmann, *J. Am. Chem. Soc.* **96:**5466 (1974).

16. A. Bienvenue, *J. Am. Chem. Soc.* **95:**7345 (1973).

17. T. Varnali, V. Aviyente, B. Terryn, and M. F. Ruiz-Lopez, *THEOCHEM* **280:**169 (1993).

18. G. Mehta and N. Mohal, *J. Chem. Soc., Perkin Trans. 1* **1998:**505.

19. D. Seebach B. Lamatsch R. Amstutz, A. K. Beck, M. Dobler, M. Egli, R. Fitzi, M. Gautschi, B. Herradon, P. C. Hibder, J. J. Irwin, R. Locher, M. Maestro, T. Maetzke, A. Mourino, E. Pfammatter, D. A. Plattner, C. Schickli, W. B. Schweizer, P. Seiler, G. Stucky, W. Petter, J. Escalarte, E. Juaristi, D. Quintana, C. Miravitlles, and E. Molins, *Helv. Chim. Acta* **75:**913 (1992).

20. E. Ghera, Y. Gaoni, and S. Shoua, *J. Am. Chem. Soc.* **98:**3627 (1976).

21. D. K. Dalling and D. M. Grant, *J. Am. Chem. Soc.* **94:**5318 (1972).

22a. H. C. Brown, J. H. Kawakami, and S. Ikegami, *J. Am. Chem. Soc.* **92:**6914 (1970).

b. B. Waegell and C. W. Jefford, *Bull. Soc. Chim. Fr.* **1964:**844.

c. H. C. Brown and J. H. Kawakami, *J. Am. Chem. Soc.* **92:**1990 (1970).

d. L A. Spurlock and K. P. Clark, *J. Am. Chem. Soc.* **94:**5349 (1972).

e. K. B. Wiberg and K. A. Saegebarth, *J. Am. Chem. Soc.* **79:**2822 (1957).

Chapter 4

1. W. Kusters and P. deMayo, *J. Am. Chem. Soc.* **96:**3502 (1974).

2a. A. Streitwieser, Jr., H. A. Hammond, R. H. Jagow, R. M. Williams, R. G. Jesaitis, C. J. Chang, and R. Wolf, *J. Am. Chem. Soc.* **92:**5141 (1970).

b. M. T. H. Liu and D. H. T. Chien, *J. Chem. Soc., Perkin Trans. 2* **1974:**937.

3. W. E. Billups, K. H. Leavell, E. S. Lewis, and S. Vanderpool, *J. Am. Chem. Soc.* **95:**8096 (1973).

4. O. Exner, in *Correlation Analysis in Chemistry,* N. B. Chapman and B. Shorter, eds. Plenum Press, New York, 1978, Chapter 10.

5. P. R. Wells, *Linear Free Energy Relationships,* Academic Press, New York, 1968, pp. 12, 13.

6a. O. R. Zaborsky and E. T. Kaiser, *J. Am. Chem. Soc.* **92:**860 (1970).

b. W. M. Schubert and J. R. Keefe, *J. Am. Chem. Soc.* **94:**559 (1972).

c. D. H. Rosenblatt, L. A. Hull, D. C. DeLuca, G. T. Davis, R. C. Weglein, and H. K. R.Williams, *J. Am. Chem. Soc.* **89:**1158 (1967).

7a. W. K. Kwok, W. G. Lee, and S. I. Miller, *J. Am. Chem. Soc.* **91:**468 (1969).

b. H. C. Brown and E. N. Peters, *J. Am. Chem. Soc.* **95:**2400 (1973).

c. W. M. Schubert and D. F. Gurka, *J. Am. Chem. Soc.* **91:**1443 (1969).

d. J. Rocek and A. Riehl, *J. Am. Chem. Soc.* **88:**4749 (1966).

8. D. E. Applequist and R. D. Gdanski, *J. Org. Chem.* **46:**2502 (1981).

9a. V. J. Shiner, Jr., and J. O. Stoffer, *J. Am Chem. Soc.* **92:**3191 (1970).

b. M. H. Davies, *J. Chem. Soc., Perkin Trans. 2* **1974:**1018.

c. J. E. Baldwin and J. A. Kapecki, *J. Am. Chem. Soc.* **91:**3106 (1969).

d. H. G. Bull, K. Koehler, T. C. Pletcher, J. J. Ortiz, and E. H. Cordes, *J. Am. Chem. Soc.* **93:**3002 (1971).

e. R. B. Timmons, J. deGuzman, and R. E. Varnerin, *J. Am. Chem. Soc.* **90:**5996 (1968).

f. Y. Pocker and J. H. Exner, *J. Am. Chem. Soc.* **90:**6764 (1968).

g. K. D. McMichael, *J. Am. Chem. Soc.* **89:**2943 (1967).

h. R. J. Cvetanovic, F. J. Duncan, W. E. Falconer, and R. S. Irwin, *J. Am. Chem. Soc.* **87:**1827 (1965).

10. J. J. Eisch and S.-G. Rhee, *J. Am. Chem. Soc.* **96:**7276 (1974).

11. C. G. Swain, A. L. Powell, W. A. Sheppard, and C. R. Morgan, *J. Am. Chem. Soc.* **101:**3576 (1979).

12. R. S. Shue, *J. Am. Chem. Soc.* **93:**7116 (1971).

13. D. N. Kevill and G. M. L. Lin, *J. Am. Chem. Soc.* **101:**3916 (1979).

14. C. G. Swain, J. E. Sheats, and K. G. Harbison, *J. Am. Chem. Soc.* **97:**783 (1975).

15. G. E. Hall and J. D. Roberts, *J. Am. Chem. Soc.* **93:**2203 (1971).

16. R. K. Lustgarten and H. G. Richey, Jr., *J. Am. Chem. Soc.* **96:**6393 (1974).

17. C. A. Bunton and C. Stedman, *J. Chem. Soc.* **1958:**2420; C. A. Bunton and E. A. Halevi, *J. Chem. Soc.* **1952:**4917.

18. P. G. Gassman and K. Sato, *Tetrahedron Lett.* **22:**1311 (1981).

19. T. B. McMahon and P. Kebarle, *J. Am. Chem. Soc.* **99:**2222 (1977).

20. J. Bromilow, R.T. C. Brownlee, V. O. Lopez, and R. W. Taft, *J. Org. Chem.* **44:**4766 (1979).

21a. S. Irle, T. M. Krygowski, J.E. Niu, and W. H. E. Schwarz, *J. Org. Chem.* **60:**6744 (1995).

21b. D. J. Craik, G. C. Levy, and R. T. C. Brownlee, *J. Org. Chem.* **48:**1601 (1983).

22. A. Fischer, W. J. Galloway, and J. Vaughan, *J. Chem. Soc.* **1964:**3591; C. L. Liotta, E. M. Perdue, and H. P. Hopkins, Jr., *J. Am. Chem. Soc.* **96:**7308 (1974).

23. E. S. Lewis, J. T. Hill, and E. R. Newman, *J. Am. Chem. Soc.* **90:**662 (1968).

24. E. M. Arnett and D. R. McKelvey, *J. Am. Chem. Soc.* **88:**2598 (1966).

25. D. N. Rogers, F. J. McLafferty, W. Fang, Q. Yang, and Y. Zhao, *Struct. Chem.* **3:**53 (1992).

26. T.-T. Wang and T.-C. Huang, *Chem. Eng. J.* **53:**107 (1993).

Chapter 5

1a. H. C. Brown and E. N. Peters, *J. Am. Chem. Soc.* **99:**1712 (1977).

b. X. Creary, M. E. Mehrsheikh-Mohammadi, and M. D. Eggers, *J. Am. Chem. Soc.* **109:**2435 (1987).

c. R. J. Blint, T. B. McMahon, and J. L. Beauchamp, *J. Am. Chem. Soc.* **96:**1269 (1974).

d. J. L. Fry, E. M. Engler, and P. v. R. Schleyer, *J. Am. Chem. Soc.* **94:**4628 (1972).

e. F. G. Bordwell and W. T. Brannen, Jr., *J. Am. Chem. Soc.* **86:**4645 (1964).

2a. R. K. Crossland, W. E. Wells, and V. J. Shiner, Jr., *J. Am. Chem. Soc.* **93:**4217 (1971).

b. R. L. Buckson and S. G. Smith, *J. Org. Chem.* **32:**634 (1967).

c. C. J. Norton, Ph.D. Thesis, Harvard University, 1955, cited by P. v. R. Schleyer, W. E. Watts, R. C. Fort, Jr., M. B. Comisarow, and G. A. Olah, *J. Am. Chem. Soc.* **86:**5679 (1964).

d. E. N. Peters and H. C. Brown, *J. Am. Chem. Soc.* **97:**2892 (1975).

e. B. R. Ree and J. C. Martin, *J. Am. Chem. Soc.* **92:**1660 (1970).

f. F. G. Bordwell and W. T. Brannen, Jr., *J. Am. Chem. Soc.* **86:**4645 (1964).

g, h. D. D. Roberts, *J. Org. Chem.* **34:**285 (1969).

i. K. L. Servis and J. D. Roberts, *J. Am. Chem. Soc.* **87:**1331 (1965).

j. R. G. Lawton, *J. Am. Chem. Soc.* **83:**2399 (1961).

k. G. M. Bennett, F. Heathcoat, and A. N. Mosses, *J. Chem. Soc.* **1929:**2567.

l. S. Kim, S. S. Friedrich, L. J. Andrews, and R. M. Keefer, *J. Am. Chem. Soc.* **92:**5452 (1970).

3b. D. H. Ball, E. D. Eades, and L. Long, Jr., *J. Am. Chem. Soc.* **86:**3579 (1964).

c. H. G. Richey, Jr., and D. V. Kinsman, *Tetrahedron Lett.* **1969:**2505.

d. P. v. R. Schleyer, W. E. Watts, and C. Cupas, *J. Am. Chem. Soc.* **86:**2722 (1964).

e. P. E. Peterson and J. E. Duddey, *J. Am. Chem. Soc.* **85:**2865 (1963).

f. A. Colter, E. C. Friedrich, N. J. Holness, and S. Winstein, *J. Am. Chem. Soc.* **87:**378 (1965).

g, h. M. Cherest, H. Felkin, J. Sicher, F. Sipos, and M. Tichy, *J. Chem. Soc.* **1965:**2513.

i. J. C. Martin and P. D. Bartlett, *J. Am. Chem. Soc.* **79:**2533 (1957).

j. S. Archer, T. R.Lewis, M. R. Bell, and J. W. Schulenberg, *J. Am. Chem. Soc.* **83:**2386 (1961).

k. D. A. Tomalia and J. N. Paige, *J. Org. Chem.* **38:**422 (1973).

l. P. Wilder, Jr., and W.-C. Hsieh, *J. Org. Chem.* **40:**717 (1975).

m. C. W. Jefford, J.-C. Rossier, J. A. Zuber, S. C. Suri, and G. Mehta, *Tetrahedron Lett.* **1980:**4081.

4. H. L. Goering and B. E. Jones, *J. Am. Chem. Soc.* **102:**1628 (1980).

5. K. Banert and W. Kirmse, *J. Am. Chem. Soc.* **104**:3766 (1982).

6. Y. Pocker and M. J. Hill, *J. Am. Chem. Soc.* **93**:691 (1971).

7. R. Baird and S. Winstein, *J. Am. Chem. Soc.* **85**:567 (1963).

8. J. B. Lambert and S. I. Featherman, *J. Am. Chem. Soc.* **99**:1542 (1977).

9a. K. Yano, *J. Org. Chem.* **40**:414 (1975).

b. N. C. Deno, N. Friedman, J. D. Hodge, and J. J. Houser, *J. Am. Chem. Soc.* **85**:2995 (1963).

c. I. Lillien and L. Handloser, *J. Am. Chem. Soc.* **93**:1682 (1971).

d. J. C. Barborak and P. v. R. Schleyer, *J. Am. Chem. Soc.* **92**:3184 (1970); P. Ahlberg, C. Engdahl, and G. Jonsäll, *J. Am. Chem. Soc.* **103**:1583 (1981).

10. S. Winstein and E. T. Stafford, *J. Am. Chem. Soc.* **79**:505 (1957).

11a M. Brookhart, A. Diaz, and S. Winstein, *J. Am. Chem. Soc.* **88**:3135 (1966).

b. G. A. Olah, J. M. Bollinger, C. A. Cupas, and J. Lukas, *J. Am. Chem. Soc.* **89**:2692 (1967).

c. G. A. Olah and R. D. Porter, *J. Am. Chem. Soc.* **92**:7627 (1970).

d. G. A. Olah and G. Liang, *J. Am. Chem. Soc.* **97**:2236 (1975).

e, f. G. A. Olah and G. Liang, *J. Am. Chem. Soc.* **93**:6873 (1971).

12. L. A. Paquerte, I. R. Dunkin, J. P. Freeman, and P. C. Storm, *J. Am. Chem. Soc.* **94**:8124 (1972).

13. C. Paradisi and J. F. Bunnett, *J. Am. Chem. Soc.* **103**:946 (1981).

14. S. Kobayashi, T. Matsumoto, H. Taniguchi, M. Mishima, M. Fujio, and Y. Tsuno, *Tetrahedron Lett.* **34**:5903 (1993).

15. M. Fujio, M. Ohe, K. Nakata, Y. Tsuji, and Y. Tsuno, *Bull. Chem. Soc. Jpn.* **70**:929 (1997).

16. J. P. Richard and W. P. Jencks, *J. Am. Chem. Soc.* **104**:4689, 4691(1982).

17. J. W. Wilt and P. J. Chenier, *J. Am. Chem. Soc.* **90**:7366 (1968); S. J. Cristol and G. W. Nachtigall, *J. Am. Chem. Soc.* **90**:7132, 7133 (1968).

b. H. C. Brown and E. N. Peters, *J. Am. Chem. Soc.* **97**:1927 (1975).

c. P. G. Gassman and W. C. Pike, *J. Am. Chem. Soc.* **97**:1250 (1975).

d. I. Tabushi, Y. Tamura, Z. Yoshida, and T. Sugimoto, *J. Am. Chem. Soc.* **97**:2886 (1975).

e. H. Weiner and R. A. Sneen, *J. Am. Chem. Soc.* **87**:287 (1965).

f. J. R. Mohrig and K. Keegstra, *J. Am. Chem. Soc.* **89**:5492 (1967).

g. E. N. Peters and H. C. Brown, *J. Am. Chem. Soc.* **96**:265 (1974).

h. L. R. C. Barclay, H. R. Sonawane, and J. C. Hudson, *Can. J. Chem.* **50**:2318 (1972).

i. C. D. Poulter, E. C. Friedrich, and S. Winstein, *J. Am. Chem. Soc.* **92**:4274 (1970).

j. J. M. Harris, J. R. Moffatt, M. G. Case, F. W. Clarke, J. S. Polley, T. K. Morgan, Jr., T. M. Ford, and R. K. Murray, Jr., *J. Org. Chem.* **47**:2740 (1982).

k. F. David, *J. Org. Chem.* **47**:2740 (1982).

l. M. Xie and W. J. le Noble, *J. Org. Chem.* **54**:3839 (1989).

m. E. Fritz-Langhals, *Tetrahedron Lett.* **35**:1851 (1994).

18. D. T. Stoelting and V. J. Shiner, Jr., *J. Am. Chem. Soc.* **115**:1695 (1993).

19. W. Franke, H. Schwarz, and D. Stahl, *J. Org. Chem.* **45**:3493 (1980); Y. Apeloig, J. B. Collins, D. Cremer, T. Bally, E. Haselbach, J. A. Pople, J. Chandrasekhar, and P. v. R. Schleyer, *J. Org. Chem.* **45**:3496 (1980).

20. J. S. Haywood-Farmer and R. E. Pincock, *J. Am. Chem. Soc.* **91**:3020 (1969).

21. G. A. Olah, A. L. Berrier, M. Arvanaghi, and G. K. S. Prakash, *J. Am. Chem. Soc.* **103**:1122 (1981).

22. J. J. Tufariello and R. J. Lorence, *J. Am. Chem. Soc.* **91**:1546 (1969); J. Lhomme, A. Diaz, and S. Winstein, *J. Am. Chem. Soc.* **91**:1548 (1969).

23. C. Lim, S.-H. Kim, S.-D. Yoh, M. Fujio, and Y. Tsuno, *Tetrahedron Lett.* **38**:3243 (1997).

24. F. David, *J. Org. Chem.* **46**:3512 (1981).

25. K. Nakata, M. Fujio, Y. Saeki, M. Mishima, Y. Tsuno, and K. Nishimoto, *J. Phys. Org. Chem.* **9**:561 (1996).

26. J. P. Richards, T. L. Amyes, and T. Vontor, *J. Am. Chem. Soc.* **113**:5871 (1991).

29. R. Fuchs and L. L. Cole, *J. Am. Chem. Soc.* **95**:3194 (1973).

Chapter 6

1a. K. Yates, G. H. Schmid, T. W. Regulski, D. G. Garratt, H.-W. Leung, and R. McDonald, *J. Am. Chem. Soc.* **95**:160 (1973).

b. J. F. King and M. J. Coppen, *Can. J. Chem.* **49**:3714 (1971).

c. N. C. Deno, F. A. Kish, and H. J. Peterson, *J. Am. Chem. Soc.* **87**:2157 (1965).

d. M. A. Cooper, C. M. Holden, P. Loftus, and D. Whittaker, *J. Chem. Soc., Perkin Trans. 2* **1973**:665.

e. R. A. Bartsch and J. F. Bunnett, *J. Am. Chem. Soc.* **91**:1376 (1969).

f. K. Yates and T. A. Go, *J. Org. Chem.* **45**:2377 (1980).

g. C. B. Quinn and J. R. Wiseman, *J. Am. Chem. Soc.* **95**:1342 (1973).

2a. H. Wong, J. Chapuis, and I. Monkovic, *J. Org. Chem.* **39**:1042 (1974).

b. P. K. Freeman, F. A. Raymond, and M. F. Grostic, *J. Org. Chem.* **32**:24 (1967).

c. T. I. Crowell, R. T. Kemp, R.E. Lutz, and A. A. Wall, *J. Am. Chem. Soc.* **90**:4638 (1968).

d. L. C. King, L. A. Subluskey, and E. W. Stern, *J. Org. Chem.* **21**:1232 (1956).

e. M. L. Poutsma and P. A. Ibarbia, *J. Am. Chem. Soc.* **93**:440 (1971).

f. S. P. Acharya and H. C. Brown, *J. Chem. Soc. Chem. Commun.* **1968**:305.

g. A. Lewis and J. Azoro, *J. Org. Chem.* **46**:1764 (1981).

h. K. Yates and T. A. Go, *J. Org. Chem.* **45**:2377 (1980).

3. D. Y. Curtin, R. D. Stolow, and W. Maya, *J. Am. Chem. Soc.* **81**:3330 (1959).

4. W. H. Saunders and A. F. Cockerill, *Mechanisms of Elimination Reactions,* John Wiley & Sons, New York, 1973, pp. 79–80.

5. M. Anteunis and H. L. Peeters, *J. Org. Chem.* **40**:307 (1975).

6a, b. D. S. Noyce, D. R. Hartter, and R. M. Pollack, *J. Am. Chem. Soc.* **90**:3791 (1968).

c. C. L. Wilkins and T. W. Regulski, *J. Am. Chem. Soc.* **94**:6016 (1972).

7. R. A. Bartsch, *J. Am. Chem. Soc.* **93**:3683 (1971).

8. K. Oyama and T. T. Tidwell, *J. Am. Chem. Soc.* **98**:947 (1976).

9a. K. D. Berlin, R. O. Lyerla, D. E. Gibbs, and J. P. Devlin, *J. Chem. Soc, Chem. Commun.* **1970**:1246.

b. R. J. Abraham and J. R. Monasterios, *J. Chem. Soc. Perkin Trans. 2* **1975**:574.

c. H. C. Brown and K.-T. Liu, *J. Am. Chem. Soc.* **97**:2469 (1975).

d. D. J. Pasto and J. A. Gontarz, *J. Am. Chem. Soc.* **93**:6902 (1971).

10a. J. F. Bunnett and S. Sridharan, *J. Org. Chem.* **44**:1458 (1979); L. F. Blackwell, A. Fischer, and J. Vaughan, *J. Chem. Soc., B* **1967**:1084.

b. J. E. Nordlander, P. O. Owuor, and J. E. Haky, *J. Am. Chem. Soc.* **101**:1288 (1979).

c. G. E. Heasley, T. R. Bower, K. W. Dougharty, J. C. Easdon, V. L. Heasley, S. Arnold, T. L. Carter, D. B. Yaeger, B. T. Gipe, and D. F. Shellhamer, *J. Org. Chem.* **45**:5150 (1980).

d. P. Cramer and T. T. Tidwell, *J. Org. Chem.* **46**:2683 (1981).

e. R. C. Fahey and C. A. McPherson, *J. Am. Chem. Soc.* **91**:3865 (1969); R. C. Fahey, M. W. Monahan, and C. A. McPherson, *J. Am. Chem. Soc.* **92**:2810 (1970).

f. C. H. DePuy and A. L. Schultz, *J. Org. Chem.* **39**:878 (1974).

11a. T. T. Coburn and W. M. Jones, *J. Am. Chem. Soc.* **96**:5218 (1974).

b. R. B. Miller and G. McGarvey, *J. Org. Chem.* **44**:4623 (1979).

c. P. F. Hudrlik and D. Peterson, *J. Am. Chem. Soc.* **97**:1464 (1975).

d. G. Bellucci, A. Marsili, E. Mastrorilli, I. Morelli, and V. Scartoni, *J. Chem. Soc., Perkin Trans 2* **1974**:201.

e. K. J. Shea, A. C. Greeley, S. Nguyen, P. D. Beauchamp, D. H. Aue, and J. S. Witzeman, *J. Am. Chem. Soc.* **108**:5901 (1986).

12. G. H. Schmid, A. Modro, and K. Yates, *J. Org. Chem.* **45**:665 (1980); A. D. Allen, Y. Chiang, A. J. Kresge, and T. T. Tidwell, *J. Org. Chem.* **47**:775 (1982).

13. G. Bellucci, R. Bianchini, and S. Vecchiani, *J. Org. Chem.* **52**:3355 (1987).

14. T. Hasam, L. B. Sims, and A. Fry, *J. Am. Chem. Soc.* **105**:3987 (1983).

15. B. Badat, M. Julia, J. N. Mallet, and C. Schmitz, *Tetrahedron Lett.* **24**:4331 (1983).

16. D. S. Noyce, D. R. Hartter, and F. B. Miles, *J. Am. Chem. Soc.* **90**:3794 (1968).

17. Y. Pocker, K. D. Stevens, and J. J. Champoux, *J. Am. Chem. Soc.* **91**:4199 (1969).

18. M. F. Ruasse, A. Argile, and J. E. Dubois, *J. Am. Chem. Soc.* **100**:7645 (1978).

19. W. K. Chwang, P. Knittel, K. M. Koshy, and T. T. Tidwell, *J. Am. Chem. Soc.* **99**:3395 (1977).

20. D. J. Pasto and J. F. Gadberry, *J. Am. Chem. Soc.* **100**:1469 (1978).

21. J. F. Bunnett and C. A. Migdal, *J. Org. Chem.* **54**:3037 (1989).

Chapter 7

2b. G. L. Closs and L. E. Closs, *J. Am. Chem. Soc.* **85**:2022 (1963).

f. R. Breslow, *J. Am. Chem. Soc.* **79**:1762 (1957).

g. K. Ogura and G. Tsuchihashi, *Tetrahedron Lett.* **1971:**3151.

h. G. L. Closs and R. B. Larabee, *Tetrahedron Lett,* **1965:**287.

i. R. B. Woodward and C. Wintner, *Tetrahedron Lett.* **1969:**2689.

j. J. A. Zoltewicz, G. M. Kauffman, and C. L. Smith, *J. Am. Chem. Soc.* **90:**5939 (1968).

3a. T.-Y. Luh and L. M. Stock, *J. Am. Chem. Soc.* **96:**3712 (1974).

b. H. W. Amburn, K. C. Kaufman, and H. Shechter, *J. Am. Chem. Soc.* **91:**530 (1969); F. G. Bordwell, J. C. Branca, C. R. Johnson, and N. R. Vanier, *J. Org. Chem.* **45:**3884 (1980).

c. F. G. Bordwell, G. E. Drucker, and H. E. Fried, *J. Org. Chem.* **46:**632 (1981); E. M. Arnett and K. G. Venkatasubramanian, *J. Org. Chem.* **48:**1569 (1983).

d. T. B. Thompson and W. T. Ford, *J. Am. Chem. Soc.* **101:**5459 (1979).

4a. G. A. Abad, S. P. Jindal, and T. T. Tidwell, *J. Am. Chem. Soc.* **95:**6326 (1973).

b. A. Nickon and J. L. Lambert, *J. Am. Chem. Soc.* **84:**4604 (1962).

6. G. B. Trimitsis and E. M. Van Dam, *J. Chem. Soc. Chem. Commun.* **1974:**610.

7. A. Streitwieser, Jr., W. B. Hollyhead, A. H. Pudjaatmaka, P. H. Owens, T. L. Kruger, P. A. Rubenstein, R. A. MacQuarrie, M. L. Brokaw, W. K. C. Chu, and H. M. Niemeyer, *J. Am. Chem. Soc.* **93:**5088 (1971).

8. F. G. Bordwell and F. J. Cornforth, *J. Org. Chem.* **43:**1763 (1978).

9a. P. T. Lansbury, *J. Am. Chem. Soc.* **83:**429 (1961).

b. D. W. Griffiths and C. D. Gutsche, *J. Am. Chem. Soc.* **93:**4788 (1971).

c. T. D. Hoffman and D. J. Cram, *J. Am. Chem. Soc.* **91:**1000 (1969).

d. G. Büchi, D. M. Foulkes, M. Kurono, G. F. Mitchell, and R. S. Schneider *J. Am. Chem. Soc.* **89:**6745 (1967).

e. G. Stork, G. L. Nelson, F. Rouessac, and O. Grigone, *J. Am. Chem. Soc.* **93:**3091(1971).

10. S. Danishefsky and R. K. Singh, *J. Am. Chem. Soc.* **97:**3239 (1975); E. M. Arnett and J. A. Harrelson, Jr., *J. Am. Chem. Soc.* **109:**809 (1987).

11. A. Streitwieser, Jr., R. G. Lawler, and C. Perrin, *J. Am. Chem. Soc.* **87:**5383 (1965); J. E. Hofmann, A. Schriesheim, and R. E. Nichols, *Tetrahedron Lett.* **1965:**1745.

12. E. J. Stamhuis, W. Mass, and H. Wynberg, *J. Org. Chem.* **30:**2160 (1965).

13a. A. P. E. Carey, S. Eustace, R. A. More O'Ferrall, and M. A. Murray, *J. Chem. Soc., Perkin Trans 2* **1993:**2285.

b. J. W. Bunting and J. P. Kanter, *J. Am. Chem. Soc.* **115:**11705 (1993).

14a. J. R. Keefe, A. J. Kresge, and Y. Yin, *J. Am. Chem. Soc.* **110:**1982 (1988); S. Eldin, R. M. Pollack, and D. L. Whalen, *J. Am. Chem. Soc.* **113:**1344 (1991); S. Eldin, D. L. Whalen, and R. M. Pollack, *J. Org. Chem.* **58:**3490 (1993).

b. F. G. Borwell and H. E. Fried, *J. Org. Chem.* **56:**4218 (1991).

c. E. M. Arnett and J. A. Harrelson, Jr., *J. Am. Chem. Soc.* **109:**809 (1987).

15. T. J. Sprules and J.-F. Lavellee, *J. Org. Chem.* **60:**5041 (1995).

16. B. G. Cox, *J. Am. Chem. Soc.* **96:**6823 (1974).

17. E. L. Eliel, A. A. Hartmann, and A. G. Abatjoglou, *J. Am. Chem. Soc.* **96:**1807 (1974); J.-M. Lehn and G. Wipff, *J. Am. Chem. Soc.* **98:**7498 (1976).

18a. N. S. Mills, J. Shapiro, and M. Hollingsworth, *J. Am. Chem. Soc.* **103:**1263 (1981).

b. N. S. Mills, *J. Am. Chem. Soc.* **104:**5689 (1982).

19. R. B. Woodward and G. Small, Jr., *J. Am. Chem. Soc.* **72:**1297 (1950).

20a. E. J. Corey, T. H. Hopie, and W. A. Wozniak, *J. Am. Chem. Soc.* **77:**5415 (1955).

b. E. W. Garbisch, Jr., *J. Org. Chem.* **30:**2109 (1965).

c. F. Caujolle and D. Q. Quan, *C. R. Acad. Sci., C* **265:**269 (1967).

d. C. W. P. Crowne, R. M. Evans, G. F. H. Green, and A. G. Long, *J. Chem. Soc.* **1956:**4351.

e. N. C. Deno and R. Fishbein, *J. Am. Chem. Soc.* **95:**7445 (1973).

21a. P. L. Stotter and K. A. Hill, *J. Am. Chem. Soc.* **96:**6524 (1974).

b. B. J. L. Huff, F. N. Tuller, and D. Caine, *J. Org. Chem.* **34:**3070 (1969).

c. H. O. House and T. M. Bare, *J. Org. Chem.* **33:**943 (1968).

d. R. S. Matthews, P. K. Hyer, and E. A. Folkers, *J. Client Soc., Chem. Commun.* **1970:**38.

e. C. H. Heathcock, R.A. Badger, and J. W. Patterson, Jr., *J. Am. Chem. Soc.* **89:**4133 (1967).

f. J. E. McMurry, *J. Am. Chem. Soc.* **90:**6821 (1968).

22. M. S. Newman, V. DeVries, and R. Darlak, *J. Org. Chem.* **31:**2171 (1966).

23. H. M. Walborsky and L. M. Turner, *J. Am. Chem. Soc.* **94:**2273 (1972).

24. Y. Jasor, M. Gaudry, and A. Marquet, *Tetrahedron Lett.* **1976:**53.

1. R. P. Bell, *Adv. Phys. Org. Chem.* **4**:1 (1966).

2a. A. Lapworth and R. H. F. Manske, *J. Chem. Soc.* **1930**:1976.

 b. T. H. Fife, J. E. C. Hutchins, and M. S. Wang, *J. Am. Chem. Soc.* **97**:5878 (1975).

 c. K. Bowden and A. M. Last, *J. Chem. Soc., Chem. Commun.,* **1970**:1315.

 d. P. R. Young and W. P. Jencks, *J. Am. Chem. Soc.* **99**:1206 (1977).

 e. H. Dahn and M.-N. Ung-Truong, *Helv Chim. Acta* **70**:2130 (1987).

3a, b. M. M. Kreevoy and R. W. Taft, Jr., *J. Am Chem Soc.* **77**:5590 (1955).

 c, d. M. M. Kreevoy, C. R. Morgan, and R. W. Taft, Jr., *J. Am. Chem. Soc.* **82**:3064 (1960).

 e. T. H. Fife and L. H. Brod, *J. Org. Chem.* **33**:4136 (1968).

4. R. G. Bergstrom, M. J. Cashen, Y. Chiang, and A. J. Kresge, *J. Org. Chem.* **44**:1639 (1979).

5a. T. Maugh II and T. C. Bruice, *J. Am. Chem. Soc.* **93**:3237 (1971).

 b. L. E. Eberson and L.-A. Svensson, *J. Am. Chem. Soc.* **93**:3827 (1971).

 c. A. Williams and G. Salvadori, *J. Chem. Soc., Perkin Trans. 2* **1972**:883.

 d. G. A. Rogers and T. C. Bruice, *J. Am. Chem. Soc.* **96**:2463 (1974).

 e. K. Bowden and G. R. Taylor, *J. Chem. Soc., B* **1971**:145, 149; M. S. Newman and A. L. Leegwater, *J. Am. Chem. Soc.* **90**:4410 (1968).

 f. T. C. Bruice and S. J. Benkovic, *J. Am. Chem. Soc.* **85**:1 (1963).

6. E. H. Cordes and W. P. Jencks, *J. Am. Chem. Soc.* **85**:2843 (1963).

7. E. Anderson and T. H. Fife, *J. Am. Chem. Soc.* **95**:6437 (1973).

8. M. W. Williams and G. T. Young, *J. Chem. Soc.* **1964**:3701.

9a. V. Somayaji and R. S. Brown, *J. Org. Chem.* **51**:2676 (1986).

 b. A. J. Briggs, C. M. Evans, R. Glenn, and A. J. Kirby, *J. Chem. Soc., Perkin Trans. 2* **1983**:1637; C. B. Quin and J. R.Wiseman, *J. Am. Chem. Soc.* **95**:1342 (1973).

10a. N. S. Nudelman, Z. Gatto, and L. Bohe, *J. Org. Chem.* **49**:1540 (1984).

 b. P. Baierwick, D. Hoell, and K. Mueller, *Angew. Chem. Int. Ed. Engl.* **24**:972 (1985).

 c. C. Alvarez-Ibarra, P. Perez-Ossorio, A. Perez-Rubalcaba, M. L. Quiroga, and M. J. Santemases, *J. Chem. Soc., Perkin Trans. 2* **1983**:1645.

11. H. R. Mahler and E. H. Cordes, *Biological Chemistry,* Harper and Row, New York, 1966, p. 201.

12. J. Hajdu and G. M. Smith, *J. Am. Chem. Soc.* **103**:6192 (1981).

13. T. Okuyama, H. Shibuya, and T. Fueno, *J. Am. Chem. Soc.* **104**:730 (1982).

14. R. P. Bell, M. H. Rand, and K. M. A. Wynne-Jones, *Trans. Faraday Soc.* **52**:1093 (1956).

15. S. T. Purrington and J. H. Pittman, *Tetrahedron Lett.* **28**:3901 (1987).

16. J. T. Edward and S. C. Wong, *J. Am. Chem. Soc.* **101**:1807 (1979).

17. H. Slebocka-Tilk, A. J. Bennet, J. W. Keillor, R. S. Brown, J. P. Guthrie, and A. Jodhan, *J. Am. Chem. Soc.* **112**:8507 (1990); H. Slebocka-Tilk, A. J. Bennet, H. J. Hogg, and R. S. Brown, *J. Am. Chem. Soc.* **113**:1288 (1991).

18. K. Bowden, *Chem. Soc. Rev.* **24**:431 (1995).

19. R. Breslow and C. McAllister, *J. Am. Chem. Soc.* **93**:7096 (1971).

20. B. Capon and K. Nimmo, *J. Chem. Soc., Perkin Trans. 2* **1975**:1113.

21. D. P. Weeks and D. B. Whitney, *J. Am. Chem. Soc.* **103**:3555 (1981).

22. R. L. Schowen, C. R. Hopper, and C. M. Bazikian, *J. Am. Chem. Soc.* **94**:3095 (1972).

23. E. Tapuhi and W. P. Jencks, *J. Am. Chem. Soc.* **104**:5758 (1982).

24. C. M. Evans, R. Glenn, and A. J. Kirby, *J. Am. Chem. Soc.* **104**:4706 (1982).

25a. T. C. Bruice and I. Oka, *J. Am. Chem. Soc.* **96**:4500 (1974).

 b. R. Hershfield and G. L. Schmir, *J. Am. Chem. Soc.* **95**:7359 (1973).

 c T. H. Fife and J. E. C. Hutchins, *J. Am. Chem. Soc.* **94**:2837 (1972).

 d. J. Hine, J. C. Craig, Jr., J. G. Underwood II, and F. A. Via, *J. Am. Chem. Soc.* **92**:5194 (1970).

26. T. S. Davies, P. D. Feil, D. G. Kubler, and D. J. Wells, Jr., *J. Org. Chem.* **40**:1478 (1975).

27. L. Do Amaral, M. P. Bastos, H. G. Bull, and E. H. Cordes, *J. Am. Chem. Soc.* **95**:7369 (1973).

28. A. J. Kirby and P. W. Lancaster, *J. Chem. Soc., Perkin Trans. 2* **1972**:1206.

29. E. H. Cordes and W. P. Jencks, *J. Am. Chem. Soc.* **85**:2843 (1963).

30. M. Hässermann, *Helv. Chim. Acta* **34**:1482 (1951).

 b. A. C. Anderson and J. A. Nelson, *J. Am. Chem. Soc.* **73**:232 (1951).

 c. W. A. Kleschick, C. T. Buse, and C. H. Heathcock, *J. Am. Chem. Soc.* **99**:247 (1977); C. H. Heathcock, C. T. Buse, W. A. Kleschick, M. C. Pirrung, J. E. Sohn, and J. Lampe, *J. Org. Chem.* **45**:1066 (1980).

d. A. T. Nielsen and W. J. Houlihan, *Org. React.* **16:**1 (1968).
31. R. S. Brown and A. Aman, *J. Org. Chem.* **62:**4816 (1997).

Chapter 9

1. P. Reeves, T. Devon, and R. Pettit, *J. Am. Chem. Soc.* **91:**5890 (1969).
2a. H. L. Ammon and G. L. Wheller, *J. Am. Chem. Soc.* **97:**2326 (1975).
 b. W. v. E. Doering and C. H. DePuy, *J. Am. Chem. Soc.* **75:**5955 (1953).
 c. J. H. M. Hill, *J. Org. Chem.* **32:**3214 (1967).
 d. D. J. Bertelli, *J. Org. Chem.* **30:**891 (1965).
3a. W. v. E. Doering and F. L. Detert, *J. Am. Chem. Soc.* **73:**876 (1951).
 b. E. F. Jenny and J. D. Roberts, *J. Am. Chem. Soc.* **78:**2005 (1956).
4a. J. J. Eisch, J. E. Galle, and S. Kozima, *J. Am. Chem. Soc.* **108:**379 (1986).
 b. P. J. Garratt, N. E. Rowland, and F. Sondheimer, *Tetrahedron* **27:**3157 (1971).
 c. R. Concepcion, R. C. Reiter, and G. R. Stevenson, *J. Am. Chem. Soc.* **105:**1778 (1983).
 d. G. Jonsäll and P. Ahlberg, *J. Am. Chem. Soc.* **108:**3819 (1986).
 e. J. M. Herbert, P. D. Woodgate, and W. A. Denny, *J. Med. Chem.* **30:**2081 (1987).
 f. R. F. X. Klein and V. Horak, *J. Org. Chem.* **51:**4644 (1986).
 g. D. A. Smith and J. Bitar, *J. Org. Chem.* **58:**6 (1993).
 h. T. K. Zywietz, H. Jiao, P. v. R. Schleyer, and A. de Meijere, *J. Org. Chem.* **63:**3417 (1998).
 i. M. J. S. Dewar and N. Trinajstic, *J. Chem. Soc. A*, **1969:**1754.
5. D. Cremer, T. Schmidt, and C. W. Bock, *J. Org. Chem.* **50:**2684 (1985).
6. B. A. Hess, Jr., and L. J. Schaad, *J. Am. Chem. Soc.* **93:**305 (1971).
7. D. Bostwick, H. F. Henneike, and H. P. Hopkins, Jr., *J. Am. Chem. Soc.* **97:**1505 (1975).
8. T. Otsubo, R. Gray, and V. Boekelheide, *J. Am. Chem. Soc.* **100:**2449 (1978).
9a. Z. Yoshida, M. Shibata, and T. Sugimoto, *Tetrahedron Lett.* **24:**4585 (1983); *J. Am. Chem. Soc.* **106:**6383 (1984).
 b. K. Nakasuji, K. Yoshida, and I. Murata, *J. Am. Chem. Soc.* **105:**5136 (1983).
 c. K. Takahashi, T. Nozoe, K. Takase, and T. Kudo, *Tetrahedron Lett.* **25:**77 (1984).
 d. A. Minsky, A. Y. Meyers, K. Hafner, and M. Rabinovitz, *J. Am. Chem. Soc.* **105:**3975 (1983).
10. R. A. Wood, T. R. Welberry, and A. D. Rae, *J. Chem. Soc., Perkin Trans. 2* **1985:**451; W. E. Rhine, J. H. Davis, and G. Stucky, *J. Organomet. Chem.* **134:**139(1977); Y. Cohen, N. H. Roelofs, G. Reinhardt, L. T. Scott, and M. Rabinovitz, *J. Org. Chem.* **52:**4207 (1987); B. Eliasson and H. Edlund, *J. Chem. Soc., Perkin Trans. 2* **1983:**1837; B. C. Becker, W. Huber, C. Schneiders, and K. Müller, *Chem. Ber.* **116:**1573 (1983); A. Minsky, A. Y. Meyers, K. Hafner, and M. Rabinovitz, *J. Am. Chem. Soc.* **105:**3975 (1983).
11. A. Greenberg, R. P. T. Tomkins, M. Dobrovolny, and J. E. Liebman, *J. Am. Chem. Soc.* **105:**6855 (1983); F. Gavina, A. M. Costero, P. Gil, and S. V. Luis, *J. Am. Chem. Soc.* **106:**2077 (1984); P. J. Garratt, *Aromaticity,* Wiley-Interscience, New York, 1986, pp. 173–181.
12. H. Prinzbach, V. Freudenberg, and U. Scheidegger, *Helv. Chim. Acta* **50:**1087 (1967).
13. J. F. M. Oth, K. Müller, H.-V. Runzheimer, P. Mues, and E. Vogel, *Angew. Chem. Int. Ed. Engl.* **16:**872 (1977).
14. G. Jonsäll and P. Ahlberg, *J. Am. Chem. Soc.* **108:**3819 (1986).
15. G. R. Stevenson and B. E. Forch, *J. Am. Chem. Soc.* **102:**5985 (1980).
16. M. A. McAllister and T. T. Tidwell, *J. Am. Chem. Soc.* **114:**5362 (1992).
17. M. A. Hempenius, J. Lugtenburg, and J. Cornelisse, *Recl. Trav. Chim. Pays-Bas* **109:**403 (1990).
18. H. Jiao and P. v. R. Schleyer, *Angew. Chem. Int. Ed. Engl.*; **35:**2383 (1996); H. A. Staab, F. Diederich, C. Krieger, and D. Schweitzer, *Chem. Ber.* **116:**3504 (1983).

Chapter 10

1a. C. K. Ingold and E. H. Ingold, *J. Chem. Soc.* **1928:**2249.
 b. R. J. Albers and E. C. Kooyman, *Recl. Trav. Chim.* **83:**930 (1964).

c. J. R. Knowles and R. O. C. Norman, *J. Chem. Soc.* **1961**:2938.

d. A. Gastaminza, T. A. Modro, J. H. Ridd, and J. H. P. Utley, *J. Chem. Soc., B* **1968**:534.

e. J. R. Knowles, R. O. C. Norman, and G. K. Radda, *J. Chem. Soc.* **1960**:4885.

f. F. L. Riley and E. Rothstein, *J. Chem. Soc.* **1964**:3860.

2. T. C. van Hock, P. E. Verkade, and B. M. Wepster, *Recl. Trav. Chim.* **77**:559 (1958); A. van Loon, P. E. Verkade, and B. M. Wepster, *Recl. Trav. Chim.* **79**:977 (1960).

3. G. A. Olah, S. J. Kuhn, S. H. Flood, and J. C. Evans, *J. Am. Chem. Soc.* **84**:3687 (1962).

4. J. E. Dubois, J. J. Aaron, P. Alcais, J. P. Doucet, F. Rothenberg, and R. Uzan, *J. Am. Chem. Soc.* **94**:6823 (1972).

5. R. M. Roberts and D. Shiengthong, *J. Am. Chem. Soc.* **86**:2851 (1964).

6. R. L. Dannley, J. E. Gagen, and K. Zak, *J. Org. Chem.* **38**:1 (1973); R. L. Dannley and W. R. Knipple, *J. Org. Chem.* **38**:6 (1973).

7a. C. D. Gutsche and K. H. No, *J. Org. Chem.* **47**:2708 (1982).

b. G. D. Figuly and J. C. Martin, *J. Org. Chem.* **45**:3728 (1980).

c. K. Dey, C. Eaborn, and D. R. M. Walton, *Organomet. Chem. Synth.* **1**:151 (1970–1971).

d. S. Winstein and R. Baird, *J. Am. Chem. Soc.* **79**:756 (1957).

8. D. S. Noyce, P. A. Kittle, and E. H. Bannitt, *J. Org. Chem.* **33**:1500 (1958).

9. M. Essiz, G. Guillaumet, J.-J. Brunet, and P. Caubere, *J. Org. Chem.* **45**:240 (1980).

10. W. Nagata, K. Okada, and T. Akoi, *Synthesis* **1979**:365.

11. W. G. Miller and C. U. Pittman, Jr., *J. Org. Chem.* **39**:1955 (1974).

12. T. A. Modro and K. Yates, *J. Am. Chem. Soc.* **98**:4247 (1976).

13. A. V. R. Rao, V. H. Deshpande, and N. L. Reddy, *Tetrahedron Lett.* **1982**:4373.

14. L. M. Jackman and V. R. Haddon, *J. Am. Chem. Soc.* **96**:5130 (1974); M. Gates, D. L. Frank, and W. C. von Felten, *J. Am. Chem. Soc.* **96**:5138 (1974).

15. C. K. Ingold, *Structure and Mechanism in Organic Chemistry,* 2nd ed., Cornell University Press, Ithaca, New York, 1969, pp. 340–344; C. G. Swain and D. R. Crist, *J. Am. Chem. Soc.* **94**:3195 (1972).

16. M. L. Bird and C. K. Ingold, *J. Chem. Soc.* **1938**:918; J. D. Roberts, J. K. Sanford, F. L. J. Sixma, H. Cerfontain, and R. Zagt, *J. Am. Chem. Soc.* **76**:4525 (1954).

17. E. Baciocchi, F. Cacace, G. Ciranni, and G. Illuminati, *J. Am. Chem. Soc.* **94**:7030 (1972).

18. R. B. Moodie and K. Schofield, *Acc Chem. Res.* **9**:287 (1976).

19. P. C. Myhre, M. Beug, and L. L. James, *J. Am. Chem. Soc.* **90**:2105 (1968).

20a. L. R. C. Barclay, B. A. Ginn, and C. E. Milligan, *Can. J. Chem.* **42**:579 (1964).

b. A. A. Khalaf and R. M. Roberts, *J. Org. Chem.* **34**:3571 (1969); A. A. Khalaf, *Rev. Clim. (Bucharest)* **19**:1373 (1974).

21a. G. P. Stahly, *J. Org. Chem.* **50**:3091 (1985).

b. A. I. Meyers and P. D. Pansegrau, *Tetrahedron Lett.* **24**:4935 (1983).

c. M. Maksoza and J. Winiarski, *J. Org. Chem.* **49**:1494 (1984).

d. I. J. Anthony and D. Wege, *Aust. J. Chem.* **37**:1283 (1984).

e. Y. Konayashi, T. Nagai, I. Kumadaki, M. Takahashi, and T. Yamauchi, *Chem. Pharm. Bull.* **32**:4382 (1984).

22. A. Wang, J. A. Maguire, and E. Bichl, *J. Org. Chem.* **63**:2451 (1998).

Chapter 11

2a. E. Vogel, *Justus Liebigs Ann. Chem.* **615**:14 (1958).

b. A. C. Cope, A. C. Haven, Jr., F. L. Ramp, and E. R. Turnbull, *J. Am. Chem. Soc.* **74**:4867 (1952).

c. R. Pettit, *J. Am. Chem. Soc.* **82**:1972 (1960).

d. M. Oda, M. Oda, and Y. Kitahara, *Tetrahedron Lett.* **1976**:839.

e. K. M. Rapp and J. Daub, *Tetrahedron Lett.* **1976**:2011.

f. E. J. Corey and D. K. Herron, *Tetrahedron Lett.* **1971**:1641.

3. A. G. Anastassiou, V. Orfanos, and J. H. Gebrian, *Tetrahedron Lett.* **1969**:4491; P. Radlick and G. Alford, *J. Am. Chem. Soc.* **91**:6529 (1969).

4a. K. B. Wiberg, V. Z. Williams, Jr., and L. E. Friedrich, *J. Am. Chem. Soc.* **90**:5338 (1968).

b. P. S. Wharton and R. A. Kretchmer, *J. Org. Chem.* **33**:4258 (1968).

c. R. L. Danheiser, C. Martinez, and J. M. Morin, *J. Org. Chem.* **45**:1340 (1980); R. L. Danheiser, C.

Martinez-Davilla, R. J. Auchus, and J. T. Kadonaga, *J. Am. Chem. Soc.* **103:**2443 (1981).

d. R. K. Hill and M. G. Bock, *J. Am. Chem. Soc.* **100:**637 (1978).

e. M. Newcomb and W. T. Ford, *J. Am. Chem. Soc.* **95:**7186 (1973).

f. R. K. Hill, C. B. Giberson, and J. V. Silverton, *J. Am. Chem. Soc.* **110:**497 (1988).

5a. L. A. Paquette and M. Oku, *J. Am. Chem. Soc.* **96:**1219 (1974).

b. A. E. Hill, G. Greenwood, and H. M. R. Hoffmann, *J. Am. Chem. Soc.* **95:**1338 (1973).

c. S. W. Staley and T. J. Henry, *J. Am. Chem. Soc.* **93:**1292 (1971).

d. T. Kauffmann and E. Köppelmann, *Angew. Chem. Int. Ed. Engl.* **11:**290 (1972).

e. C. W. Jefford, A. F. Boschung, and C. G. Rimbault, *Tetrahedron Lett.* **1974:**3387.

f. M. F. Semelhack, H. N. Weller, and J. S. Foos, *J. Am. Chem. Soc.* **99:**292 (1977).

g. R. K. Boeckman, Jr., M. H. Delton, T. Nagasaka, and T. Watanabe, *J. Org. Chem.* **42:**2946 (1977).

h. I. Hasan and F. W. Fowler, *J. Am. Chem. Soc.* **100:**6696 (1978).

i. R. Subramanyan, P. D. Bartlett, G. Y. M. Iglesias, W. H. Watson, and J. Galloy, *J. Org. Chem.* **47:**4491(1982).

6. A. Anastassiou and R. P. Cellura, *J. Chem. Soc., Chem. Commun.* **1969:**1521.

7a. L. A. Feiler, R. Huisgen, and P. Koppitz, *J. Am. Chem. Soc.* **96:**2270 (1974).

b. H. H. Wasserman, J. U. Piper, and E. V. Dehmlow, *J. Org. Chem.* **38:**1451 (1973).

c. D. A. Evans and A. M. Golob, *J. Am. Chem. Soc.* **97:**4765 (1975).

d. K. Oshima, H. Takahashi, H. Yamamoto, and H. Nozaki, *J. Am. Chem. Soc.* **95:**2693 (1973).

e. N. Shimizu, M. Tanaka, and Y. Tsuno, *J. Am. Chem. Soc.* **104:**1330 (1982).

f. V. Cere. E. Dalcanale, C. Paolucci, S. Pollicino, E. Sandri, L. Lunazzi, and A. Fava, *J. Org. Chem.* **47:**3540 (1982).

g. L. A. Paquette and M. J. Wyvratt, *J. Am. Chem. Soc.* **96:**4671 (1974); D. McNeil, B. R.Vogt, J. J. Sudol, S. Theodoropulos, and E. Hedaya, *J. Am. Chem. Soc.* **96:**4673 (1974).

h. D. Bellus, H.-C. Mez, G. Rihs, and H. Sauter, *J. Am. Chem. Soc.* **96:**5007 (1974).

i. W. Grimme, *J. Am. Chem. Soc.* **95:**2381 (1973).

j. W. Weyler, Jr., L. R. Byrd, M. C. Caserio, and H. W. Moore, *J. Am. Chem. Soc.* **94:**1027 (1972).

k. M. Nakazaki, K. Naemura, H. Harada, and H. Narutaki, *J. Org. Chem.* **47:**3470 (1982).

l. F. Binns, R. Hayes, K. J. Hodgetts, S. T. Saengchantara, T. W. Wallace, and C. J. Wallis, *Tetrahedron* **52:**3631 (1996).

8. W. H. Rastetter and T. J. Richard, *J. Am. Chem. Soc.* **101:**3893 (1979).

9. R. Huisgen and W. E. Konz, *J. Am. Chem. Soc.* **92:**4102 (1970).

10. K. Maruyama, N. Nagai, and Y. Naruta, *J. Org. Chem.* **51:**5083 (1986).

11. R. Subramanyam, P. D. Bartlett, G. Y. M. Iglesia, W. H. Watson, and J. Galloy, *J. Org. Chem.* **47:**4491 (1982).

12a. L. A. Paquette and R. S. Beckley, *J. Am. Chem. Soc.* **97:**1084 (1975).

b. K. C. Nicolaou, N. A. Petasis, R. E. Zipkin, and J. Uenishi, *J. Am. Chem. Soc.* **104:**5555 (1982).

c. B. M. Trost and A. J. Bridges, *J. Am. Chem. Soc.* **98:**5017 (1976).

d. K. C. Nicolaou, N. A. Petasis, R. E. Zipkin, and J. Uenishi, *J. Am. Chem. Soc.* **104:**5555 (1982).

e. K. J. Shea and R. B. Phillips, *J. Am. Chem. Soc.* **100:**654 (1978).

f. J. G. Millar, *Tetrahedron Lett.* **38:**7971 (1997).

13. R. K. Hill, J. W. Morgan, R.V. Shetty, and M. E. Synerholm, *J. Am. Chem. Soc.* **96:**4201 (1974); H. M. R. Hoffmann, *Angew. Chem. Int Ed. Engl.* **8:**556 (1969).

14a. T. J. Brocksom and M. G. Constantino, *J. Org. Chem.* **47:**3450 (1982).

b. L. E. Overman, G. F. Taylor, K. N. Houk, and L. N. Domelsmith, *J. Am. Chem. Soc.* **100:**3182 (1978).

c. P. W. Tang and C. A. Maggiulli, *J. Org. Chem.* **46:**3429 (1981).

d. R. B. Woodward, F. Sondheimer, D. Taub, K. Heusler, and W. M. McLamore, *J. Am. Chem. Soc.* **74:**4223 (1952).

e. T. Cohen and Z. Kosarych, *J. Org. Chem.* **47:**4005 (1982).

15. S. V. Ley and L. A. Paquette, *J. Am. Chem. Soc.* **96:**6670 (1974).

16a. A. K. Cheng, F. A. L. Anet, J. Mioduski, and J. Meinwald, *J. Am. Chem. Soc.* **96:**2887 (1974).

b. J. S. McKennis, L. Brener, J. S. Ward, and R. Pettit, *J. Am. Chem. Soc.* **93:**4957 (1971).

c. W. Grimme, H. J. Riebel, and E. Vogel, *Angew. Chem. Int. Ed. Engl.* **7:**823 (1968).

d. W. Grimme, *J. Am. Chem. Soc.* **94:**2525 (1972).

e. J. J. Gajewski, L. K. Hoffman, and C. N. Shih, *J. Am. Chem. Soc.* **96:**3705 (1974).

f. D. P. Lutz and J. D. Roberts, *J. Am. Chem. Soc.* **83:**2198 (1961).

17. A. Krantz, *J. Am. Chem. Soc.* **94:**4020 (1972).

18. H.-D. Martin and E. Eisenmann, *Tetrahedron Lett.* **1975:**661.

19a. A. Viola and L. Levasseur, *J. Am. Chem. Soc.* **87**:1150 (1965).

b. S. F. Reed, Jr., *J. Org. Chem.* **30**:1663 (1965).

c. T. S. Cantrell and H. Shechter, *J. Am. Chem. Soc.* **89**:5868 (1967).

d. R. B. Woodward, R. E. Lehr, and H. H. Inhoffen, *Justus Liebigs Ann. Chem.* **714**:57 (1968).

e. R. B. Woodward and T. J. Katz, *Tetrahedron* **5**:70 (1959).

f. N. J. Turro and W. B. Hammond, *Tetrahedron* **24**:6029 (1968).

g. J. S. McConaghy, Jr., and J. J. Bloomfield, *Tetrahedron Lett.* **1969**:3719.

h. W. J. Linn and R. E. Benson, *J. Am. Chem. Soc.* **87**:3657 (1965).

i. J. K. Crandall and W. H. Machleder, *J. Am. Chem. Soc.* **90**:7292 (1968).

j. M. Jones, Jr., S. D. Reich, and L. T. Scott, *J. Am. Chem. Soc.* **92**:3118 (1970).

k. M. Jones, Jr., and B. Fairless, *Tetrahedron Lett.* **1968**:4881; R. T. Seidner, N. Nakatsuka, and S. Masamune, *Can. J. Chem.* **48**:187 (1970).

l. P. G. Gassman, J. J. Roos, and S. J. Lee, *J. Org. Chem.* **49**:717 (1984).

m. Y. N. Gupta, M. J. Don, and K. N. Houk, *J. Am. Chem. Soc.* **104**:7336 (1982).

n. V. Glock, M. Wette, and F.-G. Klärner, *Tetrahedron Lett.* **26**:1441 (1985).

o. P. A. Zoretic, R. J. Chambers, G. D. Marbury, and A. A. Riebiro, *J. Org. Chem.* **50**:2981 (1985).

p. S. Sato, K. Tomita, H. Fujita, and Y. Sabo, *Heterocycles* **22**:1045 (1984).

q. M. Nakazaki, K. Naemura, H. Harada, and H. Narutaki, *J. Org. Chem.* **47**:3470 (1982).

20. A. G. Anastassiou and E. Yakali, *J. Chem. Soc., Chem. Commun.* **1972**:92; M. Mauksch, V. Gogonea, H. Jiao, and P. v. R. Schleyer, *Angew. Chem. Int. Ed. Engl.* **37**:2395 (1998).

Chapter 12

1a. H. O. House, C.-Y. Chu, W. V. Phillips, T. S. B. Sayer, and C.-C. Yau, *J. Org. Chem.* **42**:1709 (1977).

b. K. Fujunishi, Y. Inoue, Y. Kishimoto, and F. Mashio, *J. Org. Chem.* **40**:628 (1975).

c. C. Walling and D. Bristol, *J. Org. Chem.* **37**:3514 (1972).

d. D. H. Miles, P. Loew, W. S. Johnson, A. F. Kluge, and J. Meinwald, *Tetrahedron Lett.* **1972**:3019.

e. K. Pedersen, P. Jakobsen, and S.-O. Lawesson, *Org. Synth.* **V**:70 (1973).

f. R. Dowbenko, *Org. Synth.* **V**:93 (1973).

g. W. H. Sharkey and C. M. Langkammerer, *Org. Synth.* **V**:445 (1973).

h. P. D. Klemmensen, H. Kolind-Andersen, H. B. Madsen, and A. Svendsen, *J. Org. Chem.* **44**:416 (1979).

i. R. A. Rossi, R. H. deRossi, and A. F. Lopez, *J. Am. Chem. Soc.* **98**:1252 (1976).

j. D. W. Grattan, J. M. Locke, and S. R. Wallis, *J. Chem. Soc., Perkin Trans. 1* **1973**:2264.

3. L. H. Gale, *J. Am. Chem. Soc.* **88**:4661 (1961).

4. I. Rosenthal and D. Elad, *J. Org. Chem.* **33**:805 (1968); R. Lalande, B. Maillard, and M. Cazaux, *Tetrahedron Lett.* **1969**:745.

5a. D. D. Tanner, H. Yabuuchi, and E. V. Blackburn, *J. Am. Chem. Soc.* **93**:4802 (1971).

b. E. Müller, *Tetrahedron Lett.* **1974**:1835.

c. A. L. J. Beckwith and C. J. Easton, *J. Am. Chem. Soc.* **103**:615 (1981).

d. H. C. Brown, M. S. Kharasch, and T. H. Chao, *J. Am. Chem. Soc.* **62**:3435 (1940).

e. A. Effio, D. Griller, K. U. Ingold, J. C. Scaiano, and S. J. Sheng, *J. Am. Clem. Soc.* **102**:6063 (1980).

f. N. O. Brace, *J. Am. Chem. Soc.* **86**:523 (1964).

6a. W. H. Urry, D. R. Trecker, and H. D. Hartzler, *J. Org. Chem.* **29**:1663 (1964).

b. H. Pines, N. C. Sih, and D. B. Rosenfield, *J. Org. Chem.* **31**:2255 (1966).

c. D. D. Tanner, E. V. Blackburn, and G. E. Diaz, *J. Am. Chem. Soc.* **103**:1557 (1981).

d. H. Driquez, J. M. Paton, and J. Lessard, *Can. J. Chem.* **55**:700 (1977).

e. H. Feuer, J. Doty, and J. P. Lawrence, *J. Org. Chem.* **38**:417 (1973).

f. E. A. Mayeda, *J. Am. Chem. Soc.* **97**:4012 (1975).

g. X. Creary, *J. Org. Chem.* **45**:280 (1980).

h. A. L. J. Beckwith and R. D. Wagner, *J. Am. Chem. Soc.* **101**:7099 (1979).

7. L. K. Montgomery and J. W. Matt, *J. Am. Chem. Soc.* **89**:6556 (1967).

8. P. D. Bartlett, R. E. Pincock, J. H. Rolston, W. G. Schindel, and L. A. Singer, *J. Am. Chem. Soc.* **87**:2590 (1965).

9. E. S. Huyser and B. Giddings, *J. Org. Chem.* **27**:3391 (1962).

10. N. C. Deno, W. E. Billups, R. Fishbein, C. Pierson, R. Whalen, and J. C. Wyckoff, *J. Am. Chem. Soc.* **93**:438 (1971).

11a. D. H. Levy and R. J. Myers, *J. Chem Phys.* **41**:1062 (1964).

b. J. K. Kochi and P. J. Krusic, *J. Am. Chem. Soc.* **90**:7157 (1968).

12. H. Tanida and T. Tsuji, *J. Org. Chem.* **29**:849 (1964); P. R. Story, *Tetrahedron Lett.* **1962**:401.

13. E. J. Heiba, R. M. Dessau, and W. J. Koehl, Jr., *J. Am. Chem. Soc.* **90**:5905 (1968).

14. C. L. Karl, E. J. Maas, and W. Reusch, *J. Org. Chem.* **37**:2834 (1972).

15. P. R. Story, D. D. Denson, C. E. Bishop, B. C. Clark, Jr., and J.-C. Farine, *J. Am. Chem. Soc.* **90**:817 (1968).

16. K. J. Shea and P. S. Skell, *J. Am. Chem. Soc.* **95**:6728 (1973).

17. W. A. Bonner and F. D. Mango, *J. Org. Chem.* **29**:430 (1964).

18a. N. Kornblum, H. K. Singh, and W. J. Kelly, *J. Org. Chem.* **48**:332 (1983).

b. G. A. Russell, F. Ros, J. Hershberger, and H. Tashtoush, *J. Org. Chem.* **47**:1480 (1982).

c. G. A. Russell, J. Hershberger, and K. Owens, *J. Am. Chem. Soc.* **101**:1312 (1979).

19. C. J. Easton and M. P. Hay, *J. Chem. Soc., Chem. Commun.*, 55 (1986).

20. D. J. De Frees, R. J. McIver, Jr., and W. J. Hehre, *J. Am. Chem. Soc.* **102**:3334 (1980).

21a. K. S. Feldman, A. L. Romanelli, R. E. Ruckle, Jr., and G. Jean, *J. Org. Chem.* **57**:100 (1992).

b. J. Sejbal, J. Klinot, and A. Vystrcil, *Collect. Czech. Chem. Commun.* **53**:118 (1988).

c. J. P. Vionnet and P. Renaud, *Helv. Chim. Acta* **77**:1781 (1994).

Chapter 13

1. C. C. Lee and D. Unger, *Can. J. Chem.* **50**:593 (1972).

2a. E. Vogel, W. Grimme, and E. Dinne, *Tetrahedron Lett.* **1965**:391.

b. J. Meinwald and J. W. Young, *J. Am. Chem. Soc.* **93**:725 (1971).

c. D. M. Madigan and J. S. Swenton, *J. Am. Chem. Soc.* **93**:6316 (1971).

d. E. J. Corey and A. G. Hortmann, *J. Am. Chem. Soc.* **85**:4033 (1963).

e, f. K. M. Shumate, P. N. Neumann, and G. J. Fonken, *J. Am. Chem. Soc.* **87**:3996 (1965); K. M. Shumate and G. J. Fonken, *J. Am. Chem. Soc.* **88**:1073 (1966).

g. H. E. Zimmerman and M.-L. Viriot-Villaume, *J. Am. Chem. Soc.* **95**:1274 (1973).

h. R. L. Coffin, R.S. Givens, and R. G. Carlson, *J. Am. Chem. Soc.* **96**:7554 (1974).

3a. O. Rodriguez and H. Morrison, *J. Chem. Soc., Chem. Commun.* **1971**:679.

b. W. G. Dauben and M. S. Kellogg, *J. Am. Chem. Soc.* **93**:3805 (1971).

c. D. H. R. Barton, D. L. J. Clive, P. D. Magnus, and G. Smith, *J. Chem. Soc., C* **1971**:2193.

4a. W. C. Agosta and A. B. Smith III, *J. Am. Chem. Soc.* **93**:5513 (1971).

b. W. Ferree, Jr., J. B. Grutzner, and H. Morrison, *J. Am. Chem. Soc.* **93**:5502 (1971).

c. A. Wissner, *J. Org. Chem.* **42**:356 (1977).

d. G. W. Shaffer and M. Pesaro, *J. Org. Chem.* **39**:2489 (1974).

e. D. R. Morton and N. J. Turro, *J. Am. Chem. Soc.* **95**:3947 (1973).

f. H. Hart and A. F. Naples, *J. Am. Chem. Soc.* **94**:3256 (1972).

g. M. Pomerantz and G. W. Gruber, *J. Am. Chem. Soc.* **93**:6615 (1971).

h. O. L. Chapman, G. W. Borden, R. W. King, and B. Winkler, *J. Am. Chem. Soc.* **86**:2660 (1964); A. Padwa, L. Brodsky, and S. Clough, *J. Am. Chem. Soc.* **94**:6767 (1972).

i. R. K. Russell, R. E. Wingard, Jr., and L. A. Paquette, *J. Am. Chem. Soc.* **96**:7483 (1974).

j. D. I. Schuster and C. W. Kim, *J. Am. Chem. Soc.* **96**:7437 (1974).

k. W. G. Dauben, M. S. Kellogg, J. I. Seeman, and W. A. Spitzer, *J. Am. Chem. Soc.* **92**:1786 (1970).

l. A. Padwa and W. Eisenberg, *J. Am. Chem. Soc.* **92**:2590 (1970).

m. D. H. R. Barton and G. Quinkert, *J. Chem. Soc.* **1960**:1.

n. R. S. Cooke and G. D. Lyon, *J. Am. Chem. Soc.* **93**:3840 (1971).

o. D. Bryce-Smith and J. E. Lodge, *J. Chem. Soc.* **1963**:695; E. Grovenstein, Jr., and D. V. Rao, *Tetrahedron Lett.* **1961**:148.

p. L. E. Friedrich and J. D. Bower, *J. Am. Chem. Soc.* **95**:6869 (1973).

5. T.-Y. Leong, T. Imagawa, K. Kimoto, and M. Kawanisi, *Bull. Chem. Soc. Jpn.* **46**:596 (1973).

6. W. B. Hammond and T. S. Yeung, *Tetrahedron Lett.* **1975**:1173.

7. H. E. Zimmerman, R. S. Givens, and R. M. Pagni, *J. Am. Chem. Soc.* **90**:6096 (1968).

8. P. J. Wagner, P. A. Kelso, and R. G. Zepp, *J. Am. Chem. Soc.* **94**:7480 (1972).

10. N. J. Turro and D. S. Weiss, *J. Am. Chem. Soc.* **90**:2185 (1968).

11. S. Boue and R. Srinivasan, *J. Am. Chem. Soc.* **92**:3226 (1970); J. Saltiel *et al.*, *Org Photochem.* **3**:1 (1973).

12. W. G. Dauben, K. Koch, S. L. Smith, and O. L. Chapman, *J. Am. Chem. Soc.* **85**:2616 (1963).

13. H. E. Zimmerman and R. D. Solomon, *J. Am. Chem. Soc.* **108**:6276 (1986).

14. H. E. Zimmerman, *Angew. Chem. Int. Ed. Engl.* **8**:1 (1969).

15. K. R. Huffman, M. Burger, W. A. Henderson, Jr., M. Loy, and E. F. Ullman, *J. Org. Chem.* **34**:2407 (1969).

16. F. D. Lewis, R. W. Johnson, and D. E. Johnson, *J. Am. Chem. Soc.* **96**:6090 (1974).

17. H. E. Zimmerman, R. J. Boettcher, N. E. Buehler, G. E. Keck, and M. G. Steinmetz, *J. Am. Chem. Soc.* **98**:7680 (1976).

18. H. E. Zimmerman, R. J. Boettcher, and W. Braig, *J. Am. Chem. Soc.* **95**:2155 (1973).

19. J. R. Scheffer, K. S. Bhandari, R. E. Gayler, and R. A. Wostradowski, *J. Am. Chem. Soc.* **97**:2178 (1975).

20. C. W. Jefford and F. Delay, *J. Am. Chem. Soc.* **97**:2272 (1975).

21. G. Rattray, J. Yang, A. D. Gudmundsdottir, and J. R. Scheffer, *Tetrahedron Lett.* **34**:35 (1993).

Index